Lecture Notes in Computer Science   9771

Commenced Publication in 1973
Founding and Former Series Editors:
Gerhard Goos, Juris Hartmanis, and Jan van Leeuwen

More information about this series at http://www.springer.com/series/7409

De-Shuang Huang · Vitoantonio Bevilacqua
Prashan Premaratne (Eds.)

Intelligent Computing Theories and Application

12th International Conference, ICIC 2016
Lanzhou, China, August 2–5, 2016
Proceedings, Part I

 Springer

Editors
De-Shuang Huang
Tongji University
Shanghai
China

Prashan Premaratne
University of Wollongong
North Wollongong, NSW
Australia

Vitoantonio Bevilacqua
Polytechnic of Bari
Bari
Italy

ISSN 0302-9743   ISSN 1611-3349   (electronic)
Lecture Notes in Computer Science
ISBN 978-3-319-42290-9   ISBN 978-3-319-42291-6   (eBook)
DOI 10.1007/978-3-319-42291-6

Library of Congress Control Number: 2016943868

LNCS Sublibrary: SL3 – Information Systems and Applications, incl. Internet/Web, and HCI

Printed on acid-free paper

This Springer imprint is published by Springer Nature
The registered company is Springer International Publishing AG Switzerland

Preface

The International Conference on Intelligent Computing (ICIC) was started to provide an annual forum dedicated to the emerging and challenging topics in artificial intelligence, machine learning, pattern recognition, bioinformatics, and computational biology. It aims to bring together researchers and practitioners from both academia and industry to share ideas, problems, and solutions related to the multifaceted aspects of intelligent computing.

ICIC 2016, held in Lanzhou, China, August 2–5, 2016, constituted the 12th International Conference on Intelligent Computing. It built upon the success of ICIC 2015, ICIC 2014, ICIC 2013, ICIC 2012, ICIC 2011, ICIC 2010, ICIC 2009, ICIC 2008, ICIC 2007, ICIC 2006, and ICIC 2005 that were held in Fuzhou, Taiyuan, Nanning, Huangshan, Zhengzhou, Changsha, China, Ulsan, Korea, Shanghai, Qingdao, Kunming, and Hefei, China, respectively.

This year, the conference concentrated mainly on the theories and methodologies as well as the emerging applications of intelligent computing. Its aim was to unify the picture of contemporary intelligent computing techniques as an integral concept that highlights the trends in advanced computational intelligence and bridges theoretical research with applications. Therefore, the theme for this conference was "Advanced Intelligent Computing Technology and Applications." Papers focused on this theme were solicited, addressing theories, methodologies, and applications in science and technology.

ICIC 2016 received 639 submissions from 22 countries and regions. All papers went through a rigorous peer-review procedure and each paper received at least three review reports. Based on the review reports, the Program Committee finally selected 236 high-quality papers for presentation at ICIC 2016, included in three volumes of proceedings published by Springer: two volumes of *Lecture Notes in Computer Science* (LNCS), and one volume of *Lecture Notes in Artificial Intelligence* (LNAI).

This volume of *Lecture Notes in Computer Science* (LNCS) includes 84 papers.

The organizers of ICIC 2016, including Tongji University and Lanzhou University of Technology, China, made an enormous effort to ensure the success of the conference. We hereby would like to thank the members of the Program Committee and the referees for their collective effort in reviewing and soliciting the papers. We would like to thank Alfred Hofmann, of Springer, for his frank and helpful advice and guidance throughout and for his continuous support in publishing the proceedings. Moreover, we would like to thank all the authors in particular for contributing their papers. Without the high-quality submissions from the authors, the success of the conference would not have been possible. Finally, we are especially grateful to the IEEE Computational Intelligence Society, the International Neural Network Society, and the National Science Foundation of China for their sponsorship.

May 2016

De-Shuang Huang
Vitoantonio Bevilacqua
Prashan Premaratne

Organization

General Co-chairs

De-Shuang Huang	China
Cesare Alippi	Italy
Jie Cao	China

Program Committee Co-chairs

Kang-Hyun Jo	Korea
Vitoantonio Bevilacqua	Italy
Jinyan Li	Australia

Organizing Committee Co-chairs

Aihua Zhang	China
Ce Li	China

Organizing Committee Members

Weirong Liu	China
Erchao Li	China
Xiaolei Chen	China
Hui Chen	China
Suping Deng	China
Lin Zhu	China
Gang Wang	China

Award Committee Chair

Kyungsook Han	Korea

Tutorial Co-chairs

Laurent Heutte	France
Abir Hussain	UK

Publication Co-chairs

M. Michael Gromiha	India
Valeriya Gribova	Russia
Juan Carlos Figueroa	Colombia

Workshop/Special Session Chair

Ling Wang   China

Special Issue Co-chairs

Henry Han   USA
Phalguni Gupta   India

International Liaison Chair

Prashan Premaratne   Australia

Publicity Co-chairs

Evi Syukur   Australia
Chun-Hou Zheng   China
Jair Cervantes Canales   Mexico

Exhibition Chair

Lin Zhu   China

Program Committee

Andrea F. Abate	Fengfeng Zhou	Ming Jiang
Akhil Garg	Francesco Pappalardo	Jijun Tang
Vangalur Alagar	Shan Gao	Joaquin Torres
Angel Sappa	Liang Gao	Jun Zhang
Angelo Ciaramella	Kayhan Gulez	Kang Li
Bingqiang Liu	Hongei He	Ka-Chun Wong
Shuhui Bi	Huiyu Zhou	Seeja K.R
Bin Liu	Fei Han	Kui Liu
Cheen Sean Oon	Huanhuan Chen	Min Li
Chen Chen	Mohd Helmy Abd Wahab	Jianhua Liu
Wen-Sheng Chen	Hongjie Wu	Juan Liu
Michal Choras	Indrajit Saha	Yunxia Liu
Xiyuan Chen	Ivan Vladimir Meza Ruiz	Haiying Ma
Chunmei Liu	John Goulermas	Maurizio Fiasche
Costin Badica	Jianbo Fan	Marzio Pennisi
Dah-Jing Jwo	Jiancheng Zhong	Peter Hung
Daming Zhu	Junfeng Xia	Qiaotian Li
Dongbin Zhao	Jiangning Song	Qinmin Hu
Ben Niu	Jian Yu	Robin He
Dunwei Gong	Jim Jing-Yan Wang	Wei-Chiang Hong

Preface

The International Conference on Intelligent Computing (ICIC) was started to provide an annual forum dedicated to the emerging and challenging topics in artificial intelligence, machine learning, pattern recognition, bioinformatics, and computational biology. It aims to bring together researchers and practitioners from both academia and industry to share ideas, problems, and solutions related to the multifaceted aspects of intelligent computing.

ICIC 2016, held in Lanzhou, China, August 2–5, 2016, constituted the 12th International Conference on Intelligent Computing. It built upon the success of ICIC 2015, ICIC 2014, ICIC 2013, ICIC 2012, ICIC 2011, ICIC 2010, ICIC 2009, ICIC 2008, ICIC 2007, ICIC 2006, and ICIC 2005 that were held in Fuzhou, Taiyuan, Nanning, Huangshan, Zhengzhou, Changsha, China, Ulsan, Korea, Shanghai, Qingdao, Kunming, and Hefei, China, respectively.

This year, the conference concentrated mainly on the theories and methodologies as well as the emerging applications of intelligent computing. Its aim was to unify the picture of contemporary intelligent computing techniques as an integral concept that highlights the trends in advanced computational intelligence and bridges theoretical research with applications. Therefore, the theme for this conference was "Advanced Intelligent Computing Technology and Applications." Papers focused on this theme were solicited, addressing theories, methodologies, and applications in science and technology.

ICIC 2016 received 639 submissions from 22 countries and regions. All papers went through a rigorous peer-review procedure and each paper received at least three review reports. Based on the review reports, the Program Committee finally selected 236 high-quality papers for presentation at ICIC 2016, included in three volumes of proceedings published by Springer: two volumes of *Lecture Notes in Computer Science* (LNCS), and one volume of *Lecture Notes in Artificial Intelligence* (LNAI).

This volume of *Lecture Notes in Computer Science* (LNCS) includes 84 papers.

The organizers of ICIC 2016, including Tongji University and Lanzhou University of Technology, China, made an enormous effort to ensure the success of the conference. We hereby would like to thank the members of the Program Committee and the referees for their collective effort in reviewing and soliciting the papers. We would like to thank Alfred Hofmann, of Springer, for his frank and helpful advice and guidance throughout and for his continuous support in publishing the proceedings. Moreover, we would like to thank all the authors in particular for contributing their papers. Without the high-quality submissions from the authors, the success of the conference would not have been possible. Finally, we are especially grateful to the IEEE Computational Intelligence Society, the International Neural Network Society, and the National Science Foundation of China for their sponsorship.

May 2016

De-Shuang Huang
Vitoantonio Bevilacqua
Prashan Premaratne

Emanuele Lindo Secco
Shuigeng Zhou
Shuai Li
Shihong Yue
Saiful Islam
Jiatao Song
Shuo Liu
Shunren Xia
Surya Prakash
Shaoyuan Li
Tingwen Huang
Vasily Aristarkhov
Fei Wang
Xuesong Wang
Weihua Sun

Weidong Chen
Wei Wei
Zhi Wei
Shih-Hsin Chen
Wu Chen Su
Shitong Wang
Xiufen Zou
Xiandong Meng
Xiaoguang Zhao
Minzhu Xie
Xin Yin
Xinjian Chen
Xiaoju Dong
Xingsheng Gu
Xiwei Liu

Yingqin Luo
Yongquan Zhou
Yun Xiong
Yong Wang
Yuexian Hou
Chenghui Zhang
Weiming Zeng
Zhigang Luo
Fa Zhang
Liang Zhao
Zhenyu Xuan
Shanfeng Zhu
Quan Zou
Zhenran Jiang

Additional Reviewers

Yong Chen
Peng Xie
Yunfei Wang
Selin Ozcira
Stephen Tang
Badr Abdullah
Xuefeng Cui
Lumin Zhang
Chunye Wang
Qian Chen
Kan Qiao
Yingji Zhong
Wei Gao
Tao Yi
Liuhua Chen
Faliu Yi
Xiaoming Liu
Sheng Ding
Xin Xu
Zhebin Zhang
Shankai Yan
Yueming Lyu
Giulia Russo
Marzio Pennisi
Zhile Yang
Enting Gao
Min Sun

Lingjiao Pan
Ying Bi
Chao Jin
Shiwei Sun
Mohd Shamrie Sainin
Xing He
Xue Zhang
Junqing Li
Chen Chen
Wei-Shi Zheng
Chao Wu
Tingli Cheng
Francesco Pappalardo
Neil Buckley
Bolin Chen
Pengbo Wen
Long Wen
Bogdan Czejdo
Jing Wu
Weiwei Shen
Ximo Torres
Lan Huang
Jingchuan Wang
Savannah Bell
Alexandria Spradlin
Christina Spradlin
Li Liu

Geethan Mendiz
Jingsong Shi
Lun Li
Cheng Lian
Jin-Xing Liu
Obinna Anya
Lai Wei
Yan Cui
Peng Xiaoqing
Vivek Kanhangad
Yong Xu
Morihiro Hayashida
Yaqiang Yao
Chang Li
Jiang Bingbing
Haitao Li
Wei Peng
Jerico Revote
Xiaoyu Shi
Jia Meng
Jiawei Wang
Jing Jin
Yong Zhang
Biao Xu
Vangalur Alagar
Kaiyu Wan
Surya Prakash

Yongjin Li
Changning Liu
Xionghui Zhou
Hong Wang
Gongjing Chen
Yuntao Wei
Fangfang Zhang
Jia Liu
Jing Liu
Jnanendra Sarkar
Sayantan Singha Roy
Puneet Gupta
Shaohua Li
Zhicheng Liao
Adrian Lancucki
Julian Zubek
Srinka Basu
Xu Huang
Liangxu Liu
Qingfeng Li
Cristina Oyarzun Laura
Rina Su
Xiaojing Gu
Peng Zhou
Zewen Sun
Xin Liu
Yansheng Wang
Xiaoguang Zhao
Qing Lei
Yang Li
Wentao Fan
Hongbo Zhang
Minghai Xin
Yijun Bian
Yao Yu
Vasily Aristarkhov
Qi Liu
Vibha Patel
Jun Fan
Bojun Xie
Jie Zhu
Long Lan
Phan Cong Vinh
Zhichen Gong
Jingbin Wang
Akhil Garg

Wei Liao
Tian Tian
Xiangjuan Yao
Chenyan Bai
Guohui Li
Zheheng Jiang
Li Hailin
Huiyu Zhou
Baohua Wang
Kasi Periyasamy
Li Nie
Zhurong Wang
Ella Pereira
Danilo Caceres
Meng Lei
Changbin Du
Shaojun Gan
Yuan Xu
Chen Jianfeng
Chuanye Tang
Bo Liu
Bin Qian
Xuefen Zhu
Haoqian Huang
Fei Guo
Jiayin Zhou
Raul Montoliu
Oscar Belmonte
Farid Garcia-Lamont
Alfonso Zarco
Yi Gu
Ning Zhang
Jingli Wu
Xing Wei
Shenshen Liang
Nooraini Yusoff
Yanhui Guo
Nureize Arbaiy
Wan Hussain Wan Ishak
Yizhang Jiang
Pengjiang Qian
Si Liu
Chen Aiguo
Yunfei Yi
Rui Wang
Jiefang Liu

Aijia Ouyang
Hongjie Wu
Andrei Velichko
Wenlong Hang
Lijun Quan
Min Jiang
Tomasz Andrysiak
Faguang Wang
Liangxiu Han
Leonid Fedorischev
Wei Dai
Yifan Zhao
Xiaoyan Sun
Yiping Liu
Hui Li
Yinglei Song
Elisa Capecci
Tinting Mu
Francesco Giovanni Sisca
Austin Brockmeier
Cheng Wang
Juntao Liu
Mingyuan Xin
Chuang Ma
Marco Gianfico
Davide Nardone
Francesco Camastra
Antonino Staiano
Antonio Maratea
Pavan Kumar Gorthi
Antonio Brunetti
Fabio Cassano
Xin Chen
Fei Wang
Chen Xu
Gianpaolo Francesco
    Trotta
Alberto Cano
Xiuyang Zhao
Zhenxiang Chen
Lizhi Peng
Nagarajan Raju
e Wang
Yehu Shen
Liya Ding
Tiantai Guo

Contents – Part I

Systems Biology and Intelligent Computing in Computational Biology

Intelligent Computing in Scheduling

**Machine Learning and Data Analysis for Medical
and Engineering Applications**

Signal Processing and Image Processing

A Variable Neighborhood Search Approach for the Capacitated m-Ring-Star Problem

Carlos Franco[1(✉)], Eduyn López-Santana[2],
and Germán Mendez-Giraldo[2]

[1] Universidad Católica de Colombia, Bogotá, Colombia
cafranco@ucatolica.edu.co
[2] Universidad Distrital Francisco José de Caldas, Bogotá, Colombia
{erlopezs,gmendez}@udistrital.edu.co

Abstract. In this paper, we proposed an algorithm based on variable neighborhood search (VNS) for the capacitated m-Ring-Star problem. This problem has several real applications in communications networks, rapid transit system planning and optical fiber networks. The problem consists in design m rings or cycles that begins of a central depot and visits a set of customers and transition or steiner nodes. While the nodes don't belong to a ring these must be allocated or assign to a customer or steiner node that belongs to a ring. The number of customers allocated or visited in each ring must not exceed the maximum capacity. The goal is to minimize the visiting and allocation cost. For solving the problem, we propose a VNS approach based on random perturbation for escaping from the local optimal solutions. Our method reached the optimal solution in a reasonable amount of time in a set of instances from the literature.

Keywords: m-Ring-Star problem · Variable neighborhood search · Network design · Combinatorial optimization

1 Introduction

The capacitated m-Ring-Star problem (CmRSP) was introduced by Baldacci et al. [1]. This problem is a variant of the classical capacitated vehicle routing problem with one depot. The CmRSP consists in designing a set of rings (with size m) passing throw a central depot and visiting a subset of customers and a subset of steiner nodes. Rings may include transit nodes (Steiner nodes) so that star connections can be established between a customer and a ring through a transit node. Each ring and its star connections (ring-star) is limited by a maximum number of customers. A feasible solution is represented by a set of m rings. Each node is a visited node if is in a ring, if not, is an allocated node. If in a node is assigned an allocated node, it is call a connecting node. Figure 1 gives an example of a feasible CmRSP solution, where there are three rings, and nodes assigned with the dotted line are the allocated nodes. The goal is to minimize the total cost of visiting and assigning the customers to the routes that comes from the depot. This problem has many real-world applications like optical fiber networks, rapid transit system planning and telecommunication systems.

© Springer International Publishing Switzerland 2016
D.-S. Huang et al. (Eds.): ICIC 2016, Part I, LNCS 9771, pp. 3–11, 2016.
DOI: 10.1007/978-3-319-42291-6_1

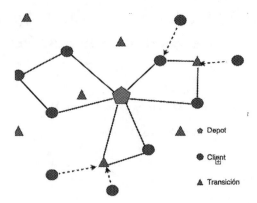

Fig. 1. An example of a feasible CmRSP solution

There are two variations of this problem, the ring star problem and the steiner ring star problem. In the ring star problem the steiner nodes are not available and the capacity constrain are not imposed. The problem consist in design a cycle through a subset of know customers looking for minimize the cycle length and the cost of assign a non-visited customer to the nearest customer visited. There are several algorithmic development for solving this problem, see [2–5].

On the other hand, the steiner ring star problem is a variation that has application in designing of digital data service networks. The problem consist in design a cycle over a set of steiner nodes and assign each customer to a node visited by a cycle. The objective is minimize the cycle and allocation costs. A branch and cut algorithm has been proposed to solve the problem [6].

The first work on this problem proposed two integer programming models inspired in classical routing formulations and introduced valid inequalities for improve the quality of the lower bounds, the problem was inspire in the design of a fiber optic communication network [1]. For solving the linear problem, authors develop a branch and cut approach and embebbed two heuristics procedure to speed up the convergence of the branch and cut algorithm.

Naji-Azimi et al. [7] proposed another heuristic based on linear programming and column generation scheme. The linear programming model selects the best point when a customer can be insert, and with the column generation scheme a reinsertion problem is solved. Some other methods are proposed for solving the problem, see [8–11].

The remainder of this paper is organized as follows: Sect. 2 present the description of the proposed algorithm based on VNS. Section 3 shows some numerical results from a set of instances of literature. Finally, Sect. 4 concludes this work and provides possible research directions.

2   Description of the Proposed Algorithm

Our algorithm is based on variable neighborhood search (VNS), for implementing the algorithm, we use the following premises:

(1) For representing the solution we use a vector that indicates the nodes on the ring and the assignation (2) We don't allow infeasible solutions in any iteration of the algorithm (3) In every moment of the algorithm, new solutions that improve the actual solution are accepted. For escaping to local optimal solutions we use a perturbation for begin in a new neighborhood (4) The stop criterion is when the algorithm can not find new best solutions.

The outline of the proposed algorithm is described in Table 1. In the following subsections, we give the details of each step.

Table 1. Proposed VNS algorithm for the CmRSP

```
current solution = initialization procedure ()
Best cost = cost (current solution)
Best solution = current solution
Stop criterion = 0
while Stop criterion=0
      current solution = perturbation within rings ()
      current solution = improve solution ()
      current solution = perturbation between rings ()
      current solution = improve solution ()
      current solution = addition of transition nodes ()
        current solution = resizing operator ()
        current solution = improve solution ()
        New cost = cost (current solution)
        New solution = current solution
        If New cost < Best cost then
              Best cost= New cost
              Best solution = New solution
              else
                 current solution = random perturbation()
                 New cost = cost (current solution)
                 New solution = current solution
                      If New cost < Best cost then
                         Best cost= New cost
                         Best solution = New solution
                         Else
                            Stop criterion=1
                      end if
      end if
end while
```

2.1 Initialization Procedure

In this procedure we look for create a feasible solution using only the customers (not the transition points or steiner nodes) that generate the structure of the ring. To do so, we select the m customer as close as possible one from each other and then we define m rings by connecting each (selected) customer to the depot. After, each one of the remaining customer is added to the best feasible position, this is the closest customer that belongs to a ring.

2.2 Perturbation Operator Within Rings

Each one of the rings is consider separately and transform each ring as the traveling salesman problem (TSP). The algorithm consists in applied variable neighborhood search (VNS) to the nodes that belongs to a ring, in this case a TSP for finding a better solution. If a better solution is find, the change is accepted. In Fig. 2 an illustrative example is presented. The node 0 represents the depot, the swap is between the customer one and three, if the cost of the new ring is lower than the actual ring then the swap is accepted. Table 2 shows the pseudocode of the perturbation operator.

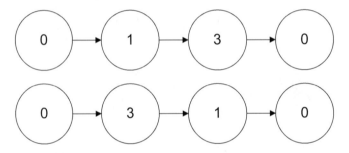

Fig. 2. An illustrative example: perturbation of a ring

2.3 Perturbation Operator Between Rings

Once the process of find better independent solutions for each ring is finished, we generate a perturbation or exchange between two rings. The exchange is made between two closest rings, if the new solution improves the oldest one, the new solution is updated. An illustrative example is presented in Fig. 3. Nodes 1 and 3 belongs to the first ring while nodes 4 and 5 belongs to the second ring. A pair of nodes is exchanged between the two rings generating a new solution that contains the nodes 4 and 3 in the first ring and nodes 1 and 5 are in the second ring. Table 3 shows the pseudocode of the perturbation operator.

Table 2. Pseudocode of swap method

```
Set m as the number of rings
Set nm as the number of nodes in each ring
For i→ 1 to m
  For j→ 1 to nm
      c→cost of the ring
      Swap two nodes without including the depot
      cnew→cost of the new ring
           if cnew<c then
              accept the new solution and save the new ring
           end
    end
end
```

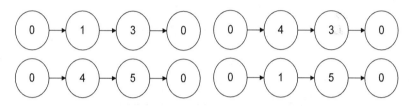

Fig. 3. An illustrative example: perturbation between rings

Table 3. Pseudocode of swap method between rings

```
Set m as the number of rings
For i→ 1 to m
  For j→ 1 to m
      c1→cost of the ring i
      c2→cost of the ring j
      Swap two nodes between the rings i and j
      cnew→cost of the new rings
      if cnew<c1+c2 then
          accept the new solution and save the new rings
      end
    end
end
```

2.4 Addition Operator of Transition Nodes

In each step of the iteration is evaluated if is convenient to replace a node of the ring for a transition node including the assigning of a customer to this transition node.

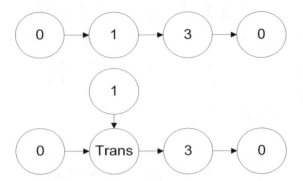

Fig. 4. An illustrative example: addition of transition nodes

Table 4. Pseudocode of adition of transition nodes

```
Set m as the number of rings
For i→ 1 to m
    For j→ 1 to m
        c1→cost of the ring i
        c2→cost of the ring j
        Swap two nodes between the rings i and j
        cnew→cost of the new rings
        if cnew<c1+c2 then
            accept the new solution and save the new rings
        end
    end
end
```

Figure 4 presents an illustrative example. Nodes 1 and 3 belongs to the ring and it is evaluated if a transition node can replace the node 1 and after assign the node 1 to the transition node. Table 4 shows the pseudocode of the perturbation operator.

2.5   Resizing Operator

The aim of this operator is to find a node in a ring that has the major cost of assignation, then extract the node of the actual ring, and reinsert in another ring, which present a less total cost. An illustrative example is presented in Fig. 5. Nodes 1, 3 and 4 belongs to ring 1 while node 5 belong to ring 2. After analyzing ring 1 the node 3 is the one with major cost of assigning, then is prove if this node can be assign in any position of ring 2. After evaluating the total cost this change can be done and the new ring 1 contains nodes 1 and 4 while ring 2 contains nodes 5 and 3. Table 5 shows the pseudocode of the resizing operator.

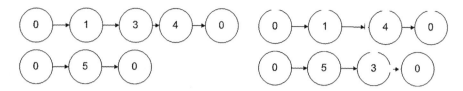

Fig. 5. An illustrative example: resizing operator

Table 5. Pseudocode of resizing operator

```
Set m as the number of rings
Set n_m as the number of nodes in each ring
For i→ 1 to m
  For j→ 1 to n_m
    Find the node with the highest cost
    Find a ring where the node has the lowest cost
    cnew→cost of the new rings
    if cnew<original cost then
       accept the new solution and save the new rings
    end
  end
end
```

2.6   Random Perturbation Operator

In order to avoid a local optimal solution, two random strategies are used for create a new solution. The first one generates a random number and select a random ring, with the random number is selected a random node and it is extract and reinsert in another random ring in a random position. The second one selects a random node of those which can be re assign in a transition node, then another random node is selected and is re assign the first node generating new rings. An illustrative example is presented in Fig. 6. Nodes 1, 4 and 5 belongs to ring 1 while nodes 3 and 2 belong to ring 2. After generate a random number on ring 1, node 4 is selected and reinserted in ring 2 between nodes 3 and 2 that is obtained by another random number.

3   Computational Experiments and Results

In order to test our algorithm we have used a well-known instances used in [1]. The results are presented in Table 6. The table is organized as follows; the first column presents the instance's name. The second column shows the best-known solution (BNS) reported in the literature. The Best solution (BSF) found by our algorithm is presented in column third, as our algorithm use a random perturbations, we run it twenty independent times and obtain the average solution on these runs (column four).

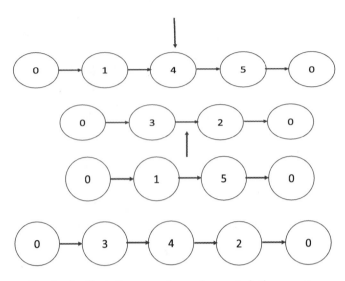

Fig. 6. An illustrative example: random perturbation operator

The standard deviation is presented in column five. The amount of computational time spend by our algorithm is presented in column six while the gap for the best solution obtained and the average gap of the twenty runs are on column six and seven respectively.

As we can see in Table 6, our algorithm is able to find the best solution know in the literature in all the instances except in the last one proving that our algorithm is efficient. For the instance that our algorithm cannot find the optimal solution, the gap is 9.75 %. The average GAP obtained in the twenty runs shows that is less than 3 %. Finally, the computational time of our algorithm is 2.96 s on average proving that it doesn't spend high computational times.

Table 6. Results of proposed method

Instance	BNS	BSF	Average	SD	Time (seconds)	GAP	Average GAP
eil26.tsp.3.12.5.A.BDS.cmrsp	242	242	242	0.00	1.11	0.00 %	0.00 %
eil26.tsp.3.25.10.B.BDS.cmrsp	2251	2251	2323.6	22.44	8.10	0.00 %	2.23 %
eil26.tsp.4.12.4.A.BDS.cmrsp	261	261	261	0.00	1.01	0.00 %	0.00 %
eil26.tsp.4.12.4.B.BDS.cmrsp	1827	1827	1827	0.00	0.79	0.00 %	0.00 %
eil26.tsp.4.18.5.A.BDS.cmrsp	339	339	343.4	3.16	1.20	0.00 %	1.30 %
eil26.tsp.4.18.5.B.BDS.cmrsp	2370	2370	2422	22.98	1.10	0.00 %	1.19 %
eil26.tsp.5.12.3.A.BDS.cmrsp	292	292	307.8	12.16	4.12	0.00 %	5.41 %
eil26.tsp.5.25.6.B.BDS.cmrsp	2674	2696	2745	31.44	5.20	0.82 %	2.66 %
eil51.tsp.3.12.5.A.BDS.cmrsp	242	242	242.4	0.69	1.92	0.00 %	0.17 %
eil51.tsp.5.50.12.B.BDS.cmrsp	3404	3736	3801.6	53.43	5.01	9.75 %	11.68 %
			Average	14.63	2.96	1.06 %	2.46 %

4   Conclusions

We have proposed a variable neighborhood search adaptation for solving the capacitated m-Ring-Star Problem. In order to avoid local optimal solutions, we have used random perturbations for avoid these local solutions and find new solutions close to another optimal local solutions or the global optimal solution.

We have used instances from the literature for testing our algorithm. The result shows the effectiveness of our algorithm in finding the optimal solution or closes one in a reasonable amount of computational time.

Future work should focus in to improve the decision-aid tool to allow speeding up the method. In addition, to decrease the computational time is possible to combine other techniques in order to decrease the computation time. Another real world constraints and characteristics can be explored as stochastic travel times, other objective functions, among others.

References

1. Baldacci, R., Dell'Amicoy, M., González, J.S.: The capacitated m-Ring-Star problem. Oper. Res. **55**, 1147–1162 (2007)
2. Calvetea, H.I., Galéy, C., Iranzo, J.A.: MEALS: a multiobjective evolutionary algorithm with local search for solving the bi-objective ring star problem. Eur. J. Oper. Res. **250**, 377–388 (2016)
3. Calvetea, H.I., Galéy, C., Iranzo, J.A.: An efficient evolutionary algorithm for the ring star problem. Eur. J. Oper. Res. **231**, 22–33 (2013)
4. Liefooghe, A., Jourdany, L., Talbi, E.G.: Metaheuristics and cooperative approaches for the bi-objective ring star problem. Comput. Oper. Res. **37**, 1033–1044 (2010)
5. Labbé, M., Laporte, G., Martíny, I., González, J.S.: The ring star problem polyhedral analysis and exact algorithm. Networks **43**, 177–189 (2004)
6. Lee, Y., Chiu, S.Y., Sanchez, J.: A branch and cut algorithm for the steiner ring star problem. Int. J. Manag. Sci. **4**, 21–34 (1998)
7. Naji-Azimi, Z., Salariy, M., Toth, P.: An integer linear programming based heuristic for the capacitated m-Ring-Star problem. Eur. J. Oper. Res. **217**, 17–25 (2012)
8. Azimi, Z.N., Salariy, M., Toth, P.: A heuristic procedure for the capacitated m-Ring-Star problem. Eur. J. Oper. Res. **207**, 1227–1234 (2010)
9. Hoshinoay, E.A., de Souza, C.C.: A branch-and-cut-and-price approach for the capacitated m-Ring-Star problem. Discrete Appl. Math. **160**, 2728–2741 (2012)
10. Hoshinoay, E.A., de Souza, C.C.: A branch-and-cut-and-price approach for the capacitated m-Ring-Star problem. Electron. Notes Discrete Math. **35**, 103–108 (2009)
11. Berinskyy, H., Zabala, P.: An integer linear programming formulation and branch-and-cut algorithm for the capacitated m-Ring-Star problem. Electron. Notes Discrete Math. **37**, 273–278 (2011)

Convolutional Neural Network Application on Leaf Classification

Yan-Hao Wu[1](✉), Li Shang[2], Zhi-Kai Huang[3], Gang Wang[1],
and Xiao-Ping Zhang[1]

[1] Institute of Machine Learning and Systems Biology,
College of Electronics and Information Engineering,
Tongji University, Caoan Road 4800, Shanghai 201804, China
wuyanhao@yeah.net
[2] Department of Communication Technology,
College of Electronic Information Engineering,
Suzhou Vocational University, Suzhou 215104, Jiangsu, China
[3] College of Mechanical and Electrical Engineering,
Nanchang Institute of Technology,
Nanchang 330099, Jiangxi, China

Abstract. Plants are everywhere in our lives, we can classify them by observing their features. But for ordinary people, the species we don't know are much more than we know. So, for amateurs who are interested in botany, a system which can classify different species of leaves must be very useful, a system like that will also help students recognize the leaves they don't know. This paper describes a system for leaf classification, which is developed with convolutional neural network technique. Previous researches in leaf identification usually use grayscale images. The main reason is that these samples mostly are green leaves. This system is trained by 1500 leaves to classify 50 kinds of plants. Compared to other research, our net use RGB images for input. And in convolutional neural network, we use PReLU instead of traditional ReLU. The experimental result shows that our method for classification gives accuracy of 94.8 %.

Keywords: Convolutional neural network · Leaf classification · Image recognition · Prelu

1 Introduction

This paper describes a convolutional neural network (CNN) for leaf classification. CNN is an effective identification method developed in recent years, and caused widespread attention. It was found in the 1960s. Now, CNN has become one of the most efficient methods in the field of pattern classification.

The invention of convolutional neural network [1] follows the discovery of visual mechanisms in animals. LeCun, et al. [2], put the back-propagation algorithm [3] into CNN in 1988. Recently, CNN has been used more widely in the field of image processing [5, 6], and it can reach a better performance than traditional methods [4] through wide verification.

© Springer International Publishing Switzerland 2016
D.-S. Huang et al. (Eds.): ICIC 2016, Part I, LNCS 9771, pp. 12–17, 2016.
DOI: 10.1007/978-3-319-42291-6_2

The convolutional neural networks are multi-layer supervised networks which can learn features automatically from datasets. For the last few years, CNNs have achieved state-of-the-art performance in almost all important classification tasks. It can perform both feature extraction and classification under the same architecture. In recent years, we believe that its main drawback is that it requires very large amounts of training data. However, latest studies have shown that state of the art performance can be achieved with networks trained using "generic" data.

AlexNet is a good convolutional neural network model for image processing. We used it for classify our leaves, got an error rate of 13.6 %. For leaf classification, 13.6 % is not enough. So we made some changes on AlexNet, and the precision became batter.

2   Related Works

In the traditional leaf classification, someone with rich taxonomic knowledge is needed, and it is very inefficient.

In the first step, leaf image is preprocessed to enhance the important features in a leaf. This step usually includes grayscale conversion, image segmentation and so on. Preprocessing is designed to improve image data and enhances the features which are relevant for further processing.

In the next step, feature extraction, is the most important part in a classification system. It extracts features for classification. The convolutional neural networks can learn features automatically from datasets. People always wanted to design some efficient feature for classification, but now it turns out that unsupervised feature extraction is more efficient.

The final step, features extracted by last step is used for classification.

Researchers around the world have used various classification techniques to classify the plants leaves for greater accuracies. Most of them extract features manually. Table 1 show the accuracies of some researches.

Table 1. Previousproducts

Research paper	Accuracy	Dataset size
Leaf Classification using shape, color, and texture features [7].	93.75 %	32
Edge and texture fusion for plant leaf classification June [8].	85.93 %	132
Shape and texture based leaf classification [9].	81.1 %	18

In this paper, we use the leaf data of dataset ICL for experiment. ICL is a database established by Institute of Machine Learning and Systems Biology, College of Electronics and Information Engineering, Tongji University. It contains 200 kinds of labeled plant leaves with different colors and shapes. There are 30 images being collected for every kind of the plants in the dataset. The images in it have high but not the same resolution, and haven't been cut. Textures and other characteristics of the leaves are completely preserved in the images of ICL dataset.

Our net is trained by 50 kinds of plant leaves. Their resolution ranged from 117*117 to 819*819.

3   Classification Based on CNN

3.1   Image Preprocessing

Images from ICL have different resolution; first of all, we must resize them to a fixed size. In this paper, we use images with 256*256 resolutions for input. In this step we choose 50 kinds of plant. So we resize all images to 256*256. Our experimentuse1500 images, 1250 for train and 250 for test.

3.2   Data Augmentation

The easiest and most common method to reduce over fitting is data augmentation.

The first way of data augmentation is adding noise to images. In this step we add 3 types of noise: Gaussian noise, Salt and pepper noise and Poisson noise, so there are 5000 images for training now. The second way of data augmentation is to flip image horizontally. After that, we will have a train set of 10000 images. The data augmentation is a very important step. Without it, the over fit problem will appear (Fig. 1).

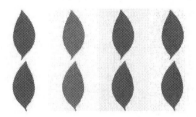

Fig. 1. Illustration of the data augmentation, from one image to eight images

3.3   The Architecture

Our net contains four layers with weights; the first five are convolutional layers and the remaining three are fully connected layers.

The output of the last fully-connected layer is fed to a 50-way softmax which produces a distribution over the 50 class labels. Our network maximizes the multinomial logistic regression objective, which is equivalent of maximizing the average across training cases of the log-probability of the correct label under the prediction distribution.

In the first layer, we use 96*11*11 convolutional kernels to conv the input images and give 96 output maps. The second layer conv the maps came from last layer with 265*5*5 convolutional kernels and give 256 output maps. The kernels of the convolutional layers are connected only to those kernel maps in the previous layer. The next two pooling layers have the same method, max pooling, with 3*3 kernels. Pooling layer can reduce the computation and maintain the stability of the structure of net at the same time.

When the feature extract layer was completed, the fully-connected layer will come. Different from the AlexNet model, our net only has two fully-connected layers with a 50-way final layer for prediction.

Actually, our net is a reduced AlexNet model.

The PReLU non-linearity is an advanced ReLU non-linearity, which is proved to be helpful for better accuracy. So we apply PReLU non-linearity to the output of every convolutional and fully-connected layer.

3.4   PReLU

The rectifier neuron Rectified Linear Unit (ReLU) [10], is one of several important points to the recent success of deep networks.

ReLU defined as:

$$f(y_i) = \max(y_i, 0) \tag{1}$$

Parametric Rectified Linear Unit (PReLU) [11]. PReLU defined as:

$$f(y_i) = \begin{cases} y_i, & \text{if } y_i > 0 \\ a_i \cdot y_i, & \text{if } y_i \le 0 \end{cases} \tag{2}$$

Here y_i is the input of the nonlinear activation f on the channel i, and a_i is a coefficient controlling the slope of the negative part. The subscript i in a_i indicates that we allow the nonlinear activation to vary on different channels.

When $a_i = 0$, it becomes ReLU.

In ReLU, all the weights less than zero will be abandoned. There must be information in the weights less than zero. So, in PReLU, we retain the weights less than zero, and regularize the function with a learnable parameter a.

Now we can describe the entire architecture of our net. Four layers with weights; the first two are convolutional layers and the remaining two are fully connected layers. And the ReLU in traditional CNN was replaced by PReLU. The architecture of our net is illustrated in.

4   Experiment and Results

Based on the ICL dataset, 25 images per species were used to train the network, and 5 images per species were used to test performance of the system (Fig. 2).

The Table. 2 shows parameters in our experiment.

There is no standard value for original learning rate. After several experiments, we found a suitable value for leaf classification based on CNN. Our original learning rate is 0.0006. After every 500 times of iteration, the rate became one tenth before. The max iteration is 2500. The learning rate is going to be 0.000000006 in the end.

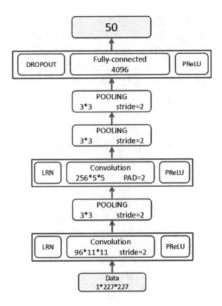

Fig. 2. Illustration of our net

Table 2. Parameters

Original learning rate	Momentum	Weight decay	Iteration	Batch-size
0.0006	0.9	0.0005	2500	64

The convolutional neural network we proposed has been trained using parameters in Table 2.

We did four experiments. The pure AlexNet is the first; and it got 86.40 % accuracy. Then we use the PReLU instead of ReLU, got 89.60 % accuracy. It seems that the PReLU bring 2.8 %accuracy for the classification net. In the third experiment, we reduced the AlexNet to make it simpler. The Accuracy was improved 7.6 %. It proves that the reduction is an effective way to enhance the performance of leaf classification. Using PReLU instead of ReLU is helpful for our classification; we got the tactic from the second experiment. So in the last experiment, we use PReLU instead of ReLU. And we got 94.8 % accuracy. Compared to the third experiment, it is not a big progress, but it is a progress after all.

We reduce the architecture and use PReLU instead of ReLU. Compared to the result of other methods, our model has a better performance. Our results, 94.8 %, show that we reached a higher precision than Pure-AlexNet. These changes help improvement on the performance of leaf classification system.

The accuracy of four methods is shown in Table 3.

Table 3. Accuracy

Method	Pure-AlexNet	PReLU-AlexNet	Reduced-AlexNet	Our net
Accuracy	86.40 %	89.60 %	94 %	**94.80 %**

5   Conclusion

This paper proposes a structure of reduced AlexNet with PReLU. Our method is better than pure AlexNet on leaf classification. For classifying leaves, it is not necessary to make the net too complex because of leaves have almost the same shapes. But their details are different, so we propose this reduction of AlexNet; it can improve about 10 % accuracy compared with pure AlexNet. And the PReLU also helps a lot, compared with pure AlexNet, the PReLU-AlexNet have a 3 % improvement. When the dataset is not big enough and objects have almost same shapes, we can use this reduced net with PReLU.

Acknowledgments. This work was supported by the grants of the National Science Foundation of China, Nos. 61520106006, 61532008, 31571364, 61303111, 61411140249, 61402334, 61472280, 61472173, 61572447, 61373098, and 61572364, China Postdoctoral Science Foundation Grant, Nos. 2014M561513 and 2015M580352.

References

1. Hinton, Geoffrey E., Salakhutdinov, Ruslan R.: Reducing the dimensionality of data with neural networks. Science **313**(5786), 504–507 (2006)
2. LeCun, Y., et al.: Backpropagation applied to handwritten zip code recognition. Neural Comput. **1**(4), 541–551 (1989)
3. Rumelhart, D.E., Hinton, G.E., Williams, R.J.: Learning representations by back-propagating errors. Nature **323**(6088), 533–536 (1986)
4. Ghodrati, A., Diba, A., Pedersoli, M., Tuytelaars, T., Van Gool, L.: DeepProposal: Hunting Objects by Cascading Deep Convolutional Layers. arXiv:1510.04445
5. Lecun, Y., Bengio, Y., Hinton, G.: Deep learning. Nature **521**(7553), 436–444 (2015)
6. Krizhevsky, A., Sutskever, I., Hinton, G.: ImageNet classification with deep convolutional neural networks. In: Advances in Neural Information Processing Systems (NIPS 2012) (2012)
7. Rashad, M.Z., El-Desouky, B.S., Khawasik, M.S.: Plants images classification based on textural features using combined classifier. Int. J. Comput. Sci. Inf. Technol. **3**(4), 93–100 (2011)
8. Sumathi, C.S., Kumar, A.V.S.: Edge and texture fusion for plant leaf classification. Int. J. Comput. Sci. Telecommun. **3**(6), 6–9 (2012)
9. He, K., Zhang, X., Ren, S., et al.: Delving deep into rectifiers: surpassing human-level performance on imagenetclassification. arXiv preprint arXiv:1502.01852 (2015)
10. Glorot, X., Bordes, A., Bengio, Y.: Deep sparse rectifier neural networks, pp. 315–323 (2011)
11. Bengio, Y.: A connectionist approach to speech recognition. Int. J. Pattern Recogn. Artif. Intell. **7**(04), 647–667 (1993)

Detecting Ventricular Fibrillation and Ventricular Tachycardia for Small Samples Based on EMD and Symbol Entropy

Yingda Wei[1,2], Qingfang Meng[1,2(✉)], Qiang Zhang[3], and Dong Wang[1]

[1] School of Information Science and Engineering,
University of Jinan, Jinan 250022, China
ise_mengqf@ujn.edu.cn
[2] Shandong Provincial Key Laboratory
of Network Based Intelligent Computing, Jinan 250022, China
[3] Institute of Jinan Semiconductor Elements Experimentation,
Jinan 250014, China

Abstract. In this paper, we proposed a new method based on Symbol Entropy and Empirical Mode Decomposition (EMD) to detect ventricular fibrillation (VF) and ventricular tachycardia (VT). Initially, we applied the EMD to decompose VF and VT signals into five sub-bands respectively. And then, we calculated the Symbol Entropy of each sub-bans as the feature to detect VT and VF. We employed the public data set to assess the proposed method. Experimental results showed that, using classification of support vector machine (SVM), the proposed method can successfully distinguish VF from VT with the classification accuracy up to 100 % based on small samples. The duration of each sample was 2 s. Moreover, the classification accuracy of the proposed method is far higher than the classification accuracy of the original signals using Symbol Entropy directly.

Keywords: Symbol entropy · EMD · VF · VT · SVM · Small samples

1 Introduction

The morbidity of cardiovascular disease has increased day by day. Heart rate of normal people is sinus rhythm, while it will become arrhythmia when there is functionality lesions on heart. Sudden cardiac death (SCD) is the most serious symptoms of arrhythmia. After couple of years the researchers indicated that in the vast majority of SCD cases was due to that the ventricular fibrillation or ventricular tachycardia had deteriorated. Once heart rate is regarded as VF, high-energy defibrillation is required for patient [1]. Nevertheless, a patient with VT demands the low-energy cardio-version instead. However, if a normal sinus rhythm or VT is misinterpreted as VF, the patient will suffer an unnecessary shock that may damage the heart. Conversely, if VF is incorrectly interpreted as VT, the result is also life-threatening [2]. Therefore, an effective detection method to distinguish VF from VT has clinical research significance.

© Springer International Publishing Switzerland 2016
D.-S. Huang et al. (Eds.): ICIC 2016, Part I, LNCS 9771, pp. 18–27, 2016.
DOI: 10.1007/978-3-319-42291-6_3

In the past, physicians distinguished VF and VT mainly depending on cardiac rhythm, while there was the overlap of two regions on the cardiac rhythm at certain degree. Therefore, mistakes may happen when distinguish VF from VT by cardiac rhythm. To avoid this situation, researchers came up with many methods about quantitative analysis of arrhythmia signal [3–7], but those methods also has limitations. At present, there are many ECG research methods based on entropy, such as Approximate Entropy analysis (ApEn) [8], Kolomogorov entropy, time-dependent entropy (TDE) [9], multi-resolution entropy (MRE) [10]. Traditional Shannon Information Entropy based on BGS statistical mechanics describe a breadth of system, it has a correlation in time and space (especially nervous system), namely, non-extensive entropy. The [11] puts forward non-breadth statistical mechanics, and defines non-breadth entropy. MRE combines with Tsallis entropy in the forms of MRET [10]. The [12] described the multi-fractal characteristics of signals. But these methods can class two signals by high accuracy base on large simples. Wang and Chen [13] proposed a method based on Symbol dynamics to research VF and VT. The study showed that a sudden drop of Symbol sequence's entropy value indicated that the patients most likely entered the sample of ventricular tachycardia and this was a crucial sample for the clinical treatment of patients. But the samples they used was also larger.

In this paper, we presented the novel method for detecting ECG based on the Symbol Entropy and EMD. The method displayed a high accuracy rate in the classification of VF and VT. Furthermore, the proposed method suited small samples.

EMD is first proposed in 1998. With defined instantaneous frequency, a finite set of band-limited signals that is termed intrinsic mode functions (IMFs) are decomposed from original signal. It's a new technique to analysis nonlinear and non-stationary signals.

2   The Method of Symbol Entropy and EMD

2.1   Data Description

The data were selected from MIT-BIH Malignant Ventricular Ectopy Database (MIT-BIH Database) and Creighton University Ventricular Tachyarrhythmia Database (CU Database). The sampling frequency of VF and VF are all 250 Hz. We extracted 100 VT samples and 100 VF samples from CU Database and MIT-BIH Database respectively. Each sample contains 500 sampling points. The sample's duration was 2 s. All the samples were normalized firstly.

2.2   Symbol Entropy

Symbol Entropy is a kind of time series analysis method developed from the Symbol dynamic theory, the chaotic time series analysis and the information theory. The method can react fundamental characteristics of signal. Main information of VF and VT are in low frequency. So Symbol Entropy suit for processing VF and VT.

Supposed the VF or VT sample is X, define $X = (x_1, x_2, \ldots, x_i, \ldots, x_N)$, and sample length was N.

Step 1: We converted into Symbol sequence $S = (s_1, s_2, \ldots, s_i, \ldots, s_N)$ according a special rule. The defined rule was as followed. If the value of larger than the mean value of, we mark as '1', else we mark as '0'. The signifying pattern was followed.

$$s(i) = \begin{cases} '0' & x(i) \leq mean(X) \\ '1' & x(i) > mean(X) \end{cases} \tag{1}$$

Here $i = 1, 2, \ldots, N$, and $x(i)$ was the value of ith point in X.

Step 2: We grouped 3 adjacent symbols as a small sequence which called "3-bit word". Supposed that S = "111100101...", then the S was coded to be 111,111,110,100,.... These were collectively called "3-bit word". A 3-bit word contained symbols of '0' and '1'. The number of different combinations is 8. We defined $S' = \{s_1', s_2' \ldots, s_i', \ldots s_M'\}$. Here is 3-bit word and M = N−3 + 1.

Step 3: We converted each 3-bit word of S' into decimal and marked D as the sequence of decimal. Therefor D = $\{d1, d2, \ldots, dM\}$. Then we measured the histogram of D which marked as H. H = $\{h0, h1, \ldots, h7,\}$. And calculated probability of hi which defined as pi = hi/M. The Symbol Entropy was defined as

$$En = -\sum_{i=0}^{7} p(h_i') \log_2 p(h_i') \tag{2}$$

2.3   Empirical Mode Decomposition

Empirical Mode Decomposition (EMD) [13–15] is adaptive processing method for signals. This method is especially suitable for the nonlinear analysis of non-stationary signal processing. The VT and VF are non-stationary signals, so we select EMD as decompose method for the VF and VF.

The EMD can decompose time series x(t) into nIMFs: IMF1(t), IMF2(t),..., IMFn (t), and a residue time series r. The x(t) can be expressed as

$$x(t) = \sum_{i=1}^{M} IMF_m(t) + r \tag{3}$$

Given a signal x(t):
Step 1: Set $r(t) = g(t) = x(t), j = 0, i = 1$.
Step 2: Get the ithIMF.

(1) Set $h_0(t) = r_{i-1}(t), j = 1$.
(2) Get local minimums and local maximums of $h_{j-1}(t)$.
(3) Generate the upper and lower envelopes $e_l(t)$ and $e_u(t)$ by connecting the minimums and the maximums with cubic spline interpolation respectively.
(4) Get the mean of $e_l(t)$ and $e_u(t)$:

$$m(t) = \frac{e_l(t) + e_u(t)}{2} \tag{4}$$

(5) the $h_j(t)$ is defined:

$$h_j(t) = h_{j-1}(t) - m_{j-1}(t) \tag{5}$$

(6) If $h_j(t)$ satisfies two basic conditions, set $IMF_i(t) = h_j(t)$;
else $j = j+1$, return Step2-(2).

Step 3: Get $r_i(t) = r_{i-1}(t) - IMF_i(t)$.

Step 4: If the sum of local minimums and local maximums larger than 2, set $i = i+1$ and return *step 2*, else stop.

2.4   The Algorithm Based on Symbol Entropy and EMD

Our proposed algorithm is defined as follows:

Step 1: Given a set X, $X = [X_1, X_2, \ldots, X_i \ldots, X_n]$, where i represents the ith sample. Defined $X_i = [x_{i1}, x_{i2}, \ldots, x_{ij}, \ldots, x_{im}]^T$, where m represents the length of X_i.

Step 2: Normalize X_i the by the followed formula $X_i' = (X_i - \bar{x})/\sigma$.

Where $\bar{X}$ and σ represent the mean and standard deviation of the th sample.

Step 3: Repeat *step 2* until all samples are normalized.

Step 4: Using EMD to get X_IMFi of X (Fig. 1).

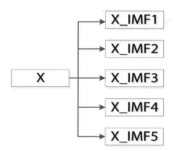

Fig. 1. IMFs of X

Step 5: Calculate the Symbol Entropy of the X_IMFi, get $En_i = [e_1, e_2, \ldots, e_n]$, where $i = 1, \ldots, 5$.

Step 6: Design classifier to calculate the classification accuracy for performance evaluation. We selected LIBSVM as the classifier, whose kernel function was RBF. Because the RBF has good performance than others in this article.

3   Experimental Results and Analysis

A sample of ECG sample from CU Database and MIT-BIH Database were plotted in Fig. 2. The IMFs of VF and VT were showed in Fig. 4.

First, we calculated the Symbol Entropy of the X directly. The result was showed in Fig. 3. The Fig. 3(a) was Symbol Entropy of 200 samples. The Fig. 3(b) were boxplot of VF and VT. From the Fig. 3(a), we can know that the original signals can't discriminate the VT and VF. And we discriminate the Symbol Entropy with SVM whose kernel function was RBF, the classification accuracy was 45.83 %.

Then we calculated the Symbol Entropy of each sample with the method described previously. The results of was presented in Fig. 5. The Fig. 5(a) was Symbol Entropy of 200 samples. The Fig. 5(b) was boxplot of VF and VT. The Fig. 5 showed that there was a good discrimination between VF and VT. Using the SVM, the result of classification accuracy was 100 %. This result was better than the former.

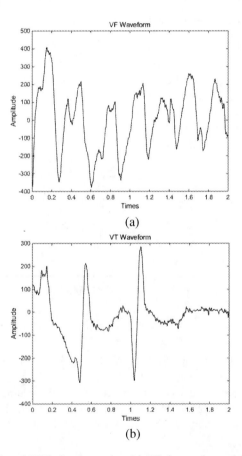

(a)

(b)

Fig. 2. Examples of ECG signal samples: (a) VF time series and (b) VT time series

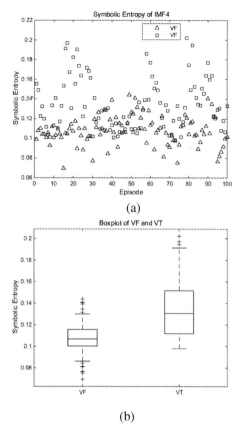

(a)

(b)

Fig. 3. (a) The Symbol Entropy of each sample.(b) The box-plot of VF and VT

We calculated five IMFs of Symbol Entropy respectively. The 3/4 of samples were used to train SVM and the rest were used to test. The test result was showed in Table 1. Table 1 showed that IMF1–5 had a higher classification accuracy than the original. From the sensitivity and specificity, we can know that VF can be regarded as VT but VT is regarded as VF hardly.

Xia et al. [16] proposed a method based on the Lempel-Ziv complexity and EMD to detect VF and VT. The sample of ECG sample also from CU Database and MIT-BIH Database. The sample length was 1000(4 s), in other word, the sample duration is larger than 2 s. The classification accuracy [16] was showed in Table 2. We can know

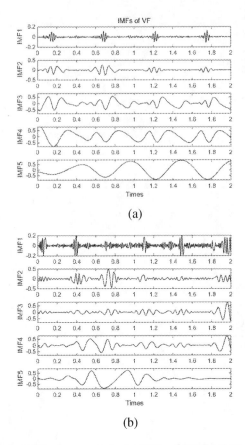

Fig. 4. IMFs of one sample: (a) IMFs of VF (b) IMFs of VT

that the classification accuracy of the proposed method in this paper was higher than [16], and the sample duration was shorter than [16].

From Tables 1 and 2, the IMF4 and IMF5 have higher Sensitivity, Specificity and Accuracy than IMF1, IMF2, and IMF3. Because IMF4 and IMF5 contains main information of signal, while IMF1–3 contain less main information.

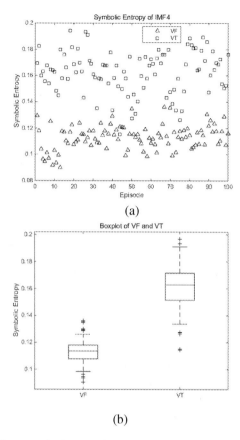

(a)

(b)

Fig. 5. (a) Symbol Entropy of 4th IMF of VF and VT.(b) The box-plot of VF and VT for 4th IMF

Table 1. Classification of EMD and Symbol Entropy with SVM.

	Sensitivity	Specificity	Accuracy
Original	94.05 %	81.9 %	45.83 %
IMF1	96.15 %	100 %	91.67 %
IMF2	99.01 %	100 %	97.92 %
IMF3	98.04 %	100 %	95.83 %
IMF4	100 %	100 %	100 %
IMF5	100 %	100 %	100 %

(Regarded the VF samples as positive samples and the VT samples as negative samples.)

Table 2. The comparison of different methods for CU and MIT-BIH data.

	Sensitivity	Specificity	Accuracy
L-Z complexity [16]	64.11 %	62.09 %	63.10 %
L-Z complexity and EMD-IMF1	87.12 %	100 %	93.56 %
L-Z complexity and EMD-IMF2	82.04 %	68.16 %	75.10 %
L-Z complexity and EMD-IMF3	77.21 %	73.18 %	75.20 %
L-Z complexity and EMD-IMF4	88.13 %	90.28 %	89.21 %
L-Z complexity and EMD-IMF5	98.15 %	96.01 %	97.08 %

4 Conclusions

An effective detection method to distinguish VF from VT has clinical research significance. In this paper, the proposed method can detect VF and VT for small samples. The timeliness of this method is better than others. The method can be improved and embedded into portable ECG monitoring wireless communication equipment to detect of VF and VT in real time. This can contribute to save the patient's life in time. There is a weakness of this paper, the proposed method should be assessed by other data-set of VF and VT.

Acknowledgments. This work was supported by the National Natural Science Foundation of China (Grant Nos. 61201428, 61302128), the Natural Science Foundation of Shandong Province, China (Grant Nos. ZR2010FQ020, ZR2013FL002), the Shandong Distinguished Middle-aged and Young Scientist Encourage and Reward Foundation, China (Grant Nos. BS2009SW003, BS2014DX015).

References

1. Othman, M.A., Safri, N.M., Ghani, I.A.: A new semantic mining approach for detecting ventricular tachycardia and ventricular fibrillation. Biomed. Sig. Process. Control **8**(2), 222–227 (2013)
2. Kong, D.R., Xie, H.B.: Use of modified sample entropy measurement to classify ventricular tachycardia and fibrillation. Measurement **44**(3), 653–662 (2011)
3. Thakor, N.V., Zhu, Y.S., Pan, K.Y.: Ventricular tachycardia and fibrillation detection by a sequential hypothesis testing algorithm. IEEE Trans. Biomed. Eng. **37**(9), 837–843 (1990)
4. Zhang, X.S., Zhu, Y.S., Thakor, N.V., Wang, Z.Z.: Detecting ventricular tachycardia and fibrillation by complexity measure. IEEE Trans. Biomed. Eng. **46**(5), 548–555 (1999)
5. Zhang, H.X., Zhu, Y.S., Xu, Y.H.: Complexity information based analysis of pathological ECG rhythm for ventricular tachycardia and ventricular fibrillation. Int. J. Bifurc. Chaos **12** (10), 2293–2303 (2002)
6. Zhang, H.X., Zhu, Y.S.: Qualitative chaos analysis for ventricular tachycardia and fibrillation based on symbol complexity. Med. Eng. Phys. **23**(8), 523–528 (2001)
7. Owis, M.I., Abou-Zied, A.H., Youssef, A.B.M.: Study of features based on nonlinear dynamical modeling in ECG arrhythmia detection and classification. IEEE Trans. Biomed. Eng. **49**(7), 733–736 (2002)

8. Fleisher, L.A., Pincus, S.M.S., Rosenbaum, H.: Approximate entropy of heart rate as a correlate of postoperative ventricular dysfunction. Anesthesiology **78**(4), 683–692 (1993)
9. Tong, S., Bezerianos, A., Paul, J., Thakor, N.: Nonextensive entropy measure of EEG following brain injury from cardiac arrest. Phys. A **305**(3), 619–628 (2002)
10. Gamero, L.G., Plastino, A., Torres, M.E.: wavelet analysis and nonlinear dynamics in a non extensive setting. Phys. A **246**(3), 487–509 (1997)
11. Tsallis, C.: Possible generalization of Boltzmann-Gibbs statistics. J. Stat. Phys. **52**, 479–487 (1988)
12. Sherman, L.D., Callaway, C.W., Menegazzi, J.J.: Ventricular fibrillation exhibits dynamical properties and self-similarity. Resuscitation **47**(2), 163–173 (2000)
13. Wang, J., Chen, J.: Symbol dynamics of ventricular tachycardia and ventricular fibrillation. Phys. A **389**(10), 2096–2100 (2010)
14. Owis, M.I., Abou-Zied, A.H., Youssef, A.B.M., Kadah, Y.M.: Study of features based on nonlinear dynamical modeling in ECG arrhythmia detection and classification. IEEE Trans. Biomed. Eng. **49**, 733–736 (2002)
15. Pachori, R.B., et al.: Analysis of normal and epileptic seizure EEG signals using empirical mode decomposition. Comput. Methods Programs Biomed. **104**, 373–381 (2011)
16. Xia, D., Meng, Q., Chen, Y., Zhang, Z.: Classification of ventricular tachycardia and fibrillation based on the Lempel-Ziv complexity and EMD. In: Huang, D.-S., Han, K., Gromiha, M. (eds.) ICIC 2014. LNCS, vol. 8590, pp. 322–329. Springer, Heidelberg (2014)

Coronary Heart Disease Recognition Based on Dynamic Pulse Rate Variability

Aihua Zhang[1(✉)], Boxuan Wei[1], and Yongxin Chou[2]

[1] College of Electrical and Information Engineering,
Lanzhou University of Technology, Lanzhou 730050, China
zhangaihua@lut.cn
[2] School of Electrical and Automatic Enginnering,
Changshu Institute of Technology, Changshu 215500, China
lutchouyx@163.com

Abstract. Objective: In order to improve the accuracy and real-time of coronary heart disease (CHD) recognition, we propose a new method to analyze the pulse signal with the idea of sliding window iterative. **Methods:** Firstly, the principle of the feature extraction method(including time domain method, Poincare plot and information entropy) that combined with the idea of sliding window iterative is described. Secondly, The continuous blood pressure signals from the website database PhysioNet are chosen to generate the dynamic pulse rate variability (DPRV) signal as experimental data, and the linear and nonlinear feature is selected for classifying the healthy people and patients with CHD. Finally, the running time and accuracy of the method in this paper are comparaed with other methods. **Result:** The pulse signal can be online analyzed by this method. The average recognizing accuracy is 97.6 %. **Conclusion:** This methods is entirely feasible. Compared with existing methods, its accuracy and real-time is higher.

Keywords: Pulse signal · Dynamic pulse rate variability (DPRV) · Coronary heart disease recognition

1 Introduction

Coronary heart disease (CHD) is one of common cardiovascular diseases. Its morbidity and mortality is increasing year by year [1]. A large number of research shows that CHD is preventable. In the early stage of CHD, some physiological signals of human body may mutate. If these mutational signals could be found, we will predict the occurrence of CHD and save precious time for treatment of patients.

At present, there are many methods of CHD recognition, either by heart sounds, or heart rate variability (HRV), or EST-ECG and so on [2–7]. Among these methods, the accuracy of method based on HRV is highest [2]. For example, Lee et al. extracted the linear and nonlinear parameters from HRV signal and used these parameters to classify the healthy people and patients with CHD based on Support vector machine (SVM). The accuracy is 90 % [3]. Dua et al. obtained the features of HRV signal by principal component analysis (PCA) and used these features classify the healthy people and

© Springer International Publishing Switzerland 2016
D.-S. Huang et al. (Eds.): ICIC 2016, Part I, LNCS 9771, pp. 28–38, 2016.
DOI: 10.1007/978-3-319-42291-6_4

patients with CHD based on multi-layer perceptron (MLP), the accuracy is 89.5 % [4]. However, HRV signal is derived from ECG signal. And multiple electrode attachments and cable connections are not convenient for portable medical instrument during ECG recording. Pulse rate variability (PRV) signal obtained from the pulse signal can reflect the small changes of heart beat cycle. It has been shown that the PRV signal can substitute HRV signal to reflect the characteristics of heart beat [8]. Compared to HRV signal, PRV signal is easier to record and apply in portable medical instrument. Thus, PRV signal can replace HRV signal to recognize the patient with CHD. Dynamic pulse rate variability (DPRV) signal is one kind of PRV signal that extracted from the dynamic pulse signal. Compared to the PRV signal extracted from off-line pulse signal, its real-time is higher [9]. Therefore, this paper will extract DPRV signal to recognize CHD and improve the real-time and measurement accuracy. At last, we expect to achieve prediction, early diagnosis and online custody of CHD.

2 Materials and Methods

2.1 Data

The data in this paper come from website database PhysioNet. Among these, the pulse signals of healthy people come from PhysioNet/fantasia database [10] and the pulse signals of patient with CHD come from PhysioNet/mghdb database [11].

PhysioNet/fantasia database contains 40 date of healthy people. During the data recording, the collected people stay in a state of rest, and keep awake by film. The signal sampling frequency is 250 Hz and the sampling time is 66 min. The data in PhysioNet/mghdb database extracts from Massachusetts General Hospital. The signal sampling frequency is 360 Hz.

Continuous blood pressure is a form of the pulse signal. So, in this paper, the continuous blood pressure signal can replace the pulse signal to generate the DPRV signal. Because the research object is mainly aimed at the peak interval of pulse signal, the influence that caused by the difference of sensor and sampling environment of two kinds of databases can be ignored. The sampling frequency of pulse signal of patients with CHD is reduced from 360 Hz to 250 Hz by down sampling to be equal to the sampling frequency of pulse signal of healthy people. 20 patients with CHD and no smoking history and 20 healthy people were selected as the experimental data. Each continuous blood pressure signal is divided into 100 successive segments, and the length of segment is 30 ms.

2.2 DPRV Signal Extraction

The continuous blood pressure signal in actual measurement is quasi periodic signal. By using discrete Fourier transform (DFT) transform, the fundamental wave can be obtained from pulse signal. In one period, the period and peak of fundamental wave is the same as the pulse signal. The fundamental wave is a narrow band signal without noise and its maximum value (i.e., peak value) is easy to record. So, the fundamental wave of pulse signal can replace the original signal to generate the DPRV signal. In this paper, the DPRV signal is extracted by the method in literature [12].

2.3 Feature Extraction

Considering the characteristics of DPRV signal is nonlinear, non-stationary and real-time. Time domain analysis, Poincare plot analysis and information entropy analysis are used to obtain the features of DPRV signal. The idea of sliding window iteration is combined with these methods for improving the real-time of algorithm.

Time Domain Analysis. Using the time domain method to analyze the DPRV signal is simple and easy. Dynamic mean (DMEAN) can reflect the overall average level of peak interval. Dynamic standard deviation (DSD) and dynamic difference standard deviation (DSDSD) can reflect the overall change of DPRV signal. Dynamic root mean square of sussessive differences (DRMSSD) can reflect the fast changing composition of DPRV signal. Dynamic percentage of adjacent RR interval difference >50 ms (DpNN50) can reflect the activity of the heart pneumogastric nerve. So above 5 kinds of features were extracted for CHD recognition. The iterative process of DNEAN is as follow.

$$DMEAN(i) = \begin{cases} \frac{S_1(i)}{i} & i \leq N_w \\ \frac{S_1(i)}{N_w} & i > N_w \end{cases} \tag{1}$$

Where, the N_w is sliding window width, $S_1(i)$ means the mean of DPRV signal.

$$S_1(i) = \begin{cases} S_1(i-1) + PP(i) & 2 \leq i \leq N_w \\ S_1(i-1) + PP(i) - PP(i-N_w) & i > N_w \end{cases} \tag{2}$$

Where, the $PP(i)$ means DPRV signal, $S_1(1) = PP(1)$.

In the same way, DSD, DSDSD, DRMSSD and DpNN50 can be obtained.

Poincare Plot Analysis. The Poincare plot is a method to represent the nonlinear variation law of in the form of graph. It has two kinds of drawing modes.

The one mode uses the adjacent PP interval as the x-and y-values to draw the scatter plot, and get its ellipse fitting pattern. The relationship between adjacent PP intb ervals can be reflected by it. Its quantitative indicators are the length of dynamic fitting ellipse major axis (recorded as DSD1), the length of dynamic fitting ellipse short axis (recorded as DSD2), The area of dynamic fitting ellipse (recorded as DA), Dynamic vectors length index (DVLI) and dynamic Porta index (DPI), dynamic Guzik index (DGI), dynamic Ehler index (DEI) that reflected the asymmetry of the pulse rate. The iterative process of DSD1 is as follow

$$DSD1(i) = \begin{cases} \sqrt{\dfrac{S_5(i)}{2(i-1)}} & 2 \leq i \leq N_w \\[3mm] \sqrt{\dfrac{S_5(i)}{2(N_w-1)}} & i > N_w \end{cases} \tag{3}$$

Where, $S_5(i)$ means the length of fitting ellipse major axis, $S_5(1) = 0$.

$$S_5(i) = \begin{cases} S_5(i-1) + (PP(i) - PP(i-1))^2 & 2 \leq i \leq N_w \\[2mm] S_5(i-1) + (PP(i) - PP(i-1))^2 - (PP(i-N_w+1) - PP(i-N_w))^2 & i > N_w \end{cases} \tag{4}$$

In the same way, DSD2, DA, DVLI, DPI, DGI, DEI can be obtained.

The other mode uses the adjacent data of the first order difference of the PP interval as the x-and y-values to draw the scatter plot and divided it into four quadrant. Its quantitative indicators are the positive feedback index (DR_{PF}), the negative feedback index (DR_{NF}) and the total feedback index (DR_{TF}) based on the relationship between the number of scattered points in each quadrant and the total number of points.

$$DR_{PF} = \frac{(DN_I + DN_{III})}{DN} \tag{5}$$

$$DR_{NF} = \frac{(DN_{II} + DN_{IV})}{DN} \tag{6}$$

$$DR_{TF} = \frac{DR_{PF}}{DR_{NF}} \tag{7}$$

Where, the first quadrant of the scattered points DN_I is:

$$DN_I(i) = \begin{cases} DN_I(i-1) + 1 & 2 \leq i, diffPP(i) > 0 \text{ and } diffPP(i-1) > 0 \\[2mm] DN_I(i-1) - 1 & i > N_w, diffPP(i-N_w+1) < 0 \text{ and } diffPP(i-N_w) < 0 \end{cases} \tag{8}$$

In the same way, DN_{II}, DN_{III}, DN_{IV} can be obtained.

Information Entropy. The information entropy is an effective tool to reflect the signal complexity. The larger entropy is, the more complex signal is. So, the basic scale entropy and the symbol sequence entropy, what are used to analyze short time signal, are selected for extracting the feature of DPRV signals [13].

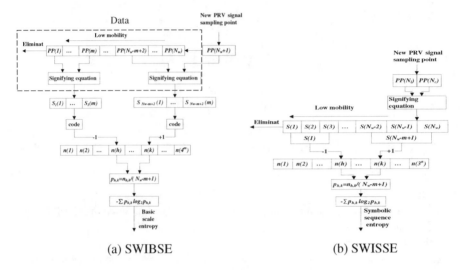

(a) SWIBSE   (b) SWISSE

Fig. 1. The process of information entropy

The basic scale entropy analysis, which is combined with the idea of sliding window iteration, is called the sliding window iteration basic scale entropy (SWIBSE) analysis. Its principle is shown as Fig. 1(a):

Firstly, a series of vectors is constructed according to time series. Secondly, the vectors are symbolized and classified based on the basic scale. The symbol of $X(1)$ is:

$$S_1(j) = \begin{cases} 0 & \mu_1 < PP(j) \leq \mu_1 + \alpha \times BS_1 \\ 1 & PP(j) > \mu_1 + \alpha \times BS_1 \\ 2 & \mu_1 - \alpha \times BS_1 < PP(j) \leq \mu_1 \\ 3 & PP(j) \leq \mu_1 - \alpha \times BS_1 \end{cases} \qquad (9)$$

Finally, the probability of each case is counted and sliding window iterative basic scale entropy (SWIBSE) is calculated.

The symbolic sequence entropy analysis method, which is combined with the idea of sliding window iteration, is called the sliding window iteration symbolic sequence entropy (SWISSE) analysis method. Its principle is similar to the SWIBSE. The difference of two methods is that the SWISSE ignores the change of amplitude of DPRV signal but uses the direction change to reflect the complexity of the DPRV signal. Its principle is shown as Fig. 1(b).

Firstly, the DPRV signal is symbolized. Here, three symbols are employed to represent the variation directions of pulse beat intervals:

$$s(i) = \begin{cases} 0 & PP(i+1) < PP(i) \\ 1 & PP(i+1) = PP(i) \qquad i = 1, 2, \ldots, N-1 \\ 2 & PP(i+1) > PP(i) \end{cases} \qquad (10)$$

Secondly, the buffer that length is N_w is opened to store the symbolic results and vector symbol sequence is constructed. Thirdly, the probability of each pattern is calculated.

In the symbol sequence vector, $s(1)$ and $s(N_w)$ have be changed. Only the probability of these two modes needs to update, recorded as $n(h)$ and $n(k)$, abbreviated as $n_{h,k}$. When each sliding window update data, $n(h) = n(h) - 1, n(k) = n(k) - 1$, The probability of their corresponding to is $p_{h,k}$. Finally, the symbol sequence entropy is calculated.

2.4 Probabilistic Neural Network

In 1989, Dr. D. F. Specht of the United States proposed a probabilistic neural network (PNN) based on the Bayesian rule and the probability density function estimation method. It is a feed-forward neural network. In essence, it is a kind of classifier, which can be used to make the Bayesian decision based on the unbiased estimation of probability density function [14]. PNN, which is on the basis of statistical principles, is need not to back propagate error to adjust the weights and biases. Meanwhile, only using the training samples can assign all the adjustable parameters. These advantages can speed up the training process, so the PNN is employed in this paper.

The PNN consists of four parts, respectively: input layer, pattern layer, summation layer and competition layer. In this paper, the input layer just receives the feature vectors of pulse signal. The pattern and summation layer has a number of neurons for classifying these input signals by multi-dimensional kernels. The output layer output the results the people is healthy or have CHD. Here, we choose the multi-dimensional Gaussian function as the multi-dimensional kernels:

$$\emptyset_{kl}(x) = \frac{1}{(2\pi)^{\frac{d}{2}}\sigma^d} exp[-\frac{(X - X_{kl})^T(X - X_{kl})}{2\sigma^2}] \tag{11}$$

Where the X_{kl} is the neuron vector, $k = 0, 1, \sigma$ is the smoothing parameter, and d is the dimension of the input vector X.

3 Experiments and Results

3.1 System Processes

When the pulse signal is input the system, its DPRV signal begin to be extracted by means of the sliding window iterative DFT. After DPRV signal is generated form one section of pulse signal width of one fixed window, the window slides, the next section of DPRV signal width begin to be generated. Meanwhile, the features of the last output DPRV signal are extracted. While the features of the former DPRV signal are output, the features of the later DPRV signal begin to be extracted. The output features are composed to feature vectors after they are preprocessed. Then the feature vectors are input PNN. Finally, input pulse signal is judged whether belong to healthy people or patient with CHD.

3.2 Characteristic Selection

The features of DPRV signal are tested by t-test and fisher discriminant analysis (FDA) (sliding window: 500 points, the length of symbol sequence vector: m = 3), and the results are shown in Tables 1 and 2:

Table 1. t-test results of DPRV features

Time domain	DMEAN	DSD	DRMSSD	DpNN50	DSDSD
	0	1.267e-60	1.305e-160	2.009e-71	1.303e-160
Poincare plot	DSD1	DSD2	DA	DVLI	DPI
	1.292e-149	4.678e-16	3.339e-91	2.088e-22	1.595e-123
	DEI	DGI	DR_{PF}	DR_{NF}	DR_{TF}
	4.449e-21	7.501e-65	8.612e-251	1.261e-44	2.462e-175
Information entropy	SWIBSE	SWISSE			
	0	0.018			

Table 2. FDA results of DPRV feature

Time domain	DMEAN	DSD	DRMSSD	DpNN50	DSDSD
	2.513	0.140	0.400	0.167	0.400
Poincare plot	DSD1	DSD2	DA	DVLI	DPI
	0.370	0.033	0.216	0.048	0.300
	DEI	DGI	DR_{PF}	DR_{NF}	DR_{TF}
	0.045	0.150	0.662	0.101	0.441
Information entropy	SWIBSE	SWISSE			
	4.141	0.001			

The results of t-test in Table 1 showed that the features of DPRV signal of healthy people and patients with CHD have significant differences. The results of Fisher discriminant rate in Table 2 showed that the features of DPRV signal of healthy people and patients with CHD were significantly different in DMEAN, DRMSSD, DSDSD, DSD1, DPI, DRPF, DRTF, and SWIBSE. There exists linear correlation between DSD1 and DRMSSD. So, DMEAN, DSDSD, DSD1, DPI, DRPF, DRTF, SWIBSE are selected to be constructed feature vectors to recognize the patients with CHD.

3.3 Coronary Heart Disease Recognition

The collected 7 kinds of features are normalized to construct feature vectors, and then the obtained feature vectors are input into PNN to classify healthy people and patients with CHD. Each feature contains 4000 samples, 3500 samples were selected as the training set, and the other 500 samples were selected as test set. In order to further verify the rationality of the feature selection, the selected features are composed to different feature vectors to input PNN. The feature vectors are shown as Table 3:

Table 3. The feature vectors consist of features

	1	2	3	4	5	6	7	8	9	10	11	12	13	...	28
DMEAN	*	*	*	*	*	*	*							...	
DSDSD		*	*	*	*	*	*	*	*	*	*	*	*		
DSD1			*	*	*	*	*		*	*	*	*	*		
DPI				*	*	*	*			*	*	*	*		
DR$_{PF}$					*	*	*				*	*	*		
DR$_{TF}$						*	*					*	*		
SWIBSE							*						*		*

Note:'*' indicates that the feature is selected.

When the above 28 feature vectors are input into PNN, the accuracy rate of one time experiment is shown as Fig. 2.

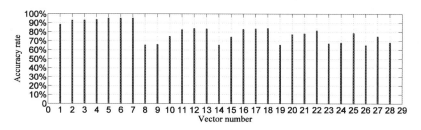

Fig. 2. The accuracy of CHD recognition using different feature vectors

It can be seen from the above that the highest accuracy rate is feature vector with number 7, reaching 97 %. So it can be concluded that using the feature vectors consisting of DMEAN, DSDSD, DSD1, DPI, DR$_{PF}$, DR$_{TF}$, and SWIBSE to recognize patients with CHD is feasible and has high accuracy.

4 Discussion and Conclusion

Because the selected training set and the test set are random when the simulation program is running at each time, the samples are different at each time. After 100 times experiments, the accuracy rate of the method in this paper is 95.4 %–98.7 %, and the average accuracy rate is 97.6 %.

The linear and nonlinear method of off-line PRV signal extraction and the method combined with the ideal of sliding window iterative are used to extract the feature of DPRV signal. The run time of two kinds of methods are shown in Table 4:

Table 4. The run time of two methods

Feature name	MEAN	SDSD	SD1	PI	R_{PF}	R_{TF}	BSE
Original method (ms)	72.8	191.2	23.8	46.0	41.0	76.2	51067.2
Method in this paper (ms)	1.3	2.6	1.6	1.3	1.3	1.9	506.6

By comparison, under equal conditions, the run time of the method combined with the ideal of sliding window iterative is much faster than the original method. Meanwhile, compared to the original method, the method in this paper can achieve parameter extraction that the precision is single data point. And the results of parameter extraction can adjust with sliding window width and achieve effect of analysis of short-time or long-time PRV signal. So the method proposed in this paper can be used to real time analysis of pulse signal.

PNN in this paper and three typical machine learning algorithms, Back-Propagation neural network (BP), generalized regression neural network (GRNN) and support vector machine (SVM), are used to recognize the patients with CHD. The accuracy rate of recognition by using of different feature vectors in Table 3 is shown as Fig. 3.

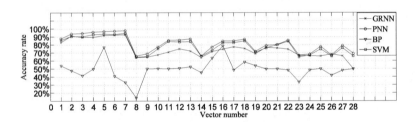

Fig. 3. The accuracy of recognition of four methods

The feature vector with number 7 is selected to make the four methods be run 100 times respectively. The accuracy rate of recognition is shown in Table 5. By comparison, PNN has the highest accuracy in four methods.

Table 5. The accuracy of four methods to run 100 times

Method name	Accuracy		
	Maximum	Minimum	Average
PNN	98.7 %	95.4 %	97.6 %
BP	80.2 %	27.4 %	47.2 %
GRNN	93.6 %	89.6 %	91.2 %
SVM	96.2 %	91.5 %	94.3 %

To sum up, based the method in this paper, it is completely feasible to recognize CHD by pulse signal, and the accuracy rate and real-time are higher.

In this paper, Firstly, based on sliding-window DFT, DPRV is extracted from pulse sign. Secondly, the methods of time domain, Poincare polt and information entropy are employed to extract the feature of DPRV. Finally, these features are input into the PNN to recognize the patients with CHD. This method obtained satisfactory simulation results. Experiment results showed that the method is feasible, high accuracy rate and real-time. However, this paper is an exploratory study, and the method is needed to further verify by a large number of clinical data. Therefore, in future research, we plan to take a large number of clinical pulse signals, and to achieve deeper data mining in order to get the better pulse signal features that reflect the CHD. And we hope to achieve prediction, early diagnosis and online custody of CHD at last.

Acknowledgments. This work was supported by the National Natural Science Foundation (grant 81360229) of China, the National Key Laboratory Open Project Foundation (grant 201407347) of Pattern Recognition in China and the Gansu Province Basic Research Innovation Group Project (1506RJIA031).

References

1. Yu, E., He, D., Su, Y., et al.: Feasibility analysis for pulse rate variability to replace heart rate variability of the healthy subjects. In: 2013 IEEE International Conference on Robotics and Biomimetics (ROBIO), pp. 1065–1070 (2013)
2. Kim, W.-S., Jin, S.-H., Park, Y.K., Choi, H.-M.: A study on development of multi-parametric measure of heart rate variability diagnosing cardiovascular disease. In: Magjarevic, R., Nagel, J.H. (eds.) World Congress on Medical Physics and Biomedical Engineering 2006. IFMBE Proceedings, vol. 14, pp. 3480–3483. Springer, Berlin (2007)
3. Lee, H.G., Noh, K.Y., Ryu, K.H.: Mining biosignal data: coronary artery disease diagnosis using linear and nonlinear features of HRV. In: Washio, T., et al. (eds.) PAKDD 2007. LNCS (LNAI), vol. 4819, pp. 218–228. Springer, Heidelberg (2007)
4. Dua, S., Du, X., Sree, S.V., et al.: Novel classification of coronary artery disease using heart rate variability analysis. J. Mech. Med. Biol. **12**(4), 1240017 (2012)
5. Karimi, M., Amirfattahi, R., Sadri, S., et al.: Noninvasive detection and classification of coronary artery occlusions using wavelet analysis of heart sounds with neural networks. In: 3rd IEE International Seminar on Medical Applications of Signal Processing, pp. 117–120 (2005)
6. Babaoglu, İ., Findik, O., Ülker, E.: A comparison of feature selection models utilizing binary particle swarm optimization and genetic algorithm in determining coronary artery disease using support vector machine. Expert Syst. Appl. **37**(4), 3177–3183 (2010)
7. Babaoğlu, I., Fındık, O., Bayrak, M.: Effects of principle component analysis on assessment of coronary artery diseases using support vector machine. Int. J. Expert Syst. Appl. **37**(3), 2182–2185 (2010)
8. Yu, E., He, D., Su, Y., et al.: Feasibility analysis for pulse rate variability to replace heart rate variability of the healthy subjects. In: 2013 IEEE International Conference on Robotics and Biomimetics (ROBIO), pp. 1065–1070 (2013)

9. Yongxin, C., Zhang, A., Jiqing, O.U., et al.: Dynamic pulse signal processing and analyzing in mobile system. Chin. J. Med. Instrum. **05**, 313–317 (2015)
10. The Physionet/Fantasia database. http://www.physionet.org/physiobank/database/fantasia
11. The Physionet/The MGH/MF waveform database. http://www.physionet.org/physiobank/database/mghdb/
12. Chou, Y., Zhang, A., Yang, X.: Dynamic pulse rate variability extraction method based on improved sliding window iterative DFT. Chin. J. Sci. Instrum. **36**(4), 812–821 (2015)
13. Bian, C.H., Ma, Q.L., Si, J.F., et al.: Entropy analysis method of short time heart rate variability symbol sequence. Chin. Sci. Bull. **03**, 340–344 (2009)
14. Wang, J.S., Chiang, W.C., Hsu, Y.L., Yang, Y.T.: ECG arrhythmia classification using a probabilistic neural network with a feature reduction method. Neuro Comput. **116**, 38–45 (2013)

Multi-dictionary Based Collaborative Representation for 3D Biometrics

Anqi Yang[1], Lin Zhang[1,2(✉)], Lida Li[1], and Hongyu Li[1]

[1] School of Software Engineering, Tongji University, Shanghai, China
{1353005, cslinzhang, 1336300, hyli}@tongji.edu.cn
[2] Shenzhen Institute of Future Media Technology, Shenzhen, China

Abstract. 3D palmprint and 3D ear based systems, two representative 3D biometric systems, have recently led to a proliferation of studies. Previous works mainly concentrated on solving one-to-one verification problems, but they cannot deal with the one-to-many identification problems quite well. Quite recently, collaborative representation (CR) framework has been exploited to solve such identification problems. The original CR based method used the whole range data for classification. However, apart from the discriminative information, the whole range data also inevitably contains confusing noises, which adversely affect the classification result. To solve this problem, we propose a multi-dictionary based collaborative representation method to separate the discriminative information from the noises. Specifically, we divide the testing image into several blocks, compute reconstruction residuals for each block using CR framework, and finally fuse the residuals to predict the class label. Experiments on benchmark datasets demonstrate that our method greatly outperforms previous one-to-many identification methods both in classification accuracy and computational complexity.

Keywords: 3D palmprint · 3D ear · Multi-dictionary · Collaborative representation

1 Introduction

Recent developments in security applications, such as access control, aviation security and e-banking, have heightened the need for recognizing the identity of a person with high confidence. As an effective solution to this problem, biometrics based approaches have received full attention. Compared with 2D ones, 3D data based systems have recently drawn increasing interest due to their robustness to illumination variations, non-intrusiveness of acquisition devices and high identification accuracy. Among the family of 3D biometric identifiers [1–3], 3D palmprint and 3D ear have proven to have the merits of high distinctiveness, robustness and user friendliness [4, 5]. Moreover, as they share common identification process (3D range data acquisition, ROI detection and extraction, feature extraction, and classification), we can easily render them under the same matching framework. In this paper, we assume that the ROIs of the range data is given and feature descriptors are available. We only focus on exploiting a universal classification method effective and efficient for both 3D palmprint and 3D ear using collaborative representation framework.

© Springer International Publishing Switzerland 2016
D.-S. Huang et al. (Eds.): ICIC 2016, Part I, LNCS 9771, pp. 39–48, 2016.
DOI: 10.1007/978-3-319-42291-6_5

1.1 Related Works

In the past decade, researchers have made tremendous efforts in developing automatic identification methods based on 3D palmprint. And 3D ear, though a relatively new member of the family of 3D biometric identifiers, has also gained significant attention in its matching schemes recently. And certain common classification methods have been applied to both of these two identifiers. For 3D palmprint matching, different schemes have been proposed. In [6], mean curvature images (MCI), Gaussian curvature images (GCI), and surface type (ST) maps were computed and for matching the authors defined two metrics, which are similar as Hamming distance [7]. In [8], Li et al. performed alignment refinement to the feature maps by using ICP (iterative closest point) [9], which has extremely heavy computational burdens. For 3D ear matching, ICP [9] and its variants are adopted as classification methods. In [10], Chen and Bhanu used the ICP algorithm to iteratively refine the transformation to bring model ears and the test ear into the best alignment. Then the root mean squares (RMS) distance is computed as the matching error criterion. In later works [11–15], ICP was also employed and modified for 3D ear matching.

The abovementioned classification methods, however, share a common paradigm: when matching two 3D palmprints or ears, two corresponding feature maps will be extracted from the range data and their matching score will be computed by using the pre-defined metric. These methods are quite suitable for one-to-one verification problem, but when facing large-scale one-to-many identification problem, they would not be computationally efficient. To solve such one-to-many identification problem, in our previous work [4, 5], collaborative representation based classification (CRC) was introduced into 3D palmprint and 3D ear matching. We extracted one feature vector from each training sample and concatenated them into one gallery dictionary. And the testing sample was also computed into a feature vector. We then represented the testing feature vector as a linear combination of a small number of vectors in the gallery dictionary and labeled the testing sample as the class that has the minimum residuals with respect to the dictionary. This method has been proved to be highly efficient and effective for large-scale identification problems based on 3D palmprint and 3D ear [4, 5].

1.2 Our Motivation and Contribution

However, different blocks of a test image contribute differently to the final identity, and consequently, simply using the whole range data to construct one dictionary ignores these differences, which may result in unsatisfying classification results. For example, some blocks contain densely-distributed biometric features that are highly discriminative and these blocks obviously add to the similarity between the testing sample and the samples in the ground-of-truth class. In contrast, there may exist some other blocks which contain useless noises that do not have large inter-class variance or even worse, resemble the features of false classes. Evidently, such blocks could not differentiate the true class from false classes and even worse, may adversely affect the classification result.

The objective of this work is to explore a novel encoding method that can not only fully utilize the discriminative information for matching but keep the computational complexity at relatively low level at the same time. We propose a multi-dictionary based collaborative representation method. Specifically speaking, we partition each training sample into the same overlapping blocks. By concatenating feature vectors extracted from blocks of the same location in each training sample, multiple dictionaries are constructed and the number of dictionaries equals the number of blocks in one training sample. Then the testing sample is divided into blocks in the same way. Each block of the testing sample is coded over corresponding dictionary and the residual is computed. We finally identify the testing sample to the class with the minimum residual summed by all the sub block residuals.

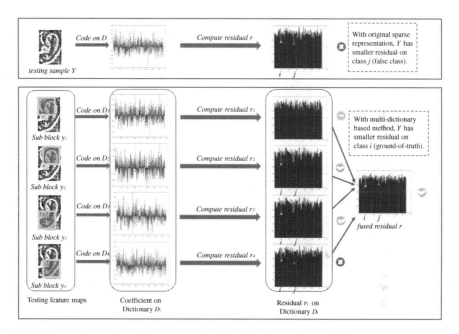

Fig. 1. Illustration of a 3D ear sample identification using collaborative representation (CR) classification in the upper panel and multi-dictionary based collaborative representation (MDCR) classification in the lower panel. (Color figure online)

The main idea of our work is illustrated in Fig. 1, which shows why multi-dictionary based collaborative representation (MDCR) classification outperforms the original collaborative representation based (CR) classification. Figure 1 shows one testing 3D ear sample. The diagrams for reconstruction coefficients are shown in the middle and the residuals are shown on the right. There exist class i and class j, which are the ground-of-truth class and a false class, respectively. In both classes, discriminative features are concentrated in certain regions; however, most regions of the two classes share greatly similar biometric features and some minor regions contain background noises, which makes the two classes difficult to be differentiated. If we use the whole feature map of the testing sample for classification, discriminative

information will be greatly interfered by non-discriminative and misleading information. As is shown in the upper block, the original collaborative representation based
classification classifies the testing sample into the false class j, which is quite
unpleasing. Therefore, we need a block-division strategy to separate an image into
different regions so that discriminative features can be isolated from interferential
information. Thus the adverse impact of non-discriminative information is constraint to
a minor block and the discriminative features are utilized to the most. As shown in the
lower panel, the feature map of the testing sample is partitioned into several blocks and
collaborative representation based classification is conducted over each block. It needs
to be noted that the misleading information affects only the result of the last block and
the right conclusion can still be reached. In this way, discriminative information is fully
exploited and makes the thoroughly contribution to the classification.

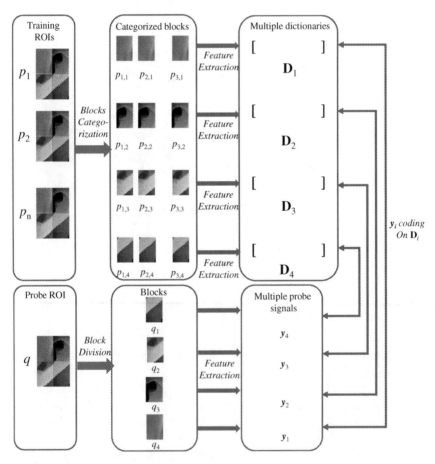

Fig. 2. Illustration for the proposed multi-dictionary based collaborative representation
(Color figure online)

2   MDCR for Classification

2.1   Multiple Dictionaries Construction

Multiple dictionaries are constructed as follows: We first partition each training sample into overlapping blocks in the same way and then extract LHST features from each block. We then combine all the feature vectors extracted from the blocks of the same location in all training samples together and construct several dictionaries. Suppose each training sample is split into C blocks and thus we can obtain C dictionaries.

$$\mathbf{D} = (\mathbf{D}_1, \mathbf{D}_2, \ldots, \mathbf{D}_C) \tag{1}$$

Suppose there are n training samples in total, thus we will have n feature vectors in each dictionary. We denote i-th dictionary as

$$\mathbf{D}_i = (\boldsymbol{d}_{i,1}, \boldsymbol{d}_{i,2}, \ldots, \boldsymbol{d}_{i,n}) \tag{2}$$

where each feature vector $\boldsymbol{d}_{i,j}$ (j = 1,2, ... ,n) is an m-dimensional vector (m is the dimension of LHST feature) and thus each dictionary is represented by an $m \times n$ matrix. Note that n is very large, which makes D_i an overcomplete description space of the n classes. Therefore, any feature vector extracted from the n classes can be linearly represented in terms of the dictionary D_i. According to the theory of compressed sensing (CS), a sparse solution is possible for an overcomplete dictionary [16]; therefore, we can express any feature vector from a testing image by a sparse linear combination of the dictionary D_i.

2.2   Multitask CR Based Classification

Given a testing 3D palmprint or 3D ear image, we partition it into C blocks and extract LHST features from each block,

$$\mathbf{Y} = (\boldsymbol{y}_1, \boldsymbol{y}_2, \ldots, \boldsymbol{y}_C) \tag{3}$$

Then, the feature vector $\boldsymbol{y}_i$ for *i*-th block of the testing image can be coded as a linear combination of all the feature vectors in dictionary $\mathbf{D}_i$, which can be expressed as follows,

$$\boldsymbol{y}_i = \mathbf{D}_i \boldsymbol{x}_i \tag{4}$$

In [17], Zhang *et al.* proposed a collaborative representation model using l_2-norm as regularization term

$$\hat{\boldsymbol{x}}_i = \arg \min_{\boldsymbol{x}_i} \left\{ \|\boldsymbol{y}_i - \mathbf{D}_i \boldsymbol{x}_i\|_2^2 + \lambda \|\boldsymbol{x}_i\|_2^2 \right\} \tag{5}$$

It can be easily verified that Eq. (5) has a close-form solution as,

$$x_i = \left(\mathbf{D}_i^{\mathrm{T}}\mathbf{D}_i + \lambda\mathbf{I}\right)^{-1}\mathbf{D}_i^{\mathrm{T}}\mathbf{y}_i \tag{6}$$

Let $\mathbf{P}_i = \left(\mathbf{D}_i^{\mathrm{T}}\mathbf{D}_i + \lambda\mathbf{I}\right)^{-1}\mathbf{D}_i^{\mathrm{T}}$. Clearly, $\mathbf{P}_i$ is independent of $\mathbf{y}_i$ and can be pre-computed totally based on the dictionary $\mathbf{D}_i$.

Inspired by Wright *et al.* [18], we classify the testing sample into class t, which has the minimum average residual with respect to all the C sub dictionaries,

$$\text{identity}(\mathbf{Y}) = \arg\min_t r_t(\mathbf{Y}) = \frac{1}{C}\sum_{i=1}^{C} ||\mathbf{y}_i - \mathbf{D}_i\delta_t(\hat{\mathbf{x}}_i)||_2^2, \tag{7}$$

where $\delta_t(\cdot)$ is a function which selects only the coefficients corresponding to class t, $||\mathbf{y}_i - \mathbf{D}_i\delta_t(\hat{\mathbf{x}}_i)||_2^2$ $(i = 1,2,\ldots,C)$ computes the reconstruction residual on class t using the i-th dictionary and thus C residuals on class t can be obtained in total. Equation (7) averages all the C reconstruction residuals and finds out the class t that has the minimum mean residual. The testing sample, therefore, is classified to class t. The resulting algorithm is what we call multi-dictionary collaborative representation (MDCR). We illustrate our proposed method in a 3D ear identification task in Fig. 2 and we further summarize the algorithm of MDCR in Table 1.

Table 1. Algorithm for multi-dictionary based collaborative representation

---

Training phase:
Input: A training set.
Output: The multiple dictionaries $(\mathbf{D}_1,\mathbf{D}_2,\ldots,\mathbf{D}_C)$.
    1. For each training sample
      Divide it into C blocks, each of which corresponds to one dictionary;
        For i-th block, extract from it a feature vector $h^{(i)}$;
    2. Concatenate all $h^{(i)}$s to construct dictionary $\mathbf{D}_i$ $(i=1,2,\ldots,C)$;
Testing phase:
Input: A testing sample $\mathbf{Y}$.
Output: Identity t of the testing sample.
    1. Divide it into C blocks, for i-th block
      (1) Extract from it a feature vector $\mathbf{y}_i$;
      (2) Code $\mathbf{y}_i$ over $\mathbf{D}_i$ as follows

$$\hat{x}_i = \arg\min_{x_i}\left\{|| \ y_i - \mathbf{D}_i x_i \ ||_2^2 + \lambda || \ x_i \ ||_2^2\right\}$$

    2. Compute the average reconstruction residual of class t on different dictionaries

$$r_t(\mathbf{Y}) = \frac{1}{C}\sum_{i=1}^{c}|| \ y_i - \mathbf{D}_i\delta_t(\hat{x}_i) \ ||_2^2, \text{ where } y_i \text{ represents the } i\text{-th block of the testing}$$

      image and $\delta_t(\hat{x}_i)$ selects from $\hat{x}_i$ all the coefficients only corresponding to
      the class t.
    3. Identity $(\mathbf{Y}) = \arg\min r_t(\mathbf{Y})$.

---

3 Experiments and Results

We evaluated our purposed method on UND Collection J2 dataset [19] and PolyU 3D Palmprint database [20], respectively. As we only focus on the effectiveness of matching methods, we uniformly adopted LHST (local histogram surface type) [4, 5] as feature descriptor, which is not only highly discriminative for 3D biometric identifies, but also quite robust even when small misalignments exist among the extracted ROIs. With respect to comparing metrics, we computed recognition rate and time cost of one identification operations including feature extraction and feature matching for each matching method. Experiments were performed on a standard HP Z620 workstation with a 3.2 GHz Intel Xeon E5-1650 CPU and an 8 GB RAM. The software platform was MATLAB 2015a.

3.1 3D Ear Identification

Dataset and Experiment Protocols. In this section, we used the UND Collection J2 dataset [19], which contains 2346 3D side face scans captured from 415 persons, making it the largest 3D ear scan dataset so far. To evaluate the performance of our method, however, we cannot simply conduct experiments on the whole dataset since some classes in UND-J2 have only 2 samples and classification schemes based on collaborative representation need sufficient samples for each class in the gallery. Consequently, we virtually created four subsets from UND-J2 for experiments. Specifically, we require that each class should have more than 6, 8, 10, and 12 samples, respectively. For subset 1, we randomly selected from each class 6 samples to form the gallery set and the test samples were used to form the test set. Similar strategies were applied to the rest subsets. To make it clear, major information about the four subsets is summarized in Table 2.

Table 2. Experiment settings on 3D palmprint and 3D ear dataset. (1) #Classes denotes the number of subjects in the gallery, (2) #Training denotes the total number of images in the training set, (3) #Testing denotes the total number of images in the testing set.

Database	UND J2 collection	PolyU 3D palmprint databases			
		Subset #1	Subset #2	Subset #3	Subset #4
#Classes	400	127	85	62	39
#Total	4000	1477	1141	911	636
#Training	2000	762	680	620	468
#Testing	2000	715	461	291	168

Performance Evaluation. In Table 3, we list the recognition rate obtained by each method on each subset. It can be seen that MDCR greatly outperforms ICP. And MDCR also reaches better recognition rate compared with SRC, although they both exploit the

Table 3. Recognition rate by using different methods on 3D ear dataset (%).

	Subset #1	Subset #2	Subset #3	Subset #4
ICP	83.22	90.02	94.09	95.83
SRC_LHST	92.17	94.36	96.56	98.81
MDCR_LHST	**93.29**	**96.31**	**97.94**	**99.40**

collaborative representation framework. This explicitly indicates the effectiveness of our multi-dictionary strategy. In Table 4, we list the time cost consumed by one identification operation by each method on all four subsets. It is evident that ICP has the highest computational complexity, which makes it not suitable for large-scale identification. Moreover, MDCR apparently speeds up the identification compared with SRC. This can be attributed to that in SRC the solution has to be reached by iterative optimization, while in MDCR the label can be easily estimated by using a linear predictive classifier, which is quite time-efficient.

3.2 3D Palmprint Identification

Dataset and Experiment Protocols. In this section, we conducted experiments on the PolyU 3D palmprint database [20]. The database contained 8,000 samples collected from 400 different palms, belonging to 200 volunteers. 20 samples from each of these palms were captured in each session, respectively. The average time interval between the two sessions was one month. In the following experiments, we took samples collected at the first session as the training set and samples collected at the second session as the testing set. Under such experimental setting, it is easy to know that for the gallery set, there are 400 classes and for each class there are 10 samples.

Performance Evaluation. We list the recognition rate and the time cost for one identification by using each method. Based on the experimental results, we could have the following findings. First, among all matching methods, CR based ones- CR_L1_LHST, Zhang et al.'s method [4], and MDCR_LHST- greatly outperform other matching methods. Second, compared to the other two CR based classification algorithms, MDCR is the best trade-off between the matching accuracy and the speed. Specifically, MDCR reaches the highest recognition rate of 99.48 % and at the same time, keep the same computational complexity as Zhang et al.'s method [4], though a bit slower than [4] Table 5.

Table 4. Time cost for one identification operation (s).

	Subset #1	Subset #2	Subset #3	Subset #4
ICP	5.356×10^5	3.763×10^5	1.876×10^5	1.287×10^5
SRC_LHST	0.074	0.070	0.066	0.056
MDCR_LHST	**0.054**	**0.055**	**0.052**	**0.047**

Table 5. Recognition rate and time cost for one identification by using various methods.

	Recognition rate (%)	Time cost for 1 identification (s)
MCI [6]	91.88	9.403
GCI [6]	91.87	9.403
ST [6]	98.78	63.276
ChiSquare_LHST	95.93	0.056
CR_L_1_LHST	99.48	0.547
Zhang et al. [20]	99.15	0.023
MDCR_LHST	**99.48**	**0.317**

4 Conclusion

In this paper, we proposed a multi-dictionary based collaborative representation method for 3D palmprint and 3D ear classification. Under multi-dictionary based scheme, the discriminative features contained in the range data are fully exploited and the adverse impact of interferential information is suppressed. Representative experiments on benchmark datasets show that our classification framework outperforms the state-of-the-art methods in both classification accuracy and computational complexity.

References

1. Drira, H., Ben Amor, B., Srivastava, A., Daoudi, M., Slama, R.: 3D face recognition under expressions, occlusions, and pose variations. IEEE Trans. PAMI **35**, 2270–2283 (2013)
2. Zhang, L., Ding, Z., Li, H., Shen, Y., Lu, J.: 3D face recognition based on multiple keypoint descriptors and sparse representation. PLoS ONE **9**(e100120), 1–9 (2014)
3. Islam, S.M., Davies, R., Bennamoun, M., Mian, A.S.: Efficient detection and recognition of 3D ears. IJCV **95**, 52–73 (2011)
4. Zhang, L., Shen, Y., Li, H., Lu, J.: 3D palmprint identification using block-wise features and collaborative representation. IEEE Trans. PAMI **37**, 1730–1736 (2015)
5. Li, L., Zhang, L., Li, H.: 3D ear identication using LC-KSVD and local histograms of surface types. In: Proceedings of IEEE ICME, pp. 1–6 (2015)
6. Zhang, D., Lu, G., Li, W., Zhang, L., Luo, N.: Palmprint recognition using 3-D information. IEEE SMC-C **39**, 505–519 (2009)
7. Daugman, J.G.: High condence visual recognition of persons by a test of statistical independence. IEEE Trans. PAMI **15**, 1148–1161 (1993)
8. Li, W., Zhang, L., Zhang, D., Lu, G., Yan, J.: Efficient joint 2D and 3D palmprint matching with alignment refinement. In: Proceedings of IEEE CVPR, pp. 795–801 (2010)
9. Besl, P.J., McKay, N.D.: A method for registration of 3-D shapes. IEEE Trans. PAMI **14**, 239–256 (1992)
10. Chen, H., Bhanu, B.: Contour matching for 3D ear recognition. In: Proceedings of IEEE WACV Workshops, pp. 123–128 (2005)
11. Chen, H., Bhanu, B.: Human ear recognition in 3D. IEEE Trans. PAMI **29**, 718–737 (2007)
12. Yan, P., Bowyer, K.: Empirical evaluation of advanced ear biometrics. In: Proceedings of IEEE CVPR Workshops, p. 41 (2005)

13. Passalis, G., Kakadiaris, I.A., Theoharis, T., Toderici, G., Papaioannou, T.: Towards fast 3D ear recognition for real-life biometric applications. In: Proceedings of IEEE AVSS, pp. 39–44 (2007)
14. Cadavid, S., Abdel-Mottaleb, M.: 3-D ear modeling and recognition from video sequences using shape from shading. IEEE TIFS **3**, 709–718 (2008)
15. Liu, H.: Fast 3D ear recognition based on local surface matching and ICP registration. In: Proceedings of IEEE INCoS, pp. 731–735 (2013)
16. Tibshirani, R.: Regression shrinkage and selection via the lasso. J. Roy. Stat. Soc. Series B **58**, 267–288 (1994)
17. Zhang, L., Yang, M., Feng, X.: Sparse representation or collaborative representation: which helps face recognition? In: Proceedings of IEEE ICCV, pp. 471–478 (2011)
18. Wright, J., Yang, A.Y., Ganesh, A., Sastry, S.S., Ma, Y.: Robust face recognition via sparse representation. IEEE Trans. PAMI **31**, 210–227 (2009)
19. CVRL datasets. http://www3.nd.edu/cvrl
20. Zhang, L., Ding, Z., Li, H., Shen, Y.: 3D ear identification based on sparse representation. PLoS ONE **9**(e95506), 1–9 (2014)

An Adaptive Multi-algorithm Ensemble
for Fingerprint Matching

Kamlesh Tiwari[1(✉)], Vandana Dixit Kaushik[2], and Phalguni Gupta[3]

[1] Department of CSIS, Birla Institute of Technology and Science,
Pilani 333031, Rajasthan, India
[2] Department of CSE, Harcourt Butler Technological Institute,
Kanpur 208002, India
[3] National Institute of Technical Teachers' Training & Research,
Kolkata 700106, WB, India
pg@cse.iitk.ac.in

Abstract. Any well known fingerprint matching algorithm cannot provide 100% accuracy for all databases. One should explore the possibility of fusion of multi-algorithms to achieve better performance on such databases. One of the major challenges is to design a fusion strategy which is both adaptive and improving with respect to the candidate database. This paper proposes an adaptive ensemble using statistical properties of two well known state-of-the-art minutiae based fingerprint matching algorithms to achieve (1) improvement on fingerprint recognition benchmark, (2) outperform on multiple databases. Experiments have been conducted on two databases containing multiple fingerprint impressions of 140 and 500 users. One of them is widely used publicly available databases and another one is our in-house database. Experimental results have shown the significant gain in performance.

Keywords: Fingerprint · Matching · Minutiae · ROC curve · Multi-algorithm · Adaptive

1 Introduction

Fingerprint is one of the most accepted biometric traits for automatic human authentication because of its invariance to changes over time, easy to acquire data and high acceptance to law [15] and society. The fingerprint is a pattern appearing on a surface touched by the upper part of the finger skin. It can be acquired digitally by using a special type of device called fingerprint scanner. Fingerprint comprises of a pattern of black and white lines called ridges and valleys respectively. The point where the ridge terminates or joins with another two ridges is termed as minutiae point. A fingerprint with its minutiae points marked is shown in Fig. 1. There can be many minutiae points in a single fingerprint. Most of the fingerprint matchers make use of minutia points as the fingerprint features for matching. Determining mapping of minutiae of one fingerprint to those of another fingerprint is important for matching. This is often a difficult task because it becomes a combinatorial problem. It is also possible that a minutiae point in one fingerprint may not have any corresponding minutia point in another fingerprint because of its spurious nature.

© Springer International Publishing Switzerland 2016
D.-S. Huang et al. (Eds.): ICIC 2016, Part I, LNCS 9771, pp. 49–60, 2016.
DOI: 10.1007/978-3-319-42291-6_6

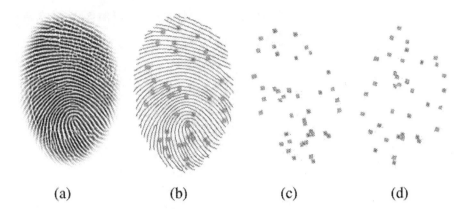

(a)   (b)   (c)   (d)

Fig. 1. Example of (a) fingerprint, (b) minutiae points on thinned image and (c) and (d) templates extracted from two fingerprints taken at different point of time from same finger.

Evaluation of a matching algorithm on different databases provides different performance figures. This is because of the variability in the database quality and the kind of users in the database [19]. It is quite possible that an algorithm being best in terms of performance on one database may perform poorly on another database. Therefore, it is beneficial to explore the suitability of multiple algorithms that can unitely contribute towards the higher accuracy on various databases. It is important to have an adaptive fusion strategy for their multi-algorithm ensemble.

There exist several minutia based matching algorithms in the literature [6]. An approach, proposed in [3], uses Delaunay triangulation for matching. Ridge patterns are used in [11] to align minutia. Its high computational cost has been reduced in [9, 10, 14] by establishing better minutiae correspondence. In [10], a minutia simplex containing a pair of minutiae and their textures has been built. It uses the ridge-based nearest neighborhood among minutiae to represent the features and to compute their similarity. In [20], a feature called polyline which calculates three transformation invariant features and performs the matching has been used. Local level matching in [18] is preferred over implicit global level matching in [8, 12] or explicit global level matching in [16, 17].

This paper proposes an adaptive multi-algorithm ensemble for fingerprint based human recognition system. It uses two fingerprint matching algorithms. It has also shown that a single fingerprint matching algorithm cannot perform well on all databases. One algorithm performing better on a database may perform disorderly on another one. It has shown that the performance of the state-of-the-art fingerprint matching algorithm can further be improved by applying suitable fusion [13]. As a result, these multi-algorithms can be used to fuse to achieve a better performance on all databases. Experiments have been conducted on two databases viz. FVC2006-DB2_A which contains 1680 fingerprint images of 140 users and InHouse500FP having 1000 fingerprint images of 500 subjects.

This paper consists of 5 Sections Next section provides a brief overview of the two algorithms used to design the multimodal system. Section 3 presents the proposed approach. Section 4 explains the experimental setup, the database and analysis of the experimental results. Conclusions are presented in the last section.

2 Background

An automatic fingerprint recognition system acquires fingerprint samples by using a digital scanner. These samples are further processed for recognition. Feature extraction and matching are the two important modules of any such system. Standard features of a fingerprint are the minutiae points that can be extracted by using mindtct [7] of NIST. Every minutiae is represented by its location and direction with respect to the fingerprint image. A fingerprint can have many distinct minutiae points.

Two different interactions with fingerprint scanner of the same user may not be identical due to variation in his behavior. User may apply different amount of pressure on the fingerprint scanner at different instant of time, his finger conditions may not be same all the time, noise in the scanner can also affect the quality of image, etc. This leads to variation in two fingerprint images and their minutiae templates even of the same user. This results in intra-class variation in the fingerprint impression as shown in Fig. 1(c), (d). This is why the two fingerprints of same finger despite being very similar do not match exactly.

Minutiae points of a fingerprint form the feature vector called template. A fingerprint database contains the feature vector of all users registered with the system. The system may accept multiple fingerprint impressions of the same user during the process of registration. Registering more fingerprints of the same person helps to improve identification accuracy of the system. A recognition system can be used in two ways. either for identification or for verification. In identification mode, the system compares features of a query fingerprint with those of all fingerprints stored in the database. Expected outcome of the identification is an ordered list of users with whom the query fingerprint matches the most. In verification mode, the system compares features of a query fingerprint with that of the claimed identity only.

This section provides details of two fingerprint matching algorithms which are used in the proposed system. First one is the Minutia Cylinder-Code (MCC) [5] of Biometric System Laboratory of University of Bologna [1]. The Second algorithm is the rotation and translation invariant fingerprint matching algorithm bozorth3 [7] provided by National Institute of Standards and Technology (NIST) [2].

2.1 BOZORTH Matcher

Fingerprint matcher bozorth3 [7] of National Institute of Standards and Technology (NIST) uses location and rotation angle of each minutiae point. To compute matching score between two minutia templates, it constructs intra-fingerprint minutia comparison tables for probe and gallery fingerprints and then determines their. compatibility by constructing Inter-Fingerprint Compatibility table. It obtains the overall matching score

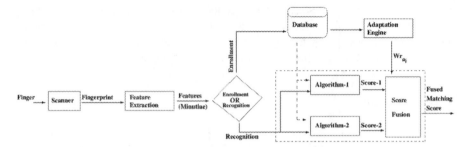

Fig. 2. Block diagram of proposed system

by traversing the Inter-Fingerprint compatibility table to link table entries into clusters and then by combining the compatible clusters. The algorithm is rotation and translation invariant.

2.2 Minutia Cylinder Code (*mcc*) Matcher

Minutia Cylinder Code (mcc) defines a 3-D local structure, called cylinder, to represent the local neighborhood of a minutia. This 3-D structure is a translation and rotation invariant [5]. These cylinders are built using distances and angle differences between minutiae points. Cylinders from the feature vector for the corresponding minutiae point. The matching score is generated based on the similarity between two cylinders. Minutia Cylinder Code (mcc) matcher verifies to what extent local matches hold at the global level and compute an overall matching score between two fingerprints. The global score is used to decide whether two minutiae templates are similar or not.

3   Proposed Approach

Any recognition system has four major components (1) Data Acquisition, (2) Feature Extraction, (3) Matching and (4) Score Fusion. Data acquisition in our setting uses a digital scanner to acquire fingerprint images. Feature extraction is done using mindtct [7] of NIST. It produces a feature template that contains specification of extracted minutiae points. Subsequently based upon the operating mode (registration or recognition) the template is either saved in the database or is used for matching in the multi-algorithms ensemble. The proposed ensemble fuses two independent matchers. The two fingerprint templates are matched by mcc and bozorth3 in parallel. Individual matching scores obtained by the two matchers are fused using threshold alignment and range compression weights obtained by adaptation engine. It is explained in detail in next subsection. The matching decision is obtained by applying a threshold on the fused score. Block diagram of the proposed system is shown in Fig. 2.

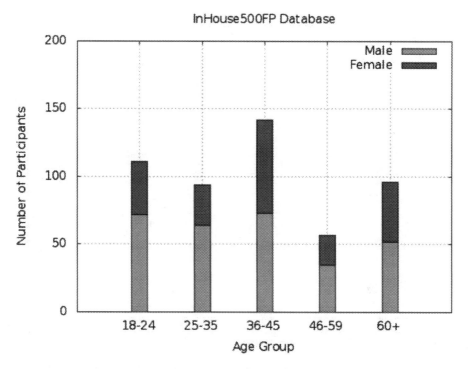

Fig. 3. InHouse500FP database, age group and gender wise distribution (Color figure online)

3.1 Normalization Scheme

Score normalization and fusion are very important component of a personal identification system. The score fusion is complex as the range of matching score produced by the two matchers may be different and they may also follow different distributions. Also, different weights may need to be assigned to different matching. Score distribution of genuine and imposter matchings of a matcher have different statistical properties. These two distributions should be well separated ideally, but in practice, they do have overlaps. A decision threshold is chosen such that it separates the two distributions and achieves maximum possible accuracy. Weights of the matching scores are necessary when the involved systems have different recognition accuracies to restrict the adverse effect of low confidence system. The normalization strategy proposed in this paper is adaptive and uses statistical properties for threshold alignment followed by assigning weights to obtain the fused matching score.

Threshold Alignment. Let the decision thresholds that typically equalizes FAR and FRR in an ensemble using n different algorithms A1, A2,..., An be Th_{A1}, Th_{A2},..., Th_{An} respectively. The system designer arbitrarily chooses a single and fixed pivot

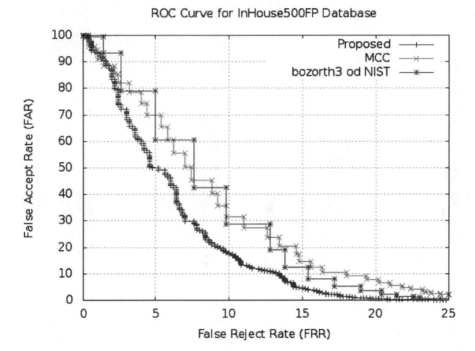

Fig. 4. ROC curve of *mcc*, *bozorth3* and proposed system on InHouse500FP database

threshold Th_p that is used for threshold alignment. Threshold aligned score v'_{Ai} for a min-max normalized matching score v_{Ai} of A_i can be obtained by

$$v'_{Ai} = v_{Ai} + \left(Th_p - Th_{Ai}\right) \tag{1}$$

In our experiment we have utilized two algorithms say $A_{bozorth}$ and A_{mcc}. Let their decision thresholds be $Th_{bozorth}$ and Th_{mcc} respectively. Threshold aligned score $v'_{A_{bozorth}}$ for a min-max normalized matching score $v_{A_{bozorth}}$ of $A_{bozorth}$ is obtained by

$$v'_{A_{bozorth}} = v_{A_{bozorth}} + \left(Th_p - Th_{A_{bozorth}}\right) \tag{2}$$

Similarly, threshold aligned scores $v'_{A_{mcc}}$ for $v_{A_{mcc}}$ of $A_{bozorth}$ can be obtained by

$$v'_{A_{mcc}} = v_{A_{mcc}} + \left(Th_p - Th_{A_{mcc}}\right) \tag{3}$$

Purpose of the initial min-max normalization is to bring the scores of different modalities in same range. This step helps to align appropriate thresholds.

Adaptation Engine For Adaptive Weighting Parameters. An adaptation engine is used to determine relative multiplicative weights of the matchers. It uses statistical properties of matching score distribution to optimize false acceptance and rejection rates of the unimodal biometric system. It is typically related to equal error rate (EER) of the unimodal biometric system. Since EER is inversely related to the accuracy; a polynomial of the form $c_1 \times (EER)^{-1} + c_2$ is used to obtain relative fusion weights where c_1, c_2 are constants. The values of constants are empirically determined by using regression on small subset of the database. Let EER_{A_i} be the equal error rate of an individual biometric sub-system using algorithm A_i then its relative multiplicative weight wr_{a_i} s determined by

$$wr_{a_i} = c_1 \times (EER_{A_i})^{-1} + c_2 \tag{4}$$

Absolute weight w_i of score from algorithm A_i is determined with the help of relative multiplicative weights of all participating algorithms and is given by

$$w_i = \frac{wr_{a_i}}{\sum_{k=1}^{n} wr_{s_k}} \tag{5}$$

The fused score $v_{s_1 s_2 \ldots s_n}$ for the scores of n biometric traits is obtained by combining $v'_{A_1}, v'_{A_2}, \ldots, v'_{A_n}$ and $w_1 w_2, \ldots, w_n$ as

$$v_{s_1 s_2 \ldots s_n} = w_1 \times v'_{A_1} + w_2 \times v'_{A_2} + \ldots + w_n \times v'_{A_n} \tag{6}$$

4   Experimental Results

This section describes the experimental setup and analyzes the result. Subsection 4.1 elaborates the two databases used in this experiment. Standard performance metrics used to measure the system performance are explained in Subsect. 4.2. Experimental results are analyzed in Subsect. 4.3.

4.1   Database

We have used two databases in our experiments viz. InHouse500FP database and a publicly available database FVC2006 to test the performance of the proposed system.

FVC2006-DB2_A Database. This is a public database created for Fourth International Fingerprint Verification Competition [4]. It is a sequestered part of a larger dataset acquired within the European Project BioSec. It does not contain deliberately introducing difficulties such as exaggerated distortion, large amounts of rotation and displacement, wet/dry impressions, etc. It contains fingerprints of 140 subjects. For every subject, there are 12 fingerprint images in the database. There are 1680 images of BMP image format having 256 gray-levels acquired through optical sensor of size

400×560 and 569 dpi resolution. This database contains 6.1 %, 63.6 %, 26.8 %, 3.0 % and 0.5 % images of NIST quality of 1, 2, 3, 4 and 5 respectively.

InHouse500FP Database. It is an in-house database that contains fingerprint images of 500 persons of age group from 18 to 75 year. Every person has provided two fingerprints in a gap of two month. This database has been collected from the persons who are not very careful about their skin condition and are involved in laborious tasks. Persons are divided into five age groups viz. 18–24, 25–34, 35–44, 45–59 and 60+. Figure 3 shows age group and gender-wise distribution of the persons in the database. According to NIST quality assessment, this database contains 34.7 %, 19 %, 23.6 %, 11.2 % and 11.5 % images with quality 1, 2, 3, 4 and 5 respectively where 1 stands for the highest quality and 5 for the poorest quality.

4.2 Standard Performance Metrics

Experimental results are analyzed using standard performance metrics like false accept rate, false rejection rate, equal error rate and correct recognition rate. They are defined as below.

False Accept Rate (FAR) refers to the rate at which unauthorized individual (imposers) are accepted as a valid user.

$$FAR = \frac{Number\ of\ Imposter\ Accepted}{Total\ Number\ of\ Imposter\ Matchings} \times 100\%$$

False Reject Rate (FRR) refers to the probability that a biometric system fails to identify a genuine subject.

$$FAR = \frac{Number\ of\ Genuine\ Rejected}{Total\ Number\ of\ Genuine\ Matchings} \times 100\%$$

Equal Error Rate (EER) is the rate at where FAR and FRR are equal. Lower the value of EER, better is the system. Receiver operating characteristic (ROC) curve is a plot between FAR and FRR. Lower the area under the curve better is the system.

Correct Recognition Rate (CRR) represents rank-1 accuracy and signifies the percentage of the best match which pertains to the same subject. It can be measured as $CRR = (N_1/N_2) \times 100$ where N_1 is the number of correct (Non-False) recognitions as top best matches and N_2 is the total number of images in the testing set. Higher the CRR, better is the system.

4.3 Results

FVC2006 DB2_A database contains multiple images of each subject. The testing strategy is based on leave one out. One fingerprint is matched with all other fingerprints

Table 1. Performance comparison. Falsely Reject (FR) and Falsely Accept (FA) are out of 18480 genuine and 2802240 imposer matchings for FVC2006-DB2_A database and are out of 500 genuine and 249500 imposer matchings for InHouse database respectively.

	FVC2006-DB2_A. Database				InHouse Database			
	CRR %	EER %	FR	FA	CRR %	EER %	FR	FA
Individual algorithm								
BOZORTH [7]	99.88	1.89	325	56658	74.60	13.08	69	30845
MCC [5]	100	1.57	292	44226	68.60	14.77	74	36810
Fusion strategy								
Maximum	99.94	1.64	3.33	41409	71.20	14.70	74	36444
Summation	100	1.22	226	34250	78.00	12.04	60	30144
Product	100	1,22	226	34264	78.00	12.02	60	30086
tanh and summation	100	1.22	226	34213	78.00	12.02	60	30086
Median and MAD and summation	100	1.09	202	30607	75.60	12.81	65	31513
Double sigmoid and summation	100	1.19	221	32785	77.00	12.04	60	30145
Minimum	100	1.22	226	34110	75.60	12.58	63	31379
z-score and summation	99.94	1.23	2228	34560	78.40	12.03	60	30100
Proposed	**100**	**0.99**	**182**	**27742**	**79.80**	**11.89**	**58**	**27579**

except itself in the database. It has been observed that the correct recognition rate (CRR) of mcc system is 100 % and the equal error rate (EER) is 1.57 %. CRR and EER of bozorth3 matcher are 99.88 % and 1.89 % respectively on this database.

CRR and EER values for different fusion schemes are shown in Table 1. A comparison of score normalization and fusion strategies like minimum, maximum, summation, product, tanh normalization, median and MAD, double Sigmoid and z-score has also been presented in Table 1.

It has been observed that CRR of the system is 99.94 % and EER is 1.64 % when maximum score among the two matchers is considered as fused score. It has falsely rejected 333 genuine out of 18480 and falsely accepted 41409 imposter out of 2802240 attempts. z-Score has same rank-1 accuracy but slightly better error rate as compared to the maximum. It can be observed that the EER of the proposed approach is better than any other fusion method. The value of EER is found to be 0.99 %. It has falsely rejected 192 genuine out of 18480 that is the least among all other fusion schemes. Further, it has falsely accepted 29037 imposters out of 2802240 attempts that is also minimum among all other fusion schemes. ROC curves of mcc, bozorth3 and the proposed system have been shown in Fig. 4. It can clearly be seen that the proposed system is better among the three.

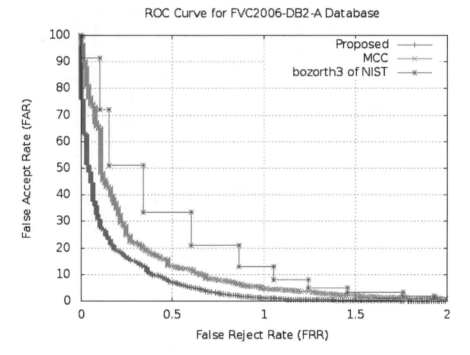

Fig. 5. ROC curve of *mcc*, *bozorth3* and proposed system on FVC2006-DB2_A database

InHouse500FP database contains two fingerprints per subject that are acquired with a gap of two months time. The testing strategy is matching fingerprint images of Phase-1 against those of Phase-2. Correct recognition rate (CRR) of bozorth3 system is found to be 74.60 % with the equal error rate (EER) of 13.08 %. But CRR and EER of mcc are 68.60 % and 14.78 % respectively. CRR, EER and DI values for the different weighted fusion of the normalized matching scores with different weights are shown in Table 1.

It can also be observed that fusion strategy which uses maximum value performs badly on this database. It has CRR of 71.20 % and EER of 14.70 %. Further, it has falsely rejected 74 genuine out of 500 and falsely accepted 36444 imposters out of 249500 attempts. Median and MAD are slightly better fusion strategies with CRR and EER being 75.60 % and 12.815 % respectively. Minimum fusion strategy as shown in the Table 1 rejects even less genuine and is marginally better with Median and MAD.

False rejections by all other methods, except the proposed one, are same but differ with respect to the number of false acceptances. Product and tanh both have same number of false acceptances that is 30086 out of 249500 attempts.

One can clearly see the superiority of the proposed fusion strategy as it rejects the lowest number of genuine (58) and accepts the less number of imposters i.e. 27579 out of 249500 attempts. CRR and EER of the proposed fusion scheme are found to be

77.80 % and 11.89 % respectively that is better than any other fusion scheme on InHouse500FP database. ROC curves of mcc, bozorth3 and the proposed system are shown in Fig. 5. It can clearly be seen that the proposed system has the lower error rate than the other two.

5 Conclusions

This paper proposes an adaptive multi-algorithm ensemble for matching in fingerprint based recognition system. It proposes a novel adaptation engine to determine parameters for threshold alignment and range compression of matching score fusion. Experimental setup has been constructed which uses two well-known state-of-the-art matching algorithms mcc and bozorth3. Further, the experiments have been conducted on two databases FVC2006-DB2_A and InHouse500FP that contain 1680 and 1000 fingerprints obtained from 140 and 500 users respectively. For FVC2006-DB2_A database, it has been observed that mcc performs better than bozorth3 whereas same is not the case for InHouse500FP database.

It has also been observed that the proposed adaptive ensemble has improved the system performance for both the databases. Comparisons with other fusion strategies such as z-score, median and MAD, tanh, double sigmoid have shown the superiority of proposed adaptation scheme. This signifies that the proposed adaptive multi-algorithm ensemble improves the fingerprint recognition benchmark and this improvement extends upon multiple databases.

References

1. Biometric System Laboratory of University of Bologna. http://biolab.csr.unibo.it/Home.asp
2. NIST Biometric Image Software. http://www.nist.gov/itl/iad/ig/nbis.cfm
3. Bebis, G., Deaconu, T., Georgiopoulos, M.: Fingerprint identification using delaunay triangulation. In: International Conference on Information Intelligence and Systems, pp. 452–459 (1999)
4. Cappelli, R., Ferrara, M., Franco, A., Maltoni, D.: Fingerprint verification competition 2006. Biometric Technol. Today **15**(7), 7–9 (2007)
5. Cappelli, R., Ferrara, M., Maltoni, D.: Minutia cylindercode: a new representation and matching technique for fingerprint recognition. IEEE Trans. Pattern Anal. Mach. Intell. **32** (12), 2128–2141 (2010)
6. Chikkerur, S., Cartwright, A.N., Govindaraju, V.: K-plet and coupled BFS: a graph based fingerprint representation and matching algorithm. In: Zhang, D., Jain, A.K. (eds.) ICB 2005. LNCS, vol. 3832, pp. 309–315. Springer, Heidelberg (2005)
7. Garris, M.D., Watson, C.I., McCabe, R.M., Wilson, C.L.: User's guide to NIST fingerprint image software (NFIS) (2001)
8. Goshtasby, A.: Piecewise linear mapping functions for image registration. Pattern Recogn. **19**(6), 459–466 (1986)
9. He, Y., Tian, J., Li, L., Chen, H., Yang, X.: Fingerprint matching based on global comprehensive similarity. IEEE Trans. Pattern Anal. Mach. Intell. **28**(6), 850–862 (2006)

10. He, Y., Tian, J., Luo, X., Zhang, T.: Image enhancement and minutiae matching in fingerprint verification. Pattern Recogn. Lett. **24**(9), 1349–1360 (2003)
11. Jain, A., Hong, L., Bolle, R.: On-line fingerprint verification. IEEE Trans. Pattern Anal. Mach. Intell. **19**(4), 302–314 (1997)
12. Jea, T.-Y., Govindaraju, V.: A minutia-based partial fingerprint recognition system. Pattern Recogn. Lett. **38**(10), 1672–1684 (2005)
13. Khalifa, A.B., Gazzah, S., BenAmara, N.E.: Adaptive score normalization: a novel approach for multimodal biometric systems. World Acad. Sci. Eng. Technol. Int. J. Comput. Sci. Eng. **7**(3), 882–890 (2013)
14. Luo, X., Tian, J., Wu, Y.: A minutiae matching algorithm in fingerprint verification. Int. Conf. Pattern Recogn. **4**, 833–836 (2000)
15. Maltoni, D., Maio, D., Jain, A.K., Prabhakar, S.: Handbook of Fingerprint Recognition. Springer, Heidelberg (2009)
16. Nilsson, K., Bigun, J.: Localization of corresponding points in fingerprints by complex filtering. Pattern Recogn. Lett. **24**(13), 2135–2144 (2003)
17. Ramo, P., Tico, M., Onnia, V., Saarinen, J.: Optimized singular point detection algorithm for fingerprint images. Int. Conf. Image Process. **3**, 242–245 (2001)
18. Ratha, N.K., Bolle, R.M., Pandit, V.D., Vaish, V.: Robust fingerprint authentication using local structural similarity. In: IEEE Workshop on Applications of Computer Vision, pp. 29–34 (2000)
19. Ross, A., Rattani, A., Tistarelli, M.: Exploiting the doddington zoo effect in biometric fusion. In: IEEE 3rd International Conference on Biometrics: Theory, Applications, and Systems 2009, BTAS2009, pp. 1–7. IEEE (2009)
20. Wang, X., Li, J., Niu, Y.: Fingerprint matching using orientation codes and polylines. Pattern Recogn. Lett. **40**(11), 3164–3177 (2007)

Stereo Matching with Improved Radiometric Invariant Matching Cost and Disparity Refinement

Jinjin Shi, Fangfa Fu, Yao Wang, Weizhe Xu, and Jinxiang Wang[(⊠)]

Microelectronics Center, Harbin Institute of Technology, Harbin 150000, China
jxwang@hit.edu.cn

Abstract. Accurate and real-time stereo correspondence is a pressing need for many computer vision applications. In this paper, an improved radiometric invariant matching cost algorithm is proposed. It effectively combines modified census transform with relative gradients measures. Although it is very simple, comparison results on Middlebury stereo testbed demonstrate that it has much lower error rates than many existing algorithms and is very close to the ANCC algorithm which represents the current state of the art under extreme luminance condition but outperforms the ANCC algorithm greatly when there are small radiometric distortions. In addition, we also develop a disparity refinement method with computational complexity invariant to the disparity range. Experimental results on Middlebury datasets show those artifacts near object boundaries are reduced using the proposed disparity refinement method.

Keywords: Stereo matching · Radiometric invariant · Census transform · Disparity refinement

1 Introduction

Stereo matching is one of the most useful technologies and challenging tasks in computer vision with wide applications such as autonomous navigation and 3-D reconstruction. It is commonly assumed that matching pixels in a stereo pair have similar color values. However, color consistency may not hold in real scenarios due to illumination variations, non-Lambertian surfaces and camera device changes etc.

In recent years, a variety of matching cost functions have been developed to tackle this issue. Heo et al. [1] proposed a new measure called Adaptive Normalized Cross-Correlation (ANCC), which utilizes color formation model explicitly to cope with complex radiometric variations. Kim et al. [2] converted pixels within each support window into Mahalanobis distance transform space. Although accuracy is an important factor, many applications require real-time processing. A host of fast similarity measures also have been developed. For example, a simple matching cost function based on relative gradients was introduced in [3] to handle the radiometric variation. Mutual information (MI) was used to model complex radiometric relationships in [4]. Fife and Archibald [5] proposed a number of sparse census transforms to reduce the resource requirements of census-based stereo systems.

© Springer International Publishing Switzerland 2016
D.-S. Huang et al. (Eds.): ICIC 2016, Part I, LNCS 9771, pp. 61–73, 2016.
DOI: 10.1007/978-3-319-42291-6_7

Disparity refinement also plays a significant role in promoting depth map's quality for stereo matching. Over the years, many refinement algorithms have been presented. For example, plane-fitting was popularly used in many top ranking algorithms [6, 7] While this approach requires segmented image as the input, which is a timing bottleneck. Local methods [8–10] utilized adaptive weights based on color and distance to refine disparities. A non-local disparity propagation on minimum spanning tree (MST) was developed by Yang [11]. Other approaches, such as region voting [12], vertical voting [13], cost spectrum peak analysis and removal [14] and sub-pixel enhancement [12], have also been developed over the past few years. Theses algorithms prevent the necessity of image segmentation, but they are still very slow due to high computational complexity.

In light of previous works, there are two main contributions in this paper: **(1)** An improved radiometric invariant matching cost algorithm called Relative Gradient Census Transform (RGCT) is proposed. It effectively combines modified census transform with relative gradients measures to take both their advantages. The relative gradient algorithm allows us to extract the "lost" contents that often appear at the over exposed or illuminated regions of the analyzed image, and the census transform is used to extract the local features. Although the RGCT algorithm is very simple, comparison results demonstrate that it has much lower error rates than many existing algorithms and is very close to the ANCC algorithm which represents the current state of the art under extreme luminance condition but outperforms the ANCC algorithm greatly when there are small radiometric distortions. More importantly, the proposed RGCT keeps very low computational amount. **(2)** We develop a disparity refinement approach with computational complexity invariant to the disparity range. Our work is inspired by [15] but with the following differences: First, the method in [15] needs MST tree construction, while the proposed method treats each scanline as a separate tree, thus it is more GPU-friendly. Second, we introduce a new technique to reduce the loss of disparity belief caused by sampling noise or unwanted local pattern. Experimental results show that artifacts near object boundaries are reduced using the proposed refinement method. The remainder of this paper is organized as follows. The proposed RGCT and disparity refinement algorithms are described in Sects. 2 and 3 respectively. Experimental results are presented in Sect. 4 followed by some conclusions in Sect. 5.

2 Matching Cost

In this section, the relative gradient algorithm in [3] is first briefly reviewed, then the proposed RGCT algorithm is presented. To handle radiometric variation problem, the following lighting model is considered in [3].

$$L = \underbrace{m_b e b + \alpha}_{\text{view independent}} + \underbrace{m_s e}_{\text{view dependent}} \tag{1}$$

where m_b and m_s are the weights in relation to view independent and dependent contributions respectively, e represents illumination energy which maybe different between corresponding pixels, b denotes surface albedo, α models diffuse lights.

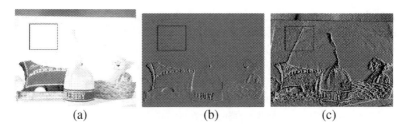

(a)   (b)   (c)

Fig. 1. A comparison between absolute and relative gradient magnitudes. (a) Middlebury dataset "Midd2". (b) Map of absolute gradient magnitudes. (c) Map of relative gradient magnitudes.

Radiometric variation occurs when $m_s e$ changes. Although α is view independent, it is unknown and therefore should be eliminated as well. As explained in [13], these variations can be eliminated by normalizing the gradient $G(x, y)$ with the largest gradient value $G_{\max}(x, y)$:

$$RG(x, y) = \frac{G(x, y)}{1 + G_{\max}(x, y)} \qquad (2)$$

where $RG(x, y)$ is the relative gradient of the pixel $p(x, y)$. $G_{\max}(x, y)$ is the largest gradient value within a 3×3 window centered at $p(x, y)$.

$$G_{\max}(x, y) = \max(|G(x+m, y+n)|), \ -1 \leq m, n \leq 1 \qquad (3)$$

Based on the assumption that the lighting geometries [13] are constant in a 3×3 neighborhood, m_s and a are excluded by the gradient operation (the numerator), and the only difference between matching pixels is the illumination energy e which is then eliminated by the division operation of (2). This algorithm allows us to extract the "lost" contents that often appear at the over exposed or illuminated regions of the analyzed image. An example is shown in Fig. 1. Relative gradients based stereo matching cost shows impressive results when there are small radiometric distortions but would have problems under extreme luminance/exposure conditions [18].

Census transform is a popular method for extracting local features. Classical census transform is defined as the raw intensity comparisons of the center pixel p with its neighboring pixels within a defined neighborhood $(q \in N(p))$. To describe the features in the computed relative gradient maps, the following modified census transform is proposed

$$T(p, q) = \begin{cases} 0, & \text{if } RG(q) < \overline{RG}(p) \\ 1, & \text{otherwise} \end{cases} \qquad (4)$$

where $\overline{RG}(p)$ denotes the average relative gradient value within a 3×3 window centered at p. Note that instead of comparing raw intensities, the modified census transform compares the relative gradients with the mean relative gradient of the patch. The results of these comparisons are then concatenated into a bit string $C(p) = \otimes_{q \in N(p)} T(p, q)$ with $\otimes$ representing the concatenation operation and $N(p)$ is a neighborhood of p. The transform is performed once both on the left and the right images prior to cost volume calculation. Finally, the census cost $C_{CT}(p,d)$ is given by the Hamming distance of the bit strings of pixel p in the left image and $(p-d)$ in the right image: $C_{CT}(p,d) = \sum_{q \in N(p)} \text{Hamming}(C_l(p), C_r(p-d))$ where C_l and C_r denote the left and the right transformed images, and d is a disparity candidate. Next, to take both the advantages of the relative gradient and census transform algorithms, the following stereo matching cost function is proposed.

$$C_{RGCT}(p,d) = \rho(\frac{C_{CT}(p,d)}{L_{CT}}, \lambda_{CT}) + \rho(C_{rg}(p,d), \lambda_{rg}) \tag{5}$$

where L_{CT} is the length of concatenated bit string ($L_{CT} = 9$), $C_{rg}(p,d)$ is the absolute difference between the pixels p and $p-d$ on the left ($RG_l(p)$) and the right ($RG_r(p-d)$) relative gradient images.

$$C_{rg}(p,d) = |RG_l(p) - RG_r(p-d)| \tag{6}$$

$\rho(c, \lambda)$ is a nonlinear function $\rho(c, \lambda) = 1 - \exp(-c/\lambda)$ that maps the census and relative gradient costs to the range [0, 1] and allows controlling their influence on outliers with only two parameters λ_{CT} ($= 55$) and λ_{rg} ($= 95$) [12].

3   Disparity Refinement

To find the disparity map, cost aggregation is performed using the method in [16]. This algorithm minimizes matching cost by recursively applying local smoothness constraint to weighted cost aggregation process.

$$C_{(p,p_d)}^A = C_{(p,p_d)} + \sum_{q \in v(p)} \min_{d \in D}(V(d_p, d_q) + C_{(q,q_d)}^A W_{p,q}(I)) \tag{7}$$

where $C_{(p,p_d)}$ denotes the cost of assigning disparity d to pixel p. $V(d_p, d_q)$ is a discontinuity penalty used to measure the cost of assigning disparity d_p and d_q to the pixels p and q (see [16] for more details). The edge weight $W_{p,q}(I)$ is defined as below in this paper:

$$W_{p,q}(I) = \exp(-\frac{\max_c |I_c(p) - I_c(q)|}{\delta}) \tag{8}$$

where $c \in \{R, G, B\}^1$, δ is a constant that adjusts the similarity between neighboring pixels. To compute the disparity map, matching cost is first aggregated independently from two inverse horizontal directions (i.e., left $\rightarrow$ right and left $\leftarrow$ right) over the image. The horizontal aggregated cost C_{hoz}^A is obtained by summing up the two intermediate results C_{lr}^A and C_{rl}^A and subtracting the initial matching cost

---

[1] For gray input image, $W_{p,q}(I) = \exp(-|I(p) - I(q)|/\delta)$.

$C_m : C_{hoz}^A = C_{lr}^A + C_{rl}^A - C_m$. Then the same operations are applied for the vertical aggregations but using C_{hoz}^A as the input. Finally, a raw disparity map is obtained from a winner-take-all operation $D_r = \arg\min_{d \in D} C_{ver}^A$, where C_{ver}^A is the vertical (final) aggregated matching cost.

Algorithm 1. Proposed Disparity Refinement

Input:

   Initial disparity map D_r and reference image I with $m \times n$ pixels. The edge weight between two adjacent pixels p (parent node) and q (child node) is $W_{p,q}(I)$.

.

Output:

   Refined disparity map D_f.

```
 1. For each pixel  p∈I,  set disparity belief  B(p)←0  if
        D_r(p) = unstable ,  otherwise  B(p)←1 .
 2. Left to right disparity belief propagation.
 3. for  i←0  to  m−1  do
 4.     for  j←0  to  n−1  do
 5.         if  D_r(p) = unstable  and  B_{lr}^A(p)≠0  then
 6.             B_{lr}^A(p) ← (W_{p,q}(I)+φ_p^{lr})B_{lr}^A(q), D_f(p) ← D_f(q)
 7.         else
 8.             B_{lr}^A(p) ← B(p), D_f(p) ← D_r(p)
 9.         end if
10.     end for
11.     end for
12.     Right to left disparity belief propagation.
13.     for  i←0  to  m−1  do
14.         Initialize:  B_{rl}^A(p) ← B_{lr}^A(p)  for all  p∈{row i}
15.         for  j←n−1  to 0  do
16.             if  D_f(p) = D_f(q)  then
17.                 B_{rl}^A(p) ← B_{lr}^A(p)+(W_{p,q}(I)+φ_p^{rl})B_{rl}^A(q) ,   D_f(p) ← D_f(q)
18.             else
19.                 if  B_{lr}^A(p) < (W_{p,q}(I)+φ_p^{rl})B_{rl}^A(q)  then
20.                     B_{rl}^A(p) ← W_{p,q}(I)B_{rl}^A(q) ,   D_f(p) = D_f(q)
21.                 end if
22.             end if
23.         end for
24.     end for
25. The same operations as lines 13-24 are applied for
    the vertical scans. Finally, a disparity map that
    maximizes total belief at each pixel is obtained.
```

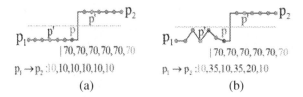

Fig. 2. A 1-D example of (a) an ideal edge and (b) an edge with sampling noise or unwanted local pattern. The numbers denote intensities for pixels from p_1 to p_2.

Since visibility is not modeled for occluded pixels, disparity refinement is often used afterwards to improve the initial results. In this paper, we propose an approach with computational complexity invariant to the disparity range. The algorithm propagates disparities from stable pixels to unstable pixels based on disparity belief propagation and loss compensation. Pixels passing left right consistency (also called LRC) check are regarded as stable, and the concept of loss compensation could be understood in the sequel. The disparity belief is defined as:

$$B(p) = \begin{cases} 1, & D_r(p) \text{ is stable} \\ 0, & \text{otherwise} \end{cases} \tag{9}$$

Similar to the cost aggregation process, the proposed disparity refinement is guided by reference image I with edge weight defined in (8). Before post-processing, mismatched pixels are detected by left-right consistency check and labeled unreliable. Let D_f denote refined left disparity map, and B_{lr}^A represent the map of accumulated disparity belief in left to right refinement. Refinement is performed via a four-pass aggregation. The proposed algorithm is summarized in Algorithm 1. For the first pass from left to right, parent node p receives belief $(W_{p,q}(I) + \varphi_p^{lr})B_{lr}^A(q)$ from child node q if $D_r(p) =$ unstable and $B_{lr}^A(p) \neq 0$ (Lines 5–6 of Algorithm 1), where φ_p^{lr} is a left-right belief compensation.

$$\varphi_p^{lr} = \begin{cases} 0, & \max_c |I_c(p) - I_c(p_1)| \geq \max_c |I_c(p) - I_c(p_2)| \\ 1 - W_{p,q}, & \text{otherwise} \end{cases} \tag{10}$$

p_1 and p_2 are boundary stable pixels (i.e., for any $p \in (p_1, p_2)$, $D_r(p) =$ unstable).

To investigate the effect of φ_p^{lr}, let us first consider the case shown in Fig. 2(a) by assuming a unit belief propagated from p_1 to p. In this case, the unit belief can reach p losslessly even without any compensation $(B_{lr}^A(p) = W_{p,q}(I)B_{lr}^A(q))$. However, for the edge with sampling noise or unwanted local pattern (Fig. 2(b)), only 0.053 belief propagates to p, which is less than that of pixel p receiving from p_2 (0.099), therefore p receives a wrong disparity from p_2. With loss compensation, the unit belief can propagate to p without a loss. This is because $W_{p'q'} + \varphi_{p'}^{lr} = 1$ for any $p', q' \in (p_1, p]$ when the absolute intensity difference between p' and p_1 is less than that between p' and p_2. On the opposite, $W_{p'q'} + \varphi_{p'}^{lr} = W_{p'q'}$ means that belief propagation across an edge is not affected, thus the edge is still preserved. Figure 3 shows an example for the

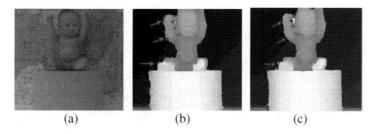

| (a) | (b) | (c) |

Fig. 3. Comparison of the proposed disparity refinement method with (b) and without (c) belief loss compensation.

proposed disparity refinement algorithm with and without loss compensation. It is clear that outliers near object boundaries are reduced with the technique.

In right to left refinement, disparity belief of current node $(B_{rl}^A(p))$ should include propagation from left to right pass $(B_{lr}^A(p))$ if $D_f(p) = D_f(q)$ (Lines 16–18). This indicates that disparity of current node is more reliable. Otherwise, current node p receives belief $W_{p,q}(I)B_{lr}^A(q)$ from child node q if belief of current node is less than child node (Lines 20–21), which indicates that disparity of current node is unreliable. Two vertical scans require the same operations as lines 13–24. After post-processing, a disparity map that maximizes total belief at each pixel is obtained.

4 Experimental Results

The proposed matching cost algorithm is evaluated using seven Middlebury image pairs (Aloe, Baby 1, Cloth 4, Art, Dolls, Laundry and Moebius)[2] taken under three different illumination conditions (indexed as 1, 2, 3) and three different exposure times (indexed as 0, 1, 2). We considered two test conditions in the experiments. For testing the effect of illumination changes, we fixed the index of exposure to 1 and varied only

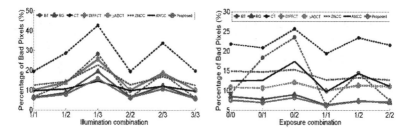

Fig. 4. Average error rates for different illumination (a) and exposure (b) conditions. (Color figure online)

---

[2] Note that since this paper is not to evaluate cost aggregation algorithm, the Middlebury 2014 datasets which contain several new features are not used for evaluation.

Fig. 5. Visual comparison of various matching cost algorithms on 'Aloe' image pair under different illumination (illu) and exposure (exp) combinations. (a, b) Left and right images with (illu-1, exp-1) and (illu-3 and exp-1) conditions respectively. (k, l) Left and right images with (illu-2, exp-0) and (illu-2 and exp-2) conditions respectively. (c–j) and (m–t) are the results using the image pair (a, b) and (k, l) separately.

the index of illumination from 1 to 3. To evaluate its robustness against camera exposure variations, we fixed the index of illumination to 2 and changed only the index of exposure from 0 to 2. The proposed algorithm was compared with seven matching cost algorithms: BT^3 ($\tau_1 = 7/255$, $\tau_2 = 2/255$, $\alpha = 0.89$) [17], RG [3], CT (7×9 window) [19], DIFFCT (7×9 window, $\lambda_{CT} = 55$, $\lambda_{Diff} = 55$) [20], ΔADCT (7×9 window, $\lambda_{CT} = 90$, $\lambda_{\Delta AD} = 90$) defined in (11), ZNCC (9×9 window), and ANCC (31×31 window, $\gamma_g = 392$ and $\delta_s = 28.8$) on RGB color channels. The parameters of each method were set with reference to the original works.

$$C_{\Delta ADCT}(p,d) = \rho(C_{CT}(p,d), \lambda_{CT}) + \rho(C_{\Delta AD}(p,d), \lambda_{\Delta AD}) \tag{11}$$

Table 1. Runtime of matching cost algorithms on 'Aloe' image pair

Method	BT	RG	CT	DIFFCT	ΔADCensus	ZNCC	ANCC	Proposed
Time(s)	0.769	0.953	0.578	0.934	0.782	38.12	230+	**0.563**

---

[3] For implementing BT, we used the code provided in [11].

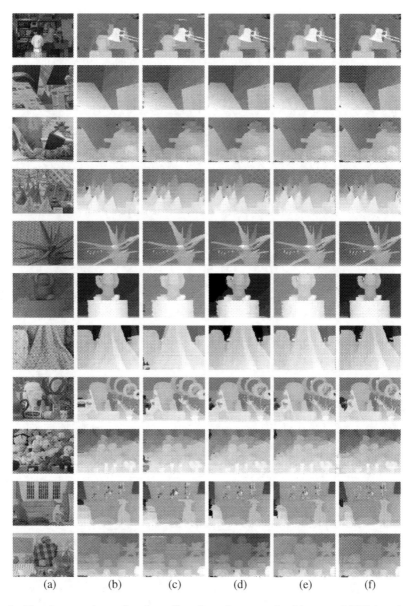

(a) (b) (c) (d) (e) (f)

Fig. 6. Visual comparison of various disparity refinement algorithms on Middlebry dataset. (a) Left images (Tsukuba, Vensus, Teddy, Cones, Aloe, Baby 1, Cloth 4, Art, Dolls, Laundry and Moebius). (b–f) are the results using [10] (b), [16] (c), [11] (d), [15] (e), and the proposed (f) disparity refinement methods respectively.

Figure 4 shows the error rates for different illumination and exposure combinations. It is clear that our algorithm has much lower error rates than the other algorithms and is very close to the ANCC algorithm under extreme luminance condition but outperforms the ANCC algorithm greatly when there are small radiometric distortions.

Table 2. Comparison of three disparity refinement methods.

Test image	Method [10]		Method [16]		Method [11]		Method [15]		Proposed	
	Nonocc	All	Nonocc	All	Nonocc	All	Nonocc	All	Nonocc	All
Tsukuba	4.19	5.38	5.88	7.64	**3.89**	**5.17**	4.17	5.63	4.30	5.46
Vensus	**0.75**	**1.55**	2.15	3.58	0.76	1.76	0.94	2.03	0.79	1.71
Teddy	**4.71**	**10.87**	6.30	13.35	5.03	11.67	5.09	12.09	4.84	11.36
Cones	3.24	**9.27**	4.17	10.84	**3.00**	10.06	3.36	10.56	3.34	9.38
Avg. Err.	3.22	**6.77**	4.63	8.85	**3.17**	7.17	3.39	7.58	3.32	6.98

Table 3. Comparison of three disparity refinement methods.

Test image	Method [10]		Method [16]		Method [11]		Method [15]		Proposed	
	Nonocc	All	Nonocc	All	Nonocc	All	Nonocc	All	Nonocc	All
Aloe	**3.28**	**6.88**	3.65	9.73	4.27	10.14	3.63	9.63	3.41	7.65
Baby 1	2.17	**3.68**	2.92	6.78	5.98	8.93	2.30	5.65	**2.15**	3.93
Cloth 4	1.22	**10.48**	1.29	12.25	1.20	11.03	1.31	11.28	**1.09**	11.10
Art	7.33	**18.44**	8.33	20.41	9.65	22.08	7.18	20.93	**7.04**	18.53
Dolls	3.86	**11.01**	4.08	11.64	4.46	12.78	3.92	12.71	**3.74**	**11.01**
Laundry	13.33	**23.38**	15.06	25.99	**12.62**	24.62	13.83	25.96	13.59	23.90
Moebius	8.58	**13.72**	8.83	15.93	9.32	16.74	8.62	15.94	**8.48**	13.91
Avg. Err.	5.68	**12.51**	6.31	14.68	6.79	15.19	5.83	14.59	**5.64**	12.86

Table 4. Runtime of different disparity refinement algorithms on 'Aloe' image pair

Method	Method [10]	Method [16]	Method [11]	Method [15]	Proposed
Time(s)	10.72	0.956	0.274	0.017	**0.013**

Figure 5 visualizes the disparity maps for the test image 'Aloe' with the '1/3' illumination and '0/2' camera exposure. All testings are performed on a 2.0 GHz Intel Dual laptop with 2 GB RAM and no explicit parallelism technique is utilized. We implemented them in C and tried to make them efficient but without paying too much effort on optimization. The computational times of the algorithms are summarized in Table 1 for the Aloe image pair[4], which has 427×370 pixels and 70 disparity levels. Note that the RG algorithm has three channels as suggested in [3] for color image, while the proposed RGCT using intensity image as the input. Comparison results indicate that the RGCT has the best overall performance.

Figure 6 visually compare the proposed disparity refinement method with the local disparity refinement method [10], the smoothness enforced disparity refinement method

---

[4] For the runtime of ANCC, we direct use the results reported in [21].

[16], the non-local disparity refinement method [16] and the nonlocal and disparity range invariant refinement method [15]. To better evaluate our algorithm's performance, two Middlebury datasets [Tsukuba, Vensus, Teddy, Cones] and [Aloe, Baby 1, Cloth 4, Art, Dolls, Laundry and Moebius] were used for comparison. Note that the comparison was made using the RGCT matching cost function and the illumination and exposure indexes were fixed to 2/2 and 1/1 respectively for the second dataset for all of the algorithms. Compared to the first dataset, the second one is much more challenging for refinement. Table 2 summarizes the error rates for the first dataset, which indicates that they are very close to each other. However, Table 3 shows that, for the second dataset, the proposed method has the *lowest* error rate for the 'non-occluded' pixels and ranks *second* among the five algorithms for the 'all' pixels. From Fig. 6, one can find that artifacts near object boundaries are reduced using the proposed belief loss compensated disparity refinement method (especially the Teddy, Baby 1, and Art images). Besides, we also measured their efficiency using the Aloe image pair. The results are summarized in Table 4. The proposed algorithm even has less computational amount than [15]. This is because the method [15] needs MST tree construction. Finally, it is worth mentioning that both the proposed matching cost and disparity refinement algorithms are parallelizable and can be further speed up by parallel computing devices such that they can meet real-time requirements.

5 Conclusion

In this paper, we proposed a novel radiometric invariant stereo matching algorithm called RGCT and a disparity refinement method based on disparity belief propagation and loss compensation. The RGCT algorithm effectively combines modified census transform with relative gradients measures to take both their advantages. The relative gradient algorithm allows us to extract the "lost" contents that often appear at the over exposed or illuminated regions of the analyzed image, and the census transform is used to extract the local features. Comparison results demonstrated that it has much lower error rates than many existing algorithms and is very close to the ANCC algorithm which represents the current state of the art under extreme luminance conditions but outperforms the ANCC algorithm greatly when there are small radiometric distortions. More importantly, the proposed RGCT algorithm keeps very low computational amount. In addition, a new technique was introduced in this paper to reduce the loss of disparity belief caused by sampling noise or unwanted local pattern. The proposed disparity refinement method is invariant to the disparity range. Experimental results showed that artifacts near object boundaries are reduced using the proposed refinement method.

Acknowledgments. This work was supported by a grant from National Natural Science Foundation of China (NSFC, No. 61504032).

References

1. Heo, Y.S., Lee, K.M., Lee, S.U.: Robust stereo matching using adaptive normalized cross-correlation. IEEE Trans. Pattern Anal. Mach. Intell. **33**, 807–822 (2011)
2. Kim, S., Ham, B., Kim, B., Sohn, K.: Mahalanobis distance cross-correlation for illumination-invariant stereo matching. IEEE Trans. Circ. Syst. Video Technol. **24**(11), 1844–1859 (2014)
3. Zhou, X., Boulanger, P.: Radiometric invariant stereo matching based on relative gradients. In: 19th 9th IEEE International Conference on Image Processing, pp. 2989–2992. IEEE (2001)
4. Viola, P., Wells, W.M.: Alignment by maximization of mutual information. Int. J. Comput. Vis. **24**(2), 137–154 (1997)
5. Fife, W., Archibald, J.: Improved census transforms for resource-optimized stereo vision. IEEE Trans. Circ. Syst. Video Technol. **23**(1), 60–73 (2013)
6. Sinha, S., Scharstein, D., Szeliski R.: Efficient high-resolution stereo matching using local plane sweeps. In: IEEE Conference on Computer Vision and Pattern Recognition, pp. 1582–1589. IEEE (2014)
7. Bleyer, M., Rother, C., Kohli, P., Scharstein D., Sinha S.: Object stereo-joint stereo matching and object segmentation. In: IEEE Conference on Computer Vision and Pattern Recognition, pp. 3081–3088. IEEE (2011)
8. Gu, Z., Su, X., Liu, Y., Zhang, Q.: Local stereo matching with adaptive support-weight, rank transform and disparity calibration. Pattern Recogn. Lett. **29**, 1230–1235 (2008)
9. Hosni, A., Bleyer, M., Gelautz, M., Rhemann, C.: Local stereo matching using geodesic weights. In: 19th IEEE International Conference on Image Processing, pp. 2093–2096. IEEE (2009)
10. Rhemann, C., Hosni, A., Bleyer, M., Rother, C., Gelautz, M.: Fast cost-volume filtering for visual correspondence and beyond. In: IEEE Conference on Computer Vision and Pattern Recognition, pp. 3017–3024. IEEE (2011)
11. Yang, Q.: A non-local cost aggregation method for stereo matching. In: IEEE Conference on Computer Vision and Pattern Recognition, pp. 1402–1409. IEEE (2012)
12. Mei, X., Sun, X., Zhou, M., Jiao, S., Wang, H., Zhang, X.: On building an accurate stereo matching system on graphics hardware. In: IEEE Conference on Computer Vision and Pattern Recognition Workshop, pp. 467–474. IEEE (2011)
13. Sun, X., Mei, X., Jiao, S., Zhou, M., Wang, H.: Stereo matching with reliable disparity propagation. In: International Conference on 3D Imaging, Modeling, Processing, Visualization, Transmission, pp. 132–139, IEEE (2011)
14. Yang, Q., Ji, P., Li, D., Yao, S., Zhang, M.: Fast stereo matching using adaptive guided filtering. Image Vis. Comput. **32**(3), 202–211 (2014)
15. Huang, X., Cui, G., Zhang, Y.: A fast non-local disparity refinement method for stereo matching. In: IEEE International Conference on Image Processing, pp. 3823–3827. IEEE (2014)
16. Yang, Q.: Local smoothness enforced cost volume regularization for fast stereo correspondence. IEEE Signal Process. Lett. **22**(9), 1429–1433 (2015)
17. Birchfield, S., Tomasi, C.: A pixel dissimilarity measure that is insensitive to image sampling. IEEE Trans. Pattern Anal. Mach. Intell. **20**(4), 401–406 (1998)
18. Xu, L., Au, O.C., Sun, W., Fang, L., Zou, F., Li, J.: Stereo matching with optimal local adaptive radiometric compensation. IEEE Signal Process. Lett. **22**(2), 131–135 (2015)

19. Zabih, R., Woodfill, J.: Non-parametric local transforms for computing visual correspondence. In: Eklundh, J.-O. (ed.) ECCV 1994. LNCS, vol. 801, pp. 151–158. Springer, Heidelberg (1994)
20. Miron, A., Ainouz, S., Rogozan, A., Bensrhair, A.: A robust cost function for stereo matching of road scenes. Pattern Recogn. Lett. **38**, 70–77 (2014)
21. Mouats, T., Aouf, N., Richardson, M.: A novel image representation via local frequency analysis for illumination invariant stereo matching. IEEE Trans. Image Process. **24**(9), 2685–2700 (2015)

A Real-Time Head Pose Estimation Using Adaptive POSIT Based on Modified Supervised Descent Method

Zhong-Qiu Zhao[1(✉)], Kewen Cheng[1], Qinmu Peng[2],
and Xindong Wu[1]

[1] College of Computer and Information,
Hefei University of Technology, Hefei, China
z.zhao@hfut.edu.cn
[2] Department of Computer Science, Hong Kong Baptist University,
Hong Kong, SAR, China

Abstract. In this paper, we proposed a real-time head pose estimation algorithm by extending Pose from Orthography and Scaling with Iterations (POSIT) (named Adaptive POSIT) method and modifying the Supervised Descent Method (SDM). Specifically, we used the modified SDM for facial landmarks detection and tracking, and adopted adaptive POSIT to estimate head pose. In the feature selection stage, we extracted different features in neighboring facial landmarks instead of a single feature. In the facial landmarks selection stage, we used partial facial landmarks instead of the whole facial landmarks. The experiments show that our method can track facial landmarks robustly with tolerance to certain illumination changes and partial occlusion, and improves the accuracy of head pose estimation.

Keywords: Head pose estimation · SDM · POSIT · Facial landmarks

1 Introduction

Head pose estimation, which determines the rotation angles involving three directions of face image in 3D space, has many applications including human-computer interaction, multi-view face recognition system [1], driving attention monitoring, and so forth. The rotation angle involves three directions including *pitch* (up and down), *yaw* (right and left), and *roll* (in-plane), as shown in Fig. 1.

The methods of head pose estimation mainly include three categories: (1) facial appearance-based; (2) classification-based; (3) models-based.

Facial appearance-based methods assume that there is a certain relationship between head pose and the features (such as gray, color, image gradient) extracted from face image, and the relationship can be established by a statistic method with a large number of samples. Extracting gray features in image space is a primitive method. However, the dimension of the gray features is very high and it requires huge computation cost. So, we generally replace image raw data with the features of lower dimensional space.

© Springer International Publishing Switzerland 2016
D.-S. Huang et al. (Eds.): ICIC 2016, Part I, LNCS 9771, pp. 74–85, 2016.
DOI: 10.1007/978-3-319-42291-6_8

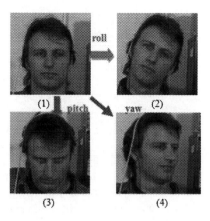

Fig. 1. Head pose in different angles. (1) The original image; (2–4) Subfigures show the images after the *roll*, *pitch*, and *yaw* rotations, respectively.

Classification-based methods mainly include manifold embedding methods and nonlinear regression methods. The former replaces face image with a high dimensional space vector and extract linear features from testing sample by mapping from the high dimensional space to its subspace. Then nearest neighbor classifier is used for classification. The latter learns the nonlinear functional mapping from the global features (such as Histogram of Oriented Gradient) extracted from face image to one or more pose directions.

Model-based methods usually apply facial geometric relationship or facial feature points for head pose estimation. They firstly use a certain geometric model or structure to represent face geometry and shape. Then some facial features are extracted to establish the corresponding relationship between the model and the images. Finally head pose is estimated by geometric model or other face model. Many model-based methods apply facial feature points for face modeling. They firstly localize the facial landmarks, and then make it match the projective landmarks of 3D face model which is parameterized by the rotation and translation. Finally, the head pose is estimated according to the matching result.

POSIT [2] method can be used to estimate 3D object pose with respect to a camera when 3D object model and the corresponding location of key points in 2D space is given. Its advantages lie in its efficiency and accuracy when facial landmarks are localized correctly. SDM [3] is a regression method of localizing facial landmarks iteratively. Its advantages lie in its robustness to certain illumination and partial occlusion, and capable of high accuracy localization. Our work is inspired partially by POSIT and SDM.

The contributions of the paper mainly include two aspects: (1) An adaptive method is proposed to select accurate facial landmarks, and thereby the accuracy of head pose estimation is improved. (2) Feature pool is proposed to extract the features in neighboring facial landmarks, and thereby the recognition performance is improved.

The paper includes five sections: Sect. 2 reviews the related works about head pose estimation. Section 3 details our algorithm of head pose estimation. Section 4 shows our experimental results. Finally, some conclusions are drawn in Sect. 5.

2   Related Works

In the early years, the work on head pose estimation mainly focused on training a model which constructing a relationship between head pose and image features. Lablack et al. [4] proposed Gabor features for constructing different linear feature space in different training image. The methods based on Gabor features achieved good performance. But the multi-scale and multi-direction transformation on the image largely increase the calculation cost and decrease the computation speed. Voit et al. [5] used neural networks for head pose estimation. However, they all assumed that the face region had already been precisely localized. But localization error had an impact on the accuracy of the head pose estimation.

Recently, many classification-based methods for head pose estimation have been proposed. Xin and Yu [6] proposed to associate a multivariate label distribution (MLD) to each image for head pose estimation. Fanelli et al. [7] used random regression forests for real-time head pose estimation. Generally, the accuracy of classification-based methods is usually not high because classification is a discrete problem which is hard to simulate the continuous process of head pose changing. On the other hand, the false classification always occurs.

Nowadays, more and more researchers focus on model-based methods for head pose estimation. Using a dimensionality reduction technique such as PCA on training facial shape results in an Active Shape Model (ASM) [8], capable of representing the primary modes of shape variation. Active Appearance Model (AAM) [9] learns the primary modes of variation in facial shape and texture from a 2D perspective. Constrained Local Model (CLM) [10] inherits the characteristic of ASM and AAM, improving the performance. Recently, several new facial landmarks location methods have been proposed. Cao et al. [11] proposed Explicit Shape Regression (ESR) utilizing a two-level boosted regressor, and combining local shape-index features and correlation features selection. Ren et al. [12] proposed Local Binary Features (LBF) to learn the locality principle. Chen et al. [13] jointed cascade face detection and alignment. The SDM learns a sequence of descent directions that minimizes the mean of Non-linear Least Squares functions sampled at facial landmarks, and extracts a single feature at each landmark. Its robustness and efficiency is currently outstanding in facial landmarks localization. However, SDM will introduce many overlaps in neighboring feature blocks due to the density distribution of facial landmarks.

To sum up, these model-based methods also have the limitation of poor landmarks localization problem, which always occurs in partial occlusion or low-resolution images. And the accuracy of facial landmarks location largely influences the head pose estimation. Our proposed method will show that it can improve the performance for head pose estimation by removing partial inaccurate landmarks adaptively. In the next section, we will detail our method for head pose estimation.

3 The Proposed Algorithm

3.1 Initialize Face Shape

During detection, we adopt face detection algorithm based on Haar features and AdaBoost cascade classifier [17] to obtain face region; During tracking, we obtain face region by face shape at the previous frame; Then face shape is initialized by current face region.

Here we need to explain the difference between Detection Model (DM) and Tracking Model (TM): The mainly difference in training lies in the initial face shape selection. The former's initial face shape is generated from face region according to face detection. The latter's initial face shape is generated from face region according to true shape. Therefore, in the detection stage, face region is obtained by face detection. However, in the tracking stage, face region is obtained by the previous shape.

3.2 Regress Face Shape

This paper applies SDM based on shape regression for facial landmarks localization. And facial landmarks tracking based on SDM is very stable, especially compared with the simple geometric model based on five feature points (two centers of eyes, nose tip, and two corners of mouth). Moreover, we can more flexibly choose feature points. Even though there is partial occlusion during tracking, head pose can be well estimated robustly.

SDM is based on machine learning and solves non-linear least squares problem. It learns the directions of gradient descent from training data and establishes corresponding regression model. During testing, we can adopt the model to predict the directions of gradient descent. Newton method is a common approach to solve least squares. However, its computational cost is too high when computing Hessian matrix directly. Therefore, SDM adopts the learned directions of gradient descent for approximation to solve Hessian matrix and gradient:

$$x_k = x_{k-1} + R_{k-1}\phi_{k-1} + b_{k-1} \tag{1}$$

Where x_k is the location of facial landmarks after the kth iteration, ϕ_{k-1} is the feature vector extracted from facial landmarks after the (k − 1)th iteration, R_{k-1} and b_{k-1} represent descent directions and bias after learning, respectively. Because SDM is common to objective function, we can apply it to predict the location of facial landmarks. During training, SDM mainly updates two parameters including shape difference $\Delta x_*^k = x_* - x_k$ and feature vector ϕ_k, and then generates new training data. The whole training process minimizes

$$\underset{R_k,b_k}{\arg\min} \sum_{x_k} \left\| \Delta x_*^k - R_k\phi_k - b_k \right\|^2. \tag{2}$$

During testing, SDM firstly updates face shape iteratively by the learned R_k and b_k, and then updates the corresponding feature vector ϕ_k, finally generates the predicted shape. However, in the practical video, SDM suffers poor landmarks localization and overlaping in neighboring feature blocks. Therefore, this paper presents two aspects of improvements:

(1) Adaptive facial landmarks selection: Facial landmarks usually include the contour of eyebrows, eyes, nose, mouth, and face. When the facial landmarks location is poor, the accuracy of head pose estimation will be largely influenced accordingly. Therefore, we propose an adaptive method to select the accurate facial landmarks. The mainly steps are as follows: Firstly, we use SDM to localize 66 key points (shown in Fig. 2) when frontal face appears and extract the corresponding feature ϕ_0 as the comparative standard. During tracking, we extract the corresponding feature ϕ_i at facial landmarks and compare the similarities between different feature blocks. The adaptive facial landmarks include 10 constant landmarks and 12 most similar landmarks. The former are selected from 49 internal facial landmarks. The latter are selected from the rest 39 internal facial landmarks. We use σ to represent the threshold of the similarity. $s_i (1 \leq i \ll 39)$ is used for representing the similarity of the corresponding 39 facial landmarks and is sorted by ascending. The threshold of the similarity is set to $s_{11} < \sigma < s_{10}$. ϕ_{ij} is the feature extracted from the jth key points in the ith frame, ϕ_{0j} is the feature extracted from the jth key points in the initialized frame f_0, $s(\phi_{0j}, \phi_{ij})$ is the similarity between the two corresponding features, when $s(\phi_{0j}, \phi_{ij}) > \sigma$, we retain the corresponding key point; Otherwise, we remove the corresponding key point. Finally, the final key points are used for head pose estimation and the corresponding 3D face model updates accordingly.

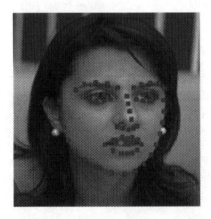

Fig. 2. 66 facial landmarks distribution.

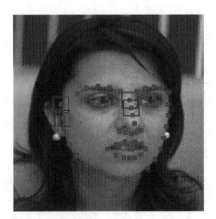

Fig. 3. Overlaps in feature blocks.

(2) Feature selection: Because we need to localize facial landmarks more accurate, feature with better identification performance is needed. However, feature blocks extracted from facial landmarks usually have a dense distribution which has impact on the identification performance. Therefore, there are probably redundant overlaps in neighboring feature blocks when we select a single feature (such as Fig. 3), which usually have an negative on the identification performance between neighboring feature blocks. Therefore, we present the concept of feature pool including many different features (such as SIFT [14], HOG [15], LBP [16], Haar [17]) extracted from neighboring feature points. We extract different features in neighboring feature points. We mainly use HOG and SIFT features with robustness to certain illumination changes.

3.3   Head Pose Estimation

POSIT is a method estimating 3D object pose according to given 3D object pose and the corresponding location of key points in 2D space. Its advantages lie in its efficiency and accuracy when facial landmarks are localized correctly. Below we will describe POSIT in details.

Object coordinate system: $b = (x,y,z)$, which is defined by the center of object. Camera coordinate system: $c = (X,Y,Z)$, which is defined by the center of camera. Through rigid transformation M (including rotation matrix R and translation matrix T), object coordinate can be transformed into camera coordinate as follows: $c = M(b, 1)^T$, where $M = (R,T)$. Screen coordinate system: $u = (x_0, y_0)$, which is the 2D coordinate in image. Through projection, camera coordinate can be transformed into screen coordinate as follows: $u = \prod_Q (c) = (\frac{x}{z}f + u_0, \frac{y}{z}f + v_0)$, $Q = \begin{bmatrix} f & 0 & u_0 \\ 0 & f & v_0 \\ 0 & 0 & 1 \end{bmatrix}$, (u_0, v_0) is the image center, f is the focal length, $\prod_Q$ represents the projection function of transformation from 3D coordinate into 2D coordinate.

In face image, facial landmarks location U corresponds to screen coordinate. According to given 3D face model S corresponding to object coordinate, POSIT solves for rotation matrix R and translation matrix T by matching predicted facial landmarks iteratively. It can be formulated as follows:

$$\arg\min_{R,T} \left\| \prod_Q (SR + T) - U \right\|^2 \tag{3}$$

The relationship between rotation matrix R and corresponding 3D angle (*roll, pitch, yaw*) is defined in Eq. (4):

$$R = R_x R_y R_z \tag{4}$$

Where

$$R_x = \begin{bmatrix} 1 & 0 & 0 \\ 0 & \cos(yaw) & -\sin(yaw) \\ 0 & \sin(yaw) & \cos(yaw) \end{bmatrix} \qquad (5)$$

$$R_y = \begin{bmatrix} \cos(pitch) & 0 & \sin(pitch) \\ 0 & 1 & 0 \\ -\sin(pitch) & 0 & \cos(pitch) \end{bmatrix} \qquad (6)$$

$$R_z = \begin{bmatrix} \cos(roll) & -\sin(roll) & 0 \\ \sin(roll) & \cos(roll) & 0 \\ 0 & 0 & 1 \end{bmatrix} \qquad (7)$$

According to the above transformation, we can easily transform from rotation matrix R into 3D angle (*roll, pitch, yaw*). Our algorithm is described as follows:

(1) Input video frame;
(2) Initialize face shape: Given current frame, face detection is done. If face detection fails, then face detection is continued; Otherwise the tracking stage is initialized, and face region are recorded. During tracking, current face region is obtained by previous face shape and recorded; According to mean shape, the face shape is initialized by current face region;
(3) Regress face shape: During detection, DM is applied for regressing face shape; During tracking, TM is applied for regressing face shape.
(4) Head pose estimation: POSIT is applied for head pose estimation. Given the focal length and image center, we can estimate 3D head pose including *roll, pitch*, and *yaw* using the given 3D face model and predicted face shape.

Figure 4 also shows the flow chart of our algorithm.

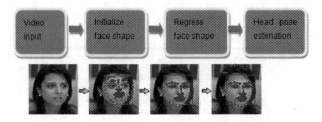

Fig. 4. The flow chart of our algorithm.

4   Experiments

Our experiments are implemented on a PC computer, i7 core CPU, 3.4 GHz, memory 32 G, and an ordinary camera is used. LFW-A&B&C [18] dataset and Helen dataset [19] are used for training and the whole dataset includes 4481 face images. LFPW [20] is only used for testing. Below the three datasets will be introduced.

LFW-A&B&C dataset includes the partial LFW dataset whose name is begin with 'A', or 'B', or 'C'. They use 66 landmarks containing 17 face contour points and 49 internal face points (including the feature points around eyebrows, eyes, nose, and mouth) marking. Due to the annotated difference between LFW and Helen, we separately mark face contour points and internal face points. The face contour points keep invariant and the internal face points use manual annotation. Each image is low-resolution with a fixed size of 250 * 250.

Helen dataset includes two parts of training and testing data. The former includes 2000 face images and the latter includes 330 face images. They use 66 landmarks marking. Each image is high-resolution with an unfixed size.

LFPW dataset includes training data and testing data. The former includes 811 face images and the latter includes 224 face images. They use 68 landmarks marking. Each image is high-resolution with an unfixed size.

4.1   Head Pose Estimation

For detected face region, we uniformly normalize the image size to 250 * 250. The block features are extracted from each key point and the size of each block is set to 32 * 32. The SDM has fast convergence and the iteration number is set to 4. For the training of DM and TM, the disturbed number is set to n = 1 and n = 5, respectively. During training TM, the translation variance is set to 20 and the scale variance is set to 0.0005. DM initializes the shape using face region which is obtained by opencv face detection [21]. TM initializes the shape using face region which is obtained by true shape. According to the predicted facial landmarks, we can apply POSIT for head pose estimation and the 3D face model can change adaptively, and the focal length is set to 1000. Figure 5 shows 10, 20, 49 facial landmarks distribution, respectively. Our adaptive points mainly include facial landmarks around eyebrows, eyes, nose, and mouth. And our adaptive landmarks selection results are shown in Fig. 6. The experimental results with different numbers of facial landmarks are shown in Table 1.

To evaluate the performance of different methods for head pose estimation, the standard videos whose ground truth are provided by the Boston University head pose database [22] are used. To measure the pose estimation error, the root-mean-square error (RMSE) value is used. We used 20 videos including three persons (Jam, Jim, Llm) in our experimental data. The person Jam and Llm all have 9 short videos. The person Jim has 2 short videos. And each video has 200 frames. The experimental results with different methods of head pose estimation are shown in Table 2. Our experiments show that our method can track facial landmarks with much higher accuracy and it is implemented in C++ at over 22 fps.

4.2   Fatigue Driving Detection

Fatigue driving detection as an important application in head pose estimation can play the role of alarming drivers. Figure 7 shows the process of downing one's head, and the corresponding head pose pitch will increase. When the driver falls in the fatigue

Table 1. Head pose estimation errors (in RMSE) with different numbers of landmarks. Adaptive refers to our adaptive facial landmarks selection.

Numbers of landmarks \ Head Pose	Roll	Pitch	Yaw
10	2.02	**3.02**	3.75
20	**1.95**	3.24	3.84
49	1.98	3.20	3.88
Adaptive	2.05	3.03	**3.66**

Table 2. Head pose estimation errors (in RMSE) with different head pose methods.

Method \ Head Pose	Roll	Pitch	Yaw
Sung et al.[23]	3.1	5.6	5.4
An et al.[24]	3.22	7.22	5.33
Roberto et al.[25]	3.00	6.10	5.26
Gritti et al.[26]	2.4	9.9	6.7
Kim et al.[27]	2.45	7.22	4.01
Adaptive	**2.05**	**3.03**	**3.66**

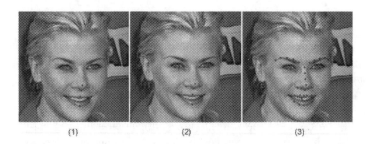

(1)   (2)   (3)

Fig. 5. Different numbers of facial landmarks distribution. (1) 10 landmarks, (2) 20 landmarks, (3) 49 landmarks.

driving, he/she possibly happens to down his/her head and we judge the fatigue driving state once the pitch is more than a setted threshold in a certain continuous time (such as in continuous five frames). Figure 9 shows that we judge the fatigue driving state by the range of head pose. Figure 8 shows the distribution of eyes landmarks which are marked in letters A–L, respectively. AD (GH, BF, HL, CE, IK) represents the distance of two corresponding landmarks. When closing one's left (right) eye, AD (GJ) is relative constant and BF (HL) or CE (IK) will decrease with assuming the head hasn't other movements or only minor movement. Therefore, BF/AD (HL/GJ) will decrease and we judge the fatigue driving state once BF/AD or HL/GJ is less than a setted

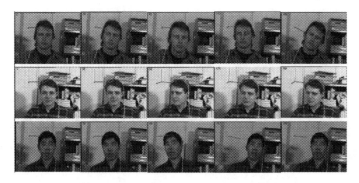

Fig. 6. Our adaptive landmarks selection.

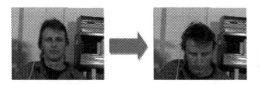

Fig. 7. The process of downing one's head.

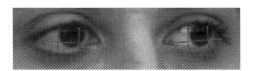

Fig. 8. Eyes landmarks distribution.

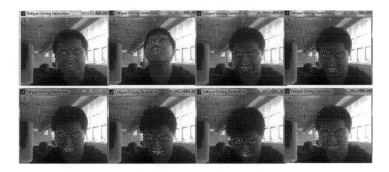

Fig. 9. We judge the fatigue driving state by the range of head pose, where Y represents the normal state, N represents the fatigue driving state.

Fig. 10. We judge the fatigue driving state by closing one's eyes, where Y represents the normal state, N represents the fatigue driving state.

threshold in a certain continuous time. Figure 10 shows that we judge the fatigue driving state by closing one's eyes. In short, we can judge fatigue driving state based on head pose estimation and closing one's eyes. When either of both is happened, we will give a warning; If not, we assume that it is a normal driving state.

5   Conclusion

This paper proposes a real-time head pose estimation using an adaptive POSIT based on modified SDM. In facial landmarks detection, we adopt the modified face alignment algorithm based on SDM which shows its robustness in a certain illumination variation and partial occlusion. In head pose estimation, we select adaptive POSIT based on 3D face model projection. It shows the promising results in terms with accuracy and efficiency for the head pose estimation.

References

1. Tu, J., Huang, T., Xiong, Y.: Calibrating head pose estimation in videos for meeting room event analysis. In: ICIP, pp. 3193–3196 (2006)
2. DeMenthon, D.F., Davis, L.S.: Model-based object pose in 25 lines of code. Int. J. Comput. Vis. **15**, 123–141 (1995)
3. Xiong, X.H., Fernando, D.L.T.: Supervised descent method and its applications to face alignment. In: CVPR, pp. 532–539 (2013)
4. Lablack, A., Zhang, Z., Djeraba, C.: Supervised learning for head pose estimation using SVD and gabor wavelets, pp. 592–596. IEEE (2008)
5. Voit, M., Nickel, K., Stiefelhagen, R.: Multi-view head pose estimation using neural networks. In: International Conference on Computer and Robot Vision, IEEE Computer Society, pp. 347–352 (2005)
6. Xin, G., Yu, X.: Head pose estimation based on multivariate label distribution. In: CVPR, pp. 1837–1842 (2014)

7. Fanelli, G., Gall, J., Van, Gool, L.: Real time head pose estimation with random regression forests. In: CVPR, pp. 617–624 (2011)
8. Cootes, T., Taylor, C., Cooper, D., Graham, J.: Active shape models-their training and application. Comput. Vis. Image Underst. **61**(1), 38–59 (1995)
9. Cootes, T., Edwards, G., Taylor, C.: Active appearance models. IEEE Trans. Pattern Anal. Mach. Intell. **23**(6), 681–685 (2001)
10. Cristinacce, D., Cootes, T.F.: Feature detection and tracking with constrained local models. In: BMVC (2006)
11. Cao, X., Wei, Y., Wen, F., et al.: Face alignment by explicit shape regression. Int. J. Comput. Vis. **107**(2), 2887–2894 (2014)
12. Ren, S., Cao, X., Wei, Y., et al.: Face alignment at 3000 FPS via regressing local binary features. In: CVPR, pp. 1685–1692 (2014)
13. Chen, D., Ren, S., Wei, Y., Cao, X., Sun, J.: Joint cascade face detection and alignment. In: Fleet, D., Pajdla, T., Schiele, B., Tuytelaars, T. (eds.) ECCV 2014, Part VI. LNCS, vol. 8694, pp. 109–122. Springer, Heidelberg (2014)
14. Lowe, D.: Distinctive image features from scale-invariant key points. Int. J. Comput. Vis. **60**(2), 91–110 (2004)
15. Dalal, N., Triggs, B.: Histograms of oriented gradients for human detection. In: Proceedings of IEEE Conference on Computer Vision and Pattern Recognition, pp. 886–893 (2005)
16. Ahonen, T., Hadid, A., Pietikäinen, M.: Face Recognition with Local Binary Patterns. In: Pajdla, T., Matas, J. (eds.) ECCV 2004. LNCS, vol. 3021, pp. 469–481. Springer, Heidelberg (2004)
17. Felzenszwalb, P.F., Girshick, R.B., McAllester, D.A., Ramanan, D.: Object detection with discriminatively trained part-based models. IEEE Trans. Pattern Anal. Mach. Intell. **32**(9), 1627–1645 (2010)
18. Saragih, J.: Principal regression analysis. In: CVPR, pp. 2, 3, 5, 6 (2011)
19. Le, V., Brandt, J., Lin, Z., Bourdev, L., Huang, T.S.: Interactive facial feature localization. In: Fitzgibbon, A., Lazebnik, S., Perona, P., Sato, Y., Schmid, C. (eds.) ECCV 2012, Part III. LNCS, vol. 7574, pp. 679–692. Springer, Heidelberg (2012)
20. Belhumeur, P.N., Jacobs, D.W., Kriegman, D.J., Kumar, N.: Localizing parts of faces using a consensus of exemplars. In: CVPR, pp. 2, 3, 5, 6 (2011)
21. Bradski, G.: The OpenCV library. Dr. Dobbs J. Softw. Tools **25**, 120–126 (2000)
22. La Cascia, M., Sclaroff, S., Athitsos, V.: Fast, reliable head tracking under varying illumination: an approach based on registration of texture-mapped 3D models. IEEE Trans. Pattern Anal. Mach. Intell. **22**(4), 322–336 (2000)
23. Tu, J., Huang, T., Tao, H.: Accurate head pose tracking in low resolution video. In: AFGR, pp. 573–578 (2006)
24. An, K.H., Chung, M.: 3D head tracking and pose-robust 2D texture map-based face recognition using a simple ellipsoid model. In: Proceedings Intelligent Robots System, pp. 307–312 (2008)
25. Roberto, V., Nicu, S., Theo, G.: Combining head pose and eye location information for gaze estimation. IEEE Trans. Image Process. **21**(2), 802–815 (2012)
26. Gritti, T.: Toward fully automated face pose estimation. In: Proceedings of the 1st International Workshop on Interactive Multimedia for Consumer Electronics, pp. 79–88. ACM (2009)
27. Kim, W.W., Park, S., Hwang, J., et al.: Automatic head pose estimation from a single camera using projective geometry. In: 2011 8th International Conference on Information Communications and Signal Processing (ICICS), pp. 1–5. IEEE (2011)

Tempo-Spatial Compactness Based Background Subtraction for Vehicle Detection and Tracking

Zubair Iftikhar$^{(\boxtimes)}$, Prashan Premaratne, Peter Vial, and Shuai Yang

School of Electrical Computer and Telecommunications Engineering,
University of Wollongong, North Wollongong, NSW, Australia
zi770@uowmail.edu.au

Abstract. Background modelling techniques use the time, spatial, intensity and image plane information to detect the objects. These features are integrated to extract the maximum information. The utilization of background techniques are mostly dependent on various parameters that can be learning rate or threshold. High dependency on parameters increase the complexity and make it difficult to control in changing weather conditions. Parameters based techniques do not provide the high efficiency in outdoor computer vision applications where illumination conditions are difficult to predict. This paper presents an algorithm that is based on background modelling with less dependency on parameters and robust to illumination changes. Camera jitter causes the major effect in modelling techniques so camera jitter is also addressed. A new way of separation of shadow from object is also implemented. Performance of the algorithm is compared with other state-of-the-art methods.

Keywords: Background modelling · Illumination conditions · Camera jitter

1 Introduction

Vision-based surveillance has become a fast growing trend in recent times. A good surveillance system has the capability to inform any irregular behavior or movement that occurs on surveillance region. Extraction of information such as traffic flow, traffic density, speed detection and classification of vehicles for a traffic surveillance system is helpful for traffic management authorities to analyze the traffic. These approaches do not need any human assistance for continuous monitoring of video stream. Intelligent transportation systems use the vision and non-vision based sensors for traffic monitoring. Computer vision-based techniques are more popular when compared to the other traditional sensors due to its versatile features [1, 2]. Apparatus of these processes do not require pavement modification of highways during installations. Multiple detection zones and lanes are monitored at the same time. It links information that is gathered from different locations to turn into a wide area surveillance. Vision-based systems have been successfully used in many research applications [3–12]. Sensors of camera and computational power have been improved a lot that results in high reliability and robust performance of the systems [13]. Many techniques are proposed in

© Springer International Publishing Switzerland 2016
D.-S. Huang et al. (Eds.): ICIC 2016, Part I, LNCS 9771, pp. 86–96, 2016.
DOI: 10.1007/978-3-319-42291-6_9

literature for detection and segmentation of moving objects, such as inter-frame differencing, edge detection, optical flow, thresholding and background subtraction. [14, 15] present the detailed reviews of vision based methodologies for traffic surveillance. Background modelling approach is widely considered in the computer vision systems for traffic surveillance applications [16–19]. The compelling reason to select the background modelling approach in vision based surveillance is that the road conditions are generally remain static so the background is modelled by observing the invariant pixels with regards to time in image plane. All variant pixels represent foreground or moving object which are segmented or categorized based on connected regions. To identify and classify the pixels as variant or invariant is the basic function of background modelling techniques. In general, detection of moving objects from a scene are on just based on intensity has limited scope as compared to spatial domain and texture. Constraints related to spatial domain and texture are helpful to differentiate the information or to perform the comparisons in temporal domain. Background modelling algorithms are highly utilized to provide clues of temporal and spatial domains to manipulate the data at pixels or frame levels. Pixel intensity, correlation with neighboring pixels and edges information are mostly integrated through temporal and spatial domains for defining a pixel as background or foreground. This classification can be effected by various factors. These factors can occur in temporal domain, spatial domain or in both due to any irregular behavior. It is compulsory for parametric techniques of background modelling to update the information in temporal domain which enhances the efficiency of the system. Results of temporal filtering are wrongly estimated if illumination changes are not addressed. Computer vision techniques are enormously used in many indoor and outdoor systems. As compared to indoor, outdoor vision under sunlight and varying climate is more challenging for extracting the information through image and video processing. One of the difficult task is to tackle the poor light or illumination conditions which do not remain constant on image plane due to gradual movement of sun's position or it is possible that illumination can suddenly change in case of cloud's movement, fog or rain. These changes of illumination is the one example of that factor which can affect in temporal domain. Camera for capturing the image plane can be installed on the specific height. This height of camera determinants that how much area will be covered for traffic surveillance. If camera is fixed on pole or overhead bridge then wind can shake the pole or high weighted movement on bridge can cause the vibration of bridge. In both cases, camera jitter will occur that disturbs the information related to spatial domain. This problem of camera jitter mainly affect background modelling techniques because these methods rely on update of parameters for each location of pixel. Vibrations result in wrongly updating parameters between two frames for a particular location. Appearance of shadows with moving objects can muddle the information of both temporal and spatial domains. As a result, it will be difficult to categorize the pixels as background or foreground in shaded regions. A robust vision-based surveillance has the capability to detect and track the moving or stopped vehicles under illumination changes in outdoor, camera jitter and shadow appearances (Fig. 1).

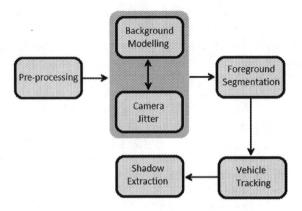

Fig. 1. Flow diagram

2 Algorithm Description

This paper presents a pixel based Tempo-Spatial Compactness Based Background Subtraction (TSCBS) algorithm which is more robust to gradual and sudden illumination. The major contributions of this paper are: (1) a new approach of background subtraction by using the temporal domain with spatial constraints; (2) a new way of determining and normalizing the camera jitter; (3) a simple geometric relations-based shadow extraction approach for vehicles.

3 Background Modeling

This TSCBS method is a pixel based approach which processes each candidate pixel and define as a background pixel or foreground pixel. Two pre-processing steps are also performed before the operation of TCBS which are discussed in Sect. 4. *0–255* range of gray scale is adequate for bit mapping in our approach. A 9×9 matrix is used for this purpose which is centered at candidate pixel and adjacent pixels which are positioned according to their location as shown in (1). Current image or frame (*Curr* Im) and background image (*Back* Im) is used in this operation of TSCBS. Following matrix for current image is named as Current Candidate Matrix (*CCM*) and Background Candidate Matrix (*BCM*) is named to matrix of background image.

$$\begin{bmatrix} i-4,j-4 & . & & . & & . & i-4,j+4 \\ & . & . & & . & . & \\ & & . & Candidate\ Pixel & . & & \\ & & & i,j & . & & \\ & . & . & & . & . & \\ i+4,j-4 & . & & . & & . & i+4,j+4 \end{bmatrix} \quad (1)$$

All those pixels which are considered as background pixels are updated in background image so this image is used as background modelling. *CCM* is multiplied with a two dimensional Gaussian function (*standard deviation = 1.2, size = 9*) which has named *GCCM*. Same process is repeated for *BCM* and named as *GBCM*. Purpose of this Gaussian multiplication is given weight to positions in the matrix so center position has more weight as compared to adjacent positions. Following steps make our approach to robust against gradual illumination change and sudden illumination change. When any illumination change occurs then intensity values of pixels are increased or decreased so only rely on intensity in temporal domain does not lead to efficient result. Relationship between adjacent pixels remains invariant (approximately) so following rule is applied to deal with illumination changes. So, derivation is applied on *GCCM* according to (2) which is named as *DGCM* and same process is applied on GBCM which is named as DGBM (3). Limited scale (0–255) of RGB and the subsequent conversion to gray scale, along with other external factors make some minor change in assumed invariant relationship (spatial domain) of adjacent pixels between two consecutive frames.

$$DGCM = derivative\,(GCCM) \tag{2}$$

$$DGBM = derivative\,(GBCM) \tag{3}$$

To address this issue, gray scale is further reduced to *1–51* range (*255/5*) so if spatial relationship between two adjacent pixels varies within five intensity values at current frame as compared to previous frame due to aforementioned factors then reduced scale assigns the same value to these two adjacent pixels. Then, *DGCM* is divided by *DGBM* and named as *DGM* as shown in (4).

$$DGM = \frac{DGCM + \alpha}{DGBM + \alpha} \tag{4}$$

Constant $\alpha = 0.1$ is used to avoid the indeterminate form. Those values in *DGM* which are between the limit of *0.98* to *1.02* are reassigned to *1* (foreground pixel) and remaining are *0* (background pixel), according to (5–6). This limit allows little variation of background pixel between two consecutive frames.

$$minL = \frac{\|DGM\|}{0.98}, \qquad maxL = \frac{\|DGM\| + \alpha}{1.02} \tag{5}$$

$$DGM = \frac{\lfloor e(-\lfloor e(-\lfloor minL \rfloor)\rfloor)\rfloor + \lfloor e(-\lfloor maxL \rfloor)\rfloor}{2} \tag{6}$$

If *76 (95 %)* out of *81* (9 × 9 Matrix) values in *DGM* are *1* then the candidate pixel will be considered as background pixel otherwise foreground pixel, according to (7–8). Background pixel is represent with *0* and foreground pixel with *1* in classified image.

$$VAR1 = \frac{\sum\sum DGM}{76} \tag{7}$$

$$Classified\_Image(i, j) = \lfloor e(-\lfloor e(-\lfloor VAR1 \rfloor)\rfloor)\rfloor \tag{8}$$

This process is repeated for every pixel. Stopped vehicle does not become the part of background model because this approach does not use the learning parameters which change the static foreground region to background region after a specific time. So, a stopped vehicle will again and again appear as foreground region for upcoming frames.

$$Gauss2D(x,y) = \frac{\left(\frac{e^{-(x,y)^2}}{2c^2}\right)}{\sum (Gauss2D(x,y))} \qquad (9)$$

4   Pre-processing and Camera Jitter

A 2D normalized Gaussian (9) with standard deviation 4 and size 9 is applied as a pre-processing step to remove the high frequency noise (10).

$$CurrGaussIm = Conv2D(CurrIm, Gauss2D(x,y)) \qquad (10)$$

This approach is based on temp-spatial relation so if the camera jitter is occurred between two consecutive frames then it disturbs the utilizing features of temporal domain and spatial domain. This jittering of camera between two frames remains very little for a capturing video which has frame rate *25 frames/second* so disturbance between adjacent pixels in spatial relation can be measured. Spatial disturbance in camera jitter is a process of shifting of pixels to new locations.

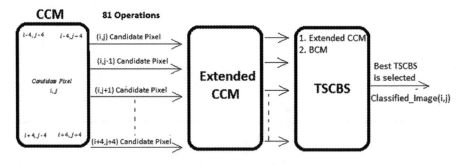

Fig. 2. Camera jitter handling

These new positions of pixels are not located as too far due to minor impact of camera jitter between two consecutive frames. To cope with this issue, each position of CCM is used as candidate pixel and new CCM is created for that candidate pixel to perform the operation of TSCBS. So, there will be 81 operations of TSCBS and position which gives the maximum result in DGM is selected as candidate pixel and result of equation is updated in background image. In this way, best relation is found in nearby locations to deal with camera jitter. Figure 2 further elaborates the procedure to deal with camera jitter.

5   Foreground Segmentation

Classified image according to (8) classifies the background pixels as 0 and foreground pixels as 1. In this section, all the connected regions are segmented together by using the Matlab built-in functions. Morphological operation is applied to fill the holes in the segmented regions. Minimum number of pixels are defined to consider a vehicle by using the learning period in start so if a foreground segmented region has less than pixels as compared to defined criteria then region is considered as ghost foreground and discarded. Also, background image is updated in this discarded region. Figure 3 describes the process of foreground segmentation. A true foreground region after vehicle tracking is shown in rectangular box for final detected image (Fig. 4).

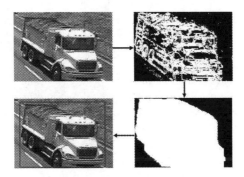

Fig. 3. Foreground segmentation

Fig. 4. Entrances of freeway

6   Vehicle Tracking

All the foreground regions which are expanded in 60 pixels of length, 40 pixels of width and assigned the value of 1 in tracking image. This is done in order to deal with the possibility of vehicle or object movement in any direction within two consecutive frames. Also, all the entrances of image plane are assigned the value 1 so that the new vehicle can be allowed in scene, as shown in Fig. 5. Classified image of (8) is multiplied with this tracking image, according to (11). In this way, ghost foregrounds are removed and vehicle is tracked in predicted region. Main reasons of this process is to eliminate the ghost regions and searching of objects in surroundings of those regions which were detected as vehicles in previous frame.

$$Tracked\_Image = Tracking\_Image \times Classified\_Image \qquad (11)$$

7 Shadow Extraction

A freeway video is used to apply and evaluate the shadow removal technique. Following procedure is applied on each detected foreground region for determination and extraction of shadow. Shadows which appear vertically along the side windows of vehicle are extracted by using lane marking of the road. Three conditions are used to detect the shadow at front or back side of the vehicle. Here, we assumed that only one side out of front and back will be under the shadow. If front side of the vehicle is shadow free then maximum intensity value in initial five rows of evaluating foreground region is named as *Threshold1* and used for comparison, *as shown* in Fig. 6. In contrast to this, if back side of the vehicle is shadow free then maximum intensity value in last five rows of evaluating foreground region is used as *Threshold1*.

$$Shadow\_Searching\ Region = [Vehicle\_Length : -1 : 0.5 \times Vehicle\_Length, Vehicle\_Width]$$
$$(12)$$

$$Shadow\_Searching\ Region = [1 : 0.5 \times Vehicle\_Length, Vehicle\_Width] \qquad (13)$$

Foreground pixels are checked from bottom to top in case 1 (12) and top to bottom in case 2 (13). In a detected region which have vehicle and possible shadow, region of shadow does not occupy the more than 50 % area so therefore searching region is limited to half by using the 0.5 in (12–13). Following two possibilities are used to determine the associated shadow.

Possibility 1: Intensity value same or higher than threshold1 in Shadow_SearchingRegion
Possibility 2: Change of intensity value in Shadow_SearchingRegion

If any possibility of these two is satisfied than that point where this possibility is occurred used as cut-off point and all the searched area discarded from foreground region. Gray scale is reduced to *1–51* range (*255/15*) in this step so that the major change in intensity values are determined which leads to aforementioned possibilities. At this stage, our approach cannot separate the occluded vehicles and considers as one object (Fig. 7).

Fig. 5. Threshold and black car and truck

Fig. 6. Shadow extraction of black car

Fig. 7. White car, black car and truck

8 Experiment Results

This paper presents a new background subtraction technique which processes and classifies each pixel either as background pixel or foreground pixel and then updates the background model. Decision is based on that how strong compactness of a candidate pixel with adjacent pixels exist in temporal domain and spatial domain. Spatial relationship of pixels makes it possible to handle both gradual illumination change and sudden illumination change without the dependency of pixel intensity. Algorithm also handles the camera jitter by selecting the best optimum result. A geometric based technique extracts the shadow from foreground region by using the features of vehicle. This algorithm was developed by using a 1440 × 1080 size freeway video but scope is global so applied for various sizes as shown in Table 1. ChangeDetection.net (CDNet 2014) benchmark dataset [20] is used to compare the result of our method in terms of F-Measure [21] with different types of MoG based background subtraction algorithms. F-Measure is the weighted average of recall and precision. Precision is the fraction of detected pixels that belong to the object by comparison with the ground truth, while recall is associated with the fraction of pixels missed. In this process, detected result is compared with ground truth which is provided in [20]. F-Measure is computed by using the software named as BMC Wizard [22]. MoG (S&G) – Stauffer and Grimson's MoG algorithm [23], MoG (ZZ) – Zoran Zivkovic's MoG algorithm [24], MoG (K&B) – KaewTraKulPong and Bowden's MoG algorithm [25], Regularized RMoG [21]. Following five datasets are selected from [20] to evaluate the performance of our algorithm.

Table 1. First Row: Baseline (highway); Second Row: Camera Jitter (traffic); Third Row: Bad Weather (snowFall); Fourth Row: Low Framerate (tramCrossroad_1fps); Fifth Row: Night Videos (tramStation).

	MoG (S&G)	MoG (ZZ)	MoG (K&B)	Regularized RMoG	TSCBS (Proposed)	Image Size
	0.82	0.84	0.71	0.80	0.87	320x240
	0.60	0.57	0.58	0.72	0.77	320x240
	0.74	0.74	0.61	0.71	0.81	720x480
	0.54	0.51	0.37	0.53	0.59	640x350
	0.41	0.40	0.34	0.41	0.43	480x295

We believe that our approach shows the good potential to deal with traffic surveillance applications. Numerous challenges which can affects the true segmentation of the vehicles are sorted-out in our suggested approach. Requirements and utilization of this system are independent especially it does not require any pre-defined threshold or learning time which makes it fast, reliable and robust for all kinds of traffic roads.

References

1. Kanhere, N.K., Pundlik, S.J., Birchfield, S.T.: Vehicle segmentation and tracking from a low-angle off-axis camera. In: Proceeding IEEE Conference Computer Vision Pattern Recognition, vol. 2, pp. 1152–1157 (2005)
2. Pang, C.C.C., Lam, W.W.L., Yung, N.H.C.: A method for vehicle count in the presence of multiple-vehicle occlusions in traffic images. IEEE Trans. Intell. Trans. Syst. **8**(3), 441–459 (2007)
3. Premaratne, P., Ajaz, S., Premaratne, M.: Hand gesture tracking and recognition system for control of consumer electronics. In: Huang, D.-S., Gan, Y., Gupta, P., Gromiha, M. (eds.) ICIC 2011. LNCS (LNAI), vol. 6839, pp. 588–593. Springer, Heidelberg (2012)
4. Premaratne, P., Nguyen, Q., Premaratne, M.: Human computer interaction using hand gestures. In: Huang, D.-S., McGinnity, M., Heutte, L., Zhang, X.-P. (eds.) ICIC 2010. CCIS, vol. 93, pp. 381–386. Springer, Heidelberg (2010)
5. Premaratne, P., Safaei, F., Nguyen, Q.: Moment invariant based control system using hand gestures. In: Huang, D.-S., Li, K., Irwin, G.W. (eds.) ICIC 2006. LNCIS, vol. 345, pp. 322–333. Springer, Heidelberg (2006)
6. Premaratne, P., Premaratne, M.: Image Matching using Moment Invariants. Neurocomputing **137**, 65–70 (2014)
7. Premaratne, P., Ajaz, S., Premaratne, M.: Hand gesture tracking and recognition system using Lucas-Kanade algorithm for control of consumer electronics. Neurocomputing **116**(20), 242–249 (2013)
8. Premaratne, P., Nguyen, Q.: Consumer electronics control system based on hand gesture moment invariants. IET Comput. Vis. **1**, 35–41 (2007)
9. Yang, S., Premaratne, P., Vial, P.: Hand gesture recognition: an overview. In: 5th IEEE International Conference on Broadband Network and Multimedia Technology (2013)
10. Zou, Z., Premaratne, P., Premaratne, M., Monaragala, R., Bandara, N.: Dynamic hand gesture recognition system using moment invariants. In: ICIAfS, IEEE Computational Intelligence Society, Colombo, Sri Lanka, pp. 108-113 (2010)
11. Herath, D.C., Kroos, C., Stevens, C.J., Cavedon, L., Premaratne, P.: Thinking head: towards human centred robotic. In: 2010 11th International Conference on Control, Automation, Robotics and Vision (ICARCV), Singapore, pp. 2042–2047 (2010)
12. Minge, E.: Evaluation of Non-intrusive Technologies for Traffic Detection. Minnesota Department of Transportation, Office of Policy Analysis, Research and Innovation, SRF Consulting Group, US Department of Transportation, Federal Highway Administration (2010)
13. Morris, B., Trivedi, M.: Robust classification and tracking of vehicles in traffic video streams. In: IEEE Conference Intelligent Transportation Systems, pp. 1078–1083 (2006)
14. Kastrinaki, V., Zervakis, M., Kalaitzakis, K.: A survey of video processing techniques for traffic applications. Image Vis. Comput. **21**, 359–381 (2003)
15. Buch, N., Velastin, S., Orwell, J.: A review of computer vision techniques for the analysis of urban traffic. IEEE Trans. Intell. Transp. Syst. **12**, 920–939 (2011)
16. Mandellos, N.A., Keramitsoglou, I., Kiranoudis, C.T.: A background subtraction algorithm for detecting and tracking vehicle. Expert Syst. Appl. **38**, 1619–1631 (2011)
17. Lima Azevedo, C., Cardoso, J., Ben-Akiva, M., Costeira, J.P., Marques, M.: Automatic vehicle trajectory extraction by aerial remote sensing. Presented at the 16th Euro Working Group Transportation, Procedia Soc. Behav. Sci., Porto, Portugal (2013)
18. Sánchez, A., Nunes, E., Conci, A.: Using adaptive background subtraction into a multilevel model for traffic surveillance. Integr. Comput. Aided Eng. **19**(3), 239–256 (2012)

19. Unzueta, L., et al.: Adaptive multicue background subtraction for robust vehicle counting and classification. IEEE Trans. Intell. Transp. Syst. **13**(2), 527–540 (2012)
20. Wang, Y., Jodoin, P.-M., Fatih, P., Janusz, K., Yannick, B., Prakash, I.: Change Detection 2014 Benchmark (2014). http://wordpress-jodoin.dmi.usherb.ca/results2014/
21. Varadarajan, S., Wang, H., Miller, P., Zhou, H.: Fast convergence of regularised region-based mixture of Gaussians for dynamic background modelling. Comput. Vis. Imaging Underst. (CVIU) **136**, 45–58 (2015)
22. http://bmc.univ-bpclermont.fr/
23. Stauffer, C., Grimson, W.E.L.: Adaptive background mixture models for real-time tracking. In: IEEE Computer Society Conference on Computer Vision and Pattern Recognition, pp. 2246–2252 (1999)
24. Zivkovic, Z.: Improved adaptive gaussian mixture model for background subtraction. In: ICPR 2004, Proceedings of the 17th International Conference on Pattern Recognition, vol. 2, pp. 28–31 (2004)
25. KaewTraKulPong, P., Bowden, R.: An improved adaptive background mixture model for real-time tracking with shadow detection. In: Remagnino, P., Jones, G.A., Paragios, N., Regazzoni, C.S. (eds.) Advanced Video Based Surveillance Systems, pp. 135–144. Springer, New York (2001)

Information Security, Knowledge Discovery and Data Mining

Research on Universal Model of Speech Perceptual Hashing Authentication System in Mobile Environment

Qiu-Yu Zhang[1(✉)], Wen-Jin Hu[1], Yi-Bo Huang[2], and Si-Bin Qiao[1]

[1] School of Computer and Communication,
Lanzhou University of Technology, Lanzhou 730050, China
zhangqylz@163.com, {295105067,742718264}@qq.com
[2] College of Physics and Electronic Engineering,
Northwest Normal University, Lanzhou 730070, China
Huangyibo1982@163.com

Abstract. To address the problem that there is no universal model for the speech perception hash algorithm in the mobile computing environment, the authentication model is studied and a speech perception hash authentication universal model for the mobile computing environment is proposed. By studying the general model for the multimedia perception hash authentication, the proposed model relies on speech perception signature and uses the multimedia perception authentication algorithm to analyze the characteristics of speech signal processing and transmission in the mobile computing environment. In this way, a complete model for the speech perception hash algorithm in the mobile computing environment is developed, providing the theoretical foundation for the subsequent design of the algorithm.

Keywords: Mobile computing environment · Multimedia information security · Speech authentication · Perceptual hashing · Perception feature extraction · Tamper detection

1 Introduction

With constant progress in modern speech communication technologies, mobile speech communication is playing an important role as a means of communication for the people. During the transmission in the mobile environment, the speech data is prone to be attacked by the noise or maliciously tampered. The speech transmission channel in the mobile computing environment is open and the transmission resources are limited. Furthermore, the processing capability and the endurance of the mobile terminals are insufficient. The speech cannot be transmitted safely due to these problems. Therefore, it is necessary to study safety certification of speech contents in the mobile computing environment [1–5]. Speech signals are informative, highly redundant and the existing speech authentication algorithms are no longer suitable for mobile computing environment [6–8].

Speech perception hash [9] is a new kind of speech authentication technology, through the hash mapping of speech features, to achieve speech content authentication,

© Springer International Publishing Switzerland 2016
D.-S. Huang et al. (Eds.): ICIC 2016, Part I, LNCS 9771, pp. 99–111, 2016.
DOI: 10.1007/978-3-319-42291-6_10

speaker authentication technology [10]. The application direction of perceptual hash function [11, 12] can be used to verify the integrity of the information content and the reliability of the information source. At present, perceptual hash functions are known to have some special properties, such as collision resistance, robustness, compactness and so on. In addition, the performance evaluation index [13, 14] of algorithm to achieve the degree of difficulty is also important. To meet the needs of multimedia information certified applications, only the algorithm is easy to implement and perception hash function of high efficiency certification. Random hash function discrimination and compactness are better than its security. Therefore, security can also rely on the setting of the key to ensure that these properties make the hash function very suitable for application in the field of information security [15, 16].

In 2009, Jiao *et al.* [17] established a speech perceptual hash authentication model. On the basis of this model, a speech content authentication algorithm based on perceptual hash is proposed. Algorithm has higher collision resistance and security and operation of the content remains the same good robustness. For malicious attacks, it can be detected correctly in a certain extent on the location of the attack position, but positioning accuracy is not high. Li *et al.* [18] proposed a high efficiency and security speech authentication algorithm based on correlation coefficient of Mel-Frequency Cepstral Coefficients (MFCC), in the process of speech authentication separately calculated the similarity between the hashing sequence using the technology of the hamming distance and similarity measuring function, this speech authentication algorithm could achieve the function as position of tampering, well security and higher sensitivity to detect and locate the speech tampering, but the robustness of normal operation of resample and re-quantization is poor.

Through the existing research of speech perceptual hashing authentication algorithm and the research of literature, it can be known that the speech perceptual hashing authentication algorithms have presented some research results [1, 10, 11, 13, 15, 17, 18], but there are still many problems needed to solve:

1. It should be further researched the original speech perceptual hashing authentication algorithm, in order to build the original speech perceptual hashing algorithm in the mobile environment;
2. It should be researched the sensitivity of the attack in a small scale of speech tampering, and the algorithm must be could searched the position of the tampering content;
3. Speech signal of compressed domain needs special speech hashing algorithm;
4. Perceptual hashing algorithm needs to be researched in the mobile speech code;
5. Evaluation method of speech perceptual distance needs to be studied.

In order to solve the above problems, this paper based on the mobile computing environment, the speech perception hashing authentication system is designed, and combined with the request of performance of mobile computing environment authentication and perceptual hashing authentication algorithm, and then a model of perceptual hashing authentication is established. The model combines with the property of human auditory perception, and optimize the current existing algorithm of speech eigenvalue extraction, causative the speech eigenvalue that is extracted in the model can better reflect the properties of the original speech signal, and the model is convenient to solve the abstract from the speech perceptual hashing function, it makes the perceptual hashing sequence satisfy the request of the performance of hashing function, and it also fulfills the request of real-time, robustness and security of the speech information in the mobile computing environment.

2 Mobile Computing Environment Analysis

2.1 Characteristics

Mobile computing environment, from a certain point of view, can be considered to be Internet, mobile user terminals and mobile data transmission technology integration [19, 20]. But the nature of the mobile computing environment is not a simple super-position of several technical properties, but rather has a lot of unique characteristics. Such as: (1) In the process of mobile network data transmission, the transmission of data as small as possible; (2) Since the mobile data transmission channel is completely open, the security of the transmission information remains to be improved; (3) Limited communication bandwidth in mobile computing environments; (4) When constructing algorithm, it needs to take into account the performance of the mobile terminal; (5) Computing and storage capacity of mobile terminal have difference, the system platform is more diverse. This brings difficulty to the application development of mobile platform. Therefore, when developing the mobile terminal application, the application compatibility issues need to be considered.

Along with constant the advance of modern speech communication technologies, mobile speech communication technology plays an important role in the way of human information communication [21]. Speech information transmission in the mobile computing environment, it may be attacked by noise and malicious tamper etc. The problem that the open of speech transmission channel, the limit of transmission resources and the lack of battery life become the prominent contradiction with the requirement of security of speech transmission. Therefore, it is very necessary to research the speech content security authentication in the mobile computing environment.

2.2 The Issue of Data Security in the Mobile Environment

When people feel the convenience by the mobile data transmission, the security of the mobile data transmission is obtained more and more attention. The data transport through the mobile channel is facing various attacks. Therefore, the problem of the real and reliability of the data in the process of the mobile data transmission need resolved urgently.

Through applying the speech perceptual hashing function in the mobile data transmission, it can be achieved the content authentication of speech and speaker authentication in the mobile environment [17], effectively improved the security and reliability of the speech in the mobile channel. When constructing authentication algorithm, it should satisfy the need of speech authentication and it also need to solve existing following problem in the mobile environment: the sources of transmission channel is limited, open is easy attacked, and the capacity of terminal handling is limited. Therefore, it should be researched the constitution of the speech perceptual algorithm, such as the robustness, discrimination, complexity, real-time and ability of tamper detection etc. The speech perceptual hashing authentication algorithm in the mobile environment should be effectively achieved through the reasonable construction.

3　Speech Content Authentication System

3.1　Basic Principle

Speech content authentication is the technology of matching speech content, which can test the authenticity and integrity of speech signal. The technology can detect whether the speech signal in the process of transmission is attacked and tampered or not, to ensure that speech information received is same as voice information before transmission. For human auditory perception system, the human can accept a certain degree of distortion, but could not accept semantic change. So the technology of speech content authentication protects semantic information rather than speech data information. Speech signal processing, which can keep hearing quality, no semantic change, will be acceptable. Similar content to keep processing and matching should not trigger detector. The channel transmitting speech signal in mobile computing environment are generally open, which often is affected by kinds of attacks, including volume increase, decrease, echo, re-sampling, low-pass filtering, compression, etc. These means of attacking will not change semantics of speech, but only affect person's auditory sense, clear degree of speech signal. Therefore, speech distortion that often through operation keeping content belongs to reasonable distortion. For speech content authentication, soft certification is needed, rather than hard certification.

Figure 1 is a diagram of speech content authentication.

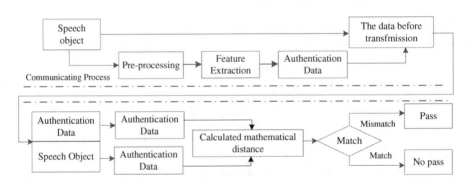

Fig. 1. Content based speech authentication system

Authentication process can be divided into four parts: speech signal preprocessing, speech feature extraction, computing feature distance and the certification of judging distance. The part of feature extraction in speech signal is the research emphasis of certification process. Speech feature selection should combine with human auditory characteristics, and choose the features that could represent the perception of ear for speech. The closer semantic level the feature extracting is obtained, the better the results of certification are. The result of feature extraction will directly affect the real-time of entire authentication algorithm. The authentication data is compressed by linear or nonlinear transformation to reduce the authentication data, at the same time, to improve the efficiency of algorithm and reduce the running time. The method of calculating mathematical distance of authentication data has a lot of kinds, such as bit error rate, Euclidean distance, information entropy, etc. The method of computing mathematics distance can be chosen according to requirements and application background.

3.2   The Inherent Fuzziness of Speech Content Authentication

Although the speech signal is similar to the audio signal, the both sounds are different, and the vibration frequencies are different. Therefore, characteristics and the audio signal are not the same, and the speech authentication methods are also different from the audio authentication method.

The system can authenticate the operation of content holding, and distinguish malicious attacks is called speech content integrity authentication system. It is easy to distinguish reasonable distortion, such as MP3 compression, it is not affected the speech quality in the sense of hearing. However, different application environments distinguish the standard of various operations is different, and in the different application environments the operation of content holding may be also recognized as a malicious attacks. Obviously, accurate authentication is unable to tolerate the distortion of speech information, and the authentication of speech information semantic is had the highest tolerance of distortion, only need ensure the integrity of semantic content that convey by speech information. As described above, it is no obvious boundaries to distinguish the attack if or not is the operation of content holding or the malicious attacks and the rule to distinguish are only related the application environments of the speech authentication system. The operation of content holding and the malicious attacks under the usual environment is as follows:

1. Operation of content holding: the volume size, echo, audio transcoding, equalization, resampling, remove noise, the frequency of low-pass filter at 3 kHz, compression, the process of audio signal, the time domain that maintain constant pitch is synchronous;
2. Malicious operations: including malicious substitution and tampering in a small range, the time axis reorder the fragment that has been disrupted the original sequence, the lossy compression of low bit rate, the frequency of low-pass filter which is damaged to the original semantic information is less than 3 kHz.

In the speech authentication system, the problem of inherent ambiguity is mainly refers to the uncertainty to distinguish the operation of content holding and malicious attack. If a secure and reliable speech authentication system is wanted to obtain, it should be known the problem of inherent ambiguity of speech authentication.

3.3 The Nature of the Speech Content Authentication System

The properties of a well-designed authentication system about the integrity of digital speech content should be satisfied as follows:

1. Authentication data should be much smaller than the original speech;
2. The introduced noise shouldn't affect certification;
3. The authentication system which is based on audio watermarking must be realized the blind detection;
4. The system is real-time and easy;
5. The system has robustness, and the operation is not sensitive to content in the course of the transmission;
6. Tamper detection: the voice and the position of tamper can be effectively detected;
7. Security: the encryption to the feature sequence is presented using the secret key in the algorithm. If the authentication information is intercepted, restore the original order or forge feature sequence are not feasible.

3.4 The Performance Evaluation of Speech Perceptual Contents

The performance evaluation of speech perceptual contents is manly used to evaluate the similar degree of the speech content to keep operating, malicious attacks with the original pronunciation.

The semantic of speech message should not be changed with the keep operating of content, but it will affect the human auditory system on speech perception. Therefore, the content to keep operating changes the content of speech perception to some degree. When changing the quality and effect of the perception, it could not be easily detected by human perceptual system, and the effect to the perceptual content is slowly changed, and less than the perception threshold (Table 1).

Table 1. The operating means and corresponding level

Operating means	Operation description
Volume adjustment	Decrease or increase the volume
Resample	Change the sample frequency
Coding	Change the bit rate or format of the coding
D/A-A/D conversion	Digital analog conversion or analog - digital conversion
Imnoise	Add noise
Filtering	through the filter process the speech
Equalization	through the equalizer process the speech
Echo	Add the echo effect
Reverberation	Add the reverberation effect
Time warp	Stretched or shortened the time of the speech

The operation means of malicious attacks cannot be tolerated by speech information, and malicious attacks may change the semantic information of the original speech. It can lead to semantic changes, which has purpose and intention to modify speech perceptual of the content. The malicious attacks contain increasing, shear, malicious tampering and replacement.

Because the speech signal is a no stationary random signal, and the perceptual hash value is a random sequence by the phonetic characteristics after processing. The speech perceptual hashing authentication system may be abstract modeling, using the analysis method of information theory random source to evaluate objectively the performance of the system. Based on information theory, reasoning and analysis of the perceptual hashing model is presented, it could be obtained a series quantitative description mathematics index of the performance evaluation. The index mainly has the bit error rate (BER), robustness and discrimination [1, 10, 11, 13, 15, 17, 18]. The robustness and discrimination are the basis performance of the perceptual hashing algorithm, the evaluation index is presented by false accept rate (FAR) and false reject rate (FRR), the essence of them is the conditional probability of random events.

4   The Universal Model of Speech Perceptual Hashing Authentication

Based on speech perceptual signature and multimedia perceptual authentication, the model framework of speech perceptual hashing authentication is established. It can be seen Fig. 2.

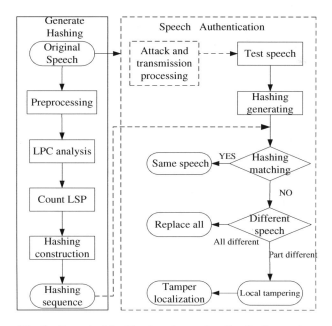

Fig. 2. Perceptual hashing based speech authentication system

Firstly, the one-way mapping process form speech message to the abstract of speech feature is analyzed. Unlike the speech signature, the abstract of speech perceptual is decided by the psychoacoustics of human auditory perceptual system. In general, the process framework of speech perceptual needs to be divided into four parts: the extraction of speech perceptual feature, the structure of perceptual hash function, the process of encryption and transfer of hash function. The technology of authentication mainly verifies the truth of the source of information, whether the information content is under attack, and the information is deleted and tampered. Identity authentication of the speaker and the information integrity authentication are what we are talking about. Therefore, authentication system should be designed in the mobile computing environment to improve the performance of the speech authentication system.

4.1   Extract the Framework of Speech Perceptual Hashing

The extraction framework of speech perceptual hashing is shown in Fig. 3. It mainly includes three parts: preprocessing, perceptual feature extraction and the hash value structure.

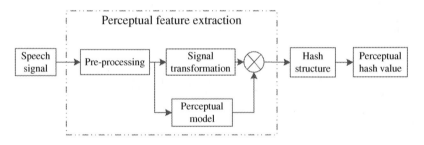

Fig. 3. Flow chart of perceptual hashing values extraction

Feature extraction of speech directly affects the consequent and efficiency of the authentication. Extraction methods include time domain and frequency domain of the feature, such as resonance peak frequency, linear prediction coefficient (LPC), linear spectral pair (LSP), linear prediction cepstral coefficient (LPCC), Mel-frequency cepstral coefficient (MFCC) based on the characteristic of the human ear hearing, linear predictive residual (LPR), Hilbert-Huang transform (HHT) [22] etc. Hash structure is divided into the abstract of eigenvalue and hash reconstruction, the abstract of eigenvalue can effectively reduce the data scale and the overhead of time. The common methods include principal component analysis (PCA), independent component analysis (ICA), singular value decomposition (SVD), vector quantization (VQ) and nonnegative matrix decomposition (NMF) etc.

The efficiency of authentication algorithm mainly relates to the scale of feature extraction, algorithm calculation and the complexity of the feature extraction.

The extracted feature will directly affect the performance of algorithm. To get the least calculation amount and data quantity of authentication, we should extract features related to semantic rather than signal features.

4.2   The Matching Framework of Speech Perceptual Hashing

It can be seen from the Fig. 4 that the matching framework of speech perceptual hashing is presented. The receiver is mainly composed of extracting the phonetic hash value under test and the distance of matching hash.

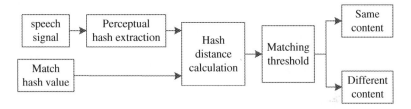

Fig. 4. Flow chart of perceptual hashing values matching

4.3   Transmitted Framework as Watermarking of Perceptual Hashing Value

At present, using the audio watermarking the picture as the watermark information is embedded in the carrier of speech. Speech perceptual hashing as the coefficient of the feature also has the characteristic of the watermark information. Speech perceptual hashing values as the watermark information have two advantages: the first one is that the hash value is related to the content of speech, the hash value as the watermark information is embedded in the speech. The second one is that audio watermarking has the robustness of itself, which may solve the transmission problem of the perceptual hashing value in the mobile channel. Therefore, speech perceptual hashing value as the watermarking can be embedded in the speech signal [23].

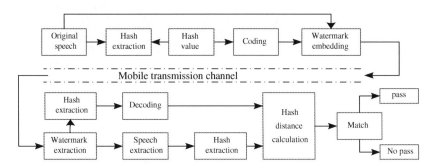

Fig. 5. Flow chart of speech watermarking based perceptual hashing values transmission in mobile environment

4.4   Flow Chart of Speech Perceptual Hashing Authentication System in Mobile Environment

According to Figs. 1, 2, 3, 4 and 5, the flow chart of speech perceptual hashing authentication system in mobile environment is shown in Fig. 6.

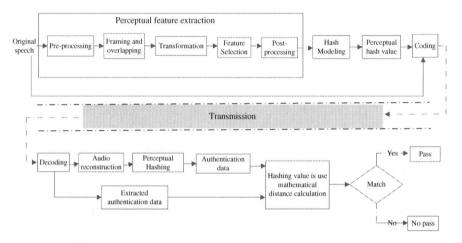

Fig. 6. Flow chart of speech perceptual hashing authentication system in mobile environment

It can be seen from Fig. 6, the pre-process of pretreatment uses the methods of A/D conversion, the conversion of single sound channel, sampling rate, normalized, bandpass filter and GSM codec to deal. The process of framing and overlapping is applied the method of the size of the frame, overlap rate and windows function to solve. The transform methods of DFT, Haar, Hadamard, Zernike moment, and wavelet, etc. can be used for the process of conversion. The process of feature selection can use the method introduced in Sect. 4.1. The process of post-processing can select the methods of normalized, decorrelation, differential and quantitative. Perceptual hashing value is the data of authentication.

In the Fig. 6 the authentication purpose is to distinguish the same speech perceptual and different speech perceptual by using the τ decision thresholds. Assuming the perceptual hashing values of speech signal a_1 and a_2 are ph_{a1} and ph_{a2}, the common method of evaluating the distance of two perceptual hashing value is presented by bit error rate (BER), if the speech information a_1 and a_2 has two speeches of the same perceptual content, $BER < \tau$, if the speech information a_1 and a_2 has two speeches of the different perceptual content, $BER \geq \tau$.

Because of the unidirectional of hashing algorithm, it could not restore the contents of the original speech, but the extraction method can extract the eigenvalue based on the speech frame, and there is one-to-one relationship between the speech eigenvalue and the speech frame. When extracting the perceptual hashing value, if the feature value is not reduced processing in the speech feature value to optimize and quantify, it can be obtained a perceptual hashing sequence which is one-to-one relationship to the original speech content. In sequence when the number of consecutive tamper is more than a cardinal number, it can be located the tampering position.

In order to verify the accuracy of tampering localization using the speech perceptual hashing algorithm, it may compare the similarity of hashing sequences. In the process of the speech authentication, the methods include Euclidean distance, hamming distance, dynamic time warping (DTW) and similarity measuring function etc. [1, 10, 11, 13, 15, 17, 18]. The common methods are the hamming distance and the similarity measuring function. Through the research of the real algorithm, compared to the hamming distance, the similarity measuring function can sensitively detect the minute changes in the speech, locate the position precise for small scale tampering, effective verify the integrity of speech contents.

A small range of tampering may reverse the semantics of the speech signal. A small scale malicious operation generally only tampers with local audio, and small scope of tampering, the lower bit error rate. Therefore BER alone could not judge whether the audio is tampered or not. The errors of audio with illegal malicious operation could cause important impact in local area. Yet the errors of audio with keeping content are often uniform distribution. In order to distinguish between the operation of keeping content and a small scale illegal malicious operation, local amplification of the audio signal is needed to achieve tamper detection after amplifying local BER. In consequence, the model introduces a new measurement method: distortion distance of frame—that the defined process can be found in [24].

In the model, in order to reduce the complexity of time of the hashing algorithm and increase the efficiency of authentication, the most essential problem is how to search a more effective method in perceptual feature extraction. In the process of research combined with the human auditory perception the algorithm of speech feature extraction is carried on optimization, choice the speech perceptual feature algorithm should be had the advantage of better express the inherent characteristic of original speech and higher efficiency of extraction. So that the algorithm can not only better reflect the characteristic of original speech, also satisfy the request of speech perceptual security authentication [11, 22–24].

5 Conclusions

This paper focuses on the speech communication in the mobile computing environment where the bandwidth resources and the terminals' processing capability are limited, a universal model of speech security authentication which is suitable the mobile computing environment is proposed. Combined with the property of human auditory perception, the model of media perceptual hashing is extended to the authentication model of speech perceptual hashing which bases on the audio watermarking. Through the Refs. [11, 22–24], it can be known that the speech perceptual hashing authentication algorithms is researched and verified, the speech eigenvalue that is extracted in the model can better reflect the properties of the original speech signal, and the model is convenient to solve the abstract from the speech perceptual hashing function, it makes the perceptual hashing sequence satisfying the request of the performance of hashing function, and it also fulfills the request of real-time, robustness and security of the speech information in the mobile computing environment.

Acknowledgments. This work is supported by the National Natural Science Foundation of China (No. 61363078), the Natural Science Foundation of Gansu Province of China (No. 1310RJYA004). The authors would like to thank the anonymous reviewers for their helpful comments and suggestions.

References

1. Wang, H., Wang, W., Chen, M., et al.: Quality-driven secure audio transmissions in wireless multimedia sensor networks. Multimed. Tools Appl. **67**(1), 119–135 (2013)
2. Wang, Z., Li, W., Zhu, B.: Audio authentication based on music content analysis. J. Comput. Res. Dev. **49**(1), 158–166 (2012)
3. Gu, J.: Research on Key Technologies of Speech Perceptual Authentication. University of Science and Technology of China, Hefei (2009)
4. Yang, R., Qu, Z., Huang, J.: Detecting digital audio forgeries by checking frame offsets. In: Proceedings of the 10th ACM Workshop on Multimedia and Security, pp. 21–26. ACM, New York (2008)
5. Zhou, L., Wu, D., Zheng, B., et al.: Joint physical-application layer security for wireless multimedia delivery. IEEE Commun. Mag. **52**(3), 66–72 (2014)
6. Zmudzinski, S., Munir, B., Steinebach, M.: Digital audio authentication by robust feature embedding. In: Proceedings of the SPIE on Media Watermarking, Security, and Forensics, vol. 8303, pp. 1–7 (2012)
7. Tomar, V.S., Rose, R.C.: Efficient manifold learning for speech recognition using locality sensitive hashing. Acoust. Speech Sig. Process. **2013**(5), 6995–6999 (2013)
8. Pathak, M.A., Raj, B., Rane, S.D., et al.: Privacy-preserving speech processing: cryptographic and string-matching frameworks show promise. Sig. Process. Mag. **30**(2), 62–74 (2013)
9. Niu, X.M., Jiao, Y.H.: An overview of perceptual hashing. Acta Electronica Sinica **36**(7), 1405–1411 (2008)
10. Chen, N., Xiao, H.D.: Perceptual audio hashing algorithm based on Zernike moment and maximum-likelihood watermark detection. Digit. Sig. Process. **23**(4), 1216–1227 (2013)
11. Huang, Y.B., Zhang, Q.Y., Yuan, Z.T., et al.: The hash algorithm of speech perception based on the integration of adaptive MFCC and LPCC. J. Huazhong Univ. Sci. Tech. **43**(2), 124–128 (2015). (Natural Science Edition)
12. Yang, G., Chen, X.O., Yang, D.S.: Efficient music identification by utilizing space-saving audio fingerprinting system. In: 2014 IEEE International Conference on Proceedings of the Multimedia and Expo (ICME), pp. 1–6. IEEE, Chengdu (2014)
13. Grutzek, G., Strobl, J., Mainka, B., et al.: Perceptual hashing for the identification of telephone speech. In: Proceedings of the Speech Communication, 10. ITG Symposium, pp. 1–4 (2012)
14. Gupta, S., Cho, S., Kuo, C.C.J.: Current developments and future trends in audio authentication. IEEE Multimed. **19**(1), 50–59 (2012)
15. Jiao, Y.H., Ji, L.P., Niu, X.M.: Perceptual speech hashing and performance evaluation. Int. J. Innovative Comput. Inf. Control **6**(3), 1447–1458 (2010)

16. Nickel, C., Zhou, X., Liu, M., et al.: Content identification and quality-based ranking. In: Wahlster, W., Grallert, H.-J., Wess, S., Friedrich, H., Widenka, T. (eds.) Towards the Internet of Services: The THESEUS Research Program, pp. 101–110. Springer, Heidelberg (2014)
17. Jiao, Y.H., Ji, L.P., Niu, X.M.: Robust speech hashing for content authentication. IEEE Sig. Process. Lett. **16**(9), 818–821 (2009)
18. Li, J.F., Wu, T., Wang, H.X.: Perceptual hashing based on correlation coefficient of MFCC for speech authentication. J. Beijing Univ. Posts Telecommun. **38**(2), 89–93 (2015)
19. Zhang, C.: Research on Key Techniques of Mobile Audio Oriented 3G Communication. Wuhan University, Wuhan (2010)
20. Xiang, K., Hu, R.M.: Research progress in frame loss concealment for mobile audio coding. J. Chin. Comput. Syst. **36**(5), 1133–1137 (2015)
21. Zhang, D.F.: Research on the Key Technologies of Identity Authentication and Content Security of 3G Network. Beijing University of Posts and Telecommunications, Beijing (2010)
22. Zhang, Q.Y., Yang, Z.P., Huang, Y.B., et al.: Efficient robust speech authentication algorithm for perceptual hashing based on Hilbert-Huang transform. J. Inf. Comput. Sci. **11**(18), 6537–6547 (2014)
23. Zhang, Q.Y., Yu, S., Xing, P.F., et al.: An improved phase coding-based watermarking algorithm for speech perceptual hashing authentication. J. Inf. Hiding Multimed. Sig. Process. **6**(6), 1231–1241 (2015)
24. Xing, P.F.: Research on Speech Perceptual Hashing Authentication Technology Based on Wavelet Transform. Lanzhou University of Technology, Lanzhou (2015)

SIPSO: Selectively Informed Particle Swarm Optimization Based on Mutual Information to Determine SNP-SNP Interactions

Wenxiang Zhang[1], Junliang Shang[1,2(✉)], Huiyu Li[1], Yingxia Sun[1], and Jin-Xing Liu[1]

[1] School of Information Science and Engineering,
Qufu Normal University, Rizhao 276826, China
{zhangwenxiang94,shangjunliang110,huizilovereaing,
sunyingxia026}@163.com, sdcavell@126.com
[2] Institute of Network Computing,
Qufu Normal University, Rizhao 276826, China

Abstract. Interactive effects of Single Nucleotide Polymorphisms (SNPs), namely, SNP-SNP interactions, have been receiving increasing attention in understanding the mechanism underlying susceptibility to complex diseases. Though many works have been done for their detection, the algorithmic development is still ongoing due to their computational complexities. In this study, we apply selectively informed particle swarm optimization (SIPSO) to determine SNP-SNP interactions with mutual information as its fitness function. The highlights of SIPSO are the introductions of scale-free networks as its population structure, and different learning strategies as its interaction modes, considering the heterogeneity of particles. Experiments are performed on both simulation and real data sets, which show that SIPSO is promising in inferring SNP-SNP interactions, and might be an alternative to existing methods. The software package is available online at http://www.bdmb-web.cn/index.php?m=content&c=index&a=show&catid=37&id=99.

Keywords: SNP-SNP interactions · Particle swarm optimization · Selectively informed · Mutual information · Scale-free network

1 Introduction

SNP-SNP interactions have been receiving increasing attention in understanding the mechanism underlying susceptibility to complex diseases, such as heart disease, Alzheimer's disease, cancer, type 2 diabetes and many others [1]. Detection of SNP-SNP interactions is therefore becomes an urgent, front-burner problem in the field of Bioinformatics. Though lots of algorithms have been proposed for inferring SNP-SNP interactions, the algorithmic development is still ongoing due to their computational and mathematical complexities, including the fitness function that decides how well an SNP combination contributes to the phenotype, the limitation of prior knowledge in interpreting complex genetic architecture of a disease, the intensive computational burden imposed by the enormous search space resulted from the problems of "high

D.-S. Huang et al. (Eds.): ICIC 2016, Part I, LNCS 9771, pp. 112–121, 2016.
DOI: 10.1007/978-3-319-42291-6_11

dimensional and small sample size" and "combinatorial explosion", which obviously hinder further applications of current algorithms in genome-wide association studies.

Recently, several particle swarm optimization (PSO) based methods have been proposed for determining SNP-SNP interactions [2–13]. Yang et al. [2] used a binary PSO to evaluate the risk of breast cancer, with odds ratio as its fitness function, which is known as OR-BPSO for short. Based on the OR-BPSO, Chang et al. [3] presented a odds ratio discrete binary PSO (OR-DBPSO) for inferring SNP-SNP interactions with the quantitative phenotype. Chuang et al. [4] also combined PSO and odds ratio to explore combinations of SNP-SNP interactions in breast cancer. They then developed a chaotic PSO (CPSO) for breast cancer association studies and indeed identified a SNP combination [5]. In order to enhance the reliability of the PSO in determining SNP-SNP interactions associated with breast cancer, they improved the PSO (IPSO) and proved that the IPSO is highly reliable than the OR-BPSO [6]. Recently, they presented a gauss chaotic map PSO (Gauss-PSO) and used it to detect the best association with breast cancer [7]. They confirmed that the Gauss-PSO was capable of identifying higher difference values between cases and controls than both the PSO and the CPSO. By analyzing the performance of the PSO and the CPSO, Yang et al. [8] presented a double-bottom chaotic map PSO (DBM-PSO) to overcome their respective limitations. Later, the DBM-PSO is used to infer SNP-SNP interactions based on *chi*-square test [9]. Hwang et al. [10] developed a complementary-logic PSO (CLPSO) to increase the efficiency of determining significant SNP-SNP interactions in case-control study. Wu et al. [11] proposed a PSO based algorithm and found SNP interactions of rennin-angiotensin system genes against hypertension. Nevertheless, above mentioned methods mainly focus on identifying a best genotype-genotype of a SNP-SNP inter-action among possible genotype combinations of SNP combinations, but not several SNP-SNP interactions among possible SNP combinations, which may lose important clues for the exploration of causative factors of complex diseases. In addition, limited sample size of SNP data obviously affects computational accuracies of these methods. More recently, Ma et al. [12] proposed PSOMiner for the detection of SNP-SNP interactions, which is a generalized PSO algorithm based on the fitness function of mutual information. Shang et al. [13] presented an improved method IOBLPSO for detecting SNP-SNP interactions by using opposition-based learning PSO. IOBLPSO has ensured the ability of global searching and prevented premature convergence. However, above methods treated all particles equally and adopt a single learning strategy, overlooking the heterogeneity of particles. Gao et al. [14] employed scale-free networks to represent the population structure of swarms and proposed a selectively informed PSO (SIPSO), where particles select different learning strategies based on their degrees: a densely connected hub particle gets full information from all of its neighbors (fully-informed) while a non-hub particle with low degree can only follow a single yet best performed neighbor (single-informed). Experiment results show that SIPSO is able to significantly improve the optimization performance.

In this paper, we combine SIPSO and mutual information to determine SNP-SNP interactions. The highlights of SIPSO are the introductions of scale-free networks as its population structure, two different learning strategies as its interaction modes, and mutual information as its fitness function. Experiments are performed on several real and simulation data sets. Results demonstrate that SIPSO is promising in inferring SNP-SNP interactions, and might be an alternative to existing methods.

2 Methods

2.1 Scale-Free Network

A scale-free network is a network whose degree distribution follows a power law, at least asymptotically. So far, many networks have been reported to be scale-free, for instance, world wide web, internet, neural networks, and so on. Liu et al. studied the influence of scale-free population structure on the performance of PSO, showing that the scale-free PSO outperforms the traditional PSO, due to only a few particles being densely connected hubs and most particles being low degree non-hubs, resulting in high heterogeneity of degrees of particles [14, 15].

In SIPSO, particles are considered as nodes of a scale-free network. This network is generated by Barabasi-Albert model [16], which incorporates two ingredients including growth and preferential attachment. Specifically, at first m_0 fully connected particles are constructed as a complete network; at each time step a new particle is added to the network, which is connected to m ($m < m_0$) nodes with a specified probability. This probability p_i of the new particle being connected to an existing particle i depends upon degree of i, i.e., k_i, which can be written as

$$p_i = \frac{k_i}{\sum_j k_j},\tag{1}$$

where j runs over all the existing particles. Here m_0 is set to 4 and m is set to 2.

2.2 Mutual Information

The SIPSO introduces mutual information to decide which SNP combinations are SNP-SNP interactions, and to measure how much the effects of captured SNP-SNP interactions to the phenotype are, since mutual information has widely been used as a promising measure for measuring dependence of two variables without a complex modeling. The formula of mutual information can be written as

$$MI(S; Y) = H(S) + H(Y) - H(S, Y),\tag{2}$$

where $H(S)$ is the entropy of S, $H(Y)$ is the entropy of Y, and $H(S, Y)$ is the joint entropy of both S and Y, S is a position of a particle indicating a SNP combination, Y is the phenotype.

The entropy and the joint entropy are defined as,

$$H(S) = -\sum_{j_1=1}^{3} \cdots \sum_{j_K=1}^{3} \left(p\left(s_{j_1}, \cdots, s_{j_K}\right) \cdot \log p\left(s_{j_1}, \cdots, s_{j_K}\right) \right), \tag{3}$$

$$H(Y) = -\sum_{j=0}^{1} \left(p\left(y_j\right) \cdot \log p\left(y_j\right) \right), \tag{4}$$

$$H(S,Y) = -\sum_{j_1=1}^{3} \cdots \sum_{j_K=1}^{3} \sum_{j=0}^{1} \left(p\left(s_{j_1}, \cdots, s_{j_K}, y_j\right) \cdot \log p\left(s_{j_1}, \cdots, s_{j_K}, y_j\right) \right), \tag{5}$$

where K is the considered order of SNP-SNP interactions, i.e., the number of SNPs in a SNP combination. s is the genotype of a SNP coded as $\{1, 2, 3\}$, corresponding to homozygous common genotype, heterozygous genotype, and homozygous minor genotype. y is the label of a sample coded as $\{0, 1\}$, corresponding to control and case. $p(\cdot)$ is the probability distribution function.

Obviously, higher mutual information value indicates stronger association between the phenotype and the SNP combination.

2.3 Selectively Informed Particle Swarm Optimization (SIPSO)

The SIPSO, first proposed by Gao *et al.* [14] that takes into account both the scale-free population structure and selectively informed learning strategies, is a typical swarm intelligence algorithm that derives the inspiration from the self-organization and adaptation in flocking phenomena. Experimental comparisons showed that SIPSO outperforms several peer algorithms in terms of solution quality, convergence speed, and success rate. We combine SIPSO and mutual information to determine SNP-SNP interactions in this paper.

For SIPSO, let $X_i^t = \left[x_{i1}^t, \cdots, x_{iK}^t\right]$ and $V_i^t = \left[v_{i1}^t, \cdots, v_{iK}^t\right]$ be the position and the velocity of particle i at iteration t, where $t \in [1, T]$, $i \in [1, I]$, T is the number of iterations, I is the number of particles, x_{iK}^t is the selected K_{th} SNP index of particle i at iteration t, and $x_{iK}^t \in [1, M]$, M is the number of SNPs in the data set. v_{iK}^t is the velocity of x_{iK}^t. At each iteration, each particle update its position and velocity according to the following equations:

$$\tilde{V}_i^{t+1} = \begin{cases} \eta \cdot \left(V_i^t + \frac{1}{|\mathrm{N}(i)|} \sum_{j \in \mathrm{N}(i)} \left(U(0, c_1 + c_2) \cdot \left(P_j^t - X_i^t \right) \right) \right) & |\mathrm{N}(i)| > \lambda \\ \eta \cdot \left(V_i^t + U(0, c_1) \cdot \left(P_i^t - X_i^t \right) + U(0, c_2) \cdot \left(G_i^t - X_i^t \right) \right) & |\mathrm{N}(i)| \leq \lambda \end{cases}, \tag{6}$$

$$V_i^{t+1} = \begin{cases} \tilde{V}_i^{t+1} & \tilde{V}_i^{t+1} \in [1 - M, M - 1] \\ U(1 - M, M - 1) & \tilde{V}_i^{t+1} \notin [1 - M, M - 1] \end{cases}, \tag{7}$$

$$\tilde{X}_i^{t+1} = X_i^t + V_i^t, \tag{8}$$

$$X_i^{t+1} = \begin{cases} \text{int}(\tilde{X}_i^{t+1}) & \tilde{X}_i^{t+1} \in [1,M] \\ \text{int}(U(1,M)) & \tilde{X}_i^{t+1} \notin [1,M] \end{cases}, \tag{9}$$

$$\eta = \frac{2}{\left| 2 - (c_1 + c_2) - \sqrt{(c_1 + c_2)^2 - 4 \cdot (c_1 + c_2)} \right|}. \tag{10}$$

Here, P_i^t is the best historical position of particle i until iteration t, G_i^t is the best historical position of neighbors of particle i at iteration t. c_1 and c_2 are the acceleration coefficients controlling how far a particle moves at an iteration. η is the learning rate. $U(a,b)$ is a random number drawn at each iteration from the uniform distribution $[a,b]$. $N(i)$ is the neighbors of particle i and $|N(i)|$ is the degree of particle i (e.g. the number of neighbors). $\text{int}(\cdot)$ is an integral function. λ is the threshold to determine a particle fully-informed or single-informed.

For constructing a positive feedback mechanism, P_i^t and G_i^t should be updated at every iteration. Specifically, whether P_i^{t+1} is updated to X_i^t or P_i^t depends on their mutual information values, which can be defined as,

$$P_i^{t+1} = \begin{cases} P_i^t & MI(P_i^t; Y) = \max(MI(P_i^t; Y), MI(X_i^t; Y)) \\ X_i^t & MI(X_i^t; Y) = \max(MI(P_i^t; Y), MI(X_i^t; Y)) \end{cases}. \tag{11}$$

Similarly, by evaluating mutual information values of $N(i)$ and G_i^t, G_i^{t+1} is updated to one of them with the higher value, which can be written as,

$$G_i^{t+1} = \begin{cases} G_i^t & MI(G_i^t; Y) = \max(MI(G_i^t; Y), MI(N(i); Y)) \\ N(i) & MI(N(i); Y) = \max(MI(G_i^t; Y), MI(N(i); Y)) \end{cases}. \tag{12}$$

While completing the iteration process, P_i^T with descending mutual information values are recorded as the detected SNP-SNP interactions of SIPSO.

3   Results and Discussion

3.1   Simulation and Real Data Sets

We exemplify 4 benchmark models of SNP-SNP interactions for the experiments [17–21], detailed parameters of which are recorded in Table 1. Specifically, Model 1 is a model that display both marginal effects and interactive effect, the penetrance of which increases only when both SNPs have at least one minor allele [17, 18]; Model 2 is a model showing both marginal effects and interactive effect, the additional minor allele at each locus of which does not further increase the penetrance [17]. Model 3 is also a model displaying both marginal effects and interactive effect, which assumes that the minor allele in one SNP has the marginal effect, however, the marginal effect is inversed while minor alleles in both SNPs are present [17]; Model 4 is a model that shows only interactive effects, which is directly cited from the reference [20].

Model 4 is exemplified here since it provides a high degree of complexity to challenge ability of a method in detecting SNP-SNP interactions. For each model, 50 data sets are simulated by *epi*SIM [22], each containing 4000 samples with the ratio of cases and controls being 1. For each data set, random SNPs are set with their MAFs chosen from [0.05, 0.5] uniformly.

A real data set of age-related macular degeneration (AMD) is used for testing the practical ability of SIPSO. AMD, refers to pathological changes in the central area of the retina, is the most important cause of irreversible visual loss in elderly populations, and is considered as a complex disease having multiple SNP-SNP interactions [18]. The AMD data set contains 103611 SNPs genotyped with 96 cases and 50 controls, which has been widely used as a benchmark data set in the field of testing methods of detecting SNP-SNP interactions [12, 13, 17, 18, 21, 23–25].

Table 1. Details of 4 models of SNP-SNP interactions. Prevalence is the proportion of samples that occur a disease. Penetrance is the probability of the occurrence of a disease given a particular genotype. MAF(α) is the minor allele frequency of α. *AA*, *Aa*, and *aa* are homozygous common genotype, heterozygous genotype, and homozygous minor genotype.

Model	MAF(a)	MAF(b)	Prevalence	Penetrance function			
				Genotypes (SNP A)	Genotypes (SNP B)		
					BB	*Bb*	*bb*
Model 1	0.300	0.200	0.100	*AA*	0.087	0.087	0.087
				Aa	0.087	0.146	0.190
				aa	0.087	0.190	0.247
Model 2	0.400	0.400	0.050	*AA*	0.042	0.042	0.042
				Aa	0.042	0.061	0.061
				aa	0.042	0.061	0.061
Model 3	0.400	0.200	0.010	*AA*	0.009	0.009	0.009
				Aa	0.013	0.006	0.006
				aa	0.013	0.006	0.006
Model 4	0.400	0.400	0.171	*AA*	0.068	0.299	0.017
				Aa	0.289	0.044	0.285
				aa	0.048	0.262	0.174

3.2   Experiments on Simulation Data Sets

Three PSO based methods for inferring SNP-SNP interactions are used for this comparison study with the SIPSO. They are DBM-PSO, PSOMiner and IOBLPSO. For a fair comparison, parameter settings of these methods are the same. Specifically, the number of particles I and the number of iterations T are respective set to 200 and 20; acceleration factors c_1 and c_2 are chosen their appropriate settings 2.05 based on previous extensive analysis [26]; the inertia weights for compared methods are set to 0.65; the threshold to determine a particle fully-informed or single-informed λ is set to 4.

Detection power is used to evaluate the performance of SIPSO by applying SIPSO on four groups of simulation data sets, each of which consists of a SNP-SNP interaction model. Power of these compared methods on simulation data sets is reported in Fig. 1.

It is seen that performance of the SIPSO is comparable, and sometimes outperforms to those of compared methods. Specifically, power of the SIPSO is much higher than that of DBM-PSO and PSOMiner; nevertheless, its power is behind that of IOBLPSO in three models except the Model 3, since the IOBLPSO introduces the strategy of opposition-leaning, which implies that introducing opposition-learning to SIPSO might also improve the performance of SIPSO, which is a direction; the DBM-PSO has the worst performance on all models because it only focus on identifying genotype combinations among SNP combinations; the PSOMiner is indeed a general PSO that used in determining SNP-SNP interactions, and hence has poor power.

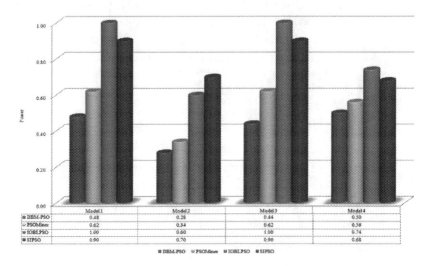

Fig. 1. Power of compared methods on simulation data sets.

3.3   Application to Real Data Set

The SIPSO is applied on AMD data set four times with parameter settings (I, T) being $(10000, 500)$, $(10000, 1000)$, $(20000, 500)$ and $(20000, 1000)$. The top 10 captured SNP-SNP interactions that might be associated with the AMD are reported in Table 2 with descending mutual information values.

It is interesting that all reported SNP-SNP interactions contain either rs380390 or rs1329428. This is because that these two SNPs have strongest main effects among all tested SNPs, and have already proofed to be significantly associated with AMD [18]. Both of them are in the *CFH* gene, and there are biologically plausible mechanisms for the involvement of *CFH* in AMD. *CFH* is a regulator that actives the alternative pathway of the complement cascade, the mutations in which can lead to an imbalance in normal homeostasis of the complement system [24]. This phenomenon is thought to account for substantial tissue damage in AMD [23, 25, 27].

Other SNPs listed in the second column of the table might be identified due to their partners, i.e., rs380390 and rs1329428, since strong main effects of them leads to their combinations with other SNPs almost displaying strong interactive effect. This also

Table 2. Top 10 captured SNP-SNP interactions associated with AMD. *CFH*: complement factor H. *MPP7*: palmitoylated membrane protein 7. *PEBP4*: phosphatidylethanolamine binding protein 4. *R3HDM1*: R3H domain (binds single-stranded nucleic acids). N/A: no gene is available. Chr: Chromosome.

SNP 1				SNP 2				Mutual information value
Index	Name	Gene	Chr	Index	Name	Gene	Chr	
43748	rs380390	*CFH*	1	80178	rs1363688	N/A	5	0.2966
54108	rs1329428	*CFH*	1	30550	rs10489076	N/A	4	0.2650
43748	rs380390	*CFH*	1	51222	rs10508731	*MPP7*	10	0.2647
43748	rs380390	*CFH*	1	8764	rs1345488	N/A	18	0.2640
54108	rs1329428	*CFH*	1	51987	rs2466215	*PEBP4*	8	0.2636
43748	rs380390	*CFH*	1	49642	rs10521121	N/A	17	0.2608
43748	rs380390	*CFH*	1	54213	rs2829015	N/A	21	0.2593
54108	rs1329428	*CFH*	1	57441	rs1917173	N/A	10	0.2470
43748	rs380390	*CFH*	1	34447	rs961360	*R3HDM1*	2	0.2468
54108	rs1329428	*CFH*	1	64573	rs717246	N/A	1	0.2435

indicates that SIPSO is sensitive to those SNPs displaying strong main effects. Further studies with the use of large scale case-control samples are needed to confirm whether these SNPs have true associations with AMD. We hope that several clues could be provided from this paper for the exploration of causative factors of AMD.

4   Conclusions

SNP-SNP interactions, also known as epistatic interactions or epistasis, is nonlinear interactive effects of SNPs, which is now believed to be one of the causative patterns of complex diseases, and is recognized fundamentally important for understanding the mechanism of disease causing genetic variation. Though many works have been done for inferring SNP-SNP interactions, the algorithmic development is still ongoing due to their mathematical and computational challenges. For instance, the limited fitness functions, the complex genetic architecture, and the intensive computational burden. By considering the advantages of the PSO including simplicity, effectiveness and low computational cost, we introduce the SIPSO with the mutual information to determine SNP-SNP interactions in this manuscript. The SIPSO improves the traditional PSO by modifying the population structure as a scale-free network and altering the interaction modes to different learning strategies. That is to say, learning strategy of each particle depends on its degree: a densely connected hub particle gets full information from all of its neighbors while a non-hub particle with low degree can only follow a single yet best performed neighbor. Experiments are performed on both simulation and real data sets, which show that SIPSO is promising in inferring SNP-SNP interactions, and might be an alternative to existing methods. The Matlab version of SIPSO software package is available online at http://www.bdmb-web.cn/index.php?m=content&c=index&a=show&catid=37&id=99.

There are several merits and demerits for SIPLO introducing to identify SNP-SNP interactions. The first advantage is that the SIPSO taking into account the individual's heterogeneity, could balance the exploration and the exploitation in the optimization process. The second advantage is that mutual information is an effective and efficient fitness function in measuring SNP-SNP interactions. Nevertheless, it still has lots of limitations. Firstly, it is only a beneficial exploration of SIPSO applying to determine SNP-SNP interactions, its performance is simply comparable to compared methods, and sometime inferior to compared methods, which indicates that the SIPSO needs more improvement while applying to the field of Bioinformatics. Secondly, SIPSO is sensitive to those SNPs displaying strong main effects. These limitations will inspire us to continue working in the future.

Acknowledgments. This work was supported by the National Natural Science Foundation of China (61502272, 61572284, 61572283), the Scientific Research Reward Foundation for Excellent Young and Middle-age Scientists of Shandong Province (BS2014DX004), the Science and Technology Planning Project of Qufu Normal University (xkj201410), the Opening Laboratory Fund of Qufu Normal University (sk201416), the Scientific Research Foundation of Qufu Normal University (BSQD20130119), The Innovation and Entrepreneurship Training Project for College Students of China (201510446044), The Innovation and Entrepreneurship Training Project for College Students of Qufu Normal University (2015A058, 2015A059).

References

1. Maher, B.: The case of the missing heritability. Nature **456**(7218), 18–21 (2008)
2. Yang, C.-H., Chang, H.-W., Cheng, Y.-H., Chuang, L.-Y.: Novel generating protective single nucleotide polymorphism barcode for breast cancer using particle swarm optimization. Cancer Epidemiol. **33**(2), 147–154 (2009)
3. Chang, H.-W., Yang, C.-H., Ho, C.-H., Wen, C.-H., Chuang, L.-Y.: Generating SNP barcode to evaluate SNP–SNP interaction of disease by particle swarm optimization. Comput. Biol. Chem. **33**(1), 114–119 (2009)
4. Chuang, L.-Y., Lin, M.-C., Chang, H.-W., Yang, C.-H.: Analysis of SNP interaction combinations to determine breast cancer risk with PSO. In: 2011 IEEE 11th International Conference on Bioinformatics and Bioengineering (BIBE), pp. 291–294. IEEE (2011)
5. Chuang, L.-Y., Chang, H.-W., Lin, M.-C., Yang, C.-H.: Chaotic particle swarm optimization for detecting SNP–SNP interactions for CXCL12-related genes in breast cancer prevention. Eur. J. Cancer Prev. **21**(4), 336–342 (2012)
6. Chuang, L.-Y., Lin, Y.-D., Chang, H.-W., Yang, C.-H.: An improved PSO algorithm for generating protective SNP barcodes in breast cancer. PLOS One **7**(5), e37018 (2012)
7. Chuang, L.-Y., Lin, Y.-D., Chang, H.-W., Yang, C.-H.: SNP-SNP interaction using gauss chaotic map particle swarm optimization to detect susceptibility to breast cancer. In: 2014 47th Hawaii International Conference on System Sciences (HICSS), pp. 2548–2554. IEEE (2014)
8. Yang, C.-H., Tsai, S.-W., Chuang, L.-Y., Yang, C.-H.: An improved particle swarm optimization with double-bottom chaotic maps for numerical optimization. Appl. Math. Comput. **219**(1), 260–279 (2012)

9. Yang, C.-H., Lin, Y.-D., Chuang, L.-Y., Chang, H.-W.: Double-bottom chaotic map particle swarm optimization based on chi-square test to determine gene-gene interactions. BioMed Res. Int. **2014**, 10 (2014)
10. Hwang, M.-L., Lin, Y.-D., Chuang, L.-Y., Yang, C.-H.: Determination of the SNP-SNP interaction between breast cancer related genes to analyze the disease susceptibility. Int. J. Mach. Learn. Comput. **4**(5), 468–473 (2014)
11. Wu, S.-J., Chuang, L.-Y., Lin, Y.-D., Ho, W.-H., Chiang, F.-T., Yang, C.-H., Chang, H.-W.: Particle swarm optimization algorithm for analyzing SNP–SNP interaction of renin-angiotensin system genes against hypertension. Mol. Biol. Rep. **40**(7), 4227–4233 (2013)
12. Ma, C., Shang, J., Li, S., Sun, Y.: Detection of SNP-SNP interaction based on the generalized particle swarm optimization algorithm. In: 2014 8th International Conference on Systems Biology (ISB), pp. 151–155. IEEE (2014)
13. Shang, J., Sun, Y., Li, S., Liu, J.-X., Zheng, C.-H., Zhang, J.: An improved opposition-based learning particle swarm optimization for the detection of SNP-SNP interactions. BioMed Res. Int. **2015**, 12 (2015)
14. Gao, Y., Du, W., Yan, G.: Selectively-informed particle swarm optimization. Scientific reports 5 (2015)
15. Liu, C., Du, W.B., Wang, W.X.: Particle swarm optimization with scale-free interactions. PLOS One **9**(5), e97822 (2014)
16. Barabasi, A.L., Albert, R.: Emergence of scaling in random networks. Science **286**(5439), 509–512 (1999)
17. Zhang, Y., Liu, J.S.: Bayesian inference of epistatic interactions in case-control studies. Nat. Genet. **39**(9), 1167–1173 (2007)
18. Tang, W., Wu, X., Jiang, R., Li, Y.: Epistatic module detection for case-control studies: a Bayesian model with a Gibbs sampling strategy. PLoS Genet. **5**(5), e1000464 (2009)
19. Frankel, W.N., Schork, N.J.: Who's afraid of epistasis? Nat. Genet. **14**(4), 371–373 (1996)
20. Li, W., Reich, J.: A complete enumeration and classification of two-locus disease models. Hum. Hered. **50**(6), 334–349 (2000)
21. Shang, J., Zhang, J., Sun, Y., Liu, D., Ye, D., Yin, Y.: Performance analysis of novel methods for detecting epistasis. BMC Bioinform. **12**(1), 475 (2011)
22. Shang, J., Zhang, J., Lei, X., Zhao, W., Dong, Y.: EpiSIM: simulation of multiple epistasis, linkage disequilibrium patterns and haplotype blocks for genome-wide interaction analysis. Genes Genomics **35**, 305–316 (2013)
23. Shang, J., Zhang, J., Lei, X., Zhang, Y., Chen, B.: Incorporating heuristic information into ant colony optimization for epistasis detection. Genes Genomics **34**(3), 321–327 (2012)
24. Shang, J., Zhang, J., Sun, Y., Zhang, Y.: EpiMiner: a three-stage co-information based method for detecting and visualizing epistatic interactions. Digit. Signal Process. **24**, 1–13 (2014)
25. Shang, J., Sun, Y., Fang, Y., Li, S., Liu, J.-X., Zhang, Y.: Hypergraph supervised search for inferring multiple epistatic interactions with different orders. In: Huang, D.-S., Jo, K.-H., Hussain, A. (eds.) ICIC 2015. LNCS, vol. 9226, pp. 623–633. Springer, Heidelberg (2015)
26. Clerc, M., Kennedy, J.: The particle swarm-explosion, stability, and convergence in a multidimensional complex space. IEEE Trans. Evol. Comput. **6**(1), 58–73 (2002)
27. Adams, M.K., Simpson, J.A., Richardson, A.J., Guymer, R.H., Williamson, E., Cantsilieris, S., English, D.R., Aung, K.Z., Makeyeva, G.A., Giles, G.G.: Can genetic associations change with age? CFH and age-related macular degeneration. Hum. Mol. Genet. **21**(23), 5229–5236 (2012)

A Clustering Based Feature Selection Method Using Feature Information Distance for Text Data

Shilong Chao, Jie Cai, Sheng Yang$^{(\boxtimes)}$, and Shulin Wang

College of Computer Science and Electronic Engineering,
Hunan University, Changsha, China
Yangsh0506@sina.com

Abstract. Feature selection is a key point in text classification. In this paper a new feature selection method based on feature clustering using information distance is put forward. This method using information distance measure builds a feature clusters space. Firstly, K-medoids clustering algorithm is employed to gather the features into k clusters. Secondly the feature which has the largest mutual information with class is selected from each cluster to make up a feature subset. Finally, choose target number features according to the mRMR algorithm from the selected subset. This algorithm fully considers the diversity between features. Unlike the incremental search algorithm mRMR, it avoids prematurely falling into local optimum. Experimental results show that the features selected by the proposed algorithm can gain better classification accuracy.

Keywords: Text classification · Feature selection · Cluster · Diversity

1 Introduction

Text classification is one of the important branches of data mining. Its target is to classify the text sets according to a certain standard or system automatically [1, 2]. In text data, Vector Space Model (VSM) [3–5] is widely used to represent the text. The VSM of text data often has high dimensionality. In order to exactly classify the text, it is necessary to reduce the dimensionality of text data by feature selection [6]. Feature selection can remove the irrelevant features, at the same time redundant features also are effectively eliminated. Finally the feature subset composed of features with strong distinguishing ability can be selected, and the accuracy and speed of classification are improved.

Feature selection based on information theory has always been a hot research topic, and a number of classical mutual information (MI) based algorithms have been proposed. Battiti proposed an algorithm called MIFS [7] which takes into account both feature-feature and feature-class mutual information for feature selection. MIFS works well when the penalty parameter is set appropriately. Peng and Ding put forward the classic mRMR algorithm [8]. The idea of mRMR is to maximize the correlation between feature and class, while minimizing the redundancy between the selected

© Springer International Publishing Switzerland 2016
D.-S. Huang et al. (Eds.): ICIC 2016, Part I, LNCS 9771, pp. 122–132, 2016.
DOI: 10.1007/978-3-319-42291-6_12

features. In mRMR, the multivariate joint probability density estimation problem of high dimensional space is replaced by the probability densities of couples. FCBF algorithm [9] uses the symmetrical uncertainty (SU) to measure the correlations of feature-class and feature-feature. In the original feature set, the features whose *SU* values with the class are less than the given threshold are deleted, and then the redundancy analysis is carried out in the remaining features. The final feature subset is obtained after the redundant features are eliminated. The CMIM algorithm [10] proposed by Fleuret takes the conditional mutual information as the measure to select features. Fleuret thinks that the more conditional mutual information the feature has in the case of selected subset, the more information about class the feature carries. The idea of CMIM is to select the feature with maximal conditional mutual information in the case of selected subset. Literature [11] made a more comprehensive and detailed summary of the feature selection based on mutual information.

Information measurements can be used to measure the uncertainty and the non-linear relationship between the features in quantitative form [12, 13]. As a result the information measures are widely used in feature selection. These above algorithms employ relevant metrics to measure the relationships of feature-class and feature-feature. Their common basic idea is to select those features relevant with class and irrelevant between selected features. These algorithms take advantage of the greedy algorithm which is easy to reach a local optimum. From the overall situation, clustering analysis on the original feature set can be considered. If the clustering analysis is applied to feature selection, the prematurely local optimal solution may be avoided. Firstly, clustering is performed on the original feature set. The features with high correlation and high redundancy are clustered together, and features in different clusters have larger diversity. The feature strongly associated with the class is selected into a feature subset from each cluster. In this way, the redundant degree of the selected feature subset is relatively low, and the correlation with class is strong. The process of feature selection based on feature clustering almost starts with calculating the corresponding measure of the specified data set. Features are clustered according to specified clustering algorithm. At last the representative features of each cluster are selected to compose the final feature subset.

Au and Chan proposed the ACA [14] algorithm, it chooses the information measurement R to measure the correlation between the features. They use K-means to cluster the features, and then select the representative feature of each class. The algorithm is very perfect in the classification of gene data. The disadvantage of this algorithm is that sometimes the clustering may enter a dead cycle, and it is needed to specify a certain number of iterations to terminate the clustering process. Song developed an algorithm FAST [15] which uses hierarchical clustering method. In this algorithm, many minimum spanning trees are built with the measure *SU* and each tree is treated as a class cluster. FAST is suitable for high dimensional data. The disadvantage is that the number of selected features is determined by the data. Liu also introduced a feature clustering feature selection algorithm MFC [16] based on minimum spanning tree. Different from FAST the measure of MFC is Variation of Information (VI). Obviously, using the diversity between features in feature selection has already become a hot spot.

2   Information Distance Measure Based Feature Clustering

2.1   Basic Idea

The algorithm proposed in this paper mainly considers the diversity between the features. The diversity metric namely the information distance is used to measure the redundancy of the feature subset. The greater diversity the feature subset has, the lower redundancy it has.

Information distance [17–19] is a kind of diversity measurement based on information theory. In the literature [17], the paper listed the most commonly used measures based on information theory. Here we choose the measure D:

$$
\begin{aligned}
D &= \frac{1}{2}[H(X) + H(Y)] - I(X;Y) \\
&= \frac{1}{2}[H(X|Y) + H(Y|X)] \\
&= \frac{1}{2}\left(\sum_{x_i}\sum_{x_j} p(x_ix_j)\log\frac{1}{p(x_i|x_j)} + \sum_{x_i}\sum_{x_j} p(x_ix_j)\log\frac{1}{p(x_j|x_i)}\right) \\
&= \frac{1}{2}\left(\sum_{x_i}\sum_{x_j} p(x_ix_j)\log\frac{1}{p(x_i|x_j)p(x_j|x_i)}\right)
\end{aligned}
\tag{1}
$$

In formula (1), X and Y are random variables, $H(X)$ and $H(Y)$ are the information entropies of X and Y, and $I(X;Y)$ is the mutual information between X and Y. The conditional entropy $H(X|Y)$ represents the conditional uncertainty of X given Y. $p(x_ix_j)$ is the joint probability density between x_i and x_j. $p(x_i|x_j)$ is the conditional probability density of x_i given x_j.

The standard distance measure must satisfy the three properties: non-negativity, symmetry and triangle inequality. It had been proved that the information distance D is in line with the three properties in the literature [17].

$$D(X,Y) \geq 0 (\text{non} - \text{negativity}) \tag{2}$$

$$D(X,Y) = D(Y,X)(\text{symmetry}) \tag{3}$$

$$D(X,Y) \leq D(X,Z) + D(Y,Z)(\text{triangle inequality}) \tag{4}$$

Essentially, it has been recognized that feature selection needs to select a subset with strong discriminatory power. The features have high relevance with class should be chosen. However, experiments had proved that the m best features are not the best m features [20, 21]. There is redundancy among these features. Therefore, the selected subset should achieve a balance point between the relevance about class and the redundancy among features. If a feature subset S has been selected, the relevance between S and class is Rel, and the redundancy among S is Red.

$$Rel = \frac{1}{|S|} \sum_{x_i \in S} I(x_i; c) \tag{5}$$

$$Red = \frac{1}{|S|^2} \sum_{x_i, x_j \in S} I(x_i; x_j) \tag{6}$$

$$MRMR : \max(Rel - Red) \tag{7}$$

In order to reach the balance, on the one hand the selected subset of features should guarantee the most relevance with class, while on the other hand the degree of redundancy among the feature subset must be lowest. The criterion combing the formulas (5) and (6) is MRMR. In practice, incremental search algorithm mRMR was used to find the near-optimal feature subset. Assuming that the original feature set is F, the subset S_{m-1} of the m-1 features is selected. Now the target should be to choose the mth feature from the feature subset $\{F - S_{m-1}\}$ which satisfies formula (8). Here, mRMR is employed to select features from a refined feature subset that produced by feature clustering using information distance measure.

$$\max_{f_j \in F - S_{m-1}} [I(f_j; c) - \frac{1}{m-1} \sum_{f_i \in S_{m-1}} I(f_j; f_i)] \tag{8}$$

2.2   K-medoids Based Feature Clustering

Clustering is a commonly used unsupervised learning method of data analysis. It can divide the data according to the diversity between each other. Similarly, cluster analysis can also be applied to the features. Actually feature clustering has been widely used in feature selection methods.

K-means and K-medoids are two classic unsupervised clustering algorithms based on partitioning. According to distance between the objects in data set, K-means could divide the objects into several clusters. The distance measure generally use geometric distance such as Minkowski distance, Euclidean distance, Manhattan distance. Suppose the Euclidean distance is chosen as the distance measure. After deciding the k center randomly, the non-center points of data set are divided to the nearest center point. Clusters are formed after every object has been assigned. The average geometric coordinate value of each cluster is set to be the new center point of the cluster. Repeat the process until the clusters are stable. However, Features are a set of discrete points in the feature information distance space, and the features have no coordinate. Therefore, K-means algorithm can't be used for feature clustering, and K-medoids clustering algorithm is adopted for feature clustering. The measure employed is the information distance D mentioned above. In the iteration process, the values of the D between the features are constant and just need calculate once, which greatly reduces the computational complexity of the algorithm.

Unlike K-means, K-medoids algorithm selects actual object as the cluster center, rather than using the mean as the center in the cluster. K-medoids clustering algorithm is based on the principle that minimizes the degree of diversity between cluster center and all the objects in data set. Corresponding to the feature cluster, the information distance sum within the cluster should be smallest. An error criterion is designed in formulas (9) and (10) to measure the cost of replacing the original center by any of the non-center features in the cluster.

$$E = \sum_{i=1}^{k} \sum_{f \in C_i} D(f, o_i) \tag{9}$$

$$T = E_f - E_{o_i} \tag{10}$$

E is the sum of the information distance between all the features and corresponding center in the data set. $D(f, o_i)$ is the information distance between feature f and nearest center o_i. T is the cost of the cluster center o_i replaced by a non-center feature f. At the initial stage of the K-medoids k features are selected randomly as the center, and then the remaining features are assigned to the nearest center in information distance. In the iterations, for each non-center feature calculate the cost T of replacing the center in each cluster. If the cost T is negative, the actual error E must be reduced, and the current center feature can be replaced by this feature. Here the original center should be replaced by the one with the minimal T in which the convergence speed can be accelerated. Conversely the center won't be changed. If all the centers don't change any more, clustering process is over.

3   Algorithm and Steps

According to the above mentioned, we propose a text data feature selection method called Information Distance Measure based Clustering for Feature Selection (IDMCFS). Its process framework is described as Fig. 1.

IDMCFS is divided into two stages: clustering stage and selecting stage. The clustering stage: assuming that the size of ultimate target features is m, the original feature set is clustered into k clusters using K-medoids. k is much larger than m. From each cluster select the feature whose mutual information with the class is largest to form the candidate feature subset S'. In the selecting stage, m features are selected from subset S' according to mRMR. The redundancy of selected features are further reduced and the diversity between the features are guaranteed. The final m features is the solution of IDMCFS. Algorithm is described as follows.

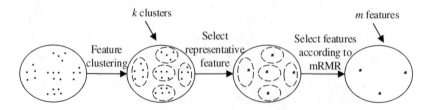

Fig. 1. The process framework of IDMCFS

Input: $F = \{f_1, f_2, ..., f_n\}$: original feature set
       m:target feature number
       k: cluster number, $k > m$
Output: S: selected feature subset

Begin
  $Center := \Phi$, Subset $S' := \Phi$, $S := \Phi$;
  For each $f_i \in F, f_j \in F$
    calculate $H(f_i)$, $I(f_i, c)$ and D_{ij} ;
  Repeat
    Randomly select a $f \in F$;
    $Center := Center + f$;
    $F := F - f$;
  Until $|Center| \geq k$
  Repeat
    For each $f \in F$
      Divide f to nearest center;
    For each $f \in F$
      Compute T_f according to formula(10);
    For each cluster
      Select f in cluster with minimal T_f to replace
center;
  Until Center not change
  Repeat
    Select f in each cluster with maximal $I(f, c)$;
    $S' := S' + f$;
  Until all clusters have been processed
  Select $f \in S'$ with maximal $I(f, c)$;
  $S := S + f$;
  Repeat
    Choose $f \in S'$ according to formula(8);
    $S := S + f$;
    $S' := S' - f$;
  Until $|S| \geq m$
  Return S;
end

The complexity of each iteration in the clustering phase of IDMCFS algorithm is $O(k(n - k)^2)$ (k is the number of clusters and n is the number of features in data set). Assuming that the total times of iterations is l, and then clustering complexity is $O(lk(n - k)^2)$. Because the feature with minimal T value is selected as the new center in iteration, thus the convergence speed is greatly accelerated and the number of iteration is very small. The complexity of selecting representative features and running mRMR are $O(n)$ and $O(km)$ respectively. Overall speaking, the complexity of IDMCFS algorithm is $O(lk(n - k)^2)$. When the cluster number k is much smaller than the original feature number n, the complexity will be $O(n^2)$.

4   Experiment Analysis

The experimental data sets are derived from the open data set (http://www.tunedit.org/repo/Data/Text-wc). These data sets are designed to test the classification performance of text data. In order to facilitate the calculation of mutual information, the data are discretized by MDL [22] discretization method. The information of data sets after discretized is displayed in Table 1. All the experiments are carried out on the experimental platform of Matlab2014a and Weka3.7. Weka is used to discretize the data and gain the classification accuracy rate of the final feature subset. The main procedure of IDMCFS is realized by Matlab2014a.

Table 1.   Information of datasets

Dataset	Features	Instances	Classes
tr12.wc	126	313	8
tr11.wc	250	414	9
wap.wc	545	1560	20
la1s.wc	1438	3204	6
la2s.wc	1438	3075	6
fbis1.wc	1138	2463	17

In order to eliminate the influence of generating initial center randomly in K-medoids algorithm to the clustering results, the experiment was repeated 50 times and took the average value as the final result. The performance of IDMCFS was compared to mRMR, CMIM and ReliefF algorithms on the same data sets. To check the performance of selected feature subset, the Naive Bayes classifier and 10-fold cross validation were adopted to test the accuracy rates of classification.

Table 2 shows the analogy of IDMCFS algorithm in different data sets and different parameters. It can be seen that in each data set there is an identical tendency. With the number of selected features m growth, the classification accuracy rates have increased. As we all know, more features often have greater distinguishing ability. Moreover, if the number of target features m is determined, the accuracy rates of classification don't

Table 2. Classification accuracy (%) of IDMCFS in different situations (m number of selected features, k number of clusters)

m	k	tr11.wc	tr12.wc	wap.wc	la1s.wc	la2s.wc	fbis1.wc
10	20	80.19	79.36	49.85	57.18	58.86	59.31
	30	81.37	**83.28**	51.36	58.65	60.35	60.09
	40	**82.56**	83.18	52.20	**60.74**	61.58	61.59
	50	82.29	83.16	**53.06**	60.46	**61.59**	**61.66**
20	30	82.31	84.02	58.68	63.97	65.84	65.63
	40	83.38	86.91	**61.07**	64.93	66.16	66.22
	50	84.23	87.82	59.84	65.66	**66.76**	**66.78**
	60	**85.53**	**88.12**	60.31	**66.25**	65.96	66.15
30	40	83.67	85.43	64.41	68.78	69.98	69.76
	50	85.24	86.55	65.01	68.80	70.21	69.89
	60	**85.85**	**87.23**	**65.10**	69.00	70.39	70.22
	70	85.58	87.20	64.77	**69.26**	**70.44**	**70.44**
40	50	85.26	86.18	63.52	70.44	70.51	68.89
	60	86.33	86.35	64.57	71.16	70.77	69.13
	70	**87.66**	**87.65**	65.69	**71.91**	70.94	69.97
	80	87.46	88.20	**66.25**	71.90	**71.83**	**70.97**
50	60	85.12	86.20	69.45	71.24	70.87	68.93
	70	85.55	87.96	**70.40**	72.64	72.56	69.54
	80	**86.61**	88.17	70.15	72.99	73.70	70.56
	90	86.16	**88.37**	70.09	**75.57**	**76.12**	**71.44**

increase as the increasement of cluster number k. The performance of IDMCFS is not related to the cluster number k. As a result it is more likely to get the global optimal solution.

Figure 2 represents the comparisons of the proposed algorithm IDMCFS and mRMR, ReliefF, CMIM algorithms performance in different data sets. Overall, the trend curves of four algorithms are similar. The classification accuracies are improved with the increase of the selected feature number. When m is determined, CMIM and mRMR have their own advantages in different data sets, and the results of ReliefF are slightly worse. The classification accuracies of IDMCFS on most data sets are higher than that of mRMR, CMIM and ReliefF. This shows that the IDMCFS algorithm can select a feature subset which has greater ability to distinguish the text. According to the theory of IDMCFS, if cluster number k equals to the feature number of the original data set IDMCFS becomes mRMR. However, the experimental results demonstrate that the performance of IDMCFS is better than mRMR and CMIM. It's believable that better feature subset can be selected by fully considering the diversity between the features. Of course, the results in the table are just the best results got in the experiment, as long as the parameter k set properly, the actual effect of IDMCFS may be further improved.

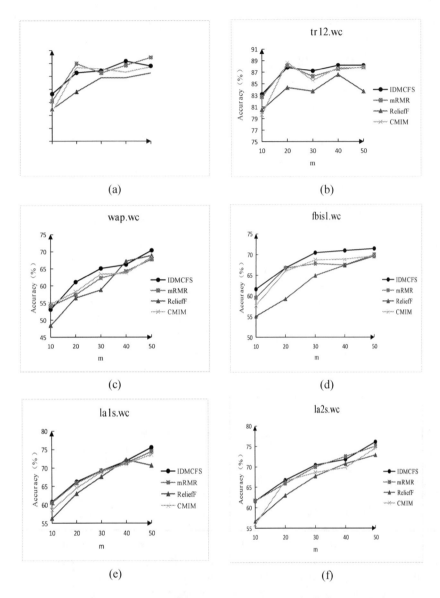

Fig. 2. The accuracies of IDMCFS, mRMR, CMIM and ReliefF on each data set

5 Conclusion

In this paper, a feature selection algorithm based on feature clustering using information distance has been proposed. This algorithm mainly emphasizes the diversity
between features, using an information distance measure to cluster the features. Highly
redundant features are clustered into a cluster, and the features of different clusters are
low relevant, so it is possible to obtain the global optimal solution. IDMCFS algorithm

uses the K-medoids clustering algorithm which only needs the distance between features. At the same time, the number of clusters k is much larger than target feature number m, which assures the high redundancy between features in each cluster. The clustering process converges quickly. The overall performance of IDMCFS is better than mRMR, CMIM and ReliefF.

There is a challenge about IDMCFS that how to choose the proper cluster number. In this paper k was set as several constants which are larger than m. Hence the value of k may be not optimal. How to find the optimal k value will be the research direction in the future.

Acknowledgments. This research was supported by the National Natural Science Foundation of China (Grant No. 61472467).

References

1. Lan, M., Tan, C.L., Su, J., Lu, Y.: Supervised and traditional term weighting methods for automatic text categorization. IEEE Trans. Pattern Anal. Mach. Intell. **31**, 721–735 (2009)
2. Xu, J., Croft, W.B.: Improving the effectiveness of information retrieval with local context analysis. ACM Trans. Inf. Syst. **18**, 79–112 (2000)
3. Chen, Z., Lü, K.: A preprocess algorithm of filtering irrelevant information based on the minimum class difference. Knowl.-Based Syst. **19**, 422–429 (2006)
4. Sebastiani, F.: Machine learning in automated text categorization. ACM Comput. Surv. **34**, 1–47 (2002)
5. Song, F., Liu, S., Yang, J.: A comparative study on text representation schemes in text categorization. Pattern Anal. Appl. **8**, 199–209 (2005)
6. Fragoudis, D., Meretakis, D., Likothanassis, S.: Best terms: an efficient feature-selection algorithm for text categorization. Knowl. Inf. Syst. **8**, 16–33 (2005)
7. Battiti, R.: Using mutual information for selecting features in supervised neural net learning. IEEE Trans. Neural Netw. **5**, 537–550 (1994)
8. Peng, H., Long, F., Ding, C.: Feature selection based on mutual information criteria of max-dependency, max-relevance, and min-redundancy. IEEE Trans. Pattern Anal. Mach. Intell. **27**, 1226–1238 (2005)
9. Yu, L., Liu, H.: Efficient feature selection via analysis of relevance and redundancy. J. Mach. Learn. Res. **5**, 1205–1224 (2004)
10. Fleuret, F.: Fast binary feature selection with conditional mutual information. J. Mach. Learn. Res. **5**, 1531–1555 (2004)
11. Vinh, N.X., Epps, J., Bailey, J.: Effective global approaches for mutual information based feature selection. In: International Conference on Knowledge Discovery and Data Mining, pp. 512–521. ACM (2014)
12. Forman, G.: An extensive empirical study of feature selection metrics for text classification. J. Mach. Learn. Res. **3**, 1289–1305 (2003)
13. Liu, H., Liu, L., Zhang, H.: Feature selection using mutual information: an experimental study. In: Ho, T.-B., Zhou, Z.-H. (eds.) PRICAI 2008. LNCS (LNAI), vol. 5351, pp. 235–246. Springer, Heidelberg (2008)
14. Au, W.H., Chan, K.C.C., Wong, A.K.C., Wang, Y.: Attribute clustering for grouping, selection, and classification of gene expression data. IEEE/ACM Trans. Comput. Biol. Bioinform. (TCBB) **2**, 83–101 (2005)

15. Song, Q., Ni, J., Wang, G.: A fast clustering-based feature subset selection algorithm for high-dimensional data. IEEE Trans. Knowl. Data Eng. **25**, 1–14 (2013)
16. Liu, Q., Zhang, J., Xiao, J., Zhu, H., Zhao, Q.: A supervised feature selection algorithm through minimum spanning tree clustering. In: IEEE 26th International Conference on Tools with Artificial Intelligence (ICTAI), pp. 264–271 (2014)
17. Vinh, N.X., Epps, J., Bailey, J.: Information theoretic measures for clusterings comparison: variants, properties, normalization and correction for chance. J. Mach. Learn. Res. **11**, 2837–2854 (2010)
18. Vinh, N.X., Epps, J., Bailey, J.: Information theoretic measures for clusterings comparison: is a correction for chance necessary? In: 26th AI Conference, pp. 1073–1080 (2009)
19. Vinh, N.X, Epps, J.: A novel approach for automatic number of clusters detection in microarray data based on consensus clustering. In: 9th IEEE International Conference on Bioinformatics and BioEngineering, pp. 84–91 (2009)
20. Jain, A.K., Duin, R.P., Mao, J.: Statistical pattern recognition: a review. IEEE Trans. Pattern Anal. Mach. Intell. **22**, 4–37 (2000)
21. Herman, G., Zhang, B., Wang, Y., Ye, G., Chen, F.: Mutual information-based method for selecting informative feature sets. Pattern Recogn. **46**, 3315–3327 (2013)
22. Fayyad, U., Irani, K.B.: Multi-interval discretization of continuous valued attributes for classification learning. In: 13th IJCAI, pp. 1022–1027 (1993)

Implementation of Microcontroller Arduino in Irrigation System

Štefan Koprda$^{(\boxtimes)}$, Martin Magdin, and Michal Munk

Faculty of Natural Sciences, Department of Computer Science,
Constantine the Philosopher University in Nitra,
Tr. a. Hlinku 1, 949 74 Nitra, Slovakia
{skoprda,mmagdin,mmunk}@ukf.sk

Abstract. The aim of this article is the design and implementation of intelligent irrigation devices using the Arduino microcontroller. The control interface of the irrigation system was created as a mobile application. The currently existing various similar solutions are all from the perspective of money unprofitable. The present solution is very cheap and effective especially for home usage. Following the extension it can also be applied on a larger scale e.g. for buildings greenhouse, intelligent garden and others. We used these hardware components for realization: the microcontroller Arduino Yun, real-time clock DS1302, two humidity sensors, relays and solenoid valves. The application for controlling irrigation device using mobile technology was created in the programming language Java. This easy smart irrigation system can be controlled by smartphone or tablet with operating systems Android 4 and above. Reactions and all functions of proposed intelligent system were verified with statistical surveyes. The results were evaluated by using technics of explorating analysis and non parametric Levene's test. By analysis we can determine the reliability of the irrigation equipment and so clarify the behavior of the system for the existence of systematic and random errors.

Keywords: Automatization · Intelligent system · Arduino YUN · PHP · Java · Reliability evaluation

1 Introduction

All human beings, animals and plants need water for survival. It is one of the basic needs for everyone. Most of the agriculture systems face water wastage as the major problem [1]. Agriculture, "The backbone of Indian economy" as quoted by MK Gandhi is defined as an integrated system of techniques to control the growth and harvesting of animal and vegetables. This statement is true and they should take an example from it and other countries [2]. Today agriculture uses for irrigation purposes 85 % of water [1].

Monitoring soil water content at high spatio-temporal resolution and coupled to other sensor data is crucial for applications oriented towards water sustainability in agriculture, such as precision irrigation or phenotyping root traits for drought tolerance [3]. The paper [4] shows that there are some technologies that benefit both farmers and the industry. ICT has certainly had a big impact on agriculture. One example that shows the successful use

© Springer International Publishing Switzerland 2016
D.-S. Huang et al. (Eds.): ICIC 2016, Part I, LNCS 9771, pp. 133–144, 2016.
DOI: 10.1007/978-3-319-42291-6_13

of ICT in agriculture are mobile telephones, in currently – smartphones [5]. This has been used to access information on weather and many other aspects. According to the work [6], ICT provide farmers with the opportunities for communication with other farmers, or options controlling their production. In current development and implementation of ICT in the form of mobile devices, more of them support greater processing power, more vivid display, and higher efficient information collecting method, which enhances interaction possibilities between human and machines. This technology can be used not only as consumer electronics products, but also considered as great assistants in the industry area. These products are then part of intelligent systems. A number of local studies [7–9] have agreed with findings by [4] which accentuated on factors such as education, negative perceptions, lack of capital, small land areas, ineffective infrastructure facilities, and limited capacity of extension workers as the main drivers that led to low technology adoption. Therefore we in this paper propose a design for home automation system using ready-to-use, cost effective and energy efficient devices including Arduino microcontrollers, and relay boards. Today we are living in 21st century where automation is playing important role in human life. Automation allows us to control appliances using automatic control. It not only provides comfort but also reduces energy, efficiency and time saving [10]. The requirement of building an automation system for an office or home is increasing day-by-day. Industrialist and researchers are working to build efficient and economic automatic systems to control different machines like lights, fans, air conditioners based on the requirement. Automation makes an efficient use of the electricity and water and reduces much of the wastage. Drip irrigation system makes the efficient use of water and fertilizer. Water is slowly dripped to the roots of the plants through narrow tubes and valves.

2   Automatic Irrigation System Using Arduino and OS Android

The concept of intelligent system is often used in research literature and popularizing magazines.

Realization of intelligent systems is a complicated task [11, 12]. Main problems is monitoring of measured variables, static values as well as dynamic processes with different Dynamics [13]. Each variable has the ability to detune the character of the measured system and thereby put the system in a constant regulatory process or even non-controllable state [14–18]. In this paper we want point out the creation of a smart irrigation system, which creates one of the research areas of intelligent systems. Determining factors in the development of the concept of a higher level of automatization intelligent systems (similarly as in industrial processes) are: properties of object control, properties of information, control and communication systems, aims of control and optimization [19]. This simple irrigation system allows the user to control irrigation in the home. The system is fully controlled using online interface and requires an active internet connection. In case of failure of internet connections, irrigation system works on the basis of the previously saved settings. Irrigation system consists of several modules, which can be divided into three parts: Control part, Regulation part and Server part. The control part consists of an Android application which is the front end of the complete irrigation system and can be implemented on any smartphone or tablet with OS Android 4.0.3 or higher. The regulation part consists of hardware elements, where the basis is the core of microcontroller Arduino Yun and ensures switching solenoid valves according to the requirements coming from the control unit. The server part is used as an interface between the control and regulation part, thus allowing communication across the Internet without the need for IP public addresses (or in case if the regulation part at some has been part of the local network).

3   Regulation Part of Smart Irrigation System

The regulation part of the irrigation system includes a description of the hardware design as the design and programming of software part. Module of regulation part is responsible for the correct interpretation of the values which are contained in the database, processing and representation at the physical layer. It consists of the control electronics and power electrical parts [20] (Fig. 1).

The core of the regulation module is a microcontroller Arduino Yun, which is based

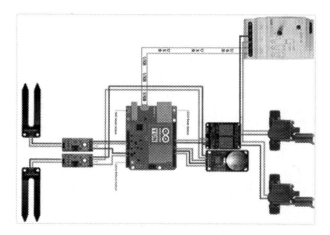

Fig. 1. Control irrigation system

on the chip ATmega32U4 and processor Atheros AR9331. Arduino is ideal for similar types of non-industrial automation [21, 22]. Primarily its modularity and sophistication facilitates the design and implementation of the final solution [18]. Arduino requires a supply voltage of 7–12 V (or 5 V in case of programming via USB) for it to run properly. USB also serves as a means for transmitting information to the terminal via UART communication [23]. Processor Atheros supports Linux distribution based on OpenWrt called as OpenWrt-Yun.

4 Concept of Communications

In design of the concept of communication, we have taken into account the requirements that have been required for realization of ideal irrigation system. The main requirement is wirelessly control of irrigation system, not only from local network, but virtually anywhere on Earth, in case if a smartphone (as control unit) has available internet connections. For this reason we designed for the solution Client-Server-Client. The microcontroller Arduino and the smartphone represent in this case clients who are connecting to a central server. This style of communication is a slightly complex but brings us benefit, if we want to create multiple irrigation systems for households, greenhouse and other. Data would be stored not only in irrigation units but also in the central database. The following figure shows the communication between microcontroller Arduino and OS Linux (first client), web server and Android application (second client). When communicating, the Linux OS sends queries in the time interval of 5 s to the Web server. In the case that in the database has been a change, it gets a response in the form of list of changes. Consequently, is realized a synchronization of the local database with the database on the Web server. Microcontroller Arduino in periodically intervals measures the current humidity and sends this data to the database on the server in time interval, which is stored in the settings. Microcontroller Arduino sends a per second request to local database and reads the current state of manual and planned irrigation and evaluates which irrigation loop is on or off. An Android application is used for creating an irrigation plan and entering commands for irrigation unit. The application saves the commands to the database using PHP scripts. Any changes to the database are recorded in the table *new_events*. With using the table irrigation unit detects whether is a database synchronization needed. Update and deleting data works on the same principles.

5 Description of the Main Features

The basic scheme of code for microcontroller Arduino consists of two main functions. After the start is automatically launched initialization method *setup()*. This method is launched only once, thereafter followed by method *loop()*. The method *loop()* is designed as infinity loop (main thread of program). The method *setup()* in our case initiallizes basic variables and sets actual time for OS Linux (after power interruption Linux does not remember this settings). The function *setCurrentTimeToLinux()* reads actual set time in RTC module and sets time for OS Linux. In method *loop()* is repeated measurement of humidity of land (*measureHumidity(sensor)*), settings state of solenoid valves in accordance of values, which is stored in database (*getValuesFromDB(valve)*), sending humidity values to the database on a Web server (*sendHumidityToDatabase()*) and synchronization of time by Internet (*synchronizeTime()*).

Other important method is *measureHumidity(sensor)*. This method measures humidity by using sensor, which is connected to the analog input port of microcontroller Arduino. Value that is read from input port is needed to be converted to the percentual units.

The method *getValuesFromDB(valve)* requires values from local database. These values determine the states of manual and planned irrigation. After is conducted evaluation on the basis of priorities – whether the system is in the mode active or no active. Top priority is manual irrigation, this priority can put the system into forced off or on state. In this state is realized the measurement of humidity of land and also sending the data to the web server, but when crossing the humidity settings, this parameter is ignored. If the manual mode is not activated, then irrigation unit works by using the planned mode. The planned mode of irrigation takes into account and also previous settings of humidity. If this value is exceeded, the system automatically switches off the irrigation loop (independently of the irrigation cycle). Irrigation is again automatically activated in the decrease of humidity.

The method *sendHumidityToDatabase()* connects to a web server and sends the actual measured value of humidity. This is realized in regular cycles according to the value of the interval that is specified in the configuration settings. The method processes with humidity for each humidity sensor in the system separately. In the database is the record containing the ID number of the sensor, the measure value of humidity and time when the measurements were realized.

The method *synchronizeTime()* is realized every 24 h and requires the current time from a Web server. This time is setting of the module of real time clock (RTC module). At the same it updates the time in OS Linux. In this way, we removed time deviation in module of real-time clock. Therefore is not necessary to adjust the time manually. If replacing the batteries in the clock module, you only need to restart the Arduino. After restart the time is synchronized.

6   Control Part

As the regulation part has no display and no keyboard, we decided that for control of the whole system is a suitable an application for smartphones. The solution we have chosen also because of low financial demands. Another advantage of this solution is the fact that most users have the smartphone or tablet always at hand.

Datamodel: This class is programmed according to the design pattern "Singleton". This solution ensures that in memory is always only one instance of the class for data consistency. Classes "*PlanItem*", "*ManualItem*", "*SettingItem*" are connected to the class *DataModel* and they are the images of SQL database on the server.

DBHandler: Another important part of the application is class DBHandler that provides communication between application and server. In order to obtain fast communication, subclasses are used: "*GetManualDataFromServer*", "*GetPlanDataFromServer*", "*Get SettingDataFromServer*", "*PostDataToServer*". These classes have its own thread. This thread is working in the background of the main thread. Main thread provides correct rendering of the GUI.

User Interface: User interface of application consists of 4 basic screens: Screen of manual control, Screen of week schedule, Screen of settings and Screen for control week of schedule.

Manual Control: On-screen for manual control is a slider bar to activate manual mode. In this mode, the system ignores the weekly schedule and is working according to the values that you entered manually. In this screen are also two buttons for control of solenoid valves.

Screen of Week Schedule: This screen consists of individual items of week schedule and button to add a new item. Individual items of the week schedule shows the day and time when the irrigation system is activated for a particular irrigation loop. In the system we can set the desired time and amount of cycles of the day (Fig. 2).

Fig. 2. Screen for settings of manual control (left figure) and week schedule (right figure).

Screen for Control Week of Schedule: Screen for control week of schedule includes a combobox on the selected day of the week, one combobox to select the start of watering, and one combobox for setting the ending of irrigation and combobox for select the irrigation loop.

Settings: Screen of settings is very simple and contains only two sliders. First slider is for settings of humidity at which the irrigation system is not active and the other to set how often information on the current humidity will be saved in the database (Fig. 3).

Server Part: We rented a Web server as free hosting (hostinger.sk). Hosting provides us with PHP and MySQL without restrictions. PHP engine has all the features enabled. We can use any version of PHP. For managing the databases is used phpMyAdmin.

MySQL Database: The database consists of four basic tables: manual, plan, settings and humidity. Table with a name *Manual* consists of two columns (*valve, valve_status*) and record data of the manual control. Column valve represents the identification of the specific valve and *valve_status* indicates its current status. Table with a name *plan* consists of five columns (*id, day, time_from, time_to and valve*) and provides to record the planning cycles. Column *day* identifies the relevant day of the week, columns

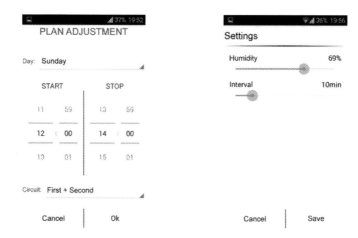

Fig. 3. Screen for control of week schedule and settings for humidity and interval.

time_from and *time_to* reserve the time period when the system is active and column *valve* provides a particular solenoid valve. Table *settings* consists of three columns (*id, setting_name, settings_value*) and contains system settings.

Column *setting_name* denotes a particular item of settings and column *setting_value* its value. Table *humidity* is determined for collection of information about humidity of irrigation land and it contains four columns (*id, timestamp, value, and sensor*). Column *timestamp* is time of record, column *value* contains the measured percentage of humidity and column *sensor* is a label of given irrigation loop. Table *new_events* provides as notification table for microcontroller Arduino. From this table Arduino receives information about which table has been changed. On the basis of the records in this table can microcontroller Arduino synchronize its local database. For complete administration of database, its integrity and control of the parameters of incoming requests were used PHP scripts. Android application uses custom scripts for loading data that is stored in the database. An integral part is the script *logger.php*, which allows you to record all operations and store the messages in the log table. On the basis on data from the log table we can create time-based graphs. These graphs show the performance of the entire system and system accesses to the central database. The simple script *time.php* is used for irrigation unit to synchronizate every 24 h or in case when we are changing battery.

7 Reliability Evaluation of the Irrigation System

For evaluation of the reliability and behavior of the irrigation system was used technics of explorating analysis and non parametric Levene's test for the purpose of observation of humidity in individual week schedule irrigation. For correct function of the irrigation system, we examined system behavior whether in the measurement process occurred errors. In the measuring, we focused on determining of systematic and random errors. Smart irrigation system has been tested in the laboratory conditions at the Department of Computer Science (FNS CPU in Nitra). The irrigation system was tested for one month. During testing we determined correct functions of solenoid valves (on/of switching) depending on the value set in the system. For analysis of correct function irrigation system we chose short time interval from irrigation period between 13.11.2015 to 16.11.2015. The irrigation system was turned on automatically without intervention during the day in three time intervals.

When using the system, the following conditions can occur:

- Solenoid valve 1 and solenoid valve 2 is closed – humidity at this time exceeded the set value,
- Solenoid valve 1 and solenoid valve 2 is opened - humidity at this time has been below the set value,
- Solenoid valve 1 is opened and solenoid valve 2 is closed,
- Solenoid valve 2 is opened and solenoid valve 1 is closed,

For evaluation of the reliability of the system we have created a data matrix. The matrix of data we have created on the basis of recorded values of measurement. On the basis of the measured data, we were able to determine the correct function of closing and opening of solenoid valve according the settings in the week schedule. In the following text, we present the results for one watering day (13.11. 2015) with a closed solenoid *valve 1*. For the following days (closed valve 1 and valve 2), the data was processed in the same way.

The irrigation plan (ID PLAN: 1#2, 7#8, 9#10) was:

- (1#2) to 60 % humidity, the valve is open, if humidity is exceeded, the valve 1 is closed,
- (7#8) to 65 % humidity, the valve is open, if humidity is exceeded, the valve 1 is closed,
- (9#10) to 70 % humidity, the valve is open, if humidity is exceeded, the valve 1 is closed.

On the basis of values from Levene's test were identified statistically significant differences in variability of humidity between observed irrigation plan (*MS Effect = 3693.626; MS Error = 44.218; F = 83.532; p = 0.0000*).

Table 1. Descriptive characteristics of irrigation plan (13.11.2015) with closed valve 1

ID PLAN	Valid N	Median	Minimum	Maximum	25th percentile	75th percentile	Average deviation	Dispersion coefficient	Range	Quartile Range
1#2	384	82.0	81.0	82.0	81.5	82.0	0.375	0.457	1.0	0.5
7#8	264	68.0	4.0	75.0	68.0	71.0	6.516	9.582	71.0	3.0
9#10	282	74.0	73.0	74.0	74.0	74.0	0.126	0.170	1.0	0.0

From the table of descriptive characteristics (Table 1), we can see that the median of humidity exceeds specified limits, but the variability in the case of the plan 7#8 is high, it reaches almost 10 % variability, whereas in case of the remaining plans is variability less than 0.5 %. The lower limit of quartile ranges, goes over the specified limits (60 %, 65 %, and 70 %) for all three irrigation plans. The high value of variability in case of the plan 7#8 could be caused by only a few extreme values.

The Box plot (Fig. 4) shows the state when the valve 1 is closed. This Graph shows the median, quartile range - middle 50 % of the values and the minimum and maximum of humidity level. On the axis y is illustrated the humidity and on the axis x specific irrigation plan for the day and time.

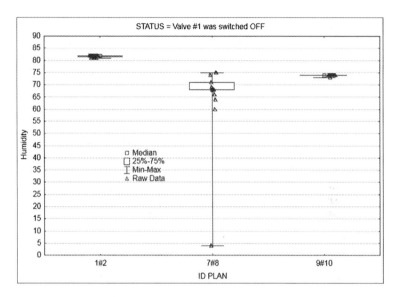

Fig. 4. Box plot of irrigation plan (13.11. 2015) with closed solenoid valve 1

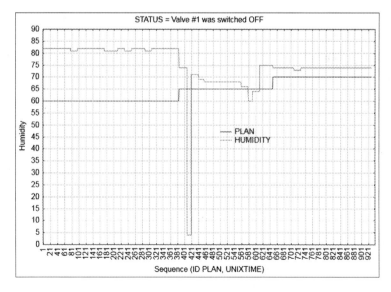

Fig. 5. Sequence graph of irrigation plan (13.11. 2015) with closed solenoid valve 1 (Color figure online)

From this graph we can clearly show greater variability of humidity in case of the irrigation plan 7#8. Moreover, in this irrigation plan we can identify only three cases with humidity below 65 % in the moment, when the valve 1 is closed. Following graph (Fig. 5) represent process of behavior solenoid valve 1, which is closed in the given sequence of irrigation plan. On the basis of this graph, we can determine whether in the system occured errors and if so whether it is a system error or a random error.

On the axis y is illustrated the humidity and on the axis x is sequence of irrigation plan (in this case, when solenoid valve 1 is closed). The blue curve represents irrigation plan, which was set using application on smartphone. The red curve represent real values of humidity.

8   Conclusion

In the two years have been developed similar systems of irrigation.

The low-cost and energy efficient drip irrigation system serves as a proof of concept by [24]. The design can be used in big agriculture fields as well as in small gardens via just sending an email to the system to water plants. The use of ultrasound sensors and solenoid valves make a smart drip irrigation system.

Designed irrigation system allows the user to control irrigation in the household. The system is fully controlled using the online interface and requires an active internet connection. In case of internet connection failure, the system works on the basis of the previous saved settings.

The advantage of the irrigation system is autonomously operated solenoid valves that ensure delivery of media to irrigation.

The accuracy of the irrigation system was greatly influenced by the used type of sensors for solenoid valve 1. For this reason, in sequence to 60 % humidity solenoid valve 1 (Fig. 5) worked reliably and was closed. The sequence to 70 % humidity solenoid valve 1 (Fig. 5) worked correctly and was closed. In a sequence 65 % humidity (Fig. 5) an error occurred, which could be caused by unpredictable changes in humidity in irrigation loop and solenoid valve 1 hasn't been opened. In an irrigation system occurred error, which can be considered after comparison of individual watering days as random, in regularly the watering sequence this error was not repeated. Random errors can be eliminated when we increase the number of humidity sensors in irrigation loop, or by replacing analog sensors for digital. The user can fully control the entire irrigation system using his/her mobile device.

Other advantages of the irrigation system are:

- possibility of operation in the event of internet connection failure,
- management of irrigation equipment using the web interface,
- possibility of supplementing the system with additional sensor elements.

On the other hand, disadvantages of the irrigation system are:

- created application is supported only on the Android OS 4 and above,
- higher input costs,
- the need to calibrate the humidity sensors, to achieve the highest possible efficiency of the irrigation system.

References

1. Nallani, S., Hency, V.B.: Low power cost effective automatic irrigation system. Indian J. Sci. Technol. **8**(23), 1 (2015)
2. Narechania, A.: An android-Arduino system to assist farmers in agricultural operations. In: IRF International Conference, New Delhi, India, pp. 19–26 (2015)
3. Bitella, G., Rossi, R., Bochicchio, R., Perniola, M., Amato, M.: A novel low-cost open-hardware platform for monitoring soil water content and multiple soil-air-vegetation parameters. Sensors **14**(10), 19639–19659 (2014). (Switzerland)
4. Truong, T.N.C.: Factors affecting technology adoption among rice farmers in the mekong delta through the lens of the local authorial managers: an analysis of qualitative data. Omonrice **16**, 107–112 (2008)
5. Pickernell, D.G., Christie, M.J., Rowe, P.A., Thomas, B.C., Putterill, L.G., Griffith, J.L.: Farmers market in Wales. Making the network? J. Brit. Food **106**, 194–210 (2004)
6. Abdullah, F.A., Samah, B.A.: Factors impinging farmers' use of agriculture technology. Asian Soc. Sci. **9**(3), 120 (2013)
7. Sadaf, S., Javed, A., Luqman, M.: Preference of rural women for agriculture information sources: a case study of District Faisalabad. Pak. J. Agric. Soc. Sci. **2**, 145–149 (2006)
8. Shaffril, H.A.M., Hassan, M.S., Samah, B.A.: Level of agro-based website surfing among Malaysian agricultural entreprenuers: a case study of Malaysia. J. Agric. Soc. Sci. **5**, 55–60 (2009)
9. Samah, B.A., Shaffril, H.A.M., Hassan, Hassan, M.A., Ismail, N.: ICT contribution in increasing agro-based entrepreneurs productivity in Malaysia. J. Agric. Ext. Soc. Sci. **5**, 93–98 (2009)
10. Kumar Sahu, C., Behera, P.: A low cost smart irrigation control system. In: 2nd International Conference on Electronics and Communication Systems, ICECS, pp. 1146–1151 (2015)
11. Pies, M., Hajovsky, R., Ozana, S.: Autonomous monitoring system for measurement of parameters of heat collection technology at thermal active mining dumps. Elektronika Ir Elektrotechnika. **19**(10), 62–65 (2013)
12. Pies, M., Hajovsky, R., Latocha, M., Ozana, S.: Radio telemetry unit for online monitoring system at mining dumps. Appl. Mech. Mater. **548–549**, 736–743 (2014)
13. Behan, M., Krejcar, O.: Modern smart device-based concept of sensoric networks. EURASIP J. Wirel. Commun. Netw. **1**, 155 (2013)
14. Luo, H., Yang, P., Li, Y., Xu, F.: An intelligent controlling system for greenhouse environment based on the architecture of the Internet of Things. Sens. Lett. **10**(1–2), 514–522 (2012)
15. Olvera-Olvera, C.A., Duarte-Correa, D., Ramírez-Rodriguez, S.R., Alaniz-Lumbreras, P.D., Lara-Herrera, A., Gómez-Meléndez, D., Herrera-Ruiz, G.: Development of a remote sensing and control system for greenhouse applications. Afr. J. Agric. Res. **6**(21), 4947–4953 (2011)
16. Lee, R.O., Suh, B.Y., Bae, Y., Yang, J.Y., Jeong, Y.Y., Nakaji, K.: Development of an integrated greenhouse monitoring and control system at province level. J. Fac. Agric. Kyushu Univ. **56**(2), 379–387 (2011)
17. David, J., Jančíkoyá, Z., Frischer, R., Vrožina, M.: Crystallizer's desks surface diagnostics with usage of robotic system. Arch. Metall. Mater. **58**(3), 907–910 (2013)
18. Bajer, L., Krejcar, O.: Design and realization of low cost control for greenhouse environment with remote control. IFAC Proceedings Volumes (IFAC-PapersOnline) **48**(4), 368–373 (2015)

19. Kachňák, A., Holiš, M., Belanský, J.: Optimization of continuous processes using hybrid neuro-fuzzy systems. In: 7th World Multiconference on Systemics, Cybernetics and Informatics, ORLANDO, pp. 107–111 (2003)
20. Koprda, Š., Magdin, M.: New trends and developments in automation in agriculture. Compusoft: Int. J. Adv. Comput. Technol. **4**(2), 1492–1494 (2015)
21. Fuentes, M., Vivar, M., Burgos, J.M., Aguilera, J., Vacas, J.A.: Design of an accurate, low-cost autonomous data logger for pv system monitoring using Arduino™ that complies with IEC Standards. Sol. Energy Mater. Sol. Cells **130**, 529–543 (2014)
22. Agudo, J.E., Pardo, P.J., Sánchez, H., Pérez, Á.L., Suero, M.I.: A low-cost real color picker based on arduino. Sensors (Switzerland)**14**(7), 11943–11956 (2014)
23. Horalek, J., Horalek, J., Sobeslav, V.: Datanetworking Aspects of Power Substation Automation. Communication and Management in Technological Innovation and Academic Globalization, pp. 147–153. World Scientific and Engineering Academy and Society, Athens (2010)
24. Agrawal, N., Singhal, S.: Smart drip irrigation system using raspberry pi and Arduino. In: International Conference on Computing, Communication and Automation, ICCCA, pp. 928–932 (2015)

TOPSIS and AHP Model in the Application Research in the Evaluation of Coal

Guoying Yang, Qingling Wang, and Jianqing Liu[✉]

Department of information processing and control engineering,
LanZhou Petrochemical College of Vocational Technology,
730060 Lanzhou, China
79546245@qq.com

Abstract. Aimed at the question of how to evaluate the coal quality, east Shenhua produced sales at shenhua group coal quality index data as the research object, using TOPSIS and AHP and combined two methods, through constructing the relative deviation matrix to determine the index weight, and gives the comprehensive sequencing of various kinds of coal.Structure deviation matrix to determine index weight not only effectively eliminate the dimensional, make the indicators are comparable, and make due to on the basis of rational scheme to build, to achieve the optimal.On the standardiZation and normaliZation of data put forward the improvement plan, gives the computing method of interval attribute data, improving the application range of the model.

Keywords: TOPSIS · AHP · Coal quality index · Relative deviation matrix

1 Introduction

The coal industry in the modern industrial development plays an indispensable role in the process, evaluation and distinguishes between coal and coal quality is a very important topic.

Several important indexes of coal: coal calorific value is calorific value of coal; It is volatile coal combustion can be volatile component; All the water is contained in coal moisture; Ash is rest after coal burning ash; Sulfur content is all the sulfur content in coal (pollution indicators) [1].

The important evaluation method has two kinds big, the main difference is the method of determining the weight. One is subjective values. The method, adopt comprehensive evaluation to determine the weight information, such as the comprehensive index method, AHP (analytic hierarchy process), the efficacy coefficient method;, another kind is objectively according to the correlation between each index or variation in Chengdu to determine weights, such as factor analysis, principal component analysis (pca), TOSPSIS (ideal solution), etc.

Relative deviation matrix constructed in TOSPSIS model by using AHP, through data related to determine the weight among indicators. Shenhua east produced sales at Table 1. Shenhua group the east Shenhua produced coal quality index of coal sales, and judging all kinds of coal quality and coal sort [2].

© Springer International Publishing Switzerland 2016
D.-S. Huang et al. (Eds.): ICIC 2016, Part I, LNCS 9771, pp. 145–152, 2016.
DOI: 10.1007/978-3-319-42291-6_14

Table 1. The east Shenhua produced coal quality index of coal sales

Category		Breed name	Z1 Qnet,ar	Z2 Vd (%)	Z3 Mt (%)	Z4 Ad (%)	Z5 St,d (%)
A	A1	Fine piece 2	≥5600	24–33	<16	<12	<0.6
	A2	Fine piece 3	≥5500	24–33	<16	<12	<0.6
	A3	Fine piece 4	≥5900	24–33	<16	<10	<0.6
	A4	Very low ash	≥5900	30–37	15-16	5-6.5	0.3-0.5
	A5	SH optimal 2	≥5650	30–37	16-17	6-7	0.3-0.6
B	B1	SHmix1-5500	≥5500	23–32	<17	<12	0.3-0.8
	B2	SHmix12-5200	5200-5499	21–30	<20	<16	0.3-1.0
	B3	SHmix 5000	5000-5199	21–30	20	20	0.3-1.0
	B4	SHmix 3-4800	4800-4999	21–30	20	20	0.3-1.0
	B5	SHmix 4-4500	4500-4799	21–30	25	20	0.3-1.5
	B6	SHmix 5-4200	4200-4499	21–28	28	15	0.3-1.5
	B7	SHmix 6-3800	⟨4200	21–28	28	20	0.3-1.5
C	C1	outsourcing1-5500	≥5500	23–32	<17	<12	0.3-0.8
	C2	outsourcing2-5200	5200-5499	21–30	<20	<16	0.3-1.0
	C3	outsourcing 5000	5000-5199	21–30	20	20	0.3-1.0
	C4	outsourcing3-4800	4800-4999	21–30	20	20	0.3-1.0
	C5	outsourcing4-4500	4500-4799	21–30	25	20	0.3-1.5
	C6	outsourcing5-4200	4200-4499	21–28	28	15	0.3-1.5
	C7	outsourcing6-3800	⟨4200	21–28	28	20	0.3-1.5
D	D1	anthracites 1-5500	≥5500	22–31	<16	<15	0.3-1.0
	D2	anthracites 2-5200	5000-5499	22–31	<15	<25	0.3-1.0
	D3	anthracites 3-4800	4800-4999	22–31	<15	<30	0.3-1.0
	D4	anthracites 4-4500	4500-4799	22–31	<15	<35	0.3-1.0
	D5	anthracites 5-4200	4200-4499	22–31	<15	<38	0.3-1.0
	D6	anthracites 6-3800	⟨4200	22–31	<15	<38	0.3-1.0

2 TOPSIS Model

TOPSIS Algorithm steps:

1. With vector planning method specification decision matrix is obtained. A multiple attribute decision making problems of decision matrix $A = (a_{ij})_{m \times n}$, Standardization of decision matrix $B = (b_{ij})_{m \times n}$,Among them

$$b_{ij} = a_{ij} / \sqrt{\sum_{i=1}^{m} a_{ij}^2}, i = 1, 2, \ldots m; j = 1, 2 \ldots n. \tag{1}$$

2. To construct weighted array $C = (c_{ij})_{m \times n}$.Each attribute weights vector

$$w = [w_1, w_2, \cdots w_n]^T, c_{ij} = w_j \cdot b_{ij}. \tag{2}$$

3. Determine the positive ideal solution is C^* and The negative ideal solution is C^0. Any of the positive ideal solution C^* of attribute values is c_j^*, The negative ideal solution C^0 of attribute values is c_j^0.

The positive ideal solution

$$c_j^* = \begin{cases} \max_i c_{ij}, j & \text{is kf property} \\ \min_i c_{ij}, j & \text{is Cost typetype} \end{cases} \tag{4}$$

The negative ideal solution

$$c_j^0 = \begin{cases} \min_i c_{ij}, j & \text{is kf property} \\ \max_i c_{ij}, j & \text{is cost type} \end{cases} \tag{5}$$

4. Solution is calculated the distance to the positive ideal solution and negative ideal solution. Sample solution to the positive ideal solution:

$$s_i^* = \sqrt{\sum_{j=1}^n \left(c_{ij} - c_j^* \right)^2} \tag{6}$$

Distance of sample solution to the negative ideal solution:

$$s_i^0 = \sqrt{\sum_{j=1}^n \left(c_{ij} - c_j^0 \right)^2} \tag{7}$$

5. Calculate the comprehensive evaluation index of each scheme

$$f_i^* = s_i^0 / (s_i^0 + s_i^*) \tag{8}$$

6. f_i^* By the order of the order of solution.

3   AHP Weighting Algorithm Is Improved

TOPSIS In the model algorithm,Attribute weights assignment has a variety of methods, common methods of artificial assignment need relevant experts to give the assignment to the various index, this method has certain subjectivity and arbitrariness [3]. In the relative deviation matrix is proposed to determine the attribute weights vector.

1. Construct ideal scheme that is, maximizing efficiency indicators and minimizing type cost index. In the evaluation index in Table 1, Z1 calorific value range, Z2 antioxidant1010 belongs to quality-benefit type indicators, Z3 all water, Z4 ash,

Z5 sulfur content type belongs to cost index. So the ideal solution for:Z = (Z1,Z2, Z3,Z4,Z5) = (5900,33.5,15,6,0.4).

2. Relative deviation matrix $R = [r_{ij}]$,

$$r_{ij} = \frac{|a_{ij} - z_j|}{\max\{a_{ij}\} - \min\{a_{ij}\}} \tag{9}$$

Not only eliminates the dimension deviation matrix structure, make the indicators are comparable, because on the basis of rational scheme to build at the same time, bring it to an optimum [4].

3. Determine the index weight. Calculate each index weight coefficient:

$$v_j = \frac{s_j}{|\overline{x_j}|} \tag{10}$$

Among them the first average indicators:

$$\overline{x_j} = \frac{1}{m} \sum_{i=1}^{m} a_{ij} \tag{11}$$

The variance:

$$s_j^2 = \frac{1}{m-1} \sum_{i=1}^{m} (a_{ij} - \overline{x_j})^2 \tag{12}$$

The normalized processing:

$$w_j = \frac{v_j}{\sum_{j=1}^{n} v_j} \tag{13}$$

The weight:

$$w = (w_1, w_2, w_3, w_4, w_5,) \tag{14}$$

4 Application Instance

According to the TOPSIS model method for integrated sorting of the data in Table 1,
    First of all indicators data standardizing. Z1 variety name assignment. Z2, Z5 interval average, Z3, Z4 take the minimum or average interval. The initial data are shown in Table 2. Using Matlab to standardize data processing A = Zscore(A),

Table 2. Shenhua group sales east Shenhua produced coal coal quality scheduling

Category		Z1	Z2	Z3	Z4	Z5	Z5	$S_{正}$	$S_{伤}$	f	P_1
A	A1	5600	28.5	16	12	0.45	0.45	0.0605	0.2546	*0.8081*	4
	A2	5500	28.5	16	12	0.45	0.45	0.0632	0.2520	0.7994	5
	A3	5900	28.5	16	10	0.45	0.45	0.0526	0.2680	0.8359	3
	A4	5900	33.5	15.2	6	0.4	0.4	0.0028	0.2984	0.9908	1
	A5	5650	33.5	16.5	9	0.45	0.45	0.0328	0.2691	0.8913	2
B	B1	5500	27.5	17	12	0.55	0.55	0.0813	0.2325	0.7409	6
	B2	5200	25.5	20	16	0.65	0.65	0.1316	0.1819	0.5802	13
	B3	5000	25.5	20	20	0.65	0.65	0.1453	0.1643	0.5306	20
	B4	4800	25.5	20	20	0.65	0.65	0.1513	0.1594	0.5131	21
	B5	4500	25.5	25	20	0.9	0.9	0.2246	0.1001	0.3083	7
	B6	4200	24.5	28	15	0.9	0.9	0.2580	0.1092	0.2975	14
	B7	3800	24.5	28	20	0.9	0.9	0.2674	0.0851	0.2414	22
C	C1	5500	27.5	17	12	0.55	0.55	0.0813	0.2325	0.7409	8
	C2	5200	25.5	20	16	0.65	0.65	0.1316	0.1819	0.5802	15
	C3	5000	25.5	20	20	0.65	0.65	0.1453	0.1643	0.5306	9
	C4	4800	25.5	20	20	0.65	0.65	0.1513	0.1594	0.5131	16
	C5	4500	25.5	25	20	0.9	0.9	0.2246	0.1001	0.3083	23
	C6	4200	24.5	28	15	0.9	0.9	0.2580	0.1092	0.2975	24
	C7	3800	24.5	28	20	0.9	0.9	0.2674	0.0851	0.2414	25
D	D1	5500	26.5	16	15	0.65	0.65	0.1018	0.2266	0.6899	10
	D2	5200	26.5	15	25	0.65	0.65	0.1332	0.2126	0.6149	17
	D3	4800	26.5	15	30	0.65	0.65	0.1593	0.1996	0.5562	11
	D4	4500	26.5	15	35	0.65	0.65	0.1845	0.1929	0.5111	18
	D5	4200	26.5	15	38	0.65	0.65	0.2056	0.1901	0.4804	12
	D6	3800	26.5	15	38	0.65	0.65	0.2113	0.1898	0.4732	19

$$A = \begin{bmatrix} 1.0516 & 0.7433 & -0.7303 & -0.8233 & -1.3119 \\ 0.8704 & 0.7433 & -0.7303 & -0.8233 & -1.3119 \\ 1.5953 & 0.7433 & -0.7303 & -1.0572 & -1.3119 \\ \cdots & \cdots & \cdots & \cdots & \cdots \\ -1.4856 & -0.1014 & -0.9409 & 2.2172 & -0.0625 \end{bmatrix}$$

Tectonic relative deviation matrix:

$$R = \begin{bmatrix} 0.1765 & 0.5556 & 0.0769 & 0.1875 & 0.1000 \\ 0.2353 & 0.5556 & 0.0769 & 0.1875 & 0.1000 \\ 0 & 0.5556 & 0.0769 & 0.1250 & 0.1000 \\ \cdots & \cdots & \cdots & \cdots & \cdots \\ 1.0000 & 1.0000 & 1.0000 & 1.0000 & 1.0000 \end{bmatrix}_{25 \times 5}$$

The weight $w = = (0.1893, 0.1058, 0.3210, 0.1980, 0.1859)$.

Z3 entire water 0.3210 as the greatest weight, Z2 volatilization of 0.1058 for minimum weight. Illustrate Z3 data can clearly separate samples, evaluation resolution information [5].

$$c_j^* = (0.0442, 0.0264, 0.0481, 0.0114, 0.0219)$$

$$c_j^0 = (0.0315, 0.0193, 0.0898, 0.0724, 0.0493)$$

Sample solution to the positive ideal solution

$$S_{\text{positive}} = s_i^* = [0.0605 \ 0.0632 \ 0.0526 \ldots 0.2113]_{1 \times 25}$$

Distance of sample solution to the negative ideal solution

$$S_{\text{negative}} = s_i^0 = [0.2546 \quad 0.2520 \quad 0.2680 \ldots 0.1898]_{1 \times 25}.$$

Comprehensive evaluation index of each scheme

$$f_i^* = [0.8081 \ 0.7994 \ 0.8359 \ldots 0.4732]_{1 \times 25}$$

Shenhua group sales east Shenhua produced coal coal quality scheduling of P1 as shown in the Table 2, the higher the comprehensive evaluation index, coal quality, the better. What do we learn from this model, the top 10 high quality coal respectively: A4, A5, A3, A1, A2, B1 and B5, C1, C3, D1; Number five is relatively inferior coal as B4, B7, C5, C6, C7. A kind of special varieties of A4, low ash 0.9908 evaluation index for the high quality coal, class C series C7 outsourcing evaluation index 0.2414 6-3800 as the inferior coal [6].

5   TOPSIS Algorithm to Improve

Common types are interval attribute values, efficiency and cost, the attribute value standardization can effectively improve the accuracy of the model. The interval type attribute is a certain range, the best efficiency properties was the bigger the better, cost type attributes as small as possible. The attribute specification can be used to solve nonlinear change certain target level and the nonlinear relationship between the attribute values and the target of not fully compensated.

1. Interval attribute type is neither cost nor efficiency data, the optimal attribute set for a given interval $[a_j^0, a_j^*]$, a_j' is can't stand the lower limit, a_j'' is can't stand upper limit,

$$b_{ij} = \begin{cases} 1 - (a_j^0 - a_{ij})/(a_j^0 - a_j'), & a_j' \leq a_{ij} < a_j^0 \\ 1, & a_j^0 \leq a_{ij} \leq a_j^* \\ 1 - (a_{ij} - a_j^*)/(a_j'' - a_j^*), & a_j^* < a_{ij} \leq a_j'' \\ 0, & \text{other} \end{cases} \tag{15}$$

After transformation of attribute values b_{ij} With the original attribute values a_{ij} Function graphic is commonly between trapezoid. The optimal interval attribute value equals the upper and lower, the optimal range for a point, degradation of triangle function graphics.

As Table 1 shows the Z1, Z2, Z3 coal index data are interval display, such as set a certain interval optimal attribute index, can enhance the accuracy evaluation model.

If the Z2 to determine the optimal interval value for (30, 32), cannot tolerate lower limit for the 21st, cannot tolerate cap of 33, using the Matlab arithmetic, get coal quality P2 is: [4 5 3 1 2 20 6 13 21 7 14 22 8 15 23 9 16 24 25 10 17 11 1812 19]. Coal optimal for A2, worst for C7.

If the Z5 to determine the optimal interval value is [0.5, 0.6], cannot tolerate lower limit of 0.3, can't stand up to 1, and get the quality of coal P3: [4 5 3 1 2 20 6 13 21 7 14 22 8 15 23 9 16 24 25 10 17 11 1812 19] The coal quality optimal for A4, worst for C7.

2. Vector normalization. The cost and efficiency are available under the type of transformation. $b_{ij} = a_{ij}/\sqrt{\sum_{i-1}^{m} a_{ij}^2}$ Canonical,Sum of squares is 1, each scheme attribute value is often used in computing solutions with some virtual the occasion of the Euclidean distance.

3. Standardized processing. The actual problem, the measurements of different variables are often not consistent, eliminate the dimensional effect of variables, each variable all have the same expressive force, Standardization of data processing $b_{ij} = (a_{ij} - \overline{a_j})/s_j$.

6 Conclusion

The coal industry in our country has a very important role in the process of industrialization, the merits of the objective evaluation of coal resources, give full play to the efficiency of coal resources, can improve the utilization value of all kinds of coal resources, avoid high energy using high or low energy, low, to make reasonable allocation.

This paper sales at Shenhua group east Shenhua produced coal coal resources index data as the research object, using TOPSIS model for comprehensive evaluation, using AHP to avoid subjective weight assignment, improve the effect of the model evaluation objective. Interval attribute data processing in the model are given, and put forward the vector normalization and standardization of data to avoid dimensional effect, and improve the evaluation precision of the model.

The objective of this paper only for coal resources attribute data for the comprehensive evaluation analysis, if can be combined with the efficiency of input and output data, comprehensive evaluation of coal output, coal can be the reasonable distribution of the efficiency of coal, coal and other practical problems [7, 8].

References

1. Li, D.Y., Li, H.M., Zhou, Y., et al.: Scientific evaluation method of mining of coal resources study. J. Coal **543**, 543–547 (2012)
2. Bai, X.M., Zhao, S.S.: The Pros and cons of a variety of comprehensive evaluation method to judge the research. J. Stat. Res. 45–48 (2000). (cutflower production potentials)
3. Gong, J., Hu, N.L., Cui, X., et al.: Based on AHP And TOPSIS evaluation model of rock burst tendency prediction. J. Rock Mech. Eng. 1442–1448 (2014). (cutflower production potentials)
4. Hu, Z.Y., Tan, S.M., Peng, Y.: TOPSIS evaluation method based on fuzzy AHP study. Microcomput. Inf. Practices **261**, 228–230 (2006)
5. Peng, X.Y., Zhang, Q.H.: The coal resources city sustainable development of the comprehensive evaluation method research. Pract. Underst. Math. 22–27 (2009)
6. Qin, Y., Tang, D.Z., Liu, D.M., et al.: Coal reservoir development dynamic geological evaluation theory and technology progress. J. Coal Sci. Technol. **1**, 80–88 (2014)
7. Wan, M., Hu, M.Y., Tian, X.: Ningwu basin coal reservoir evaluation. J. Inner Mongolia, Petrochemical Ind. **120**, 120–121 (2006)
8. Li, C.W., Xie, B.J.: Evaluation model of coal and gas outburst intensity energy. J. Coal. **9**, 1547–1552 (2012)

An Improved Context-Aware Recommender Algorithm

Huiyu Miao, Bingqing Luo, and Zhixin Sun$^{(\boxtimes)}$

Key Laboratory of Broadband Wireless Communication
and Sensor Network Technology,
Nanjing University of Posts and Telecommunications, Nanjing 210003, China
sunzx@njupt.edu.cn

Abstract. The paper proposes a kind of improved algorithm for context-aware recommender algorithm which is based on matrix factorization. The dot product of items and users is calculated for modeling the factors in paper. And then the average scores of items, different users-baseline and items-baseline on different contexts are added as users-items bias terms. Finally the evaluation scores as the recommender results for each items are the summation of users-items bias terms, interactions on items-contexts, users-contexts and users-items-contexts, and the dot products. The experimental results on test sets show that this improved algorithm has better accuracy of recommendation.

Keywords: Context-aware recommender system · Recommender system · Matrix factorization

1 Introduction

Context-aware was firstly put forward by Schilit [1] in 1994. He categories contexts into location, identities of nearby people and objects, and changing to those objects. Dey [2] defined context that "Context is any information that can be used to characterize the situation of an entity. An entity is a person, place, or object that is considered relevant to the interaction between a user and an application, including the user and applications themselves."

Recommender System (RS) [3, 4], as a part of personalized recommender service, tries to excavate the relationships between users and items, to acquire useful items informations (like information in Web, commodities and services) and to generate recommender results for users. Common recommender algorithms are collaborative filtering, content-based filtering and hybrid filtering.

Context-aware Recommender Systems, also called CARS, are firstly proposed by Adomavicius and Tuzhilin [5] in 2011. The contexts which affected the recommender results are added into recommender system, and the accuracy of recommendation are increased. Adomavicius [5, 6] raised three kinds of context-filters which are based on different effects of contexts: Contextual Pre-filtering, Contextual Post-filtering and Contextual Modeling.

The first part of this paper introduces the definitions of context-aware, recommender system and CARS. The next part analyzes related papers of CARS filed. The

© Springer International Publishing Switzerland 2016
D.-S. Huang et al. (Eds.): ICIC 2016, Part I, LNCS 9771, pp. 153–162, 2016.
DOI: 10.1007/978-3-319-42291-6_15

third part addresses the details of the improved algorithm. The fourth part shows the results and comparisons in test sets. And the last part discusses the prospects of CARS and the future works on it.

2 Related Research

There are many studies and reports about context-aware. Paper [7] proposes an architecture of context awareness in Ubiquitous computing, which contains a lot of modules like data collections, data classifications, data representations and ratiocinates, and databases updating. Paper [8] discusses context-aware mobile p2p social network, which consists of system architecture, aggregation model and discovery algorithm. The aggregate model intelligently aggregates potential p2p social network and discovers the suited relationship according to users' demands.

Algorithms and applications of Recommender System (RS) are hot research topics nowadays. Paper [9] proposes a way solves the data sparseness problem effectively. It put forwards a multi-level association and some classification rules. The multi-level association is based on users' information contents and classification rules used to integrate items project. Paper [10] enhances the accuracy of collaborative filtering and models users' or items' timing behaviors. Paper [11] proposes a new algorithm based on collaborative filtering, which replaces nulls by predictive values produced by content-based filtering in matrices. Paper [12] is based on a mixed filtering algorithm. And top-model-based clustering and the TF-IDE Cosine correlation algorithm are used to provide the potential topics for microblog users.

Context aware Recommender System (CARS) has caused a wide discussion in the field of context-aware and RS since 2011. Paper [13] proposes a model modified classical matrix factor, and the model builds the fuzzy matching relations between contexts and potential factors. Paper [14] proposes MCRS model, which is based on hybrid-ε-greed algorithm. The MCRS model is a combination of standard of ε-greedy algorithm and content-based and collaborative filtering technology projects. Paper [15] proposes a new kind of text mining technique which automatically extracts specific areas of knowledge for context-aware recommendation, so the algorithm has many advantages such as high recommended effects, self-learning abilities and interpretation capabilities. Paper [16] proposes a new approach that combines contexts into recommender system. This approach offers a framework which can identify related context variables and products a serious of context features which describes similar rating-pattern for each class of entities. Paper [17] analyzes the limitations of current CARS and proposes a new algorithm combined pre-filter with post-filter, and the algorithm is based on the importance of contexts which depends on the dynamic behaviors of the user uniquely.

Paper [18, 19] all use topic hierarchy to optimize the context-aware recommender model. Paper [18] proposes two item-recommender algorithms which generalize from literature recommended score prediction algorithm. And paper [19] improves the accuracy of CARS using the context from topic hierarchy structure. And the paper discusses three different themes hierarchies based on LUPI respectively. Paper [20, 21] proposed some algorithm for CARS based on matrix factorization. Paper [20] models the interactions of the contextual factors with item ratings introducing additional model parameters. Paper [21] adds many parameters which can influence the accuracy of recommender based on paper [20], and experimental results show that the improved algorithm in paper [21] has better results.

Through the analysis of relevant reports and papers, studies of context-aware, RS or CARS are all wide and deep. Depend on current reports mentioned before, a large number of algorithms emphasis the interactions between items and users, but ignored the interactions of users-items-contexts. However, this paper is based on those interactions. Experiences results show that the improved algorithm has better recommender effects.

3 Details of Algorithm

3.1 The Improved Algorithm (ICARA)

A Context-aware Recommender Algorithm based on the matrix factorization is proposed by Baltrunas [20]. The algorithm in [20] is expended into three conditions by different assumptions. The first CAMF-C is based on one contextual condition; the second is CAMF-CI, it has one parameter per contextual condition and item pair; the last one CAMF-CC introduces one model parameters for each contextual condition and item category. The unify equation is as follow:

$$r_{\widehat{uic_1c_2...c_k}} = \bar{i} + b_u + \sum_{l=1}^{k} B_{ic_l} + I_i^T U_u \tag{1}$$

In Eq. 1, $r_{\widehat{uic_1c_2...c_k}}$ represents the evaluations from different users for the items under contexts among $c_1 - c_k$. $\bar{i}$ represents the average scores of items i ratings in the data set R, b_u represents the baseline parameter for users, B_{ic_l} represents the interaction of the contextual conditions and the items.

The algorithms of paper [21] are improved algorithms. It proposed two equations: ICAMF-I and ICAMF-II. The equations are showed as follows:

$$r_{\widehat{uic_1c_2...c_k}} = \bar{i} + b_u + b_i + I_i^T U_u + \sum_{l=1}^{k} I_i^T C_{lc_l} + \sum_{l=1}^{k} U_u^T C_{lc_l} \tag{2}$$

$$r_{\widehat{uic_1c_2...c_k}} = \bar{i} + b_u(c_1,...c_k) + b_i(c_1,...c_k) + I_i^T U_u + \sum_{l=1}^{k} I_i^T C_{lc_l} + \sum_{l=1}^{k} U_u^T C_{lc_l} \tag{3}$$

Though adding context information into recommender algorithm, there are several limitations on paper [21]. In Eq. 1, it ignores the baseline of items; in Eqs. 2 and 3, they all ignore different influences for users or items by different contexts. So the improved algorithm is introduced in this paper and called ICARA:

$$r_{\widehat{uic_1c_2\ldots c_k}} = \bar{i} + b_u(c_1,\ldots c_k) + b_i(c_1,\ldots c_k) + I_i^T U_u + B_1 \tag{4}$$

$$B_1 = \sum_{l=1}^{k} B_{ic_l} + \sum_{l=1}^{k} B_{uc_l} + \sum_{l=1}^{k} B_{uic_l} \tag{5}$$

$$b_u(c_1,\ldots c_k) = b_u + \frac{1}{k}\sum_{l=1}^{k} b_{uc_l} \tag{6}$$

$$b_i(c_1,\ldots c_k) = b_i + \frac{1}{k}\sum_{l=1}^{k} b_{ic_l} \tag{7}$$

In Eq. 4, $r_{\widehat{uic_1c_2\ldots c_k}}$ represents the evaluated scores for users. $\bar{i}$ represents average scores for each items. The next two parameters represent baseline evaluated factors, which mean the users' or items' baselines with different contexts. The next parameter builds the model of factors for users on different items. The B_1 is showed in Eq. 5 and explanation in follow paragraph.

In Eq. 5, B_{uc_l}, B_{ic_l} and B_{uic_l} represent the interactions of items-contexts, users-contexts and users-items-contexts.

In Eq. 6, b_u represents the users' baselines, $\frac{1}{k}\sum_{l=1}^{k} b_{uc_l}$ represents the affections on baseline of users.

In Eq. 7, b_i represents the items' baselines, $\frac{1}{k}\sum_{l=1}^{k} b_{ic_l}$ represents the affections on baseline of items.

The Minimum Squared-Error (MSE) is calculated for getting those parameters in equation:

$$\text{Min} \sum_{r_{uic_1c_2\ldots c_k}\in R} [L_1(r_{uic_1c_2\ldots c_k} - \overline{r_{uic_1c_2\ldots c_k}}) + \frac{1}{2}\lambda_1(B_2 + \|I_i\|^2 + \|U_u\|^2)] \tag{8}$$

$$B_2 = b_u^2 + b_i^2 + \sum_{l=1}^{k} \|b_{uc_l}\|^2 + \sum_{l=1}^{k} \|b_{ic_l}\|^2 + B_3 \tag{9}$$

$$B_3 = \sum_{l=1}^{k} \|B_{uc_l}\|^2 + \sum_{l=1}^{k} \|B_{ic_l}\|^2 + \sum_{l=1}^{k} \|B_{uic_l}\|^2 \tag{10}$$

3.2   The Training Algorithm

The following is learning algorithm, we assumes that the objective function is F_1:

Input: test set (R), dimension of factor space (d),
parameter(γ), velocity of learning (η)

Output: all $b_u(c_1,\cdots c_k)$, $b_i(c_1,\cdots c_k)$, b_{uc_l}, b_{ic_l}, B_{uc_l}, B_{ic_l}, B_{uic_l} and
U, I, $c_{1,\cdots}c_k$

Initialize: the number of all $b_u(c_1,\cdots c_k)$, $b_i(c_1,\cdots c_k)$,

b_{uc_l}, b_{ic_l}, B_{uc_l}, B_{ic_l}, B_{uic_l} are zero, initialize $U \epsilon R^{|U|*d}$, $I \epsilon R^{|I|*d}$,
$C_1 \epsilon R^{|C_1|*d}$, $\cdots$, $C_k \epsilon R^{|C_k|*d}$ as a random small number.

For j = 1: k **do**

While $(u,\ i,\ C_1,\ C_k)$ **in R do**

$b_u \leftarrow b_u - \eta \frac{\partial F_1}{\partial b_u}$

$b_i \leftarrow b_i - \eta \frac{\partial F_1}{\partial b_i}$

$b_{uc_l} \leftarrow b_{uc_l} - \eta \frac{\partial F_1}{\partial b_{uc_l}}$

$b_{ic_l} \leftarrow b_{ic_l} - \eta \frac{\partial F_1}{\partial b_{ic_l}}$

$B_{uc_l} \leftarrow B_{uc_l} - \eta \frac{\partial F_1}{\partial B_{uc_l}}$

$B_{ic_l} \leftarrow B_{ic_l} - \eta \frac{\partial F_1}{\partial B_{ic_l}}$

$B_{uic_l} \leftarrow B_{uic_l} - \eta \frac{\partial F_1}{\partial B_{uic_l}}$

$U_u \leftarrow U_u - \eta \frac{\partial F_1}{\partial U_u}$

$I_i \leftarrow I_i - \eta \frac{\partial F_1}{\partial I_i}$

$C_{lc_l} \leftarrow C_{lc_l} - \eta \frac{\partial F_1}{\partial C_{lc_l}}$

end while

j ← j + 1

end for

4   Experiments and Results

Two kinds of test sets are used in this paper: the first kind of test set is true movie-scored, also called LDOS-CoMoDa-set, which is supplied by Odic [22]. Except the data of users and items, the set offers 12 contexts among 30 informational variables. The lost data is set 0 in training process.

Another kind of sets are three half-simulation movie-scored test sets, which are built on Yahoo! Webscope. In those sets, age and sex are considered as two simulated information, and age is divided into 3 categories: under 18 years old, between 18 and 50 years old, and above 50 years old and $c \in \{0, 1\}$ are used to represent males or females. From raw data-set, the number for extracting scores is randomly. According to this rules, 3 half-simulation data-sets are get, which are corresponding to the rule of $\alpha \in \{0.1, 0.5, 0.9\}$, $\beta = 0.9$.

In our paper, we use MAE (Mean Absolute Error) as the judgment for ICARA.

$$\text{MAE} = \frac{\sum_{r_{uic_1c_2...c_k} \in R_{test}} \left| r_{uic_1c_2...c_k} - \overline{r}_{uic_1c_2...c_k} \right|}{N(R_{(test)})} \tag{11}$$

In Eq. 8, $R_{(test)}$ represents test set, $r_{uic_1c_2...c_k}$ represents the true scores for test sets, and $N(R_{(test)})$ represents the numbers of the data in test sets. So the smaller MAE is, the better the performance algorithm has.

4.1   Number of Contexts Variables for the True Sets

We use information gain algorithm to get every variables' information gains which in LDOS-CoMoDa.

Table 1. Context variables' information gains in LDQS-CoMoDa

Context variable	Information gain
endEmo	0.17263
dominantEmo	0.13699
Interaction	0.02727
Mood	0.02488
Social	0.02053
Weather	0.01432
Time	0.00901
Season	0.00732
Physical	0.00721
Decision	0.00621
DayType	0.00582
Location	0.00503

From Table 1, the first two variables' information gains are bigger than the others and the last six variables' information gains are smaller. So the threshold is range from 0.01432 to 0.13699. To get best range of threshold, the best number of variables is studied. Figure 1 shows the first five variables are best number for contexts.

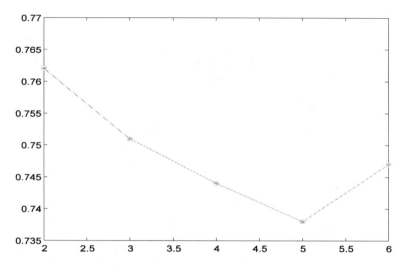

Fig. 1. The relationship between N of context variable and MAE

4.2 Experiments on Set of LDOS-CoMoDa Set

The comparison is showed in Fig. 2. From Fig. 2, CAMF-CC has a better performance than CAMF-C and CAMF-CI because contexts have same influence in same groups of data-set. And among those algorithms which are mentioned, ICARA has best performance due to more specific evaluations into recommender algorithm. From the figure, ICAMF-I, ICAMF-II and ICARA all get lower MAE and have better performance in recommendation with this parameter, so the baseline b_{ic_l} is important in recommender system. And then with the interaction among items, users and contexts, the recommender performance is specific and individual. Due to the limitation of true date-set, the tests in Half-Simulation are implemented.

4.3 Experiments on Sets of Half-Simulation

From Fig. 3, ICARA has a better performance than the others. In this part of comparison, CAMF-CC is not included in testing for no groups in half-simulation sets. When $\alpha = 0.1$ all algorithms get higher MAE while lower when MAE is $\alpha = 0.9$. It shows that when contexts have a big percentage in comparisons, the performance of recommender algorithm is better. And because context variations which affect recommender performance in ICARA, no matter what α is, it always get the lowest MAE.

When α from 0.1 to 0.5, MAE of ICAMF-I reduces 5.79 %, ICMF-II reduces 4.47 %, and ICARA reduces 9.37 %. When α from 0.5 to 0.9, MAE of ICAMF-I

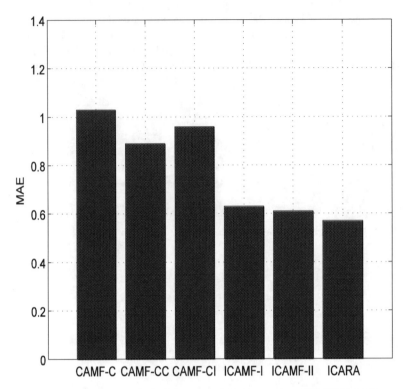

Fig. 2. Different MAE for different algorithm in test set of LDOS-CoMoDa

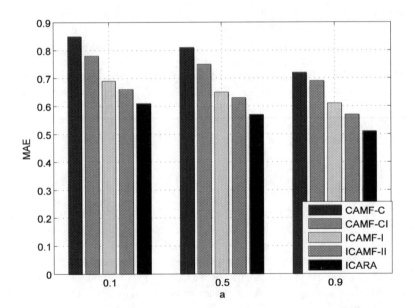

Fig. 3. Different MAE for different algorithm in test set of half-simulation

reduces 9.23 %, ICMF-II reduces 10.93 %, and ICARA reduces 15.51 %. Those data show that with the percentage of context bigger, the MAE of each algorithm reduces quickly.

ICARA collects information from different users, contexts and items and calculate the interactions among them. Though ICARA has better performance in recommender, the computational cost is higher and computational space is bigger than others. This algorithm is suited for recommender system with strong and fast computing power. Cloud computing is required for small size of recommender system or mobile devices.

5 Prospects of CARS and Future Works

Nowadays, the big data exists in network, websites and every lifetime. CARS becomes famous for the systems in mobile devices with related sciences flourishing. In this paper, An improved algorithm which is based on matrix factorization are put forward. More parameters about contexts are added in this algorithm (ICARA), like $b_i(c_1, \ldots c_k)$, B_{uc_l}, B_{ic_l} and B_{uic_l}. Tests in one true data set and three half-simulation data sets show that ICARA has better performance in recommender system, with more computational cost and space.

Though ICARA has some advantages for recommendation, there are several limitations. Firstly, ICARA does not deal with the cold-start problem, which exists widely in the field of CARS and traditional RS. Secondly, it only tests on disperse data-sets but not on continuation data-sets. Third, more cost and space are required in computations are waited for improving. So in the future study, those works above will be covered.

Acknowledgment. This work is supported by the National Natural Science Foundation of China (60973140, 61170276, 61373135), The research project of Jiangsu Province (BY2013011), The Jiangsu provincial science and technology enterprises innovation fund project (BC2013027), the High-level Personnel Project Funding of Jiangsu Province Six Talents Peak and Jiangsu Province Blue Engineering Project. The major project of Jiangsu Province University Natural Science (12KJA520003).

References

1. Schilit, B., Adams, N., Want, R.: Context-aware computing applications. In: IEEE Workshop on Mobile Computing Systems and Applications. 85–90 (1994)
2. Dey, A.K.: Understanding and using context. Pers. Ubiquit. Comput. **5**(1), 4–7 (2001)
3. Adomavicius, G., Tuzhilin, A.: Toward the next generation of recommender systems: a survey of the state-of-the-art and possible extensions. IEEE Trans. Knowl. Data Eng. (TKDE). **17**(6), 734–749 (2005)
4. Ricci, F., Rokach, L., Shapira, B., Kantor, P.B.: Recommender Systems Handbook. Springer, Berlin (2011)
5. Adomavicius, G., Tuzhilin, A.: Context-aware recommender systems. In: Ricci, F., Rokach, L., Shapira, B., Kantor, P.B. (eds.) Recommender Systems Handbook, pp. 217–253. Springer, New York (2011)

6. Adomavicius, G., Sankaranarayanan, R., Sen, S., Tuzhilin, A.: Incorporating contextual information in recommender systems using a multidimensional approach. ACM Trans. Inf. Syst. (TOIS) **23**(1), 103–145 (2005)
7. Mao, H.Y., Jiang, N.K., Su, W., et al.: A context aware modeling framework for pervasive applications. In: Cloud and Service Computing, pp. 40–44 (2012)
8. Cao, H.H., Zhu, J.M., Pan, Y., et al.: Context-aware p2p mobile social network structure and discovery algorithm. Chin. J. Comput. **35**, 1223–1234 (2012)
9. Paireekreng, W.: Mobile content recommendation system for re-visiting user using content-based filtering and client-side user profile. In: International Conference on Machine Learning and Cybernetics (ICMLC), pp. 1655–1660 (2013)
10. Sun, G.F., Wu, L., Liu, Q., et al.: Recommendations based on collaborative filtering by exploiting sequential behaviors. J. Softw. **24**(11), 2721–2733 (2013)
11. Fan, J.Q., Pan, W.M., Jiang, L.S.: An improved collaborative filtering algorithm combining content-based algorithm and user activity. In: International Conference on Big Data and Smart Computing, pp. 88–91 (2014)
12. Dong, K.J., Shen, Y.: A hybrid content-based filtering approach: recommending microbloggers for web-based communities. In: IEEE International Conference on Green Computing and Communications (GreenCom), pp. 1254–1258 (2013)
13. Fang, Y.N., Guo, Y.F.: A context-aware matrix factorization recommender algorithm. In: 4th IEEE International Conference on Software Engineering and Service Science (ICSESS), pp. 914–918 (2013)
14. Bouneffouf, D., Bouzeghoub, A., Gancarski, A.L.: Following the user's interests in mobile context-aware recommender systems: the hybrid-ε-greedy algorithm. In: Advanced Information Networking and Applications Workshops (WAINA), pp. 657–662 (2012)
15. Zhang, W.P., Lau, R., Tao, X.H.: Mining contextual knowledge for context-aware recommender systems. In: 9th International Conference on e-Business Engineering (ICEBE), pp. 356–360 (2012)
16. Rana, S., Jain, A., Panchal, V.K.: Most influential contextual-features [MICF] based model for context-aware recommender system. In: 2013 International Conference on Emerging Trends in Communication, Control, Signal Processing and Computing Applications (C2SPCA), pp. 1–6, 10–11 October 2013
17. Katarya, R., Verma, O.P., Jain, I.: User behaviour analysis in context-aware recommender system using hybrid filtering approach. In: 4th International Conference on Computer and Communication Technology (ICCCT), pp. 222–227 (2013)
18. Domingues, M.A., Manzato, M.G., Marcacini, R.M., et al.: Using contextual information from topic hierarchies to improve context-aware recommender systems. In: 22nd International Conference on Pattern Recognition (ICPR), pp. 3606–3611 (2014)
19. Sundermann, C.V., Domingues, M.A., Marcacini, R.M., Rezende, S.O.: Using topic hierarchies with privileged information to improve context-aware recommender systems. In: Brazilian Conference on Intelligent Systems (BRACIS), pp. 61–66 (2014)
20. Baltrunas, L., Ludwig, B., Ricci, F.: Matrix factorization techniques for context aware recommendation. In: 5th ACM Conference on Recommender Systems, pp. 301–304 (2011)
21. Feng, P.C.: Research on Context Aware Personal Recommendation Algorithm. Donghua University, Shanghai (2014)
22. Odic, A., Tkalcic, M., Tasic, J.F., et al.: Relevant context in a movie recommender system: users opinion vs. statistical detection. ACM RecSys (2011)

Effectively Classifying Short Texts via Improved Lexical Category and Semantic Features

Huifang Ma[(✉)], Runan Zhou, Fang Liu, and Xiaoyong Lu

College of Computer Science and Engineering,
Northwest Normal University, Lanzhou, China
mahuifang@yeah.net

Abstract. Classification of short text is challenging due to its severe sparseness and high dimension, which are typical characteristics of short text. In this paper, we propose a novel approach to classify short texts based on both lexical and semantic features. Firstly, the term dictionary is constructed by selecting lexical features that are most representative words of a certain category, and then the optimal topic distribution from the background knowledge repository is extracted via Latent Dirichlet Allocation. The new feature for short text is thereafter constructed. The experimental results show that our method achieved significant quality enhancement in terms of short text classification.

Keywords: Short text classification · Latent Dirichlet allocation · Lexical features · Semantic features · Optimal topic distribution

1 Introduction

With the rapid development of the social network, we are now dealing with much more short texts in various applications. Examples are snippets in search results, tweets, status updates, news comments, and reviews from various social platforms. There is an urgent demand to interpret these short texts efficiently and effectively. However, short texts do not provide sufficient word occurrences, traditional methods [1] such as "Bag-Of-Words" fail to represent these short texts accurately due to their high-dimension and severe sparsity. One crucial question, for conventional classifying methods like support vector machine (SVM) [2] and k nearest neighbors (k-NN) [3], is how to select representative features in such short documents. The characteristics of short text will hinder the application of conventional machine learning and text mining algorithms.

Most existing approaches have attempted to enrich short text to get more features, which combine more semantic, contexts and associations. Existing work in the literature tried to address the aforementioned challenges from two directions. One was to employ search engines and utilize the search results to expand related contextual content [4, 5] while the other was to utilize external repositories as background knowledge [6, 7]. These two types of methods can solve somewhat the problem, but still leave much space for improvement. Both methods tried to obtain more semantic

© Springer International Publishing Switzerland 2016
D.-S. Huang et al. (Eds.): ICIC 2016, Part I, LNCS 9771, pp. 163–174, 2016.
DOI: 10.1007/978-3-319-42291-6_16

information by retrieving terms in search engines, Wikipedia or other resources, therefore it will not only go against some irrelevant information but also consume much more time.

In the opposite direction of enriching short text representation, some researchers [8] proposed to trim a short text representation to get a few most representative words for short text classification. Others attempted to represent short text via effective feature selection method to reduce the dimension of feature space. Short texts can then be resented by these selected features. Several feature selection measures have been put forward to reduce dimensionality in the past years, such as term frequency-inverse document frequency (TF-IDF), information gain (IG), mutual information (MI) and expected cross entropy (ECE), etc. [9].

In recent years, topic modeling approaches, such as Latent Dirichlet distribution (LDA) [10], are widely applied for text classification. One typical example is that Blei et al. [11] derived a set of hidden topics obtained by LDA as new features from a large existing Web corpus. Phan et al. [12] put forward an improved method which exploits topics with multi-granularity based on LDA, modeling the short text more precisely to certain extent. Vo and Ock [5] proposed methods for enhancing features using topic models, which make short text seem less sparse and more topic-oriented for classification. These approaches provided new methods for enhancing features by combining external texts from topic models that make documents more effective for classification.

This paper presents a short text classification method based on lexical category features and optimal topic distribution. Firstly, a mechanism is introduced to build a term dictionary by selecting lexical features which are the most representative words of a category. And then the optimal topic distribution of the background knowledge repository is obtained via LDA. Then the original features are extended by combination of both the selected features and the optimal topic distribution on which the SVM classifier and k-NN are trained. Finally, experimental evaluation on real text data is conducted.

The remainder of this paper is organized as follows. We describe the relevant theoretical knowledge in Sect. 2. Section 3 introduces the proposed general framework combining lexical category features and semantics for short text classification. Experimental designs and findings are presented in Sect. 4. Section 5 concludes the proposed work and points out our future work.

2   Problem Preliminaries

In this paper, we mainly focus on news titles which appear frequently on social networks, and the background knowledge is extracted from news content. Let $D = \{d_1, d_2, ..., d_m\}$ be the short text corpus, where m is the number of texts in D. $W = \{t_1, t_2, ..., t_n\}$ denotes the vocabulary of D, where n is the number of unique words in D. $C = \{c_1, c_2, ..., c_k\}$ is the collection of class labels, where k is the number of class labels. The term dictionary $W(t) = \{t_1, t_2, ..., t_l\}$ is constructed by selecting lexical features that are the most distinctive words with regard to a certain category, where l denotes the number of selected distinctive words from all categories and $l \ll n$. Next, we will introduce relevant theoretical knowledge.

2.1 Expected Cross Entropy

Expected cross entropy (*ECE*) is a kind of feature selection measure based on the information theory, which considers both word frequency and the relationship between word and category. The bigger the *ECE* value, the more informative a feature has for the purpose classification. *ECE* value of word t_i is usually calculated as follows:

$$ECE(t_i) = P(t_i) \sum_j P(C_j|t_i) \log_2 \frac{P(C_j|t_i)}{P(C_j)} \tag{1}$$

Where, t_i represents for a word, C_j represents for category j, $P(C_j | t_i)$ represents that when contains the word t_i the probability of belong to the C_j.

2.2 Correlation Weight

Correlation weight (*COW*) is an efficient way of term weighting for short text, which considers the correlation of terms within a short text [8]. Concretely, the conditional probability is used to model the probability that terms appear together in a short text and the probabilistic correlation of terms is defined in a symmetric way as follows:

$$cor(t_i, t_j) = p(t_i|t_j) \times p(t_j|t_i) \tag{2}$$

which represents the probabilistic degree that words t_i and t_j belong to the same short texts.

The correlation weight denotes the reliability and importance of the word t_i in the short text. A higher correlation weight implies a higher probability that if word t_i appears in the short text, other words t_j will also appear in the short texts. In other words, the more words t_j that show high correlation with word t_i, the higher the probability word t_i is relevant to the short text. Given a short text d_l with an initial weight w_i (term frequency) of each word t_i, the correlation weight of word t_i in the short text d_l is defined as:

$$COW(t_i) = w_i + \frac{\sum_{t_j \in d_l} w_j \cdot cor(t_j, t_i)}{|d_l|} \tag{3}$$

Where, $|d_l|$ means the total number of words in the short text d_l.

2.3 Latent Dirichlet Allocation

Latent Dirichlet Allocation (LDA), first introduced by Yang et al. [9], is a generative probability model of a corpus that can be used to estimate the multinomial observations by unsupervised learning. It can be used to model and discover underlying topic structures of any kind of discrete data in which text is a typical example. The basic idea is that documents are characterized as multinomial distributions over latent topic, meanwhile each topic is represented by a multinomial distribution over words.

3   General Framework

The whole framework consists of four major steps, namely constructing term dictionary, extracting optimal topic distribution, forming the new feature and training the classifier as shown in Fig. 1.

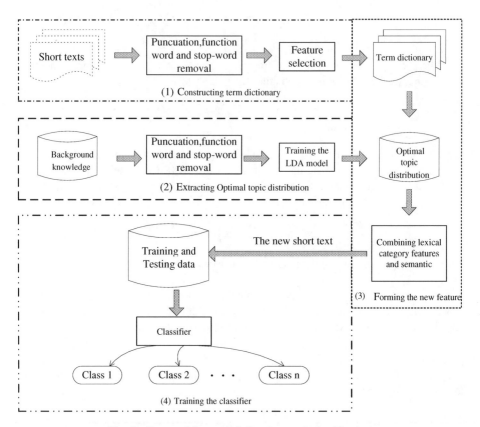

Fig. 1. Proposed framework for short text classification

The main process of our approach is as follows:

Step 1. Constructing term dictionary by tokening, removing stop-words and selecting lexical features which are the most of distinctive words with regard to a category;

Step 2. Extracting optimal topic distribution from the background knowledge repository via Latent Dirichlet Allocation;

Step 3. Forming the new feature for short text by combining both lexical category features with optimal topic distribution as external features;

Step 4. Training the classification model on labeled data.

3.1 Constructing Term Dictionary

The most critical step in the proposed approach is to select the most representative words from a short text as features to construct the term dictionary. Traditionally, whether a word is chosen as the feature or not depends on two ideal conditions, the features should be (i) well representing the main content of the short text, and (ii) topically indicative. *COW* weighting scheme is sufficient for the first requirement. For the second requirement, we prefer to use *ECE′* (an improved *ECE* method) value to measure the topical-specificity of a word.

In Eq. (1), a word t_i has an overall weight in all the categories. However, in most instances, a representative word of category A may be not of great importance in category B. Accordingly, we prefer to assign different weights to terms in different categories. So the weight of a word t_i in different categories C_j is calculated separately as following:

$$ECE(t_i, C_j) = P(t_i|C_j)P(C_j|t_i) \log_2 \frac{P(C_j|t_i)}{P(C_j)} \tag{4}$$

From Eq. (4), we can see that if t_i has a strong relationship with category C_j, or the category is of small size, the word has higher possibility to have high weight with regard to category C_j. Ideally, we wish that most of distinctive words would have strong relationship with one category and less coherent with others. Equation (5) is used to measure the weight of a word with regard to a certain category.

$$ECE'(t_i, C_j) = ECE(t_i, C_j) - \sum_{k \neq j} ECE(t_i, C_k) \tag{5}$$

In order to make the word features as discriminative as possible, the final weight of a word t_i with regard to a category C_j is defined using the equation below.

$$t_{i,j} = COW(t_i) \times ECE'(t_i, C_j) \tag{6}$$

The top-l distinctive words for each category are selected to construct the feature dictionary.

3.2 Extracting Optimal Topic Distribution

This subsection provides a detailed description of topic model analysis. When applied to text classification, the topics are considered as features for classification and each topic is characterized by a multinomial distribution over words. Short texts are in general much shorter, nosier, and sparser, which makes the potential structure matrix very large. Therefore, LDA model is adopted in the background knowledge, which is considered as an external repository. How to select the optimal number of adapted topics becomes our primary concern. Intuitively, the quality of topics depends on their capability in helping discriminate short text with different class labels. Table 1 shows the definition of each symbol in extracting optimal distribution.

Table 1. Notations for extracting topic distribution

Notation	Definition
N	A set of numbers
$T = \{T_1, T_2, ..., T_{|N|}\}$	The topic set with all topic distributions
$T_i = \{z_{i_1}, z_{i_2}, ..., z_{iK_i}\}$	A topic set with topic distribution T_i
z_{ij}	One of topic in topic distribution T_i
K_i	The total number of topics in topic distribution T_i
$W = \{w(T_1), w(T_2), ..., w(T_{|N|})\}$	The weight vector with different topic distribution
$nh(d_i)$	The text in the same class for text d_i
$nm(d_i)$	The text in the different class for text d_i
$d_i^{T_i}$	A topic distribution for d_i
$dis(d_i^{T_i}, d_j^{T_i})$	The distance between d_i and d_j on a topic T_i

Given a set of numbers empirically denoted by N, with the size $|N|$, we run LDA over the background knowledge repository to generate the topics with respect to each item in N. As a result, we can obtain $|N|$ different sets of topics. For example, $N = \{10, 30, 50, 70, 90, 120, 150\}$, and $|N| = 7$. We present $|N|$ different sets of topics as $T = \{T_1, T_2, ..., T_{|N|}\}$, where each entry T_i is a topic set in the form of $T_i = \{z_{i_1}, z_{i_2}, ..., z_{iK_i}\}$, K_i is the number of topics and z_{ij} denotes a topic, which is a probability distribution over words. Aiming at the topic's ability to distinguish between different categories, we can construct a topic distribution of the weight vector $W = \{w(T_1), w(T_2), ..., w(T_{|N|})\}$, where $w(T_i)$ is the weight indicating the importance of topic set T_i.

To get W, an algorithm based on the key idea of Relief [14] is proposed. Specifically, for each long text d_i in the knowledge repository, finding n-nearest neighbors $nh(d_i)$ from the same class and the n-nearest neighbors $nm(d_i)$ from different classes. And then the distance between d_i and d_j over a certain topic T_i is calculated, which becomes the basis on which the weight $w(T_i)$ is updated. Conventionally, KL-divergence can be used to measure the distance between two probability distributions, it's an asymmetric measure distance. In this paper, we use the symmetric KL-divergence value as distance. Then the weight $w(T_i)$ is updated as follows.

$$w(T_i) = w(T_i) + \sum_{i=1}^{n} dis(d_i^{T_i}, nm(d_i)^{T_i}) - \sum_{i=1}^{n} dis(d_i^{T_i}, nh(d_i)^{T_i}) \qquad (7)$$

Where, $dis(d_i^{T_i}, nm(d_i)^{T_i}) = \frac{1}{2} \sum_{z_k \in T_i} \{p(z_k|d_i) \log \frac{p(z_k|d_i)}{p(z_k|nm(d_i))} + p(z_k|nm(d_i)) \log \frac{p(z_k|nm(d_i))}{p(z_k|d_i)}\}$ So is the value $dis(d_i^{T_i}, nh(d_i)^{T_i})$.

Equation (7) shows that for the topic distribution T_i, the smaller the distance within the same class, or the bigger the distance within the different class, the higher the weight $w(T_i)$ is. Finally, topic T^* with biggest weight value $w(T_i)$ is taken as optimal topic, namely, $T^* = \arg\max_{T_i \in T}\{w(T_i)\}$, The algorithm for selecting optimal topic distribution is summarized as follows:

```
Algorithm 1: Selecting Optimal Topic Distribution
1: Input: background knowledge repository;
2: Output: optimal topic distribution Tᵢ;
3: Let T = {T₁ ,T₂ ,…,T|N|}, W = {w(T₁) ,w(T₂),…,w(T|N|)};
4: Initialize W(Tᵢ)    0 for all topic distribution;
5: for each text dᵢ ∈ D do
6:       Find N-nearest neighbors: nm(dᵢ) and nh(dᵢ);
7:       for i    1 to |N| do
8:             Update W(Tᵢ) using (7);
9:       end for
10: end for
11: return  T* = arg max{w(Tᵢ)} ;
                  Tᵢ∈T
```

3.3 Form New Feature for Short Text

Our primary idea of establishing the classification framework is to combine external words and optimal number of adapted topics as representation model for short text. Given a short text d_i and optimal topic distribution T_i, $d_i^{T_i}$ is a topic distribution for d_i. Finally the new feature for short text d_i is obtained by appending the topic feature $d_i^{T_i}$ to the selected term dictionary $\vec{w}_{\underline{d}}$ as follows:

$$\overrightarrow{F_{\underline{d}}} = [\vec{w}_{\underline{d}}, \alpha \cdot d_i^{T_i}] \tag{8}$$

After obtaining the new features, we can train classifiers in conventional ways. We choose SVM and k-NN as the classifiers for the experiments, and the improved algorithm has advantages on performance compared to the traditional ones.

4 Experiment

In this section, we conduct a series of experiments to evaluate the classification performance of our algorithm and describe these experiments and the results. The classification algorithms selected are SVM and k-NN. We parameterize k-NN by choosing the value $k = 30$, which is known to perform well in [5]. For SVM, a linear kernel is preferred due to its good balance between execution time and accuracy.

4.1 Data Set

Experiments are conducted on two dataset collected from different platforms:

Sougou Dataset [15] is a collection of news articles provided by Sougou Lab, including 18 categories such as International, Sports, Society, and Entertainment etc. Each article document involve page URL, page ID, news title and page content. The html tags have been removed in the page content. Our experiment selects 8 categories from the dataset and each category contains 4000 documents, with the ratio of training documents to test documents as 7:3.

Search Snippets are very short in comparison with normal documents. It has been used in [4, 12, 13]. Each snippet consists of three parts: a URL, a very short title and a short text description. To prepare the labeled training and test data, we perform Web search transactions using various phrases belonging to different domains. For each search transaction, we select top 20 snippets from the results to ensure that most of them belong to the same domain. The dataset consists of 10,060 train and 2,280 test snippets from 8 categories.

Short text refers to the title, the news contents and the description of short text are used for background knowledge extraction. All Chinese documents were pre-processed by word segmentation using ICTCLAS [16].

4.2   Results and Analysis

We present experimental results on three tasks. First, we evaluate our feature extraction schemes and compare the performances with other classical methods. Second, we study the performance of our method under different topic settings. At last, we conduct the impact analysis on parameter α and compare the classification performance of our algorithm with different combinations of category features and topic distribution.

4.2.1   Evaluation of $COW \times ECE'$

We evaluate 6 schemes to select features: $TF \times IDF$, $TF \times ECE$, $TF \times ECE'$ (ECE' refers to an improved ECE method), $COW \times IDF$, $COW \times ECE$ and $COW \times ECE'$. The evaluation is conducted on both datasets. We select different size of features ranging from 50 to 350 for both traditional methods and our measure and then apply SVM and k-NN for short text classification. The classification accuracy is shown in Fig. 2.

From the results, we make the following observations. First, more words generally lead to better classification accuracy. The improvement becomes very minor when more than 250 words are used. In almost all cases of different feature sizes, it is observed that our approach $COW \times ECE'$ achieves the highest performance on both datasets. Besides, $COW \times ECE'$ achieves stable accuracy with the growing size of features. This demonstrates the robustness of our method to noisy data. Hence, we take $COW \times ECE'$ as an effective method for selecting features for short text classification.

4.2.2   The Different Number of Topics

Let the set of numbers $N = \{10, 30, 50, 70, 90, 110, 130\}$, $|N|$ sets of topics from the background knowledge is extracted and topics are selected via LDA based on our proposed algorithm. All the experimental results are obtained by averaging 20 runs. $W(T_i)$ is updated based on Eq. (7). The results of classification accuracy of test data are shown in Table 2.

As we can see in Table 2, for both datasets, SVM classifier performs generally better than k-NN. More topics (from 10 topics to 70 topics) generally lead to better classification accuracy. It is clear that the more numbers of topic is set in the LDA

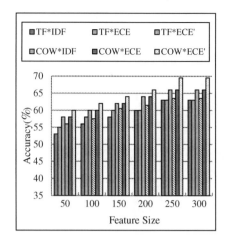

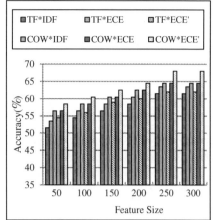

(a)SVM on Sougou Dataset

(b)k-NN on Sougou Dataset

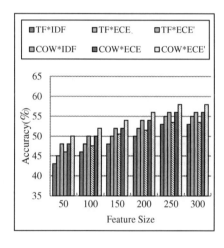

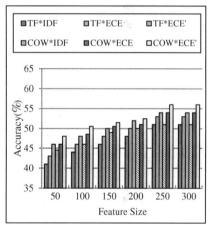

(c)SVM on Search Snippets

(d)k-NN on Search Snippets

Fig. 2. Classification accuracy of traditional feature selection measure

model, the higher the computational cost. And from Table 2 we can see that the improvement becomes very minor when more than 90 topics are used. Therefore we select $T(90)$, which achieves remarkable improvement in classification accuracy over any topics.

4.2.3 The Impact Analysis on the Parameter α

Parameters α is the most important parameter in our method, which is a user specified parameter indicating the importance of topic features. We set parameters α varied from 0 to 200. In Fig. 3, classification accuracy is reported to depict the performance of our

Table 2. Classification accuracy of SVM and k-NN with different number of topics

Data sets	Method	T10	T30	T50	T70	T90	T110	T130
Sougou	SVM	67.34	71.36	74.69	73.83	**75.82**	73.98	72.89
Dataset	k-NN	65.56	70.96	73.89	72.76	**74.92**	72.84	70.98
Search	SVM	65.12	69.45	71.56	72.89	**73.58**	71.03	72.56
Snippets	k-NN	63.45	69.03	70.06	70.89	**71.12**	70.36	69.89

proposed approach. As for constructing short term vector, top 200 words are chosen as the word feature in each category.

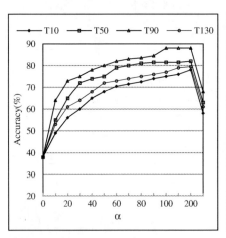

(a)SVM on Sougou Dataset

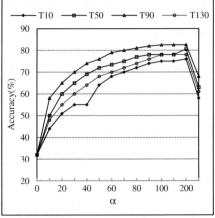

(b)k-NN on Sougou Dataset

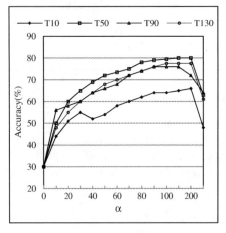

(c)SVM on Search Snippets

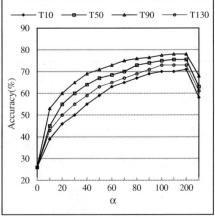

(d)k-NN on Search Snippets

Fig. 3. Accuracy of SVM and k-NN on various α

From the Fig. 3, we can see that as the enriched representation is a combination of word feature from the short text and topic feature from a topic space derived from background knowledge, it demonstrates its superior performance to either considering word feature or topic feature in isolation, indicating that an optimal performance comes from an appropriate combination of both. Besides, word-only approach (corresponding to $\alpha = 0$) in Fig. 3 performs very poor. When optimal topic distribution is adopted, the classification algorithm shows its best performance, which means for classifying short text, the proposed general framework could improve the classification performance remarkably and is superior to word-only methods.

5   Conclusions and Future Work

The main challenges for short text classification are the high dimensionality of feature space, data sparseness, and the lack of context information. In this paper, we put forward a new method for short text classification by combining both lexical and semantic features. Specifically, a term dictionary is built by selecting the most of distinctive words with regard to a category and the optimal topic distribution from the background knowledge is extracted by LDA. New feature for short text can therefore be generated. With the new features, we can train classifiers in conventional ways. Regarding future work, we are also interested in how to extend classical LDA model to the correlated topic models to extract not only semantic features, but also the correlations between these features for representing short texts.

Acknowledgement. This work is supported by the National Natural Science Foundation of China (No. 61363058), Youth Science and technology support program of Gansu Province (145RJZA232, 145RJYA259), 2016 undergraduate innovation capacity enhancement program and 2016 annual public record open space Fund Project 1505JTCA007.

References

1. Gupta, V., Lehal, G.S.: A survey of text mining techniques and applications. J. Emerg. Technol. Web Intell. **1**(1), 60–76 (2009)
2. Sebastiani, F.: Machine learning in automated text categorization. ACM Comput. Surv. **34** (1), 1–47 (2002)
3. Cheng, Q.Q., Wang, L.L., Zheng, T., et al.: Microblog friend recommendation based on multi-feature classification. Comput. Eng. **41**(4), 65–69 (2015)
4. Sun, A.: Short text classification using very few words. In: Proceedings of the 35th International ACM SIGIR Conference on Research and Development in Information Retrieval, New York, USA, pp. 1145–1146 (2012)
5. Vo, D.T., Ock, C.Y.: Learning to classify short text from scientific documents using topic models with various types of knowledge. Expert Syst. Appl. **42**(3), 1684–1698 (2015)
6. Hu, X., Zhang, X., Lu, C., et al.: Exploiting Wikipedia as external knowledge for document clustering. In: Proceedings of the 15th ACM SIGKDD International Conference on Knowledge Discovery and Data Mining, Paris, France, pp. 389–396 (2009)

7. Hu, J., Fang, L., Cao, Y.: Enhancing text clustering by leveraging Wikipedia semantics. In: Proceedings of the 31th Annual International ACM SIGIR Conference on Research and Development in Information Retrieval, Singapore, pp. 179–186 (2008)
8. Song, S., Zhu, H., Chen, L.: Probabilistic correlation-based similarity measure on text records. Inf. Sci. **289**(1), 8–24 (2014)
9. Yang, L.L., Li, C.P., Ding, Q., et al.: Combining lexical and semantic features for short text classification. In: Proceedings of the 17th International Conference in Knowledge Based and Intelligent Information and Engineering Systems, KES, pp. 78–86 (2013)
10. Cheng, H., Qin, Z., Qian, W., et al.: Conditional mutual information based feature selection. In: International Symposium on Knowledge Acquisition and Modeling, pp. 103–107 (2008)
11. Blei, D.M., Ng, A.Y., Jordan, M.I.: Latent Dirichlet allocation. J. Mach. Learn. Res. **3**, 993–1022 (2003)
12. Phan, X.H, Nguyen, L.M, Horiguchi, S.: Learning to classify short and sparse text & web with hidden topics from large-scale data collections. In: Proceedings of the 17th International Conference on World Wide Web, pp. 91–100. ACM, New York, USA (2008)
13. Chen, M., Jin, X., Shen, D.: Short text classification improved by learning multi-granularity topics. In: Proceedings of the 22th International Joint Conference on Artificial Intelligence, pp. 1776–1781 (2011)
14. Kononenko, I.: Estimating attributes: analysis and extensions of relief. In: Bergadano, F., De Raedt, L. (eds.) ECML 1994. LNCS, vol. 784, pp. 171–182. Springer, Heidelberg (1994)
15. Sogou Labs: Text Categorization Dataset [EB/OL]. http://www.sogou.com/labs/dl/c.html. Accessed 01 Sept 2008
16. ICTCLAS, ICTCLAS2012-SDK-0101, rar [EB/OL]. http://www.nlpir.org/download/. Accessed 18 Aug 2014

Selection of Optimal Cutting Parameters in Parallel Turnings Using Genetic Heuristics

Lifang Pan[1], Shutong Xie[2(✉)], Kunhong Liu[3], and Jiangfu Liao[2]

[1] School of Science, Jimei University, Xiamen 361021, China
[2] School of Computer Engineering, Jimei University, Xiamen 361021, China
stxie@126.com
[3] School of Software, Xiamen University, Xiamen 361005, China

Abstract. Optimal cutting parameters can lead to considerable savings in manufacturing fields. In this paper, to deal with the optimization problem of cutting parameters which aims to minimize the unit production cost (UC) in parallel turnings, we propose a novel optimization approach which divides this complicated problem into several sub-problems. Then a genetic algorithm (GA) is developed to search the optimal results for each sub-problem. Simulations show that the corresponding approach can find better results than previous approach to significantly reduce the production cost.

Keywords: Parallel turning · Cutting parameters · Optimization · Genetic algorithm

1 Introduction

The desirable objective in machining operations is to produce high-quality products with low cost and high productivity. In order to minimize the machining cost for machining economics problem, the optimization of cutting parameters is one of the most important issues since these parameters strongly affect the cost, productivity and quality [1]. The optimization problem in multi-pass turnings becomes very complicated when plenty of practical constraints have to be involved [2]. Therefore, meta-heuristic algorithms have been introduced in solving machining economics problems because of their power in global searching and robustness as well. Chen and Tsai [3] have employed the combination of simulated annealing (SA) and pattern search (PS) for solving these problems. Later Chen [4] proposed a scatter search to tackle the optimization problems again and get better results. Genetic algorithm (GA), Ants colony optimization (ACO) [5–7] and particle swarm optimization (PSO) [8, 9] were developed by many researchers to optimize the machining parameters in multi-pass turnings.

Most research focus on optimization of machining parameters for singe-tool turnings. However, parallel turning, i.e., turning operations that utilize multiple cutting tools, are being utilized more and more to increase operation productivity, increase part quality by decreasing setups. So process planning of parallel machining has recently begun to receive attention, but the corresponding research is scarce yet. Tang et al. [10] proposed a PSO algorithm to optimize the cutting parameters in parallel turning for both single pass and multi-pass in order to minimize production time. Xie and Pan [11]

© Springer International Publishing Switzerland 2016
D.-S. Huang et al. (Eds.): ICIC 2016, Part I, LNCS 9771, pp. 175–182, 2016.
DOI: 10.1007/978-3-319-42291-6_17

developed an EDA-based approach to find the optimal cutting parameters in parallel turning. On the other hand, most researchers combined various heuristic algorithms with local improvement methods to deal with the optimization problem in turnings as a whole in order to further reduce the unit production cost (UC). These approaches are useful to some extent. But they are generally very complicated. Different from the previous approaches which tackled the optimization problem as a whole, the complicated optimization problem was firstly divided into a number of smaller optimization sub-problems, and each of them was conquered one by one in this paper. By doing so, this complex problem becomes simple enough to be solved. In order to tackle the sub-problems, the genetic heuristics was proposed which is efficient and effective.

2   Mathematical Model of Optimization in Parallel Turnings

In this paper, the mathematical model [3] was extended to describe the parallel turnings, which was used to optimize cutting parameters. The parallel turning process is divided into multi-pass roughing and single pass finishing. And there are four cutters to realize parallel turnings, the symmetrical two of them is used for rough machining, and other symmetrical two of them is used for finish machining. The objective of the optimization problem was to find the optimal cutting parameters including cutting speed, feed rate and depth of cut for both rough and finish machining in order to minimize production cost (UC).

2.1   Cost Function

The UC for parallel turning operations can be divided into four basic cost elements.

(1) Cutting cost by actual time in cut (C_M).
(2) Machine idle cost for loading/unloading operations and idling tool motion (C_I).
(3) Tool replacement cost (C_R).
(4) Tool cost (C_T).

Hence UC can be represented as

$$
\begin{aligned}
UC &= C_M + C_I + C_R + C_T \\
&= \left[\frac{\pi DL}{2000 V_r f_r} \left(\frac{d_t - d_s}{d_r} \right) + \frac{\pi DL}{2000 V_s f_s} \right] k_0 \\
&\quad + \left[t_c + (h_1 L + h_2) \left(\frac{d_t - d_s}{d_r} + 1 \right) \right] k_0 \\
&\quad + \left[\frac{\pi DL}{1000 V_r f_r} \left(\frac{d_t - d_s}{d_r} \right) \frac{t_e}{T_r} + \frac{\pi DL}{1000 V_s f_s} \frac{t_e}{T_s} \right] k_0 \\
&\quad + \left[\frac{\pi DL}{1000 V_r f_r} \left(\frac{d_t - d_s}{d_r} \right) \frac{k_r}{T_r} + \frac{\pi DL}{1000 V_s f_s} \frac{k_s}{T_s} \right]
\end{aligned}
\tag{1}
$$

Where

UC:	Unit production cost excluding material cost (\$/piece)
k_0:	Direct labor cost, including overhead (\$/min)
k_r, k_s:	Cutting edge cost for rough cutter and finish cutter (\$/edge)
D, L:	Diameter and length of work-piece (mm)
d_t:	Depth of material to be removed (mm)
h_1, h_2:	Constants relating to tools travel and approach/departure time (min)
t_c:	Preparation time for loading and unloading time (min)
t_e:	Time required to exchange a tool (min)
T_r, T_s:	Tool life, expected tool life for rough and for finish machining (min)
V_r, V_s:	Cutting speeds in rough and finish machining (m/min)
f_r, f_s:	Feed rates in rough and finish machining (mm/rev)
d_r, d_s:	Depths of cut for each pass of rough and finish machining (mm)
n:	Number of rough cuts, an integer

$$n = \frac{d_t - d_s}{d_r}, \text{ and } n \in Z_+ \tag{2}$$

For the minimization of UC, practical constraints that present the states of machining process including rough and finish operations are to be considered, which are listed as following.

(1) Parameter bounds; (2) Tool-life constraint; (3) Operating constraints consisting of surface finish constraint, force and power constraint; (4) Stable cutting region constraint; (5) Chip-tool interface temperature constraint; (6) Roughing and finishing parameter relations.

2.2 Calculation of the Number of the Sub-problems

The parameters to be optimized include multiple rough cuts and a single finish cut. The possible number of rough cuts is restricted by

$$n_L \leq n \leq n_U \tag{3}$$

where $n_L = (d_t - d_{sU})/d_{rU}$, $n_U = (d_t - d_{sL})/d_{rL}$

Thus the number of possible value of n can be expressed as

$$m = (n_U - n_L + 1) \tag{4}$$

In general, m is not a big integer. The optimization problem in parallel turnings can be divided into m sub-problems, in each of which the number of rough cuts n is fixed. Thus the solution of the whole optimization problem is divided into searching the optimal results of m sub-problems and the minimum of them is the objective of whole optimization problem.

3   GA-Based Optimization Approach

Genetic algorithm (GA) is one of popular evolutionary algorithm, which has been applied extensively in solving kinds of optimization problems. Due to its power in solving combinatorial optimization problems, we use the GA to search for the optimal cutting parameters of the optimization problems.

3.1   The Framework of GA-Based Optimization Approach

The optimization problem in parallel turnings can be divided into m sub-problems according to Eq. (4), in each of which the number of rough cuts n is fixed. Thus the solution of the whole optimization problem is divided into searching the optimal results of m sub-problems and the minimum of them is the objective of whole optimization problem. The flow chart of the optimization approach is shown in Fig. 1, in which the MaxGen is the maximum of iterations of the GA, and m is the number of sub-problems. The realization of individual representation, genetic operators, handling of constraints, and fitness assignment are described in the following subsections.

The optimization approach is an enumerative method incorporated with a GA heuristic. We employed it to search for the optimal result in each sub-problem, starting from the one with the smallest number of rough cuts, i.e. one pass. And the first optimal result UC_{1o} in first sub-problem will be found. Then the algorithm goes on searching process in the rest of sub-problems. Finally, the algorithm finishes and the optimal result must be the minimum of results found in all searches $UC_{1o}, \dots, UC_{io}, \dots, UC_{mo}$.

3.2   Individual Representation

The individual chromosome is consists of five variables to be optimized, i.e., cut speed, feed rates of both rough and finish machining, and the depths of cut of finish machining. Each variable is represented by a 20-bit binary string, as shown in Table 1. The depth of cut of rough d_r can be obtained by $n = (d_t - d_s)/d_r$ when n is fixed.

The chromosome encoding is different from the chromosome encoding method in standard GA. The difference is that encoding bound of the variables, i.e. f_r, f_s, d_s are adjusted reasonably according to the specific constraints and number of rough cuts. During evolution, the chromosome encoding will be different for each sub-problem because of the changing constraints such as n. The encoding method can not only reduce the number of infeasible individuals in the population, but also eliminates some constraints to avoid the corresponding penalty functions in order to reduce the complexity of the problems.

3.3   Genetic Operators

The crossover operator performs single-point crossover between pairs of individuals; the mutation operator performs inversion of binary bit string; the decoding performs the

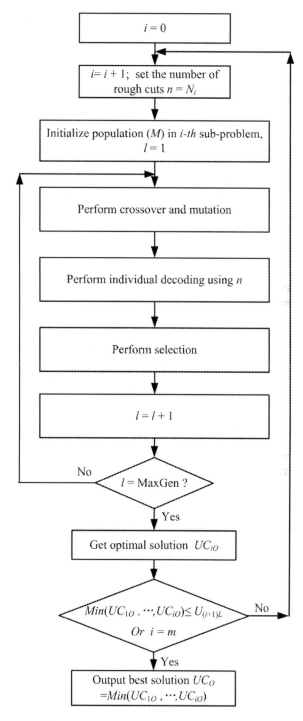

Fig. 1. Flow chart of the optimization approach

Table 1. Representation of individual chromosome

Bit	1–20	21–40	41–60	61–80	81–100
Variant	V_r	f_r	V_s	f_s	d_s

calculation of fitness for the individuals. Finally the selection operator performs the selection of individuals in terms of the fitness.

3.4   Handling of Constraints

The penalty function is used to penalize the individuals who violate the constraints. The more the constraints are violated, the heavier the penalty will be needed. As a result, the fitness of them will be small. So the penalty function is expressed as

$$Penalty(X) = \sum_{i=1}^{k} a_i \times h_i, \text{ where } a_i = \begin{cases} 0 & \text{Satisfy constraints} \\ 1 & \text{Violate constraint(s)} \end{cases} \tag{5}$$

where k is the number of constraints, h_i is a penalty positive constant.

The individual that obtains lower UC gets higher fitness value. Thus the fitness function can be expressed as

$$fitness(X) = \begin{cases} 1/UC(X) & \text{Satisfy constraints} \\ 1/(UC(X) + Penalty(X)) & \text{Violate constraint(s)} \end{cases} \tag{6}$$

4   Simulation and Discussion

In all the simulations, algorithms were run in a computer with a Core Duo T2080 CPU. Each problem was tested 20 times with different random initial populations but only the best results are given in this paper. The running time of each algorithm is the total of 20 runs. As for GA, the population and offspring sizes are both set to 200 and the algorithm stops after 400 generations. For each individual, the crossover and the mutation rates are set to 0.6 and 0.1, respectively. We used the machining experiment example shown in Table 2, which is an extended version from [3].

Table 2. Data for machining case

$D = 50$ mm	$L = 300$ mm	$d_t = 6$ mm	$V_{rL} = 50$ m/min
$f_{rL} = 0.1$ mm/rev	$d_{rL} = 1$ mm	$V_{rU} = 500$ m/min	$f_{rU} = 0.9$ mm/rev
$d_{rU} = 3$ mm	$V_{sL} = 50$ m/min	$f_{sL} = 0.1$ mm/rev	$d_{sL} = 1$ mm
$V_{sU} = 500$ m/min	$f_{sU} = 0.9$ mm/rev	$d_{sU} = 3$ mm	$p = 5$
$q = 1.75$	$r = 0.75$	$\mu = 0.75$	$v = 0.95$

(Continued)

Table 2. (*Continued*)

$\eta = 0.85$	$\lambda = 2$	$\upsilon = -1$	$\tau = 0.4$
$\phi = 0.2$	$\delta = 0.105$	$R = 1.2$ mm	$C_0 = 6 \times 10^{11}$
$T_L = 25$ min	$T_U = 45$ min	$F_U = 200$ Kgf	$P_U = 10$ kW
$SC = 140$	$Q_U = 1000°$ C	$SR_U = 10$ μm	$h_1 = 7 \times 10^{-4}$
$h_2 = 0.3$	$t_e = 1.5$ min/edge	$t_c = 0.75$ min/piece	$k_r = 2.5$ \$/edge
$k_s = 2.5$ \$/edge	$k_0 = 0.5$ \$/min	$k_1 = 108$	$k_2 = 132$
$k_3 = 1$	$k_4 = 2.5$	$k_5 = 1$	

The optimization approaches were tested in the machining problems with the total depth of cuts varying from 6 mm to 12 mm. Table 3 show the optimal results obtained by our algorithms are smaller than those obtained by EDA [11], which means the proposed GA-based approach can further reduce the UC. Furthermore, the standard deviations of the results obtained by our approach are much smaller than those obtained by EDA [11]. But, compared to EDA, the GA-based approach consumes much time to get optimal results. On the other hand, despite different total depth of cuts, the running time is about 220 s, which means that the proposed approach do not increase the computational complexity with the scale of the optimization problems. An interesting phenomenon is that the optimal solutions are always found in the minimum number of rough cuts, i.e. 1, 2, and 3. Our algorithms had a larger probability to find the optimal result and to avoid being converged into a local optimum. Therefore, the proposed approaches in this paper tackled the problem reasonably to achieve satisfying results.

Table 3. Comparison of the performance between different algorithms

Total depth of cuts d_t(mm)	Algorithms	UC (\$)	Standard deviation	Running time (S)	Optimal n
6	GA	1.5741	0.0001	227.6	1
	EDA	1.5926	0.0112	108.7	
8	GA	1.9737	0.0006	267.7	2
	EDA	2.0263	0.0170	107.4	
10	GA	2.3837	0.0011	220.9	3
	EDA	2.4618	0.0265	107.8	
12	GA	2.5739	0.0002	195.3	3
	EDA	2.6037	0.0125	108.1	

5 Conclusion

An optimization approach was proposed to search optimal cutting parameters in parallel turnings. In the optimization approach, by dividing this complicated optimization problem into several sub-problems, the solution of this hard problem was converted to some simple ones. A genetic heuristic (GA) was also developed to solve the sub-problems. By using the chromosome encoding method, the number of infeasible individuals in the evolutionary population was greatly reduced. Simulations showed that the optimization approach combining pass enumerating method with heuristic algorithms was efficient and effective.

In this paper, we only studied the parallel turning with two symmetrical cutters, which is the easy one in multi-tool turning operations. However, the optimization approach, which incorporated pass numerating method with meta-heuristic algorithms, can be extended to investigate other parallel turning systems with two cutters operating in different cutting parameters, or with more than two cutters.

Acknowledgments. This work was supported by the Natural Science Foundation of Fujian Province of China (No. 2016J01735, JK2015025), and the Foundation for Young Professors of Jimei University, China (No. 2011C002).

References

1. Mukherjee, I., Ray, P.K.: A review of optimization techniques in metal cutting processes. Comput. Ind. Eng. **50**, 15–34 (2006)
2. Shin, Y.C., Joo, Y.S.: Optimization of machining conditions with practical constraints. Int. J. Prod. Res. **30**, 2907–2919 (1992)
3. Chen, M., Tsai, D.: A simulated annealing approach for optimization of multi-pass turning operations. Int. J. Prod. Res. **34**, 2803–2825 (1996)
4. Chen, M.: Optimizing machining economics models of turning operations using the scatter search approach. Int. J. Prod. Res. **42**, 2611–2625 (2004)
5. Sankar, R., Asokan, P., Saravanan, R., Kumanan, S., Prabhaharan, G.: Selection of machining parameters for constrained machining problem using evolutionary computation. Int. J. Adv. Manuf. Technol. **32**, 892–901 (2007)
6. Vijayakumar, K., Prabhaharan, G., Asokan, P., Saravanan, R.: Optimization of multi-pass turning operations using ant colony system. Int. J. Mach. Tools Manuf. **43**, 1633–1639 (2003)
7. Wang, Y.: A note on 'optimization of multi-pass turning operations using ant colony system'. Int. J. Mach. Tools Manuf. **47**, 2057–2059 (2007)
8. Srinivas, J., Giri, R., Yang, S.: Optimization of multi-pass turning using particle swarm intelligence. Int. J. Adv. Manuf. Technol. **40**, 56–66 (2009)
9. Yildiz, A.R.: A novel particle swarm optimization approach for product design and manufacturing. Int. J. Adv. Manuf. Technol. **40**, 617–628 (2009)
10. Tang, L., Landers, R. Balakrishnan, S.N.: Parallel turning process parameter optimization based on a novel heuristic approach. J. Manuf. Sci. Eng.-Trans. ASME **130**, 031002-031001-031012 (2008)
11. Xie, S., Pan, L.: Optimization of machining parameters for parallel turnings using estimation of distribution algorithms. In: 3rd International Conference on Advanced Engineering Materials and Technology, vols. 753–755, pp. 1192–1195. Trans Tech Publications Ltd., Switzerland (2013)

Balanced Tree-Based Support Vector Machine for Friendly Analysis on Mobile Network

Han Wu, Bingqing Luo, and Zhixin Sun[✉]

Key Laboratory of Broadband Wireless Communication
and Sensor Network Technology,
Nanjing University of Posts and Telecommunications, Nanjing 210003, China
sunzx@njupt.edu.cn

Abstract. In this paper, a balanced tree-based Support Vector Machine is proposed. Based on the traditional Support Vector Machine algorithm, Support Vector Machine classifier is trained by comparing the dissimilarity matrix. The samples which have minimum dissimilarity will be trained first, and the less their dissimilarities are, the earlier they will be trained. These samples are merged in accordance with the training sequence and generate a binary balanced tree gradually. The classification algorithm is used in the friendly analysis on the mobile network, which makes the analysis more convinced. The new proposed SVM algorithm analyze on different kinds of basic energy consumption data collected by the mobile terminal and use these data to classify the applications, and after the related judge process, the mobile application will get the friendly analysis results. Experimental results show that the proposed balanced tree-based SVM algorithm can realize the classification and friendly analysis more accurately.

Keywords: Support Vector Machine · Friendly analysis on mobile network · Classifier

1 Introduction

Support Vector Machine (SVM) was first proposed by Vapnik, it is a supervised learning classification algorithm. It has many advantages in small sample, nonlinear and high-dimension pattern identification, and it can be extended to solve the functional fitting problem. SVM is a machine learning algorithm based on the statistical learning theory, which is famous during the 1990's, and it can improve the generalization ability of the learning machine.

With the development of technology, the application of SVM is very wide, such as face detection, speech recognition, image processing and so on. Among them, the friendly analysis on mobile network is influential. In order to improve the classification accuracy, it analyzes the basic energy consumption data to classify the application. Based on the classification progress, the related evaluation algorithm is chosen to get the final friendly analysis results. The chosen of SVM algorithm makes a great effect on the whole analysis.

© Springer International Publishing Switzerland 2016
D.-S. Huang et al. (Eds.): ICIC 2016, Part I, LNCS 9771, pp. 183–191, 2016.
DOI: 10.1007/978-3-319-42291-6_18

The traditional SVM classification algorithm is used to find an optimal hyper plane as a standard in the training sample, and the formal sample is divided into a positive part and a negative one. As the result, the distance between them would be the longest. In order to achieve multiclass classification, the combination of two-value classification is the common strategy. So this paper focus on how to choose the appropriate algorithm.

The paper proposes a new balanced tree-based SVM algorithm. In the first section, we introduce the related works. The second section describes the balanced tree-based SVM classification algorithm. The third section is the experiments and discussion, the algorithm is used to compare with other related algorithms. The fourth section is conclusion, and the prospect of this research are presented in this section.

2 Related Works

Support vector machine (SVM) algorithm has been widely studied. SVM is an identification algorithm for two-class problem and can not be directly applied to multiclass problem itself. Five kinds of SVM classification algorithms for multiclass problem are widely used: One versus rest (OVR), One versus one (OVO), error correcting output coding (ECOC) and binary tree-based or directed a cyclic graph.

OVR and OVO algorithm are proposed in the early time, One versus one (OVO) performs very well in the classification of 20 sets of data, significantly better than Bias and Rocchio algorithm [1].

ECOC is widely used in speech recognition, [2] and the classification accuracy and coding length are closely contact with the Hamming distance between the codes. In the field of speech recognition, four kinds of error correcting coding, which are proposed in Reference [3], can improve the overall fault-tolerance rate effectively. In the case of the limited length of encoding, a good algorithm is determined by an encoding matrix A [4], and it has error correction ability and good generalization. The choice of coding itself will be totally different and the algorithm is not stable to reveal all the classification situations. In the field of friendly analysis on mobile network, we prefer other algorithms.

The binary tree-based multiclass classification is a traditional method to transfer multiclass classification problems into the two-class classification. Reference [5] proposes a binary tree-based multiclass SVM algorithm, in order to solve the establishment of the group sub-swarm. So as to form a multiclass SVM classification, it distributes the classifier at each node. The algorithm is realized in software, and the experiment uses related data to train the classifier. The experiment gets a favorable grouping result by adjusting SVM parameters. An partial binary tree-based twin support vector machine algorithm is proposed in reference [6], In this paper, it shows the absolute advantage in the training time, and when nonlinear kernel is used, it has a better classification effects than OVR and traditional binary tree SVM method and overcomes the disadvantage that the classical OVR algorithm may exist the problem that it has unclassifiable regions in imbalanced data situation. Compared with the proposed partial binary tree-based, the more close to a normal binary tree it is, the more likely it will get the best training and classification speed. In order to construct a normal feature of tree, a balanced binary decision tree algorithm is proposed in reference [7]. It selects out two classes that have the maximum distance in the interior of class firstly, and then, the other classes are classified according to the principle of proximity, and finally, it can get two equal-number group. In the aspects of classification accuracy, it is superior than the unbalanced decision tree-based algorithm proposed in reference [8]. Huffman tree-based multiclass classification algorithm is proposed in reference [9], and it is constructed from the leaf to the root. The algorithm chooses two classes which have the maximum distance as the classifier, merge them and put them into the original data sets. After a certain number of merging and replacing actions, the whole Huffman tree is constructed. Compared with fixed tree-based algorithm, reference [10] proposes a feature selection algorithm, and it can reduce the number of classifiers and improve the classification accuracy by using the new feature selection method. But reference [11] proposes that the selection of the entire process is based on the feature, the selection may just reveal one part of the whole data information. Reference [12] is similar to reference [7], it is based on a balanced binary decision tree, as well. What's the difference is, in the process of construction, it considers the effective distance between the classes instead. The accuracy of the classification algorithm is related to the selection of data sets and the parameters of training model [13]. About the complexity of the method, Least Square Support Vector Machine (LS-SVM) is simpler than traditional SVM algorithm, reference [14] proposes that the improved LS-SVM algorithms can effectively improve the overall performance. Particle Filter Method for Improved Flooding Classification [15, 16] is popular in a big sample situation, and it can prove the efficiency outstandingly.

The overall idea of the friendly analysis on mobile network is to collect many kinds of basic information from the server. It evaluates the application and scores it according to many measure standards, including the user experience, power consumption in the terminal, signal consumption, traffic flows consumption, and so on. Then, all the scores are added to get the final analysis results. While, different applications will need the different evaluations, so it is necessary to do appropriate classification before the evaluation. This paper hopes to collect sufficient basic data when the application is used, using SVM to train them and find the difference between the application, and finally, it can judge what kind of category the application belongs to.

It is relatively complex to use SVM to solve multiclass classification directly. So, it is common to adopt a strategy of two value classification. How to improve the classification accuracy has become the focus of study. Although the classification accuracy of the OVO is high, it has too much classifier. Multiclass classification method based on tree or graph can be a new solution, but small errors may accumulate and become a big one. In addition, error correcting output coding can also solve it, but the classification accuracy is related to the chosen encoding, the method is not stable. In the traditional tree-based algorithm, the maximum-distance class will be divided into two category, While it may exist the situation that the maximum distance is not an effective one.

This paper proposes a balanced tree-based SVM for friendly analysis on mobile network. It is an improvement to the traditional tree-based algorithm, using the new combination strategy to achieve the classification of two-value class. The algorithm can reduce not only the number of SVM classifier, but also the prediction time of mobile applications, and improve the efficiency of multiclass classification.

3 Balanced Tree-Based Support Vector Machine

3.1 Dissimilarity Calculation

Given n train sets $\{X_1, X_2, X_3, \ldots, X_n\}$

In the i-th set, there are n training samples:

$$\{X_{i1}, X_{i2}, X_{i3} \ldots, X_{in}\}$$

A set of super spherical surfaces can be generated by creating a hyper sphere for each class of training samples.

$$\text{Dissimilarity matrix } D = \begin{pmatrix} D_{11} & \cdots & D_{1n} \\ \vdots & \ddots & \vdots \\ D_{n1} & \cdots & D_{nn} \end{pmatrix}$$

$$D_{ij} = \frac{N_{ij}(D_{ij}) + N_{ji}(D_{ij})}{n_i + n_j}$$

$$d_{ij} = \frac{\sum_{k=0}^{k=n_i}(X_k^i - C_i)}{n_i + n_j}$$

D_{ij} is the dissimilarity between X_i and X_j, $i, j = 1, 2, 3, \ldots$, n; d_{ij} is the distance from X_i to the center of X_j, Corresponding super sphere; $N_{ij}(D_{ij})$ is all the Xi sample that is not less than d_{ij}.

n_i, n_j, is the number of sample X_i and X_j.

$D_{ij} = \frac{N_{ij}(D_{ij}) + N_{ji}(D_{ij})}{n_i + n_j}$ shows the sample distribution in the class.

If D_{ij} is relatively big, the training sample trends to exist in the two-side space, otherwise, it distribute in the center zone. So the more D_{ij} is, the more difference X_i and X_j should be.

3.2 Construction of Balanced Tree

The construction method is from bottom to top, and it is shown in Fig. 1. It trains from the leaf nodes to the root, viewing the distance as the ability to distinguish, and selects out two classes which are the most difficult to distinguish as the train sample each time and use two-class SVM to train them. The training will get a node in Balanced Tree-Based Support Vector Machine model, and remove these two points out of the train set, and so on until all of the set are finished training, then merge the formal points two-by-two, and gradually, it forms a new set. This is the construction process.

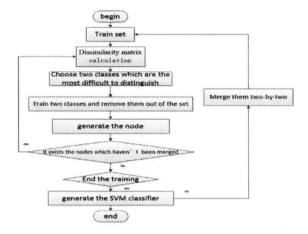

Fig. 1. Method of classifier construction

The train process is shown in Fig. 2:

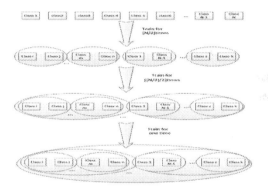

Fig. 2. Process of training all the train samples

After the construction of the classifier, it is necessary to test the classifier, and it is opposite to process of construction, which means from top to bottom. The whole process is shown in Fig. 3. It starts from the root node firstly, and uses the two-class classifier to test the set, according to the classification results, the next level's node are decided, and so on until it reaches to the leaf node.

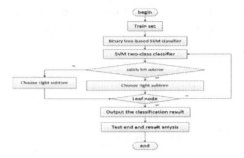

Fig. 3. The method of test the classifier

As shown in Fig. 4, the non-leaf nodes in the tree represent the two-class SVM, and the leaf nodes represent all samples. The top-to-bottom model can reduce the errors.

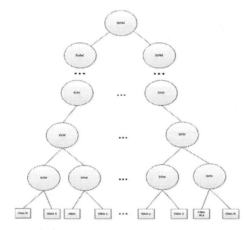

Fig. 4. The balanced tree-based multiclass classifier

4 Experiments and Discussion

4.1 Data Sets

Data record the basic energy consume when mobile is running on certain applications. The experiments collect 800 sets of data as the sample. In the experiment, the applications are divided into eight classes:

1. Instant communication
2. Game
3. Office
4. Web browsing
5. Music and video
6. Shopping
7. Mapping
8. Photoing.

The average number of sample are from 90 to 110, and use the proportion of 3:1 for training and testing.

4.2 Analysis on the Result

The System try to compare with other three algorithms:

Algorithm 1: OVR;
Algorithm 2: OVO;
Algorithm 3: improved binary tree-based SVM;
Algorithm 4: balanced tree-based SVM.

The results are shown in the next Table 1:

Table 1. The data comparison of four kinds of algorithms

The number of sample			The number of right classification			
	Train sets	Test sets	Algorithm 1	Algorithm 2	Algorithm 3	Algorithm 4
Class 1	70	20	13	12	16	17
Class 2	80	30	21	20	26	27
Class 3	80	30	22	20	26	27
Class 4	70	20	13	12	17	17
Class 5	70	20	13	12	16	17
Class 6	80	30	21	20	25	26
Class 7	80	30	22	21	27	28
Class 8	70	20	13	12	16	17
Total	600	200	138	129	170	176

Now define: accrancy $= \dfrac{\text{correctly predicted data}}{\text{total data}} \times 100\%$, X axis means the class; and Y axis means the accuracy, and get the Fig. 5:

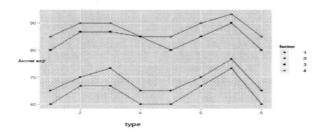

Fig. 5. The accuracy comparison of four kinds of algorithms

In Table 2, it lists the comparison among these four algorithm. The results show that the proposed algorithm has the optimal accuracy, and its cost time is quite close to other algorithms.

Table 2. The accuracy and time comparison of four kinds of algorithms

Algorithm	Accuracy/%	Training time/s	Test time/s
OVR SVM	69	3.63	2.56
OVO SVM	64.5	23.32	2.54
Improved binary tree-based SVM	85	4.53	0.54
Balanced tree-based SVM	88	5.87	0.51

5 Summary

The application of SVM is a quite complex but important problem. Until now, it doesn't exist a optimal method to solve the classification problem. While, the use of different methods are determined by the situations.

Friendly analysis on mobile network has a bright prospect. Facing with the mess application in the smart terminal, the user experience decline a lot. In this situation, a high-accuracy classification can analyze whether the application is friendly, and improve the user experience effectively. Future work on the friendly analyze on mobile network is focus on that the evaluation standard is too simple, cannot reveal all the situation, and the complex scoring should relay on the high-accuracy classification. In this case, the proposed algorithm can effectively improve the scoring accuracy.

Some improvements still exist in the proposed method. Currently, train set is chosen from the basic information collected from the weighted application, in the future work, we should choose better methods for concrete application instead of some representative ones.

References

1. Zhang, Q., Luo, M., Xue, Y., Tan, J.: Multi-class text categorization based on immune algorithm. In: International Workshop on Education Technology and Training, 2008, and 2008 International Workshop on Geoscience and Remote Sensing, ETT and GRS 2008, vol. 1, pp. 749–752, 21–22 December 2008
2. Ye, Q., Liang, J., Jiao, J.: Pedestrian detection in video images via error correcting output code classification of manifold subclasses. IEEE Trans. Intell. Transp. Syst. **13**(1), 193–202 (2012)
3. Xiao-feng, L., Xue-ying, Z., Ji-kang, D.: Speech recognition based on support vector machine and error correcting output codes. In: 2010 First International Conference on Pervasive Computing Signal Processing and Applications (PCSPA), pp. 336–339, 17–19 September 2010
4. Jian, B., Liu, R.: M-ary support vector machine (M-ary SVM) for multi-category classification. J. Comput. Appl. **03**, 661–664 (2012)

5. Duan, T., Zhang, D.-N.: Application of multiclass SVM based on binary tree in targeting grouping. Radio Eng. **06**, 88–91 (2015)
6. Juan-Ying, X., Bing-Quan, Z., Wan-Zi, W.: A partial binary tree algorithm for multiclass classification based on twin support vector machines. J. Nanjing Univ. **47**(4), 354–363 (2011)
7. Xue, S., Jing, X., Sun, S., Huang, H.: Binary-decision-tree-based multiclass Support Vector Machines. In: International Symposium on Communications and Information Technologies, pp. 85–89 (2014)
8. Guan, T., Frey, C.W.: Using ensemble of decision trees with SVM nodes to learn the behaviour of a transmission control software. In: 2014 IEEE 17th International Conference on Intelligent Transportation Systems (ITSC), pp. 1323–1328, 8–11 October 2014
9. Li, J.J., Wang, Y., Liu, R.Q.: Selection of materialized view based on information weight and using Huffman-tree on spatial data warehouse. In: First International Conference on Innovative Computing, Information and Control, ICICIC 2006, vol. 2, pp. 71–74, 30 August 2006–1 September 2006
10. Xu, Q., Zhang, X.: Multiclass feature selection algorithms base on R-SVM. In: 2014 IEEE China Summit and International Conference on Signal and Information Processing (ChinaSIP), pp. 525–529, 9–13 July 2012
11. Freeman, C., Kulic, D., Basir, O.: Feature-selected tree-based classification. IEEE Trans. Cybern. **43**(6), 1990–2004 (2013)
12. Yuncu, E., Hacihabiboglu, H., Bozsahin, C.: Automatic speech emotion recognition using auditory models with binary decision tree and SVM. In: 2014 22nd International Conference on Pattern Recognition (ICPR), pp. 773–778, 24–28 August 2014
13. Gu, Y., Zhao, W., Wu, Z.: Least squares support vector machine algorithm. J. Tsinghua Univ. (Sci. Technol.) **07**, 1063–1066+1071 (2010)
14. Wang, C., Chen, S.: An improved LS-SVM based on SSOR-PCG. In: 2013 Ninth International Conference on Natural Computation (ICNC), pp. 28–33, 23–25 July 2013
15. Insom, P., Cao, C., Boonsrimuang, P., Liu, D., Saokarn, A., Yomwan, P., Xu, Y.: A support vector machine-based particle filter method for improved flooding classification. IEEE Geosci. Remote Sens. Lett. **PP**(99), 1–5 (2015)
16. Zeng, L.-M., Wu, X.-B., Liu, P.: Sample reduction strategy for SVM large scale training data set using PSO. Comput. Sci. **09**, 215–217 (2009)

Feature Combination Methods for Prediction of Subcellular Locations of Proteins with Both Single and Multiple Sites

Luyao Wang[1,2], Dong Wang[1,2(✉)], Yuehui Chen[1,2(✉)],
Shanping Qiao[1,2], Yaou Zhao[1,2], and Hanhan Cong[3]

[1] School of Information Science and Engineering,
University of Jinan, Jinan 250022, China
897112057@qq.com,
{ise_wangd,yhchen,ise_qiaosp,ise_zhaoyo}@ujn.edu.cn
[2] Shandong Provincial Key Laboratory of Network Based Intelligent
Computing, Jinan 250022, China
[3] University of Jinan, Jinan 250022, China
ado_conghh@ujn.edu.cn

Abstract. Effective feature extraction methods play very important role for prediction of multisite protein subcellular locations. With the progress of many proteome projects, more and more proteins are annotated with more than one subcellular location. However, compared with the problems of single-site protein, the problems of multiplex protein subcellular localizations are far more difficult and complicated to deal with. To improve the multisite prediction quality, it is necessary to incorporate different feature extraction methods. In this paper, a version of feature combination method which is to make use of the 20 dimensions of entropy density instead of the former 20 dimensions of amphiphilic pseudo amino acid composition (AmPseAAC), is used in two different datasets. It is different from the way of simple dimensions additive feature fusion. On base of this novel feature combination method, we adopt the multi-label k-nearest neighbors (ML-KNN) algorithm and setting different weights into different attributes' ML-KNN, which is called wML-KNN, to predict multiplex protein subcellular locations. The best overall accuracy rate on dataset S1 from the predictor of Virus-mPLoc is 61.11 % and 82.03 % on dataset S2 from Gpos-mPLoc, respectively.

Keywords: Multisite protein subcellular localizations · The entropy density · AmPseAAC · Multi-label k-nearest neighbors algorithm · wML-KNN

1 Introduction

The success of the human genome project has stimulated the emergence of a new and far more challenging frontier: proteomics. Although proteins are generated according to their DNA code, they are far more complex and varied than DNA. While the function of a protein is closely correlated with its subcellular attributes, and the knowledge of protein subcellular attributes plays a vitally important role in molecular biology, cell biology, pharmacology, and medical science [1]. So predicting protein subcellular

© Springer International Publishing Switzerland 2016
D.-S. Huang et al. (Eds.): ICIC 2016, Part I, LNCS 9771, pp. 192–201, 2016.
DOI: 10.1007/978-3-319-42291-6_19

locations have become a hot topic in bioinformatics. Although various experimental technologies have been developed for determining the protein subcellular locations, almost every available approach is costly and time consuming [2]. Therefore, computational methods as alternative choices are developed to predict multisite protein subcellular locations. Moreover, some studies have discovered that more and more proteins exist or move between two or more subcellular locations, and only located in the specific and accurate subcellular locations, can the protein play its functions normally. With the avalanche of protein sequences generated in the post-genomic age, it is highly desired to develop automated methods for efficiently identifying various attributes of uncharacterized proteins [3]. However, for the proteins with multiple subcellular locations, this issue becomes more complicated and challengeable than before.

In this paper, the particular feature fusion method which with the 20 dimensions of entropy density replace to the former 20 dimensions of amphiphilic pseudo amino acid composition (AmPseAAC) is adopted. On base of this novel feature combination representation, we adopt the multi-label k-nearest neighbors (ML-KNN) algorithm and setting different weights into different attributes' ML-KNN, which is called wML-KNN to predict multiplex protein subcellular locations. After that, several different performance measures for multilabel classifications have been used to evaluate this algorithm.

2 Datasets

We choose to use the dataset S1 from the predictor of Virus-mPLoc [4] and the dataset S2 from the predictor of Gpos-mPLoc [5] as the benchmark datasets for the current study. The reasons why doing so are as follows. (1) None of proteins included in S1 and S2 has ≥25 % sequence identity to any other in the same subset (subcellular location). (2) They also contain proteins with more than one location and hence can be used to deal with proteins having both single and multiple location sites.

The dataset S1 contains 252 locative protein sequences (207 different proteins), of which 165 belong to one subcellular location, 39 to two locations, 3 to three locations, and none to four or more locations. The dataset S1 is classified into 6 subcellular locations, and hence can be formulated as:

$$S1 = S_1 \cup S_2 \cup S_3 \cup S_4 \cup S_5 \cup S_6 \tag{1}$$

where S_1 represents the subset for the subcellular location of "viral capsid", S_2 for "host cell membrane", S_3 for "host endoplasmic reticulum", and so forth. Because some proteins may simultaneously occur in two or more locations, so different subcellular locations have different proteins. If a protein coexistes at two different subcellular location sites, it will be counted as two locative proteins; if it coexistes at three different subcellular location sites, it will be counted as three locative proteins, and so forth. It can be seen from Table 1 that different subcellular locations locate different proteins.

Table 1. Detail of the viral protein benchmark dataset S1 taken from Virus-mPloc

Subset	Subcellular location	Number of proteins
S_1	Host viral capsid	8
S_2	Host cell membrane	33
S_3	Host endoplasmic reticulum	20
S_4	Host cytoplasm	87
S_5	Host nucleus	84
S_6	Secreted	20
Total number of locative proteins		252
Total number of different proteins		207

The dataset S2 contains 523 gram-positive bacterial protein sequences (519 different proteins), of which 515 belong to one location and 4 to two locations. The dataset S2 is classified into 4 subcellular locations, and hence can be formulated as:

$$S2 = S_1' \cup S_2' \cup S_3' \cup S_4' \tag{2}$$

where S_1' represents the subset for subcellular location "cell membrane", S_2' for "cell wall", S_3' for "cytoplasm", and so forth. It can be seen from Table 2 different subcellular locations include different amount of proteins.

Table 2. Detail of the gram-positive bacterial protein benchmark dataset S2 taken from Gpos-mPLoc

Subset	Subcellular	Number of proteins
S_1'	Cell membrane	173
S_2'	Cell wall	18
S_3'	Cytoplasm	208
S_4'	Extracell	123
Total number of locative proteins		523
Total number of different proteins		519

3   Methods

3.1   Entropy Density

Zhu [6] proposed entropy density method to express the sequence of DNA and determine the exon in the gene sequences. The definition of shannon entropy method [7] can be formulated as:

$$H(S) = -\sum_{i=1}^{20} f_i \log f_i \qquad i = 1, 2, 3, \ldots, 20 \tag{3}$$

where $f_i (i = 1, 2, \ldots, 20)$ are the normalized occurrence frequencies of the 20 amino acids in the protein S. Hence the entropy density function can be expressed as:

$$s_i(S) = -\frac{1}{H(s)} f_i \log f_i \qquad i = 1, 2, 3, \ldots, 20 \tag{4}$$

So protein sequence S can be represented as:

$$s(S) = (s_1(S), s_2(S), s_3(S), \ldots, s_{20}(S)) \tag{5}$$

where the 20 elements are associated with the 20 elements completely in Eq. (4) or the 20 amino acid components of the protein S.

3.2 Amphiphilic Pseudo Amino Acid Composition

To develop a powerful method for statistically predicting protein subcellular localizations according to the sequence information, one of the most important things is to formulate the protein sequences with an effective mathematical expression that can truly reflect the intrinsic correlation with their subcellular localizations [8]. Chou [9] has also proposed an effective way of representing protein character sequence by some of its physiochemical properties, which is called amphiphilic pseudo amino acid composition (AmPseAAC).

Suppose a protein sequence P with L amino acid resides, which can be formulated as:

$$P = R_1 R_2 R_3 R_4 R_5 \ldots \ldots R_L \tag{6}$$

where $R_i(i = 1, 2, \ldots, L)$ represents the residue at sequence position i. The sequence order effect along with a protein chain is approximated by a set of sequence order correlation factor which is defined as:

$$\begin{cases} \tau_{2\lambda-1} = \dfrac{1}{L-\lambda} \sum_{i=1}^{L-\lambda} H_{i,i+\lambda}^1 \\[4mm] \tau_{2\lambda} = \dfrac{1}{L-\lambda} \sum_{i=1}^{L-\lambda} H_{i,i+\lambda}^2 \end{cases} , (\lambda < L) \tag{7}$$

$$\begin{cases} H_{i,j}^1 = h^1(R_i) \cdot h^1(R_j) \\ H_{i,j}^2 = h^2(R_i) \cdot h^2(R_j) \end{cases} \tag{8}$$

where $H^1(R_i)$ and $H^2(R_i)$ are respectively the hydrophobicity and hydrophilicity values for the i-th (i = 1, 2,..., L) amino acid in Eq. (8).

After incorporating the sequence-order correlated factors from Eq. (8) into the classical 20-D (dimensional) amino acid composition, we obtain a pseudo amino acid composition with $(20 + 2\lambda)$ components. In other words, the representation for a protein sample P is now formulated as:

$$P = [p_1, p_2, \ldots, p_{20}, p_{20+1}, \ldots, p_{20+\lambda}, p_{20+\lambda+1}, \ldots, p_{20+2\lambda}]^T, (\lambda < L) \qquad (9)$$

$$p_k = \begin{cases} \dfrac{f_k}{\sum\limits_{i=1}^{20} f_i + \omega \sum\limits_{j=1}^{2\lambda} \tau_j}, (1 \le k \le 20) \\[4mm] \dfrac{\omega \tau_{k-20}}{\sum\limits_{i=1}^{20} f_i + \omega \sum\limits_{j=1}^{2\lambda} \tau_j}, (20 + 1 \le k \le 20 + 2\lambda) \end{cases} \qquad (10)$$

where $f_i (i = 1, 2, \ldots, 20)$ are the normalized occurrence frequencies of the 20 amino acids in the protein P, j the j-tier sequence-correlation factor computed according to Eq. (7), and in the above formula, ω is the weight factor and its value usually is 0.05. As we can see from Eqs. (9)–(10), the first 20 components reflect the effect of the classical amino acid composition, while the components from 20 + 1 to $20 + 2\lambda$ reflect the amphipathic sequence-order pattern.

3.3   Multi-label K Nearest Neighbor (ML-KNN)

ML-KNN proposed by Zhang and Zhou is a simple non-parametric multilabel classifier, which uses the k-nearest neighbor algorithm to statistics the category tag information of neighbor samples, and exploits the principle of maximum posterior probability to inference the no example of label set [10]. Let $D = \{(x_i, Y_i)|1 \le i \le p\}$ be multilabel training set and let x be no example. Suppose $N(x)$ denote the set of k nearest neighbors of x identified in the training set. For the j-th category of y_j, ML-KNN algorithm will calculate the following statistics,

$$C_j = \sum_{(x,Y) \in N(x)} \{y_j \in Y\} \qquad (11)$$

where C_j counts the number of neighbors of x belonging to the $N(x)$ class. Let H_j represents this event about x containing the category of y_j, then let $P(H_j|C_j)$ represents the posterior probability established H_j that when $N(x)$ includes the number samples of C_j with a category label y_j. Multi-label classifier required can be expressed as:

$$h(x) = \{y_j | \frac{P(H_j|C_j)}{P(\neg H_j|C_j)} > 0.5, 1 \le j \le q\} \qquad (12)$$

the formula is when the posterior probability $P(H_j|C_j)$ is greater than $P(\neg H_j|C_j)$, the mark y_j belongs to the example of x. The function based on Bayes' theorem can be rewritten as:

$$h(x) = \{y_j | \frac{P(H_j)P(C_j|H_j)}{P(\neg H_j)P(C_j|\neg H_j)} > 0.5, 1 \leq j \leq q\} \tag{13}$$

However, objectively uneven training set results in a relatively low accuracy. To overcome this problem, an attribute weighted based improving algorithm was proposed [11]. The weighted factor w_t is defined as:

$$w_t = \frac{\log(a + \frac{AvgNum}{Num(C_t)})}{\log(a+1)} (1 < t < N) \tag{14}$$

$$a = \frac{MaxNum}{AvgNum} \tag{15}$$

where $AvgNum$ refers to the average number of samples which are different categories, $Num(C_t)$ refers to the C_t class containing the numbers of sample. In the ML-KNN algorithm, the prior probability is weighted, which is named wML-KNN.

4 Evaluation Functions

Due to the comprehensiveness and understandability of these metrics, it is very meaningful and interesting to make efforts to apply these five metrics in our future study. Evaluation metrics for multi-label learning include the following five items [12].
   Absolute-true:

$$T(h) = \frac{1}{p}\sum_{i=1}^{p}[h(x_i) = y_i] \tag{16}$$

The evaluation index is used to show the proportion of predicted labels consistent with real tag set. It has been increasingly used and widely recognized by investigators to examine the accuracy of various prediction methods. The bigger the value of $T(h)$, the better the performance.
   Hamming Loss:

$$H(h) = \frac{1}{p}\sum_{i=1}^{p}\frac{1}{q}|h(x_i)\Delta y_i| \tag{17}$$

Hamming Loss is used to evaluate how many times an instance-label pair is mis-classified. The smaller the value of $H(h)$, the better the performance.
   One-error:

$$O(h) = \frac{1}{p} \sum_{i=1}^{p} [[\arg\ \max f(x_i, y)] \notin y_i] \tag{18}$$

One-error evaluates how many times the top-ranked label is not in set of proper labels of the instance. The smaller the value of $O(h)$, the better the performance.

Coverage:

$$C(h) = \frac{1}{p} \sum_{i=1}^{p} \max\ \ rank_f\ (x_i, y) - 1 \tag{19}$$

Coverage is used to evaluate how far we need to go down the list of labels in order to cover all the proper labels of the instance on the average. The smaller the value of $C(h)$, the better the performance.

Average Precision:

$$A(h) = \frac{1}{p} \sum_{i=1}^{p} \frac{1}{|y_i|} \sum_{y \in y_i} \frac{\left|\{y' | rankf(x, y') \le rankf(x_i, y), y' \in y_i\}\right|}{rankf(x_i, y)} \tag{20}$$

Average precision evaluates the average fraction of labels ranked above a particular label $y \in Y$ which actually are in Y. The bigger the value of $A(h)$, the better the performance.

5 Experiments and Results

In this paper, in order to enhance the accuracy for prediction of multiplex protein subcellular localizations, we adopt the simple dimension additive feature fusion method and the special fusion method to compare results of ML-KNN and wML-KNN by many experiments. A large number of experiments show that when k take the value of 1, the system works best and get highest accuracy. The best experimental results on each metrics are showed in the following four tables. In every table, using the letters of T, H, O, C and A replace Absolute-true, Hamming Loss, One-error, Coverage and Average Precision respectively.

In this thesis, the two different feature extraction methods of the entropy density and AmPseAAC are combined specially. The sequences of protein sample P can be expressed as:

$$P = (s_1(P), s_2(P), \ldots, s_{20}(P), p_{20+1}, p_{20+2}, \ldots, p_{20+2\lambda}) \tag{21}$$

where $s_i(P)(i = 1, 2, \ldots, 20)$ represents the entropy density, and $p_{20+2\lambda}$ $(\lambda = 1, 2, \ldots, 10)$ is made up of after AmPseAAC's 20 dimensions. Suppose the value of λ is 10, and thus the feature vector from this model is 40 dimensions. While the simple dimension additive feature fusion method can be defined as:

$$P = (s_1(P), s_2(P), \ldots, s_{20}(P), p_1, p_2, \ldots, p_{20+2\lambda}) \tag{22}$$

Table 3. Using the simple dimensions additive feature fusion way to obtain results of ML-KNN in two different datasets

Dataset	Evaluation criterion				
	T	H	O	C	A
S1	0.5675	0.1369	0.3968	1.0794	0.7634
S2	0.6310	0.1764	0.3556	0.5813	0.7902

Table 4. Using the particular feature fusion way to obtain results of ML-KNN in two different datasets

Dataset	Evaluation criterion				
	T	H	O	C	A
S1	0.5913	0.1263	0.3611	1.0159	0.7846
S2	0.6577	0.1635	0.3289	0.5615	0.8069

Table 3 shows that the sample dimensions additive feature combination of the entropy density and AmPseAAC yields the best performance on every evaluation measurement. Table 4 shows that the special combination of the above two methods of feature extraction achieves rather superior performance than Table 3. What is more, the result using exceptional fusion method significantly outperform that using sample fusion way on every evaluation measurement no matter which dataset is tested. Although using the special combination method along with ML-KNN is superior to using the simple dimensions additive feature fusion way, the accuracy for prediction is not high than that of others studies.

Table 5. Using the simple dimensions additive feature fusion way to obtain results of wML-KNN in two different datasets

Dataset	Evaluation criterion				
	T	H	O	C	A
S1	0.5992	0.1276	0.3889	1.0357	0.7761
S2	0.7648	0.1162	0.2352	0.3078	0.8728

Table 6. Using the particular feature fusion way to obtain results of wML-KNN in two different datasets

Dataset	Evaluation criterion				
	T	H	O	C	A
S1	0.6111	0.1210	0.3254	0.9762	0.8098
S2	0.8203	0.0884	0.1797	0.2390	0.9028

Table 5 shows that although we use the simple dimensions additive feature fusion as algorithm input, adopting wML-KNN as prediction algorithm can obtain better the result of prediction. Table 6 shows that using the particular feature fusion way achieves rather competitive performance where on all evaluation metrics no algorithm has ever outperformed wML-KNN. From Tables 5 and 6, we can see that using the special combination method along with wML-KNN still shows superior performance, which achieves 61.11 % and 82.03 % accuracy in two different datasets, and the best result on each metric is better than Tables 3 and 4 that show. So wML-KNN approach is useful to predict subcellular locations.

6 Conclusion

Prediction of protein subcellular localization is a challenging problem, particularly when the system concerned contains both single and multiplex proteins. In this paper, using the particular feature fusion way can obtain higher accurate rate on dataset S1 from the predictor of Virus-mPLoc and on dataset S2 from Gpos-mPLoc, respectively. On base of this novel feature combination representation, when adopt wML-KNN we can achieve better result than adopting ML-KNN algorithm by kinds of evaluation methods.

On the whole, it deserves to notice that one combination does not necessary yield the best performance on different datasets in every metric, so exploring whether better performance can be received by using other method of feature extraction or other predict algorithm is still a meaningful task. Hence, we shall make efforts in our future work to study multisite protein subcellular localization.

Acknowledgment. This research was partially supported by the Science and Technology Foundation of University of Jinan (Grant No. XKY1402), Shandong Provincial Natural Science Foundation, China, under Grant ZR2015JL025, the Youth Project of National Natural Science Fund (Grant No. 61302128), the Youth Science and Technology Star Program of Jinan City (201406003), the Natural Science Foundation of Shandong Province (ZR2011FL022, ZR2013FL002), the Scientific Research Fund of Jinan University (XKY1410, XKY1411), the Program for Scientific research innovation team in Colleges and Universities of Shandong Province (2012–2015), and the Shandong Provincial Key Laboratory of Network Based Intelligent Computing.

References

1. Chou, K.C.: Prediction of protein cellular attributes using pseudo amino acid composition. Proteins: Struct. Funct. Genet. **43**, 246–256 (2001)
2. Du, P.F., Xu, C.: Predicting multisite protein subcellular locations: progress and challenges. Proteomics **10**(3), 227–237 (2013)
3. Chou, K.C.: Pseudo amino acid composition and its applications in bioinformatics, proteomics and system biology. Curr. Proteomics **6**, 262–274 (2009)
4. Chou, K.C., Cai, Y.D.: Predicting protein localization in budding yeast. Bioinformatics **21** (7), 944–950 (2005)
5. Su, C.Y., Lo, A., Lin, C.C., et al.: A novel approach for prediction of multi-labeled protein subcellular localization for prokaryotic bacteria. In: Proceedings of the 2005 IEEE Computational Systems Bioinformatics Conference Workshops, Stanford, California, 8–12 August, pp. 79–80. IEEE, Piscataway (2005)

6. Zhu, H.Q., She, Z.S., Wang, J.: An EDP-based description of DNA sequences and its application in identification of exons in human genome. In: The Second Chinese Bioinformatics Conference Proceedings, Beijing, pp. 23–24 (2002)
7. Shannon, C.E.: The mathematical theory of communication. Bell Syst. Tech. **27**, 623–656 (1948)
8. Chou, K.C., Wu, Z.C., Xiao, X.: iLoc-virus: a multi-label learning classifier for identifying the subcellular localization of virus proteins with both single and multiple sites. J. Theor. Biol. **284**, 42–51 (2011)
9. Chou, K.C.: Using amphiphilic pseudo amino acid composition to predict enzyme subfamily classes. Bioinformatics **21**, 10–19 (2005)
10. Zhang, M.L., Zhou, Z.H.: ML-KNN: a lazy learning approach to multi-label learning. Pattern Recogn. **40**(7), 2038 (2007)
11. Shen, Z.B., Bai, Q.Y.: KNN text classification method based on weight modify. Comput. Sci. **35**(10), 123–126 (2008)
12. Qu, X., Chen, Y., Qiao, S., Wang, D., Zhao, Q.: Predicting the subcellular localization of proteins with multiple sites based on multiple features fusion. In: Huang, D.-S., Han, K., Gromiha, M. (eds.) ICIC 2014. LNCS, vol. 8590, pp. 456–465. Springer, Heidelberg (2014)

Systems Biology and Intelligent
Computing in Computational Biology

Performance and Improvement of Tree-Based Methods for Gene Regulatory Network Reconstruction

Ming Shi, Yan-Wen Chong$^{(\boxtimes)}$, and Shao-Ming Pan

State Key Laboratory of Information Engineering in Surveying, Mapping and
Remote Sensing, Wuhan University, Wuhan, China
{shiming,ywchong,pansm}@whu.edu.cn

Abstract. Computational reconstruction of gene regulatory networks (GRNs)
from gene expression data is of great importance in systems biology. Dialogue
for Reverse Engineering Assessments and Methods (DREAM) challenge aims
to evaluate the success of computational GRN inference algorithm on benchmarks of simulated data. Tree-based methods, such as Random Forest, infer true
regulators of a target gene in a feature selection way andexhibitcompeti-
tiveperformance. GENIE3 algorithm is a Random Forest-based algorithm and
was winner of the DREAM4 *InSilico Multifactorial* challenge. In this paper, we
further investigated the performance of tree-based algorithms for GRN inference. Experimental results showed that GENIE3 loses robustness on small-scale
heterozygous knock-down datasets, and a slightly modified version of GENIE3
algorithm mGENIE3 was provided. Experiments conducted on simulation and
real gene expression datasets show superior performance of mGENIE3.

Keywords: Gene expression data · Gene regulatory network · DREAM
challenge · Tree-based method · Random Forest · Modified version

1 Introduction

Inferring causal relationships among genes is one of the fundamental problems in
understanding cell behaviors [1, 2]. With the development of biological techniques, a
large amount and various types of biological datasets are available, these
high-throughput datasets can be considered as the outputs of the biological systems and
contain important biological information [3]. Computationally deciphering the complex
structure of transcriptional regulations among genes from gene expression data is a
challenging task. For the target of GRNs inference, the DREAM3 and DREAM4
challenge was held to encourage researchers to develop new efficient computation
methods to infer robust GRNs.

   Many methods have been proposed for reverse engineering of GRNs. These
methods differ in two ways: state representation of nodes/genes and the types of
interactions between genes/nodes. Boolean networks which were first introduced as a
model of a GRN by Kauffman in 1970s are discrete dynamic systems [4]. When a GRN
is modeled into a Boolean network, each gene is considered to be either "on" or "off".

© Springer International Publishing Switzerland 2016
D.-S. Huang et al. (Eds.): ICIC 2016, Part I, LNCS 9771, pp. 205–213, 2016.
DOI: 10.1007/978-3-319-42291-6_20

The "on" state means an active or over-expressed state, and the "off" state means an inactive or under-expressed state of genes. Regulatory interactions among genes are described by Boolean functions, which are combinations of logic operators: AND, OR and NOT. For the target of Boolean GRN reconstruction, both the underlying topology as well as the Boolean functions of genes/nodes should be inferred from gene expression data. The Boolean networks can be inferred from both time-course gene expression data and perturbation data. A Bayesian network (BN), which belongs to the family of probabilistic graphical models (GMs), is a directed acyclic graph. In particular, each node of BN in the graph represents a random variable, while the edges between the nodes represent probabilistic dependencies among the corresponding random variables. As a stochastic model, the Bayesian network can deal with noisy data and stochastic aspects of GRNs in a natural way [5, 6]. Though BNs have a number of advantages as a modeling framework, the main disadvantage of Bayesian networks is that they require the network structure to be acyclic and the dynamic aspects are not considered in the models. Since feedback loops [7] and dynamics are important features of GRNs, these disadvantages limit the application of Bayesian network models. Gaussian graphical models (GGMs) that are undirected probabilistic graphical models that allow the identification of conditional independence relations among the nodes under the assumption of a multivariate Gaussian distribution of the data. GGMs offer a realistic way to represent complex gene networks due to its interpretation in terms of conditional correlations. Some GGM-based approaches infer the network structure based on low-order partial correlations [7–9], where the order of partial correlation is the number of variables conditioned on. To estimate the full-order partial correlations, the connection between partial correlation and ordinary least squares regression has been implemented [10, 11].

Various types of methods has been proposed for GRN inference from gene expression data, however they are notorious for high false positive rate, especially when the number of genes is large and the number of samples is small. Tree-based methods such as Random Forest [12], extra trees [13], boosting trees [14] et al., are effective and efficient for GRN inference [15]. In these methods, candidate regulators of a target gene are considered as features, the true regulators are identified in feature-selection procedures. Although they are competitive with other methods such as Boolean network-based methods, Bayesian network-based methods et al., experimental results have shown that outliers and noise in datasets will reduce their performance [15].

In this article, we first introduce a tree-based GRN inference algorithmGENIE3, which was first proposed by Huynh-Thu in DREAM4 challenge [14]. This method was the best performer in the DREAM4 *InSilico* Multifactorial challenge. Its main feature and advantage over existing technique is that it makes very few assumptions about the nature of the relationships between genes. We first introduce the framework of GENIE3, show how it works for GRNs inference from various types of gene expression data. Then experiments conducted on DREAM3 and DREAM4 datasets show that GENIE3's performance is related to the type of samples as well as the size of networks. Analysis shows that GENIE3's performance loses robustness on heterozygous knock-down samples, and finally a corresponding method for remedy is proposed.

2   Algorithm and Performance

2.1   Problem Definition

Our goal is to reconstruct directed gene regulatory network using gene expression data. Suppose that a gene expression dataset contains the expression profiles of p genes across n samples, and in the corresponding GRN we aim to reconstruct, there are totally p nodes representing the p genes respectively. The directed edges in a GRN represent regulatory relationships between regulators and their target genes. In this work, we focus on inferring GRNs from static steady-state gene expression data, *i.e.* expression profiles resulting from the systematic knockout or knockdown of genes, multifactorial perturbation data in GRNs, and so on [15].

Assume that vector $\mathbf{x}_i = \left(\mathbf{x}_i^1, \mathbf{x}_i^2, \cdots, \mathbf{x}_i^n\right)^T$, $i = 1, 2, \cdots, p$ denote the expression values of the i-th gene across n samples, The basic assumption for GRN inference is that expression values of a target gene are function of the other genes in the network. For example, the expression level of the i-th gene is related with the expression levels of all the other genes:

$$\mathbf{x}_i = f_i\left(\mathbf{x}_1, \mathbf{x}_2, \cdots, \mathbf{x}_{i-1}, \mathbf{x}_{i+1}, \cdots, \mathbf{x}_p\right) + \epsilon_i \tag{1}$$

where $\left(\mathbf{x}_1, \mathbf{x}_2, \cdots, \mathbf{x}_{i-1}, \mathbf{x}_{i+1}, \cdots, \mathbf{x}_p\right)$ is the expression values all the genes in a GRN except the i-*th* gene. These genes can be considered as the candidate regulators of the i-*th* gene; ϵ_i is a random noise with zero mean. Based on the hypothesis above, we can find that reconstructing a GRN from expression datasets is equivalent to finding an appropriate function f_i, $i = 1, 2, \cdots, p$ that best fits (1).

2.2   GENIE3 Algorithm

The basic thoughts of Tree-based algorithms are that within the framework of supervised learning, the problems of identifying true regulators of target genes, *i.e.* finding the function f_i, $i = 1, 2, \cdots, p$ best fit (1) is equivalent to a feature selection problem. Candidate regulators of a target gene are considered as features.

GENIE3 algorithm, which was first proposed by Huynh-Thu in DREAM4 *InSilico Multifactorial* challenge, is essentially a tree-based ensemble method for GRN inference [14]. As is well known, performance of a single regression tree will be greatly improved by ensemble methods. GENIE3 mainly employ two kinds of tree-based ensemble algorithms: Random Forest and Extra-Trees. As a tree-based method, GENIE3 algorithm aims to find the function $\tilde{f}_i$, *i.e.* a collection of regression trees, which minimizes the following error construct regression trees:

$$\min_{\tilde{f}_i}\|\epsilon\|_2^2 = \min_{f_i}\left\|\mathbf{x}_i - f_i\left(\mathbf{x}_1, \mathbf{x}_2, \cdots, \mathbf{x}_{i-1}, \mathbf{x}_{i+1}, \cdots, \mathbf{x}_p\right)\right\|_2^2 \tag{2}$$

One key characteristic of regression trees is that importance of the input variables, *i.e.* the expression levels of candidate regulators, can be measured using their relevance for predicting outputs, *i.e.* the expression levels of a target gene. In our context, higher importance score of a candidate regulator indicates higher probability of regulatory effect imposed on the target gene. The flow chart of GENIE3 is shown below (Fig. 1).

Input:

x_i, the expression value of the i-th gene, $i = 1, 2, \cdots, p$

$\mathbf{X}_{-i} = \left(\mathbf{x}_1, \mathbf{x}_2, \cdots, \mathbf{x}_{i-1}, \mathbf{x}_{i+1}, \cdots, \mathbf{x}_p \right)$, the expression values of all genes except the i-th gene, these genes are considered as candidate regulators of target gene i, $i = 1, 2, \cdots, p$

Output: p ensembles of regression trees corresponding to p genes, the i-th ensembles of regression trees is equivalent to function $\tilde{f}_i, i = 1, 2, \cdots, p$, so that

$$\left\| \mathbf{x}_i - \tilde{f}_i \left(\mathbf{x}_1, \mathbf{x}_2, \cdots, \mathbf{x}_{i-1}, \mathbf{x}_{i+1}, \cdots, \mathbf{x}_p \right) \right\|_2^2 = \min_{f_i} \left\| \mathbf{x}_i - f_i \left(\mathbf{x}_1, \mathbf{x}_2, \cdots, \mathbf{x}_{i-1}, \mathbf{x}_{i+1}, \cdots, \mathbf{x}_p \right) \right\|_2^2$$

Repeat until $i = p$

  Initialize $i = 1$

  Construct ensembles of regression trees with tree-based methods, such as random forests and extra trees, where $\mathbf{x}_i$ is a dependent variable and

$\mathbf{x}_1, \mathbf{x}_2, \cdots, \mathbf{x}_{i-1}, \mathbf{x}_{i+1}, \cdots, \mathbf{x}_p$ are regressors. These constructed trees represent a approximate solution of equation (2), i.e. $\tilde{f}_i$.

  Variable scores of regressors are measured according to their relevance for predicting the dependent variable in the regression trees.

  Set $i = i + 1$

Fig. 1. Flowchart of GENIE3 algorithm

2.3   Performance of GENIE3

Experiments with both synthetic and real data show that GENIE3 performs well. Datasets used for inferring GRNs collected from DREAM3 and DREAM4 *InSilico* Network Inference Challenge. Accuracy of the inferred GRNs is measured by the area under the Receiver Operating Characteristic curve, or AUROC. Higher AUROC scores represent higher accuracy level of the inferred GRNs.

DREAM3 *InSilico* Network Challenge. The goal of the DREAM3 *InSilico* Network Challenge is to predict the directed unsigned network topology from the given *InSilico* generated gene expression datasets. DREAM3 is divided into 3 sub-challenges: *InSilico*10-gene network sub-challenge, *InSilico*50-gene network sub-challenge and *InSilico*100-gene network sub-challenge. Each sub-challenge contains noisy time series and static steady-state datasets from 5 different benchmark networks. The three types of gene expression data for each challenge are: heterozygous knock-down data containing the steady state levels for the wild-type and the heterozygous knock-down strains for each gene, null-mutants data contain the steady state levels for the wild-type and the null-mutant strains for each gene, trajectories data contain time courses of the network recovering from several external perturbations. However, as a tree-based method, GENIE3 can only deal with steady state levels, *i.e.* heterozygous data, knock-down data and null-mutants data.

Table 1 exhibits the performance of GENIE3 on these datasets. From the contents of the table, we can make the following conclusions. First, it is obvious that the performance of GENIE3 algorithm on null-mutants data is better than that on heterozygous knock-down, no matter the size of networks is 10, 50 or 100. Second with the sizes of networks increase from 10, 50 to 100, the average AUROC score of null-mutants data increases from 0.663 to 0.801. But the average AUROC score of heterozygous knock-down data stay around 0.55. It means that the GENIE3's GRN reconstruction quality on null-mutants data improves while the number of genes in GRN increases to a properly number. But regardless of the size of GRN changes, GENIE3's reconstruction quality on heterozygous knock-down gene expression data stays undesirable.

Table 1. AUROC scores of GENIE3 in DREAM3

Type of data	AUROC scores					Mean
	NET1	NET2	NET3	NET4	NET5	
InSilico Size 10 sub-challenge						
Heterozygous knock-down datasets	0.3391	0.4580	0.6189	0.7248	0.5163	0.5314
Null-mutants datasets	0.6854	0.6408	0.7189	0.6197	0.6503	0.6630
InSilico size 50 sub-challenge						
Heterozygous knock-down datasets	0.6049	0.5679	0.5849	0.5896	0.5571	0.5808
Null-mutants datasets	0.7944	0.7805	0.8474	0.7316	0.6938	0.7695
InSilico Size 100 sub-challenge						
Heterozygous knock-down datasets	0.5893	0.6107	0.5490	0.5662	0.5270	0.5684
Null-mutants datasets	0.7982	0.8767	0.8417	0.7813	0.7059	0.8007

DREAM4 *InSilico* Network Challenge. The goal of the *InSilico* network challenge is to reverse engineer gene regulation networks from simulated steady-state and time-series data. There are three *InSilico* sub-challenge called: *InSilico* Size10, *InSilico* Size100 and *InSilico* Size100 Multifactorial respectively. Each sub-challenge consists of five networks. Though totally 5 types of datasets are provided, GENIE3 algorithm can only deal with the static steady-state data, i.e. the knockouts, the knockdowns as well as the multifactorial perturbations expression data. AUROC scores of GENIE3 algorithm on these datasets are shown in Table 2. The same phenomenon that was observed in the DREAM3 Challenge above also appears in these experiments too: with the size of network growing, the average AUROC scores of each type increase. So we can make a conclusion that, when the number of genes in GRN is less than 100, the reconstruction quality of GENIE3 is positive correlation with it. Also, it can be found that GENIE3's performance on multifactorial perturbations data is almost as well as that on knockouts. In consideration of the fact that, GENIE3 is the best performer in DREAM4 InSilico Size100 Multifactorial sub-challenge, we can make the conclusion that GENIE3 algorithm is excellent for GRN inference on multifactorial perturbation data.

Table 2. AUROC scores of GENIE3 inDREAM4

Type of data	AUROC scores					Mean
	NET1	NET2	NET3	NET4	NET5	
InSilico Size 10 sub-challenge						
Knock-down data	0.7255	0.6071	0.6729	0.7082	0.6828	0.6793
Knock-out data	0.7341	0.6257	0.6894	0.7489	0.7008	0.6998
Multi-factorial perturbation data	0.7098	0.7411	0.6675	0.5756	0.7424	0.6873
InSilico Size 100 sub-challenge						
Knock-down data	0.7078	0.667	0.6678	0.6734	0.7094	0.6851
Knock-out data	0.8113	0.7623	0.8139	0.8144	0.7895	0.7983
Multi-factorial perturbation data	0.7439	0.728	0.773	0.7908	0.8003	0.7672

3   Detailed Analysis of GENIE3 Algorithm

3.1   Robustness on Small-Scale Datasets

Each time we repeat the experiment on DREAM3 and DREAM4 datasets, the results are not exactly the same. This is because that GENIE3 is a Random Forest based method, and randomness is one of its key characteristic. For example, in the construction of Random Forests, the technique of bootstrap resampling is needed, this bring uncertainty to the reconstructed GRNs [16].

We conducted repeated experiments on different types of datasets and computed the variance and mean of AUROC scores for further analysis. The mean of AUROC scores reflected the average accuracy of inferred GRNs while variance reflected the variability of the distribution for AUROC scores. We repeatedly inferred GRNs from the small-scale heterozygous knock-down datasets in DREAM3 and DREAM4 challenge using GENIE3 algorithms for 1000 times, and exhibit the statistical analysis of the AUROC scores in Table 3.

Table 3. Statistical analysis of AUROCs on heterozygous knock-down datasets from DREAM3 and DREAM4

Network size	Interval of AUROC scores	Interval length	Mean	Standard deviation
DREAM3 *InSilico* network inference challenge				
10	[0.32, 0.70]	0.38	0.66	0.068
50	[0.59, 0.62]	0.03	0.60	0.005
100	[0.57, 0.60]	0.03	0.58	0.006
DREAM4 *InSilico* network inference challenge				
10	[0.68, 0.76]	0.08	0.69	0.012
100	[0.70, 0.72]	0.02	0.71	0.004

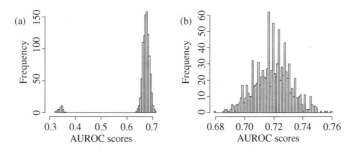

Fig. 2. Distribution of AUROC scores for the GENIE3 algorithm on heterozygous knock-down data. Gene expression data used for inferring GRNs were collected from DREAM3 *InSilico* Size 10 challenge (A) and DREAM4 *InSilico* Size 10 challenge (B).

It can be seen from Table 3 that when the size of network is 10, the AUROC scores vary between 0.32 and 0.70, *i.e.*, the length of the distribution interval is 0.38; this reflects a drastic change in network reconstruction accuracy. But when the size of networks grows up to 50 and 100, the phenomena of instability become weaker and the according interval length reduce to 0.3 and 0.3 respectively. It indicates that when GENIE3 algorithm is applied to small-scale heterozygous knock-down datasets, it loses robustness. This may due to the small sample size. Figure 2 shows the histogram of these AUROC scores. As can be seen from Fig. 2(A), there is a probability of 3 % that the tree-based algorithm GENIE3 inferred GRNs with very low accuracy for which the AUROC scores are below 0.35, even less than 0.5. This may due to the small sample size. Efforts should be made to eliminate this instability and our method for correcting the defect will be depicted in Sect. 3.2.

3.2 Methods to Eliminate Instability and Improve Accuracy

A natural idea to eliminate instability is to average the results of GENIE3 algorithm, and this is based on the fact that the variance of the sample mean is smaller than the sample by a factor of the square root of the sample size. The new algorithm is called mGENIE3. Suppose that mGENIE3 algorithm haverepeatedGENIE3 algorithm N times on one dataset, and for a target gene, and correspondingly the j-th candidate regulator gets N different importance scores in these N experiments. Then the average of the importance scores is set to be the final score of the j-th regulator.

Experimental results have verified the effectiveness of this method. For example, when this slightly modified algorithm, *i.e.* mGENIE3, is applied to reconstruct gene networks using datasets from DREAM3 *InSilico* Size10 sub-challenge, the AUROC score is about 0.68 and the estimated standard deviation is 10^{-4}. Another notable point is that the AUROC score of mGENIE3 is 0.68, which is very near the highest AUROC score (0.70) of GENIE3 on the same dataset (Table 3). This demonstrates the effectiveness and efficiency of mGENIE3. A detailed exhibition of the performance of mGENIE3 in the small-scale datasets from DREAM3 and DREAM4 challenge is shown in Table 4.

212 M. Shi et al.

Table 4. AUROC scores of mGENIE3 inDREAM3 and DREAM4 challenge

Type of data	AUROC scores					Mean
	NET1	NET2	NET3	NET4	NET5	
DREAM3 *InSilico* Size 10 sub-challenge						
Heterozygous knock-down datasets	0.6749	0.4571	0.6229	0.7201	0.5039	0.6749
Null-mutants datasets	0.7017	0.6376	0.7022	0.6069	0.6235	0.7017
DREAM4 *InSilico* Size 10 sub-challenge						
Knock-down data	0.7231	0.6079	0.6698	0.7153	0.6913	0.6815
Knock-out data	0.7467	0.6265	0.6886	0.7569	0.7188	0.7075
Multi-factorial perturbation data	0.7020	0.7359	0.6776	0.5623	0.7652	0.6886

4 Conclusion

In this paper, we first introduce GENIE3 algorithm, a tree-based ensemble method for gene regulatory network reconstruction. Then we evaluate GENIE3's performance on the datasets collected from DREAM3 and DREAM4 *InSilico* Network Inference challenge. Finally, noticing that the performance of GENIE3 may become poor due to the small sample size, we use an average method to modify the original algorithm. Experiments conducted on *InSilico* datasets verified the efficiency and effectiveness of our method. Furthermore, when dealing with instability of recovery results, average may be effective but there is still space to improve, the Z-statistic may be another way to improve the recovery quality and it would be studied in the future.

Acknowledgement. This work has been partially supported by the National Natural Science Foundation of China (Grant Nos. 61572372 and 41271398), LIESMARS Special Research Funding, and also partially supported by the Fund of SAST (Project No. SAST201425). The funders had no role in study design, data collection and analysis, decision to publish, or preparation of the manuscript.

Conflict of Interest.
The authors declare that there is no conflict of interest regarding the publication of this article.

References

1. Zhou, T., Wang, Y.L.: Causal relationship inference for a large-scale cellular network. Bioinformatics **26**(16), 2020–2028 (2010)
2. Zhang, X.J., Liu, K.Q., Liu, Z.P., Duval, B., Richer, J.M., Zhao, X.M., Hao, J.K., Chen, L. N.: NARROMI: a noise and redundancy reduction technique improves accuracy of gene regulatory network inference. Bioinformatics **29**(1), 106–113 (2013)
3. Wang, Y.Y., Li, Z.G., Chen, T., Zhao, X.M.: Understanding the aristolochic acid toxicities in rat kidneys with regulatory networks. IET Syst. Biol. **9**(4), 141–146 (2015)
4. Dover, G.A.: The origins of order-self-organization and selection in evolution- Kauffman, SA. Nature **365**(6448), 704–706 (1993)

5. De Jong, H.: Modeling and simulation of genetic regulatory systems: a literature review. J. Comput. Biol. **9**(1), 67–103 (2002)
6. McAdams, H.H., Arkin, A.: Stochastic mechanisms in gene expression. Proc. Natl. Acad. Sci. USA **94**(3), 814–819 (1997)
7. Hasty, J., McMillen, D., Isaacs, F., Collins, J.J.: Computational studies of gene regulatory networks: in numero molecular biology. Nat. Rev. Genet. **2**(4), 268–279 (2001)
8. Wille, A., Buhlmann, P.: Low-order conditional independence graphs for inferring genetic networks. Stat. Appl. Genet. Mol. Biol. Mol. **5**(1) (2006)
9. De La Fuente, A., Bing, N., Hoeschele, I., Mendes, P.: Discovery of meaningful associations in genomic data using partial correlation coefficients. Bioinformatics **20**(18), 3565–3574 (2004)
10. Tenenhaus, A., Guillemot, V., Gidrol, X., Frouin, V.: Gene association networks from microarray data using a regularized estimation of partial correlation based on PLS regression. IEEE/ACM Trans. Comput. Biol. Bioinform. **7**(2), 251–262 (2010)
11. Peng, J., Wang, P., Zhou, N.F., Zhu, J.: Partial correlation estimation by joint sparse regression models. J. Am. Stat. Assoc. **104**(486), 735–746 (2009)
12. Liaw, A., Wiener, M.: Classification and regression by Random Forest. R news **2**(3), 18–22 (2002)
13. Geurts, P., Ernst, D., Wehenkel, L.: Extremely randomized trees. Mach. Learn. **63**(1), 3–42 (2006)
14. Drucker, H., Cortes, C.: Boosting decision trees. Adv. Neural Inf. Process. Syst. **1059**, 479–485 (1996)
15. Marbach, D., Costello, J.C., Kuffner, R., Vega, N.M., Prill, R.J., Camacho, D.M., Allison, K.R., Kellis, M., Collins, J.J., Stolovitzky, G., DREAM5 Consortium: Wisdom of crowds for robust gene network inference. Nat. Methods **9**(8), 796–804 (2012)
16. James, G., Witten, D., Hastie, T., Tibshirani, R.: An Introduction to Statistical Learning. Springer, New York (2013)

A Hybrid Tumor Gene Selection Method with Laplacian Score and Correlation Analysis

Bo Li[1,2,3(✉)], Xiao-Hui Lei[1,2], Yang Hu[1,2], and Xiao-Long Zhang[1,2]

[1] School of Computer Science of Technology,
Wuhan University of Science of Technology, Wuhan, Hubei 430065, China
liberol@126.com
[2] Hubei Province Key Laboratory of Intelligent Information
Processing and Real-Time Industrial System, Wuhan, Hubei 430065, China
[3] School of Electronics and Information Engineering,
Tongji University, Shanghai 201804, China

Abstract. In the proposed method, Laplacian criteria is firstly introduced to sort the genes as their descending scores. And then, correlation analysis is applied to select those pathogenic genes from the sorted sequence to reduce the redundancy. At last, SVM classifier is used to predict the class labels of the optimal gene subset. Compared to some other related gene selection methods such as Fisher score and Laplacian score, Experimental results on four standard datasets have shown the stability and efficiency of the proposed method.

Keywords: Correlation analysis · Laplacian score · Tumor gene expressive data · Gene selection

1 Introduction

A DNA microarray is a collection of microscopic DNA spots attached to a solid surface, from which tumor gene expressive data have been achieved. Based on tumor gene expressive data, studies have been conducted to measure the expression levels of large numbers of genes simultaneously or to explore the genotype of a genome, which will contribute to making insight into both biological processes and mechanisms of human cancer diseases. However, how to interpret those tumor gene expressive data still needs further demonstration. Many scholars have reported their research results on tumor gene expressive data, where more attentions are paid to both molecular classification of cancer and key tumor genes selection.

For cancer prediction, because tumor gene expressive data are always characterized by a large amount of variables (genes or features), some dimensionality reduction methods have to be adopted to extract low dimensional discriminant features before making classification to them. Linear models such as linear discriminant analysis(LDA) [1], principal component analysis (PCA) [2], partial least squares (PLS) [3] and independent component analysis (ICA) [4] have been successfully applied to tumor gene expressive data classification. Later, Pochet et al. systematically proved that nonlinear models outperform those linear ones on many tumor gene expressive data sets [5]. So how to nonlinearly mine the tumor gene expressive data has been

© Springer International Publishing Switzerland 2016
D.-S. Huang et al. (Eds.): ICIC 2016, Part I, LNCS 9771, pp. 214–223, 2016.
DOI: 10.1007/978-3-319-42291-6_21

concentrated on and some nonlinear models were presented. Alexandridis et al. put forward a nonlinear method using finite mixture distribution for tumor gene data analysis [6]. Meanwhile, Martella proposed a nonlinear factor mixture model with both factor factorization and normal mixture [7]. Moreover, other nonlinear feature extraction methods such as kernel transformation and manifold learning have also been advanced for tumor gene expressive data analysis [8, 9]. It must be noted that tumor gene expressive data are also indicated with a small amount of observations (samples), so sparse representation technique is also introduced to these traditional linear and nonlinear methods for gene tumor gene expressive data classification, for example, Cui et al. proposed a sparse maximum margin discriminant analysis for tumor gene expressive data feature extraction [10]. In this method, a linear transformation will be explored to maximize the margin with sparse representation. Juan et al. used a sparse manifold clustering and embedding method to make discriminant feature extraction and classification for gene expression profiles of glioblastoma and meningioma tumors [11]. However, it is still difficult to make a reasonable biological explanation for the results by these feature extraction or subspace learning approaches.

The methods mentioned above benefit to predict cancer genre with high efficiency and low computational cost. However, they also fail to explore those key tumor pathogenic genes, which shows heavy impact on the final cancer diagnosis and treatment. Under such circumstance, some methods aiming to select key tumor genes from numerous genes have shown their superiority.

Tumor gene selection, also known as feature subset selection, is the process of selecting a subset of key tumor genes for use in tumor classification, which will simplify learning models to make them easy to be interpreted by researchers and users, shorten training times and enhance generalization by reducing over-fitting. As a result, how to select the pathogenic gene subset has attracting many attentions. Researchers have tried their attempts to make gene selection before carrying out tumor data classification. Based on the strategies applied to search key tumor genes, they can be divided into complete mode, heuristic mode and random mode. When conducting complete selection, both exhaustive and non-exhaustive techniques are all contained. In order to enhance the gene subset selection efficiency, heuristic mode is also introduced, where forward selection, backward selection, combine forward and backward selection and instance based selection are all involved. The third is random selection including random generation plus sequential selection, simulated annealing, genetic algorithms, random forest and rotation forest. However, in these methods, much high computational cost will be expended, which will lead to difficulty in real-world data application.

In addition, tumor gene selection methods can also be further categorized into either wrapper method or filter one by whether or not classifier models are adopted to justify the performance of the subset selection. In the wrapper methods, classifiers must be used to evaluate the performance of the selected tumor gene subset, thus support vector machine (SVM) based wrapper methods are always proposed. However, these methods work well only with binary classification. In order to overcome the problem, by incorporating mRMR, a SVM-REF was put forward by Mundra and Rajapakse to improve SVM based method's performance [12], but it still remains unclear that how SVM-REF can be extended to multiclass classification. In general, the wrapper method can obtain high performance for key tumor gene selection. However, the computational expense for wrapper methods is significantly high. Another disadvantage of the wrapper method is that the gene subset is heavily depended on the corresponding classifier. For instance the gene subset selected by SVM model may be distinct from that subset selected by the k nearest neighbor method.

Instead of classifier, the intrinsic characteristics of the original tumor gene expressive data, such as locality preserve capacity [13], similarity preserve capacity [14] or structural sparsity [15], is employed to choose genes in filter methods. By contrast to the wrapper method, the filter is also highly efficient because no classifier is involved in for gene subset selection, which results in it more favored in classification problems, especially for high dimensional tumor gene expressive data label prediction. Some famous tumor gene selection methods, i.e. relief, T-test, Fisher score and Laplacian score, are all filter models, which have been widely applied to tumor gene expressive data. But these classical filter methods directly use the corresponding criterion to select key tumor genes from the original data, where redundant genes and even noise are always involved. Moreover, to reduce the redundancy and to remove the noise in tumor gene expressive data will contribute to selecting key tumor genes accurately and even to improving the performance of final data classification.

So in this paper, a hybrid tumor gene selection method is presented to select relevant biological correlations or "molecular logic" genes from tumor gene expression data, by which their genres will be easily predicted with high performance. In the proposed method, Laplacian criterion will be also adopted to sort the genes, where those with high scores will be of high possibility to key tumor genes. Although high scored, some of them may be also heavily correlated. Thus, in the light of Laplacian scores, those genes from the original tumor gene expressive data are ranked as a descending order. And then correlation matrixes can be constructed to measure the redundancy between the sorted genes, according to which a gene subset can be selected to reduce the redundancy existing in the sorted gene sequence. Finally, SVM is introduced to predict the class labels of tumor gene expressive data using the selected optimal gene subset.

2   Method

2.1   Motivation

It is well-known that tumor gene expressive data are always characterized by a large amount of features (genes) with a small amount of observations (samples), where any feature dimension in all samples actually represents a gene. For example, there are 72 samples in Leukemia data set, where 7129 genes are contained. On the one hand, in the original tumor gene expressive data, genes are ranked as their biological characteristics, where no discriminant information for classification is hidden. That is to say, those genes contributing more to tumor gene expressive data classification are randomly distributed or their locations in the samples are unrelated to the final classification performance, based on which it is unreasonable or difficult to select key tumor genes. Thus it occurs to gene selection a problem that which criterion will be adopted. Combining to the characteristics that the filter model searches feature subset by using the intrinsic characteristics of the original data, manifold learning based method can be well performed for key tumor gene selection because manifold learning method is good at exploring inherent geometry information of the original data. In this paper, Laplacian score will be introduced to rearrange the genes in all samples.

On the other hand, there are many genes or even noise contained in a tumor gene expressive sample. Firstly, some show no relation or few relations to the corresponding tumor. In other words, only small number of genes will result in cancer and other numerous genes are redundant. Secondly, some will be high possibility to key tumor genes, however, there are also heavy correlations in them. Thus how to reduce the redundancy between genes is necessary to key tumor gene selection because it will make classification tasks more difficult if unnecessary or redundancy genes are involved in. So in the proposed method, correlation matrix will be recommended to quantify the correlations between genes. Those genes with high correlation scores will be considered to be more interrelated. On the basis of the correlation matrix, the genes with both less Laplacian scores and high correlation coefficients can be removed and the rest are composed of the optimal gene subset. Therefore the redundancy of the tumor gene expressive data will be reduced as much as possible. At last, SVM will be applied to classify the optimal gene subset.

2.2 Laplacian Score

Originating from manifold learning method, i.e. Laplacain eigenmaps and locality preserving projection, Laplacian score is presented by He et al., which is modeled to evaluate the features by their locality preserving ability [13].

Let L_r denote the Laplacian score of the r-th feature. Let f_{ri} denote the i-th sample of the r-th feature, where $i = 1, 2, \ldots, n$. Laplacian score can be stated as the following steps.

Firstly, a nearest neighbor graph G will be constructed with n nodes, where the i-th node is sample X_i. And then, an edge is put between two points X_i and X_j if point X_i is among the k nearest neighbors of point X_j or point X_j is among the k nearest neighbors of point X_i. Moreover, when the class label information are available, an edge can also be put between two points with the same class.

Secondly, if there is an edge between two points X_i and X_j, set the similarity between them as $S_{ij} = \exp(-\|X_i - X_j\|^2/t)$, otherwise, $S_{ij} = 0$.

Thirdly, for the r-th feature, we define f_r is the r-th feature of samples and $f_r = [f_{r1}, f_{r2}, \ldots, f_{rm}]$, $D = diag(S1)$, $1 = [1, 1, \ldots, 1]^T$, $L = D - S$, where L is a Laplacian matrix. Construct the function:

$$\tilde{f}_r = f_r - \frac{f_r^T D1}{1^T D1} 1 \tag{1}$$

At last, calculate the following Laplacian score.

$$L_r = \frac{(\tilde{f}_r^T L \tilde{f}_r)}{(\tilde{f}_r^T D \tilde{f}_r)} \tag{2}$$

2.3   Correlation Analysis

For high dimensional tumor gene expressive data, there exist some genes with high correlations, which show much effects on the final performance. On the contrary, those genes with low correlations will be make more contributions to data classification. Thus, to remove the redundant genes from tumor gene expressive data will help to improve the efficiency of tumor genre classification. Thus a correlation matrix is constructed to measure the similarity between genes for all samples $X = [X_1, X_2, \ldots, X_n]$, which can be stated below.

$$R = \begin{bmatrix} \rho_{11} & \rho_{12} & \cdots & \rho_{1n} \\ \rho_{21} & \rho_{22} & \cdots & \rho_{2n} \\ \cdots & \cdots & \cdots & \cdots \\ \rho_{n1} & \rho_{n2} & \cdots & \rho_{nn} \end{bmatrix} \tag{3}$$

where R denotes the correlation matrix between samples and ρ_{rs} means the correlation coefficient between any two features f_r and f_s. Moreover, the correlation coefficient ρ_{rs} is calculated as follows.

$$\rho_{rs} = \frac{E((f_r - E(f_r)) \cdot (f_s - E(f_s)))}{\sqrt{D(f_r)} \cdot \sqrt{D(f_s)}} \tag{4}$$

where $E(\bullet)$ and $D(\bullet)$ represent mean and variance of features or genes, respectively.

From the above equation, it can find that the correlation matrix is symmetrical, where the value of any element is located between 0 and 1 except that the diagonal elements are 1.

Generally speaking, the larger of the element ρ_{rs} in correlation matrix, the higher correlation between genes f_r and f_s. So in this paper, a parameter ρ, named correlation coefficient threshold, is set to determine whether or not two genes are correlated, based on which the uncorrelated key tumor genes will be selected. In details, those with correlation coefficient are larger than ρ, they will be correlated, otherwise, they will be uncorrelated.

2.4   Outline of the Proposed Method

Based on the above analysis, tumor genes can be reorganized as its descending Laplacian scores, from which those with high Lapalcian scores and low correlation coefficients will be chosen as key tumor genes.

Input: Tumor gene expressive data $X = [X_1, X_2, ..., X_n] \in \mathbb{R}^{D \times n}$ and their labels $C = [C_1, C_2, ..., C_c]$, $C_i = \pm 1, i = 1, 2, ..., c$, numbers of nearest neighbors k, correlation coefficient threshold ρ.

Output: Selected key tumor gene subset F and label classification

Algorithm:

Step 1 Tumor gene sorting

Sort the tumor gene expressive data X according to the descending Laplacian scores and obtain the sorted matrix X';

Step 2 Correlation analysis

Make correlation analysis to features in the sorted matrix X' using correlation matrix and obtain the corresponding correlation matrix R;

Step 3 Key tumor gene selection

From the features with high Laplacian score in the sorted matrix X', select those correlation coefficient smaller than ρ as key gene subset;

Step 4 Data classification

Adopt SVM classifier to classify the key tumor gene subset.

3   Experiments

In this Section, the proposed method is compared with several related gene selection methods including Fisher score and Laplacian score. After employing Fisher score, Laplacian score and the proposed method for key gene subset selection, SVM classifier is adopted for key tumor gene label prediction. At the same time, SVM is also taken as another comparison method for tumor gene expressive data classification.

For Laplacian score and the proposed method, k nearest neighbors criterion is also used to construct the corresponding local graph, where parameter k is introduced. In the experiments, we train k using leave-one-out-cross-validation (LOO-CV) on tumor gene expressive samples.

Moreover, when making correlation analysis on tumor gene expressive data, another parameter, i.e. correlation coefficient threshold ρ also appears. In the experiment, LOO-CV is also introduced to adjust parameter ρ.

3.1   Experiment Data

In this paper, four benchmark tumor gene expressive data sets including Acute leukemia, High-grade gliomas, Diffuse large B-cell lymphomas (DLBCL) and Colon are all applied in the experiments to validate the proposed method.

An conclusion for the above mentioned four benchmark tumor gene expressive data has been drawn, where samples for each class and genes are all described. It can be found from the following table (Table 1).

Table 1. Summary of four benchmark tumor gene expressive data sets

Data sets	Training set		Test set		Genes
	Class1	Class2	Class1	Class2	
Leukemia	11	27	14	20	7129
Gliomas	14	7	14	15	12625
DLBCL	20	12	38	7	5469
Colon	14	8	26	14	2000

3.2　Experiment Results

When carrying out experiment, it must be noted that there are two parameters including k nearest numbers and correlation coefficient threshold ρ are involved in the experiments. So LOO-CV method is adopted to tune them.

For benchmark four tumor gene expressive data, we first using LOO-CV to train nearest neighbors number k when constructing the local graph. Figure 1 shows the experimental results. In Fig. 1, it can be found that LOO-CV will climb to the top for High-grade gliomas, Acute leukemia, Colon and DLBCL data sets when k is set to 3, 2 or 3, 2 and 5, respectively.

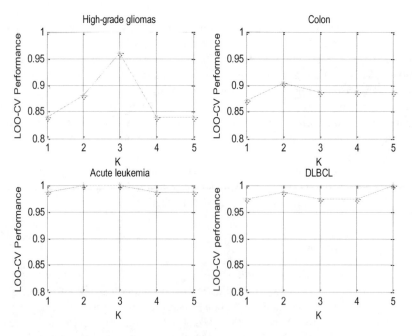

Fig. 1. LOO-CV performance with varied k for High-grade gliomas, Acute leukemia, Colon and DLBCL data sets

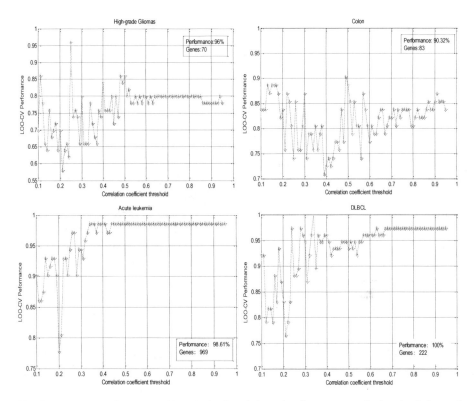

Fig. 2 LOO-CV performance with varied ρ for High-grade gliomas, Acute leukemia, Colon and DLBCL data sets

After trained parameter k, LOO-CV is also used again to adjust correlation coefficient threshold ρ. For High-grade gliomas, Acute leukemia, Colon and DLBCL data sets, Fig. 2 illustrates LOO-CV performance curves with varied parameter ρ, where ρ will be tuned to be 0.25, 0.38, 0.49 and 0.34 companying with the fixed k, i.e. 3, 2, 2 and 5, respectively.

Table 2. LOO-CV Performance comparisons on four data sets

Methods	Performance (Selected genes)			
	Leukemia	Gliomas	Colon	DLBCL
Fisher score	90.67 % (300)	97.12 % (125)	97.12 % (125)	95.72 % (20)
Laplacian score	75.11 % (300)	96.64 % (300)	96.64 % (300)	95.19 % (300)
SVM	78.56 % (12625)	97.24 % (7129)	97.24 % (7129)	95.5 % (5469)
Proposed method	96.00 % (70)	98.61 % (969)	98.61 % (969)	100 % (222)

At last, with the fixed parameters k and ρ, performance comparisons can be made to Fisher score, Laplacian Score, SVM and the proposed method. The experimental results on four tumor gene expressive data sets are displayed in the following Table. From Table 2, an conclusion can be drawn that the proposed method outperforms the other three methods.

4 Conclusions

In this paper, a hybrid feature selection method is presented for tumor gene expressive data classification. In the proposed method, firstly, Laplacian score is introduced to rearrange the original tumor expressive data as descending order, then correlation analysis is applied to the sorted gene sequence, where those with high Laplacain scores and low correlation coefficients are selected as an optimal gene subset. At last, SVM is also used to predict the genre of the optimal gene subset. Experiments on four benchmark tumor gene expressive data sets have been conducted by making comparisons to some other related gene selection methods such as Fisher score and Laplacian score, by which the proposed method can be validated to be efficiency and efficient.

Acknowledgement. This work was partly supported by the grants of Natural Science Foundation of China (61273303, 61273225, 61373109 and 61572381), China Postdoctoral Science Foundation (20100470613 and 201104173), Natural Science Foundation of Hubei Province (2010CDB03302), the Research Foundation of Education Bureau of Hubei Province (Q20121115).

References

1. Martinez, A.M., Kak, A.C.: PCA versus LDA. IEEE Trans. Pattern Anal. Mach. Intell. **23** (2), 228–233 (2001)
2. Jolliffe, I.T.: Principal Component Analysis. Springer, Heidelberg (2002)
3. Fort, G., Lacroix, S.L.: Classification using partial least squares with penalized logistic regression. Bioinformatics **21**(7), 1104–1111 (2005)
4. Huang, D.S., Zheng, C.H.: Independent component analysis based penalized discriminate method for tumor classification using gene expression data. Bioinformatics **22**, 1855–1862 (2006)
5. Pochet, N., Smet, F.D., Suykens, J.A., Moor, B.D.: Systematic benchmarking of microarray data classification: assessing the role of non-linearity and dimensionality reduction. Bioinformatics **20**(3), 3185–3195 (2004)
6. Alexandridis, R., Lin, S., Irwin, M.: Class discovery and classification of tumor samples using mixture modeling of gene expression data—a unified approach. Bioinformatics **20** (16), 2545–2552 (2004)
7. Martella, F.: Classification of microarray data with factor mixture models. Bioinformatics **22** (3), 202–208 (2006)

8. Pillati, M., Viroli, C.: Supervised locally linear embedding for classification: an application to gene expression data analysis. In: 29th Annual Conference of the German Classification Society (GfKl 2005), pp. 15–18 (2005)
9. Li, B., Tian, B.B., Zhang, X.L., Zhang, X.P.: Locally linear representation Fisher criterion based tumor gene expressive data classification. Comput. Biol. Med. **44**(10), 48–54 (2014)
10. Cui, Y., Zheng, C.H., Yang, J., Sha, W.: Sparse maximum margin discriminant analysis for feature extraction and gene delection on gene expressive data. Comput. Biol. Med. **43**, 933–941 (2013)
11. Juan, M.G.G., Juan, G.S., Pablo, E.M., Elies, F.G., Emilio, S.O.: Sparse manifold clustering and embedding to discriminant gene expression profiles of glioblastoma and meningioma tumors. Comput. Biol. Med. **43**, 1863–1869 (2013)
12. Mundra, P.A., Rajapakse, J.C.: SVM-RFE with MRMR filterfor gene selection. IEEE Trans. Nanobiosci. **9**(1), 31–37 (2010)
13. He, X., Cai, D., Niyogi, P.: Laplacian score for feature selection. In: Advances in Neural Information Processing Systems, pp. 507–514 (2005)
14. Zhao, Z., Wang, L., Liu, H., Ye, J.: On similarity preserving feature selection. IEEE Trans. Knowl. Data Eng. **25**(3), 619–632 (2011)
15. Nie, F., Huang, H., Cai, X., Ding, C.H.: Efficient and robust feature selection via joint l2, 1-norms minimization. In: Advances in Neural Information Processing Systems, pp. 1813–1821 (2010)

Analysis of Mitochondrial Hsp70 Homolog Amino Acid Sequences of Amitochondriate Organisms Using Apriori and Decision Tree

Jiwon Song[(✉)] and Taeseon Yoon

Hankuk Academy of Foreign Studies/Natural Science,
Yong-in, Republic of Korea
sjw_512@naver.com,
tsyoon@hafs.hs.kr

Abstract. In the endosymbiont hypothesis, the host cell was initially believed to have developed a nucleus and other major characteristics of a eukaryotic cell before taking in the bacteria which in time became mitochondria. This explanation is mainly grounded on existence of early-branching lineages of eukaryote that lack mitochondria. However, an alternative view has gained more support after evolutionary remnants of mitochondria were found in these organisms, where the host cell was yet to evolve the intricacy of eukaryote when the endosymbiosis first began. In this research, we examined the amino acid sequences of mitochondrial Hsp70 found in *Giardia intestinalis*, *Trichomonas vaginalis*, and *Vairimorpha necatrix* from each of the three major amitochondriate lineages. Analyzing the sequences with apriori and decision tree algorithm, we will compare the data with the sequence of the mitochondriate *Mus musculus*, and therefore provide grounds to evaluate the scenarios of endosymbiosis.

Keywords: Mitochondria · Endosymbiosis · Symbiogenesis · Archezoa · Giardia intestinalis · Trichomonas vaginalis · Vairimorpha necatrix · Mitochondrial Hsp70 · mtHsp70 · Heat shock protein 70 · Heat shock protein 70kDa · HSPA9 · Decision tree · Apriori

1 Introduction

The endosymbiont hypothesis assumes that the organelles such as mitochondria and chloroplast evolved from prokaryotes taken in as an endosymbiont. After Russian botanist Mereschowsky presented the basic idea of symbiosis in 1905 [1], Lynn Margulis further advanced the theory with microbiological evidences and presented it in a 1967 paper [2].

Originally, the endosymbiosis resulted in mitochondria was believed to take place after the nucleus evolved. Research on SSU rRNA sequences showed that there were several early branching eukaryote lineages, collectively termed "archezoa", which included protists such as microsporidians, diplomonads, and parabasalids. The main characteristic of these organisms was that they lacked recognizable mitochondria [3, 4],

D.-S. Huang et al. (Eds.): ICIC 2016, Part I, LNCS 9771, pp. 224–233, 2016.
DOI: 10.1007/978-3-319-42291-6_22

and therefore was concluded to evolve out of the common eukaryote ancestor before the endosymbiosis.

However, further phylogenetic analysis of the archezoan sequences showed a considerably high divergence, indicating that the early branching position of the taxa was rather a methodological artifact [5, 6]. These species are now considered as fast-evolving rather than early-branching [7], and the current eukaryotic tree has no single lineage that is clearly the earliest to diverge [8, 9].

Furthermore, evolutionary remnants of mitochondria were found in every seemingly amitochondriate organisms thoroughly inspected [10]. They contained mitochondrion-related organelles (MROs) such as hydrogenosomes [11] and mitosomes [12], along with the genes and proteins typical of mitochondria [13–16].

A different theory gained momentum from these studies, also known as the symbiogenesis scenario [17], supposing that the endosymbiosis of which later became the mitochondria took place before the evolution of the complex characteristics found in existent eukaryotes. Further examination of eukaryotes that lack mitochondria showed a variety of mitochondrial genes, especially the ones related with iron-sulfur (Fe-S) clusters (ISCs). Mitochondrial ISC assembly machinery is engaged in the maturation of all cellular iron-sulfur proteins, and it is speculated that the ISC proteins are accommodated in MROs in eukaryotes that lack mitochondria [18].

70 kD heat shock protein, one of the ISC proteins, are commonly found in Archaebacteria, Eubacteria, and Eukaryota [19]. It functions in correct folding of mature protein, and in the translocation of newly synthesized polypeptides across membranes as well [20]. Among the families of Hsp70, mitochondrial Hsp70, also known as Hspa9, is a result of transfer from mitochondria DNA to nucleus [21]. The mtHsp70 is one of the evolutionary remnants found in amitochondriate eukaryotes of various phylums.

In this work, we compared mitochondrial-like Hsp70 amino acid sequences for eukaryotes with and without mitochondria using apriori and decision tree algorithm. These include 3 eukaryotes that lack mitochondria: *Giardia intestinalis, Trichomonas vaginalis, and Vairimorpha necatrix,* along with *Mus musculus* as a representative of mitochondriate species. By studying their phylogenetic relationship with algorithmic methods, we aim to find the similarities between Hsp70 of amitochondriate and mitochondriate eukaryotes and therefore provide the grounds to evaluate the scenarios of mitochondria endosymbiosis.

2   Materials and Methods

There are mainly three amitochondriate lineages: diplomonads, trichomonads, and microsporidia [14]. We used three amitochondriate species, one from each lineage, which were found to have mitochondrial Hsp70 and have its mRNA fully sequenced. Each mRNA was converted to amino acid sequence then was analyzed using apriori and decision tree algorithms.

2.1   Species

A. *Giardia intestinalis. Giardia intestinalis*, also known as *Giardia lamblia*, is a flagellated protozoan parasite which colonizes and reproduces in the small intestine, and leads to giardiasis. Classified as diplomonad, it contains mitosomes instead of mitochondria. This organelle functions in iron-sulphur protein maturation and is considered as an evolutionary remnant of mitochondria [22]. The phylogenetic analysis of its Hsp70 homolog gene found that the sequence belonged to the eukaryotic mitochondrial clade [23].

B. *Trichomonas vaginalis. Trichomonas vaginalis* is also a flagellated protozoan parasite, and causes trichomoniasis when infected. It is classified as trichomonad, but it differs with *Giardia* in that it contains hydrogenosome as a remnant of mitochondria. Due to this characteristic, the organism produces energy via glycolysis, succeeded by the conversion of pyruvate and malate to hydrogen and acetate [24]. Hsp70 homolog gene was cloned and analyzed, to be found that the protein belonged to the mitochondrial Hsp70 [14].

C. *Vairimorpha necatrix. Vairimorpha necatrix* is a parasite of some species of moth. It belongs to microsporidia, another amitochondriate lineage. The micosporidia used to be classified as protists, but are now considered as fungi. Like the *G. intestinalis,* it doesn't possess mitochondria and has mitosomes instead [25]. Research found out that its Hsp70 gene is an ortholog of the mitochondrial Hsp70, also providing grounds that the species is secondly amitochondriate [26].

D. *Mus musculus. Mus musculus,* commonly known as the house mouse, is studied widely as an important model organism. The researches involving the house mouse include the sequencing of complete mouse reference genome in 2002. This work uses the reference sequence of mtHsp70 mRNA of *Mus musculus* as a representative of mitochondriate species.

2.2   Algorithms

A. Apriori. Apriori is an algorithm used in frequent item set mining to find the general trends of the data, and therefore discover association rules over transactional databases. The algorithm uses a "bottom up" approach, starting from finding frequent individual items and then extending them to larger item sets. These sets, called 'candidates', are each tested against the data. The candidates are removed if they don't appear sufficiently often, while others are extended further to generate candidates with longer lengths. This process continues until no more sets appear with enough frequency. Breadth-first search and hash tree structure are used to find the sets accurately and efficiently.

Used in the analysis of mRNA in this study, the apriori algorithm provides the information of which amino acid appears on each position when divided by window of certain size. If the results of different organisms share a significant similarity, it can be concluded that the proteins of the organisms have a lot in common in the amino acid sequence.

B. Decision Tree. Traditionally, decision tree is one of the decision support systems manually created to show the decisions and their possible consequences as a tree-like diagram. When adopted in machine learning, it is used as a predictive model. The goal of this decision tree is to create a model that can predict the output variable based on several input variables. The overall structure of this model resembles the traditional tree. Each internal node corresponds with an input feature, and the arcs coming from a node corresponds with each one of the possible values of the feature. Learning of decision tree is the process of finding the appropriate rules of classification which can split the given set into proper subsets.

The algorithm is used in this work to reveal the characteristics of the amino acid sequences of each species. If the classification of the data does not appear clearly, it can be inferred that the sequences are highly similar with each other that there isn't a clear distinguishing line between the data.

3 Results

3.1 Apriori

The experiment was done under 9-window, 11-window, and 13-window, and the most dominant amino acids in each position are listed by their frequency in the tables below (Tables 1, 2, 3, 4 and 5).

In all windows, the amino acid sequences of each species appeared to be highly divergent. Glycine and arginine were the most frequent in *G. intestinalis* sequence. In *T. vaginalis* no amino acid was remarkably frequent. V. *necatrix*'s mainly consisted of lysine, leucine, and isoleucine. *M. musculus* mtHsp70 protein was distinctive in that leucine was the most dominant in almost all positions. The patterns of each position also appeared to have no specific commonality.

Table 1. 9-window apriori experiment results (amino acid/frequency)

Species	Position 1		Position 2		Position 3		Position 4		Position 5	
G. intestinalis	R	14	G	7	R	11	G	13	R	9
	S	7	R	7	P	8	D	7	Q	7
							S	7		
T. vaginalis	K	7	V	10	E	8	G	9	E	8
					L	8			T	8
					A	7			K	7
V. necatrix	K	19	L	11	K	16	L	12	I	15
	H	8	I	9	I	10	I	9	K	10
					R	9	T	8	L	10
M. musculus	L	12	K	15	L	17	L	17	L	23
			L	12						

Table 2. 9-window apriori experiment results (amino acid/frequency) (2)

Species	Position 6		Position 7		Position 8		Position 9	
G. intestinalis	S	9	R	9	R	11	G	10
	A	8	A	7	G	10		
	G	8	G	7				
	R	8						
T. vaginalis	K	10	V	9	A	8	G	8
	A	7	I	8	K	8	D	7
			A	7	D	7		
					I	7		
V. necatrix	L	12			K	12	K	14
	K	11			L	8	M	8
	I	8						
M. musculus	L	21	L	20	L	19	L	18
	S	14	R	11	K	12	V	12
	R	12	S	11			S	11

Table 3. 11-window apriori experiment results (amino acid/frequency)

Species	Position 1		Position 2		Position 3		Position 4		Position 5		Position 6	
G. intestinalis	R	8	S	8	R	9	G	9	G	6	G	10
	H	6			H	8	R	7	S	6	R	7
							S	6				
T. vaginalis	K	7	S	8	G	8	I	7	A	7	D	7
			Q	7	E	7	T	7	T	7	E	7
			T	6	N	6	D	6				
			V	6								
V. necatrix	L	11	K	8	K	12	I	8	K	12	R	9
	K	9			I	8	K	7	N	8		
	I	7			L	7	L	7				
							R	7				
M. musculus	L	15	L	16	L	14	L	19	L	11	L	13
	K	9	S	9	S	9	S	9			K	12
							V	9			R	12

3.2   Decision Tree

The experiment was done in 10 folds, also under 9-window, 11-window, and 13-window. Rules of frequency over 0.800 were selected and listed by their positions in the tables below. Class 1, 2, 3, 4 each represents G. intestinalis, T. vaginalis, V. necatrix, and M. musculus.

For all the amitochondriate species, the most rules were found in 9-window experiment. Therefore, it can be inferred that the 9-window best reveals the distinctive

Table 4. 11-window apriori experiment results (amino acid/frequency) (2)

Species	Position 7		Position 8		Position 9		Position 10		Position 11	
G. intestinalis	P	6	G	9	R	11	R	10	G	7
	R	6	A	7			G	6	R	7
							V	6	Q	6
									S	6
T. vaginalis	K	8	I	6	I	7	I	8	A	9
	V	8	V	6	V	7	G	7	V	6
	A	7			K	6	K	6		
V. necatrix	I	14	K	9	L	11	K	11	I	8
	K	13	R	9	K	7	L	9		
M. musculus	L	12	L	19	L	12	L	12	L	16
					R	12	Q	10	P	9
					S	10				
					K	9				

Table 5. 13-window apriori experiment results (amino acid/frequency)

Species	Position 1		Position 2		Position 3		Position 4		Position 5		Position 6	
G. intestinalis	G	9	G	6	G	7	G	6	Q	5	R	12
	S	6	Q	5	R	6	A	5				
	R	5	R	5			R	5				
T. vaginalis	A	5	V	6	G	6	D	7	A	9	V	8
	E	5	G	5	T	6	I	5	T	6	G	7
	K	5			K	5			E	5		
	S	5							F	5		
	V	5										
V. necatrix	I	6	K	10	K	7	I	9	I	6	I	9
	K	6			L	7	K	7	K	6	K	7
	R	6			R	7	L	6			V	6
M. musculus	L	13	L	13	L	11	S	9	L	15	L	12
	S	10	R	10			L	8	K	8	I	9
	T	8									R	9

characteristics the proteins of each species have. For the exception of *G. intestinalis* in 9-window, all sequences of amitochondriate species appeared to have few distinguishing aspects. On contrary, the sequence of *M. musculus* was found to have much more unique patterns. The results shown in Table 6 also fit with the data of Table 7. Since the organisms lacking mitochondria had fewer classification rules, their sequences tended to be classified as *Mus musculus* rather than other amitochondriate species, and in some cases, even themselves. The *M. musculus* sequence tended to be more distinctive and was mostly classified as itself (Table 8).

Table 6. 13-window apriori experiment results (amino acid/frequency) (2)

Species	Position 7	Position 8	Position 9	Position 10	Position 11	Position 12	Position 13
G. intestinalis	A 6	S 9	R 8	R 8	R 8	R 8	G 8
	S 6	E 5	G 6	A 5	H 5	P 5	L 7
	R 5		L 5	V 5		S 5	
T. vaginalis	K 7	K 8	D 6	K 8	D 5	A 6	L 7
	T 6	Q 6	I 6		K 5	E 6	T 6
		N 5	K 6		Q 5	I 5	A 5
			A 5			L 5	I 5
			E 5				
V. necatrix	K 8	I 6	L 9	K 11	K 10	K 10	K 7
		K 6	R 6	I 7	L 8	Y 6	L 6
		R 6					
M. musculus	L 14	L 20	V 9	L 9	L 19	R 10	L 9
	S 10		L 8	T 8		K 8	
			R 8			L 8	

Table 7. Decision tree experiment results

Class 1	Class 2	Class 3	Class 4
9-window			
Position1 = R	Position3 = A	Position1 = V	Position1 = L
Position9 = C	Position9 = Q	Position1 = K	Position6 = K
Position4 = S	Position6 = K	Position6 = I	Position1 = S
Position6 = G	Position7 = K	Position3 = K	Position9 = A
Position5 = R	Position6 = I	Position6 = E	Position4 = W
Position9 = H	Position7 = V	Position9 = C	Position5 = W
Position5 = G		Position9 = M	Position5 = L
Position9 = G			Position6 = C
Position6 = S			Position6 = G
Position7 = P			Position9 = S
Position6 = S			Position6 = L
Position7 = G			Position9 = P
11-window			
Position6 = G	Position3 = E	Position3 = K	Position1 = S
Position10 = G	Position4 = K	Position4 = I	Position10 = L
	Position7 = A	Position4 = I	Position3 = W
	Position10 = I	Position8 = R	Position3 = L
		Position10 = K	Position10 = E
			Position3 = L
			Position10 = V
			Position4 = L
			Position10 = L
			Position4 = L
			Position10 = T
			Position5 = L
			Position10 = V
			Position10 = W
13-window			
Position9 = H	Position5 = A	Position2 = K	Position1 = L
	Position9 = A	Position8 = I	Position1 = S
		Position2 = K	Position8 = S
		Position9 = R	Position1 = T
			Position9 = L
			Position4 = K
			Position8 = K
			Position7 = W
			Position9 = F
			Position13 = W

Table 8. decision tree experiment results (2)

(a)	(b)	(c)	(d)	←classified as
19	15	9	29	(a): class 1
10	16	16	27	(b): class 2
5	7	36	33	(c): class 3
9	10	19	76	(d): class 4

4   Conclusion

By analysis using apriori and decision tree, we compared the amino acid sequence of mitochondrial Hsp70 in species with and without mitochondria. In apriori experiment results, no specific similarity was found in all sequences. This can be interpreted as a result of long time evolution and divergence. However, in decision tree results, the sequence of those that lack mitochondria tended to have few distinctive characteristics and be classified as *M. musculus* as well as itself. This seems to be because after diverging from the common ancestor, *M. musculus* has developed a lot of different characteristics as it evolved into a more complex organism while others didn't.

In short, the mitochondrial Hsp70 of *G. intestinalis*, *T. vaginalis*, *V. necatrix*, and *M. musculus* are all clearly different from each other. However, the three amitochondriate species do share certain similarities with *M. musculus* while *M. musculus* tends to have more unique traits that others do not have.

When and how endosymbiosis exactly happened still remains a mystery, and many scenarios are being proposed and evaluated. We believe that further research on mitochondrial Hsp70 using other algorithm such as support vector machine will lend meaningful results as this work did. Furthermore, algorithmic approach on other evolutionary remnants of mitochondria will be able to provide more accurate understanding of these genes and proteins, shedding light on the evolution of eukaryotes and the truth about mitochondrial evolution.

References

1. Mereschkowsky, C.: Über natur und ursprung der chromatophoren im pflanzenreiche (1905)
2. Sagan, L.: On the origin of mitosing cells. J. Theoret. Biol. **14**(3), 225–IN6 (1967)
3. Cavalier-Smith, T.: A 6-kingdom classification and a unified phylogeny. In: Endocytobiology II, pp. 1027–1034 (1983)
4. Cavalier-Smith, T.: Kingdom protozoa and its 18 phyla. Microbiol. Rev. **57**(4), 953–994 (1993)
5. Hirt, R.P., et al.: Microsporidia are related to Fungi: evidence from the largest subunit of RNA polymerase II and other proteins. Proc. Nat. Acad. Sci. **96**(2), 580–585 (1999)
6. Keeling, P.J., Luker, M.A., Palmer, J.D.: Evidence from beta-tubulin phylogeny that microsporidia evolved from within the fungi. Mol. Biol. Evol. **17**(1), 23–31 (2000)
7. Philippe, H.: Early–branching or fast–evolving eukaryotes? An answer based on slowly evolving positions. Proc. Roy. Soc. Lond. B: Biolog. Sci. **267**(1449), 1213–1221 (2000)

8. Keeling, M.J., Eames, K.T.: Networks and epidemic models. J. Roy. Soc. Interface **2**(4), 295–307 (2005)
9. Koonin, E.V.: The origin and early evolution of eukaryotes in the light of phylogenomics. Genome Biol. **11**(5), 209 (2010)
10. Embley, T.M., Hirt, R.P.: Early branching eukaryotes? Curr. Opin. Genet. Dev. **8**(6), 624–629 (1998)
11. Müller, M., Lindmark, D.G.: Respiration of hydrogenosomes of Tritrichomonas foetus. II. Effect of CoA on pyruvate oxidation. J. Biolog. Chem. **253**(4), 1215–1218 (1978)
12. Tovar, J., Fischer, A., Clark, C.G.: The mitosome, a novel organelle related to mitochondria in the amitochondrial parasite Entamoeba histolytica. Mol. Microbiol. **32**(5), 1013–1021 (1999)
13. Bui, E.T., Bradley, P.J., Johnson, P.J.: A common evolutionary origin for mitochondria and hydrogenosomes. Proc. Nat. Acad. Sci. **93**(18), 9651–9656 (1996)
14. Germot, A., Philippe, H., Le Guyader, H.: Presence of a mitochondrial-type 70-kDa heat shock protein in Trichomonas vaginalis suggests a very early mitochondrial endosymbiosis in eukaryotes. Proc. Nat. Acad. Sci. **93**(25), 14614–14617 (1996)
15. Horner, D.S., et al.: Molecular data suggest an early acquisition of the mitochondrion endosymbiont. Proc. Roy. Soc. Lond. B: Biolog. Sci. **263**(1373), 1053–1059 (1996)
16. Roger, A.J., Clark, C.G., Doolittle, W.F.: A possible mitochondrial gene in the early-branching amitochondriate protist Trichomonas vaginalis. Proc. Nat. Acad. Sci. **93** (25), 14618–14622 (1996)
17. Gray, M.W.: Mitochondrial evolution. Cold Spring Harb. Perspect. Biol. **4**(9), a011403 (2012)
18. Lill, R., Mühlenhoff, U.: Iron–sulfur-protein biogenesis in eukaryotes. Trends Biochem. Sci. **30**(3), 133–141 (2005)
19. Gupta, R.S., Singh, B.: Phylogenetic analysis of 70 kD heat shock protein sequences suggests a chimeric origin for the eukaryotic cell nucleus. Curr. Biol. **4**(12), 1104–1114 (1994)
20. Hartl, F.-U., Hlodan, R., Langer, T.: Molecular chaperones in protein folding: the art of avoiding sticky situations. Trends Biochem. Sci. **19**(1), 20–25 (1994)
21. Boorstein, W.R., Ziegelhoffer, T., Craig, E.A.: Molecular evolution of the Hsp70 multigene family. J. Mol. Evol. **38**(1), 1–17 (1994)
22. Tovar, J., et al.: Mitochondrial remnant organelles of Giardia function in iron-sulphur protein maturation. Nature **426**(6963), 172–176 (2003)
23. Arisue, N., et al.: Mitochondrial-type Hsp70 genes of the amitochondriate protists, Giardia intestinalis, Entamoeba histolytica and two microsporidians. Parasitol. Int. **51**(1), 9–16 (2002)
24. Steinbüchel, A., Müller, M.: Anaerobic pyruvate metabolism of Tritrichomonas foetus and Trichomonas vaginalis hydrogenosomes. Mol. Biochem. Parasitol. **20**(1), 57–65 (1986)
25. Vávra, J.: "Polar vesicles" of microsporidia are mitochondrial remnants ("mitosomes"). Folia Parasitol. **52**(1/2), 193–195 (2005)
26. Hirt, R.P., et al.: A mitochondrial Hsp70 orthologue in Vairimorpha necatrix: molecular evidence that microsporidia once contained mitochondria. Curr. Biol. **7**(12), 995–998 (1997)

Identifying miRNA-mRNA Regulatory Modules Based on Overlapping Neighborhood Expansion from Multiple Types of Genomic Data

Jiawei Luo$^{(\boxtimes)}$, Bin Liu, Buwen Cao, and Shulin Wang

College of Computer Science and Electronics Engineering & Collaboration
and Innovation Center for Digital Chinese Medicine of 2011 Project of Colleges
and Universities in Hunan Province, Hunan University, Changsha 410082, China
luojiawei@hnu.edu.cn

Abstract. MicroRNA (miRNA)-mRNA regulatory modules are key entities to disorders. Several computational methods are developed to identify miRNA-mRNA modules. Although these methods have achieved ideal performance, the number of modules needed to be predefined. Therefore, identification of modules is still computationally challenging. In this study, a new algorithm called MiRMD (miRNA-mRNA Regulatory Modules Detection) is presented to identify miRNA-mRNA modules, which do not need to predefine the number of modules. Firstly, a miRNA-mRNA regulatory network is constructed, then core structures are detected in this network by merging cohesive modules. Next, some overlapping neighbor nodes are added into the cores according to the density. Finally, some overlap modules are filtered. The experimental results based on three cancers datasets show that modules identified by MiRMD are more coherent and functional enriched than the other two methods according to MiMEC and GO enrichment. Particularly, modules identified by our method are strongly implicated in cancer.

Keywords: miRNA-mRNA regulatory modules · Cores · Merging · Density

1 Introduction

MicroRNAs (miRNAs) are small (~ 22 nucleotides) non-protein-coding RNAs that cause mRNA degradation and translational inhibition at the 3 untranslated region by base pair with target mRNAs [1]. In the past ten years, some studies showed that miRNAs are involved in the process of development, cell differentiation, proliferation, apoptosis [2–4]. Recent researches have reported miRNA expressed differentially in many cancers [5, 6]. Therefore, the dysregulation of miRNA may lead to diseases.

Several researchers have put focus on how miRNA, genes and protein interact with each others on a system level. For example, miRNA regulates their targets in cellular networks globally [7, 8] and miRNA regulates in cellular pathways together with other miRNAs [9, 10]. However, a better understanding of the mechanisms of miRNA regulation is still needed.

© Springer International Publishing Switzerland 2016
D.-S. Huang et al. (Eds.): ICIC 2016, Part I, LNCS 9771, pp. 234–246, 2016.
DOI: 10.1007/978-3-319-42291-6_23

The modular organization of biological networks is crucial for the understanding of biological pathways and complex cellular system [11]. More important, the module structures have greatly stimulated the development of miRNA-based drugs [12]. Moreover, the increasing expression files of miRNA and mRNA helps further research on the regulatory relationship between miRNA and mRNA. For example, Joung et al. [13] presented a method to identify the highest probabilities functional modules from the candidate modules which is decided by the target sites of miRNAs and the respective miRNA-mRNA expression correlation (MiMEC) [14]. However, their algorithm needs prior setting of several parameters of the fitness function, which would substantially affect the efficiency. Liu et al. [15] proposed a method based on proba-bilistic graphics model called Correspondence Latent Dirichlet Allocation to discover miRNA-mRNA functional modules by integrating multiple types of expression pro-files. However, this method requires a predefined number of modules.

More recently, Zhang et al. [16] proposed a framework called SNMNMF which factorizes multiple non-negative matrix to reconstruct miRNA regulatory modules by using three types of data consisting of the expression profiles, the target sites infor-mation and gene-gene interaction network. SNMNMF also needs to predefine the number of modules, and cannot guarantee all of the identified modules contain both miRNA and mRNA. Pio et al. [17] proposed a novel biclustering algorithm which can automatically determine the number of biclusters. The algorithm starts with an initial set of bicliques and then it iteratively get the highly cohesive biclusters with a bottom-up strategy. Nonetheless, it requires confirmed interactions, does not consider miRNA, mRNA expression profiles and the target site information. In addition, Li et al. [18] proposed a new model called Mirsynergy to detect synergistic miRNA-mRNA modules by two-stage clustering. However, the effectiveness of Mirsynergy is sharply decreased when it is used in a relatively dense miRNA regulatory network.

In this article, we propose a novel framework called MiRMD (miRNA-mRNA Regulatory Modules Detection) which integrates miRNA expression profiles, miRNA target site information, mRNA expression profiles and gene-gene interaction networks to identify miRNA-mRNA regulatory modules, and the number of modules do not have to be predefined. Firstly, we get several high quality cores by merging pair of modules iteratively with maximum cohesiveness score, then we will modify each core structure by adding neighbor nodes if the density is greater than a threshold value, finally filter the overlap modules. The proposed method is tested on three cancer datasets including ovarian, breast and thyroid cancer. Comparing with some state-of-the-art algorithms, we find that the miRNA-mRNA regulatory modules iden-tified by our method are more functionally enriched and negative MiMEC. Moreover, the modules identified by MiRMD are strongly implicated in cancer.

2   Method

We propose an algorithm to detect miRNA-mRNA regulatory modules based on overlapping neighborhood expansion, called MiRMD, which can be divided into four major phases: (1) Constructing the miRNA-mRNA regulatory network; (2) Detecting the cores in the miRNA-mRNA regulatory network; (3) Extending the core structures by adding their neighbor nodes; (4) Filtering the overlap modules.

2.1   Construction of the miRNA-mRNA Regulatory Network

Several computational prediction algorithms have been proposed to get targets of miRNA, such as TargetScan [19], PicTar [20], Pearson correlation coefficient (PCC) [16], GenMiR++ [21] and LASSO [22]. According to Ref. [18], LASSO can achieve the better performance than the other methods, which is employed to predict the targets of miRNAs in this study. Based on the LASSO linear regression model which is described as Eq. (1), we can achieve the Eqs. (2) and (3).

$$\hat{\beta} = \arg \ \min_\beta \left\{ \sum_{i=1}^n \left(y_i - \sum x_{ij}\beta_j \right)^2 + \lambda \sum_{j=1}^p |\beta_j| \right\} \tag{1}$$

$$x_{it} \sim w_{i0} + \sum_j w_{ij}(z_{jt}c_{ij}) \tag{2}$$

$$\sum_j |w_{ij}| \le \lambda \tag{3}$$

where n means the number of all miRNAs, y_i means the expression level of the i-th miRNA, j means the j-th mRNA, x_{ij} represents the expression level of the j-th potential targeting mRNA of the i-th miRNA, p represents the number of all potential targeting mRNAs of the i-th miRNA, t is the index of samples, $x_{i,t}$ is the expression profile of mRNAs, $z_{j,t}$ is the expression profile of miRNAs which have one or more target sites, $w_{i,0}$ presents the cutoff value, $w_{i,j}$ means the fitted linear coefficients, and $c_{i,j}$ is the number of target sites in mRNA i corresponding to miRNA j.

LASSO is implemented with R package glmnet and the tune parameter λ is set by using cross-validation [18]. Since miRNAs suppress the expression of mRNAs, all of the positive coefficients of W are set to zero, then interactions corresponding to negative values are regarded as the interactions between miRNAs and mRNAs.

2.2   Detecting the Cores

In the second phase, the cores are detected in the miRNA regulatory network. Algorithm 1 presents the detailed process. First, each miRNA and the corresponding target mRNAs are regarded as initial module (step 1). Second, for each module, we measure the closeness between pairs of modules (V_i, V_j). Base on the condition of a good core according to Ref. [17], we consider those pairs of modules which satisfy the condition Eq. (4) as the candidate modules for the merging process (steps 2–6).

$$\left| V_i^{mir} \cap V_j^{mir} \right| \le \alpha \quad \left| V_i^m \cup V_j^m \right| \ge \theta \tag{4}$$

where V_i^{mir} is the set of miRNAs in module V_i, V_i^m is the set of mRNAs in module V_i.

Third, we compute the cohesiveness between pairs of candidate modules to generate new modules (step 4 and 13), the cohesiveness score function is described as:

$$q(V_i, V_j) = \frac{\left|V_i^m \cap V_j^m\right|}{\left|V_i^m \cup V_j^m\right|} \times \frac{\sum_{x \in M} \sum_{y \in N} A(x, y)}{|M| \times |N|} \tag{5}$$

where A is the adjacency matrix, N is the set of mRNAs, and M is the set of miRNAs in the module V_{new}. Among those pairs of candidate modules (V_i, V_j), the one with maximum cohesiveness score is selected to merge at each iteration (steps 8–10), the merging process is described as follows:

$$V_{new}^{mir} = V_i^{mir} \cup V_j^{mir} \quad V_{new}^m = V_i^m \cap V_j^m \tag{6}$$

```
Algorithm 1: Detecting the cores;
Input: miRNA-mRNA matrix A; the threshold α, θ;
Output: cores V;
1  V'←{V₁, V₂, . . ., Vₘ}; M_can←ϕ; V←V'
2  for pairs of modules Vᵢ, Vⱼ ∈ V do
3     if |Vᵢᵐ ∪ Vⱼᵐ| ≥ θ then
4        M_can←M_can ∪ {<Vᵢ, Vⱼ, q(Vᵢ, Vⱼ)>}
5     end
6  end
7  while |M_can| > 0 do
8     <Vₘ, Vₙ, q(Vₘ, Vₙ)> ← getBest(M_can) according to score q
9     V_new←merge(Vₘ, Vₙ); V←V\(Vₘ, Vₙ)
10       for each module Vₓ ∈ V do
11          if |Vₓᵐⁱʳ ∩ V_newᵐⁱʳ| ≤ α and |Vᵢᵐ ∪ Vⱼᵐ| ≥ θ then
12  M_can←M_can ∪ {<Vₓ, V_new, q(Vₓ, V_new)>}
13          end
14       end
15 V←V ∪ {V_new};
16 M_can←M_can\ <Vₘ, Vₙ, q(Vₘ, Vₙ)>
17 end
18 V←V\(V ∩ V')
19 return V
```

When a new module is generated, we recheck whether it satisfy the merging condition and recalculate cohesiveness score with other modules in V (steps 11–14). If a module cannot be merged with others, the algorithm would be keep it as the final core. The iterative process stops when there is no new module.

2.3 Extending Core Structures to Co-regulated Modules

Prior to extending each core structure to miRNA-mRNA regulatory module, we present the density formula of the different node type, which is described as follows:

$$dens(V_i) = \frac{\sum_{x \in M} \sum_{y \in N} A(x,y) + \sum_{i \in N} \sum_{j \in N} H(i,j)}{|N| \times (|M| + |N| - 1)} \tag{7}$$

where N is the set of mRNAs, M is the set of miRNAs in the module, and H is the adjacency matrix of the gene-gene interaction network. For H_{ij}, 0 means no interaction and 1 means the reliable interaction between gene i and j ($i \neq j$).

```
Algorithm 2: Extending core structures
Input: interaction matrix A, H, cores V, threshold t;
Output: the set of modules LC;
1  LC←φ
2  for each V_i∈V do
3     lc←V_i; flag←0
4     if dens(V_i)<t then
5        insert lc into LC
6     end
7     else
8        while flag==0 do
9           ms←the neighbor mRNAs of nodes in V_i
10          mis←the miRNAs which regulate mRNAs in V_i
11          v_0 ←argmaxdens(ms∪ mis)
12          V←lc∪ v_0
13          if dens(V)<t then
14             flag←1; insert lc into LC
15          end
16          else
17             lc←V
18          end
19       end
20    end
21 end
22 return LC;
```

After obtaining the cores, we extend each core by iteratively adding neighbor miRNAs and mRNAs in terms of the density. As described in Algorithm 2, in extending each module V_i, we check the density of V_i first and keep the module whose density is smaller than a threshold t as the final module (step 5). Then for others, we calculate the density of $V_i' = V_i \cup \{v_0\}$ by considering every neighbor node v_0 (steps 8–12). If the node with the maximum density ensure the density of V_i' is greater than t, then we add it to form V_i', let $V_i = V_i'$ (step 17). The entire process terminates when all of the neighbor nodes are considered (steps 13–15).

2.4 Filtering the Overlap Modules

Algorithm 1 allows some modules to be merged with different modules and Algorithm 2 allows the adding of overlapping neighbor nodes. We achieve several overlap modules, and the overlap score of every pair of modules is defined as [23]:

$$\omega(V_i, V_j) = \frac{|V_i \cap V_j|^2}{|V_i| \times |V_j|} \tag{8}$$

where V_i is the elements in module i, and V_j is the elements in module j. If the overlap score of the two modules are greater than the predefined threshold, which is set as 0.8 [24], we keep the module with a higher density as final module.

3 Experiments and Results

3.1 Data Collection and Pre-processing

In order to show the performance of our method, we applied it with SNMNMF and Mirsynergy on three cancers datasets (OV, BRCA, and THCA). The OV dataset was downloaded from Ref. [16]. In this dataset, the m/miRNAs expression profiles were originally obtained from TCGA, the target site information were originally processed from MicroCosm (v5) and GGI data were originally downloaded from TRANSFAC [25]. For the latter two datasets, the expression data were downloaded from TCGA and further log2-transformed and mean-centred. The BRCA contains 331 samples, 688 miRNAs and 9174 mRNAs, the THCA contains 551samples, 696 miRNAs and 9759 mRNAs. Since TargetScan can provide more efficiently target site information [26] besides having higher accuracy and sensitivity than others [22], the target site information for the BRCA and THCA were obtained from TargetScan database website. GGI data consist of Transcription factor binding site (TFBS) and PPI data processed from TRANSFAC and BioGrid [27] databases respectively.

3.2 Comparison with Other Methods

Since Mirsynergy can obtain better quality modules than others, and SNMNMF is one of the most cited works in the recent literatures, we compared MiRMD with those two methods. We set the initial numbers of modules to 50 which is the same as the value in SNMNMF. We got 58, 62, 68 modules in OV, BRCA, and THCA datasets, respectively, and the summary of the result can be seen in Table 1.

3.3 Verifying Co-expression of miRNA-mRNA Modules

Since miRNAs repress their target mRNAs, the negativity of MiMEC is a reasonable indicator of the quality [18]. Then we calculated the averaged pairwise MiMEC of each module on the three datasets. As seen in Table 1, the averaged MiMEC of over all of the MiRMD modules (−0.068, −0.127, −0.109) are more negative than Mirsynergy (−0.05, −0.08, −0.08), especially three times as more negative as those of SNMNMF (−0.04, −0.04) for BRCA and THCA. Therefore, modules identified by MiRMD have better quality. Moreover, some experimentally confirmed interactions [28] can only be identified by MiRMD in BRCA were shown in Table 2, that is to say, MiRMD can identify meaningful biological correlations between miRNAs and genes.

3.4   Assessing the Functional Enrichment of Modules

To investigate the biological relevance of the modules, we calculated the biological

Table 1. Comparison of MiMRD with SNMNMF and Misynergy

Cancer	Method	M#	$\overline{mi}$	$\overline{m}$	MiMEC	Sensitivity	Time
OV	MiRMD	58	4.23	12.3	−0.068	0.273	1
	Mirsynergy	84	4.76	7.57	−0.05	0.186	1
	SNMNMF	49	4.12	81.37	0.07	0.153	24+
BRCA	MiRMD	62	5.45	16.08	−0.127	0.172	1
	Mirsynergy	53	5.77	24.15	−0.08	0.165	1.5
	SNMNMF	39	2.62	71.56	−0.04	0.143	24+
THCA	MiRMD	68	5.23	24.37	−0.109	0.168	1.5
	Mirsynergy	50	7.6	32.26	−0.08	0.161	2
	SNMNMF	39	2.23	74.82	−0.04	0.156	24+

Note: M#: the number of modules; $\overline{mi}$: average miRNA per module; $\overline{m}$: average mRNA per module; MiMEC: miRNA-mRNA expression correlation; Sensitivity: the percentage of significant GO annotation terms; Time: the number of hours algorithms took to run.

functional enrichment in Gene Ontology (GO) terms or pathways (KEGG), and downloaded 2007 GO terms of biological processes (GOBPs) and 9438 genes from R package *biomaRt*. Then *enrichGO* and *enrichKEGG* function in package *clusterProfiler* were used to analysis GO and KEGG pathway enrichment by correcting the *P-values* to false discovery rates (FDR), respectively. Here, we set *pAdjustMethod* to *BH* and *pvalueCutoff* to 0.05. Then we compared the meaningful functional enrichment of the identified modules from MiRMD, Mirsynergy, and SNMNMF, the GO enrichment score was used to measure the functional enrichment, which is described as follows [18]:

$$GOES = \frac{M}{G}\sum_{g}^{G} -\log(FDR) \tag{9}$$

where g is a specific significant GOBP, M is the number of the modules, and G is the number of all of the significant GOBPs.

As shown in Fig. 1, the averaged enrichment scores over all of the modules identified by MiRMD are higher than the scores from others in all three test datasets (for OV, the *GOES* is 15.64, 7.51, 15.84, respectively; for BRCA, the *GOES* is 8.74, 5.56, 10.66, respectively; for THCA, the *GOES* is 8.04, 6.73, 11.42, respectively), which demonstrated that the modules identified by MiRMD are more functionally enriched.

Moreover, we compared the sensitivity of MiRMD with other two methods. The sensitivity proposed by Chen *et al.* [29] indicates the percentage of significant GO

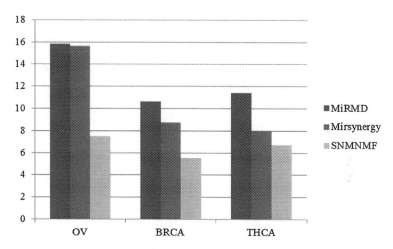

Fig. 1. *GOES* of three methods tested on three cancers. (Color figure online)

Table 2. Some experimentally confirmed interactions identified only by MiRMD in BRCA.

ID	Confirmed interactions
15	hsa-miR-22: ERBB3, FAM171A1, GBE1, GLYR1
	hsa-miR-100: C11orf52, CASZ1, CLDN3, FXN
25	hsa-miR-30a: IKBIP, ITPK1, KIAA0040
	hsa-miR-1238: ADAMTS14, ADAMTS6, C1orf135, CA10
44	hsa-miR-141: CFL2, CXCL2, EMP2, FOXA1, LHFP, MATR3

annotation terms, which is defined as Eq. (10). As shown in Table 1, the sensitivity is superior than other two methods, that is to say, MiRMD can identify modules with more affluent biological meanings.

$$Sensitivity = \frac{the\ number\ of\ significant\ GO\ annotations}{the\ number\ of\ all\ GO\ annotations} \qquad (10)$$

3.5   Parameters Analysis

According to Ref. [17], a good core should contain approximately avg_mirna miRNAs, while keeping at least min_mrna mRNAs, therefore, the threshold α was varied from 1 to $\lfloor avg\_mirna \rfloor$, meanwhile the threshold θ was changed from $\lceil min\_mrna \rceil$ to $\lceil max\_mrna \rceil$, where avg_mirna is the average number of the miRNAs which target each mRNA, min_mrna is the minimize number of the mRNAs targeted by each miRNA, max_mrna is the maximum number of the mRNAs which targeted by each miRNA. For OV dataset, $\lfloor avg\_mirna \rfloor$ is 2, $\lceil min\_mrna \rceil$ is 2, and $\lceil max\_mrna \rceil$ is 9, as shown in Table 3, when α is 2 ($\lfloor avg\_mirna \rfloor$) and θ is 2 ($\lceil min\_mrna \rceil$), the specificity Eq. (11) of MiMRD is the highest. Moreover, the results of analysis on BRCA and THCA were same as above. Therefore, we set α to avg_mirna and θ to min_mrna for all datasets. Besides, the density threshold t was set to 0.4 according to Ref. [23].

$$Specificity = \frac{the\ number\ of\ significant\ modules}{the\ number\ of\ all\ modules} \tag{11}$$

Table 3. The analysis of threshold α and θ of MiMRD on OV.

α	θ	Num.	Specificity
1	2	88	0.665
	3	97	0.628
	4	107	0.603
	5	115	0.574
	6	132	0.536
	7	145	0.484
	8	152	0.406
	9	164	0.375
2	2	70	0.828
	3	76	0.713
	4	80	0.692
	5	82	0.684
	6	86	0.677
	7	95	0.641
	8	102	0.615
	9	114	0.578

3.6   Case Study

Since experimental data consist of three cancers datasets, we expected the modules identified are closely related to cancer. To further demonstrate the power of our method, we proceeded to analyze some significance modules related to ovarian cancer. 147 miRNAs were downloaded from Ref. [30] as our cancer miRNA standard dataset and in which 41 miRNAs are closely related to ovarian cancer. The identified modules contain 123 different miRNAs, 58 of which are included in the standard set of cancer miRNAs, which is significant ($P = 1.227 \times 10^{-6}$). Particularly, 24 miRNAs are related to ovarian cancer, which is enrichment significant ($P = 2.475 \times 10^{-7}$) (Fig. 2). Therefore, we can observe that miRNAs in modules can closely related to cancer.

For example, a module (Fig. 3) contains five miRNAs, four of them belong to the cancer miRNA standard sets. mir-22, mir-199a-5p, mir-145, mir-10b play crucial roles in ovarian cancer [25]. Moreover, genes of this module are enriched in some kinds of pathways which are related to cancer such as *Rap1 signaling pathway, Pathways in cancer, and PI3K-Akt signaling pathway.*

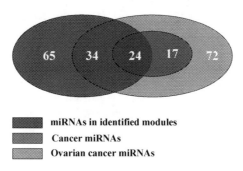

Fig. 2. Analysis of miRNA involved in modules detected by our method. (Color figure online)

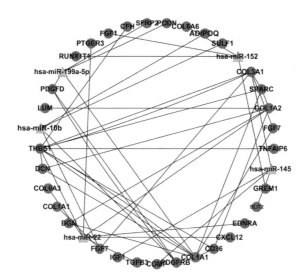

Fig. 3. An example of the detected modules in OV. The pink nodes represent mRNAs, green nodes represent miRNA, black edges present gene-gene interactions and red edges represent miRNA-mRNA regulatory interactions. (Color figure online)

4 Conclusion

Recent studies show that miRNAs are important regulators in cancers, and work with their target gene in some important biological processes. Many methods have been proposed to identify the modules, however, most of them needed to predefine the number of modules.

In this paper, we proposed a new method called MiRMD to detect miRNA-mRNA regulatory modules based on the expression profiles of m/miRNA, the target sites information and the gene-gene interactions. We detected some cores in miRNA regulatory networks by merging some cohesive modules. Firstly, then extended the cores by adding the neighbor miRNAs and mRNAs according to the density, and meanwhile keeping the number of modules constant, finally filtered some overlap modules. Compared with SNMNMF, the main advantage of our method is that MiRMD can automatically determine the number of modules.

To demonstrate the performance of MiRMD, we compared it with SNMNMF and Mirsynergy on OV, BRCA and THCA datasets. We showed that the MiMEC of modules identified by our method is more negative than other methods, which demonstrated the effectiveness of our method. Moreover, the modules identified by our method were enriched in significant GO and some of KEGG pathways. Finally, the modules identified by MiRMD are strongly implicated in cancer.

Recent research suggests that the collective relationships between groups of miRNAs and mRNAs are more readily interpreted than those between individual miRNAs and mRNAs, however, which did not be considered in MiRMD. In the future, we will develop a framework to discover modules based on the collective relationships.

Acknowledgments. The authors would like to acknowledge the assistance provided by National Natural Science Foundation of China (Grant nos. 61572180 and 61472467) and Hunan Provincial Natural Science Foundation of China (Grant no. 13JJ2017).

References

1. Bartel, D.P.: MicroRNAs: target recognition and regulatory functions. Cell **136**, 215–233 (2009)
2. Ambros, V.: The functions of animal micrornas. Nature **431**, 350–355 (2004)
3. Du, T., Zamore, P.D.: Beginning to understand microRNA function. Cell Res. **17**, 661–663 (2007)
4. Bushati, N., Cohen, S.M.: Microrna functions. Annu. Rev. Cell Dev. Biol. **23**, 175–205 (2007)
5. Iorio, M.V., Ferracin, M., Liu, C.G., Veronese, A., Spizzo, R., Sabbioni, S., Magri, E., Pedriali, M., Fabbri, M., Campiglio, M., Mnard, S., Palazzo, J.P., Rosenberg, A., Musiani, P., Volinia, S., Nenci, I., Calin, G.A., Querzoli, P., Negrini, M., Croce, C.M.: MicroRNA gene expression deregulation in human breast cancer. Cancer Res. **65**, 7065–7070 (2005)
6. Porkka, K.P., Pfeiffer, M.J., Waltering, K.K., Vessella, R.L., Tammela, T.L.J., Visakorpi, T.: microRNA expression profiling in prostate cancer. Cancer Res. **67**, 6130–6135 (2007)

7. Cui, Q., Yu, Z., Purisima, E.O., Wang, W.: Principles of microRNA regulation of a human cellular signaling network. Mol. Syst. Bio. **2**, 46 (2006)
8. Liang, H., Li, W.H.: microRNA regulation of human protein–protein interaction network. RNA **13**, 1402–1408 (2007)
9. Gusev, Y., Schmittgen, T.D., Lerner, M., Postier, R., Brackett, D.: Computational analysis of biological functions and pathways collectively targeted by co-expressed microRNAs in Cancer. BMC Bioinform. **8**, S16 (2007)
10. Xu, J., Wong, C.: A computational screen for mouse signaling pathways targeted by microRNA clusters. RNA **14**, 1276–1283 (2008)
11. Hartwell, L.H., John, J.H., Leibler, S., Murray, A.W.: From molecular to modular cell biology. Nature **402**, C47–C52 (1999)
12. Croce, C.M.: Causes and consequences of microRNA dysregulation in cancer. Nat. Rev. Genet. **10**, 704–714 (2009)
13. Joung, J.G., Hwang, K.-B., Nam, J.-W., Zhang, B.T.: Discovery of microRNA-mRNA modules via population based probabilistic learning. Bioinformatics **23**, 1141–1147 (2007)
14. Shalgi, R., Lieber, D., Oren, M., Pilpel, Y.: Global and local architecture of the mammalian microRNA transcription factor regulatory network. PLoS Comput. Biol. **3**, e131 (2007)
15. Liu, B., Liu, L., Tsykin, A., Goodall, G.J., Green, J.E., Zhu, M., Kim, C.H., Li, J.: Identifying functional miRNA-mRNA regulatory modules with correspondence latent Dirichlet allocation. Bioinformatics **26**, 3015–3111 (2010)
16. Zhang, S., Li, Q., Liu, J., Zhou, X.J.: A novel computational framework for simultaneous integration of multiple types of genomic data to identify microRNA-gene regulatory modules. Bioinformatics **27**, i401–i409 (2011)
17. Pio, G., Ceci, M., D'Elia, D., Loglisci, C., Malerba, D.: A novel biclustering algorithm for the discovery of meaningful biological correlations between microRNAs and their target genes. BMC Bioinform. **14**, S8 (2013)
18. Li, Y., Liang, C., Wong, K.C., Luo, J., Zhang, L.: Mirsynergy: detecting synergistic miRNA regulatory modules by overlapping neighbourhood expansion. Bioinformatics **30**, i2627–i2635 (2014)
19. Lewis, B.P., Burge, C.B., Bartel, D.P.: Conserved seed pairing, often flanked by Adenosines, indicates that thousands of human genes are microRNA targets. Cell **120**, 15–20 (2005)
20. Krek, A., Grun, D., Proy, M.N., Wolf, R., Rosenberg, L., Epstein, E.J., MacMenamin, P., da Piedade, I., Gunsalus, K.C., Stoffel, M., Rajewsky, N.: Combinatorial microRNA target predictions. Nat. Genet. **37**, 495–500 (2005)
21. Huang, J.C., Babak, T., Corson, T.W., Chua, G., Khan, S., Gallie, B.L., Hughes, T.R., Blencowe, B.J., Frey, B.J., Morris, Q.D.: Using expression profiling data to identify human microRNA targets. Nat. Methods **4**, 1045–1049 (2007)
22. Lu, Y., Zhou, Y., Qu, W., Minghua, D., Chenggang, Z.: A Lasso regression model for the construction of microRNA target regulatory networks. Bioinformatics **27**, 2406–2413 (2011)
23. Min, W., Li, X., Kwoh, C.-K., Ng, S.K.: A core-attachment based method to detect protein complexes in PPI networks. BMC Bioinform. **10**, 169 (2009)
24. Nepusz, T., Yu, H., Paccanaro, A.: Detecting overlapping protein complexes in protein-protein interaction networks. Nat. Methods **9**, 471–472 (2012)
25. Wingender, E., Chen, X., Hehl, R., Karas, H., Liebich, I., Matys, V., Meinhardt, T., Prss, M., Reuter, I., Schacherer, F.: TRANSFAC: an integrated system for gene expression regulation. Nucl. Acids Res. **28**, 316–319 (2000)
26. Friedman, R.C., Farh, K.K., Burge, C.B., Bartel, D.P.: Most mammalian mRNAs are conserved targets of microRNAs. Genome Res. **19**, 92–105 (2009)

27. Livstone, M.S., Breitkreutz, B.J., Stark, C., Boucher, L., Andrew, C.A., Oughtred, R., Nixon, J., Reguly, T., Rust, J., Winter, A., Dolinski, K., Tyers, M.: The BioGRID interaction database: 2011 update. Nucl. Acids Res. **39**, D698–D704 (2011)
28. Karim, S., Masud, M., Liu, L., Le, T.D., Lim, J.: Identification of miRNA-mRNA regulatory modules by exploring collective group relationships. BMC Genomics **17**, 7 (2015)
29. Chen, L., Wang, H., Zhang, L., Wan, L., Wang, Q., Yukui, S., He, Y., He, W.: Uncovering packaging features of co-regulated modules based on human protein interaction and transcriptional regulatory networks. BMC Bioinform. **11**, 392 (2011)
30. Koturbash, I., Zemp, F.J., Pogribny, I., Kovalchuk, O.: Small molecules with big effects: the role of the microRNAome in cancer and carcinogennesis. Mutat. Res. **722**, 94–105 (2011)

PFC: An Efficient Soft Graph Clustering Method for PPI Networks Based on Purifying and Filtering the Coupling Matrix

Ying Liu[1]([⊠]) and Amir Foroushani[2]

[1] Division of Computer Science, Mathematics, and Science,
College of Professional Studies, St. John's University, Queens, NY, USA
liuyl@stjohns.edu
[2] Department of Computer Science, Texas State University,
601 University Drive, San Marcos, TX 78666-4684, USA

Abstract. One of the most pressing problems of the post genomic era is identifying protein functions. Clustering Protein-Protein-Interaction networks is a systems biological approach to this problem. Traditional Graph Clustering Methods are crisp, and allow only membership of each node in at most one cluster. However, most real world networks contain overlapping clusters. Recently the need for scalable, accurate and efficient overlapping graph clustering methods has been recognized and various soft (overlapping) graph clustering methods have been proposed. In this paper, an efficient, novel, and fast overlapping clustering method is proposed based on purifying and filtering the coupling matrix (PFC). PFC is tested on PPI networks. The experimental results show that PFC method outperforms many existing methods by a few orders of magnitude in terms of average statistical (hypergeometrical) confidence regarding biological enrichment of the identified clusters.

Keywords: Protein-protein interaction networks · Purifying and filtering the coupling matrix · Overlapping clusters · Functional modules

1 Introduction

Homology based approaches have been the traditional bioinformatics approach to the problem of protein function identification. Variations of tools like BLAST [1] and Clustal [2] and concepts like COGs (Clusters of orthologous Groups) [3] have been applied to infer the function of a protein or the encoding gene from the known a closely related gene or protein in a closely related species. Although very useful, this approach has some serious limitations. For many proteins, no characterized homologs exist. Furthermore, form does not always determine function, and the closest hit returned by heuristic oriented sequence alignment tools is not always the closest relative or the best functional counterpart. Phenomena like Horizontal Gene Transfer complicate matters additionally. Last but not least, most biological Functions are achieved by collaboration of many different proteins and a proteins function is often context sensitive, depending on presence or absence of certain interaction partners.

© Springer International Publishing Switzerland 2016
D.-S. Huang et al. (Eds.): ICIC 2016, Part I, LNCS 9771, pp. 247–257, 2016.
DOI: 10.1007/978-3-319-42291-6_24

A Systems Biology Approach to the problem aims at identifying functional modules (groups of closely cooperating and physically interacting cellular components that achieve a common biological function) or protein complexes by identifying network communities (groups of densely connected nodes in PPI networks). This involves clustering of PPI-networks as a main step. Once communities are detected, a hypergeometrical p-value is computed for each cluster and each biological function to evaluate the biological relevance of the clusters. Research on network clustering has focused for the most part on crisp clustering. However, many real world functional modules overlap. The present paper introduces a new simple soft clustering method for which the biological enrichment of the identified clusters seem to have in average somewhat better confidence values than current soft clustering methods.

2   Previous Work

Examples for crisp clustering methods include HCS [4], RNSC [5] and SPC [6]. More recently, soft or overlapping network clustering methods have evolved. The importance of soft clustering methods was first discussed in [7], the same group of authors also developed one of the first soft clustering algorithms for soft clustering, Clique Percolation Method or CPM [8]. An implementation of CPM, called CFinder [9] is available online. The CPM approach is basically based on the "defective cliques" idea and has received some much deserved attention. Another soft clustering tool is Chinese Whisper [10] with origins in Natural Language Processing. According to its author, CW can be seen as a special case of the Random Walks based method Markov-Chain-Clustering (MCL) [11] with an aggressive pruning strategy.

Recently, some authors [12, 13] have proposed and implemented betweenness based [14] Clustering (NG) method, which makes NG's divisive hierarchical approach capable of identifying overlapping clusters. NG's method finds communities by edge removal. The modifications involve node removal or node splitting. The decisions about which edges to remove and which nodes to split, are based on iterated all pair shortest path calculations.

In this paper, we present a new approach, called PFC, which is based on the notion of Coupling matrix (or common neighbors). In the rest of the paper, we first describe PFC and compare its results with the best results achieved by the aforementioned soft approaches. The second part of this work aims to illustrate the biological relevance of soft methods by giving several examples of how the biological functions of overlap nodes relate to biological functions of respective clusters.

3   PFC Method

The method introduced here is based on the purification and filtering of coupling matrix, PFC. PFC is a soft graph clustering method that involves only a few matrix multiplications/manipulation. Our experimental results show that it outperforms the above mentioned methods in terms of the p-values for MIPS functional enrichment [15] of the identified clusters. The PPI net works we used in the paper are yeast PPI networks (4873 proteins and 17200 interactions).

3.1 Coupling Matrix

Bibliographical coupling is an idea from text classification: If two documents (for example two scientific papers) share a significant number of cited references, they are likely to deal with similar topics. A coupling matrix in a network describes the number of shared neighbors (or paths of length two) for each node pair. For undirected graphs like PPI networks, this matrix is symmetric and can be easily obtained from the original adjacency matrix A by: $B = A * A$. Notably, for second degree neighbors, the entry in coupling matrix is nonzero, even if there is no edge between the nodes. The importance of second degree neighbors in PPI networks has been emphasized before in the literature. For example: [16] note that "A substantial number of proteins are observed to share functions with level-2 neighbors but not with level-1 neighbors."

3.2 Purification of the Coupling Matrix

Adjacency matrices of biological networks are in general very sparse. The coupling matrix described above is slightly denser. However, not all nonzero-values are equally valuable. In the purification step, we determine the number of nonzero values (in unweighted graphs like PPI-Networks, this corresponds to the row sum), the maximum entry and the minimum non-zero value for each line of the coupling matrix. Rows in which the minimum nonzero entry and the maximum value are relatively close are considered homogenous and left unchanged. For other rows, we delete nonzero entries that don't make a significant contribution to the row sum. The Purification Process is summarized below:

```
FOREACH row i of the Coupling Matrix B
IF  min(B(i,:))   <   ⌊max(B(i,:)) * α⌋
THEN  B(i,:) = ⌊ B(i,:)./ (Bavg (i) * β) ⌋
Where:  "./" is the Matlab cell wise division operator,
⌊ ⌋ is the basic floor operation and α  and β are values
less than and  greater than 1 respectively.
```

This purification step is robust in regard to choice of values for its parameters. In particular in our experiment with a yeast PPI network, the results for $\alpha = 0.8$ and $\beta = 1.2$ did not differ from those for $\alpha = 0.7$ and $\beta = 1.3$.

3.3 Filtering of the Purified Coupling Matrix

The set of nonzero entries in each line of the Purified Coupling matrix can be considered as a candidate cluster. For a network of n nodes, this generally means n candidate clusters. However, not all rows are equally interesting. The set of nonzero entries (the information content) of many rows is likely to be very similar to, or contained largely within the sets of nonzero entries of other rows. This means that many rows are likely to represent spurious or redundant clusters. In the filtering step, we address this problem and try to select the most relevant and interesting rows of the purified coupling matrix. The set of nonzero entries in each of the selected lines of the purified coupling matrix represent our final clusters. The filtering step of PFC is a flexible step. Two alternative filtering approaches are discussed below.

Filtering by Simple, Local Criteria. The first Filtering approach is motivated by assumptions about the nature of the data and size of the target clusters. PPI data are for the most part results of high throughput experiments like yeast two hybrid and are known to contain many false positive and many false negative entries. For certain, more thoroughly studied parts of the network, additional data might be available from small scale, more accurate experiments. In PFC, the emphasis lies on common second degree neighbors and this can magnify the effects of noise. Under the assumption that Nodes with low degree belong in general to the less thoroughly examined parts of the network, it is conceivable that the current data for the graph around these low nodes contains many missing links. Missing links in these areas can have dramatic effects on the constellation of second degree neighbors. This means the Coupling data for low degree nodes is particularly unreliable. On the other hand, many extremely well connected nodes are known to be central hubs that in general help to connect many nodes of very different functionality with each other, hence, their second degree neighbors compromise huge sets that are less likely to be all functionally related. Additionally, it has been shown that most functional modules are meso-scale [Spirin, V. 2003]. There are also some fundamental physical constrains on the size and shape of a protein complex that make very large modules unlikely. Taking these considerations into account, a filter is easily constructed by the following rules:

> Discard all clusters (rows of purified coupling matrix) where the labeling node (the i.th node in the i.th row) has a particularly low (<14) or particularly high (>30) degree. Discard all clusters where the module size is too small (<35) or particularly large (>65).

The selected minimum and maximum values for degree of labeling nodes and module size are heuristically motivated. The intervals can be easily changed to obtain or discard more clusters, but the enrichment results for these intervals seem reasonably good. The peak log value for the enrichment of selected clusters is at -91.00 and the average lies at -18.99. Using this filter, by clustering yeast PPI networks, PFC yields 151 clusters from 52 different Functional categories. Figure 1 gives an example.

Filtering by Corroboration. Filtering by local criteria gives impressing results but it does not guarantee that a few of the remaining clusters do not overlap in majority of their elements. Although PFC is an overlapping clustering algorithm, very large overlaps between clusters are bound to indicate presence of redundant clusters. At the same time, repeated concurrence of large groups of proteins in different rows does reinforce the hypothesis that these groups are indeed closely related, and that the corresponding rows represent a high quality cluster. These observations can be used to construct an alternative filter that removes both low quality and redundant clusters from the coupling matrix. The main idea is that a line A is corroborated by a Line B if the majority of nonzero elements in A are also nonzero in B. The following summarizes this filter:

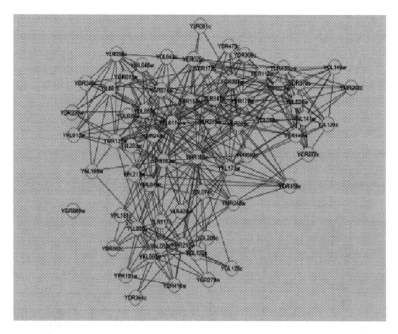

Fig. 1. This Figure shows the community for the row labeled "YKL173w" in the purified coupling matrix of yeat PPI network. It is one of the clustered selected by PFC1. Out of the 63 proteins in this community, 58 belong to MIPS Funcat 11.04.03.01.

Given the Binary version of the Purified Coupling Matrix B
Calculate Overlap Matrix $O = B * B$
Normalize $O(i, j)$ by Size of Module j
Calculate Corroboration Matrix $C = \lfloor O ./ \alpha \rfloor$
  Where: $0.5 < \alpha \le 1$; and "./" is the Matlab cellwise division.
Calculate Common Corroborator Matrix C Com= $C * C$
Rank the rows of Ccom by the sum of their entries
Interpret Ccom as description of a directed Confirmation graph between clusters, where the direction of confirmation is from lower ranked to higher ranked rows.
  Select clusters whose in-degree in the confirmation graph is higher than a threshold and whose out degree is 0.

Given the sparse nature of the involved matrices, this Corroboration based filter can be implemented very efficiently in Matlab. It discards by design redundant clusters (out-degree >0 in the confirmation graph indicates that there is a similar cluster with a higher rank) and retains only high quality clusters (clusters with a high in-degree in the confirmation graph have been confirmed by presence of many other clusters with similar structure). The ranking by row sum helps consolidate and summarize relevant parts of smaller clusters into larger ones. Figure 2 gives two examples of clusters selected by this approach on Yeast-PPI network.

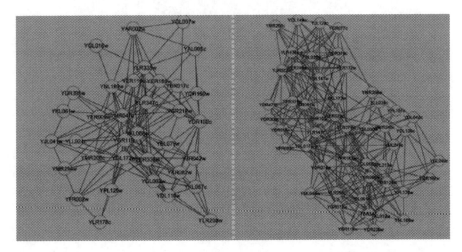

Fig. 2. Two of the clusters selected by PFC2. The left Figure shows the selected community for the row labeled "YDR335w" in the purified coupling matrix. Out of the 35 proteins in this community, 29 belong to MIPS Funcat 20.09.01(nuclear transport). The right Figure shows the selected community for the row labeled "YKL173w" in the purified coupling matrix. It is one of the clustered selected by PFC1. Out of the 63 proteins in this community, 58 belong to MIPS Funcat 11.04.03.01 (Splicing).

4 Experimental Results and Discussions

The results for two versions of PFC (PFC1: using the local criteria filter and PFC2: using the corroboration based filter) are compared with results obtained by other soft clustering methods. A PPI network of yeast with 4873 Nodes and 17200 edges is used as the test data set. The other methods are an in-house implementation of Pinney and Westhead's Betweenness Based proposal [12], Chinese Whisper [10] as available from its author's webpage, CPM as implemented in C-Finder [9]. Whenever other methods needed additional input parameters, we tried to choose parameters that gave the best values. The results from different methods are summarized in Table 1.

To ensure comparability of results from different methods, for each pair of methods it was determined which functional categories are covered by clusters from both methods. For example, Clusters obtained by Chinese Whisper fell into 32 different

Table 1. Comparison of results from different methods.

Method	Cluster count	Average cluster size	Average enrichment	Network coverage	Diversity	Remarks
Betweenness based	20	302.70	−15.11	0.58	19/20	Extended communities, size >25
Chinese whisper	38	23.45	−12.11	0.17	32/38	Communities of size >10
C finder	68	14.50	−15.70	0.19	48/68	K = 4 size >4
PFC 2	40	25.4	−19.40	0.17	36/40	Corroborated filtering, in-degree in confirmation graph >4 degree >1
PFC 1	183	44.76	−19.35	0.31	55/183	Filtering by local criteria: degree in [13,29] Module size in [33,64]

MIPS-Funcat categories and those found by C-Finder fell into 48 different categories, 18 of which were also among the 32 categories associated with results from Chinese Whisper. We then compared for each method pair and each common category the best enrichment results (Table 2).

Table 2. Pair-wise overlaps between top rated functional categories of all methods

Method	BW	CW	C-Finder	PFC1	PFC2
BW	19	6	12	12	11
CW	6	32	18	18	16
C-finder	12	18	48	25	28
PFC1	12	18	25	55	22
PFC2	11	16	28	22	36

5   Biological Functions of Overlap Nodes

The hypergeometric evaluation of individual clusters is the main pillar in assessing the quality of crisp clustering methods. For soft clustering methods, further interesting questions arise that deal with relationships between clusters. A possible conceptual disadvantage, production of widely overlapping, redundant clusters was addressed in previous sections. Figures 3, 4, 5 and 6 are the clustering results of two versions of PFC (PFC1: using the local criteria filter and PFC2: using the corroboration based filter). The results demonstrate an important advantage of soft methods against crisp ones: They show how soft clustering can adequately mirror the fact that many proteins have context dependent functions, and how in some cases overlap nodes can act as functional bridges between different modules.

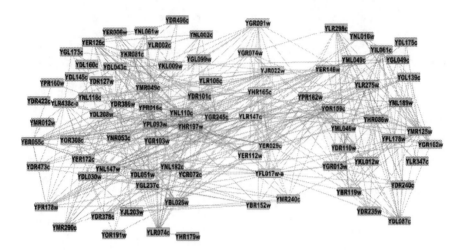

Fig. 3. PFC1 result #1: There is a relatively large overlap (yellow nodes). All 10 overlap nodes are involved in "nuclear mRNA splicing, via spliceosome-A". The same is true for ca.25 % (12 out of 45) of the green nodes to the left and 68 % (17 out of 25) of the green nodes to the right of the overlap. Furthermore, two of the overlap nodes are also involved in spliceosome assembly the total number of such nodes in the entire network is 19.

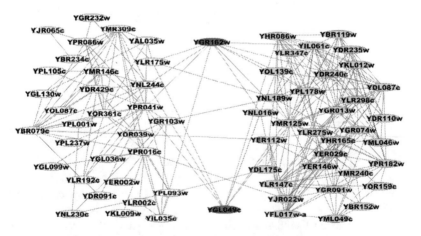

Fig. 4. PFC1 result #2: The dominant function for the left module is translation initiation (10 out of 31) for the right module, it is nuclear mRNA splicing (27 out of 33); both overlap nodes are involved in translation initiation and Protein-RNA complex assembly.

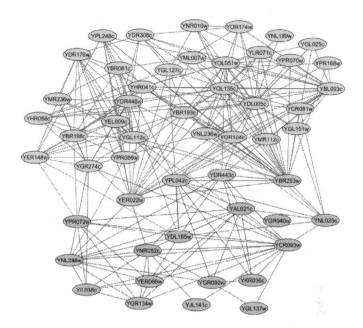

Fig. 5. PFC2 result #1: The main functions of the top and bottom clusters are identical: on both sides, over 80 % of the nodes are involved in "transcription from RNA polymerase II promoter" and this is also the main function of all of the overlap nodes. However, the bottom part also contains a specialized module for poly tail shortening: all 7 node in the entire network that are involved in poly tail shortening are gathered here

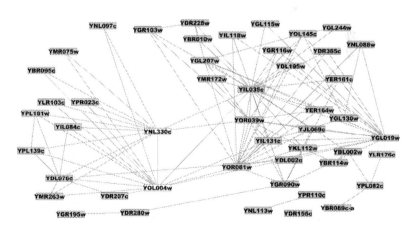

Fig. 6. PFC2 result #2.10 out of 13 yellow nodes are involved in histone deacytlation (left), 21 out of 33 green nodes are involved in transcription, DNA dependent (right); both white nodes are involved in both functions.

6 Conclusions

This paper introduced PFC, a new clustering concept based on purification and filtering of a coupling (common neighbor) matrix. It discussed two very different filtering methods resulting in two flavors of PFC. PFC consists of only a few matrix multiplications and manipulations and is therefore very efficient. Both flavors of PFC seem to outperform current soft clustering methods on PPI networks by a few orders of magnitude in terms of average statistical confidence on biological enrichment of the identified clusters.

The paper illustrated the importance of soft clustering methods in systems biology by giving a few concrete examples of how the biological function of the overlap nodes relates to the functions of the respective clusters.

References

1. Adamcsek, B.G., et al.: CFinder: locating cliques and overlapping modules in biological networks. Bioinformatics **22**(8), 1021–1023 (2006)
2. Altschul, S.F., et al.: Gapped BLAST and PSI-BLAST: a new generation of protein database search programs. Nucleic Acids Res. **25**(17), 3389 (1997)
3. Biemann, C.: Chinese whispers-an efficient graph clustering algorithm and its application to natural language processing problems. In: Proceedings of the HLT-NAACL-06 Workshop on Textgraphs-06, New York, USA (2006)
4. Chua, H.N., Sung, W.K., Wong, L.: Exploiting indirect neighbours and topological weight to predict protein function from protein-protein interactions. Bioinformatics **22**(13), 1623–1630, 1 July 2006. PMID: 16632496; btl145 [pii] (Oxford, England)
5. Derenyi, I., et al.: Clique percolation in random networks. Phys. Rev. Lett. **94**(16), 160202 (2005)
6. Girvan, M., Newman, M. E.: Community structure in social and biological networks. In: Proceedings of the National Academy of Sciences of the United States of America, vol. 99, no. 12, pp. 7821–7826, 11 June 2002. PMID: 12060727; 99/12/7821 [pii]
7. Gregory, S.: An algorithm to find overlapping community structure in networks. In: Kok, J. N., Koronacki, J., Lopez de Mantaras, R., Matwin, S., Mladenič, D., Skowron, A. (eds.) PKDD 2007. LNCS (LNAI), vol. 4702, pp. 91–102. Springer, Heidelberg (2007)
8. Hartuv, E., Shamir, R.: A clustering algorithm based on graph connectivity. Inf. Process. Lett. **76**(4–6), 175–181 (2000)
9. King, A.D., Przulj, N., Jurisica, I.: Protein complex prediction via cost-based clustering. Bioinformatics **20**(17), 3013–3020, 22 November 2004. PMID: 15180928; bth351 [pii] (Oxford, England)
10. MIPS. The functional catalogue (FunCat). Internet on-line. http://mips.gsf.de/projects/funcat (2007). Accessed 10 Feb 2008
11. Palla, G., Derenyi, I., Farkas, I., Vicsek, T.: Uncovering the overlapping community structure of complex networks in nature and society. Nature **435**(7043), 814–818, 9 June 2005. PMID: 15944704; nature03607 [pii]
12. Pinney, J.W., Westhead, D.R.: Betweenness-based decomposition methods for social and biological networks. In: Barber, S., Baxter, P.D., Mardia, K.V., Walls, R.E. (eds.) Interdisciplinary Statistics and Bioinformatics. Leeds University Press, Leeds (2006)

13. Spirin, V., Mirny, L.A.: Protein complexes and functional modules in molecular networks. Proc. Natl. Acad. Sci. **100**(21), 12123–12128 (2003)
14. Tatusov, R.L., Koonin, E.V., Lipman, D.J.: A genomic perspective on protein families. Science **278**(5338), 631 (1997)
15. Thompson, J.D., Higgins, D.G., Gibson, T.J.: CLUSTAL W: improving the sensitivity of progressive multiple sequence alignment through sequence weighting, position-specific gap penalties and weight matrix choice. Nucleic Acids Res. **22**(22), 4673–4680 (1994)
16. Van Dongen, S.: A cluster algorithm for graphs. Rep.-Inf. Syst. **10**, 1–40 (2000)

MDAGenera: An Efficient and Accurate Simulator for Multiple Displacement Amplification

Weiheng Huang[1], Hongmin Cai[1(✉)], Wei Shao[1], Bo Xu[1],
and Fuqiang Li[2]

[1] School of Computer Science and Engineering,
South China University of Technology, Guangzhou, China
hmcai@scut.edu.cn
[2] Beijing Genomics Institute, Shenzhen, China

Abstract. The advance of single cell sequencing advocates a new era to delineate intratumor heterogeneity and traces the evolution of single cells at molecular level. However, current single cell technology is hindered by an indispensable step of genome amplification to accumulate enough samples to reach the sequencing requirement. Multiple Displacement Amplification (MDA) method is the major technology adopted for genome amplification. But it suffers from a major drawback of large amplification bias, resulting in time and label consuming. To fulfill this gap, we have presented a simulation software for the MDA process in this paper. The proposed simulator was based on an original hypothesis for catering to empirical MDA process. It was implemented to achieve high efficiency with affordable computational cost, thus allowing for an individual MDA experiment to be simulated quickly. Surprising nice experiments demonstrated the simulator is promising in providing guidance and cross-validation for experimental MDA.

Keywords: Single cell sequencing · Multiple Displacement Amplification (MDA) · DNA amplification simulator

1 Introduction

The next-generation sequencing (NGS) is adopted as a popular strategy for genotyping and has included comprehensive characterization of genome by generating hundreds of millions of short reads in a single run [6]. The NGS could achieve higher coverage and resolution, thus allowing more accurate estimation of copy numbers and detection of breakpoints with high throughput, than traditional methods do. Due to the high heterogeneity between individual single cells at the molecular level, NGS has been successfully tailored to sequence entire genome at the single cell level. Current single-cell sequencing technology starts by isolating single cells from large tissues and organs and then amplifies the small amount of the whole genome of the single cells to a large amount so that the amplicons can be used for genome sequencing and analysis. Then we can reveal life secretes from the results like the cause of cellular differentiation and the process of single cell evolution [1, 7, 8, 12]. As a revolutionary technology,

© Springer International Publishing Switzerland 2016
D.-S. Huang et al. (Eds.): ICIC 2016, Part I, LNCS 9771, pp. 258–267, 2016.
DOI: 10.1007/978-3-319-42291-6_25

single cell sequencing quickly received wide spot lights and has been employed in many biological and medical fields like tumor population structure and evolution [7].

The single cell sequencing needs an extra step to amplify the genome from a single cell. However, owing to low amounts of input material, the amplification step will introduce substantial levels of technical noise. Currently there are three major amplification schemes, DOP-PCR, multiple displacement amplification (MDA) [9] and multiple annealing looping-based amplification cycle (MALBAC) [13]. DOP-PCR could achieve remarkable uniform amplifications, but its coverage is very low. MDA is an isothermal process where the sample DNA needs no purification, which makes it easier to be operated than other methods like PCR. It uses the DNA polymerase ϕ29 to amplify single cell genomes ranging from human genome like crude whole blood and tissue culture cells [2] to environmental bacteria. And the amplicons have the average length of 70, 000 bp, while the number is only 1 kbp for other amplification methods. Even though the method of MDA introduce less bias than other amplification methods [3], the bias are inevitable, with some genomic regions being amplified more than others. Such drawbacks greatly limits the power of single-cell sequencing in improving its discrimination resolution as well as its popularities. To gain a thorough understanding on the complex biochemical process of the amplification, one may need to conduct extensive experiments to enhance its purity. To alleviate the economic burden as well as quantitatively investigate the key characteristics of the amplification process, we have reported a simulated software for MDA process, named by MDAGenera, in this paper.

The main contribution of this paper is in two aspects: (1) an efficient simulation software for MDA process, based on rigorous mathematical modeling, is provided. It could efficiently produce reliable amplification results for MDA process, thus is capable of providing economically experimental design and technology improvement; (2) the primers used in MDA process was designed to be freely attached at single-stranded region, our preliminary studies disproof this heuristic claim. Our simulator assumed that the probability of the primers to anneal to a DNA strand is negatively proportional to the stand's length. Extensive experiments based on this hypothesis obtained satisfactory results and demonstrated its validness. The frequency distribution of the reads coverage after simulation nearly perfectly matched to that after real sequencing. The proposed Simulator is promising and is able to provide guidance as well as evaluation criteria for experimental MDA design and improvement.

2   Methods

2.1   Biochemical Introduction of MDA Process

MDA is a non-PCR based DNA amplification technique. The reaction takes place at the constant temperature of 30 degree. Its success is highly dependent on the DNA polymerase of ϕ29 and random hexamer primers. The DNA polymerase ϕ29 is a high-processivity enzyme that can produce DNA amplicons of 70 to 100 kbp length. Due to its high fidelity and 3'–5' ends proofer-ading activity, it can effectively reduce the amplification error rate to 1 in 10^6–10^7 bases compared to conventional Taq polymerase Tindall and Kunkel [11]. Random hexamer primer, another important material for the reaction, consists of six random nucleotides. The process of MDA is a complicated chemical reaction and the key procedures are shown in Fig. 1. Basically, it consists of six steps as follows:

1. Stabilize the reaction temperature at 30° C and pour the ϕ29 DNA polymerases, primers and DNA strand(s) into the reaction container.
2. The random hexamer primers start by looking for attachable single-stranded positions on the DNA template sequences and then attach to them.
3. ϕ29 DNA polymerases begin synthesizing DNA strands from the positions where the primers have attached to.
4. If ϕ29 DNA polymerases meet any double-stranded areas, they will displace the previously amplified sequence and continue to synthesize its own sequence.
5. A tree-like structure is generated within a period of time.
6. Put S1 nuclease into the container to dismantle the entire tree-like structure Tagliavi and Draghici [10].

Fig. 1. Illustration of the MDA process. The hexamer primers (dark gray) search for position on the single-stranded DNA templates (red) and then attach to them; (b) The ϕ29 DNA polymerases synthesize new strands from the templates; (c) Once the polymerases reach a double-stranded region, it replaces the previous strand and continues to synthesize its own strand, and the displaced strands can be attached by other primers (red); (d) Finally, a tree-like structure is formed after a period of time. This process is repeated after rounds of reactions. (Color figure online)

With the aforementioned steps, the DNA sequence is amplified until enough samples are accumulated to meet the required quantity for sequencing.

2.2   Mathematical Modelling of MDA Reaction

The previous MDA process involves complex biological reactions. Among the factors, four of them have fundamental influence on the reaction, including DNA polymerase ϕ29, random hexamer primers, DNA strands and nucleotides. While the speed of the reaction is decided by four parameters of DNA polymerases, including the average amplified strand length μ, its standard deviation σ, number of the bases synthesized within one unit time interval k and the number of available polymerases n_φ. The first three parameters solely determine the speed of the polymerases to synthesize a new strand, while the last one imposes a limit on the reaction speed when it is less than the number of positions with primer attached. The primer also has a profound influence on MDA. We shall use two parameters to quantify its effect, including the number of available primers n_p and the bounding speed of the primer α. The former value affects the concentration of primers, thus influence the overall reaction speed. The later one controls how much time a primer needs to anneal to a DNA strand. The longer it takes, the slower the whole process is. In real MDA experiment, one can reasonably assume that sufficient nucleotides for the reaction are provided, therefore the influence of the number of nucleotides on final amplification is ignored.

In summary, one may see that the number of available polymerases $\phi29$ affects the reaction speed if the number of polymerases available is not enough. The parameters of μ, σ and k mainly affect the speed of $\phi29$ DNA polymerases to synthesize new strands. The other three variables, namely α, n_p and l_s affect the number of primers bound on unit time interval v. V also indicates how fast the overall reaction speed is. Since MDA is isothermal, the value of α is proportional to v. The number of available primers n_p is positively correlated to v in that a larger number of the available primers n_p leads to higher concentration of primers, thereby accelerates the MDA process. The length of all single-stranded DNA sequences l_s is also positively correlated to v since a longer single-stranded DNA fragment provides more positions for a primer to be attached to.

Mathematically, the number of primers bound on unit time interval, i.e., the overall reaction speed, can be formulated as

$$v = \alpha \beta n_p l_s \tag{1}$$

where β is a trade-off parameters indicating the correlation between v and $n_p l_s$. In our experiments, it is set to be 1 for simplicity.

Although MDA process was designed and assumed that the primers attach to single-stranded region at fully random position. Our preliminary simulation experiments strongly suggested that this assumption may be disputable. Therefore, we proposed a new hypothesis that the primers are inclined to attach to the short and dissociative single sequences instead of attaching to positions randomly. Thus we updated the MDA algorithm by assuming that the probability of the primers to be attached is negatively proportional to its length. Mathematically,

$$\alpha = \exp(-c * \mu) \tag{2}$$

where c is a trade-off parameter quantifying the influence of averaged strand length over the reaction speed α of the primes.

The pseudocode for the proposed model is summarized in Table 1.

Table 1. MDA Simulation Algorithm

MDA Simulation Algorithm

Require: The DNA sequence that needs to be amplified(fasta format)

1: Average length of newly synthesized strands μ

2: Standard deviation of average amplified length σ

3: Number of bases synthesized in unit time interval k

4: Number of available polymerase n_φ

5: Number of available primers n_p

6: Bounding speed of primers α

7: Expected reaction time T

8: Expected coverage of the reaction C

Ensure: Amplified DNA sequences in fasta format

9: Initialization. The input DNA strand is initialized as the first sequence in *HashMap<Integer, Sequence> sequence*. *t=0, c=0*.

10: **while** $t < T$ and $c < C$ and $n_p > 0$. **do**

11: $t \leftarrow t \ 1$, update *sequenceAttachable*

12: Calculate *v*, delete *v* polymerases if available, otherwise delete all polymerases from *idleEnzyme*, add them to *activateEnzyme*.

13: **for** every polymerase in *activateEnzyme* **do**

14: **If** not attached to any DNA sequence **then**

MDA Simulation Algorithm

15:      find an attachable position in
sequenceAttachable , dissociative and short strand have
higher possibility to be attached to

16:   **else**

17:      do nothing

18: Sort all the polymerases in *activateEnzyme* using the
EnzymeComparator

19: All polymerases in *activateEnzyme* add *k* bases to
their corresponding sequence

20: **for** every polymerase in *activateEnzyme* **do**

21:      **if** length of synthesized length reaches
amplifyLength **then**

22:        delete it from *activateEnzyme*, add it to the
idleEnzyme

23:     else

24:     do nothing

25: Update current coverage *c*

26: Update the number of single-stranded bases of each
DNA sequence.

2.3   Implementation

We use an object-oriented programming approach with four Java classes, *Sequence*, *Enzyme, MDA* and *EnzymeComparator*, respectively. *MDA* serves as an overall control class, Sequence represents a single-stranded DNA sequence. It has an unique *sequenceNumber* to number the sequence and an array *DNASequence* to store the nucleotides information of the strand. Enzyme acts the role as the polymerase. *TemplateSequenceNumber* is used to record the *sequenceNumber* of the sequence that is attached by the polymerase, and *nontemplateSequenceNumber* stores the *sequence*

Number of the newly synthesized sequences. *EnzymeComparator* is used as a comparator class to sort the polymerases.

3   Results and Discussion

Our simulation process starts with the DNA sequence of E.coli. Its input file was fed into MDAGenera with parameters $\alpha = 2.6 * 10^{-11}$, k = 1000, n_p = 250000, $\mu = 80000$, $\sigma = 2300$, T = 10000. We set the target coverage of E. coli to $600X$ to approximate the quantity of the experimental data. When we obtained the amplicons of E. coli from MDAGenera, we applied an Illumina read generator, ART [4], to generate sequenced Illumina reads for our outputs. Then we used Fastq Quality Converter in the Fastx Toolkit to convert the quality format of our results from Sanger to Illumina. Finally, the obtained sequence is then aligned to the reference genome by a sequence alignment tool of BWA [5]. For performance comparatione, a recently reported compute simulator for MDA, named by MDAsim, was employed with default setting [10]. The MDAsim was implemented to simulate the MDA reactions under ideal conditions.

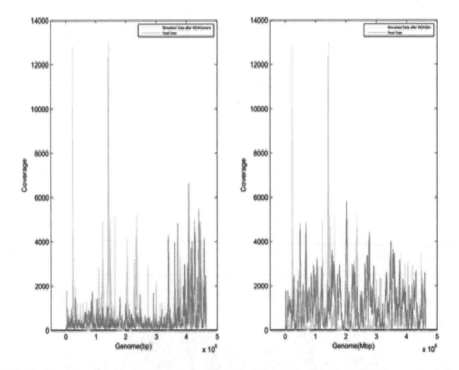

Fig. 2. Experimental comparison between the simulated result after the MDAGenera (the left figure) and MDAsim (the right figure) on real sequenced data of E. coli (lane 7). The horizontal axis represents the reference DNA position of the bases along the reference DNA strand while the vertical axis represents its coverage depth. The simulated results were highlighted in green and were overlaid by its empirical counterpart in light gray for difference. The simulated result after the proposed MDAGenera demonstrated a high similarity to the real data. (Color figure online)

A typical result after the MDAGenera and MDAsim were shown in Fig. 2. In both figures, the horizontal axis represents the reference DNA position of the bases on the reference DNA strand while the vertical axis represents its coverage depth. For ease of comparison, the simulated reads were overlaid by the empirical data. The empirical sequenced result on lane 7 was highlighted in grey while the simulated one in green. One may easily notice that the results of coverage distribution by MDAGenera could fit to the real sequencing data more accurately than by MDAsim did. The MDAGenera obtained several coverage peaks as that of the real data with nearly same altitude. In comparison, the MDAsim seldom captured the peaks.

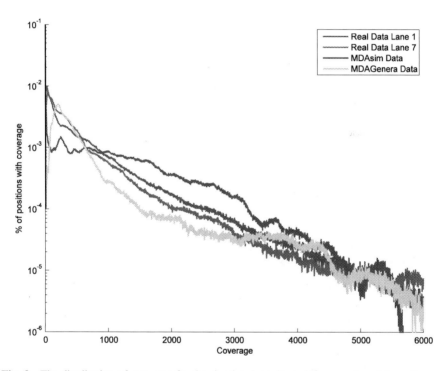

Fig. 3. The distribution of coverage for the simulated results and the experimental results were shown to be matched nearly perfect. The horizontal axis denotes the coverage while vertical axis is the percentage of positions with the corresponding coverage, i.e., a point (x,y) implies there is y percent of positions having x coverage (Color figure online)

We also plotted the coverage histogram distribution for both the simulated and real data to demonstrate their differences. The results were shown in Fig. 3. In this figure, each point (x,y) represents that there is y percent of bases have gained a coverage equal to x. The real data of E. coli Lane 1 and Lane 7 were denoted in dark green and red, respectively. The simulated results after Mdasim and MDAGenera were highlighted in blue and light green, respectively. The result by MDAGenera was shown to achieve a very nice similarity of the histogram distribution to its real peers. Under most cases of the coverage, it matches the coverage distribution of single cells very well. In comparison, the MDAsim was shown to have larger discrepancy to the real distribution. One may also notice that our simulations were less similar to the real data in the region (highlighted in light gray) where coverage is less than 200. The simulated results after MDAGenera in blue have few (around 0.01 %) number of genome positions with coverage less than 200, while in real MDA process there are more genome positions (>1 %) amplified less than 200 times. This is because that both of the real data and our simulated results amplify the DNA sample to 600×, but the real data has vary unbalanced coverage distribution where the coverage depth for several genomic positions reaches 12000 but many are barely amplified. Our simulated results has no such unbalanced coverage distribution and therefore has less positions being in low coverage depth.

Despite the minor differences mentioned above, our extensive experimental results strongly support our hypothesis and effectively demonstrated the reliability of the proposed simulator. It offered a plausible explanation for the cause of unbalanced coverage distribution of MDA. We currently are working to validate the finding through biological experiments. Moreover, the nice similarity between the proposed MDAGenera to the real data strongly support our hypothesis and effectively demonstrated the reliability of the proposed simulator. It offered a plausible explanation for the cause of unbalanced coverage distribution of MDA. We currently are working to validate the finding through biological experiments.

4 Conclusion

Highly accurate and reliable single cell sequencing technology is largely hindered by the technological bias introduced by genome amplification. MDA has been widely used to amplify entire genomes from a few cells or even a single cell, but its performance is less satisfactory due to uneven amplification rates. The complex biological reactions in MDA has not been fully comprehended and thus limited its improvements.

We have developed a computer simulator, named by MDAGenera, for MDA process. The software is coded with Java and it is intended to run on Windows system. It has user-friendly GUI and high efficiency. To amplify a DNA sample with 3 MB to a coverage of 600, MDAGenera costs 300 s with 12 GB of RAM to achieve full coverage. It is designed to be highly efficient, even working on personal computers. Under ideal conditions, the Generator can easily achieve the expected average coverage for each genome position. To resolve the distribution disagreements between the real and simulated data, we have proposed a hypothesis that the coverage peaks are caused by the inclination of primers when it is attaching to the reference sequence. It implies that the primers tend to find dissociative and short strands to attach to. The experimental results based on the hypothesis strongly support our proposition. The frequency distribution of the reads coverage after simulation nearly perfectly matched to that after real sequencing. The proposed Generator is promising and is able to provide guidance as well as evaluation criteria for experimental MDA design and improvement.

Acknowledgments. This work was supported by National Nature Science Foundation of China (61372141), Special Program for Applied Research on Super Computation of the NSFC-Guangdong Joint Fund (the second phase), Science and Technology Planning Project of Guangdong Province, and the Fundamental Research Fund for the Central Universities (2015ZZ025).

References

1. Cai, H., Ruan, P., Ng, M., Akutsu, T.: Feature weight estimation for gene selection: a local hyperlinear learning approach. BMC Bioinf. **15**(1), 70–78 (2014)
2. Dean, F.B., Hosono, S., Fang, L., Wu, X., Fawad Faruqi, A., Bray-Ward, P., Sun, Z., Zong, Q., Du, Y., Du, J.: Comprehensive human genome amplification using multiple displacement amplification. Proc. Nat. Acad. Sci. **99**(8), 5261–5266 (2002)
3. Hosono, S., Faruqi, A.F., Dean, F.B., Du, Y., Sun, Z., Wu, X., Du, J., Kingsmore, S.F., Egholm, M., Lasken, R.S.: Unbiased whole-genome amplification directly from clinical samples. Genome Res. **13**(5), 954–964 (2003)
4. Huang, W., Li, L., Myers, J.R., Marth, G.T.: Art: a next-generation sequencing read simulator. Bioinformatics **28**(4), 593–594 (2012)
5. Li, H., Durbin, R.: Fast and accurate short read alignment with burrows–wheeler transform. Bioinformatics **25**(14), 1754–1760 (2009)
6. Metzker, M.L.: Sequencing technologies the next generation. Nat. Rev. Genet. **11**(1), 31–46 (2010)
7. Navin, N., Kendall, J., Troge, J., Andrews, P., Rodgers, L., Mcindoo, J., Cook, K., Stepansky, A., Levy, D., Esposito, D.: Tumour evolution inferred by single-cell sequencing. Nature **472**(7341), 90–94 (2011)
8. Ni, X., Zhuo, M., Su, Z., Duan, J., Gao, Y., Wang, Z., Zong, C., Bai, H., Chapman, A.R., Zhao, J., et al.: Reproducible copy number variation patterns among single circulating tumor cells of lung cancer patients. Proc. Nat. Acad. Sci. **110**(52), 21083–21088 (2013)
9. Spits, C., Le Caignec, C., De Rycke, M., Van Haute, L., Van Steirteghem, A., Liebaers, I., Ser mon, K.: Whole-genome multiple displacement amplification from single cells. Nat. Protoc. **1**(4), 1965–1970 (2006)
10. Tagliavi, Z., Draghici, S.: MDAsim: a multiple displacement amplification simulator. In: 2012 IEEE International Conference on Bio informatics and Biomedicine (BIBM), pp. 1–4. IEEE (2012)
11. Tindall, K.R., Kunkel, T.A.: Fidelity of DNA synthesis by the Thermus aquaticus DNA polymerase. Biochemistry **27**(16), 6008–6013 (1988)
12. Xu, X., Hou, Y., Yin, X., Bao, L., Tang, A., Song, L., Li, F., Tsang, S., Wu, K., Wu, H., et al.: Single-cell exome sequencing reveals single-nucleotide mutation characteristics of a kidney tumor. Cell **148**(5), 886–895 (2012)
13. Zong, C., Lu, S., Chapman, A.R., Xie, X.S.: Genome-wide detection of single nucleotide and copy-number variations of a single human cell. Science **338**(6114), 1622–1626 (2012)

Construction of Protein Phosphorylation Network Based on Boolean Network Methods Using Proteomics Data

Han Yu[1,2], Yaou Zhao[1,2], Shiyuan Han[1,2], Yuehui Chen[1,2(✉)],
Wenxing He[3(✉)], and Likai Dong[1]

[1] School of Information Science and Engineering,
University of Jinan, Jinan 250022, China
yuhan_1991@163.com, yhchen@ujn.edu.cn
[2] Shandong Provincial Key Laboratory of Network Based
Intelligent Computing, Jinan 250022, China
[3] School of Biological Science and Technology,
University of Jinan, Jinan 250022, China
163.hwx@163.com

Abstract. Post-translational Modification (PTM) of Proteins is a key biological process in the regulation of protein function. This paper discusses the problem of construction of PTM network based on the reverse engineering principles, which is constructed by using PBIL and TDE algorithms. Experiments which are based on two well-known pathways by the time series data of protein phosphorylation data show that the new method can be successfully validated and further reveal the regulation of protein phosphorylation.

Keywords: PTM · Computational intelligence · Phosphorylation network · PBIL · TDE

1 Introduction

Post-translational Modification (PTM) of Proteins plays a key role in many biological processes. PTM could alter protein structure, activity, stability and interaction with other molecule. With the development of modern biological mass spectrometry technology, the high-throughput screening and quantitative analysis of PTM have been greatly facilitated, and the detection sensitivity has been greatly improved [1]. Currently, nearly 400 kinds of PTM such as ubiquitination, methylation, nitration, and other similar things have been found. Of those, there are more than 300 kinds of PTM of proteins is widely involved in many life activities [2]. However, it is very likely to cause disease, once the abnormal PTM of proteins. Phosphorylation of a protein may lead to activation or repression of its activity, alternative subcellular localization and interaction with different binding partners [3]. Through damaging the functions of kinases and phosphatases, some signaling pathways of normal biological processes may be changed and cause many diseases.

Recently, with deep learning, 'networks' has emerged as a hotspot that takes into consideration both key genes/proteins and their relationship with specific modules,

© Springer International Publishing Switzerland 2016
D.-S. Huang et al. (Eds.): ICIC 2016, Part I, LNCS 9771, pp. 268–277, 2016.
DOI: 10.1007/978-3-319-42291-6_26

pathway and processes [4]. There is a wealth of experimental data through proteomics methods. However, the data analysis methods are not perfect and this has brought about great challenges. In addition, how to construct PTM networks, analyze its spatial-temporal variance and the influence to the outside environment, is still a problem. In this research, we construct a phosphorylation network based on Boolean network method with PBIL and TDE algorithm, to discover and explore the regulation of protein phosphorylation.

2 Methods

2.1 Boolean Network Model

Boolean network consists of N Boolean variables, and each variable is defined as a binary number to describe their states. Boolean functions are used to illustrate the logical relations between variables. A large T cell receptor signal transduction Boolean network is firstly constructed by Saez-Rodriguez et al. [6]. Then, Kaufman et al. applies Boolean network for constructing a simplified T cell receptor pathway model. In addition, EGFR/ErbB signal network based on Boolean network is built by Samaga et al. details [7], and cell apoptosis Boolean network is built by Schlatter et al. [8]. Boolean networks are inherently deterministic. Probabilistic Boolean networks are also proposed by Helikar et al., and RTK, G protein coupled receptors and Integrin signaling network based on Boolean network are built [9]. Currently, Boolean work has been a significant model for biological network.

This paper used Boolean network to build phosphorylation network. Network nodes are protein phosphorylation sites, and variables are the states of modified sites (0 denotes unmodification, 1 denotes modification). Different modified sites are connected by Boolean functions to describe the logical relationship between upstream and downstream sites.

2.2 PBIL Algorithm

Population-Based Incremental Learning (PBIL) is a distributed estimation algorithm, which is proposed by Carnegie Mellon University Baluja in 1994 [10–12]. Unlike traditional Genetic Algorithm (GA), there are not corresponding crossover operators and selection operators in PBIL algorithm, which could make the algorithm run faster. The distribution of the population is described by a probability model, and the evolution of the population is accomplished through the revision of the probability model. Through learning and sampling of the sample and gradual correcting, PBIL algorithm makes the initial model of uniform distribution eventually be able to represent the spatial distribution of the samples. Compared with the evolutionary computation method, which is led by the GA, PBIL algorithm is a probabilistic model of the whole population distribution, which is a mathematical description of the "macro" level of biological evolution. What's more, PBIL algorithm describes the relationship between the various variables to be optimized through a probabilistic model, and it is more effective to solve multi variable and nonlinear problem, and it is more suitable to solve the optimization problem of high dimensional data. PBIL algorithm can be described as follow:

Step 1 initial probability model;
Step 2 uniform generate the initial probability table **P**;
Step 3 generate **N** individuals, then calculation the fitness of each individual;
Step 4 choose the fitness of the optimal individual **B** according to *step 3*;
Step 5 update the table **P** by a certain mutation probability in *step 2*;
Step 6 if meet the termination condition, then the algorithm is stopped, otherwise go to *step 1*.

P is the probability that the generated $\&X_n$ of individual equal 1 in each dimension. If the dimension of the individual is M, **P** is denoted as $\{P_1, P_2, ..., P_M\}$. While $P_i = 1$, P_i denotes $\&X_n$ (corresponding to 1 denotes modification, $\bar{\&}X_n$ corresponding to 0 denotes unmodification). Optimal individual B is a Boolean function of the output of the model.

2.3   TDE Modeling Parameters

Individuals of PTM network generated by PBIL. To obtain the optimal individual, Trigonometric Differential Evolution (TDE) is applied in this paper which proposed by Fan and Lampinen in 2003 [13]. This method improves the rate of DE convergence by embedding the TDE mutation operator onto the method of differential evolution (DE) [14]. The procedure of the TDE is shown as follow:

Step 1 initial population;
Step 2 generate **N** individuals randomly, and calculate fitness of each individual;
Step 3 performing the mutation operator, the crossover operator and select operator successively;
Step 4 if meet the termination condition, then the algorithm is stopped, otherwise go to *step 1*.

2.4   Fitness Function and Evaluation Index

Fitness function offers the TDE optimization model parameters a value, the evaluation model to solve the problem of the ability. In this research, Mean Square Error MSE, which is usually used in optimization problem, is applied to fitness function, it is described as follows:

$$fit_{MSE}(i) = \frac{1}{p}\sum (y_i^a - y_i^b)^2 \tag{1}$$

Where, $fit_{MSE}(i)$ is the i-th individual fitness and P is the total number of samples. y_i^a is a Boolean function of true phosphorylation relationship; y_i^b is a Boolean function of the output of the model.

From data, the phosphorylation network derived by data, the matching degree of the structure and the real network should be judged. And the matching is generally through the check the concordance with corresponding edge in the map. Recognition of protein phosphorylation can be seen as a binary classification problem. The positive examples indicate the presence of phosphorylation relationship between two protein locis, on the contrary modification. In this way, there is a total of 4 cases in the process of classification: True Positive (**TP**), True Negative (**TN**), False Positive (**FP**), False Negative (**FN**). And the counts and percentages of the four cases are usually used as the evaluation index of the classification. The indicators were included:

1. Sensitivity/True Positive Rate (**TPR**) and Specificity/True Negative Rate (**TNR**):

$$TPR = \frac{TP}{TP + FN} \tag{2}$$

$$TNR = \frac{TN}{TN + FP} \tag{3}$$

2. False Positive Rate (**FPR**) and False Negative Rate (**FNR**):

$$FPR = \frac{FP}{TN + FP} \tag{4}$$

$$FNR = \frac{FN}{TP + FN} \tag{5}$$

3. Precision (**P**) and Recall (**R**):

$$P = \frac{TP}{TP + FP} \tag{6}$$

$$R = \frac{TP}{TP + FN} \tag{7}$$

4. Accuracy (**Acc**):

$$Acc = \frac{TP + TN}{TP + FP + TN + FN} \tag{8}$$

5. F-score (**F**):

$$F = \frac{2P \bullet TPR}{P + TPR} \tag{9}$$

2.5   Overview of the Phosphorylation Network Workflow

The detailed flow of the algorithm is shown in Fig. 1:

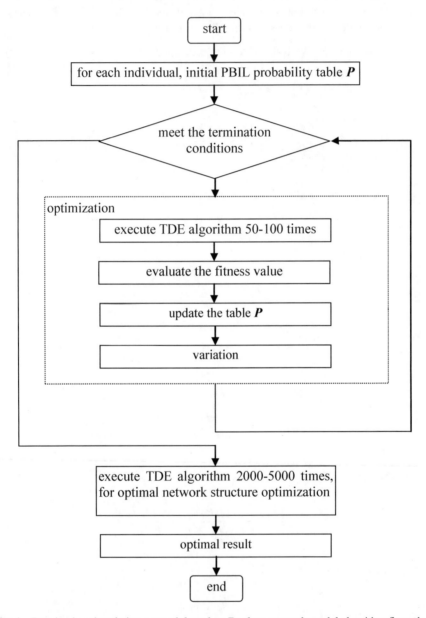

Fig. 1. Protein phosphorylation network based on Boolean network model algorithm flow chart.

3   Experimental Results

3.1   Data Sources

In this paper, the experiment data are offered by Key Laboratory of Cardiovscular Remodeling and Function Research, Chinese Ministry of Education Chinese Ministry of Health. The laboratory is equipped with MALDI-TOF-TOF, LTQ Orbitrap and LCMS-IT-TOF high-resolution mass spectrometer, which could determine the modification of each site of protein. The action of ox-LDL, which is the main pathogenic factors of atherosclerosis, on vascular endothelial cell is established as studying object. The experiment time series data from 14 different points in time is collected with changes from normality and added stimulation to stability by nano-2D-HPLC combined with high resolution mass spectrometer. Each time point of time series data is a two-dimensional table, and the rows mean the name of protein and the set serial number, the columns mean the protein phosphorylation site modification at different time points. The data in the row i, and column j means the modified situation (phosporylation or not) of one set of the protein i at the time of tj, and the content of it. The data are produced by high-resolution mass spectrometer will get rid of low reliability after a series of preprocessing.

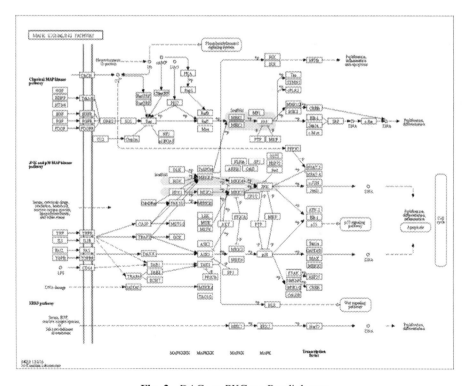

Fig. 2. DAG → PKC → Ras link way

Experimenters take sample at the 15 min, 30 min, 45 min, 1 h, 1.5 h, 2 h, 4 h, 6 h, 8 h, 12 h, 18 h, 24 h, 48 h and 72 h. Because of the larger sampling interval, the rate of modification degree is virtually unchanged. In this paper, we use the method of linear interpolation to expand 14 sets to 288 sets in every 15 min between the sampling points with little change in the late stage. This method not only does not affect the accuracy of the results, but also make the sampling data more even.

3.2   Results

To test the reliability and effectiveness of the method, one of link way (DAG → PKC → Ras) in the MAPK signaling pathway included in KEGG database (Fig. 2) and one link way (Ras → MEK → Src → AR) that contains phosphorylation reaction in Endocrinology/Hormones signaling pathway (Fig. 3) are used as experiments.

Put the time series of protein phosphorylated site in each pathway, as the input variable of experimental, and the modification relationship matrixes have shown in Fig. 4(a) and (b). Among them, the A, B, C, D, E and F represent DAG, PKC, Ras, MEK, Src and AR respectively.

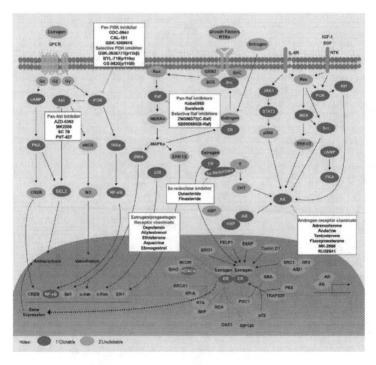

Fig. 3. Ras → MEK → Src → AR

$$
\begin{array}{c}
\begin{array}{ccc}
 & A & B & C \\
A & \begin{bmatrix} 0 & 1 & 1 \\ B & 1 & 0 & 1 \\ C & 0 & 0 & 0 \end{bmatrix}
\end{array}
\end{array}
\qquad
\begin{array}{ccccc}
 & C & D & E & F \\
C & \begin{bmatrix} 0 & 1 & 1 & 1 \\ D & 0 & 0 & 1 & 1 \\ E & 0 & 0 & 0 & 1 \\ F & 0 & 0 & 0 & 0 \end{bmatrix}
\end{array}
$$

(a)   (b)

Fig. 4. (a) DAG → PKC → Ras phosphorylation relationship matrix, means PKC and Ras phosphorylated modification by DAG, DAG and Ras phosphorylated modification by PKC. **(b)** Ras → MEK → Src → AR phosphorylation relationship matrix, means MEK, Src and AR phosphorylated modification by Ras, Src and AR phosphorylated modification by MEK, and AR phosphorylated modification by Src.

The Boolean network and state transition are inferred by phosphorylation modification relationship matrix have been shown in Figs. 5 and 6.

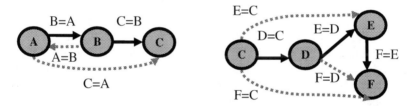

Fig. 5. The Boolean network are inferred by algorithm. The solid line shows the modified relation consistent with the real relationship, the dotted line shows the contrary reaction.

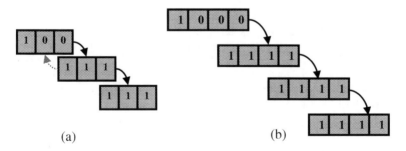

(a)   (b)

Fig. 6. (a) Is the phosphorylation process of DAG → PKC → Ras; (b) is the phosphorylation process of Ras → MEK → Src → AR.

Through training, we gain the reverse modeling for the phosphorylation network. Repeat the experiment 50 times and average the results, and the evaluation index points shown in Table 1.

Table 1. Evaluation index

	TPR	TNR	FPR	FNR	P	R	Acc	F
DAG → PKC → Ras	1	0.714	0.286	0	0.5	1	0.778	0.667
Ras → MEK → Src → AR	1	0.769	0.231	0	0.5	1	0.813	0.667

Obviously, according to the infer model and calculate the evaluation index points, the vast majority of relationships are inferred accurately. The average accuracy of the algorithm is about 80 %. It also shows that, the method proposed in this paper with higher accuracy and reliability in building phosphorylation network.

4 Conclusions

The PTM is a key process of protein function regulation and important for revealing the pathogenesis of various diseases. Nowadays, how to build effective regulation networks of PTM has become a characterization field of proteomics problem needed to be solved. According to the experimental data from proteomics technology, in this paper, we successfully build a phosphorylation network based on Boolean Network model by the method of PBIL and TDE, and prove the validity and reliability by reverse engineering methods. However, lots of experimental proof and different protein phosphorylation still has some unknown complex interactive control in the process of the protein phosphorylation modification. In further research, we hope to find more effective methods for constructing protein post-translational multi-modifications regulated networks and further reveal the inside control laws.

Acknowledgement. This research was partially supported by Program for Scientific research innovation team in Colleges and universities of Shandong Province 2012–2015, the Key Project of Natural Science Foundation of Shandong Province (ZR2011FZ001), the Natural Science Foundation of Shandong Province (ZR2011FL022, ZR2013FL002), the Youth Science and Technology Star Program of Jinan City (201406003), the Shandong Provincial Key Laboratory of Network Based Intelligent Computing and Shandong Provincial Natural Science Foundation, China, under Grant ZR2015JL025. This work was also supported by the National Natural Science Foundation of China (Grant No. 61302128) and Shandong Distinguished Middle-Aged and Young Scientist Encourage and Reward Foundation Grant BS2014DX015. Research Fund for the Doctoral Program of University of Jinan (No. XBS1604). The scientific research foundation of University of Jinan (XKY1410, XKY1411). Authors would like to thank Huixiang Xu of School of Information Science and Engineering, University of Jinan. Thanks the anonymous referees for the technical suggestions and remarks which helped to improve the contents and the quality of presentation.

References

1. Na, S., Paek, E.: Software eyes for protein post-translational modifications. Mass Spectrom. Rev. **34**(2), 133–147 (2015)
2. Witze, E.S., Old, W.M., Resing, K.A.: Mapping protein post-translational modifications with mass spectrometry. Nat. Method **4**, 798–806 (2007)
3. Ross, K.E., Arighi, C.N., Ren, J., Huang, H., Wu, C.H.: Construction of protein phosphorylation networks by data mining, text mining and ontology integration: analysis of the spindle checkpoint (2013)
4. Pawson, T., Linding, R.: Network medicine. FEBS Lett. **582**, 1266–1270 (2008)
5. Videla, S., Guziolowski, C., Eduati, F.: Learning boolean logic models of signaling networks with ASP. Theor. Comput. Sci. (2014)
6. Saez-Rodriguez, J., Simeoni, L., Lindquist, J.A.: A logical model provides insights into T cell receptor signaling. PLoS Comput. Biol. **3**(8), 163 (2007)
7. Samaga, R., Saez-Rodriguez, J., Alexopoulos, L.G.: The logic of EGFR/ErbB signaling. Theoretical properties and analysis of high-throughput data. PLoS Comput. Biol. **5**(8), 1000438 (2009)
8. Schlatter, R., Schmich, K., Vizcarra, I.A.: ON/OFF and beyond-a boolean model of apoptosis. PLoS Comput. Biol. **5**(12), 1000595 (2009)
9. Helikar, T., Kochi, N., Konvalina, J.: Boolean modeling of biochemical networks. Open Bioinform. J. **5**, 16–25 (2011)
10. Baluja, S.: Population-based incremental learning. A method for integrating genetic search based function optimization and competitive learning. Technical report CMU-CS-94-163, Carnegie Mellon University, Pittsburgh, PA (1994)
11. Baluja, S.: An empirical comparison of seven iterative and evolutionary function optimization heuristics. Technical report CMU-CS-95-193, Computer Science Department, Carnegie Mellon University (1995)
12. Baiuja, S., Caruana, R.: Removing the genetics from standard genetic algorithm. In: Proceedings of the International Conference on Machine Learning, pp. 38–46. Morgan Kaufmann, San Mateo, USA (1995)
13. Fan, H.Y., Lampinen, J.: A trigonometric mutation operation to differential evolution. J. Glob. Optim. **27**(1), 105–129 (2003)
14. Storn, R., Price, K.: Differential evolution-a simple and efficient heuristic for global optimization over continuous spaces. J. Glob. Optim. **11**(4), 341–359 (1997)

Analysis of MicroRNA and Transcription Factor Regulation

Wei-Li Guo[1], Kyungsook Han[2], and De-Shuang Huang[1(✉)]

[1] Institute of Machine Learning and Systems Biology, College of Electronics
and Information Engineering, Tongji University, Shanghai 201804, China
guoweili_henu@126.com, dshuang@tongji.edu.cn
[2] Department of Computer Science and Engineering, Inha University,
Incheon, South Korea
khan@inha.ac.kr

Abstract. Gene regulatory networks in different tissues offer insight into the mechanism of tissue identity and function. Here we construct regulatory networks in 10 human tissues including regulations among miRNAs, transcription factors and genes. The results reveal that TS miRNAs are regulated largely by non-tissue specific TFs. TS miRNAs connect with more TFs compared with trivial miRNAs, inferring tight co-regulation of gene expression for TS miRNAs and TFs. Both TS miRNAs and TSTFs tend to regulate broad sets of genes involved in tissue specific functions. In particular, we identified tissue specific regulations instrumental to defining tissue specific functions, and some pathways important to tissue identity or disease, which cannot be explained by only tissue specific genes, can be captured in our tissue specific regulations.

Keywords: Gene regulatory network · miRNA · Transcription factor · Tissue specific regulation · Tissue specificity

1 Introduction

There are hundreds of cell types in human body, but they behave diversity in morphology and perform different functions despite harboring the same genetic information, which is called tissue specificity. Tissue specificity is developed through tissue-dependent mechanisms of gene regulation, including transcriptional regulation and post-transcriptional regulation [1–3]. Tissue specificity has significant consequence, for example, tissue specific genes are more likely to be drug target, and the disordered regulations usually result in disease. Understanding these regulatory underpinnings of complex tissues is crucial for uncovering human physiology.

In recent years, Identification of gene regulatory networks that integrate both the transcriptional and post-transcriptional regulation has been extensively studied, for example, Shalgi et al. constructed mammalian miRNA-transcriptional factor regulatory network and studied the global and local architecture of regulating [4]. More recently, Gersten et al. constructed an integrated gene regulatory network in human by combining high-throughput ChIP-seq data with miRNA regulatory data as well as protein interaction data and studied the general principles of regulation without considering

© Springer International Publishing Switzerland 2016
D.-S. Huang et al. (Eds.): ICIC 2016, Part I, LNCS 9771, pp. 278–284, 2016.
DOI: 10.1007/978-3-319-42291-6_27

tissue specificity [5]. The main limitations of these approaches are that: (1) Few TFs were included. (2) The regulatory networks are studied only in a single cell type, or various tissues were considered as a whole. Systematically mapping integrated regulatory networks among TFs and miRNAs across different tissues is largely missing.

In this paper, we constructed large-scale gene regulatory networks in 10 human tissues by integrating TF and miRNA regulation database which cover almost all the publicly curated and experimentally predicted regulations, with gene expression data and protein expression data. We analyze the regulatory features of miRNAs and TFs through analysis of topological properties in the network as well as well as how different regulators cooperate in regulation to achieve tissue function. Furthermore, we identify tissue specific regulations for each tissue, and investigate the role they play in tissue identity and disease. The results show that TS miRNAs tend to connect with more TFs compared with trivial miRNAs, indicating tight co-regulation of gene expression for TS miRNAs and TFs. Functional analysis of TSRNs shows that specific regulations are often instrumental to tissue specific functions. Our regulatory networks tend to provide new insight into how different regulators and TSRNs contribute to tissue identity.

2 Methods

2.1 Construction of Tissue Regulatory Network

To construct the regulatory network for each tissue, first, a background regulatory network was constructed using literature reported and experimentally predicted regulatory interactions among TFs, miRNAs and genes, without considering tissue condition, for TF-gene regulatory interactions, a large number of TF-gene curated regulations were collected from databases TRED [6], PAZAR [7], TRRUST [8]. Besieds, experimentally predicted TF target relations were obtained from Encyclopedia of DNA elements (ENCODE) ChIP-seq data [9] and JASPAR [10]. we mapped the binding sites to human genome, and identified the TF target genes by identifying the TFBSs within the region from 1 kb upstream and 500b downstream of the transcription start site (TSS) for all human genes. Human miRNAs were from miRBase [11], for the TF-miRNA regulations, we obtained from TRansmiR [12]. For miRNA-gene interactions, the experimentally confirmed interactions were obtained from public databases Tarbase [13] and miRTarBase [13], and the potential interactions were the ones predicted by PicTar [14]. Finally, a reference regulatory network for human was built with 356436 edges, consisting of 925 TFs, 1602 miRNAs and 19855 genes.

Based on the background regulatory network, and the expression data for TFs, genes and miRNAs in each tissue, the tissue regulatory network for a particular tissue was built by extracting TFs that were expressed both at protein level and transcript level, miRNAs and genes that ware expressed at transcript level in the respective tissue as well as regulatory relations between them from the background regulatory network.

2.2   Tissue-Specific Genes and Housekeeping Genes

Tissue specific genes were identified from GSE2361 using method [15], HK genes were collected from Chang et al. [15]. In total, 3378 tissue specific genes and 2064 HK genes were recognized. Tissue specific TFs were identified using protein expression data from [16] with the same method as TSGs, and general TFs were considered as those expressed at in at least 90 % tissues at protein level. As a result, 128 tissue specific transcript factors and 120 general TFs were got in 10 tissues. For miRNAs, we considered it as tissue-specific miRNAs if it is specifically expressed in one tissue, 103 tissue-specific miRNAs were extracted from paper [17].

3   Results

3.1   General Property of Tissue Regulatory Networks

The tissue regulatory networks for 10 tissues (i.e. heart, ovary, kidney, pancreas, prostate, placenta, brain, testis, liver and lung) were constructed based on the background regulatory network. The size of regulatory network for each tissue is quite different, the number of TFs varies from 104 to 276, the miRNAs from 64 to 165, and regulatory interactions vary from 14985 to 57100.

3.2   miRNA and TF Regulation

TFs and miRNAs are crucial in maintaining tissue functions, besides tissue-specific TFs and miRNAs are likely to be involved in complex disease [18–20]. We analyze the regulatory feature of TFs and miRNAs in the network and how TFs and miRNAs cooperate to regulate gene expression in different tissues.

To investigate the regulatory relationship between miRNAs and TFs and how they cooperate to control gene expression, we did the following studies. By studying the proportion of TSTFs across tissues for regulating TS miRNAs (Fig. 1(a)), we found that in most tissues, TS miRNAs are regulated widely by non-TS TFs. This result is consistent with previous reports that TFs involved in tissue-specific regulatory networks were usually non-TSTFs in the corresponding tissue [21]. To further investigate the activity of miRNAs and TFs, we studied the connectivity with TFs for both TS miRNAs and trivial miRNAs (Fig. 1(b)). The result shows that TS miRNAs tend to connect with significantly higher number of TFs compared with trivial miRNAs (p-value <0.01; Wilcoxon rank-sum test). The close interaction between TS miRNAs and TFs suggests strong cooperation of them in regulating gene expression.

MiRNAs achieve their function through their target genes. To further determine the function of target genes of miRNAs, we performed functional enrichment analysis using DAVID [22], for example, in heart, the target genes of TS miRNAs are enriched with functions like "muscle contraction", "blood circulation" and "circulatory system process", which is important for heart function and development. This indicates that TS miRNAs tend to regulate genes involved in tissue specific functions.

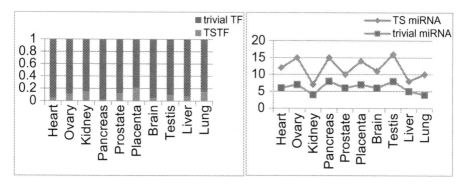

Fig. 1. (a) Proportion of TSTFs and trivial TFs regulating TS miRNAs in each tissue. (b) The number of TFs that miRNAs connect in each tissue. (Color figure online)

By investigating the degree distribution of TFs, we found that most of TFs have a low degree, whereas a few TFs are high connected, indicating the presence of hubs. The hubs of TFs were identified in each tissue and the number and degree distributions of hubs in each tissue (Fig. 2) reveal the quite different activity of TFs across tissues. For example, in lung and ovary, there tend to be more TFs as hubs. In some tissue, like kidney, the number of hubs is very low, while for these hub TFs, the average degree are higher compared with other tissues. This indicates that for each tissue, there are some specific regulatory modes for its own.

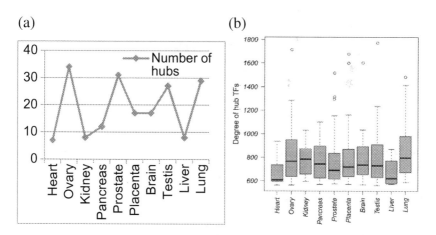

Fig. 2. Hubs of TFs in each tissue. (a) Number of hubs in each tissue. (b) The degree of hub TFs in each tissue.

3.3   Tissue-Specific Regulations

Similar to genes, some regulatory interactions appear in only one tissue, while most others appear in many tissues. These that are only existed in one tissue are called tissue specific regulations. These specific regulations provide important insight into complex biological system and disease.

Our specific regulatory network can not only capture disease genes, but also offer clues about important pathways to the pathogenesis of some disease. Figure 3 illustrate the brain specific regulatory interactions are involved in glioma pathway. We extracted the genes enriched for glioma as well as their related regulators in our brain specific regulatory network. In the network, POU3F2 specifically regulate TBP and TCF7, and these two factors regulate OLIG2, indicating the tight cooperation of POU3F2 and OLIG2. Prior research proved that POU3F2, SOX2 and OLIG2 are core transcription factors in regulation of glioblastoma tumorigenicity [23]. Gliomas often comes with activation of HIF1A pathway, in which HIF1A could activate angiogenesis and invasion through upregulating its target genes crucial for these functions [24]. These specific regulatory interactions captured genes involved in glioma, also offer how genes are controlled in glioma and provide clues to pathogenesis of glioma.

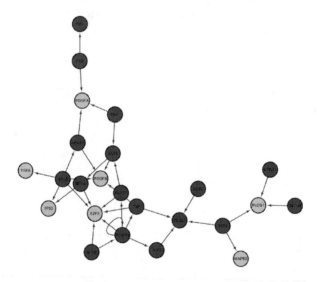

Fig. 3. The brain specific regulatory network in the pathway glioma gene set. The genes enriched in glioma pathway together with their regulators as well as the regulatory interactions from the brain specific regulatory network were analyzed. TFs and genes are in green and cyan, respectively. (Color figure online)

4 Conclusion

In this paper, we constructed integrated specific regulatory networks in 10 human tissues including TF and miRNA regulation. Then we conducted a detailed analysis of the regulatory networks, exploring regulatory principles underlying the diversity of tissue functions.

Through analyze topological property of miRNAs and TFs in the network, we find different classes of regulators behave quite different and act different roles in these networks, TS miRNAs tend to connect with more TFs compared with trivial miRNAs, inferring the tight co-operation of TS miRNAs and TFs in regulating gene expression. The difference of number of hubs as well as the degree of hubs in each tissue shows the diversity and specificity of TF regulation. In particular, tissue specific regulations prefer to be enriched for pathways crucial to maintain tissue specific functions, and some of the pathway information cannot be captured from only tissue specific genes, indicating that TSRNs provide complementary insight into understanding tissue specificity. Together, our work provides new insight into how miRNAs, TFs as well as TSRNs contribute to tissue identity from system level, and the results uncovered here are important for understanding tissue specific regulation.

Acknowledgements. This work was supported by the grants of the National Science Foundation of China, Nos. 61133010, 61520106006, 31571364, 61532008, 61572364, 61373105, 61303111, 61411140249, 61402334, 61472282, 61472280, 61472173, 61572447, and 61373098, China Postdoctoral Science Foundation Grant, Nos. 2014M561513 and 2015M580352.

References

1. Greene, C.S., Krishnan, A., Wong, A.K., Ricciotti, E., Zelaya, R.A., Himmelstein, D.S., Zhang, R., Hartmann, B.M., Zaslavsky, E., Sealfon, S.C.: Understanding multicellular function and disease with human tissue-specific networks. Nat. Genet. **47**, 569–576 (2015)
2. Pierson, E., Koller, D., Battle, A., Mostafavi, S., Consortium, G.: Sharing and specificity of co-expression networks across 35 human tissues (2015)
3. Zhu, L., Guo, W.-L., Deng, S.-P., Huang, D.-S.: ChIP-PIT: enhancing the analysis of ChIP-seq data using convex-relaxed pair-wise tensor decomposition
4. Shalgi, R., Lieber, D., Oren, M., Pilpel, Y.: Global and local architecture of the mammalian microRNA–transcription factor regulatory network. PLoS Comput. Biol. **3**(7), e131 (2007)
5. Gerstein, M.B., Kundaje, A., Hariharan, M., Landt, S.G., Yan, K.-K., Cheng, C., Mu, X.J., Khurana, E., Rozowsky, J., Alexander, R.: Architecture of the human regulatory network derived from ENCODE data. Nature **489**(7414), 91–100 (2012)
6. Jiang, C., Xuan, Z., Zhao, F., Zhang, M.Q.: TRED: a transcriptional regulatory element database, new entries and other development. Nucleic Acids Res. **35**(suppl 1), D137–D140 (2007)
7. Portales-Casamar, E., Arenillas, D., Lim, J., Swanson, M.I., Jiang, S., McCallum, A., Kirov, S., Wasserman, W.W.: The PAZAR database of gene regulatory information coupled to the ORCA toolkit for the study of regulatory sequences. Nucleic Acids Res. **37**(suppl 1), D54–D60 (2009)
8. Han, H., Shim, H., Shin, D., Shim, J.E., Ko, Y., Shin, J., Kim, H., Cho, A., Kim, E., Lee, T.: TRRUST: a reference database of human transcriptional regulatory interactions. Sci. Rep. **5** (2015)
9. Consortium, E.P.: An integrated encyclopedia of DNA elements in the human genome. Nature **489**(7414), 57–74 (2012)

10. Mathelier, A., Zhao, X., Zhang, A.W., Parcy, F., Worsley-Hunt, R., Arenillas, D.J., Buchman, S., Chen, C.-Y., Chou, A., Ienasescu, H.: JASPAR 2014: an extensively expanded and updated open-access database of transcription factor binding profiles. Nucleic Acids Res. gkt997 (2013)

11. Griffiths-Jones, S., Grocock, R.J., Van Dongen, S., Bateman, A., Enright, A.J.: miRBase: microRNA sequences, targets and gene nomenclature. Nucleic Acids Res. **34**(suppl 1), D140–D144 (2006)

12. Wang, J., Lu, M., Qiu, C., Cui, Q.: TransmiR: a transcription factor–microRNA regulation database. Nucleic Acids Res. **38**(suppl 1), D119–D122 (2010)

13. Hsu, S.-D., Lin, F.-M., Wu, W.-Y., Liang, C., Huang, W.-C., Chan, W.-L., Tsai, W.-T., Chen, G.-Z., Lee, C.-J., Chiu, C.-M.: miRTarBase: a database curates experimentally validated microRNA–target interactions. Nucleic Acids Res. gkq1107 (2010)

14. Krek, A., Grün, D., Poy, M.N., Wolf, R., Rosenberg, L., Epstein, E.J., MacMenamin, P., da Piedade, I., Gunsalus, K.C., Stoffel, M.: Combinatorial microRNA target predictions. Nat. Genet. **37**(5), 495–500 (2005)

15. Chang, C.-W., Cheng, W.-C., Chen, C.-R., Shu, W.-Y., Tsai, M.-L., Huang, C.-L., Hsu, I. C.: Identification of human housekeeping genes and tissue-selective genes by microarray meta-analysis. PLoS ONE **6**(7), e22859 (2011)

16. Kim, M.-S., Pinto, S.M., Getnet, D., Nirujogi, R.S., Manda, S.S., Chaerkady, R., Madugundu, A.K., Kelkar, D.S., Isserlin, R., Jain, S.: A draft map of the human proteome. Nature **509**(7502), 575–581 (2014)

17. Landgraf, P., Rusu, M., Sheridan, R., Sewer, A., Iovino, N., Aravin, A., Pfeffer, S., Rice, A., Kamphorst, A.O., Landthaler, M.: A mammalian microRNA expression atlas based on small RNA library sequencing. Cell **129**(7), 1401–1414 (2007)

18. Ravasi, T., Suzuki, H., Cannistraci, C.V., Katayama, S., Bajic, V.B., Tan, K., Akalin, A., Schmeier, S., Kanamori-Katayama, M., Bertin, N.: An atlas of combinatorial transcriptional regulation in mouse and man. Cell **140**(5), 744–752 (2010)

19. Esquela-Kerscher, A., Slack, F.J.: Oncomirs—microRNAs with a role in cancer. Nat. Rev. Cancer **6**(4), 259–269 (2006)

20. Huang, D.-S.: Systematic Theory of Neural Networks for Pattern Recognition, vol. 28, pp. 323–332. Publishing House of Electronic Industry of China, Beijing (1996)

21. Yu, X., Lin, J., Zack, D.J., Qian, J.: Computational analysis of tissue-specific combinatorial gene regulation: predicting interaction between transcription factors in human tissues. Nucleic Acids Res. **34**(17), 4925–4936 (2006)

22. Dennis Jr., G., Sherman, B.T., Hosack, D.A., Yang, J., Gao, W., Lane, H.C., Lempicki, R. A.: DAVID: database for annotation, visualization, and integrated discovery. Genome Biol. **4**(5), P3 (2003)

23. Li, J., Kozono, D., Nitta, M., Sampetrean, O., Gonda, D., Kushwaha, D.S., Merzon, D., Ramakrishnan, V., Zhu, S., Zhu, K.: Dynamic epigenetic regulation of glioblastoma tumorigenicity through LSD1 modulation of MYC expression. Cancer Res. **75**(15 Supplement), 979 (2015)

24. Kaur, B., Khwaja, F.W., Severson, E.A., Matheny, S.L., Brat, D.J., Van Meir, E.G.: Hypoxia and the hypoxia-inducible-factor pathway in glioma growth and angiogenesis. Neuro-oncology **7**(2), 134–153 (2005)

Gene Extraction Based on Sparse Singular Value Decomposition

Xiangzhen Kong[1], Jinxing Liu[1,2(✉)], Chunhou Zheng[1],
and Junliang Shang[1]

[1] School of Information Science and Engineering, Qufu Normal University,
Rizhao 276826, Shandong, China
{kongxzhen, shangjunliang110}@163.com,
{sdcavell, zhengch99}@126.com
[2] Bio-Computing Research Center, Shenzhen Graduate School,
Harbin Institute of Technology, Shenzhen 518055, Guangdong, China

Abstract. In this paper, we develop a new feature extraction method based on sparse singular value decomposition (SSVD). We apply SSVD algorithm to select the characteristic genes from Colorectal Cancer (CRC) genomic dataset, and then the differentially expressed genes obtained are evaluated by the tools based on Gene Ontology. As a gene extraction method, SSVD is also compared with some existing feature extraction methods such as independent component analysis (ICA), the p-norm robust feature extraction (PREE) and sparse principal component analysis (SPCA). The experimental results show that SSVD method outperforms the existing algorithms.

Keywords: Singular value decomposition · Gene extraction · Sparse constraint · Gene Ontology

1 Introduction

Recently, feature extraction and dimensionality reduction have become fundamental tools for many data mining tasks, especially for processing high-dimension low sample size (HDLSS) data such as genomic data, which make many feature extraction methods lose effective. In the large number of genes, only a handful of them can regulate the gene expression. The minor number of genes associated with a special biological process are called differentially expressed genes. Up to now, more and more novel algorithms are designed to extract those genes which are relevant to a biological process from gene expression data. Lee et al. applied PCA to analyze gene expression data [1]. Sparse principal component analysis (SPCA) was used to analyze gene expression data by Journée et al. [2]. Liu et al. employed the robust PCA (RPCA) based method for discovering differentially expressed genes [3]. Witten et al. proposed penalized matrix decomposition (PMD), which was used to extract plant core genes by Liu et al. [4]. Huang et al. used ICA to analyze gene expression data [5, 13]. Liu et al. proposed a feature extraction method based on the Schatten p-norm and L_p-norm named p-norm robust feature extraction (PREE) [6].

© Springer International Publishing Switzerland 2016
D.-S. Huang et al. (Eds.): ICIC 2016, Part I, LNCS 9771, pp. 285–293, 2016.
DOI: 10.1007/978-3-319-42291-6_28

In the most feature extraction methods mentioned above, sparseness constraint has a great influence on the performance for analyzing data. The reason lies in that only a few number of genes regulate the gene expression. SSVD is an algorithm also based on sparseness constraint. It is first developed as an exploratory analysis tool for biclustering [7]. We established the model for gene extraction based on SSVD, and then use it to select the characteristic genes from Colorectal Cancer (CRC) genomic datasets. This dataset is from The Cancer Genome Atlas (TCGA) and has been preprocessed by us. Finally, we evaluate the selected genes through the Gene Ontology tools. By comparing with ICA, PREE, and SPCA methods, all empirical results show that the SSVD-based gene extraction method is more prominent than the competitive methods for identifying differentially expressed genes.

2   Gene Extraction: A SSVD-Based Solution

Let X be a n × p data matrix whose every row represents all of genes' expression level in one sample. The ordinary singular value decomposition (SVD) of X is defined as

$$X = UDV^T = \sum_{k=1}^{r} d_k u_k v_k^T, \tag{1}$$

In the Eq. (1), r is the rank of matrix X, $U = (u_1, \cdots, u_r)$ is a matrix composed by orthonormal left singular vectors $u_i (i = 1, \cdots, r)$, $V = (v_1, \cdots, v_r)$ is a matrix composed by orthonormal right singular vectors $v_i (i = 1, \cdots, r)$, $D = \mathrm{diag}(d, \cdots, d_r)$ is a diagonal matrix with singular values $d_i (i = 1, \cdots, r)$, and $d_1 \geq \cdots \geq d_r > 0$. SVD decomposes X into a summation of rank-one matrices $d_k u_k v_k^T$, each of which can be called a SVD layer. In applications, one usually focuses on the SVD layers corresponding to large d_k values. The rest of SVD layers corresponding to small $d_k s$ can often be recognized as noise or less useful. If we only take the first $K(K < r)$ layers in (1), and then we obtain the following rank-K approximation for X, as shown in Eq. (2) and Fig. 1.,

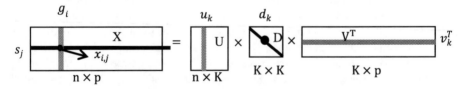

Fig. 1. Graphical depiction of SVD of a matrix X, where $s_j(j = 1,2, \cdots, p)$ is the sample expression profile, g_i is the gene transcriptional responses, $u_k(k = 1, \cdots, K)$ is a unit n-vector, $v_k^T (k = 1, \cdots, K)$ is a unit p-vector.

$$X \approx X^{(K)} = \sum_{k=1}^{K} d_k u_k v_k^T, \tag{2}$$

where $X^{(K)}$ gives the closest rank-K matrix approximation to X in the sense that $X^{(K)}$ minimizes the squared Frobenius norm, i.e.

$$X^{(K)} = \underset{X^* \in A_K}{\operatorname{argmin}} \|X - X^*\|_F^2 = \underset{X^* \in A_K}{\operatorname{argmin}} \operatorname{tr}\left\{ (X - X^*)(X - X^*)^T \right\}, \tag{3}$$

where A_K is the set of all n × p matrices of rank-K [8].

SSVD seeks a low-rank matrix approximation to X as that in Eq. (2) with the requirement that the vectors u_k and v_k are sparse. i.e., they have many zero entries. Mihee et al. first proposed SSVD as an exploratory analysis tool for biclustering [7]. They use the method of iteration to obtain the first l layers corresponding to the larger singular values such as $d_1, \cdots d_l$, each SSVD layer referred to a rank-one matrix $d_k u_k v_k^T$. Firstly, they only focus on the first layer obtained by SSVD. The best rank-one matrix approximation of X under the Frobenius norm is as,

$$(d_1, u_1, v_1) = \underset{X^* \in A_K}{\operatorname{argmin}} \|X - duv^T\|_F^2, \tag{4}$$

where d is a positive scalar, u is a unit n-vector, v is a unit p-vector. In order to obtain sparse vectors u and v, they proposed to add sparsity-inducing penalties on u and v in Eq. (4). The following penalized sum-of-squares criterion is the optimization objective

$$\|X - duv^T\|_F^2 + \lambda_u P_1(u) + \lambda_v P_2(v), \tag{5}$$

where $P_1(u)$ and $P_2(v)$ are sparsity-inducing penalty terms. λ_u and λ_v are non-negative sparse penalty parameters balancing the goodness-of-fit. When $\lambda_u = \lambda_v = 0$, the criterion (5) is reduced to (4). For $P_1(u)$ and $P_2(v)$, SSVD takes advantage of the adaptive lasso penalties proposed by Zou et al. [11],

$$P_1(u) = \sum_{i=1}^{n} w_{1,i} |u_i| \ and \ P_2(v) = \sum_{j=1}^{p} w_{2,j} |v_j|, \tag{6}$$

where $w_{1,i}$ and $w_{2,j}$ are data-driven weights. If $w_{1,i} = w_{2,j} = 1$ for every i and j, then Eq. (6) become the plain lasso penalty. Following Zou [9], the weights $w_{1,i}s$ can be chosen as $w_1 = \left(w_{1,1}, \cdots, w_{1,n} \right)^T = |\hat{u}|^{-\gamma_1}$, $|\hat{u}|^{-\gamma_1}$ is defined as an operation to each component of the vector $\hat{u}$, and $\hat{u}$ is the ordinary least squares (OLS) estimate of u. The Bayesian information criterion (BIC) [10] can be used to select an appropriate γ_1 from a finite set of candidate values. Zou Suggests using $\gamma_1 = 0.5$ or 2 [9]. $w_2 = (w_{2,1}, \cdots, w_{2,p})^T = |\hat{v}|^{-\gamma_2}$, and $\hat{v}$ is the OLS estimator of v, and γ_2 can be chosen similarly as γ_1.

According to the adaptive lasso penalty, the minimizing objective (5) is written as

$$\|X - duv^T\|_F^2 + \lambda_u \sum_{i=1}^{n} w_{1,i} |u_i| + \lambda_v \sum_{j=1}^{p} w_{2,j} |v_j|. \tag{7}$$

The alternating direction method is utilized to evaluate u and v. When fixing u, Eq. (7) is equivalent to minimizing

$$\left\|X - duv^T\right\|_{\mathrm{F}}^2 + \lambda_v \sum_{j=1}^{p} w_{2,j}|v_j|$$
$$= \|X\|_{\mathrm{F}}^2 + \sum_{j=1}^{p} \{v_j^2 - 2v_j(X^T u)_j + \lambda_v w_{2,j}|v_j|\}. \tag{8}$$

In the same manner, for fixed v, minimizing Eq. (7) is equivalent to minimizing

$$\left\|X - duv^T\right\|_{\mathrm{F}}^2 + \lambda_u \sum_{i=1}^{n} w_{1,i}|u_i|$$
$$= \|X\|_{\mathrm{F}}^2 + \sum_{i=1}^{n} \{u_i^2 - 2u_i(X^T v)_i + \lambda_u w_{1,i}|u_i|\}. \tag{9}$$

In [7], Mihee et al. proposed an efficient iterative algorithm and a lemma that effectively utilized the special structure of SSVD. The lemma gave a closed-form solution to such minimization problems as Eqs. (8) and (9). People interested in the lemma could review the literature [7]. According to the lemma, we obtained the optimizing u is $\tilde{u}_i = sign\{(Xv)_i\}(|(Xv)_i| - \lambda_u w_{1,i}/2)_+$, and then we got the unit vector by letting $u = u/\|u\|$, and the optimizing v is $\tilde{v}_j = sign\{(Xu)_j\}(|(Xu)_j| - \lambda_v w_{2,j}/2)_+$. The minimization of Eq. (7) with respect to u and v is iterated until convergence.

The degree of singular vectors u and v is closely related to the two penalty parameters λ_u and λ_v presented in Eqs. (5) and (7). As discussed above, we can conclude that the degree of sparsity of u is the number of $(Xv)_i$ s that are bigger than $\lambda_u w_{1,i}/2$. In other words, fixing v, the degree of sparsity of u depends on λ_u, the same for vector v. It means that the parameter λ_v or λ_u is larger, the u or v is more sparser. Zou et al. [11] suggested that the Bayesian information criterion (BIC) [10] can be used to select the optimal number of non-zero coefficients. SSVD adopted the BIC to selected the parameters λ_u and λ_v.

When extracting feature genes using SSVD, we only focus on the first layer which corresponding to the largest singular value, because the largest singular value highlights the most obvious characteristic of the matrix. The genes are usually grouped into up-regulated genes and down-regulated genes [12], which can be reflected by the positive items and negative items in the sparse right singular vector v. Here, we only consider the absolute value of the items in v to identify the differentially expressed genes. So the algorithm to obtain the characteristic genes can be introduced as this:

1. Get the first triplet $\{d_{old}, u_{old}, v_{old}\}$ using the standard SVD to X, here d_{old} is first singular value in the diagonal matrix D, u_{old} is a unit n-vector, i.e. the first column of left singular matrix U, v_{old} is the a unit p-vector, the first column of the right singular matrix V.
2. Update:
   (a) Set $\tilde{v}_j = sign\{(X^T u_{old})_j\}(|(X^T u_{old})_j| - \lambda_v w_{2,j}/2)_+$, $j = 1, \cdots, p$, where λ_v is the minimizer of BIC(λ_v) defined in [10]. Let $\tilde{v} = (\tilde{v}_1, \cdots, \tilde{v}_p)$, and $v_{new} = \tilde{v}/\|v\|$.
   (b) Set $\tilde{u}_i = sign\{(X^T v_{new})_i\}(|(X^T v_{new})_i| - \lambda_u w_{1,i}/2)_+$, $j = 1, \cdots, p$, where λ_u is the minimizer of BIC(λ_u) defined in [10]. Let $\tilde{u} = (\tilde{u}_1, \cdots, \tilde{u}_n)$, and $v_{new} = \tilde{u}/\|u\|$.

(c) Set $u_{old} = u_{new}$ and repeat steps (a) and (b) until convergence.

3. Set $u = u_{new}$, $v = v_{new}$, $d = u_{new}^T X v_{new}$ at convergence.

4. In the three steps above, we get the first rank-one matrix layer, i.e., the sparse left singular vector u_1 and the sparse right singular vector v_1 are obtained. The genes corresponding to non-zero entries in the vector v_1 are selected as the differentially expressed genes.

5. The absolute value on each component in v_1 is sorted in descending order. We take the top c ($c \ll p$) genes as the differentially expressed ones, where c is the number of extracted genes, and p is the total number of genes.

3 Results and Discussion

In this section, we use the CRC dataset to evaluate the SSVD-based gene extraction method, and compare it with other competing methods such as ICA [13], PRFE [6] and SPCA [2]. For every competitive method, we select their parameters corresponding to

Table 1. The GO TERMS associated with **Molecular Function** on the genes identified from CRC dataset by different methods.

ID	Name	ICA	PRFE	SPCA	SSVD	Genes in annotation
		pValue	pValue	pValue	pValue	
		Genes in query	Genes in query	Genes in query	Genes in query	
GO:0044822	Poly(A) RNA binding	1.713E−37	6.965E−54	2.039E−36	**2.204E−99**	1180
		116	138	115	**187**	
GO:0003723	RNA binding	6.491E−36	2.185E−51	1.70E−35	**8.593E−94**	1608
		134	157	134	**207**	
GO:0003735	Structural constituent of ribosome	1.815E−46	1.949E−56	**8.837E−62**	1.983E−60	214
		61	69	**73**	72	
GO:0005198	Structural molecule activity	7.096E−43	6.425E−55	**1.919E−71**	8.841E−57	748
		99	113	**130**	115	
GO:0032403	Protein complex binding	2.258E−12	4.226E−15	**1.115E−24**	7.526E−23	1053
		68	74	**91**	88	
GO:0044877	**Macromolecular complex binding**	4.752E−11	1.11E−13	2.888E−17	**3.245E−21**	1578
		85	92	100	**108**	
GO:0019899	Enzyme binding	1.949E−13	1.102E−15	1.368E−07	**9.868E−21**	1851
		101	107	85	**118**	
GO:0008135	Translation factor activity, RNA binding	-[a]	1.047E−08	0.00002158	**1.109E−14**	90
		-	15	11	**21**	
GO:0050839	Cell adhesion molecule binding	1.751E−11	2.105E−11	**6.459E−22**	1.235E−14	203
		26	26	**38**	30	
GO:0019843	rRNA binding	1.22E−08	1.298E−15	**2.994E−18**	2.391E−14	55
		12	18	**20**	17	

[a]In Table 1., the mark "-" means the pValue is too bigger or the number of "Genes in Query" is too small to list.

the optimal performance. The CRC dataset used in our experiment contains 20,502 genes in 281 samples, The SSVD-base method is firstly used to identify characteristic genes from these datasets. For fair comparison, 500 genes are selected by every competitive methods.

The differentially expressed genes extracted are evaluated by Gene Ontology (GO) tool. Many GO tools have been developed recently for ontology based gene list functional enrichment analysis [14, 15]. We choose ToppFun [16] which is available at http://toppgene.cchmc.org/enrichment.jsp. The main objectives to be analyzed are GO: "Molecular Function", "Biological Process", "Cellular Component", and "Coexpression" [17], and the minimum number of genes is set to 2 and maximum P-value is 0.01, the other parameters are default values.

Tables 1, 2, 3 and 4 list the top 10 closely related terms by different methods. In the listed Tables, "Genes in Annotation" denotes the number of genes associated with the

Table 2. The GOTERMS associated with **Biological Process** on the genes identified from CRC dataset by different methods.

ID	Name	ICA	PRFE	SPCA	SSVD	Genes in annotation
		pValue	pValue	pValue	pValue	
		Genes in query	Genes in query	Genes in query	Genes in query	
GO:0070972	Protein localization to endoplasmic reticulum	1.822E−61	2.465E−74	2.465E−74	**6.871E−92**	137
		62	70	70	**80**	
GO:0006613	Cotranslational protein targeting to membrane	5.936E−66	2.299E−78	4.739E−82	**7.282E−92**	110
		60	67	69	**74**	
GO:0006614	SRP-dependent cotranslational protein targeting to membrane	1.27E−66	3.622E−79	6.779E−83	**7.48E−91**	108
		60	67	69	**73**	
GO:0045047	Protein targeting to ER	1.143E−64	7.857E−77	2.876E−82	**4.869E−90**	114
		60	67	70	**74**	
GO:0072599	Establishment of protein localization to endoplasmic reticulum	1.885E−63	2.20E−75	1.04E−80	**2.52E−88**	118
		60	67	70	**74**	
GO:0000184	Nuclear-transcribed mRNA catabolic process, nonsense-mediated decay	1.408E−62	7.822E−78	2.216E−81	**2.856E−85**	121
		60	69	71	**73**	
GO:0000956	Nuclear-transcribed mRNA catabolic process	3.107E−53	5.653E−67	9.266E−63	**1.662E−74**	200
		65	75	72	**80**	
GO:0006612	Protein targeting to membrane	4.559E−54	1.341E−63	1.341E−63	**7.067E−74**	203
		66	73	73	**80**	
GO:0006413	Translational initiation	1.464E−47	1.009E−59	5.372E−56	**4.82E−73**	269
		68	78	75	**88**	
GO:0090150	Establishment of protein localization to membrane	2.715E−50	2.769E−59	5.232E−57	**1.051E−72**	371
		80	88	86	**99**	

Table 3. The GO TERMS associated with **Cellular Component** on the genes identified from CRC dataset by different methods.

ID	Name	ICA	PRFE	SPCA	SSVD	Genes in annotation
		pValue	pValue	pValue	pValue	
		Genes in query	Genes in query	Genes in query	Genes in query	
GO:0022626	Cytosolic ribosome	2.367E−65	9.138E−78	2.762E−83	**2.583E−85**	108
		59	66	69	**70**	
GO:0005925	Focal adhesion	4.006E−52	6.807E−67	4.074E−69	**1.395E−81**	392
		83	95	98	**108**	
GO:0030055	Cell-substrate junction	2.835E−52	1.167E−65	5.599E−68	**1.521E−81**	402
		84	97	98	**109**	
GO:0005924	Cell-substrate adherens junction	1.169E−51	4.154E−65	1.526E−68	**6.219E−81**	397
		83	95	98	**108**	
GO:0070161	Anchoring junction	1.671E−49	4.87E−63	8.57E−61	**4.018E−80**	497
		89	102	100	**117**	
GO:0005912	Adherens junction	5.767E−48	1.635E−61	2.007E−61	**1.215E−78**	478
		86	99	99	**114**	
GO:0044391	Ribosomal subunit	1.99E−53	6.097E−62	2.72E−66	**6.445E−68**	170
		61	67	70	**71**	
GO:0044445	Cytosolic part	1.62E−44	3.102E−53	4.6E−57	**2.978E−65**	220
		60	67	70	**76**	
GO:1990904	Ribonucleoprotein complex	1.894E−28	5.96E−37	2.84E−35	**3.802E−63**	719
		79	90	88	**119**	
GO:0030529	Intracellular ribonucleoprotein complex	1.894E−28	5.961E−37	2.841E−35	**3.802E−63**	719
		79	90	88	**119**	

Table 4. The GO TERMS associated with **Coexpression** on the genes identified from CRC dataset by different methods.

ID	Name	ICA	PRFE	SPCA	SSVD	Genes in annotation
		pValue	pValue	pValue	pValue	
		Genes in query	Genes in query	Genes in query	Genes in query	
M11197	Housekeeping genes dentified as expressed across 19 normal tissues.	9.072E−118	4.169E−158	3.938E−131	**1.197E−204**	389
		128	154	137	**181**	
16872506-SuppTable1	Human Leukemia Yukinawa06 2000genes	3.01E−126	4.508E−148	3.595E−136	**2.961E−199**	1505
		218	237	227	**277**	
17210682-SuppTable3	Human Colon Grade07 1102genes	3.659E−54	9.783E−64	1.683E−49	**3.837E−89**	833
		109	119	104	**143**	
12456497-Table 4	Human Leukemia Durig03 88genes	4.338E−61	6.394E−77	7.217E−77	**1.444E−85**	81
		49	57	57	**61**	
M14524	Up-regulated genes in colon carcinoma tumors compared to the matched normal mucosa samples.	8.896E−42	3.04E−52	1.55E−42	**6.625E−78**	870
		97	109	98	**135**	
18689800-TableS 7	Human EmbryonicStemCell Thomas08 1088genes	1.501E−55	1.528E−67	3.262E−50	**1.345E−76**	1023
		121	134	115	**143**	
17210682-SuppTable2	Human Colon Grade07 1950genes	4.944E−40	7.823E−49	1.768E−36	**5.107E−74**	1549
		126	138	121	**168**	
M5792	Genes with increased copy number that correlates with increased expression across six different lung adenocarcinoma cell lines.	1.009E−66	2.647E−71	1.027E−60	**5.503E−73**	178
		68	71	64	**72**	
15546871-Table 1S	Mouse Liver White05 638genes	4.597E−51	9.95E−61	6.96E−47	**3.966E−71**	501
		85	94	81	**103**	
M2328	Genes translationally regulated in MEF cells (embryonic fibroblasts) in response to serum starvation and by rapamycin (sirolimus) [PubChemID = 6610346].	2.205E−40	7.172E−57	8.458E−59	**4.195E−67**	68
		35	44	45	**49**	

term in global genome, "Genes in Query" denotes the number of genes associated with the term in query. It demonstrates that SSVD-GO results have very lower P-values and larger proportion of input genes in the annotated genes. We mark out the superior results in bold type. For example, in Table 1, there are 1180 genes in the genome with the term of "GO:0044822:poly(A) RNA binding", ICA, PREE, SPCA can identify 116, 138, 115 genes, respectively. However, SSVD can identify 187 genes and has the lowest P-value (2.204E-99). From Tables 2, 3 and 4., it can be clearly found that in all GO functional categories SSVD-based method performs best in the competitive methods including ICA, PRFE and SPCA. Only in Table 1, there are five items SPCA outperforms other methods.

4 Conclusion

In this paper, based on the SSVD, we propose a novel feature extraction method to identify differentially expressed genes from gene expression datasets. Numerous elaborate experiments on two gene expression datasets demonstrate that the proposed method has a better performance than the other gene identification methods. In future, systematic studies on other datasets will be conducted for more convincing arguments.

Acknowledgement. This work was supported in part by the grants of the National Science Foundation of China, Nos. 61572284, 61502272, 61572283; Shenzhen Municipal Science and Technology Innovation Council, No. JCYJ20140417172417174; Natural Science Foundation of Shandong Province, No. BS2014DX004.

References

1. Lee, D., Lee, W., Lee, Y., Pawitan, Y.: Super-sparse principal component analyses for high-throughput genomic data. BMC Bioinf. **11**(1), 296 (2010)
2. Journée, M., Nesterov, Y., Richtárik, P., Sepulchre, R.: Generalized power method for sparse principal component analysis. J. Mach. Learn. Res. **11**, 517–553 (2010)
3. Liu, J.X., Wang, Y.T., Zheng, C.H.: Robust PCA based method for discovering differentially expressed genes. BMC Bioinf. **14**(Suppl 8), S3 (2013)
4. Liu, J.X., Zheng, C.H., Xu, Y.: Extracting plants core genes responding to abiotic stresses by penalized matrix decomposition. Comput. Biol. Med. **42**(5), 582–589 (2012)
5. Huang, D.S., Zheng, C.H.: Independent component analysis-based penalized discriminant method for tumor classification using gene expression data. Bioinformatics **22**(15), 1855–1862 (2006)
6. Liu, J., Liu, J.X., Gao, Y.L., Kong, X.Z., Wang, D.: A p-norm robust feature extraction method for identifying differentially expressed genes. PLoSONE **10**(7), e0133124 (2015)
7. Lee, M., Shen, H.P., Huang, J.Z., Marron, J.S.: Biclustering via sparse value decomposition. Biometrics **66**, 1087–1095 (2010)
8. Eckart, C., Young, G.: The approximation of one matrix by another of lower rank. Psychometrika **1**, 211–218 (1936)
9. Zou, H.: The adaptive lasso and its oracle properties. J. Am. Stat. Assoc. **101**(475), 1418–1429 (2006)

10. Schwarz, G.: Estimating the dimension of a model. Ann. Stat. **6**, 461–464 (1978)
11. Zou, H., Hastie, T., Tibshirani, R.: On the "degrees of freedom" of the lasso. Ann. Stat. **35**, 2173–2192 (2007)
12. Kilian, J., Whitehead, D., Horak, J., Wanke, D., Weinl, S., Batistic, O.: The AtGenExpress global stress expression data set: protocols, evaluation and model data analysis of UV-B light, drought and cold stress responses. Plant J. **50**(2), 347–363 (2007)
13. Zheng, C.H., Huang, D.S., Zhang, L., Kong, X.Z.: Tumor clustering using nonnegative matrix factorization with gene selection. IEEE Trans. Inf. Technol. Biomed. **13**(4), 599–607 (2009)
14. Sartor, M.A., Mahavisno, V., Keshamouni, V.G., Cavalcoli, J., Wright, Z., Karnovsky, A., Kuick, R., Jagadish, H., Mirel, B., Weymouth, T.: ConceptGen: a gene set enrichment and gene set relation mapping tool. Bioinformatics **26**(4), 456–463 (2010)
15. Boyle, E.I., Weng, S.A., Gollub, J., Jin, H., Botstein, D., Cherry, J.M., Sherlock, G.: GO: termfinder-open source software for accessing gene ontology information and finding significantly enriched gene ontology terms associated with a list of genes. Bioinformatics **20**(18), 3710–3715 (2004)
16. Chen, J., Bardes, E.E., Aronow, B.J., Jegga, A.G.: ToppGene suite for gene list enrichment analysis and candidate gene prioritization. Nucleic Acids Res. **37**(suppl 2), W305–W311 (2009)
17. Wang, E.T., Sandberg, R., Luo, S., Khrebtukova, I., Zhang, L., Mayr, C., Kingsmore, S.F., Schroth, G.P., Burge, C.B.: Alternative isoform regulation in human tissue transcriptomes. Nature **456**(7221), 470–476 (2008)

Explore the Brain Response to Naturalistic and Continuous Music Using EEG Phase Characteristics

Jie Li[1], Hongfei Ji[1(✉)], Rong Gu[1], Lusong Hou[1], Zhicheng Zhang[1],
Qiang Wu[2], Rongrong Lu[3], and Maozhen Li

[1] Department of Computer Science and Technology, Tong Ji University,
No. 4800 CaoAn Highway, Shanghai 201804, People's Republic of China
{nijanice,gurong2001,zzc_0108}@163.com,
jhf@tongji.edu.cn
[2] School of Information Science and Engineering, Shandong University,
No. 27 Shanda Nanlu, Jinan 250100, Shandong, People's Republic of China
wuqiang@sdu.edu.cn
[3] Department of Rehabilitation, Huashan Hospital, Fudan University,
No. 12 Ulumuqi Middle Road, Shanghai 200040, China
echoll66@hotmail.com
[4] Department of Electronic and Computer Engineering, Brunel University,
Uxbridge, London UB8 3PH, UK
Maozhen.Li@brunel.ac.uk

Abstract. Although many researches attempt to extract music-related EEG activities, they usually focus on EEG amplitude characteristics. So far, there is no publication reporting naturalistic and continuous music related components based on EEG phase characteristics. In this work, we explore the brain response to long natural music using only EEG phase characteristics. Benefiting from multiway representation, the Ordered PARAFAC model decomposition, and pattern correlation analysis, related phase factors can be extracted and reveal that the alpha and theta oscillations and central and occipital area are most relevant to the music stimulus, which is consistent with not only the previous work but also the results of corresponding EEG amplitude characteristics. Moreover, phase factors can be combined to identify plausible real brain activities elicited by music. Our studies attest to the effectiveness of EEG phase characteristics in exploring the brain response to naturalistic and continuous music.

Keywords: EEG · Phase · Music · Tensor

1 Introduction

Scalp recorded Electroencephalogram (EEG) is a noninvasive measurement of electrical activities produced by the firing of neuros within the brain. It is assumed to provide a reliable and convenient window to an understanding of the human mind. Usually, the spontaneous EEG is recorded in the rest state of a subject, that is, in the absence of specific sensory stimuli, while Event-related potential (ERP) is collected by presenting auditory,

© Springer International Publishing Switzerland 2016
D.-S. Huang et al. (Eds.): ICIC 2016, Part I, LNCS 9771, pp. 294–305, 2016.
DOI: 10.1007/978-3-319-42291-6_29

visual or somatosensory stimuli. Although the spontaneous EEG is widely used for the clinical application [1], the research fields of neuroscience, cognitive science, psychology, are extensively depend on ERP [2], since it is the measured brain response that is the direct result of a specific sensory or cognitive event. Traditionally, ERP is acquired in a specially designed experiment, in which a large number of identical stimuli are presented [3] repeatedly, and then ERP features in connection with a specific stimulus, such as amplitude, latency, are extracted by averaging EEG epochs together.

There has been a growing interest in studying cognitive brain function in more practical situations, like listening to naturalistic and continuous music. In this case, analysis mechanicals must consider the continuity property of elicited activity, and relying on plenty of trials repetitions is not feasible. Recently, many researches attempt to extract music-related activity hidden in a mixture of brain activity. Jäncke et al. identified a more or less continuous increase in power relative to a baseline period, indicating strong event-related synchronization (ERS) during music listening [4]. Schaefer et al. demonstrated the amplitude envelope of some short music stimuli (3s-fragment) correlated well with the ERP component [5]. Sturm et al. proposed a multivariate regression based method extracting onset-related brain responses from the ongoing EEG, and led to a significant Cortico-Acoustic Correlation [6]. In Cong's studies [7, 8], EEG data and spectrogram of the ongoing EEG were decomposed by independent component analysis (ICA) or nonnegative tensor factorization (NTF) techniques, and extracted components significantly correlated with the temporal course of a long-term music features.

In neuroscience, phase characteristic have been proven to contain most important information about the neural activity [9]. However, up to now, the most exploited features in elicited EEG to naturalistic and continuous music are all the amplitude characteristics [4–8], phase has been ignored. In our past study [10], we defined phase interval value (PIV) and proposed a computational model based on the ordered Parallel Factors (PARAFAC) algorithm to extract feature from epoched EEG data in brain computer interface (BCI). In this work, we further extend this model to explore the brain response to naturalistic and continuous music using EEG phase characteristic, and figure out the following questions: Can brain response to music be revealed from only phase characteristics of EEG signal? Can phase characteristics be combined with amplitude characteristic to further identify real brain activities elicited by music?

The rest of the paper is organized as follows. Experiment setup and data description in our study are introduced in Sect. 2, Sect. 3 describes the method, and Sect. 4 presents corresponding result. Section 5 offers the conclusion.

2 Experiment Setup and Data Description

The experiment was carried out in the Lab for Brain Cognition and Intelligent Computing, Tong Ji University, China. Totally, ten healthy male subjects, aged from 19 to 32, took part in the experiment. All subjects had normal hearing, and none of them had musical expertise. In the data collection stage, each subject was seated in front of a notebook computer, keeping arms on the chair arms with body relaxing and eye opened. And then an 8.5-min long musical piece of modern tango was played as the stimulus (same with Cong's work [7, 8]). During the subject listened to the music, his EEG data were acquired by Brain Vision actiCHamp amplifier (Brain Products, Germany), and recorded by Brain Vision Recorder. 60-channels' EEG signals (electrodes located at standard positions of the 10–20 international system as FP1, FP2, AF7, AF3, AFz, AF4, AF8, F7, F5, F3, F1, Fz, F2, F4, F6, F8, FT9, FT7, FC5, FC3, FC1, FCz, FC2, FC4, FC6, FT8, FT10, T7, C5, C3, C1, Cz, C2, C4, C6, T8, TP7, CP5, CP3, CP1, CPz, CP2, CP4, CP6, P7, P5, P3, P1, Pz, P2, P4, P6, P8, PO7, PO3, POz, PO4, PO8, O1, Oz, and O2), 2-channels' reference signals (respectively fixed at the left and right mastoids), and 2-channels' ElectroOculogram (EOG) (horizontally EOG and vertically EOG) were also recorded in this experiment. EEG data were collected with the sampling rate of 500 Hz and then rereferenced to average references, down-sampled to 250 Hz, band-pass filtered within 4–30 Hz (using two-way least-squares FIR filtering).

3   Method

3.1   Music Feature Calculation

Five musical features usually studied in related work [7, 11], that is, fluctuation centroid, fluctuation entropy, and pulse clarity, strength of major of minor mode (i.e. mode), and Key Clarity, were calculated using the MIR toolbox [12]. In details, the 512-s long music was segmented by a 3-s sliding window, and the overlap between two adjacent windows was 2 s. Then for each feature, 510 samples were obtained at a sampling frequency of 1 Hz.

3.2   EEG Phase Characteristics Multiway Representation

In this work, we apply phase locking value (PLV) and phase interval value (PIV) to detect phase characteristics. PLV reflects the stability of the phase differences, that is, phase synchrony between signals. It was applied in EEG to evaluate the dynamic integration of distributed neural networks in the brain [13–15]. PIV reflects the degree of the phase difference directly, and was firstly proposed and effectively applied for epoched EEG data classification in our previous work [10].

Continuous EEG data were collected from the subject listening to the long music. By calculating each channel's data $x(c, t)$ (collected at channel c and time t) convolutions with a complex wavelet (the complex Morlet wavelet, with the center frequency $\Omega = 1$, and the bandwidth parameter $\sigma = 2$ as the wavelet mother), 3-way complex coefficients can be obtained (at channel c, frequency f and time t), which can be written as:

$$WTx(c, f, t) = WTx_{r(c,f,t)} + jWTx_{i(c,f,t)} \qquad (1)$$

Where WTx_r denotes its real part, and WTx_i denotes its imaginary part.

A sliding window was used to segment those temporal continuous coefficients, similar to the music feature calculation. In order to include the evoked delay time, the duration of the window was 4 s, and the windows' overlap was the same as musical feature calculation. During each window, 4-way PLV (c1, c2, f, t') and PIV (c1, c2, f, t'), which denote PIV and PLV between channel c1 and channel c2 at frequency f and time t', were calculated as follows:

$$PLV(c1, c2, f, t') = \left| \sum_{t=(t'-1)*250+1}^{(t'-1)*250+4*250} \exp(j(\text{artan}\frac{WTx_{i(c1,f,t)}}{WTx_{r(c1,f,t)}} - \frac{WTx_{i(c2,f,t)}}{WTx_{r(c2,f,t)}}) \right| \quad (2)$$

$$PIV(c1, c2, f, t') = \sum_{(t'-1)*250+1}^{(t'-1)*250+4*250} \left| \exp\left(j \text{ artan}\frac{WTx_{i(c1,f,t)}}{WTx_{r(c1,f,t)}} \right) + \exp(j \text{ artan}\frac{WTx_{i(c1,f,t)}}{WTx_{r(c1,f,t)}}) \right|$$
$$(3)$$

By this way, PLV and PIV samples were obtained at a sampling frequency of 1 Hz, which is consistent with music features at the temporal mode. Finally, for each subject, we constructed 4-way (channel × channel × frequency × time) PLV and PIV tensors at 60-channels between, the frequency of 4–30 Hz (step by 1 Hz), and 1–510 s (step by 1 s).

3.3   EEG Amplitude Characteristic Multiway Representation

In order to evaluate the performance of the phase characteristics, we calculate the amplitudes of the convolution with the wavelet function, reflecting the power at channel c, frequency f and time t, as follows

$$AMPS(c, f, t') = \sum_{t=(t'-1)*250+1}^{(t'-1)*250+4*250} \left| WTx_{r(c,f,t)} + jWTx_{i(c,f,t)} \right| \quad (4)$$

By this way, AMPS samples were obtained at a sampling frequency of 1 Hz, which is also consistent with music features at the temporal mode, and consistent with PIV and PLV at both temporal and spectral modes. For each subject, a 3-way (channel × frequency × time) AMPS tensor was constructed at 60-channels, the frequency of 4–30 Hz (step by 1 Hz), and 1–510 s (step by 1 s).

3.4   Music Related Factors Extraction

When we obtain the multiway-representation (channel × channel × frequency × time) of EEG phase characteristics, the next is how to explore it. Tensor decomposition can factorize those multi-way data without losing potential information among modalities [16]. As one of the most fundamental tensor decomposition models, parallel factor analysis-(PARAFAC) can achieve an easily interpretable model under mild conditions, without constraints in the form of orthogonality or statistical independence, and it is especially appropriate for EEG multi-modal analysis [16].

The tensor factorization under the PARAFAC mode has some problems, such as it is difficult to determine the number of components, and the number of components greatly impacts on the decomposition results. In our previous work, we have developed the Ordered PARAFAC Model [10] and the non-redundant rank-one tensor decomposition [17] based on PARAFAC, and demonstrated they could help to identify the characteristics of epoched EEG data in multi-modes. Compared with the PARAFAC Model, the Ordered PARAFAC Model is designed to find a set of rank-one tensors sequentially, and the most primary factors can be extracted in the first place. Therefore, we extend this Ordered PARAFAC Model to factorize the PIV and PLV from continuous EEG data.

In order to explore the brain response based on EEG phase characteristic, and address the questions proposed earlier, we conduct three different data processing procedures to identify the brain actives related to music, as the following:

(1) Figure 1 illustrates the flow chart of data processing and analysis for 4-way phase tensors. According to the method above mentioned, 4-way tensors PLV and PIV (channel × channel × frequency × time) are constructed based on continuous EEG data from individual subject. The Ordered PARAFAC Model is applied to decompose those 4-way tensors into rank-one tensor factors, and at the same time related phase factors are selected by correlating the temporal pattern of the factors with the temporal courses of music features.

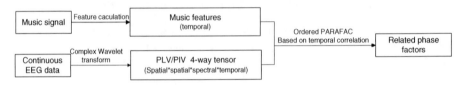

Fig. 1. Flow chart of data processing and analysis for 4-way phase tensors

(2) Figure 2 illustrates the flow chart of data processing and analysis for 5-way phase tensors. Firstly, for each subject, 4-way tensors PLV and PIV (channel × channel × frequency × time) are constructed. Then by combining the PLV and PIV of all subjects together, we can construct 5-way tensors PLV and PIV (channel × channel × frequency × time × subject). The Ordered PARAFAC Model is applied to decompose those 5-way tensors into rank-one tensor factors, and

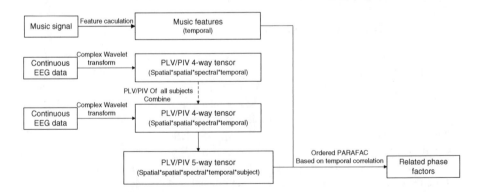

Fig. 2. Flow chart of data processing and analysis for 5-way phase tensors

related phase factors are selected by correlating the temporal pattern of the factors with the temporal courses of music features.

(3) Figure 3 illustrates the flow chart of data processing and analysis for 3-way amplitude tensor. As illustrated in Fig. 3, according to the method above mentioned, for each subject, a 3-way tensor AMPS (channel × frequency × time) is constructed. The Ordered PARAFAC Model is applied to decompose those 3-way tensors into rank-one tensor factors. Although related amplitude factors can be selected by correlating the temporal pattern of the factors with the temporal

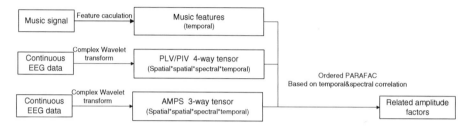

Fig. 3. Flow chart of data processing and analysis for 3-way amplitude tensors

courses of music features, it will be demonstrated that the factors could be further identified by correlating the spectral patterns with those of related phase tensors.

4 Results

In 4-way phase tensors analysis, as listed in Table 1, for almost all subjects (except Sub. 7), factors significantly correlated with the temporal course of the music features (especially the first three features, i.e. pulse clarity, fluctuation entropy, and fluctuation centroid) can be found from PIV or PLV, or both of them. For comparison, Table 1 also lists the corresponding results of 3-way tensor AMPS. As can be seen, phase characteristics can achieve close results to amplitude characteristics, and the most functional to elicit brain response is the fluctuation centroid, which is consistent with the Cong's previous finding by the EEG spectrum (amplitude characteristics) [7, 8].

Furthermore, for almost all subjects, the largest correlations always focus in the pulse clarity and fluctuation centroid. Figure 4 shows the spatial, spectral, and temporal patterns of some related PLV factors (correlative with the pulse clarity and fluctuation centroid) from different subjects, and Fig. 5 shows the spatial, spectral, and temporal patterns of some related PIV factors (correlative with the pulse clarity and fluctuation centroid) from different subjects. They reveal that the alpha and theta oscillation, and central and occipital area are most relevant to the music stimulus, which is also consistent with the Cong's previous findings by the EEG amplitude characteristics [7, 8].

Table 1. The subjects having regarding factors significantly correlative with the regarding music features.

	Pulse clarity	Fluctuation entropy	Fluctuation centroid	Mode	Key clarity
PIV	Sub. 1, 3, 5, 6	Sub. 8, 9, 10	Sub. 2, 3, 4, 6, 7, 8, 9, 10	Sub. 1	/
PLV	Sub. 1, 3, 4, 5, 8, 9	Sub. 2, 3, 8, 9, 10	Sub. 2, 3, 4, 6, 8, 9, 10	Sub. 1	/
AMPS	Sub. 1, 3, 4, 5, 6, 8, 9, 10	Sub. 1, 2, 3, 10	Sub. 1, 2, 3, 4, 5, 7, 8, 9, 10	Sub. 10	/

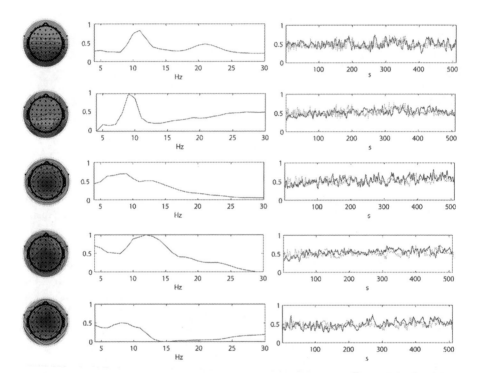

Fig. 4. The spatial, spectral, and temporal patterns of some related 4-way PLV factors from different subjects. (the purple and orange lines denote the temporal course of music feature-pulse clarity and fluctuation centroid respectively) (Color figure online)

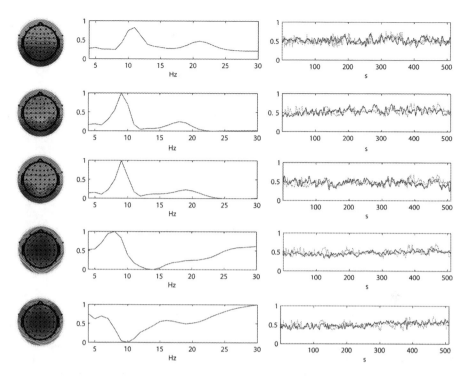

Fig. 5. The spatial, spectral, and temporal patterns of some related 4-way PIV factors from different subjects. (the purple and orange lines denote the temporal course of music feature-pulse clarity and fluctuation centroid respectively) (Color figure online)

It is concluded that brain response to music can be extracted from only EEG phase characteristics.

Although we cannot obtain related factors for all subjects, for 5-way phase tensors analysis, factors significantly correlated (p values < 0.001) with some music features (pulse clarity and fluctuation centroid) present on each subject. As illustrated in Fig. 6, the related phase factors present on each subject, and they reflect that the alpha and theta oscillation, and central and occipital area are most relevant to the music features. We conclude that the music related factors become more primary by combining all subjects' PLV and PIV together. It is confirmed that the efficiency of the proposed algorithm in finding the most primary factors.

For 3-way amplitude tensor analysis, as listed in Table 1, for all subjects, significantly correlations with the temporal course of the music features (especially the first

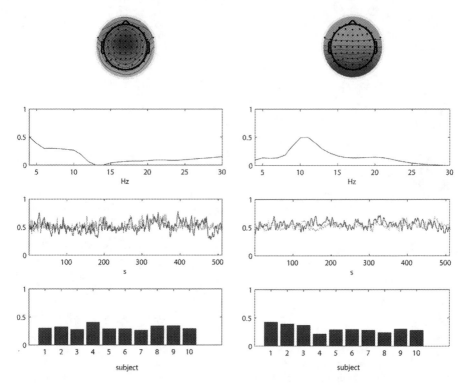

Fig. 6. The spatial, spectral, temporal, and subject patterns of related phase factors of 5-way PLV and PIV from all subjects. (the purple and orange lines denote the temporal course of music feature-pulse clarity and fluctuation centroid respectively) (Color figure online)

three features, i.e. pulse clarity, fluctuation entropy, and fluctuation centroid) can be found from individual AMPS tensors. Actually, we can obtain more related amplitude factors than phase factors. However, although some amplitude factors significant correlative with the music features (on the temporal pattern), they are not real brain activities. As illustrated in Fig. 7, all of the factors show significantly correlated (p values < 0.001) with some of music features, but the second and the fourth factors are evident eye blink artifacts. Here, the amplitude related factors can be further identified by correlating the spectral pattern with those of corresponding subjects' related phase tensors. Phase characteristics can be combined to identify plausible real brain activities elicited by music.

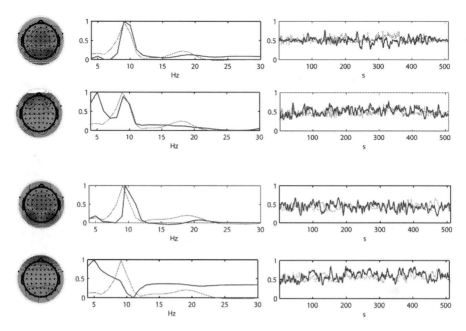

Fig. 7. The spatial, spectral, and temporal patterns of some related 3-way AMPS factors from two subjects. (The first two rows come from one subject, and the bottom two rows come from the other subject. The green lines denote spectral patterns of corresponding subjects' related phase tensors, and the purple and orange lines denote the temporal course of music feature-pulse clarity and fluctuation centroid respectively) (Color figure online)

5   Conclusion

Many researches attempt to extract music-related EEG activities, however, they usually focus on EEG amplitude characteristics. So far, there is no publication reporting naturalistic and continuous music related components based on EEG phase characteristics. In this work, we explore the brain response to long natural music using only EEG phase characteristics. 4-way (channel × channel × frequency × time) and 5-way (channel × channel × frequency × time × subject) PLV and PIV tensor are constructed, and the Ordered PARAFAC Model is applied to decompose those phase tensors into rank-one tensor factors, and then related phase factors are identified by correlating the temporal pattern of the factors with the temporal courses of music features. The phase factors reveal that the alpha and theta oscillation, and central and occipital area are most relevant to the music stimulus, which is consistent with not only the previous work but also the results of corresponding EEG amplitude characteristics. Moreover, it is demonstrated that amplitude related factors can be further selected by correlating the spectral patterns with those of corresponding subjects' related phase tensors, which will help to identify plausible real brain activities elicited by music.

In sum, our studies attest to the effectiveness of EEG phase characteristics in exploring brain response to naturalistic and continuous music.

Acknowledgments. The work was supported by the National Natural Science Foundation of China (Grant Nos. 61105122, 61305060), the Fundamental Research Funds for the Central Universities, Specialized Research Fund for the Doctoral Program of Higher Education (Grant no. 20130131120025), Jinan Youth Star of Science and Technology Plan (Grant no. 201406002), Science and Technology Commission of Shanghai Municipality (Grant Nos. 16JC1401300) and Shanghai sailing program (Grant No. 16YF1415300).

References

1. Henry, J.C.: Electroencephalography: basic principles, clinical applications, and related fields, fifth edition. Neurology **67**(11), 2092 (2006)
2. Reinvang, I.: Cognitive event-related potentials in neuropsychological assessment. Neuropsychol. Rev. **9**(4), 231–248 (1999)
3. Steven, J.: An Introduction to the Event-Related Potential Technique, vol. 18(11), p. 66 (2005). Neuroreport Brisson and Jolicur Copyright © Lippincott Williams
4. Jäncke, L., Kühnis, J., Rogenmoser, L., Elmer, S.: Time course of EEG oscillations during repeated listening of a well-known aria. Front. Hum. Neurosci. **9**, 401 (2015)
5. Schaefer, R.S., Farquhar, J., Blokland, Y., Sadakata, M., Desain, P.: Name that tune: decoding music from the listening brain. Neuroimage **56**(2), 843–849 (2011)
6. Sturm, I., Dähne, S., Blankertz, B., Curio, G.: Multi-variate EEG analysis as a novel tool to examine brain responses to naturalistic music stimuli. PLoS One **10**(10) (2015)
7. Cong, F., Alluri, V., Nandi, A.K., Toiviainen, P., Rui, F., Abu-Jamous, B., et al.: Linking brain responses to naturalistic music through analysis of ongoing EEG and stimulus features. IEEE Trans. Multimedia **15**(5), 1060–1069 (2013)
8. Cong, F., Phan, A.H., Zhao, Q., Nandi, A.K., Alluri, V., Toiviainen, P., et al.: Analysis of ongoing EEG elicited by natural music stimuli using nonnegative tensor factorization. In: Signal Processing Conference (EUSIPCO), 2012 Proceedings of the 20th European, pp. 494–498. IEEE (2012)
9. Rodriguez, E., George, N., Lachaux, J.P., Martinerie, J., Renault, B., Varela, F.J.: Perception's shadow: long-distance synchronization of human brain activity. Nature **397** (6718), 430–433 (1999)
10. Li, J., Zhang, L.: Phase interval value analysis for the motor imagery task in BCI. J. Circuits Syst. Comput. **18**(8), 1441–1452 (2009)
11. Alluri, V., Toiviainen, P., Jääskeläinen, I.P., Glerean, E., Sams, M., Brattico, E.: Large-scale brain networks emerge from dynamic processing of musical timbre, key and rhythm. Neuroimage **59**(4), 3677–3689 (2012)
12. Lartillot, O., Toiviainen, P.: MIR in matlab (II): a toolbox for musical feature extraction from audio. In: Proceedings of the International Conference on Music Information Retrieval 2007, pp. 237–244 (2007)
13. Bhattacharya, J., Petsche, H., Feldmann, U., Rescher, B.: EEG gamma-band phase synchronization between posterior and frontal cortex during mental rotation in humans. Neurosci. Lett. **311**, 29–32 (2001)
14. Mormann, F., Kreuz, T., Andrzejak, R., David, P., Lehnertz, K., Elger, C.: Epileptic seizures are preceded by a decrease in synchronization. Epilepsy Res. **53**, 173–185 (2003)

15. Gysels, E., Celka, P.: Phase synchronization for the recognition of mental tasks in a brain-computer interface. IEEE Trans. Neural Syst. Rehabil. Eng. **12**(4), 406–415 (2005). A Publication of the IEEE Engineering in Medicine and Biology Society
16. Cong, F., Lin, Q.H., Kuang, L.D., Gong, X.F., Astikainen, P., Ristaniemi, T.: Tensor decomposition of EEG signals: a brief review. J. Neurosci. Methods **248**, 59–69 (2015)
17. Ji, H., Li, J., Lu, R., Gu, R., Cao, L., Gong, X.: EEG classification for hybrid brain-computer interface using a tensor based multiclass multimodal analysis scheme. Comput. Intell. Neurosci. **2016**, 1–15 (2016)

Computer Assisted Detection of Breast Lesions in Magnetic Resonance Images

Vitoantonio Bevilacqua[1](✉), Maurizio Triggiani[1],
Maurizio Dimatteo[1], Giuseppe Bellantuono[1], Antonio Brunetti[1],
Leonarda Carnimeo[1], Francescomaria Marino[1], Michele Telegrafo[2],
and Marco Moschetta[2]

[1] Department of Electrical and Information Engineering,
Technical University of Bari, Bari, Italy
vitoantonio.bevilacqua@poliba.it
[2] Department of Radiology, University of Bari, Bari, Italy
marco.moschetta@gmail.com

Abstract. Nowadays preventive screening policies and increased awareness initiatives are up surging the workload of radiologists. Due to the growing number of women undergoing first-level screening tests, systems that can make these operations faster and more effective are required. This paper presents a Computer Assisted Detection system based on medical imaging techniques and capable of labeling potentially cancerous breast lesions. This work is based on MRIs performed with morphological and dynamic sequences, obtained and classified thanks to the collaboration of the specialists from the University of Bari Aldo Moro (Italy). A first set of 60 images was acquired without Contrast Method for each patient and, subsequently, 100 more slices were taken with Contrast Method. This article formally describes the techniques adopted to segment these images and extract the most significant features from each Region of Interest (ROI). Then, the underlying architecture of the suggested Artificial Neural Network (ANN) responsible of identifying suspect lesions will be presented. We will discuss the architecture of the supervised neural network based on the algorithm named Robust Error Back Propagation, trained and optimized so to maximize the number of True Positive ROIs, i.e., the actual tumor regions. The training set, built with physicians' help, consists of 94 lesions and 3700 regions of any interest extracted with the proposed segmentation technique. Performances of the ANN, trained using 60 % of the samples, are evaluated in terms of accuracy, sensitivity and specificity indices. In conclusion, these tests show that a supervised machine learning approach to the detection of breast lesions in Magnetic Resonance Images is consistent, and shows good performance, especially from a False Negative reduction perspective.

Keywords: Medical image classification · Artificial neural networks · Breast lesions detection · Decision support systems

© Springer International Publishing Switzerland 2016
D.-S. Huang et al. (Eds.): ICIC 2016, Part I, LNCS 9771, pp. 306–316, 2016.
DOI: 10.1007/978-3-319-42291-6_30

1   Introduction

According to the US Cancer Statistics Working Group [1] the breast cancer is the second leading cause of death among women after lung tumor. Automated diagnostic tools actually allow clinicians to perform mass screening for breast cancer and other common diseases [2, 3]. Specific clinical markers help specialists in their diagnosis of breast cancer lesions. Several data processing approaches in medical domains have already being developed [2–8], due to the effective improvement in classification and prediction networks. This trend significantly helps clinicians in their diagnosis.

Classification and prediction are two forms of data analysis that can be used to extract models describing important data classes or to predict future data trends [5]. In this work, a Computer Assisted Detection (CAD) system, based on medical imaging techniques and neural detection, is proposed. The techniques adopted to segment these images and extract the most significant features from each ROI are formally described. Then, a Neural Network (NN) capable of detecting potentially cancerous breast lesions in patients' breast MRIs is implemented and a developed case study is reported.

An important role in the diagnosis of breast cancer is played by Contrast Enhanced Magnetic Resonance Imaging (CE-MRI). This method has a reported sensitivity and specificity respectively of 95–99 % and 80 % [9–19]. Image analysis is based on morphological features and on the enhancement patterns of lesions in dynamic breast MRIs.

The enhancement kinetic includes the lesion highlighting in the early, intermediate and late post-contrast phases as depicted by means of Time-signal Intensity Curves (TIC). These curves are classified as steady (type I), plateau (type II) and wash-out (type III); type I indicates potentially cancerous regions, type II benign lesions and type III indicates the malignant ones, as in [20]. Diverging results have been published concerning diagnostic values of TICs [10, 20, 22] due to the not completely known biologic and pathophysiologic mechanisms of wash-in and wash-out and the histological variability of breast lesions. Recent studies suggest the use of modified dynamic protocols with a reduced temporal and increased spatial resolution compared with the standard dynamic protocol for detecting subtle morphologic details of key diagnostic importance in breast MRI [21]. Therefore, abbreviated MR protocols including the pre-contrast and the third post-contrast THRIVE sequence have been proposed in the field of breast MRI for lesion detection and characterization.

The third post contrast THRIVE sequence acquired 3 min after contrast material injection during the intermediate post contrast period, that is, when wash-in phase has been completed, and the not yet occurred wash-out phase has been proposed [22]. THRIVE images allow to evaluate the lesion morphologic characteristics with high spatial resolution, including their internal architecture and the contrast enhancement. Therefore, the post-contrast image analysis mainly depends on the detection of contrast-enhanced areas and on the evaluation of their morphology and internal architecture.

The paper is organized as follows: the Sect. 2 deals with all patients' characteristics, all details of the Magnetic Resonance Acquisition Protocol and image specifications; in the Sect. 3, proper preprocessing/masking techniques able to provide consistent identifications of ROIs are discussed, together with the extracted features; in the Sect. 4, the dataset for synthesizing an adequate ANN are described, together with the features required to the neural network and classification performances of the implemented neural network; in the last section all results obtained from the proposed method are presented and discussed.

2 Materials

2.1 Patients and Magnetic Resonance Acquisition Protocol

Between January 2015 and March 2015, a group of twenty patients deemed significant by the medical team, consisting of ten diseased patients and ten healthy ones underwent MR examination. The twenty patients were women with age range 28–77 years (mean age ± standard deviation (SD), 53.2 ± 8.9 years) with family history of breast cancer and dense glandular structure or with suspected breast lesions detected by mammography or ultrasonography and informed according to the Declaration of Helsinki principles.

MR examination was performed in the second week of menstrual cycle, in case of premenopausal women. In case of dynamic imaging positivity, histological examination provided by US-guided core needle biopsy was performed at second look US in all patients. In case of negativity, patients were invited to resume periodic breast US examination six months later. The acquisition protocol of the images includes two phases during which seven X-rays are performed.

MR examinations were all performed on a 1.5 T MR device (Achieva, Philips Medical Systems, Best, The Netherlands) by using a four-channel breast coil. The standard protocol consisted of:

- Transverse short TI inversion recovery (STIR) turbo-spin-echo (TSE) sequence (TR/TE/TI = 3.800/60/165 ms, field of view (FOV) = 250 × 450 mm (AP × RL), matrix 168 × 300, 50 slices with 3-mm slice thickness and without gaps, 3 averages, turbo factor 23, resulting in a voxel size of 1.5 × 1.5 × 3.0 mm^3; acquisition time: 4 min);
- Transverse T2-weighted TSE (TR/TE = 6.300/130 ms, FOV = 250 × 450 mm (AP × RL), matrix 336 × 600, 50 slices with 3-mm slice thickness and without gaps, 3 averages, turbo factor 59, SENSE factor 1.7, resulting in a voxel size of 0.75 × 0.75 × 3.0 mm^3; acquisition time: 3 min);
- Three-dimensional dynamic, contrast-enhanced (CE) T1-weighted high resolution isotropic volume (THRIVE) sequences (TR/TE = 4.4/2.0 ms, FOV = 250 × 450 × 150 mm (AP × RL × FH), matrix 168 × 300, 100 slices with 4-mm slice thickness, spacing between slices: 2 mm; turbo factor 50, SENSE factor 1.6, 6 dynamic acquisitions, resulting in 1.5−mm^3 isotropic voxels, a dynamic data acquisition time of 1 min 30 s, and a total sequence duration of 9 min). Gadobenate dimeglumine (Multihance, Bracco, Milan, Italy) was intravenously injected at a dose of 0.1 mmol/kg of body weight and flow rate of 1.5 ml/s followed by 20 ml of saline solution.

2.2 Description of the Images

The first tests were carried out on a sample of twenty patients deemed significant by the medical team, consisting of ten diseased patients and ten healthy ones. The acquisition protocol of their images included two phases during which seven analysis were performed. During the first phase, that is without Contrast Medium (CM), sixty images per patient were produced. The resolution of each image is 512×512 pixels and represents on a 11-bit grayscale a 3-mm-thick slice of the patient's torso. The second phase includes six more scans. The first one was still performed without CM, then the patient has been injected with a contrast molecule and five more evenly-timed scans took place. Eventually the second phase produced six hundreds images, only one hundred of which are significant to our analysis (third post-contrast subset). Each image produced during this phase has a (320×320)-pixel size and a 4-mm-thick slice of the torso is represented with 2048 gray levels.

3 Methods

3.1 Image Registration

As advised by our team of experts, we focused our attention on the whole first set of images (60) and on a significant subset of the second one (100 out of 600), where the CM absorption is at its peak. Following the extraction of the 100 slices we matched each slice without CM to the closest one in the subset with CM within a radius of 1 mm. This proved to be sufficient given the slice thickness of our samples. A trimming of the ends of the series might occur since the two sets of images span different lengths over the longitudinal axis.

Given that the two sets of images were captured at different times, they might not be perfectly superimposable due to unavoidable movements of the patient's body. Therefore, after upscaling the smaller images from the second set, a procedure of Image Registration is applied, so that each image with CM can be superimposed on the corresponding image without CM as best as possible.

Finally we downscale the levels of gray from 2048 to 256, by performing a min-max normalization and auto-adjustment of the contrast on each set of images, considered as a whole.

3.2 Thorax Masking

Before processing images, the thoracic area is excluded to avoid subsequent processing of uninteresting regions. The proposed CAD considers a geometric parabola to identify thorax, which approximately follows the external border of the rib cage, so as to find three interpolation points A, B, C, as shown in Fig. 1.

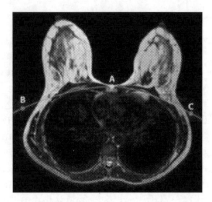

Fig. 1. Masking result: interpolation of a geometric parabola by means of points A, B, C

The Image Processing procedure is applied to each MRI as shown in Fig. 2. More in detail, it is applied to the images acquired without CM, since they are less noisy. The masking procedure is executed with Prewitt edge detection algorithm and the removal of stray pixels or isolated ones.

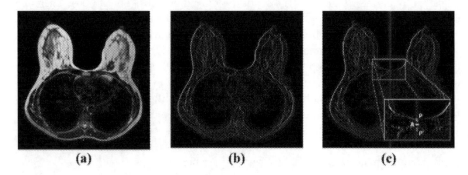

Fig. 2. Image Processing procedure applied to an MRI: (a) image acquired without CM: (b) image obtained using a Prewitt edge detection; (c) identification of point A as the midpoint of the segment P-P′ from the previous image

Thanks to the work of involved physicians in making a precise positioning of each patient, a correct image centering can be assumed. Starting from the median axis of the image, the first intersection (P) with the sternum can be searched, and then the second intersection output (P′); as a consequence, the first point A in the geometric parabola will be found as the midpoint of the segment P–P′ as shown in Fig. 2(c).

Then, every image has to be analyzed (starting from left to determine point B, and starting from right to determine point C as indicated in Fig. 1) and processed by columns, until three adjacent light pixels are found. These pixels allow to recognize a point E on the left side in the external lateral border of thorax, as shown in Fig. 3. The vertical distance between point A and point E can thus be evaluated, and the interpolation point B will be taken at 2/3 of this segment height as shown in Fig. 3. In a similar way, the external lateral point E′ will be subsequently recognized on the right side of thorax. The same procedure will be applied to determine the interpolation point C.

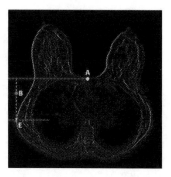

Fig. 3. Identification of point B at 2/3 of the vertical distance between points A and E

Then, the individuated area can be subtracted from all images, which have been previously equally centered and registered. The curve leaves in view some thoracic skin tissue in order to highlight any injuries. In this way, the breast area of interest results effectively isolated, in order to proceed to the segmentation phase.

3.3   Image Segmentation

The first step of this phase regards with Thresholding, being available a Region of Interest without undesired elements. Images without CM are firstly considered to find the optimal value of threshold. This value has been herein determined with an inductive procedure as equal to the 95th percentile of the distribution of the gray levels, after having excluded the range of levels (0–10) from the evaluation of the histogram. The choice of removing these very dark pixels, is due to their dominance in the histogram, that is also linked to the body of each patient: more slender a patient is, greater the number of black pixels in his histogram. Any region with a diameter less than 5 mm has not to be considered as a ROI [19]; then the values of areas have been converted from pixels to mm^2. In this way, a binary mask is obtained, to be used both for images with CM and without it.

3.4   Feature Extraction

After having determined adequate ROIs according to the previous steps, the necessary features can be extracted:

- size in mm^2 of the suspicious lesion [19];
- average value of the gray levels of images with CM and without it in ROIs, to determine areas with gray intensity different from standard ones;

- circularity, given by the following expression, where A = ROI area and p = ROI perimeter

$$4\pi\left(\frac{A}{P^2}\right)$$

- aspect ratio, intended as *major Axis/minor Axis* according to matlab documentation
- eccentricity of the ellipse, whose second order moments coincide with those of each ROI;
- solidity, defined as the proportion of the pixels in the convex hull that are also in the region;
- convexity (edge roughness), given by the ratio between the perimeter of the convex hull and the one of the ROI.

Circularity and Aspect ratio are used to differentiate blood vessels from other regions due to their elongated shape.

Since solidity and convexity are measurements of the roughness of each ROI, this feature is useful to distinguish tumor regions due to the high presence of blood vessels.

For a better comprehension of all features, a selected MRI is reported in Fig. 4, where:

A. Blood Vessel: Area 68.3, Average gray value TSE/CM 196.16/147.23, Standard Deviation TSE/CM 29.77/30.21, Circularity: 0.24, Aspect ratio: 4.92, Eccentricity: 0.97, Solidity: 0.62, Convexity: 1.12;
B. Tumor: Area 166.35, Average gray value TSE/CM 95.66/203.07, Standard Deviation TSE/CM 54.53/47.34, Circularity: 0.77, Aspect ratio: 1.41, Eccentricity: 0.70, Solidity: 0.91, Convexity: 0.99

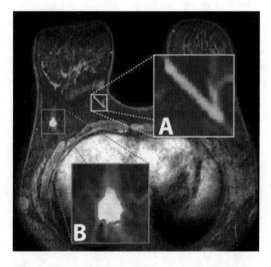

Fig. 4. Selected MRI: (A) blood vessel; (B) tumor

4   Detection and Classification

4.1   Dataset Balancement via the SMOTE Technique

It is well known that performances of machine learning algorithms in medicine are typically evaluated using predictive accuracy, that is, a measure of diagnostic accuracy [8]. Unfortunately, this does not result appropriate when data are imbalanced, that is, if the classification categories are not approximately equally represented, as well as in the classification of pixels in mammogram images as possibly cancerous [23]. In detail, a typical mammography dataset often contains 98 % normal pixels and 2 % abnormal ones. Thus, a simple default strategy of guessing the majority class gives a predictive accuracy of 98 %. However, the nature of this classification requires a fairly high rate of correct detection in the minority class and allows for a small error rate in the majority class in order to achieve this result. In this paper the considered mammography dataset is composed by 3700 negative cases and 94 positive ones (including both malignant lesions and benign ones). Due to the fact that a simple predictive accuracy is clearly not appropriate in such situations, in this paper a data balance with a Synthetic Minority Over-sampling Technique (SMOTE) has been preferred [24] to increase the number of positive patterns until the amount of negative ones achieves better classifier performance.

4.2   Neural Network Implementation and Validation

The implemented neural network has been trained by means of the 60 % of the dataset (3697 negative, 3612 positive); reserving the 20 % of data for a validation phase and the 20 % of data for a test phase. The choice of using the 60 % of the dataset for the neural training phase, in the experience of the authors, can guarantee a solid classification with no risk of overfitting. The topology of the implemented neural network involves four layers with:

- 1st layer: 212 neurons with logsig activation function
- 2nd layer: 29 neurons with logsig activation function
- 3rd layer: 1 neuron with logsig activation function
- 4th layer: 1 neuron with tansig activation function

The used fitting function is the resilient backpropagation algorithm function [25], and the threshold is dynamically adjusted using ROC curves over 100 runs.

4.3   Performance Measures

In the context of balanced datasets and equal error costs, it is reasonable to use error rate as a performance metric, but, in the presence of imbalanced datasets with unequal error costs, it is appropriate to use the ROC curve or other similar techniques [8].

The validation procedure consists in 100 elaboration with random provided Training, Validation and Test set. Obtained results are provided in Table 1, where the indices TPR, TNR, Accuracy are defined as.

- Accuracy = (T P + T N)/(T P + FP + T N + FN)
- TPR = TP/(TP + FN)
- TNR = TN/(TN + FP)

Table 1. Results of the validation procedure with the indices accuracy, TPR and TNR

	Min	Max	Mean	Standard deviation
Accuracy	0.9624	0.9849	0.9736	0.0044
TPR	0.9598	0.9958	0.9791	0.0075
TNR	0.9459	0.9892	0.9684	0.0075

5 Discussion and Conclusions

A CAD system based on medical imaging techniques and capable of marking potentially cancerous breast lesions has been presented. The development of the proposed CAD system has been based on analysis, detection and classification of ROIs, taken on MRIs performed with morphological and dynamic sequences. It has to be noticed that all considered sequences of MRI have been obtained and classified thanks to the collaboration of the specialists from the University of Bari Aldo Moro (Italy). The techniques adopted to segment all the selected images, and extract the most significant features from each ROI have been formally described in this paper. The architecture of the supervised neural network, based on the algorithm named Robust Error Back Propagation, has been then discussed, trained and optimized, so as to maximize the number of True Positive ROIs, i.e., the real regions suffering from cancer. Physicians helped to determine an adequate training set, consisting of 94 tumor regions and 3700 regions of any interest extracted with the proposed segmentation techniques. The synthesized ANN has been trained using 60 % of samples, and corresponding performances have been evaluated in terms of accuracy, sensitivity and specificity. A satisfactory behavior has been found, thus confirming that the definition of a CAD system based on a supervised machine learning approach is consistent, especially from a False Negative reduction perspective. In fact, the neural detection system has been synthesized to present a reduced number of False Negative pixels, intended as pixels belonging to breast regions which could be left out in a subsequent investigation, despite containing lesions of any possible interest.

Future work might concern with a comparison of the behavior of the implemented ANN with respect to spiking neural networks techniques, when data are collected during a different time period. A predictive model might probably be explored with this data set.

References

1. US Cancer Statistics Working Group, et al.: United States Cancer Statistics: 1999–2010 Incidence and Mortality Web-Based Report, Atlanta: US Department of Health and Human Services, Centers For Disease Control and Prevention and National Cancer Institute (2013)
2. Wolberg, W.H., Street, W.N., Mangasarian, O.L.: Image analysis and machine learning applied to breast cancer diagnosis and prognosis. Anal. Quant. Cytol. Histol. **17**(2), 77–87 (1995)
3. Wolberg, W.H., Street, W.N., Mangasarian, O.L.: Breast cytology diagnosis via digital image analysis. Anal. Quant. Cytol. Histol. **15**(6), 396–404 (1993)
4. Mangasarian, O.L., Street, W.N., Wolberg, W.H.: Breast cancer diagnosis and prognosis via linear programming. Oper. Res. **43**(4), 570–577 (1995)
5. Bevilacqua, V., Mastronardi, G., Menolascina, F.: Hybrid data analysis methods and artificial neural network design in breast cancer diagnosis: IDEST experience. In: International Conference on Computational Intelligence for Modelling, Control and Automation, 2005 and International Conference on Intelligent Agents, Web Technologies and Internet Commerce, vol. 2, pp. 373–378. IEEE (2005)
6. Wolberg, W.H., Street, W.N., Heisey, D.M., Mangasarian, O.L.: Computer derived nuclear features distinguish malignant from benign breast cytology. Hum. Pathol. **26**(7), 792–796 (1995)
7. Bevilacqua, V., Mastronardi, G., Menolascina, F., Pannarale, P., Pedone, A.: A novel multi-objective genetic algorithm approach to artificial neural network topology optimisation: the breast cancer classification problem. In: International Joint Conference on Neural Networks, 2006, IJCNN 2006, pp. 1958–1965. IEEE (2006)
8. Bevilacqua, V.: Three-dimensional virtual colonoscopy for automatic polyps detection by artificial neural network approach: new tests on an enlarged cohort of polyps. Neurocomputing **116**, 62–75 (2013)
9. Heywang-Köbrunner, S.H., Katzberg, R.W.: Contrast-enhanced magnetic resonance imaging of the breast. Invest. Radiol. **29**(1), 94–104 (1994)
10. Orel, S.G., Schnall, M.D., Livolsi, V.A., Troupin, R.H.: Suspicious breast lesions: MR imaging with radiologic-pathologic correlation. Radiology **190**(2), 485–493 (1994)
11. Huang, W., Fisher, P.R., Dulaimy, K., Tudorica, L.A., Ohea, B., Button, T.M.: Detection of breast malignancy: diagnostic MR protocol for improved specificity. Radiology **232**(2), 585–591 (2004)
12. Kuhl, C.K.: MRI of breast tumors. Eur. Radiol. **10**(1), 46–58 (2000)
13. Baum, F., Fischer, U., Vosshenrich, R., Grabbe, E.: Classification of hypervascularized lesions in CE MR imaging of the breast. Eur. Radiol. **12**(5), 1087–1092 (2002)
14. Hoffmann, U., Brix, G., Knopp, M.V., Hess, T., Lorenz, W.J.: Pharmacokinetic mapping of the breast: a new method for dynamic MR mammography. Magn. Reson. Med. **33**(4), 506–514 (1995)
15. Degani, H., Gusis, V., Weinstein, D., Fields, S., Strano, S.: Mapping pathophysiological features of breast tumors by MRI at high spatial resolution. Nat. Med. **3**(7), 780–782 (1997)
16. Orel, S.G.: High-resolution MR imaging for the detection, diagnosis, and staging of breast cancer. Radiographics **18**(4), 903–912 (1998)
17. Belli, P., Costantini, M., Bufi, E., Magistrelli, A., La Torre, G., Bonomo, L.: Diffusion-weighted imaging in breast lesion evaluation. La Radiologia Medica **115**(1), 51–69 (2010)

18. Kul, S., Cansu, A., Alhan, E., Dinc, H., Gunes, G., Reis, A.: Contribution of diffusion-weighted imaging to dynamic contrast-enhanced MRI in the characterization of breast tumors. Am. J. Roentgenol. **196**(1), 210–217 (2011)
19. Moschetta, M., Telegrafo, M., Rella, L., Capolongo, A., Ianora, A.A.S., Angelelli, G.: MR evaluation of breast lesions obtained by diffusion-weighted imaging with background body signal suppression (DWIBS) and correlations with histological findings. Magn. Reson. Imaging **32**(6), 605–609 (2014)
20. Fobben, E.S., Rubin, C.Z., Kalisher, L., Dembner, A.G., Seltzer, M.H., Santoro, E.J.: Breast MR imaging with commercially available techniques: radiologic-pathologic correlation. Radiology **196**(1), 143–152 (1995)
21. Kuhl, C.K., Schild, H.H., Morakkabati, N.: Dynamic bilateral contrast-enhanced MR imaging of the breast: trade-off between spatial and temporal resolution. Radiology **236**(3), 789–800 (2005)
22. Kuhl, C.K., Mielcareck, P., Klaschik, S., Leutner, C., Wardelmann, E., Gieseke, J., Schild, H.H.: Dynamic breast MR imaging: are signal intensity time course data useful for differential diagnosis of enhancing lesions? Radiology **211**(1), 101–110 (1999)
23. Woods, K.S., Doss, C.S., Bowyer, K.W., Solka, J.L., Priebe, C.E., Kegelmeyer Jr., W.P.: Comparative evaluation of pattern recognition techniques for detection of microcalcifications in mammography. Int. J. Pattern Recognit. Artif. Intell. **7**(06), 1417–1436 (1993)
24. Chawla, N.V., Bowyer, K.W., Hall, L.O., Philip Kegelmeyer, W.: SMOTE: synthetic minority over-sampling technique. J. Artif. Intell. Res. **16**, 321–357 (2002)
25. Kaur, H., Singh Salaria, D.: Bayesian regularization based neural network tool for software effort estimation. GJSFR-D: Agric. Vet. **13**(2), 45–50 (2013)

Using the Ranking-Based KNN Approach for Drug Repositioning Based on Multiple Information

Xin Tian[1], Mingyuan Xin[1], Jian Luo[1], and Zhenran Jiang[2($\boxtimes$)]

[1] Shanghai Key Laboratory of Regulatory Biology,
Institute of Biomedical Sciences and School of Life Sciences,
East China Normal University, Shanghai 200241, China
[2] Shanghai Key Laboratory of Multidimensional Information Processing,
Department of Computer Science and Technology,
East China Normal University, Shanghai 200241, China
zrjiang@cs.ecnu.edu.cn

Abstract. Using effective computer methods to infer potential drug-disease relationships can provide clues for the discovery new uses of old drugs. This paper introduced a Ranking-based k-Nearest Neighbor (Re-KNN) method to drug repositioning for cardiovascular diseases. The main characteristic of the Re-KNN lies in combining conventional KNN algorithm with Ranking SVM (Support Vector Machine) algorithm to get neighbors that are more trustable. By integrating the chemical structural similarity, target-based similarity, side-effect similarity and topological similarity information, Re-KNN method can obtain an improved AUC (Area under ROC Curve) and AUPR (Area under Precision-Recall curve) compared with other methods, which prove the validity and efficiency of multiple features integration.

Keywords: Cardiovascular disease · Drug reposition · Ranking-based KNN · Ranking SVM

1 Introduction

Drug discovery is an extremely time-consuming and high-risk procedure. It has been reported that developing a new drug often takes about 15 years and $800 million to $1 billion for pharmaceutical companies [1]. However, FDA approves only a relatively small number of new drugs each year. Therefore, drug repositioning (New uses of old drugs) is widely regarded as an effective solution to solve the dilemma [2].

The existing methods for drug repositioning can be categorized into drug-based method and disease-based method. (1) The drug-based method uses the drug information about the chemical or pharmaceutical perspective to predict the new function of 'old' drugs. The drug information contain chemical similarity, target-based similarity or side effect similarity and so on [3, 4]. These methods are usually applied when the drug profiling are more comprehensive and important for drug repositioning. (2) The disease-based method concerns more about the disease phenotype or pathology. This method can be applied to solve the problem of missing drug knowledge or to focus on a

© Springer International Publishing Switzerland 2016
D.-S. Huang et al. (Eds.): ICIC 2016, Part I, LNCS 9771, pp. 317–327, 2016.
DOI: 10.1007/978-3-319-42291-6_31

specific disease [5, 6]. These methods provide a variety of similarity calculation methods for reference. However, these methods cannot be used for samples with missing biological information. Recently, network-based methods combined with more biological knowledge get more applications for drug repositioning [7, 8].

Wu et al. combined a network-based precition method with the high-throughput genome-wide data for drug repositioning [9]. However, the noise of the gene expression data can influence prediction accuracy. Gottlieb et al. developed a novel algorithm, PRETICT, to infer novel drug indications [10]. The PRETICT method integrates multiple drug-drug and disease-disease similarity based on their biological nature, and get excellent prediction performance. However, the number of negative drug-disease interactions is far more than the positive ones in the drug-disease networks, which affects the accuracy of the PRETICT algorithm. Chen et al. introduced the ProbS and HeatS to predict direct drug-disease associations based only on the basic network topology [11]. The method implement naïve topology-based inference and do not take into account important features within the drug-disease domain. Zhang and Wang attempted to utilize chemical structure, protein targets, or phenotypic information to construct computational models and achieved promising results [12, 13]. This illustrates the importance of comprehensive information. Therefore, this paper attempted to integrate four types of information, including the chemical structural similarity, the sequence similarity, the side-effect similarity and the topology similarity for drug repositioning.

We introduce a new multi-label classification algorithm called Ranking-based KNN approach (denoted as Re-KNN, [14]) based on four types similarity for drug repositioning. The Re-KNN approach can be divided into two phases. (i) Using traditional KNN algorithm to identify the k-nearest-neighbors of the test drug. (ii) The selected neighbors are re-ranked by the Ranking SVM model [15] trained on their trustiness. The higher weights will be assigned to the trustable neighbors for making the weighted-voting decisions. The Re-KNN methods combined with the similarity matrix of drug-drug was applicated in the drug-disease dataset by 10-fold cross-validation tests. To illustrate the performance of the method, we compared Re-KNN method with WNN-GIP method [16], Super-Target method [17] and KNN method. The increased AUC and AUPR of the Re-KNN method illustrate a higher accuacy prediction than the three methods. Further, some interesting relationships between cardiovascular disease and drugs were found in this paper.

2   Materials

Cardiovascular disease is a type of chronic disease and a serious threat to the health of the elderly. 38 different kinds of cardiovascular diseases treated as the research objective were obtained from Human Diseases database of KEGG [18]. In addition, 104 different types of drugs, which are known as the treatment for the 38 cardiovascular diseases were drawn from the KEGG DRUG database. Including 38 cardiovascular diseases associated with these 104 kinds of drugs, we also picked out other 55 diseases. Both these diseases and cardiovascular disease at least have one same drug. Finally, we obtained 93 kinds of diseases and 540 kinds of related drugs. There are 1,060 known disease-drug relationships (Table 1).

Table 1. Statistic of the validated drug-disease association network

Attention	Number of diseases	Number of drugs	Number of known associations	Average degree of diseases
Cardiovascular disease	93	540	1060	11.4

3 Methods

3.1 Overview of the Method

Tsung-Hsien Chiang first developed the Ranking-based KNN method for multi-label classification. In the paper, the drug-repositioning problem will be relocated as a multi-label classification problem. The 'old' drugs are considered as samples to be classified while the names of diseases will be treated as the classification labels. First, all of the drugs were divided into training set and test set, then the similarity matrix of the training set was acquired by combining four categories of similarity calculation. With the given a test drug D, we need to find its k-nearest neighbors using KNN algorithm, denoted as $\{N_k (D,j)|j = 1,\ldots,k\}$. Then, we should create new instances from $\{N_k (D,j)|j = 1,\ldots,k\}$ and send all the new instances to train the Ranking SVM model for re-ranking. The re-ranking k-nearest neighbors of drug D, which denoted as $\{N_k' (D, j)|j = 1,\ldots,k\}$, can be obtained by the Ranking SVM model. Therefore, the re-ranked neighbors can be treated as trustable neighbors that possess the most similarity with the labels of drug D. Finally, a weighted voting strategy was applied to obtain the final

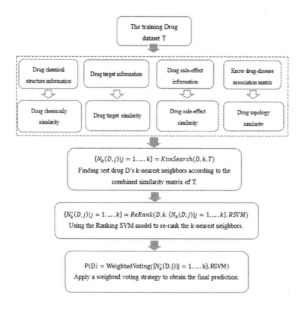

Fig. 1. The flowchart of the ranking-based KNN approach

predicted labels for the test drug D. The flowchart of the Re-KNN approach in drug-repositioning process is shown in Fig. 1. In the following section, the whole training process of Ranking SVM will be explained in details.

3.2   Similarity Measures

To take full advantage of biological prior knowledge and known network topology information, the linear combination of four kinds of similarity calculation methods, including the chemical structure similarity, target-based similarity, side-effect similarity and topology similarity were applied.

(1) Chemical similarity: We downloaded the canonical MOL files from KEGG. The chemical fingerprints were computed using PaDel-Descriptor software [19]. A chemical fingerprint contains 881 chemical substructures according to Pub-Chem database. Each drug D is described by a binary fingerprint C(D). The element of C(D) is encoded as 1 or 0 which indicates the presence or absence of corresponding PubChem substructure. The similarity score between two drugs is computed based on their fingerprints C(D) according to the Tanimoto score. Tanimoto score is equivalent to the Jaccard score of the drugs' fingerprints. We use the Tanimoto score as the chemical similarity of drugs.

$$S_{cs}(D_i, D_j) = \frac{|C(D_i) \cap C(D_j)|}{|C(D_i)| + C(D_j) - |C(D_i) \cap C(D_j)|} \tag{1}$$

Where $|C(D_i)|$ and $|C(D_j)|$ means the counts of structure fragments in drug i and drug j, respectively. The $|C(D_i) \cap C(D_j)|$ means the amount of chemical substructures shared by drug i and drug j.

(2) Target-based similarity: The targets of drugs can be obtained from DrugBank or KEGG DRUG. To compute the sequence similarity between two proteins, we use the normalized version of Smith-Waterman scores [20].

(3) Side-effect similarity: SIDER database contains information about 1430 kinds of marked drugs and 5868 kinds of recorded adverse drug reactions. Each drug D was represented by 5868-dimensional binary side-effect profile E (D). The elements of E (D) is encoded as 1 or 0 which indicate the presence or absence of each of the side-effect key words respectively. The pairwise side-effect similarity between drug i and drug j is computed as the Tanimoto score of their chemical structure similarity:

$$S_{se}(D_i, D_j) = \frac{|E(D_i) \cap E(D_j)|}{|E(D_i)| + E(D_j) - |E(D_i) \cap E(D_j)|} \tag{2}$$

(4) Topology similarity: According to the existing drugs-diseases network relation-ship, we constructed a matrix K, where the element k_{ij} of the K represents the number of shared diseases by drug i and drug j [21]. Drug i and drug j shared

more diseases, the more similar the two drugs will be. At the same time, we use Floyd algorithm to find the shortest distance between two nodes in the drug-disease network, and then calculate the Pearson correlation coefficients of the distance vectors between any two drugs to construct a similar matrix. The matrix reflects the topology similarity between the drugs.

$$S_{tp}\left(D_i, D_j\right) = \left| \frac{\sum_{k=1}^{n}(V_{ik} - \overline{V_i})(V_{jk} - \overline{V_j})}{\sqrt{\sum_{k=1}^{n}(V_{ik} - \overline{V_i})^2}\sqrt{\sum_{k=1}^{n}(V_{jk} - \overline{V_j})^2}} \right| \tag{3}$$

3.3 Ranking SVM Model

KNN algorithm is an effective learning method for solving conventional classification problems. The label of a sample is usually based on a majority vote of its k-nearest-neighbors. However, not all of the neighbors are trustable for label prediction. Thus, the Ranking SVM model is used to give the trustable ranking of these neighbors. Figure 2 shows the training process of the Ranking-SVM model, where x denotes the drug's feature vector and the Y(x) denotes the label set associated with x.

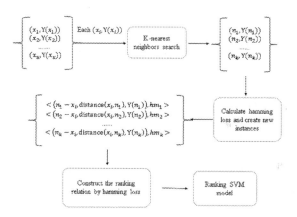

Fig. 2. The training process of the ranking SVM model

First, with KNN algorithm, the k-nearest neighbors for the sample x_i in training set will be searched.

Second, by constructing k instances $Q = \{q_1, q_2, \ldots, q_k\}$ for the sample x_i, the ranking SVM model is built. A new instance q_j can be established for $x_i's$ j-th nearest neighbors by KNN. The elements of the $q_j's$ features are listed as follows:

- The difference among each feature value of j-th nearest neighbors and sample x_i (size = dim(S)).
- The Euclidean distance between j-th nearest neighbors and sample x_i (size = 1).
- The cosine distance between j-th nearest neighbors and sample x_i (size = 1).
- The label set of j-th nearest neighbors (size = n).

Then, using Hamming loss function [22] to determine the quality of each new instance q_j, Hamming loss calculates the percentage of labels that are not consistent between two instances. It is defined as follows:

$$\text{HammingLoss}(L_1, L_2) = \frac{1}{|L|} |L_1 \Delta L_2| \tag{4}$$

Where L_1 and L_2 represent two label sets respectively. The lower Hamming loss is, the higher rank will be assigned to the new instance q_j.

3.4 Weighted Voting

According to the above training process, a Ranking SVM model is built on the training data set. With sending an un-labeled drug x to the model, we get the re-ranked trustable neighbors $\{N_k'(x, j) | j = 1, \ldots, k\}$ of drug x. In this section, we will give the corresponding weight to each trustable neighbors. Let $(w_1, w_2, \ldots, w_k)$ denote the weight scores of the re-rank neighbors $\{N_k'(x, j) | j = 1, \ldots, k\}$. We choose the $\{w_j | j = 1, \ldots, k\}$ with 10-fold cross-validation to maximize AUC, where $w_j \in (0, 1)$ follow the uniform distribution.

The probability $Pr_i(x)$ between drug x and disease-label y_i is defined as following formula. As the $Pr_i(x)$ increases, the probability that the drug x can treat the disease y_i will also arise.

$$Pr_i(x) = \frac{\sum_{j=1}^{k} w_j \cdot y_i(N_k'(x, j))}{\sum_{j=1}^{k} w_j} \tag{5}$$

4 Results

In order to find the trustable neighbors, the Re-KNN method is applied to the drug-disease network associated with cardiovascular disease. In addition, 10-fold cross validation tests is used to evaluate the performance of the method. The known drugs-diseases dataset are randomly divided into ten subsets of equal size. Each one of the ten parts will be used as test data in turn while the remaining nine parts should be treated as training data. Both AUC and AUPR are used to evaluate the results as the quality measures.

4.1 Comparison with Other Methods

Recently, several similar types of methods have been proposed for drug-disease association prediction. This study selects the WNN-GIP, the Super-Target and KNN in order to compare the performance of Re-KNN method.

The performance comparision for the Re-KNN method, WNN-GIP method, Super-Target method and KNN method on cardiovascular dataset are listed in Table 2. As shown in Table 2, we can see the Re-KNN method performed better with both AUC and AUPR.

Table 2. Comparison results for the four prediction methods

Method	Re-KNN	WNN-GIP	Super-target	KNN
AUC	0.928	0.845	0.800	0.821
AUPR	0.588	0.375	0.322	0.258
SN	0.388	0.134	0.111	0.147
SP	0.977	0.985	0.968	0.982
ACC	0.983	0.974	0.959	0.969
MCC	0.542	0.351	0.343	0.250

SN–Sensitivity; SP–Specificity; ACC–Accuracy;
MCC–Matthews Correlation Coefficient.

By comparing with WNN-GIP method, the AUC value and the AUPR value of the Re-KNN method is increased by 0.08 and 0.21 respectively. In addition, the Re-KNN method had a better accuracy by comparing with other three methods. For instance, the accuracy 0.983 of the Re-KNN method is larger than the accuracy of the WNN-GIP. Since the number of negative sample is far more than the number of the positive ones, sensitivity should be given higher priority. The sensitivity (0.388) of Re-KNN method is larger than other three methods, which justified the contribution of Ranking-SVM model. The ROC curves and the P-R curves of four methods are shown in Fig. 3.

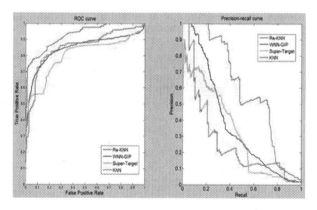

Fig. 3. ROC and P-R curve of four methods in cardiovascular diseases related datasets

4.2 Predicted New Drug-Disease Associations for Cardiovascular Diseases

After the Re-KNN method is proved to perform better than other three methods, the experiment was repeated 10 times for the drug-disease dataset associated with cardiovascular diseases. By ranking the 10 experimental results, the diseases that can be possibly treated are picked out. In particular, we found 300 new associations between 117 drugs and 35 cardiovascular diseases that can be seen on the top of Fig. 4. The red diamonds represent cardiovascular diseases, green triangles represent the drugs, black solid line represents the existing relationship and the purple dashed line represents the possible existing new associations predicted by our methods. As the figure shows, the new relationship is distributed more concentratedly in the marked Part A and Part B respectively. The gathered drugs in part A that used to be treatment for other diseases may become the therapeutic for cardiovascular. The diseases of the cardiovascular type can possibly treat another disease type in Part B.

The bottom part of Fig. 4 presents the specific examples predicted by Re-KNN method. In the left of botton part shows the drug Adomiparin (D07510) can be used as the treatment for the inherited thrombophilia (H00223) and Gordon's syndrome (H00243) simultaneously. The drugs, Digiglusin (D03819) and Quelicin (D00766), may become the treatment to H00233, which known as the therapeutic method of Gordon's syndrome. In the right, primary pulmonary hypertension (H01300) may be treated by the drugs D08953, D03658, D06414, etc. In addition, these drugs can also treat chronic myeloid leukemia (H00004), a cancer of haematopoietic and lymphoid tissues.

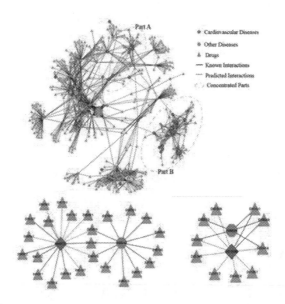

Fig. 4. The predictied associations and the known interactions between drug-diseases (Color figure online)

In order to verify the predicted new associations, we searched the literatures in the PubMed database with the predicted relationships. Table 3 lists a part of the searched results. It showed some of the drug predicted by this paper can be validated by relevant literatures and show potentiality for further study.

Table 3. The confirmed drug-disease associations predicted by Re-KNN in PubMed

Diseases(KEGG code)	Known drugs	Predicted drugs	References
Hemophilia (H00219)	Blood-coagulation factor VIII, etc.	Epsilon-Aminocaproic acid	[23]
Inherited thrombphilia (H00223)	Warfarin and Heparin	Dalteparin sodium	[24]
Beta thalassemia (H00228)	Deferoxamine, etc.	Vitamin E	[25]
Preexcitation syndrome(H01154)	Vagal maneuvers, etc.	Disopyramide phosphate	[26]
Primary pulmonary hypertension (H01300)	Imatinib	Nilotinib	[27]

5 Discussion and Conclusion

This paper proposed a Re-KNN method to the drug repositioning of cardiovascular diseases. One characteristic of the method lies in combining four types of similarity calculation methods including chemical structural similarity, target-based similarity, side-effect similarity and topological similarity information. By comparing with WNN-GIP method and Super-Target method, the increased AUC value and AUPR value proved the better predictive capability of Re-KNN method. In addition, some part of new predicted associations can be supported by the PubMed database. We expect the method can provide valuable clues for drug discovery.

Although the Re-KNN method had a better predictive capability in drug repositioning, the performance can be further improved by integrating more biological information including gene expression experiments and pathway-related information. Further, good prediction models need to integrate more prior information and constantly improve its own algorithm. In the Re-KNN method, the ranking-SVM model and the weighted vote strategy still have a room for improvement. In the future, we will devote ourselves to improve the method from the above aspects.

Acknowledgements. This work was partially supported by National Basic Research Program of China (Grants No. 2012CB910400), the Fundamental Research Funds for the Central Universities (78260026), and National Science and Technology Support Plan Project (2015BAH12F01).

References

1. Adams, C.P., Brantner, V.V.: Estimating the cost of new drug development: is it really $802 million? Health Aff. **25**(2), 420–428 (2006)
2. Ashburn, T.T., Thor, K.B.: Drug repositioning: identifying and developing new uses for existing drugs. Nat. Rev. Drug Discov. **3**(8), 673–683 (2004)

3. Yang, L., Agarwal, P.: Systematic drug repositioning based on clinical side-effects. PLoS ONE **6**(12), e28025 (2011)
4. Sanseau, P., Agarwal, P., Barnes, M.R., et al.: Use of genome-wide association studies for drug repositioning. Nat. Biotechnol. **30**(4), 317–320 (2012)
5. Clouser, C.L., Patterson, S.E., Mansky, L.M.: Exploiting drug repositioning for discovery of a novel HIV combination therapy. J. Virol. **84**(18), 9301–9309 (2010)
6. Anne, C., James, P., Alistair, B., et al.: Drug repositioning for Alzheimer's disease. Nat. Rev. Drug Discov. **11**(11), 833–846 (2012)
7. Cheng, F., Liu, C., Jiang, J., et al.: Prediction of drug-target interactions and drug repositioning via network-based inference. PLoS Comput. Biol. **8**(5), e1002503 (2012)
8. Alaimo, S., Pulvirenti, A., Giugno, R., et al.: Drug-target interaction prediction through domain-tuned network-based inference. Bioinformatics **29**(16), 2004–2008 (2013)
9. Wu, Z., Wang, Y., Chen, L.: Network-based drug repositioning. Mol. BioSyst. **9**(6), 1268–1281 (2013)
10. Gottlieb, A., Stein, G.Y., Ruppin, E., et al.: Predict: a method for inferring novel drug indications with application to personalized medicine. Mol. Syst. Biol. **7**, 496 (2011)
11. Chen, H., Zhang, H., Zhang, Z., et al.: Network-based inference methods for drug repositioning. Comput. Math. Methods Med. **2015**, 130620 (2015)
12. Zhang, P., Agarwal, P., Obradovic, Z.: Computational drug repositioning by ranking and integrating multiple data sources. In: Blockeel, H., Kersting, K., Nijssen, S., Železný, F. (eds.) ECML PKDD 2013, Part III. LNCS, vol. 8190, pp. 579–594. Springer, Heidelberg (2013)
13. Wang, Y., Chen, S., Deng, N., et al.: Drug repositioning by kernel-based integration of molecular structure, molecular activity, and phenotype data. PLoS ONE **8**(11), e78518 (2013)
14. Chiang, T.H., Lo, H.Y., Lin, S.D.: A ranking-based KNN approach for multi-label classification. JMLR W&CP. **25**, 81–96 (2012)
15. Yu, H., Kim, J., Kim, Y., et al.: An efficient method for learning nonlinear ranking SVM functions. Inf. Sci. **209**, 37–48 (2012)
16. Laarhoven, T.V., Marchiori, E.: Predicting drug-target interactions for new drug compounds using a weighted nearest neighbor profile. PLoS ONE **8**(6), e66952 (2013)
17. Shi, J.Y., Yiu, S.M., Li, Y., et al.: Predicting drug-target interaction for new drugs using enhanced similarity measures and super-target clustering. Methods **83**, 98–104 (2015)
18. Kanehisa, M., Goto, S.: KEGG: kyoto encyclopedia of genes and genomes. Nucleic Acids Res. **28**(1), 29–34 (2000)
19. Yap, C.W.: PaDEL-descriptor: an open source software to calculate molecular descriptors and fingerprints. J. Comput. Chem. **32**(7), 1466–1474 (2011)
20. Smith, T.F., Waterman, M.S.: Identification of common molecular subsequences. J. Mol. Biol. **147**(1), 195–197 (1981)
21. Xia, Z., Wu, L.Y., Zhou, X., et al.: Semi-supervised drug-protein interaction prediction from heterogeneous biological spaces. BMC Syst. Biol. **4**(Suppl 2), S6 (2010)
22. Tsoumakas, G., Katakis, I.: Multi-label classification: an overview. Depatment of Informatics, Aristotle University of Thessaloniki, Greece (2006)
23. Biggs, R., Hayton-Williams, D.S.: Epsilon-aminocaproic acid in the treatment of hemophilia. Br. Dent. J. **124**(124), 157 (1968)
24. Leduc, L., Dubois, E., Takser, L., et al.: Dalteparin and low-dose aspirin in the prevention of adverse obstetric outcomes in women with inherited thrombophilia. J. Obstet. Gynaecol. Can. **29**(10), 787–793 (2007)
25. Giardini, O., Cantani, A., Donfrancesco, A.: Vitamin E therapy in homozygous beta-thalassemia. N. Engl. J. Med. **305**(11), 644 (1981)

26. Kerr, C.R., Prystowsky, E.N., Smith, W.M., et al.: Disopyramide phosphate in wolff-parkinson-white syndrome. Am. J. Cardiol. **47**(2), 495 (1981)
27. Chaumais, M.C., Perros, F., Dorfmuller, P., et al.: Nilotinib and imatinib therapy in experimental pulmonary hypertension. Arterioscler. Thromb. Vasc. Biol. **32**(6), 1354–1365 (2012)

Depth-First Search Encoding
of RNA Substructures

Qingfeng Chen[1,2(✉)], Chaowang Lan[1], Jinyan Li[3], Baoshan Chen[2],
Lusheng Wang[4], and Chengqi Zhang[5]

[1] School of Computer, Electronic and Information, Guangxi University,
Nanning 530004, China
qingfeng@gxu.edu.cn, 295267178@qq.com

[2] State Key Laboratory for Conservation and Utilization of Subtropical
Agro-bioresources, Guangxi University, Nanning, China
chenyaoj@gxu.edu.cn

[3] Advanced Analytics Institute, University of Technology Sydney,
P.O. Box 123 Broadway, Ultimo, NSW 2007, Australia
jinyan.li@uts.edu.au

[4] Department of Computer Science, City University of Hong Kong,
Kowloon, Hong Kong
cswangl@cityu.edu.hk

[5] Centre for Quantum Computation and Intelligent Systems,
University of Technology Sydney, P.O. Box 123 Broadway,
Ultimo, NSW 2007, Australia
chengqi.zhang@uts.edu.au

Abstract. RNA structural motifs are important in RNA folding process. Traditional index-based and shape-based schemas are useful in modeling RNA secondary structures but ignore the structural discrepancy of individual RNA family member. Further, the in-depth analysis of underlying substructure pattern is underdeveloped owing to varied and unnormalized substructures. This prevents us from understanding RNAs functions. This article proposes a DFS (depth-first search) encoding for RNA substructures. The results show that our methods are useful in modelling complex RNA secondary structures.

Keywords: Data mining · RNA · Subgraph · Substructure · Support

1 Introduction

Most studied sequences in human genome are protein-coding genes. In recent years, there has been increasing evidence to indicate that the non-coding portion of the genome is of crucial functional importance: for normal development and physiology and for disease. For example, microRNAs (miRNAs) have been uncovered as key regulators of gene expression at the post-transcriptional level, and epigenetic and genetic defects in miRNAs and their processing machinery are a common hallmark of disease [1]. ncRNAs are emerging as key regulators of embryogenesis by controlling embryonic gene expression.

© Springer International Publishing Switzerland 2016
D.-S. Huang et al. (Eds.): ICIC 2016, Part I, LNCS 9771, pp. 328–334, 2016.
DOI: 10.1007/978-3-319-42291-6_32

Recent advent of high-throughput sequencing enabled genome-wide measurements of RNA structure, including de novo structure prediction and comparative structure prediction on the basis of a single sequence and multiple homologous ncRNAs, respectively. Computational structure prediction leads to an increasingly growth of RNA structure data in the last decade, such as fRNAdb [2], Rfam 12.0 [3] for short regulatory ncRNAs, and lncRNAdb for long non-coding RNAs. Identifying and validating regulatory RNA motifs involved in diverse cellular processes from the valuable data is essential to bring about comprehensive understanding of RNA function [4].

Several methods have been developed to normalize RNA structure data. A solution with sublinear running time would require index-based structure modeling [5]. However, widely used index structures like suffix trees or arrays or the FMindex have unsatisfactory performance on typical RNA sequence-structure patterns, because they cannot utilize the RNA structure information [6]. Shape-based modeling focuses on the common shape in a more abstract sense [1], but ignores the structural variation of individual RNA family member, which naturally has different lengths and sequence compositions. Traditional methods focus on finding integrally conserved secondary structures, whereas the substructure features have been largely overlooked, especially the frequent substructures shared by a collection of RNAs.

This paper proposes a novel framework to encode RNA substructures in terms of depth-first search (DFS) code. The vertex and edge in a graph represent base pairing stems and loops, respectively. Attribute data is used to indicate the substructure information, including length and sequence as well as graph data. This assists in distinguishing complicated secondary structures and facilitating the coding and ordering of RNA structure graphs.

2 Preliminaries

Labeled graph is a general data structure for modeling complicated patterns. A labeled graph composes of labels related to its edges and vertices. The vertex set and edge set of a graph g are represented by $V(g)$ and $E(g)$, respectively. A vertex or an edge can be mapped to a label by a labeling function l. A graph g_1 is said to be a subgraph of another graph g_2 if the graph g_1 is included in g_2.

Definition 3.1: Suppose l_1 and l_2 are labeling function of g_1 and g_2, respectively. A subgraph isomorphism is a mapping function $f : V(g_1) \rightarrow V(g_1)$. $\forall\ u \in V(g_1)$, $l_1 (u) = l_2(f(u))$, and $\forall\ (u, v) \in E(g_1)$, $(f(u), f(v)) \in E(g_2)$ and $l_1 (u, v) = l_2(f(u), f(v))$. $g_1 \cong g_2$ is used to represent g_1 and g_2 are isomorphic.

Given a graph dataset $D = \{G_1, G_2, \cdots, G_n\}$, $supp(g)$ represents the percentage of graphs where g is a subgraph. Thus, we have.

$$supp(g) = |f(g, G)|/|D| \qquad (1)$$

where $f(g, G) = \{G_i \in D | g \subseteq G_i\}$.

In the similar way to Apriori-like scheme, g is a frequent subgraph if its support is not less than a given minimum support threshold *minsup*, namely $supp(g) \geq minsup$.

To avoid redundant patterns in case of a small minimum support, closed frequent graph patterns CP are identified in our method. Let FP be a frequent graph pattern. Thus, we have $CP = \{g_i \mid g_i \in FP \text{ and } \not\exists\, g_j \in FP, g_i \subset g_j, i \neq j, \text{ and } supp(g_i) = supp(g_j)\}$.

Traditional approaches identify frequent graphs, including closed and non-closed, by adding an edge e to a graph g each time until all the frequent graphs FP with g are identified. In other words, if the support of a graph is less than *minsup*, the recursively extension of edge will be terminated. However, it is observed that this may result in generation of duplicated graphs and unnecessary computation for support. Thus, this article extends the *CloseGraph* method [7] by transferring ncRNA structure into graph data combining attribute data.

3   Graph-Based Model for NcRNA

3.1   Graph Representation for NcRNA Structures

Traditional dynamic programming approach can identify the optimal or suboptimal secondary structures, whereas it is unable to determine which one within the derived structures from the program is closest to the one occurring naturally [8]. Thus, a framework was proposed by us to normalize the RNA data into a collection of distance vectors, whereas the deep analysis of substructure patterns has been largely neglected. This might result in incorrect results since some RNAs that have high global similarity may show much difference in substructures.

A RNA structure graph g can be expressed as vertex and edge as usual. Vertex V comprises stems of RNA structure, and edge E includes loops, bulge and hairpin loops. Let lowercase letter a, b, c, d denote vertex identifiers, and v_i be label of vertices, such as v_0 for vertex a and v_1 for vertex b. The base-pairing region is viewed as a vertex in the graph.

Figure 1b presents the transformation from original RNA structure in Fig. 1a to structure graph, in which a, b, c, d, e, f are vertices corresponding to stems or base-pairs $(a, a'), (b, b'), (c, c'), (d, d), (e, e), (f, f)$, respectively, which are labeled with $v_0, v_1, v_2, v_3, v_4, v_5$. $l_i, 0 \leq i \leq 5$, describes the loops of RNA structure from 5' end to 3' end. Nevertheless, some vertices may connect to loop edges, such as v_2, v_4, v_5 linked to l_2, l_5, l_7. It is necessary to insert additional labels, such as v'_2, v'_4, v'_5.

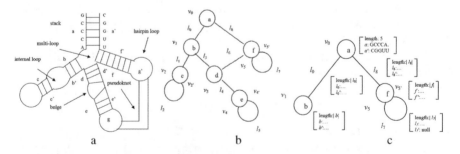

Fig. 1. (a) A RNA structure, (b) structure graph transformed from Fig. 1, (c) an instance of combination of graph and attribute data

The original graph in Fig. 1b alone cannot include all information of complex RNA structures. Attribute data, including length and sequence information of substructures are thus applied in this article. For brevity, an instance including three vertices, v_0, v_1, v_5, are presented in Fig. 1c. Attribute data are added for all vertices and edges, including hairpin loop. Nevertheless, bulge and hairpin loop do not have symmetric sequence. Thus, their symmetric sequences are recorded as *null* herein, such as null value with l_7'.

It is observed that there are actually two sequences with regular stems and loops. Their lengths may be unequal in many cases. To clarify this issue, an average length is defined. Thus, we have

$$|x| = \begin{cases} |s_x| & \text{if } x \text{ is a bulge or a hairpin loop} \\ round\left(\sqrt{|s_x| * |s_{x'}|}\right) & \text{otherwise} \end{cases} \qquad (2)$$

where x represents regular stems and loop regions or irregular structures including bulge and hairpin loop. $|s_x|$ and $|s_{x'}|$ represent the lengths of x and x', respectively.

A simple way to identify frequent subgraph patterns is to extend a graph g by adding a new edge e. However, the same graph can be extended in more than one way, and the naive graph mining was criticized due to its low efficiency and generation of duplicate graphs.

3.2 DFS Lexicographical Order

This section presents the process of mapping a graph to a DFS code, by which to detect frequently connected subgraphs that exhibit feature similarity in subsets of the dimensions including vertices and edges. A subgraph g_1' of g_1 is called a nesting of g_2 in g_1 if $g_2 \cong g_1'$.

For each labeled graph, it can be converted into an edge sequence, including subscripts of vertices and edges, namely subscripted graph G_T with DFS code, in which T denotes DFS subscripts. An order of these sequences is generated in terms of labels and subscripts, namely sequence order. In addition, edge order is built by subscripts.

Definition 3.2: Edge order can be defined as $\prec T$ in a formal way. Suppose $e_1 = (i_1, j_1)$ and $e_2 = (i_2, j_2)$. $e_1 \prec T \, e_2$ is hold if one of the following conditions is satisfied.

- $j_1 < j_2$ or
- $i_1 > i_2$ and $j_1 = j_2$
- $i_2 > i_1$ in case of $j_1 = i_1'$ or $j_2 = i_2'$

where the last condition indicates that there is a hairpin loop in the compared edges. In that case, we need to compare the first subscript i rather than j of two edges.

This article only considers forward edge since the base-paring regions (stems) are viewed as a vertex and we focus on structure similarity. For some specific ncRNA secondary structures, such as pseudoknot, the base-pairing region is divided into two vertices v and v' of graph. There is a forward edge from v to v' via a hairpin loop, and a back edge occurs from v' to v. In that case, the edge order is needed to define again.

For brevity, an edge is represented as a 6-tuples, $(i, j, x_i, l_{i,j}, x_j, \vec{\alpha})$, where i and j are the order pair to represent the edge between v_i and v_j, and x_i and x_j are the labels of v_i and v_j, respectively. $l_{i,j}$ represents the label of edge between v_i and v_j. $\vec{\alpha}$ is a vector to indicate the attribute data of the edge, including hairpin loop. Note, not all edge has connected hairpin loop. For those with hairpin loop, we have $x_i = x_j$.

Definition 3.3: *DFS code.* Suppose (e_i) denotes a 6-tuple, $(i, j, x_i, l_{i,j}, x_j, \vec{\alpha})$ for a given G_T. An edge sequence can be generated by the edge order $\prec_T$ as Definition 3.2, such that $e_i \prec_T e_{i+1}$, $0 \le i \le |E|$. C_g denotes DFS codes of graph g, which consists of a collection of (e_i).

Example 3.1: In Fig. 1b, (v_0, v_1), (v_1, v_2), (v_2, v_2'), (v_1, v_3), (v_3, v_4), (v_4, v_4'), (v_4, v_5), (v_5, v_5'), (v_0, v_5) are represented by a collection of DFS codes, namely $(0, 1, a, l_0, b, \vec{\alpha}_0)$, $(1, 2, b, l_1, c, \vec{\alpha}_1)$, $(2, 2', c, l_2, c, \vec{\alpha}_2)$, $(1, 3, b, l_3, d, \vec{\alpha}_3)$, $(3, 4, d, l_4, e, \vec{\alpha}_4)$, $(4, 4', e, l_5, e, \vec{\alpha}_5)$, $(3, 5, d, l_6, f, \vec{\alpha}_6)$, $(5, 5', f, l_7, f, \vec{\alpha}_7)$, $(0, 5, a, l_8, f, \vec{\alpha}_8)$. They can be recorded as $e_0, e_1, e_2, e_3, e_4, e_5$ according to Definition 3.3. $\vec{\alpha}_0$ is the attribute data of l_0 and is represented by $(|l_0|, S_{l0}, S_{l0'})$, where S_{l0} and $S_{l0'}$ are the sequence of the regions l_0 and l_0', respectively. $\vec{\alpha}_2$ is the attribute data of hairpin loop l_2 and is represented by $(|l_2|, S_{l2})$, where S_{l2} denotes the sequence of the hairpin region l_2.

It is necessary to consider the label information as one of the ordering criteria while searching the frequent subgraphs from the yielded DFS code data. This assists in distinguishing two edges that have the same subscripts but discrepant labels. It is normal that the order of two edges follow the priority from $\prec_T$, the vertex label x_i, the edge label l_{ij}, to the vertex label x_j.

Definition 3.4: Lexicographic order. Suppose S denotes a set of codes of all connected labeled graphs, namely $S = \{C_g \mid g$ is a labeled graph of $G\}$. Let $\prec_L$ be a linear order in the label set L, which comprises vertex labels, l_i and l_j, and edge label l_{ij}. $\prec$ is a linear order of combination of $\prec_T$ and $\prec_L$ regarding DFS codes. Suppose $\alpha = C_\alpha = \{a_0, a_1, \cdots, a_m\}$ and $\beta = C_\beta = \{b_0, b_1, \cdots, b_n\}$, $\alpha, \beta \in S$. Lexicographic order $\le$ is a linear order, and $\alpha \le \beta$ iff either of the condition is satisfied.

- $\exists\, i,\ 0 \le i \le min(m, n)$, $a_k = b_k$ for $k < i$, $a_i \prec b_i$
- $a_k = b_k$ for $0 \le k \le m$, and $m \le n$.

where the linear orders $\prec$ and $\le$ are used for comparison between DFS codes, and between labeled graphs, respectively. $\prec_T$ and $\prec_L$ are applied to distinguish edge order from vertex order.

Each labeled RNA structure graph G has only one subscripting T_G. Thus, it is viewed as a base subscripting of G. It is unnecessary to consider minimum DFS code of G herein since the completeness of mining result can be guaranteed based on the generated labeled graph by DFS codes. Suppose $D = G_1, G_2, \cdot \leftarrow \cdot \leftarrow \cdot \leftarrow, G_n$

$\Diamond \leftarrow$ represents a set of graphs. The support of g in D is the number of subgraph isomorphisms of g occurs in every $G_i \psi \in \leftarrow D$. That is

$$O(g, D) = \sum_{i=1}^{n} \psi(g, G_i) \tag{3}$$

From the observation, a graph g can have more than one subgraph isomorphism in the same graph G_i. However, if two DFS codes s and $s' \leftarrow$ are found to represent the same graph and $s \prec s'$, it is unnecessary to search any descendant of s'. Given a graph g, it can be extended as g' by adding a new edge e. If the occurrence of g' is less or equal to $O(g, D)$ and the support of g' is smaller than a support threshold specified by users, it is called a closed graph.

4 Experiments

The raw data set from fRNAdb is represented as the form of dots, and left bracket "(" and right brackets ")", which represent the stem and loop of RNA secondary structure, respectively. The initial graph is constructed from the 5′ end to the 3′ end of RNAs by storing the dots to a loop object. In contrast, if a left bracket is detected, its sequence and structure information are saved to a stem object. This is repeated until a right bracket is found. To generate the labeled graph described above, the stored brackets and dots by the initial graph are translated into vertices and edges. They are labeled with number according to their positions in the RNA graph.

The experiment is implemented on the supercomputer platform of Guangxi University, which consists of AMD 843, 8 CPUs, 128G memory, 8 * 600G SAS HD, Linux 5.6 OS. The approach to identify interesting frequent RNA substructures from the graph data is partially adapted from the FP-growth algorithm. In a similar manner, it is composed of FP tree construction and frequent itemset (substructure) generation.

Table 1 shows the number of frequent pattern, candidate closed graph, and closed graph under different minimum supports. It is observed that the number of frequent pattern decreases as minimum support grows. The number of the closed graph is far

Table 1. Number of frequent pattern in different minimum supports

Minimum support	Frequent pattern	Candidate pattern	Closed graph
0.165	307	75	54
0.17	273	69	47
0.175	229	69	40
0.18	207	64	38
0.185	189	58	32
0.19	165	58	29
0.195	149	58	30
0.2	135	58	27
0.205	121	53	21
0.21	107	53	21

less than the number of the frequent pattern. Nevertheless, the number of the candidate closed graph is a bit more than the number of the closed graph. Thus, the generation of candidate graph is able to greatly reduce those non-closed graphs.

5 Conclusions

The study of RNA secondary structures has become an important filed in recent years since structure-function correlation is an efficient way to explore its central roles in biology systems. Although, the traditional approaches such as index-based schema show good performance in finding similar structure in shape, the sequence and substructure information is ignored. This article proposes an approach to identify frequent RNA substructure patterns by DFS code. The raw RNA structure data is translated into graph data, in which the stems and loops are viewed as vertex and edge, respectively. The experimental results demonstrate our method is useful in identifying RNA substructures.

Acknowledgement. The work reported in this paper was partially supported by two National Natural Science Foundation of China projects 61363025 and 61373048, two key projects of Natural Science Foundation of Guangxi 2012GXNSFCB053006 and 2013GXNSFDA019029, and a grant from the Research Grants Council of the Hong Kong Special Administrative Region, [Project No. CityU 123013].

References

1. Rager, J.E., Smeester, L., Jaspers, I., et al.: Epigenetic changes induced by air toxics: formaldehyde exposure alters miRNA expression profiles in human lung cells. Environ. Health Perspect. **119**(4), 494500 (2011)
2. Kin, T., Yamada, K., Terai, G., et al.: fRNAdb: a platform for mining/annotating functional RNA candidates from non-coding RNA sequences. Nucleic Acids Res. **35**(Database issue), D145–D148 (2007)
3. Nawrocki, E.P., Burge, S.W., Bateman, A., et al.: Rfam 12.0: updates to the RNA families database. Nucleic Acids Res. **43**, D130–D137 (2015)
4. Wan, Y., Kertesz, M., Spitale, R.C., Segal, E., Chang, H.Y.: Understanding the transcriptome through RNA structure. Nat. Rev. Genet. **12**, 641–655 (2011)
5. Meyer, F., Kurtz, S., Beckstette, M.: Fast online and index-based algorithms for approximate search of RNA sequence-structure patterns. BMC Bioinform. **14**, 226 (2013)
6. Vyverman, M., Baets, B.D., Fack, V., Dawyndt, P.: Prospects and limitations of full-text index structures in genome analysis. Nucleic Acids Res. **40**(15), 6993 (2012)
7. Yan, X., Han, JW.: CloseGraph: mining closed frequent graph patterns. In: Proceedings of the Ninth ACM SIGKDD International Conference on Knowledge Discovery and Data Mining, pp. 286–295 (2003)
8. Le, S.Y., Owens, J., Nussinov, R., et al.: RNA secondary structures: comparison and determination of frequently recurring substructures by consensus. Comput. Appl. Biosci. CABIOS **5**(3), 205–210 (1989)

A Practical Algorithm for the 2-Species Duplication-Loss Small Phylogeny Problem

Jingli Wu[1,2(✉)] and Junwei Wang[2]

[1] Guangxi Key Lab of Multi-source Information Mining and Security,
Guangxi Normal University, Guilin 541004, China
wjlhappy@mailbox.gxnu.edu.cn
[2] College of Computer Science and Information Technology,
Guangxi Normal University, Guilin 541004, China

Abstract. In this paper, the 2-Species Duplication-Loss Small Phylogeny Problem is studied. By introducing an alignment algorithm, a labeling algorithm and three smart mutate operators, a genetic algorithm G2SP is presented. Algorithm G2SP adopts the method of combining the general operator and the smart ones. The general operator maintains the population diversity effectively, while the smart ones improve population convergence and make it evolve to the optimal solution more quickly. The tRNA and rRNA gene data of six kinds of real bacterium were used to test the performance of algorithms. The experimental results indicate that the G2SP algorithm can get fewer evolution cost than the PBLB algorithm proposed by Holloway et al., and it is an effective method for solving the 2-species duplication-loss small phylogeny problem.

Keywords: Duplication · Loss · Small phylogeny problem · Alignment · Genetic algorithm · Bioinformatics

1 Introduction

With the rapid development of sequencing technology, abundant completely sequenced and annotated genomes are available in public repositories. The inference of evolutionary histories of genomes and gene families has become one of the main topics in comparative genomics [1–5], since it can help to reveal the genomic basis of phenotypes. When comparing the genomes of related species, the mutations of interest are genome-scale changes such as rearrangements (inversions, transpositions, translocations, etc.) and content modifying operations (duplications, losses, horizontal gene transfers, etc.), which affects the overall organization of genes.

Recent works have indicated that duplication and loss operations dominate the evolution process of transfer RNA (tRNA) gene families [6–9]. Since tRNAs are essential for decoding the genome sequences into proteins, reconstructing the evolutionary history of their content and organization among species might bring us new insights into the translational machinery. In 2013, Holloway et al. [10] proposed a problem which aims to infer the gene order information of an ancestral genome from the gene order information of two given extant genomes in a simplified evolution model, including only duplication and loss operations. Because the gene order is preserved by

© Springer International Publishing Switzerland 2016
D.-S. Huang et al. (Eds.): ICIC 2016, Part I, LNCS 9771, pp. 335–346, 2016.
DOI: 10.1007/978-3-319-42291-6_33

duplication and loss operations, they further cast the problem into an alignment one. The problem has been proven to be NP-hard and APX-hard [11, 12]. Holloway et al. [10] presented a pseudo-boolean linear programming (PBLP) approach for an integer linear programming formulation of the alignment problem. Tremblay-Savard et al. [13] developed a heuristic algorithm for the gene order alignment problem in an evolution model including rearrangements (inversions and transpositions). Benzaid and El-Mabrouk [14] presented a dynamic programming based approach for the alignment of a set of gene orders related through a phylogenetic tree under a model involving a wide range of evolutionary events. In this paper, a practical algorithm G2SP based on genetic algorithm is presented for solving the Duplication-Loss Alignment Problem (DLA) [10]. An alignment algorithm, a labeling algorithm and three smart mutate operators are designed for the algorithm. Experimental results indicate that the G2SP algorithm can get fewer evolution cost than the PBLB one.

2   Definitions and Notations

Let Σ be an alphabet in which each character represents a specific gene family. A genome $X = x_1x_2...x_n$ can be represented as a string over Σ where each character may appear many times, i.e., $x_i \in \Sigma (i = 1, ..., n)$, $X \in \Sigma^n$. Let $X[i..j]$ denote the substring $x_i...x_j (1 \le i \le j \le n)$ of X. $\bar{X}$ denotes the inverse string of X. Let $\varphi = \{D, L\}$ be the operations set in an evolutionary model involving duplication (D) and loss (L) [10].

- Duplication Operation (D): A Duplication of size $k + 1$ on genome $X = x_1x_2...x_n$ is an operation that copies a substring $X[i..i + k]$ to a location j ($j < i$ or $j > i + k$) of X. The original copy $X[i..i + k]$ is called the origin, and the copied string is called the product of the duplication D.
- Loss Operation (L): A loss of size $k + 1$ on genome $X = x_1x_2...x_n$ is an operation that removes a substring $X[i..i + k]$ from genome X.

Given genomes X and Y, an operation sequence $O = O_1, O_2, ..., O_m$ ($O_i \in \varphi, i = 1, ..., m$) that can transform X into Y is called an evolutionary history, $O_{X \rightarrow Y}$. In this paper, $O_{X \rightarrow Y}$ is restricted to a visible history [10], which means no duplication in $O_{X \rightarrow Y}$ is modified subsequently by inserting (by duplication) or removing (by loss) genes from its origin or product. In this case, X is called a visible ancestor of Y [10]. Let $C(O_i)$ be the cost of the i-th operation in $O_{X \rightarrow Y}$, and the cost of $O_{X \rightarrow Y}$ is defined as follows.

$$C(O_{X \rightarrow Y}) = \sum_{1 \le i \le m} C(O_i)$$ (1)

Let $\varphi_{X \rightarrow Y}$ denote the set of all possible evolutionary histories transforming X to Y. If $\varphi_{X \rightarrow Y}$ is not empty, X is called a potential ancestor of Y [10]. The cost $C(X \rightarrow Y)$ of transforming X into Y is defined as formula (2):

$$C(X \rightarrow Y) = \min_{O_{X \rightarrow Y} \in \varphi_{X \rightarrow Y}} C(O_{X \rightarrow Y})$$ (2)

In 2012, Holloway et al. [10] proposed the two species small phylogeny problem in the duplication-loss evolution model (we abbreviate it as 2-SPP-DL): given two genomes $G1$ and $G2$, and evolutionary operations set $\varphi = \{D, L\}$, find a potential common ancestor G^* of $G1$ and $G2$ minimizing $C(G^* \rightarrow G1) + C(G^* \rightarrow G2)$.

As both duplication and loss operations belong to content-modifying ones, the 2-SPP-DL problem can be reformulated as an alignment problem [10]. Let matrix $G_{2 \times \alpha}$ record an alignment of G_1 and G_2, where $G_j^i \in (\Sigma \cup \{-\})(1 \leq i \leq 2, 1 \leq j \leq \alpha)$. There are only two kinds of relationships between elements G_j^1 and G_j^2, as follows:

(1) G_j^1 matches G_j^2: $G_j^1 \neq -$, $G_j^2 \neq -$, $G_j^1 = G_j^2$;
(2) G_j^1 mismatches G_j^2: $G_j^1 \neq -$, $G_j^2 = -$; or $G_j^1 = -$, $G_j^2 \neq -$.

"G_j^1 mismatches G_j^2" means G_j^2 (resp. G_j^1) is a loss, or G_j^1 (resp. G_j^2) is a duplication. Therefore, alignment G can be interpreted as a sequence of duplication (D) and loss (L) operations, which is called a labeling of G [10] and is denoted as $l(G) = O_1, O_2, ..., O_{|l(G)|}$ ($O_i \in \varphi$, $i = 1, ..., |l(G)|$). The cost of labeled alignment G is the sum of costs of all operations in $l(G)$, as defined in formula (3).

$$C(G) = \sum_{1 \leq i \leq l(G)} C(O_i) \tag{3}$$

It has been proven that given genomes G_1 and G_2, there is a one-to-one correspondence between their labeled alignments and visible ancestors [10]. Based on the above notations and definitions, Holloway et al. [10] reduced the 2-SPP-DL into the Duplication-Loss Alignment Problem (DLA), which is defined as follows: given genomes G_1 and G_2, determine a labeled alignment G of minimum cost.

3   G2SP Algorithm

In this section, firstly, algorithm ALIGN is presented for obtaining an alignment of two given genomes. Secondly, algorithm LABLE is proposed for labeling a given alignment. Finally, based on the ALIGN and LABLE algorithms, and three novel smart mutate operators, a genetic algorithm G2SP is proposed.

3.1   ALIGN Algorithm

Given genomes G_1 and G_2, algorithm ALIGN attempts to produce as many matches as possible by inserting characters '$-$' into them, and outputs an alignment G of G_1 and G_2. It compares G_1 and G_2 character by character. When a mismatch is met, algorithm ALIGN tries to search the longest matching strings after that mismatch position, and aligns matching strings through inserting '$-$' in terms of certain criteria. As shown in Fig. 1(a) shows two genomes, Fig. 1(b) shows an alignment obtained by performing algorithm ALIGN. The description of algorithm ALIGN is as follows.

G_1: a b b b b d G: a b b b b d

G_2: a b d   a b - - - d

(a) genomes G_1 and G_2 (b) an alignment G of G_1 and G_2

Fig. 1. An example of the ALIGN algorithm.

```
Algorithm ALIGN
```

```
Input: genomes G₁ and G₂,
Output:an alignment G of G₁ and G₂.
```
1. for(i=1;i≤$|G_1|$, i≤$|G_2|$;i++)
2.   if($G_{1i}\neq G_{2i}$)
3.   search the longest string $G_2[j..k]$ ($i<j≤k≤|G_2|$) in
        G_2 with $G_2[j..k]=G_1[i..1]$ ($i≤1≤|G_1|$);
4.   search the longest string $G_1[j'..k']$ ($i<j'≤k'≤|G_1|$)
        in G_1 with $G_1[j'..k']=G_2[i..1']$ ($i≤ 1'≤|G_2|$);
5.   if($|G_2[j..k]|$>0 && $|G_1[j'..k']|$>0)
6.   if((($(k-j+1)-(j-i)$)≥0 and (($(k'-j'+1)-(j'-i)$)≥0)
7.   if(($(k-j)>(k'-j')$)||(($(k-j)=(k'-j')$&& $j'>j$))
8.   insert ($j-i$) '-' into G_1 from the i-th
              position; $i=k+1$;
9.   else if(($(k-j)<(k'-j')$)||(($(k-j)=(k'-j')$&&$j'<j$))
10.   insert ($j'-i$) '-' into G_2 from the i-the
              position; $i=k'+1$;
11.   else if(($(k-j+1)-(j-i)$)≥0)
12.   insert ($j-i$) '-' into G_1 from the i-th
            position; $i=k+1$;
13.   else if(($(k'-j'+1)-(j'-i)$)≥0)
14.   insert ($j'-i$) '-' into G_2 from the i-th
            position; $i=k'+1$;
15.   else if($|G_2[j..k]|$>0 && (($(k-j+1)-(j-i)$))≥0)
16.   insert ($j-i$) '-' into G_1 from the i-th
          position; $i=k+1$;
17.   else if($|G_1[j'..k']|$>0 && (($(k'-j'+1)-(j'-i)$))≥0)
18.   insert ($j'-i$) '-' into G_2 from the i-th
          position; $i=k'+1$;
19. r=rand()%2+1; G_{ri}='-'; i++;
20.   recalculate$|G_1|$ and $|G_2|$;
21.if($|G_1|\neq|G_2|$)
22.   insert some '-' at the end of the shorter genome
      to make $|G_1|=|G_2|$.

Algorithm LABLE (left to right)

Input: An alignment G of G_1 and G_2

Output: A labeling $l(G)$ of G, and the cost $C(G)$

1. $i=1$; $C(G)=0$; $l(G)=`\ `$;
2. while($i\leq a$) //a is the number of columns of G
3.   if $(G_i^1=G_i^2)$
4. $G_i^*=\ G_i^1$; i++;//G^* is the ancestor of G_1 and G_2;
5.   else if$(G_i^1=\$||G_i^2=\$||G_i^1=L||G_i^2=L||0\leq G_i^1\leq a||0\leq G_i^2\leq a)$
6. i++;
7.   else if$(G_i^1\neq G_i^2$ && $G_{i+1}^1=G_{i+1}^2)$
8.   if$(G_i^1\neq-$ && $G_i^2=-)$
9. $P_1=G^1$; $P_2=G^2$;
10.   else $P_1=G^2$; $P_2=G^1$;
11.   if$(P_{1i}=P_{1j}$&&$i\neq j)$ //P_{1i} has a duplication in P_1
12. $P_{2i}=\$$; i++; /* $\$$ indicates either P_{1i} is the product of duplicating P_{1j} $(1\leq j\leq|P_1|,i\neq j)$, or P_{2i} is a loss*/
13.   else // P_{2i} is a loss
14. $G_i^*=P_{1i}$; $P_{2i}=L$; $l(G)=l(G)+L(P_{2i})$; i++; $C(G)$++;
15.   else//continuous mismatch gene block length > 1
16.   if$(G^1[i..u]\in\Sigma^{u-i+1}$ && $G^2[i..u]\in\{-\}^{u-i+1}(i\leq u\leq a))$
17. $P_1=G^1$; $P_2=G^2$;
18.   else $P_1=G^2$; $P_2=G^1$;
19.   if$(P_1[i..u]=P_1[j..k])$ and is not a cyclic duplication)//$(1<j\leq k\leq a,j>u$ or $i>k)$
20. $P_2[i..u]=$index$(P_1[j..k])$; //indexes string $l(G)=l(G)+D(P_1[j..k])$; $C(G)$++; $i=u+1$;
21.   else
22.   search $P_1[j..k]\in(\Sigma\cup\{-\})^{k-j+1}$) meeting:
            (1)$P_1[j'..k']$ obtained by removing `$-$' from $P_1[j..k]$ is a prefix of $P_1[i..u]$;
            (2)$P_{2t}(j\leq t\leq k,P_{1t'}=`-')$ has duplication origin
23.   if $(P_1[j..k]$ is found and $|P_1[j'..k']|>1$ and is not a cyclic duplication)
24. $P_2[i..i+(k'-j')]=$index$(P_1[j'..k'])$; $l(G)=l(G)+D(P_1[j'..k'])$; $C(G)$++; $i=i+(k'-j')+1$;
25.   else if$(P_{1i}=P_{1j}$ && $i\neq j)$
26. $P_{2i}=\$$; i++;
27.   else
28. $G_i^*=P_{1i}$; $P_{2i}=L$; $l(G)=l(G)+L(P_{2i})$; i++; $C(G)$++;
29. for $(i=1;i\leq a;i$++)//process $\$$, default operation is L
30.   if $(G_i^1=\$)$
31. $G_i^*=G_i^2$; $G_i^1=L$; $l(G)=l(G)+L(G_i^1)$; $C(G)$++;
32.   else if $(G_i^2=\$)$
33. $G_i^*=G_i^1$; $G_i^2=L$; $l(G)=l(G)+L(G_i^2)$; $C(G)$++.

3.2   LABLE Algorithm

A recent literature has indicated that the minimum labeling alignment problem is APX-hard [15]. In this section, a heuristic labeling algorithm LABLE is proposed. It labels an alignment sequence G from both "left to right" and "right to left", and chooses the operations sequence with fewer cost as the labeling of G. As in reference [10], $C(D(k)) = 1$ and $C(L(k)) = k$ are used to compute the costs of k-size duplication and loss operations respectively. During the labeling process, cyclic duplications are not permitted [10], for the cycle implies that it is not possible to define a history of duplications leading to extant genomes. Since the labeling methods are the same for both directions, only the "left to right" labeling method is depicted.

3.3   G2SP Algorithm

In this section, a genetic algorithm G2SP is devised to solve the DLA Problem. The input is two genomes G_1 and G_2. The output is a labeled alignment G and an ancestral sequences G^*. In the following, some key techniques in the G2SP algorithm are given.

Chromosome Presentation and the Generation of the Initial Population. An alignment of G_1 and G_2 is adopted to encode a solution, hence the chromosome length is variable. An initial chromosome is generated by executing algorithm ALIGN. The injection of randomness in algorithm ALIGN maintains the individual diversity of the initial population.

Fitness Function. Because each individual represents a viable solution to the DLA problem, we need to make an estimate of the result. Given a chromosome G, the fitness function *Fitness* (G) is defined as follows:

$$Fitness(G) = \frac{1}{C(G)}, \tag{4}$$

where $C(G)$ denotes the cost of a labeled alignment G.

Selection Operator. Roulette wheel selection [16] is adopted to generate a new population for the next generation. It evaluates an individual survival probability in terms of its relative fitness so that every individual has a chance to be selected.

Crossover Operator. In this paper, single point crossover operator is used. Given two parent chromosomes $F1$ and $F2$, firstly, $F1$ is split at a randomly determined crossover point. Denote by L_{F1} (resp. R_{F1}) as the block after removing character '−' from the left (resp. right) genome block. $F2$ is split at the corresponding crossover point so that L_{F2} (resp. R_{F2}) is exactly the same as L_{F1} (resp. R_{F1}), here L_{F2} (resp. R_{F2}) has the similar meaning as L_{F1} (resp. R_{F1}). Secondly, all data beyond the crossover point in both chromosomes are swapped between $F1$ and $F2$, and the blanks are filled with '−'. Finally, two offspring chromosomes $C1$ and $C2$ are obtained by deleting the columns with only '−'. As shown in Fig. 2, the bold column in $F1$ is a randomly determined

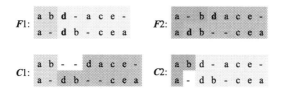

Fig. 2. An example of crossover operator.

crossover point, and the one in $F2$ is the corresponding crossover point. Two offspring $C1$ and $C2$ are got by performing the single point crossover operator.

Mutation Operator. In this paper, the smart mutate operators with guidance information are proposed according to the features of duplication and loss operations in biological evolution. It distinguishes which kind of change is a better one in terms of fitness. In the following, three smart mutate operators are described in detail. Given a parent chromosome $F_{2\times\alpha}$, F_1 and F_2 represent the rows of $F_{2\times\alpha}$ respectively.

(1)   Re-aligning gene block

   The operator tries to re-align some continuous mismatch gene block, which is chosen randomly from the parent chromosome. Given a gene block $(F_1[s..e], F_2[s..e])$ $(F_1[s..e], F_2[s..e] \in (\Sigma \cup \{-\})^{e-s+1}$, F_{1i} mismatches F_{2i}, $s \leq i \leq e)$, where s (resp. e) $(1 \leq s \leq e \leq \alpha)$ denotes the start (resp. end) position of the gene block. Let (F'_1, F'_2) be the new block after removing the characters '$-$' from $(F_1[s..e], F_2[s..e])$. Compute alignments for (F'_1, F'_2) and its inverse $(\overline{F'_1}, \overline{F'_2})$ respectively with the ALIGN algorithm, and the alignment results are still denoted by (F'_1, F'_2) and $(\overline{F'_1}, \overline{F'_2})$ respectively. Replace $(F_1[s..e], F_2[s..e])$ with (F'_1, F'_2) (resp. $(\overline{\overline{F'_1}}, \overline{\overline{F'_2}})$ that is the inverse of $(\overline{F'_1}, \overline{F'_1})$) to obtain a new individual $C1$ (resp. $C2$). Choose the one with larger fitness between $C1$ and $C2$ as the offspring C. As shown in Fig. 3(a) shows the parent chromosome F, where the shadow part represents the continuous mismatch gene block. Figure 3(b) is the new block after removing the characters '$-$' from the shadow part in Fig. 3(a). Figure 3(c) and Fig. 3(d) show two new individuals, and the one in Fig. 3(d) with larger fitness is regarded as the offspring C.

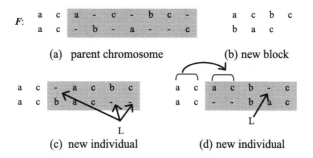

Fig. 3. An example of re-matching gene block.

(2) Smart moving gene

The operator attempts to improve the individual fitness by moving the '$-$' in it. Given a parent F, if there exist two mismatch columns (F_{1i}, F_{2i}) and (F_{1j}, F_{2j}) to meet the condition of F_{1i} matches F_{2j} (resp. F_{2i} matches F_{1j}), and substring $F_1[i+1..j-1]$ (resp. $F_2[i+1..j-1]$) contains only '$-$' values, F_{1i} and F_{2j} (resp. F_{2i} and F_{1j}) can be shifted to the same column by moving '$-$' without changing the gene order in F. The columns having only '$-$' are deleted from F and a new individual is obtained. Choose the one with larger fitness between F and the new individual as the offspring C. Figure 4 gives an example. Figure 4(a) shows a parent F, where the shadow parts indicate two mismatch columns. In Fig. 4(b), the match genes have been shifted to the same column. Figure 4(c) shows the new individual whose evolution cost is less than F.

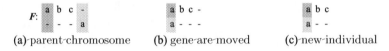

(a) parent chromosome   (b) gene are moved   (c) new individual

Fig. 4. An example of the smart moving gene.

(3) Moving gene block

The operator improves an individual fitness by moving the gene block in it. Given a parent F, find some continuous mismatch gene block $(F_t[i..j], F_{3-t}[i..j])$ $(F_t[i..j] \in \Sigma^{j-i+1}, F_{3-t}[i..j] \in \{-\}^{j-i+1}, t = 1,2, 1 \le i < j \le \alpha)$ in F. Attempt to move some substring of $F_{3-t}[1..i-1]$ (resp. $F_{3-t}[j+1..\alpha]$) rightward (resp. leftward) and make it matching or partial matching $F_t[i..j]$ without changing the gene order in F, and a new individual $C1$ (resp. $C2$) is obtained. Choose the one with larger fitness between

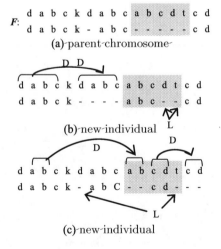

(a) parent chromosome

(b) new individual L

(c) new individual

Fig. 5. An example of moving gene block.

G2SP algorithm

Input: genomes G_1 and G_2,

Output: a labeled alignment G, and the ancestral genome G^*.

Step 0: Give proper parameter setting: population size N, crossover rate p_c, mutation rate p_m and the maximum number of population generation *maxgen*;

Step 1: Randomly generate N individuals as an initial population pop_0, *gen*=0;

Step 2: Compute the fitness of every individual in pop_{gen} in terms of formula (4) and retain the best individual G;

Step 3: If *gen*>*maxgen*, go to **Step7**, otherwise go to **Step4**;

Step 4: Select $(1-p_c) \times N$ members of pop_{gen} by using selection operator, and add them to pop_{gen+1} directly; Select $p_c \times N/2$ pairs of individuals from pop_{gen}. For each pair of individuals, produce two offspring individuals by applying crossover operator and add them to pop_{gen+1};

Step5: Select $p_m \times N$ individuals from pop_{gen+1} with uniform probability. For each, produce a new individual by randomly applying one of the three mutation operators;

Step6: *gen*=*gen*+1, go to **Step2**;

Step7: Infer the ancestral genome G^* from the labeled alignment G, return G^* and G, stop.

Fig. 6. G2SP algorithm.

$C1$ and $C2$ as the offspring C. Figure 5 shows an example. Figure 5(a) is a given parent F, where the shadow part denotes a continuous mismatch gene block. Figure 5(b) shows the new individual $C1$ obtained from moving $F_2[7..9]$ rightward. Figure 5(c) shows the new individual $C2$ got from moving $F_2[15..16]$ leftward.

Based on the above mentioned design, the G2SP algorithm is summarized in Fig. 6.

4   Experimental Results

In this section, real genetic data were used to compare the PBLP algorithm with the G2SP one. Algorithm PBLP was implemented on a Lenovo workstation with Intel(R) Core (TM) i5-3570@3.4 GHz Processors and 4 GB of RAM. The operating system was Ubuntu 12.04 and the compiler was Python 2.7.3. Algorithm G2SP also ran on a Lenovo workstation with the same configuration. The operating system was Windows XP professional and the compiler was Microsoft Visual C# 2012.

4.1   Experimental Data

The stable RNAs are primarily transfer RNAs (tRNAs) and ribosomal RNAs (rRNAs). Since the stable RNA families are rapidly evolving by duplication and loss [10], three sets of real-world instances, which contain the stable tRNA and rRNA contents of 5 Bacillus, 5 Yersinia and 5 Escherichia, were used in our experiments. All the data can be obtained from Pathosystems Resource Integration Center (PATRIC) [17] and National Center of Biotechnology Information (NCBI) [18]. For convenience of description, the genomes in each set were numbered as follows: Bacillus ((1) NC 004722 Bacillus cereus ATCC 14579 Bacteria, (2) NC 006274 Bacillus cereus E33L Bacteria, (3) NC 007530 Bacillus anthracis str.'Ames Ancestor' Bacteria, (4) NC 006322 Bacillus licheniformis ATCC 14580 Bacteria; Firmicutes, (5) NC 000964 Bacillus subtilis subsp. subtilis str. 168 Bacteria), Yersinia ((1) Yersinia pestis Pestoides F, (2) Yersinia pestis Pestoides A, (3) Yersinia pestis Nepal516, (4) Yersinia enterocolitica subsp palearctica YO527, (5) Yersinia enterocolitica subsp palearctica 6475307), Escherichia ((1) Escherichia blattae DSM 4481, (2) Escherichia coli KTE210, (3) Escherichia coli O111-H- str 11128, (4) Escherichia coli TW10509, (5) Escherichia fergusonii ATCC 35469).

4.2   Performance Evaluation

The following two measurements were used to evaluate the performance of the two algorithms: (1) the evolution cost; (2) the running time. The parameters of algorithm G2SP were set as follow: $N = 100, maxgen = 50, p_c = 0.8$ and $p_m = 0.2$. In the following tables, the genomes are denoted by their number specified above. For example, the pair of genomes {1,2} of bacillus represents genomes NC_004722 and NC_006274.

In Table 1, ten genome pairs are generated for Bacillus, Yersinia and Escherichia, respectively. The overall number of represented RNA families in Bacillus, Yersinia and Escherichia genomes is around 42, 45 and 42, and the total number of RNAs in each of their genome is between 96 and 149, 93 and 133, 87 and 103, respectively. The experimental results show that for different pairs of genomes, algorithm G2SP can get fewer evolution cost than algorithm PBLP. As shown in Table 1, take Bacillus for an example, the evolution history inferred from NC_006274 (numbered 2) and NC_007530 (numbered 3) has the minimal evolution cost, while the one inferred from NC_007530 (numbered 3) and NC_000964 (numbered 5) has the maximal evolution cost. Therefore, among the five Bacillus genomes, the phylogenetic relationship between NC_006274 and NC_007530 is closer than other genome pairs, and the one between NC_007530 and NC_000964 is more distant than other pairs.

Table 1. Comparisons of evolution cost.

Genome pair	Bacillus		Yersinia		Escherichia	
	PBLP	G2SP	PBLP	G2SP	PBLP	G2SP
{1, 2}	17	9	109	101	142	134
{1, 3}	18	10	105	92	166	147
{1, 4}	64	47	132	125	71	59
{1,5}	59	38	115	102	90	81
{2,3}	3	2	118	118	77	66
{2,4}	50	41	136	129	127	118
{2,5}	63	59	116	106	118	105
{3,4}	49	40	140	127	145	135
{3,5}	62	60	124	110	125	106
{4,5}	21	14	132	117	109	98

Table 2 shows the average running time of all pairs of each instance set. Although the running time of the G2SP algorithm is longer than that of the PBLP algorithm, the longest one is only 1044 s, which is still feasible in practical application.

The above experimental results show that the G2SP algorithm can get fewer evolution cost than the PBLP one with different pairs of genomes. Following the

Table 2. Comparisons of running time (s).

Genome	PBLP	G2SP
Bacillus	58.2	164.2
Yersinia	12.6	1044.0
Escherichia	11.7	910.3

characteristics of algorithm G2SP are analyzed. The convergence of it is analyzed at first. (1) The G2SP algorithm generates an initial population by using the ALIGN algorithm, which aims at generating as many matches as possible and ensures the high quality of initial solutions. (2) The three smart mutation operators adjust the positions of '−' by guidance in terms of the situation of sequences so as to improve the alignment, that plays a positive role in making the population evolve to the optimal solution rapidly. (3) Retaining the best individual in each generation also ensures the convergence of algorithm G2SP. In addition, some strategies are adopted to avoid the premature convergence of the G2SP algorithm. (1) Algorithm ALIGN injects random information into the initial solution and ensures the diversity of the initial population. (2) In the single point crossover operator, filling the blanks with '−' will introduce some new solution and give a good chance to maintain the population diversity during the evolution process.

5   Conclusions

The two-species small phylogeny problem in the duplication-loss model is studied based on the duplication-loss alignment problem. A genetic algorithm G2SP is presented for solving this model, which adopts both one ordinary single point crossover operator and three smart mutation operators. The three smart mutation operators can also be integrated with other algorithms, which will be studied in the future. Compared with algorithm PBLP, the G2SP algorithm can get fewer evolution cost with reasonable running time, which was tested by a number of experiments.

Acknowledgments. The authors are grateful to anonymous referees for their helpful comments. This research is supported by the National Natural Science Foundation of China under Grant No. 61363035 and No. 61502111, Guangxi Natural Science Foundation under Grant No. 2015GXNSFAA139288, No. 2013GXNSFBA019263 and No. 2012GXNSFAA053219, Research Fund of Guangxi Key Lab of Multisource Information Mining and Security No. 14-A-03-02 and No. 15-A-03-02, "Bagui Scholar" Project Special Funds, Guangxi Collaborative Innovation Center of Multi-source Information Integration and Intelligent Processing.

References

1. Hallett, M.T., Lagergren, J.: New algorithms for the duplication-loss model. In: Shamir, R., Miyano, S., Istrail, S., Pevzner, P., Waterman, M. (eds.) RECOMB 2000, pp. 138–146. ACM Press, New York (2000)

2. Chauve, C., Doyon, J.P., El-Mabrouk, N.: Gene family evolution by duplication, speciation, and loss. J. Comput. Biol. **15**(8), 1043–1062 (2008)
3. Berard, B., Szollosi, G.J., Daubin, V.: Evolution of gene neighborhoods within reconciled phylogenies. Bioinformatics **28**, i382–i388 (2012)
4. Patterson, M., Szöllösi, G., Daubin, V., Tannier, E.: Lateral gene transfer, rearrangement, reconciliation. BMC Bioinf. **14**, S4 (2013)
5. Sankoff, D.: Genome rearrangement with gene families. Bioinformatics **15**(11), 909–917 (1999)
6. Rogers, H.H., Bergman, C.M., Griffiths-Jones, S.: The evolution of tRNA genes in Drosophila. Genome Biol. Evol. **2**, 467–477 (2010)
7. Withers, M., Wernisch, L., Reis, M.D.: Archaeology and evolution of transfer RNA genes in the Escherichia coli genome. Bioinformatics **12**, 933–942 (2006)
8. Bermudez-Santana, C., Attolini, C.S., Kirsten, T., Engelhardt, J., Prohaska, S., Steigele, S., Stadler, P.: Genomic organization of eukaryotic tRNAs. BMC Genomics **11**, 270 (2010)
9. Tang, D., Glazov, E., McWilliam, S., Barris, W., Dalrymple, B.: Analysis of the complement and molecular evolution of tRNA genes in cow. BMC Genomics **10**, 188 (2009)
10. Holloway, P., Swenson, K., Ardell, D.H., et al.: Ancestral genome organization: an alignment approach. J. Comput. Biol. **20**(4), 280–295 (2013)
11. Andreotti, S., Reinert, K., Canzar, S.: The duplication-loss small phylogeny problem: from cherries to trees. J. Comput. Biol. **20**(9), 643–659 (2013)
12. Dondi, R., El-Mabrouk, N.: Aligning and labeling genomes under the duplication-loss model. In: Bonizzoni, P., Brattka, V., Löwe, B. (eds.) CiE 2013. LNCS, vol. 7921, pp. 97–107. Springer, Heidelberg (2013)
13. Tremblay-Savard, O., Benzaid, B., Lang, B.F., El-Mabrouk, N.: Evolution of tRNA repertoires in bacillus inferred with OrthoAlign. Mol. Biol. Evol. (2015). doi:10.1093/molbev/msv029
14. Benzaid, B., El-Mabrouk, N.: Gene order alignment on trees with multiOrthoAlign. BMC-Genomics **15**(Suppl. 6), S5 (2014)
15. Benzaid, B., Dondi, R., El-Mabrouk, N.: Duplication-loss genome alignment: complexity and algorithm. In: Dediu, A.-H., Martín-Vide, C., Truthe, B. (eds.) LATA 2013. LNCS, vol. 7810, pp. 116–127. Springer, Heidelberg (2013)
16. Holland, J.H.: Adaptation in natural and artificial systems. Ph.D. dissertation, University of Michigan (1975)
17. http://patricbrc.vbi.vt.edu/
18. http://www.ncbi.nlm.nih.gov/

Prediction of Phosphorylation Sites Using PSO-ANNs

Ruizhi Han[1,2], Dong Wang[1,2(✉)], Yuehui Chen[1,2(✉)],
Wenzheng Bao[3], Qianqian Zhang[1], and Hanhan Cong[4]

[1] School of Information Science and Engineering,
University of Jinan, Jinan 250022, China
501248792@qq.com, {ise_wangd,yhchen}@ujn.edu.cn
[2] Shandong Provincial Key Laboratory of Network
Based Intelligent Computing, Jinan 250022, China
[3] College of Electronics and Information Engineering,
Tongji University, Shanghai 201804, China
[4] University of Jinan, Jinan 250022, China

Abstract. Post-translational modifications (PTMs) are essential for regulating conformational changes, activities and functions of proteins, and are involved in almost all cellular pathways and processes. Phosphorylation is one of the most important post-translational modifications of proteins, which is related to many activities of life. It can regulate signal transduction, gene expression and cell cycle regulation of many cellular processes by protein phosphorylation and dephosphorylation. With the development and application of proteomics technology, researchers pay close attention on protein phosphorylation research more and more widely. In this paper, we use PSO algorithm to optimize neural network weight coefficients and classify the data which has secondary encoding according to the physical and chemical properties of amino acids for feature extraction. The experimental results compared with the result of the support vector machine (SVM) and experimental results show that the prediction accuracy of PSO-ANNs 2.44 % higher than that of SVM. And this paper at the same time, this paper also analyzes the experimental results under different window values. The results of the experiment are best when the window value is 11.

Keywords: Phosphorylation sites prediction · Particle swarm optimization (PSO) · Artificial neural network

1 Introduction

The cell is the smallest unit of life and the basic elements of cell protein. Protein post-translational modifications (PTMs) is an important mechanism for the regulation of protein function, in the biological processes and pathways plays an irreplaceable role, and reversibly determines the cell dynamics and plasticity [1]. Phosphorylation is one of the most important post-translational modifications (PTMs). This modification to amino acid side chains by covalent bond connecting a phosphate group, usually occurs on serine *S* threonine *T* and tyrosine *Y*, known as site. The process of phosphorylation regulates almost all life activities, including cell proliferation, differentiation and

© Springer International Publishing Switzerland 2016
D.-S. Huang et al. (Eds.): ICIC 2016, Part I, LNCS 9771, pp. 347–355, 2016.
DOI: 10.1007/978-3-319-42291-6_34

development, neural activity, muscle contraction, cancer and new supersedes the old. Especially in the cellular response to external stimuli, protein phosphorylation is the main signal known as transfer mode [2, 11].

In 2004, South Korea Kim et al. [3] for the first time, scholars using support vector machine (Support vector machine, SVM) algorithm to design the kinase specific phosphorylation sites prediction method. In 2005, Li et al. [4] by k- (k-Nearest Neighbor, k-NN nearest neighbor) algorithm to design specific kinase (Kinase-specific) phosphorylation site prediction method. 2007, Tang et al. [6] using genetic algorithm integrated neural network (genetic algorithm integrated neural network, GANN) design no specific phosphorylation sites prediction method, and prediction accuracy can respectively reach: serine (S) was 81.1 %, threonine (T) was 76.7 %, tyrosine (Y) was 73.3 %. In 2014, Wu et al. [5] k- neighbor algorithm is designed based on the phosphatase recognition site prediction method.

In this paper, we extract the 32 proteins related to signal transduction and constructs the data sets used in the experiment because phosphorylated modification occurs frequently in the signal transduction. And we use particle swarm optimization (PSO) algorithm to optimize the parameters of the neural network and adopt two kinds of feature extraction method in the feature extraction. Through experimental verification, we get more ideal results by using this classification model.

2   Materials and Methods

2.1   Materials

The occurrences of phosphorylation in a certain site of protein sequences is mainly influenced on site adjacent sequences. [12] So the sample is usually consist of m amino acid residues behind site and m amino acid residues in front of the site. So, each micro amino acid sequence contains 2m + 1 amino acid residues. Researchers mark modified sites' adjacent sequence as positive samples and mark could not be modified sites' adjacent sequence as negative samples. In this article, we select value (the value of m) as 11. So, each sequences covered 23 amino acid residues.

Protein phosphorylation function is common especially in the signal transduction [1]. So, when build datasets, we extract the 32 proteins which were related to signal transduction [8] from phosphorylation site database [15] (www.phosphosite.org). Such 32 proteins contained 596 amino acid residues which have been marked as modification sites clearly. There are 305 serine (S), 167 threonine (T) and 124 tyrosine (Y). At the same time, we extract not be identified sites as negative samples in this 32 protein amino acid residues and we get 1945 points which includes 929 serine (S), 634 threonine (T) and 382 tyrosine (Y). In order to ensure the datasets is positive and negative balance, we extract 596 not modified sites randomly as a negative sample dataset in 1945 not be indicated modified sites. The details of the data set can be seen in Table 1, it contains the number of three kinds of amino acid residues (S, T, Y) in positive samples and negative samples.

Table 1. Number of amino acid residues in data set.

Amino acid residues	Positive samples	Negative samples
S	305	310
T	167	164
Y	124	122

2.2   Feature Extraction

2.2.1   Encoding Based on Attribute Grouping

Previously, Fan and Zhang have detected that the serine and threonine acceptor site microenvironment is depleted in nonpolar and hydrophobic amino acids. Whereas the tyrosine acceptor site microenvironment is characterized by only one enriched property, namely the charge, and is depleted in cysteine (*C*) and proline (*P*), which are neutral residues [7, 13].

Thus, we adopted an encoding scheme of protein sequences considering the hydrophobicity and charged character of amino acid residues [13]. The encoding method based on attribute grouping (named as EBAG) divides the 20 amino acid residues and a kind of gap into five different classes according to their physicochemical property: the hydrophobic group C1 = [*A, F, G, I, L, M, P, V, W*], the polar group C2 = [*C, N, Q, S, T, Y*], the acidic group C3 = [D, E], the basic group C4 = [*H, K, R*] [9, 10], and the gaps' group C5 = [*X*]. Given a protein sequence p fragment with 2L + 1 amino acid residues, we used the above classification convert protein sequences fragments P into the following form (Table 2):

Table 2. Groups of amino acid residues.

Groups	Amino acid residues	Label
C1	*A,F,G,I,L,M,P,V,W*	Hydrophobic group
C2	*C,N,Q,S,T,Y*	Polar group
C3	*D,E*	Acidic group
C4	*H,K,R*	Basic group
C5	*X*	Gaps group

2.2.2   Profile Encoding

Profile encoding is according the frequency of each amino acid in protein sequences to encode. The frequency of each type of amino acids in a protein sequence is F_A, the number of occurrences of each amino acid is C_A, and total number of each sequence is L. The value of A range of 1, 2, 3, 4,... and 20. Then the calculated frequency of each amino acid formula is:

$$F_A = \frac{CA}{L} \tag{1}$$

In this way, we can put a protein sequence into a characteristic vector. Each amino acid residue in a certain order. PV represents a trace protein sequences, as shown in the following formula:

$$PV = [F_1, F_2, F_3, \ldots F_{20}] \tag{2}$$

A collection of all protein sequences is PV vector [14].

2.2.3   Encoding Based on Attribute Grouping and Profile Encoding

In this article, we use EBAG data encoding at first. The protein sequence fragment P can be divided into five categories and then according the frequency of the five types in a same sequence fragment P to encoding second. We can get the recoding protein sequence fragment depending on the frequency of five types. As shown in Fig. 1:

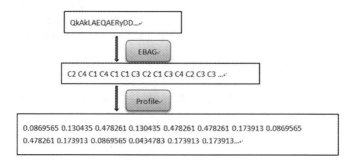

Fig. 1. Process of feature extraction.

2.3   Models

The PSO-ANNs uses PSO to optimize the connection weights of the ANNs over the training dataset. With a basic architecture similar to the BPNN, the current PSO-ANNs contains one input layer, one hidden layer and one output layer (Fig. 1). According to our preliminary optimization, the number of hidden nodes is set as 15 in this work. Only one output node (y) is needed, and the corresponding value is '1' or '0', representing a phosphorylation site or a non-phosphorylation site. W_{ij} denotes the weight from an input node to a hidden node and W_{jk} the weight from a hidden node to the only output node. The neural network uses a sigmoid function to provide a continuous activation function.

2.3.1   PSO

PSO initialized to a group of random particles (random solutions). Then through iteration to find the optimal solution. During each generation, the particles update them-selves by tracking two "extreme". The first is the particles found the optimal solution. The solution is called individual extreme value pBest. Another "extreme" value is the whole population found the optimal solution. This is global extreme value gBest. It also just need one part of the particle's neighbors only instead of the entire population.

So local extremum is the extremum among all the neighbors. After found the two optimal values, the particle will according the formula (3) and (4) to update their new position and speed of students usually.

$$
\begin{aligned}
v_{id}^{k+1} &= wv_{id}^k + c_1 rand()(p_{id} - x_{id}^k) + c_2 rand()(p_{pbest} - x_{id}^k) \\
x_{id}^{k+1} &= x_{id}^k + v_{id}^{k+1} \quad i = 1,2,\cdots,m; \quad d = 1,2,\cdots,D
\end{aligned}
\tag{3}
$$

$$
\begin{aligned}
v_{id}^{k+1} &= wv_{id}^k + c_1 rand()(p_{id} - x_{id}^k) + c_2 rand()(p_{gbest} - x_{id}^k) \\
x_{id}^{k+1} &= x_{id}^k + v_{id}^{k+1} \quad i = 1,2,\cdots,m; \quad d = 1,2,\cdots,D
\end{aligned}
\tag{4}
$$

Most evolutionary computation technique is using the same process, the particle swarm algorithm has the following optimization process:

1. The random initialization population
2. Each individual within the population on calculation of fitness (the fitness value).
3. Update population based on fitness
4. If the termination condition is met, stop, or turn to step 2.

2.3.2 ANNs

The basic structure of ANNs has three layers: input layer, output layer and hidden layer. The structure of ANNs is shown in Fig. 2.

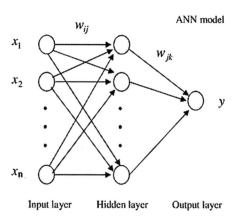

Fig. 2. Structure of ANNs

Input layer: Many neurons accept large nonlinear input information. The input information referred as the input vector.

Output layer: Information transmission, analysis, balance in the neuronal links, form the output. The output of information referred as the output vector.

Hidden layer: Hidden layer is the dimensions consist of many neurons and links between input layer and output layer. Hidden layer could have many layers but one layer is always adopted as usual.

2.3.3 Model Evaluation

Ten-fold cross validation was applied to evaluate the powers of the prediction method. Sensitivity (Sn), specificity (Sp), accuracy (Acc) and Matthew's Correlation Coefficient (MCC) were utilized to assess the performance of prediction system. All of the above measurements are defined as follows:

$$Sn = \frac{TP}{TP + FN} \tag{5}$$

$$Sp = \frac{TN}{TN + FP} \tag{6}$$

$$Acc = \frac{TN + TP}{TN + FP + FN + TP} \tag{7}$$

$$MCC = \frac{TP \times TN - FP \times FN}{\sqrt{(TN + FP) \times (TN + FN) \times (TP + FP) \times (TP + FN)}} \tag{8}$$

TP, TN, FP and FN are the number of true positives, true negatives, false positives and false negatives, respectively. Sensitivity and specificity illustrate the correct prediction ratios of positive and negative data sets, while accuracy represents the correct ratio among both positive and negative data sets. The MCC considers both the TP and the TN as successful predictions, and it is usually regarded as a balanced measure that can be used even if the classes are of very different sizes. For these reasons, the MCC is more reliable than the Acc. The value of MCC ranges from −1 to 1, with larger values standing for better predictive performance.

3   Results and Discussion

3.1   Investigation of Different Models and Features

In the prediction issue, all data sets could be regarded as the input vector common input neural network, instead of input a kind of amino acid residues to predict respectively. For example, we input the features of three kinds of amino acid residues, such as serine, threonine and tyrosine, to classification model at the same time. Let the classification model distinguish weather three kinds of amino acid residues are modified site or not based on prior extract the features of vector. This approach reduces the time complexity and space complexity to a certain extent. At the same time, we can get the overall data accuracy of the classification model's prediction instead of the prediction accuracy of three kinds of modification sites.

In this paper, the experiments are tested 32 proteins' sequence data using 10-fold cross validation. Here, four measurements, i.e. accuracy (Ac), sensitivity (Sn), specificity (Sp) and MCC, were jointly used to assess the performance of PSO-ANNs. The overall prediction accuracy (Ac) of PSO-ANNs reached 85.91 %.

Table 3. Prediction performance of models trained with different features in window size 11.

Models	Feature	Performance			
		ACC	SN	SP	MCC
PSO-ANNs	Profile	61.92 %	56.67 %	77.50 %	0.32
	EBAG + Profile	**85.91 %**	**86.40 %**	**85.42 %**	**0.72**
SVM	Profile	51.67 %	54.90 %	48.47 %	0.03
	EBAG + Profile	**83.47 %**	**84.21 %**	**82.72 %**	**0.67**

As visible in Table 3, in view of the two kinds of classification model for the same dataset of the ten cross validation experiments, show that the experimental accuracy of PSO–ANNs is higher than the SVM. We get the PSO-ANNs in EBAG + Profile prediction accuracy is 85.91 %, more superior than the 83.47 % of SVM's prediction accuracy. And the other experimental results have some extent advantages. All results are averages of ten times experiments.

As shown in Table 3: when using Profile coding only, both PSO-ANNs and SVM, and classification have not ideal results. The indicators are far below the EBAG + Profile feature extraction method. And after we add physical and chemical properties of amino acids to the Profile feature extraction group, two kinds of model prediction accuracy has improved significantly. Meanwhile, Mathews index of the two models in the case of the fusion of the two feature extraction methods have been greatly improved, especially SVM. Its Mathews index increased from 0.03 to 0.67. Therefore, before the characteristics of dataset are extracted, grouping dataset according to their own physical and chemical properties of amino acid residues first owes great influence to the result of the experiment. We can safely draw the conclusion that when feature extracting the protein microarrays, add more physical and chemical properties which related to the amino acid residues can improve accuracy of classification.

3.2 Investigation of Window Sizes

The number of residues surrounding the phosphorylation site that are taken into account is important because too few means information useful for making predictions gets ignored, while too many will decrease the signal-to-noise ratio. However, the residues contacted by the kinase may not be the same as the residues surrounding the phosphorylation site in the linear sequence [12].

In this article, we draw lessons from Huang et al. [13]. Huang used SVM to virPTM in experiment and discusses. Then he get the conclusion: when the window of the amino acid sequence is 23, the SVM classification effect better to some extent. Window values as 23, is 11 amino acid residues before and after the site, coupled with the modified site itself for a total of 23 amino acid residues. At the same time we also do the experiment on window values from 17 to 29 used the PSO-ANNs. The final experimental results show that, when the window is a value of 23 and adopts the EBAG + Profile feature extraction method, the experimental results of PSO - ANNs are also more effective. Details of experimental results are shown in Fig. 3.

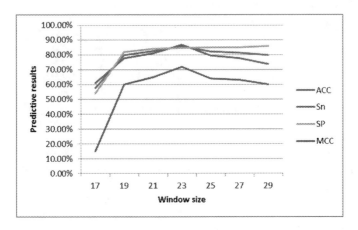

Fig. 3. Performance of different size of window. (Color figure online)

4 Conclusions

Above all, this paper used two methods of feature extraction to extract features of the original datasets twice. The results of two groups are very different whether predictive accuracy or sensitivity, or specificity. And we used neural network to classify test datasets in terms of classification model and using PSO algorithm to optimize the parameters of the neural network. The experimental results compared with SVM, which used same datasets and same feature extraction. The experimental results show that the prediction accuracy of neural network is 2.44 % higher than the SVM's prediction results.

Next, in order to improve the accuracy, we can start from the following two aspects: 1. Find a more reasonable feature extraction method for protein sequences. 2. Build a more reasonable classification model.

Acknowledgements. This research was partially supported by Program for Scientific research innovation team in Colleges and universities of Shandong Province 2012–2015, the Key Project of Natural Science Foundation of Shandong Province (ZR2011FZ001), the Natural Science Foundation of Shandong Province (ZR2011FL022, ZR2013FL002), the Youth Science and Technology Star Program of Jinan City (201406003), Shandong Provincial Natural Science Foundation, China, under Grant ZR2015JL025 and the Shandong Provincial Key Laboratory of Network Based Intelligent Computing. This work was also supported by the National Natural Science Foundation of China (Grant No. 61302128). The scientific research foundation of University of Jinan (XKY1410, XKY1411).

References

1. Xue, Y., Liu, Z.X., Cao, J., Ren, J.: Computational prediction of post-translational modification sites in proteins. Syst. Comput. Biol.-Mol. Cell. Exp. Syst. **5772**(6), 18559 (2011)
2. Huang, Z.Y., Yu, Y.L., Fang, C.Y., Yang, F.Y.: Progress in identification of protein phosphorylation by mass spectrometry. J. Chin. Mass Spectrom. Soc. **24**(4), 490–500 (2003)
3. Kim, J.H., Lee, J., Oh, B., Kimm, K., Koh, I.: Prediction of phosphorylation sites using SVMs. Bioinformatics **20**(17), 3179–3184 (2004)
4. Li, A., Wang, L.R., Shi, Y.Z., Wang, M.H., Jiang, Z.H., Feng, H.Q.: Phosphorylation site prediction with a modified k-nearest neighbor algorithm and BLOSUM62 matrix. Conf. Proc. IEEE Eng. Med. Biol. Soc. **6**, 6075–6078 (2005)
5. Wu, Z., Lu, M., Li, T.T.: Prediction of substrate sites for protein phosphatases 1B, SHP-1, and SHP-2 based on sequence features. Amino Acids **46**(8), 1919–1928 (2014)
6. Tang, Y.R., Chen, Y.Z., Canchaya, C.A., Zhang, Z.D.: GANNPhos: a new phosphorylation site predictor based on a genetic algorithm integrated neural network. Protein Eng. Des. Sel. **20**(8), 405–412 (2007)
7. Fan, S.C., Zhang, X.G.: Characterizing the microenvironment surrounding phosphorylated protein sites. Genomics Proteomics Bioinf. **3**, 213–217 (2005)
8. Wang, J.Y., Zhu, S.G., Xu, C.F.: Biochemistry, 3rd edn. Higher Education Press, Peking (2002)
9. Zhang, Z.H., Wang, Z.H., Zhang, Z.R., Wang, Y.X.: A novel method for apoptosis protein subcellular localization prediction combining encoding based on grouped weight and support vector machine. FEBS Lett. **580**, 6169–6174 (2006)
10. Nanni, L., Lumini, A.: An ensemble of reduced alphabets with protein encoding based on grouped weight for predicting DNA-binding proteins. Amino Acids **36**, 167–175 (2009)
11. Li, H., Xie, L.: Biological information method for prediction and identification of protein translation modification. Prog. Mod. Biomed. **8**, 1729–1735 (2008)
12. Trost, B., Kusalik, A.: Computational prediction of eukaryotic phosphorylation sites. Bioinformatics **27**, 2927–2935 (2011)
13. Huang, S.Y., Shi, S.P., Qiu, J.D., Liu, M.C.: Using support vector machines to identify protein phosphorylation sites in viruses. J. Mol. Graph. Model. **56**, 84–90 (2015)
14. Liu, Q.F.: Protein sequence coding and function prediction. Hunan University, May 2011
15. Hornbeck, P.V., et al.: PhosphoSitePlus, 2014: mutations, PTMs and recalibrations. Nucleic Acidc Res. **43**, D512–D520 (2015)

Predicting Subcellular Localization of Multiple Sites Proteins

Dong Wang[1], Wenzheng Bao[2], Yuehui Chen[1(✉)], Wenxing He[3(✉)],
Luyao Wang[1], and Yuling Fan[1]

[1] School of Information Science and Engineering,
University of Jinan, Jinan 250022, China
{ise_wangd,yhchen,ise_fanyl}@ujn.edu.cn,
897112057@qq.com
[2] Institute of Machine Learning and Systems Biology,
College of Electronics and Information Engineering,
Tongji University, Shanghai 201804, China
baowz55555@126.com
[3] School of Biological Science and Technology,
University of Jinan, Jinan 250022, China
163.hwx@163.com

Abstract. Accurate classification on protein subcellular localization plays an important role in Bioinformatics. An increasingly evidences demonstrate that a variety of classification methods have been employed in this field. This research adopts feature fusion method to extract the information of the protein subcellular. Several types of features are employed in this protein coding method, which include amino acid index distribution, the stereo-chemical properties of amino acids and the information for local sequence of amino acids. On base of this feature combination method, flexible neutral tree (FNT) is employed to predict multiplex protein subcellular locations. The overall accuracy rate of using flexible neutral tree as prediction algorithm may reach a better result.

Keywords: Amino acid index distribution (AAID) · Pseudo amino acid composition (PseAAC) · Stereo-chemical properties (SP) · Flexible neutral tree (FNT)

1 Introduction

The structure of a cell is composed of compartments named "subcellular fractions", which include cell membrane, nucleus, endoplasmic reticulum and so on. The function of the sub-cell is also performed by different kinds of proteins, which were localized to "protein subcellular location" and carried out biochemical function in different ranges. One protein may have one protein subcellular location, while others may have several different locations. However, proteins can be in working order only when they must be localized to their correct protein subcellular locations. Otherwise dysfunction and diseases may come out [1–3]. Without this knowledge the function of protein in a cell has cannot be fully elucidated. This information provides insight into finding novel

© Springer International Publishing Switzerland 2016
D.-S. Huang et al. (Eds.): ICIC 2016, Part I, LNCS 9771, pp. 356–365, 2016.
DOI: 10.1007/978-3-319-42291-6_35

proteins and novel protein function theoretically. Meanwhile, a more specific organellar context in which to investigate a particular protein can also be exhibited. For a long time, these related data have been difficult to produce on a large scale for higher eukaryotic organisms. However, recent advances in computational intelligence methods and high-throughput subcellular localization assays have made it possible to generate these datasets. So we can use Internet Protein Database through high-throughput methods to predict the membrane organization for the entire proteome and to determine the subcellular localization of a subset of the proteome.

Extracting more feature information from the protein sequence and the design of an efficient algorithm is the most important aspect which needs much more focus. Shortcomings are exist in most of feature extraction method, such as amino acid composition, it misses all the position information of the protein sequence; although the dipeptide model includes some position information, it is extremely easy to cause dimension disaster [4]. For algorithms, there are several algorithms which can deal with the multi-label prediction problem. Some proteins may localized at multiple sites, this kind of proteins have thus more than one subcellular locations [5, 6]. In this paper, we combined several feature extraction methods to extract features as much as possible. Then we obtained the flexible neutral tree to predict the subcellular locations of the proteins.

2   Dataset

The dataset we used is the dataset in constructing Gpos-mPloc [7]. The proteins in this dataset have one or more subcellular locations. There are 519 different protein sequences, among which 515 have only one subcellular location and 4 protein sequences have 2 subcellular locations. None of the proteins has ≥25 % sequence identify to any other in the same dataset (subcellular location). The four subcellular locations are shown in Fig. 1, which can be download at http://www.csbio.sjtu.edu.cn/bioinf/Cell-PLoc-2/ [8].

Protein numbers of each sub-cellular location pattern are shown in Table 1. The organelle markers subcellular locations are all have the probability to exist the multisite protein sequences.

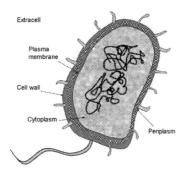

Fig. 1. The organelle markers subcellular locations of Gram-positive bacteria proteins

Table 1. Protein distribution situation of each sub-cellular location.

Sub-cellular locations	Number of protein sequences
Cell membrane	174
Cell wall	18
Cytoplasm	208
Extra-cell	123

3   Feature Extraction Method

3.1   Amino Acid Index Distribution

In the amino acid index distribution model, feature vector can be represented by the following formula [9]:

$$F_{\text{AAID}} = [x_1, x_2, \ldots x_{20}; y_1, y_2, \ldots y_{20}; z_1, z_2, \ldots z_{20}] \tag{1}$$

The first twenty dimensions vector is the combination of composition information and physicochemical values, its calculate equation is:

$$x_i = J_i f_i \quad i = 1, 2, \ldots 20 \tag{2}$$

where is the physicochemical values of the 20 amino acids, $J_A = 0.486$, $J_C = 0.2$, $J_D = 0.288$, $J_E = 0.538$, $J_F = 0.318$, $J_G = 0.12$, $J_H = 0.4$, $J_I = 0.37$, $J_K = 0.402$, $J_L = 0.42$, $J_M = 0.417$, $J_N = 0.193$, $J_P = 0.208$, $J_Q = 0.418$, $J_K = 0.262$, $J_S = 0.2$, $J_T = 0.272$, $J_V = 0.379$, $J_W = 0.462$, $J_Y = 0.161$, is the frequency of the 20 amino acids in the corresponding protein sequence. So is the product of protein composition information and physicochemical values.

Then the following 20 dimension feature vectors is 2-order center distance information, it includes statistical information, physicochemical values and sequence information. The formula is as follows:

$$y_i = \sum_{j=1}^{N_{n_i}} \left(\frac{p_{i,j} - \bar{p}_i}{T} J_i\right)^2 \quad i = 1, 2, \ldots 20 \tag{3}$$

Where $p_{i,j}$ is the position number of the i-th amino acid residue, $\bar{p}_i$ is the mean position number of the i-th amino acid. T is the total number of the protein sequence.

$$z_i = \sum_{j=1}^{N_{n_i}} \left(\frac{p_{i,j} - \bar{p}_i}{T} J_i\right)^3 \quad i = 1, 2, \ldots 20 \tag{4}$$

The equation above is 3-order center distance computing equation which is similar to 2-order center distance. Obviously, we can increase the order number easily, but we may get unsatisfied results with the increasing of the order number because it may produce much more redundant information. Therefore we use the first 40 dimension vectors.

3.2   Pseudo Amino Acid Composition

To develop a powerful method for statistically predicting protein subcellular localization according to the sequence information. Amino acid composition (AAC) is proposed to extract protein features. However, the location information may be ignored. So chou K.C proposed the pseudo amino acid composition (PseAAC) model [10], which contained both composition information and location information. The feature vector is formulated by:

$$P = \left[v_1, v_2, \ldots v_{20}, v_{20+1}, \ldots v_{20+\lambda} \right]^{\mathrm{T}} \tag{5}$$

First 20 dimension vectors are the normalized occurrence frequencies of the 20 native amino acids in protein P, and the rest λ dimension vectors are the location information. In this model, the location information of residue in a protein sequence can be represented by the following equations:

$$\Theta_\theta = \frac{1}{L-k} \sum_{i=1}^{L-k} \Omega_{i,i+k}, (k<L) \tag{6}$$

L is the length of protein sequence.

$$\Omega_{i,i+k} = \frac{1}{3} \left\{ [S_1(A_{i+k}) - S_1(A_i)]^2 + [S_2(A_{i+k}) - S_2(A_i)]^2 + [T(A_{i+k}) - T(A_i)]^2 \right\} \tag{7}$$

Here, $S_1(A_i)$, $S_2(A_i)$ and $T(A_i)$ is the value of normalized hydrophobicity, hydrophilicity and the side chain mass for amino acid residue A_i respectively. While $S_1(A_{i+k})$, $S_2(A_{i+k})$ and $T(A_{i+k})$ are those values for amino acid residue A_{i+k}. The sequence information of the k most adjacent residues are contained in the k-th correlation factor, which is coupling factor [11]. The dimension of feature vector in this model depends on the value of λ. In this paper, the value of λ is 20. So the total dimension of the feature vector in this model is 40. The value of normalized hydrophobicity, hydrophilicity and the side chain mass for amino acid residue are shown in Table 2.

3.3   Stereo-Chemical Property (SP)

Proteins have many intrinsic physicochemical properties. We could classify the amino acids according to their physicochemical property [12]. Predicting the protein sub-cellular location is a research priority in recent years. The protein's physicochemical properties have a certain relationship with their sub-cellular locations,

Table 2. Original hydrophobic parameter, hydrophilic parameter and side chain atomic weight value of the 20 amino

Residues	Hydrophobic	Hydrophilic	The weight for amino acid
A	1.8	−0.5	89.079
R	−4.5	3.0	174.188
N	−3.5	0.2	132.104
D	−3.5	3.0	133.089
C	2.5	−1.0	121.145
Q	−3.5	0.2	146.131
E	−3.5	3.0	147.116
G	−0.4	0.0	75.052
H	−3.2	−0.5	155.141
I	4.5	−1.8	131.160
L	3.8	−1.8	131.160
K	−3.9	3.0	146.17
M	1.9	−1.3	149.199
F	2.8	−2.5	165.177
P	−1.6	0.0	115.177
S	−0.8	0.3	105.078
T	−0.7	−0.4	119.105
W	−0.9	−3.4	204.213
Y	−1.3	−2.3	181.176
V	4.2	−1.5	117.133

so extracting their physicochemical properties feature to predict their sub-cellular locations is necessary and meaningful.

In this paper, the 20 amino acids were classified into ten categories according to their hydrophilic and hydrophobic properties, size, Acidity and basicity properties and polarity. The classification situation is shown in Table 3.

Table 3. Classification situation of amino acid stereo-chemical properties

Stereo-chemical properties	Amino acid residues
Small	ACDGNPSTV
Polar	CDEHKNQRSTVWY
Hydrophobic	ACFGHIKLMTVWY
Proline	P
Charged	DEHKR
Negative	DE
Positive	HKR
Aromatic	FHWY
Aliphatic	ILV
Tiny	ACGS

$\Psi_i(i = 1, 2, \ldots 10)$ was used to define the amino acid group whose attribute is i, thus

$$\psi_1 = \{ACDGNPSTV\} \tag{8}$$

$$\psi_4 = \{P\} \tag{9}$$

Then statistic the frequency of the amino acids whose attribute is i, the ten frequencies will be the feature vector of the protein sequence. It is a ten dimension feature vector:

$$P = [p_{\Psi 1}, p_{\Psi 2}, p_{\Psi 3} \ldots p_{\Psi 10}]^T \tag{10}$$

4   Algorithm

Machine learning method algorithm design is an important segment of the protein subcellular localization for an efficient algorithm is the guarantee of accurate rate. Large amount of algorithms can be used for proteins subcellular localization prediction. But these classifiers usually assign only one subcellular location for a protein sequence, they did not take the proteins which localize to more than one location into consideration. But the number of the multiple sites is increasing, so protein subcellular localization is becoming crucial and urgent. In this paper, we use flexible neutral network [13] to predict the protein subcellular locations.

The FNT model, which is a novel machine learning method, has been put forward by YH Chen [14, 15]. The output of a non-leaf node is calculated using a FNT model shown in Fig. 2. From this point of view, the instruction +i is also called a flexible neuron operator with i inputs.

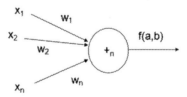

Fig. 2. Non-leaf node of flexible neural tree with a terminal instruction set T = {x_1, x_2, ..., x_n}.

The output of a flexible neuron is calculated as follows:

$$net_n = \sum_{j=1}^{n} w_j x_j \tag{11}$$

where $x_j (j = 1, 2, \ldots, n)$ are the input nodes. The output of the node n is calculated by:

$$f(a_n, b_n, net_n) = e^{-(\frac{net_n - a_n}{b_n})^2} \tag{12}$$

The model has a good performance on the classification. Because of the specialty of the alternative tree, such novel machine learning method could be used in the feature selection [16]. So the principle of the special neutral network can be described as follow:

Algorithm 1. Flexible Neutral Tree

Input: the parameter of fitness;
        the parameter of elitist;
        the features of protein tertiary structure sequence.

Output:the lable of protein tertiary structure sequence.
1. Initializing the parameters of PSO;
2. Setting the elitist & fitness;
3. Creating the initial population;
4. Constructing optimization by PSO;
5.   ***If*** Finding the better tree structure;
6.   ***Then*** optimizing parameter in the better tree
7.   ***else*** return to the step5;
8.   ***If*** Judging the termination of FNT
9.   ***then*** output the type of predicted sequence;
10. ***else*** return to the step8.

The reason for choosing this representation is that the tree can be created and evolved using the existing or modified tree-structure-based approaches in the Fig. 3. So, during this research, the PSO has been employed by the parameter optimization algorithm.

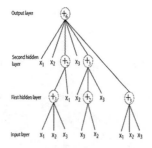

Fig. 3. Typical representation of FNT

A typical FNT model is showed in Fig. 3. Its overall output can be computed from left to right by a depth-first method recursively.

5   Evaluation Criteria

The evaluation criteria for machine learning method is effective. This research adopts absolute-true to evaluate the criteria for flexible neutral tree method.

Ten-fold cross-validation was applied to evaluate the powers of the prediction method. Accuracy (Acc) was utilized to assess the performance of prediction system. All of the above measurements are defined as follows:

$$Acc = \frac{TN + TP}{TN + FP + FN + TP} \tag{13}$$

TP, TN, FP and FN are the number of true positives, true negatives, false positives and false negatives, respectively. Sensitivity and specificity illustrate the correct prediction ratios of positive and negative data sets, while accuracy represents the correct ratio among both positive and negative data sets.

6   Results and Conclusions

The final accuracy rate is closely related to the feature we have extracted and the algorithm designed. In this paper, three feature extraction methods were used, amino acid index distribution, the pseudo amino acid composition and the stereo-chemical properties [17–19]. For a protein sequence, it contains N-terminal signal part and the mature part. There is little relationship between the two parts. So we can divide the protein sequence into two parts, and use the pseudo amino acid composition model and amino acid index distribution model to extract feature information respectively. After that, we can get an 80 dimension feature vector. At last, plus the ten stereo-chemical properties vector, a 90 dimension feature vector which can be used to represent the protein sequence was got. Then, we use the flexible neutral tree algorithm to predict the protein subcellular locations.

Feature fusion is neither simply combining several feature extraction methods, nor the more extraction methods we used, the better performance we got. Even if the extraction method is efficient in other papers, when we used for solving specific problems in this paper, it may be inappropriate.

Table 4. Comparision results of two fusion methods

Feature extraction model	Accurate rate
PseAAC + SP + AAID + FNT	60.2174 %
PseAAC + SP + FNT	68.1472 %

When this feature extraction method is added, the accurate rate is lower than before although the information is much more than before. The comparison results are shown in Table 4. From the table we can see that when the amino acid index distribution feature extraction method added, the accurate rate is only 60.2174 % which are lower than that in our method, which employed the classifier of *FNT*. So the feature fusion is not combining several feature extraction methods simply. So for different dataset, different feature fusion methods produce different effects on it. It's important to choose the appropriate methods. Especially, those ones, who have much more biological significance feature fusion method, should be paid more attention.

Acknowledgment. This research was partially supported by the Youth Project of National Natural Science Fund (Grant No. 61302128), Shandong Provincial Natural Science Foundation, China, under Grant ZR2015JL025, the Youth Science and Technology Star Program of Jinan City (201406003), the Natural Science Foundation of Shandong Province (ZR2011FL022, ZR2013FL002), the Scientific Research Fund of Jinan University (XKY1410, XKY1411), the Program for Scientific research innovation team in Colleges and Universities of Shandong Province (2012–2015), and the Shandong Provincial Key Laboratory of Network Based Intelligent Computing.

References

1. Du, P.F., Xu, C.: Predicting multisite protein subcellular locations: progress and challenges. Expert Rev. Proteomics **10**, 227–237 (2013)
2. Chou, K.C.: Some remarks on predicting multi-label attributes in molecular biosystems. Mol. BioSyst. **9**, 1092–1100 (2013)
3. Xiao, X., Wu, Z.C., Chou, K.C.: iLoc-Virus: a multi-label learning classifier for identifying the subcellular localization of virus proteins with both single and multiple sites. J. Theor. Biol. **284**, 42–51 (2011)
4. Chou, K.C.: Pseudo amino acid composition and its applications in bioinformatics, proteomics and system biology. Curr. Proteomics **6**, 262–274 (2009)
5. Zhang, M.L., Zhou, Z.H.: ML_KNN: a lazy learning approach to multi-label learning. Pattern Recogn. **40**, 2038–2048 (2007)
6. Wan, S., Mak, M., Kung, S.: mGOASVM: multi-label protein subcellular localization based on gene ontology and support vector machines. BMC Bioinform. **13**(1), 290 (2012)
7. Su, C.Y., Lo, A., Lin, C.C., et al.: A novel approach for prediction of multi-labeled protein subcellular localization for prokaryotic bacteria. In: Proceedings of the 2005 IEEE Computational Systems Bioinformatics Conference Workshops, pp. 79–80. IEEE, Stanford, California, Piscataway, 8–12 August 2005
8. Shen, H.B., Chou, K.C.: Gpos-mPLoc: a top-down approach to improve the quality of predicting subcellular localization of gram-positive bacterial proteins. Protein Pept. Lett. **16**, 1478–1484 (2009)
9. Meis, A., Andradenavarro, M.: A novel approach for protein subcellular location prediction using amino acid exposure. BMC Bioinform. **14**, 342 (2013)
10. Luo, H.: Predicted protein subcellular localization in dominant surface ocean bacterioplankton. Appl. Environ. Microbiol. **78**(18), 6550–6557 (2012)
11. Mooney, C., Wang, Y., Pollastri, G.: SCLpred: protein subcellular localization prediction by N-to-1 neural networks. Bioinformatics **27**(20), 2812–2819 (2011)

12. Yu, H., Jiang, W., Liu, Q.: Expression pattern and subcellular localization of the ovate protein family in rice. PLoS ONE **10**(3), e0118966 (2015)
13. Wu, Z., Xiao, X., Chou, K.: iLoc-Plant: a multi-label classifier for predicting the subcellular localization of plant proteins with both single and multiple sites. Mol. BioSyst. **7**(12), 3287–3297 (2011)
14. Yang, B., Chen, Y.H., Jiang, M.Y.: Reverse engineering of gene regulatory networks using flexible neural tree models. Neurocomputing **99**, 458–466 (2013)
15. Chen, Y.H., Yang, B., Dong, J.: Evolving flexible neural networks using ant programming and PSO algorithm. In: Yin, F.-L., Wang, J., Guo, C. (eds.) Advances in Neural Networks–ISNN. LNCS, vol. 3173, pp. 211–216. Springer, Heidelberg (2004)
16. Bao, W.Z., Chen, Y.H., Wang, D.: Prediction of protein structure classes with flexible neural tree. Bio-Med. Mater. Eng. **24**, 3797–3806 (2014)
17. Reyck, B.D., Degraeve, Z., Vandenborre, R.: Project options valuation with net present value and decision tree analysis. Eur. J. Oper. Res. **184**(1), 341–355 (2008)
18. Hanigovszki, N., Poulsen, J., Blaabjerg, F.: A novel output filter topology to reduce motor overvoltage. J. Electroanal. Chem. **40**(3), 845–852 (2003)
19. Hwang, D., Green, P.: Bayesian Markov chain Monte Carlo sequence analysis reveals varying neutral substitution patterns in mammalian evolution. Proc. Natl. Acad. Sci. U.S.A. **101**(39), 13994–14001 (2004)

Dynamically Heuristic Method for Identifying Mutated Driver Pathways in Cancer

Shu-Lin Wang and Yiyan Tan[✉]

College of Computer Science and Electronics Engineering,
Hunan University, Changsha 410082, Hunan, China
smartforesting@gmail.com,
walk_sunshine@foxmail.com

Abstract. Many genomics projects are bringing convenience to the research of identifying driver genes and driver pathways. However, it also brings us the biggest challenge that is how to screen functional mutation and superannuate the unfunctional mutation called as passenger mutation. In our study, we integrate dynamic ant colony optimization into genetic algorithm (DACGA) to identify driver pathways and the problem is equivalent to solve the so-called maximum weight submatrix problem in which driver pathways should satisfy two properties: high coverage and exclusivity. Integrating the two algorithms can make the most use of speed ability and global convergence of genetic algorithm (GA) and positive feedback of ant colony optimization (AC). AC is chose when it approaches stagnation in population of evolution, while GA is chose under other conditions, thus it can dynamically select AC or GA for achieving maximum weight. The proposed method is evaluated on simulated and biological datasets, respectively, and the experimental results indicate that our method is efficient and robust in identifying driver pathways.

Keywords: Dynamically heuristic method · Driver pathways · Driver mutations · Maximum weight submatrix problem

1 Introduction

As we all know, cancer is a very complex disease, and it is very difficult to cure so far. Cancer can be caused by gradually accumulated mutations in DNA replication [1]. Usually, there are two types of mutation in the progress of cancer. The first one is driver mutation which plays an important role in the development of cancer. Once driver genes are activated, it might lead to a carcinoma and promote the cancer cell to proliferate infinitely and diffuse. It results in driver genes having a high mutated frequency in the most patient, such as TP53 gene, because some driver genes are biological functional genes. But not all driver genes have high frequency or are biologically functional genes, such as the longest gene TTN [2]. The second one is passenger mutation which has less influence on cancer.

© Springer International Publishing Switzerland 2016
D.-S. Huang et al. (Eds.): ICIC 2016, Part I, LNCS 9771, pp. 366–376, 2016.
DOI: 10.1007/978-3-319-42291-6_36

The most challenge is how to distinguish driver mutation from passenger mutation in identifying driver pathways. The researchers have focused on identifying driver genes in prior work [3, 4], but the discovered driver genes obtain a simple set which include independent genes have no relationship between genes. Then those driver pathways in cancer have been focused on. Some studies prove that genes have certain relationship with each other in driver pathways [5, 6]. The standard method is that compute frequency of mutation for identifying driver genes in a large cohort of cancer genomes. However different samples from same patient rarely have same mutation. Thus, it has been known as an inefficient method for identifying driver genes. Other methods for identifying driver pathways based on the prior knowledge of which pathways have been found by biological experiment. It cannot distinguish driver mutation from randomly passenger genes quickly and efficiently. There are many other features on driver genes not limited to high frequency. Thus driver pathways should satisfy two properties: high coverage and high exclusivity, where high coverage denotes most patients have at least one mutation in the set, and high exclusivity denotes nearly all patients have no more than one mutation in the set [7]. However there are paradoxical properties between them. First of all, it is essential to discover new algorithm which is not rely on prior knowledge for identifying driver pathways. Second, it is not useful to compute background mutation rate (BMR), and it will become computational challenge. Thus we should design a new algorithm to keep the balance between high mutual exclusive and coverage.

In recent years, several approaches turn the problem into solving maximizing weight submatrix problem by the two properties: high coverage and high exclusivity. De novo Driver Exclusivity (Dendrix) [7, 8] algorithm were introduced by discovering driver pathways which is to simultaneously maximize coverage and minimize the coverage overlap. The algorithm based on MCMC to solve the maximizing weight submatrix problem, but efficiency of Dendrix is very low. Genetic Algorithm (GA) were used to identifying driver pathways [9], but GA cannot make the most use of feedback in system, and it slips into the local solution. Thus we propose dynamically heuristic algorithm which integrates genetic algorithms into ant colony optimization (DACGA) [10] to solve this problem. GA algorithm has strong robustness with the potential of parallelism, but it solves problem in a low accuracy. Although ant colony optimization (AC) solve problem in slow convergence, it makes the use of feedback by leaning [11]. The proposed algorithm DACGA combines GA and AC to discover driver pathways. The basic idea of DACGA is that AC is chose when it approaches stagnation in population of evolution, while GA is chose under other conditions, thus it can dynamically select AC or GA for achieving maximum weight [12]. Furthermore the experimental results show that our method is more efficient than GA and MCMC not only in simulated data but also in biological data.

2 Methods

We try to discover the approximately exclusive driver pathways that can cover most of patients. As Vandin proposed, there are two properties in driver pathways. Thus they define the coverage overlap as a set M of genes to satisfy exclusive property [2]. The following is exclusivity:

$$\omega(M) = \sum_{g \in M} |\Gamma(g)| - |\Gamma(M)| \tag{1}$$

And then we use $\Gamma(g) = \{i : A_{ig} = 1\}$ as coverage. We should keep trade off in high coverage and exclusive. So they define difference of coverage and coverage overlap as weight.

$$W(M) = |\Gamma(M)| - \omega(M) = 2|\Gamma(M)| - \sum_{g \in M} |\Gamma(g)| \tag{2}$$

So it is transformed to find maximum weight submatrix problem in data.

2.1 Mechanism of Ant Colony Optimization

Ants find the shortest route without any information for seeking food, and they have strong adapted ability in the complex environment. One of the fundamental reason is that ants send out a special secretion–pheromone which will evaporate with the lapse of time, when they look for food [11]. Ants choose route by the intensity of phenomenon, and ant will discover shortest pathway by some positive feedback. All ants round an obstruction and find the shortest pathway by feedback of phenomenon at the end.

The model of ant colony optimization is a good algorithm in global optimization. We define as intensity of pheromone in the (i,j) pathway at the moment. Intensity of pheromone is defined as follows:

$$\tau_{ij}(t+1) = \tau_{ij}(t) + \sum \Delta \tau_{ij}^h(t) \tag{3}$$

We define Z_h as weight of the pathway of the h-th ant. The length of pathways is k which is number of driver genes in a set. Q is constant coefficient [11].

$$\Delta \tau_{ij}^h(t) = Q/Z_h \tag{4}$$

It represents the importance of pathways. And U is the set of genes, ant h transfer with (t) which is the probability of transfer at the moment. We can define (t) [11, 13] as the following:

$$P_{ij}^h(t) = \begin{cases} \dfrac{[\tau_{ij}(t)]^\alpha}{\sum_{i \in U} [\tau_{ij}(t)]^\alpha} & j \in U \\ 0 & else \end{cases} \tag{5}$$

2.2   Mechanism of Genetic Algorithm

GA bases on the specie of evolution, and there are some steps to simulate the evolution. The basic elements of GA include encoding, fitness function, operation and parameter. Firstly, we select the initial population. Secondly we compute the weight of individual by fitness function which has the difference of the coverage and coverage overlap, more detail about fitness function we can see (2). According to the weight of individual, we choose next generation. Thirdly, we run the cross operation with a probability of P_c, and run mutation operation with probability of P_m. Finally, it produces next generation and repeats the algorithm until satisfied conditions. We decide the probability by roulette method. Eventually we define P_c and P_m as the followings:

$$P_i = \frac{f_i}{\sum_{i=1}^{D} f_i} (i = 1, 2..., D) \tag{6}$$

D is size of population. Understandably, the weight of individual is the proportional to the possibility.

2.3   Dynamically Genetic Algorithm and Ant Colony Optimization

The basic idea of combination of GA and AC is that aims at making the use of their respective advantages. We designed a combination algorithm which is superior to ant colony optimization in time and advantage over genetic algorithm in accuracy [10, 12]. Operations of combination of two algorithms make the most use of speed ability and global convergence of GA and positive feedback of AC. AC is chose when it approaches stagnation in population of evolution, while GA is chose under other conditions, thus it can dynamically select AC or GA for achieving maximum weight [14]. We define threshold to dynamically control the selection as the following:

$$\gamma_{ij} = (|\nabla f(X_i^j)| - \nabla f_j min) / (\nabla f_j max - \nabla f_j min) \tag{7}$$

Because the problem is discrete, it not exist grads.

$$\nabla f(X_i^j) = f\left(X_{pi}^j\right) - f(X_{si}^j) \tag{8}$$

$$\nabla f_j min = min_i \left\{ \left| f\left(X_{pi}^j\right) - f(X_{si}^j) \right| \right\} \tag{9}$$

$$\nabla f_j max = max_i \{ | f\left(X_{pi}^j\right) - f(X_{si}^j) | \} \tag{10}$$

$f\left(X_{pi}^j\right)$ is parent generation and $f(X_{si}^j)$ is next generation in population. We choose ant colony optimization when less than threshold (10^{-1} we have been used in our study).

To effectively understand the framework of DACGA, we describe the produce as follows:

Step 0: Defined the fitness function and give the proper parameter. Parameters included constant coefficient Q, the number of desired genes k (k is the length of driver pathways), and the maximal number of individual in the population and mutation matrix A.

Step 1: According to the Eq. (5), compute the probability of every ant form current gene to next gene. Every ant choose next gene by roulette method. Continue computing and choosing until all ants finishes discovering of the pathway that length is k.

Step 2: Select optimization of pathways by weight, and add the optimization of pathways. Update pheromone by Eq. (3) in all pathways.

Step 3: According to the Eq. (7), compute the threshold to choose algorithm. If it chooses AC, going to step 1, going to step 3.

Step 4: According to the fitness function and Eq. (6), select two individuals. Crossing genes in two individuals, and then mutate genes in two individuals. We compute the weight of each pathway.

Step 5: Continue the algorithm until reach the number of iteration or satisfy conditions.

3   Results

3.1   Simulation Data Result

We simulate mutation data starting with some gene sets [13]. Every set have k genes ($k = 3, 5$ has been used in our study). We generated the mutation data of 100 patients and change the number of genes. For each row, we set the number to 1 with probability p_i ($p_i = 0.008$ has been used in our study). The genes are mutated using a random model.

We have compared the accuracy of MCMC, GA, DACGA on solving maximum weight submatrix problem on simulate data (shown in Fig. 1) when $k = 3$. From the plot, it's obviously to know that DACGA is better accuracy in simulate data except when the number of genes is 100. The accuracy means that the approach can reach the maximum weight in datasets and how many times the method can reach when k is a certain number. We can clearly know the superiority of our method to solve the maximum weight submatrix problem when $k = 5$ (Fig. 2). The weight of our method finding is no less than the weight of GA method and MCMC method. In summary, DACGA method has competitive ability with GA and MCMC.

3.2   Biological Data Result

We collect three data sets to assess our methods. For the three data, we used the mutation data to evaluate DACGA. We obtain the first two data sets (LC and GBM1) from Vandin *et al.* [2, 7]. And the other data sets download. We apply our approach DACGA onto three data sets when we solve the maximum weight submatrix.

The Table 1 describes the information of data resource including the cancer type, the number of genes which have been summarized. We apply our method to test data sets. This analysis describe that the DACGA is more accuracy than MCMC and GA in the simulation data. In the following, we use our approach onto three datasets and discuss more driver pathways are being discovered.

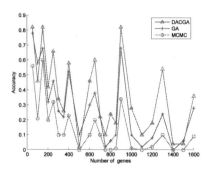

Fig. 1. Comparison of the results obtained by three methods in accuracy. In this plot, the y-axis is accuracy and x-axis is the number of genes. Red line denotes accuracy of DACGA, and rosy line (MCMC), then the blue line (GA). (Color figure online)

Glioblastoma. The glioblastoma dataset got from Vandin *et al.* [2, 7]. We can discover the pathway which gene set is (TP53, CDKN2A) with $k = 2$. This gene set belongs to p53signaling pathways. This gene set covers 69 % in patients. And we also find the gene set (CDK4, RB1) which is other cell cycle of p53 signaling pathways. MCMC is effective method not merely in reaching maximum weighted submatrix, but also in founding driver pathways. GA can reach maximum weighted submatrix with identify (CDKN2A CYP27B1), but it cannot find RB1 and TP53 which is very important genes. We identify (TP53, CDKN2A, MDM2) gene set and (CDKN2A, CDK4, RB1) gene set which is cell cycle of p53 signaling pathway when $k = 3$. We find gene set (PIK3R1, ERBB2, CDKN2A, KRAS, RB1) which is part of PI3K-Akt/MAPK signaling pathway when $k = 5$. Those cell cycle and the relationship of genes is significant in the development of cancer. Found driver pathways include much noisy by MCMC algorithm and cannot include significant genes by GA algorithm when $k = 2, 3, 5$. We compute numerical values which associated with pathways by t test. All the genes in the set were drawn in 3D plot by a tool which is Kyoto Encyclopedia of Genes and Genomes (KEGG) Color Pathway 3D [15]. The tool also called KEGG Color Pathway WebGL which is a variant of the Color Pathway tool, where the result is shown using WebGL graphics. Its height of the red bar is proportional to the numerical values mentioned in Fig. 3. We can see that the height of red bar of CDKN2A and TP53 is very high in pattern. Moreover previous studies have shown they involved in the development of cancer and disturbing signaling pathways.

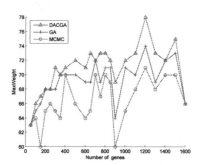

Fig. 2. Compared maximum weight among three methods. In this plot, the y-axis is maximum weight and x-axis is number of genes. (Color figure online)

Lung cancer. Lung cancer is a common cancer in this world. It is increasing in incidence rate and mortality rate. Thus, it is one of malignant tumor and threats to our health and life. We download dataset from Vandin *et al.* [2, 7] directly.

We stable discover gene set (EGFR, KRAS) which is part of MAPK signaling pathway when $k = 2$ (Fig. 4). This gene set covers 55 % patient and have high exclusivity which means all patients have no more than one mutation in the set and no co-mutation. Although the mutation matrix is very sparse, two genes (EGFR, KRAS) that can also be found by GA and MCMC are mutated in more than 20 samples. They mutated in 60 and 30 samples respectively. When $k = 3$, the optimal gene set includes this pair and STK11 which belong to cell growth pathway, so it plays important roles in proliferation of cell.

We also identify calcium signaling pathway of which the gene set is (EGFR, PTK2, ERBB4, KRAS, and STK11) with never being found driver pathways by other methods and covering 73 % sample. We drew 3D plot to easily understand the relationship of gene set we found and signaling pathways prior discovered (Fig. 3). It obviously knows that the gene (KRAS) which is high bar in pattern has an influence on signaling pathways form KEGG Color Pathway WebGL.

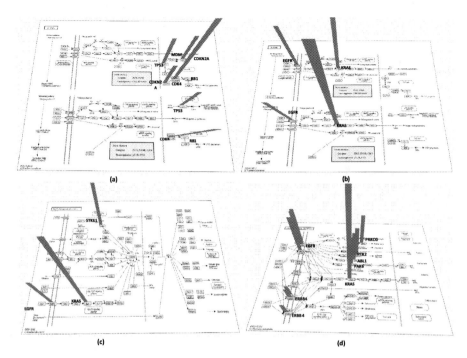

Fig. 3. Color pathway described on WebGL. Its height of the red bar is proportional to the p-values. (a) The optimal solution in GBM1 dataset when $k = 2$ or $k = 3$. (b) We identify pathway by DACGA in LC when $k = 2$. (c) When $k = 3$, we get unique gene set (EGFR, KRAS, STK11) in LC dataset. (d) Discovered the optimal gene set when $k = 5$ in LC dataset. (Color figure online)

Table 1. Brief introduction of three data sets. First column is cancer type. LC is lung cancer, and GBM1 is glioblastoma, HNSCC is head and neck squamous cell carcinoma. Samples: number of patient. Genes: number of genes. Aver: average number of mutation per sample.

Cancer type	Samples	Genes	Aver
LC	163	356	6.0
GBM1	84	178	9.6
HNSCC	74	4920	21.8

Head and neck squamous cell carcinoma. The incident of head and neck squamous cell carcinoma is increasing in patient over 50 years old. Head and neck squamous cell carcinoma is a disease which belongs to a type of skin cancer.

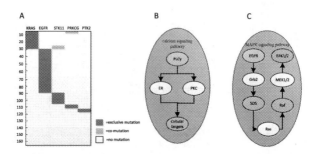

Fig. 4. Submatrix of optimal gene set. (A) The maximum weight submatrix in glioblastoma data when $k = 5$. (B) It describes calcium signaling pathway. (C) It is MAPK signaling pathway. (Color figure online)

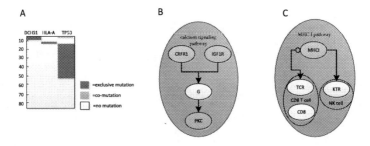

Fig. 5. Submatrix of optimal gene set in HNSCC. (A). This gene set we discovered when $k = 3$ as the plot describes which is cover 73 % patient (54/74). (B) It describes calcium signaling pathway. G gene set including GNAS. (C) MHCI genes have great influence on CD8 T cell and NK cell. It describes MHC I pathway. (Color figure online)

TP53 gene is very prevalent in mutation of dataset. The gene set (DCHS1, HLA-A, TP53) discovered when $k = 3$ which is cover 73 % patient. We identify TP53 gene and HLA-A genes which is a necessary condition to MHC I pathway. Because the MHCI gene set include HLA-A gene and HLA-B gene. Thus HLA-A is second to none in term of importance to MHC I pathway. MHCI gene set has impact on natural killer cell mediated cytotoxicity and T cell receptor signaling pathway which including CD8 T cell and CD4Tcell (Fig. 5). We also discover GNAS gene which belongs to G gene set which is a part of calcium signaling pathway. The gene set (ABL2FBN2, TP53) discovered by MCMC with including some passage genes and ABL2 is mutated in one sample, obviously, MCMC is not effective methods to find maximum weighted sub-matrix in lager scales of genes. With increasing number of genes, it is low performance for identify driver pathways in robustness with GA, such as the gene set (SFI1, SLC2A13, TP53) is found when $k = 3$ with including unless and unfunctional genes that are SFI1and SLC2A13.

4 Conclusion

This paper aims to identify driver pathways, which is of great benefit to clinical care and identify new driver genes. We designed a dynamically heuristic algorithm to solve the so-called maximum weight submatrix problem. Then, we proved our DACGA method has competitive efficiency compared with other methods on simulation and biological data.

The proposed method yields two merits. (1) We found the driver pathways without relying on prior knowledge and BMR. The driver pathways discovered have higher coverage and higher exclusivity. (2) Operation of combination of two algorithms makes the most use of speed ability and global convergence of GA and positive feedback of AC. AC is chose when it approaches stagnation in population of evolution, while GA is chose under other conditions, thus it can dynamically select AC or GA for achieving maximum weight. The experimental results indicate that our method can effectively identify many driver pathways discovered by other studies and even novel pathways.

Acknowledgement. This research was supported by the National Natural Science Foundation of China (Grant Nos. 61472467, 60973153 and 61471169) and the Collaboration and Innovation Center for Digital Chinese Medicine of 2011 Project of Colleges and Universities in Hunan Province.

References

1. Hanahan, D., Weinberg, R.A.: The hallmarks of cancer. Cell **100**(1), 57–70 (2000)
2. Vandin, F., Upfal, E., Raphael, B.J.: Algorithms and genome sequencing: identifying driver pathways in cancer. Computer **45**(3), 39–46 (2012)
3. Boca, S.M., Kinzler, K.W., Velculescu, V.E., Vogelstein, B., Parmigiani, G.: Patient-oriented gene set analysis for cancer mutation data. Genome Biol. **11**(11), R112 (2010)
4. Efroni, S., Ben-Hamo, R., Edmonson, M., Greenblum, S., Schaefer, C.F., Buetow, K.H.: Detecting cancer gene networks characterized by recurrent genomic alterations in a population. PLoS ONE **6**(1), e14437 (2011)
5. Overdevest, J.B., Theodorescu, D., Lee, J.K.: Utilizing the molecular gateway: the path to personalized cancer management. Clin. Chem. **55**(4), 684–697 (2009)
6. Swanton, C., Caldas, C.: Molecular classification of solid tumours: towards pathway-driven therapeutics. Br. J. Cancer **100**(10), 1517–1522 (2009)
7. Vandin, F., Upfal, E., Raphael, B.J.: De novo discovery of mutated driver pathways in cancer. Genome Res. **22**(2), 375–385 (2012)

8. Zhao, J.F., Zhang, S.H., Wu, L.Y., Zhang, X.S.: Efficient methods for identifying mutated driver pathways in cancer. Bioinformatics **28**(22), 2940–2947 (2012)
9. Vogelstein, B., Kinzler, K.W.: Cancer genes and the pathways they control. Nat. Med. **10**(8), 789–799 (2004)
10. Liu, X., Feng, X.U.: Hybrid ant colony genetic algorithm based on change rate of objective function. Comput. Eng. Appl. **49**(18), 41–44 (2013)
11. Dorigo, M., Bonabeau, E., Theraulaz, G.: Ant algorithms and stigmergy. Future Gener. Comput. Syst.-Int. J. Grid Comput. Escience **16**(8), 851–871 (2000)
12. Coonrod, L.A., Lohman, J.R., Berglund, J.A.: Utilizing the GAAA tetraloop/receptor to facilitate crystal packing and determination of the structure of a CUG RNA helix. Biochemistry **51**(42), 8330–8337 (2012)
13. Stutzle, T., Hoos, H.H.: MAX-MIN ant system. Future Gener. Comput. Syst. **16**(8), 889–914 (2000)
14. Hahn, W.C., Weinberg, R.A.: Modelling the molecular circuitry of cancer. Nat. Rev. Cancer **2**(5), 331–341 (2002)
15. Liu, Y., Hu, Z.: Identification of collaborative driver pathways in breast cancer. BMC Genomics **15**(1), 1–16 (2014)

System Prediction of Drug-Drug Interactions Through the Integration of Drug Phenotypic, Therapeutic, Structural, and Genomic Similarities

Binglei Wang[2], Xingxing Yu[1], Ran Wei[1], Chenxing Yuan[3],
Xiaoyu Li[2], and Chun-Hou Zheng[2(✉)]

[1] Institute of Health Sciences, School of Life Sciences, Anhui University,
Hefei 230601, Anhui, China
[2] School of Computer Science and Technology, Anhui University,
Hefei 230601, Anhui, China
zhengch99@126.com
[3] College of Electrical Engineering and Automation, Anhui University,
Hefei 230601, Anhui, China

Abstract. Prediction of drug-drug interactions (DDIs) is an essential step in both drug development and clinical application. As the number of approved drugs increases, the number of potential DDIs rapidly rises. Several drugs have been withdrawn from the market due to DDI-related adverse drug reactions recently. Therefore, it is necessary to develop an accurate prediction tool that can identify potential DDIs during clinical trials. We propose a new methodology for DDIs prediction by integrating the drug-drug pair similarity, including drug phenotypic, therapeutic, structural, and genomic similarity. A large-scale study was conducted to predict 6946 known DDIs of 721 approved drugs. The area under the receiver operating characteristic curve of the integrated models is 0.953 as evaluated using five-fold cross-validation. Additionally, the integrated model is able to detect the biological effect produced by the DDI. Through the integration of drug phenotypic, therapeutic, structural, and genomic similarities, we demonstrated that the proposed method is simple, efficient, allows the uncovering DDIs in the drug development process and postmarketing surveillance.

Keywords: Drug-drug interaction · Structural similarity · Therapeutic similarity · Genotypic similarity · Phenotypic similarity

1 Introduction

Drug-drug interactions (DDIs) are a common cause of adverse drug reactions, especially in patients on multiple drug therapy [1–3]. They occur when the pharmacologic effect of a given drug is greater or less than expected in the presence of another drug, leading to unpredictable clinical effects [4–6]. Many DDIs are not identified in the early stage and result in clinically high morbidity and mortality [7]. As the number of approved drugs increases, the number of potential DDIs rapidly rises [8, 9]. Therefore,

© Springer International Publishing Switzerland 2016
D.-S. Huang et al. (Eds.): ICIC 2016, Part I, LNCS 9771, pp. 377–385, 2016.
DOI: 10.1007/978-3-319-42291-6_37

reliable and efficient large-scale detection methods for identifying DDIs are greatly desired and urgently required.

A popular experimental technique for identifying DDIs is through testing metabolic profiles (especially Cytochrome P450 enzymes) or transporter-based pharmacokinetic interactions. However, experimental DDIs detection methods are not applicable on a large scale since they are time consuming and expensive. As a result, there is a practical need for computational inference methods that can identify potential DDIs [10, 11]. During the past decade, several methods have been developed for the prediction of potential DDIs [12–15]. In these works, several approaches based on molecular similarities were used to mine the interactions between two drugs. Therefore, development of similarity-based model is a reliable strategy to predict DDIs.

In this study, we proposed a large-scale method for new DDI prediction by using four kinds of molecular similarity, including phenotypic similarity, therapeutic similarity, chemical structural similarity, and genomic similarity. Our approach is supported by the assumption that if drug A have some similarities with drug Band obvious effect with drug C, then drug C might have the same effect with drug B [16]. For example, it has been reported in DrugBank that ritonavir is an HIV protease inhibitor that works by interfering with the reproductive cycle of HIV and itraconazole may increase the effect and toxicity of ritonavir. Itraconazole is one of the triazole antifungal agents that inhibits cytochrome P-450-dependent enzymes resulting in impairment of ergosterol synthesis. It has been used against histoplasmosis, blastomycosis, cryptococcal meningitis and aspergillosis [17]. On the basis of our assumption, the drugs similar to ritonavir or itraconazole could cause a similar effect as described above. Through the proposed methodology, we demonstrated that phenotypic, therapeutic, structural, and genomic similarities are promising for uncovering DDIs in drug development and post-marketing surveillance.

2   Materials and Methods

2.1   Data Collection

We collected DDI data from the DrugBank database [18] (V3.0; http://www.DrugBank.ca). More than 16 453 DDI pairs differs from approved and experimental drugs. We eliminated the following drugs that cannot be used to calculate four types of similarity: inorganic salts, antibody drugs, and drugs that do not have Anatomical Therapeutic Chemical (ATC) classification system codes or known target information or known ADR information. In the aggregate, 6946 high-quality and unique ligand–receptor DDI pairs connecting 721 approved drugs are applied to this research [15]. Ultimately, we retrieved SMILES data for those drugs, and then converted them into canonical SMILES using Open Babel (V2.3.1) [19].

2.2   Molecular Similarities Analysis

2.2.1   Drug Phenotypic Similarity

Each drug was coded by 13060 ADE bit vectors in MetaADEDB and each bit represents one ADE. If an ADE is associated with a given drug in MetaADEDB, the corresponding bit is set to "1" or else it is set to "0". Next, by using the drug's ADE bit vectors, Tanimoto coefficient of the phenotypic similarity $S_P(d_i, d_j)$ between drugs d_i and d_j was calculated [20].

2.2.2   Drug Structural Similarity

The structural similarity $Ss(d_i, d_j)$ between drugs d_i and d_j was calculated by the Tanimoto similarity metric using MACCS keys which is freely available from Open-Babel (v2.3.1) [19].

2.2.3   Drug Therapeutic Similarity

We collected the ATC codes for all drugs from DrugBank database [18]. The kth level drug therapeutic similarity (Sk) between two drugs is defined via the ATC codes as:

$$Sk(di, dl) = \frac{ATCk(di) \cap ATCk(dl)}{ATCk(di) \cup ATCk(dl)}$$

where $ATC_k(d)$ represents all ATC codes at the kth level of drug d. A score $S_T(d_i, d_j)$ reflects the therapeutic similarity between two drugs:

$$ST(di, dl) = \frac{\sum_{k=1}^{n} Sk(di, dl)}{n}$$

where n represents the five level ATC codes which range from 1 to 5 [21].

2.2.4   Genomic Similarity

The drug's genomic similarity based on a large drug-target interaction (DTI) network built from DrugBank [18] and the Therapeutic Target Database (TTD) [22]. By using target protein bit vectors, the corresponding bit will be set to "1" if a target protein is associated with a drug in the DTI network, otherwise "0". Finally, by inputting the drug's target protein bit vectors, the $S_G(d_i, d_j)$ is calculated using the Tanimoto coefficient [23].

2.3   Predicting New DDIs

First step, we built matrix 1 (M1) which consists of the similarity matrix and matrix 2 (M2) which consists of the established interactions. In addition, the interaction of a drug with itself is not considered so, in all matrices, the values in diagonals are 0. Next, as step 4 in Fig. 1, in order to find the highest Tanimoto coefficient (TC) value, we only retain the maximum value in the array for each entry. For considering the highest value for each pair of drugs, a symmetric transformation is carried out to obtain the final M01 matrix [16]. By this matrix operation, M01 can be obtained from structural similarity, M02 can be obtained from genotypic similarity, M03 can be obtained from therapeutic similarity, and M04 can be obtained from phenotypic similarity.

$$M05 = (M01 + M02)/2 \tag{1}$$

$$M06 = (M01 + M03)/2 \tag{2}$$

$$M07 = (M01 + M04)/2 \tag{3}$$

$$M08 = (M02 + M03)/2 \tag{4}$$

$$M9 = (M02 + M04)/2 \tag{5}$$

$$M10 = (M03 + M04)/2 \tag{6}$$

$$M11 = (M01 + M02 + M03)/3 \tag{7}$$

$$M12 = (M01 + M02 + M04)/3 \tag{8}$$

$$M13 = (M01 + M03 + M04)/3 \tag{9}$$

$$M14 = (M02 + M03 + M04)/3 \tag{10}$$

$$M15 = (M01 + M02 + M03 + M04)/4 \tag{11}$$

In the example of Fig. 1, drug3 and drug5 have interaction in DrugBank as it showed in M2. The final score retrieved in M01 with a TC = 0.9. This method can also predict new DDI such as drug1 and drug5 (TC = 0.9).

2.4   Performance Measures

To assess the performance of this model, some measures, including specificity, precision, sensitivity, enrichment factor and F1 score, have been used.

$$\Pr ecision = \frac{TP}{TP + FP}$$

$$Sensitivity = \frac{TP}{TP + FN}$$

$$Specificity = \frac{TN}{TN + FP}$$

$$F1 = \frac{2 \times Sensitivity \times \Pr ecision}{Sensitivity + \Pr ecision}$$

$$Enrichment\ Factor\ (x\%) = \frac{\frac{N\ retrieved\ known\ interactions\ in\ topx\%}{N\ interactions\ in\ topx\%}}{\frac{N\ known\ interactions\ in\ DrugBank}{N\ of\ possible\ interactions}}$$

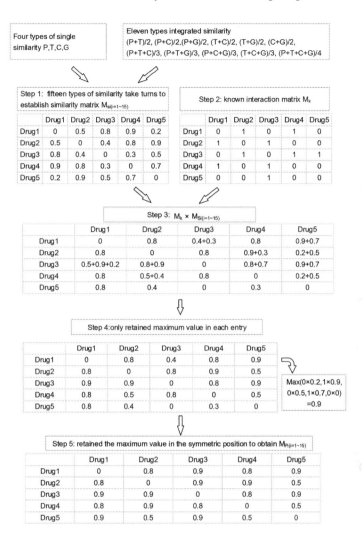

Fig. 1. Establishing a DDIs prediction model through combination of the DrugBank interaction database and four kinds of molecular similarity. In step one, the M1 is created by TC values of four kinds of molecular similarity. In step two, the M2 is created by interactions of 721 approved drugs which have present in DrugBank. "1" stands for interaction which has occurred between two drugs. "0" stands for the opposite situation. In step 4, we only retain the maximum value in each entry. In step 5, M01 is formed based on a symmetry-based transformation. Since the interaction of a drug with itself is not considered, we set the values in the diagonal of all the matrices "0".

where TP, FP, TN and FN represent true positive, false positive, true negative and false negative respectively.

3 Results and Discussions

3.1 Analysis the Performance of the Model

A sum of 6946 DDIs were acquired from DrugBank which were associated with 721 approved drugs. Through these DDI pairs, 721 drugs and four types of similarity, the model can detect new DDIs has been set up. Some measures, specificity, precision, sensitivity, enrichment factor and F1 score, have been used to estimate different cut-off values of TC. The Table 1 shows the performance of our model using different similarities and cut-off values for the TC. Based on single similarity, the structural similarity achieved highest F1 score 0.4497. But the highest F1 score 0.4987 of all matrices appear in M06 which came from structural and therapeutic similarity. The reason of this phenomenon perhaps is that the uniformly spreads data of these two kinds of similarity. As for the AUC value, the highest one 0.953 (Fig. 2B) come up in M15, while based on M06, the AUC value is 0.945 (Fig. 2A). With regard to M11-M14, the TC value is generally lower than other matrices. So, based on TC > 0.7, F1 score of the four matrices is 0.

Table 1. Comparison of model performance using different similarities and cut-off values for TC.

M06 TC value	0.6	0.65	0.7	0.75	0.8	0.85	0.9	0.95
TP	6754	6584	6380	6158	5713	4705	3613	1921
TN	158333	192898	212508	221771	231882	243203	248683	251339
FN	192	362	566	788	1233	2241	3333	5025
FP	94281	59716	40106	30843	20732	9411	3931	1275
Precision	0.0668	0.0993	0.1372	0.1664	0.2160	0.3333	0.4789	0.6011
Sensitivity	0.9724	0.9479	0.9185	0.8866	0.8225	0.6774	0.5202	0.2766
Specificity	0.6268	0.7636	0.8412	0.8779	0.9179	0.9627	0.9844	0.9950
EnFa	2.5115	3.7152	5.1286	6.2191	8.0728	12.4552	17.8965	22.4607
F1	0.1251	0.1798	0.2388	0.2802	0.3422	0.4468	0.4987	0.3788

* EnFa is short for Enrichment Factor.

3.2 Prediction of the Effect Produced by the DDI

As we expected, if drugs have the similar features, they might produce similar effect. Because biological interactions due to being metabolized of same enzymes or competition for the same transporter point which based on similarities between two molecules [16]. For example, in our prediction the structural and therapeutic similarity between clomipramine and methadone is 0.915, they are both a CYP2D6 inhibitor and may decrease the metabolism and clearance of tamsulosin. In order to testify our model having the ability of detecting the biological effect produced by the DDIs, a random selection of interactions from DrugBank was used to determine the precision of the predicted biological effect. We selected 100 interactions randomly from M06 using a

TC cut-offs $0.9 \geq TC > 0.8$ and $1 \geq TC > 0.9$, the model correctly predicted the effect in 95 % and 98 % of the interactions, respectively. This assumption must be beneficial to future prediction.

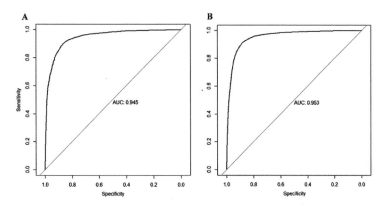

Fig. 2. AUC **value showing the performance of the model.** A is obtained from structural similarity and therapeutic similarity. B is obtained from structural similarity, genotypic similarity, therapeutic similarity and phenotypic similarity.

3.3 Limitations and Future Work

There are several limitations in this method. First of all, the performance of the model depends on the comprehensiveness of the information in DrugBank database [16]. Because we can only find the similarity drugs implicated in the interactions described previously. An additional question is that the 3D structure must be a better representation of the molecules and a very important component in the interactions [24, 25]. But the four kinds of similarity we concerned are all 2D features. Last but not least important, the combinational method of the similarities might have problems. However, we predict thousands of possible DDIs using this model and have a pleased performance. In the future, we intend to consummate our model using more representative features and databases.

4 Conclusion

In this article, we provide a really simple, efficient, accessible large-scale computational method to predict unknown DDIs. This model built by matrix algorithm predicts interactions by utilizing the structural, genotypic, therapeutic similarity and phenotypic similarity. On the basis of DrugBank database, high performance was yielded and showed a F1 score of 0.4987 and AUC value of 0.953. In summary, our study will be pretty helpful *in silico* prediction and provide a novel perspective for interactions between drugs.

Acknowledgement. This work was supported by grants from the National Natural Science Foundation of China (61272339, and 61272333), and the Anhui Provincial Natural Science Foundation (1508085QF135).

References

1. Beijnen, J.H., Schellens, J.H.: Drug interactions in oncology. Lancet Oncol. **5**, 489–496 (2004)
2. Nemeroff, C.B., Preskorn, S.H., Devane, C.L.: Antidepressant drug-drug interactions: clinical relevance and risk management. CNS Spectr. **12**, 1–13 (2007)
3. Classen, D.C., Pestotnik, S.L., Evans, R.S., Lloyd, J.F., Burke, J.P.: Adverse drug events in hospitalized patientsExcess length of stay, extra costs, and attributable mortality. JAMA **277**, 301–306 (1997)
4. Gottlieb, S.: Antihistamine drug withdrawn by manufacturer. BMJ. Br. Med. J. **319**, 7 (1999)
5. Henney, J.E.: Withdrawal of troglitazone and cisapride. JAMA, J. Am. Med. Assoc. **283**, 2228 (2000)
6. SoRelle, R.: Withdrawal of Posicor from market. Circulation **98**, 831–832 (1998)
7. Moore, T.J., Cohen, M.R., Furberg, C.D.: Serious adverse drug events reported to the food and drug administration, 1998–2005. Arch. Intern. Med. **167**, 1752–1759 (2007)
8. Bjornsson, T.D., Callaghan, J.T., Einolf, H.J., Fischer, V., Gan, L., et al.: The conduct of in vitro and in vivo drug-drug interaction studies: a pharmaceutical research and manufacturers of America (PhRMA) perspective. Drug Metab. Dispos. **31**, 815–832 (2003)
9. http://www.fda.gov
10. Cheng, F., Li, W., Liu, G., Tang, Y.: In silico ADMET prediction: recent advances, current challenges and future trends. Curr. Top. Med. Chem. **13**, 1273–1289 (2013)
11. Percha, B., Altman, R.B.: Informatics confronts drug–drug interactions. Trends Pharmacol. Sci. **34**, 178–184 (2013)
12. Gottlieb, A., Stein, G.Y., Oron, Y., Ruppin, E., Sharan, R.: INDI: a computational framework for inferring drug interactions and their associated recommendations. Mol. Syst. Biol. **8**, 592 (2012)
13. Huang, J., Niu, C., Green, C.D., Yang, L., Mei, H., et al.: Systematic prediction of pharmacodynamic drug-drug interactions through protein-protein-interaction network. PLoS Comput. Biol. **9**, e1002998 (2013)
14. Cami, A., Manzi, S., Arnold, A., Reis, B.Y.: Pharmacointeraction network models predict unknown drug-drug interactions. PLoS ONE **8**, e61468 (2013)
15. Cheng, F., Zhao, Z.: Machine learning-based prediction of drug–drug interactions by integrating drug phenotypic, therapeutic, chemical, and genomic properties. J. Am. Med. Inform. Assoc. **21**, e278–e286 (2014)
16. Vilar, S., Harpaz, R, Uriarte, E., Santana, L., Rabadan., R., et al.: Drug–drug interaction through molecular structure similarity analysis. J. Am. Med. Inform. Assoc. (2012). doi:10.1136/amiajnl-2012-000935
17. Wishart, D.S., Knox, C., Guo, A.C., et al.: DrugBank: a knowledgebase for drugs, drug actions and drug targets. Nucl. Acids Res. **36**(suppl 1), D901–D906 (2008)
18. Knox, C., Law, V., Jewison, T., Liu, P., Ly, S., et al.: DrugBank 3.0: a comprehensive resource for 'omics' research on drugs. Nucl. Acids Res. **39**, D1035–D1041 (2011)

19. OLBoyle, N.M., Banck, M., James, C.A., Morley, C., Vandermeersch, T., et al.: Open babel: an open chemical toolbox. J. Cheminf. **3**, 33 (2011)
20. Cheng, F., Li, W., Wu, Z., Wang, X., Zhang, C., et al.: Prediction of polypharmacological profiles of drugs by the integration of chemical, side effect, and therapeutic space. J. Chem. Inf. Model. **53**, 753–762 (2013)
21. Xu, K.-J., Song, J., Zhao, X.-M.: The drug cocktail network. BMC Syst. Biol. **6**, S5 (2012)
22. Zhu, F., Shi, Z., Qin, C., Tao, L., Liu, X., et al.: Therapeutic target database update 2012: a resource for facilitating target-oriented drug discovery. Nucl. Acids Res. **40**, D1128–D1136 (2012)
23. Willett, P.: Similarity-based virtual screening using 2D fingerprints. Drug Discov. Today **11**, 1046–1053 (2006)
24. Vilar, S., Karpiak, J., Costanzi, S.: Ligand and structure-based models for the prediction of ligand-receptor affinities and virtual screenings: development and application to the β2-adrenergic receptor. J. Comput. Chem. **31**, 707–720 (2010)
25. Engel, S., Skoumbourdis, A.P., Childress, J., Neumann, S., Deschamps, J.R., et al.: A virtual screen for diverse ligands: discovery of selective G protein-coupled receptor antagonists. J. Am. Chem. Soc. **130**, 5115–5123 (2008)

Predicting Transcription Factor Binding Sites in DNA Sequences Without Prior Knowledge

Wook Lee, Byungkyu Park, Daesik Choi, Chungkeun Lee,
Hanju Chae, and Kyungsook Han[(⊠)]

Department of Computer Science and Engineering,
Inha University, Incheon, South Korea
{wooklee,dschoi,lchmo4,hanju93}@inha.edu,
{bpark,khan}@inha.ac.kr

Abstract. Transcription factors are proteins involved in converting DNA to RNA by binding to specific regions of DNA. Many computational methods developed for predicting transcription factor binding sites in DNA are either tissue-specific or species-specific methods, so cannot be used without prior knowledge of tissue or species. Some prediction methods are limited to short DNA sequences only, so cannot be used to find potential transcription factor binding sites in long DNA sequences. In this study, we developed a new method that predicts transcription factor binding sites in DNA sequences of any length without prior knowledge of tissue or species. In independent testing with datasets that were not used in training the method, it achieved reasonably good performances (accuracy of 81.84 % and MCC of 0.634 in one testing, and accuracy of 71.16 % and MCC of 0.403 in another testing). Our method will be useful for finding putative transcription factor binding sites in the absence of prior knowledge of tissue or species.

Keywords: Transcription factor binding site · Protein-DNA interaction

1 Introduction

Interactions between protein and DNA play an important role in many cellular processes. In particular, transcription factors (TFs) are special proteins that are involved in transcribing DNA to RNA. TFs bind to specific regions of DNA called promoter or enhancer sequences. Some transcription factors bind to a promoter region of DNA near the transcription start site and help initiate transcription. Other TFs regulate expression of a gene by binding to a regulatory sequence such as an enhancer region [1].

Detecting TF binding sites in DNA by biochemical experiments takes a lot of time and cost. Due to recent advances in experimental techniques, many transcription factor binding sites have been identified *in vitro* or *in vivo*, but a large number of TF binding sites in a wide range of species and tissues are still unknown.

Several computational methods have been developed to predict TF binding sites in DNA. PIPES [2], for example, predicts tissue-specific binding targets of mouse TFs. PROMO [3] finds known TF binding sites in DNA sequences using species-tailored searches. DeepBind [4], which was developed to predict protein-binding regions in

© Springer International Publishing Switzerland 2016
D.-S. Huang et al. (Eds.): ICIC 2016, Part I, LNCS 9771, pp. 386–391, 2016.
DOI: 10.1007/978-3-319-42291-6_38

DNA or RNA in a specified species, cannot be used to predict TF binding sites because it only tells the user whether a specified protein binds to the input DNA sequence or not. It also has a DNA sequence length limit of 1,000 nucleotides.

We have previously developed a prediction model called PNImodeler [5] to predict protein-binding nucleotides in DNA or DNA-binding amino acids in protein. PNImodeler predicts binding sites at the nucleotide- and residue-level. In an attempt to construct a model specialized in predicting TF binding sites in DNA of any length without requiring prior knowledge of species or tissue, we have recently analyzed a large amount of TF binding sites identified by biochemical experiments and developed a new computational method based on the analysis of the data. The rest of this paper presents details of the method and results with actual data.

2   Materials and Methods

2.1   Training and Test Datasets

The data of TF binding sites in DNA were obtained from the JASPAR CORE database [6]. JASPAR provides curated and non-redundant data of experimentally identified TF binding sites. As of February 2016, the JASPAR CORE database [6] provides 519 motifs of TF binding sites, which were detected from vertebrates by biochemical experiments. Only 44 out of the 519 motif had sequence data in fasta format. For each motif, the database provides many redundant binding sites, each in both forward and reverse directions. We selected a single forward strand for each binding site and obtained a total of 4,995 non-redundant DNA sequences as a test dataset after running CD-HIT-EST [7] with the sequence similarity threshold of 80 %. The remaining 38,279 DNA sequences were used to construct training datasets.

One of the challenges in predicting TF binding sites in DNA is to find a way of representing the important information content of TF and DNA sequences. An interacting pair of TF and DNA sequences of various lengths should be transformed into a feature vector of the same length for a prediction model. TF binding sites in DNA are of various lengths, typically ranging from 11 to 21 nucleotides in the TF binding sites provided by JASPAR. The TF binding sites in DNA were used as positive data for a prediction model. For each TF binding site of length s, two non-binding sites of length s around the binding site were used as negative data. Thus, the ratio of the number of positive data to the number of negative data is 1:2 in this training dataset. A prediction model trained with this dataset is called DP_w2.

For comparative purposes, we constructed another prediction model called DP_w4. DP_w4 was trained with a dataset in which the ratio of the number of positive data to the number of negative data is 1:4. The training dataset for DP_w4 had same positive data as that for DP_w2. But, for each TF binding site of length s, four non-binding sites of length s (two non-binding sites from each side of the binding site) were used as negative data. Redundant data in the training datasets were removed by the feature-based redundancy removal method, which was developed in our previous work [8].

Our method is intended to find TF binding sites in DNA sequences of any length. Thus, we constructed two types of test datasets for models DP_w2 and DP_w4 from the 4,995 non-redundant DNA sequences and their TF sequences. The first test dataset was built in same method as the training dataset. The second test dataset was generated using a sliding window on long DNA sequences. A sliding window that contains binding nucleotides more than or equal to $n\%$ of the window size was regarded as positive.

Since the minimum length of typical TF binding sites is 11, there is a limit on the maximum size of a window. For example, with a cutoff value of 80 % for a window size ≥ 14, all windows with 11 binding nucleotides will be regarded as negative even if all the 11 TF binding nucleotides are included in the windows. So, we determine the cutoff value of a window as follows: 70 % for a window of 15 nucleotides, 80 % for a window of 13 nucleotides, and 90 % for a window of 12 nucleotides.

2.2   Feature Selection and Representation

Representing the key features TF and DNA sequences of various lengths into feature vectors of a fixed length is important for the success of a machine learning-based prediction method. To encode a feature vector, we used three types of features [9]: composition, transition and distribution. Four nucleotides of DNA are not clustered, but twenty amino acids of protein are clustered into seven groups {AGV}, {C}, {MSTY}, {FILP}, {HNQW}, {KR} and {DE} based on the dipoles and volumes of the side chains [9].

- Composition: normalized frequency of each nucleotide (or amino acid group) in the sequence
- Transition: normalized frequency of transition between each nucleotide (or each amino acid group) in the sequence
- Distribution: normalized position of the first, 25 %, 50 %, 75 % and 100 % nucleotide (or amino acid) of each type (or amino acid group) in the sequence

A single feature vector contains a total of 93 elements (4 compositions, 6 transitions and 20 distributions for the DNA sequence, and 7 compositions, 21 transitions and 35 distributions for the TF sequence).

3   Results

We built prediction models DP_w2 and DP_w4 using the library for support vector machines (LIBSVM) [10] with the radial basis function (RBF) kernel. We tried several different values between 100 and 5,000 for the parameter C of the RBF kernel. The parameter γ of RBF was set to 0.0107527, and the positive data of DP_w2 and DP_w4 were assigned the weights of 2 and 4, respectively.

The performance of the prediction models was evaluated by six measures: sensitivity, specificity, accuracy, positive predictive value (PPV), negative predictive value (NPV) and Matthews correlation coefficient (MCC) (refer to reference [5] for the definition of the six measures).

We tested the prediction models DP_w2 and DP_w4 on two different test datasets discussed earlier. Tables 1 and 2 show the performance of DP_w2 and DP_w4 with test dataset 1, which consists of DNA sequences of fixed length. As shown in Table 1, DP_w2 achieved the best performance with C of 1,000 (sensitivity of 87.69 %, specificity of 78.91 %, accuracy of 81.84 %, PPV of 67.52 %, NPV of 92.76 % and MCC of 0.634). Likewise, DP_w4 achieved the best result with C of 1,000 (sensitivity of 87.97 %, specificity of 80.85 %, accuracy of 82.27 %, PPV of 53.45 %, NPV of 96.41 % and MCC of 0.586) (Table 2).

Table 1. Performance of the model DP_w2 with test dataset 1, which consists of DNA sequences of fixed length

C	Sensitivity	Specificity	Accuracy	PPV	NPV	MCC
100	83.70 %	72.85 %	76.47 %	60.66 %	89.94 %	0.535
200	84.60 %	74.91 %	78.14 %	62.77 %	90.68 %	0.564
300	85.49 %	75.86 %	79.07 %	63.90 %	91.27 %	0.582
400	85.81 %	76.66 %	79.71 %	64.76 %	91.53 %	0.593
500	86.35 %	77.13 %	80.20 %	65.37 %	91.87 %	0.603
600	86.53 %	77.74 %	80.67 %	66.03 %	92.03 %	0.611
700	86.85 %	78.29 %	81.14 %	66.67 %	92.25 %	0.619
800	87.11 %	78.42 %	81.31 %	66.87 %	92.40 %	0.623
900	87.37 %	78.65 %	81.55 %	67.17 %	92.57 %	0.628
1,000	**87.69 %**	**78.91 %**	**81.84 %**	**67.52 %**	**92.76 %**	**0.634**

Table 2. Performance of the model DP_w4 with test dataset 1, which consists of DNA sequences of fixed length

C	Sensitivity	Specificity	Accuracy	PPV	NPV	MCC
100	84.06 %	75.86 %	77.50 %	46.54 %	95.01 %	0.499
200	85.49 %	77.41 %	79.02 %	48.61 %	95.52 %	0.527
300	86.17 %	78.19 %	79.78 %	49.69 %	95.76 %	0.541
400	86.61 %	78.97 %	80.50 %	50.73 %	95.93 %	0.553
500	86.93 %	79.28 %	80.81 %	51.20 %	96.04 %	0.559
600	87.13 %	79.60 %	81.11 %	51.64 %	96.11 %	0.565
700	87.13 %	80.01 %	81.43 %	52.14 %	96.13 %	0.569
800	87.75 %	80.30 %	81.79 %	52.69 %	96.33 %	0.578
900	87.91 %	80.63 %	82.08 %	53.15 %	96.39 %	0.583
1,000	**87.97 %**	**80.85 %**	**82.27 %**	**53.45 %**	**96.41 %**	**0.586**

We also tested DP_w2 and DP_w4 on test dataset 2, which consists of DNA sequences of various lengths. As shown in Table 3, DP_w2 achieved the best balanced performance with 70 % binding proportion of a window of 15 nucleotides and C of 1,000 (sensitivity of 73.96 %, specificity of 70.02 %, accuracy of 71.16 %, PPV of 50.03 %, NPV of 86.89 % and MCC of 0.403). Table 4 shows the performance of

DP_w4 with test dataset 2. DP_w4 also achieved the best balanced performance with 70 % binding proportion of a window of 15 nucleotides and C of 1,000 (sensitivity of 72.49 %, specificity of 77.10 %, accuracy of 76.43 %, PPV of 35.08 %, NPV of 94.26 % and MCC of 0.381). Both in DP_w2 and DP_w4, different binding thresholds resulted in a larger difference in PPV than in other performance measures (data not shown).

In test dataset 1, DP_w4 showed a slightly higher sensitivity, specificity, accuracy and NPV but a much lower PPV (about 12 %) than DP_w2. In test dataset 2, DP_w4 showed a higher specificity, accuracy and NPV but a lower sensitivity, PPV and MCC than DP_w2. In particular, the difference of PPV between the models is about 15 %.

Table 3. Performance of DP_w2 with test dataset 2 of DNA sequences of variable lengths using a window of 15 nucleotides and 70 % binding threshold. Different C values in the range 1005,000 were tried and the results with C values in the range 5001,000 are shown here.

C	Sensitivity	Specificity	Accuracy	PPV	NPV	MCC
500	73.24 %	69.13 %	70.32 %	49.05 %	86.42 %	0.388
600	73.36 %	69.41 %	70.55 %	49.32 %	86.52 %	0.391
700	73.34 %	69.85 %	70.86 %	49.68 %	86.59 %	0.396
800	73.64 %	69.83 %	70.93 %	49.76 %	86.72 %	0.398
900	73.81 %	69.94 %	71.06 %	49.91 %	86.81 %	0.401
1,000	**73.96 %**	**70.02 %**	**71.16 %**	**50.03 %**	**86.89 %**	**0.403**

Table 4. Performance of DP_w4 with test dataset 2 of DNA sequences of variable lengths using a window of 15 nucleotides and 70 % binding threshold. Different C values in the range 100-5,000 were tried and the results with C values in the range 500-1,000 are shown here.

C	Sensitivity	Specificity	Accuracy	PPV	NPV	MCC
500	72.64 %	75.82 %	75.36 %	33.90 %	94.20 %	0.369
600	72.51 %	76.16 %	75.63 %	34.18 %	94.19 %	0.372
700	72.35 %	76.43 %	75.84 %	34.39 %	94.18 %	0.373
800	72.51 %	76.64 %	76.03 %	34.64 %	94.23 %	0.377
900	72.42 %	76.93 %	76.27 %	34.90 %	94.23 %	0.379
1,000	**72.49 %**	**77.10 %**	**76.43 %**	**35.08 %**	**94.26 %**	**0.381**

4   Conclusion

Despite recent advances in experimental techniques to identify transcription factor binding sites in DNA, a large number of transcription factor binding sites in a genome are still unknown. In an effort to reduce the time and cost required for biochemical experiments, several computational methods have been developed. However, most computational methods are either tissue-specific or species-specific methods, so cannot be used without prior knowledge of tissue or species. Some methods are limited to short DNA sequences and cannot be used to find putative transcription binding sites in long DNA sequences or genomes.

In this study, we developed a new method that predicts transcription factor binding sites in DNA sequences of any length without requiring prior knowledge of tissue or species. In independent testing with datasets that were not used in training the method, it achieved reasonably good performances (accuracy of 81.84 % and MCC of 0.634 in one testing, and accuracy of 71.16 % and MCC of 0.403 in another testing). Our method will be useful for finding potential transcription factor binding sites in DNA in the absence of prior knowledge of tissue or species or in a wide range of species and tissues before conducting biochemical experiments.

Acknowledgments. This research was supported by Basic Science Research Program through the National Research Foundation of Korea (NRF) funded by the Ministry of Science, ICT & Future Planning (2015R1A1A3A04001243) and in part by the international cooperation program managed by the National Research Foundation (NRF) (2014K2A2A2000670).

References

1. Latchman, D.S.: Transcription factors: an overview. Int. J. Biochem. Cell Biol. **29**(12), 1305–1312 (1997)
2. Zhong, S., He, X., Bar-Joseph, Z.: Predicting tissue specific transcription factor binding sites. BMC Genom. **14**, 796 (2013)
3. Messeguer, X., Escudero, R., Farré, D., Nuñez, O., Martínez, J., Albà, M.M.: PROMO: detection of known transcription regulatory elements using species-tailored searches. Bioinformatics **18**(2), 333–334 (2002)
4. Alipanhi, B., Delong, A., Weirauch, M., Frey, B.: Predicting the sequence specificities of DNA- and RNA-binding proteins by deep learning. Nat. Biotechnol. **33**(8), 831–838 (2015)
5. Im, J., Tuvshinjargal, N., Park, B., Lee, W., Huang, D.S., Han, K.: PNImodeler: web server for inferring protein-binding nucleotides from sequence data. BMC Genom. **16**(Suppl 3), S6 (2015)
6. Mathelier, A., Zhao, X., Zhang, A.W., Parcy, F., Worseley-Hunt, R., Arenillas, D.J., Buchman, S., Chen, C.Y., Chou, A., Ienasescu, H., Lim, J., Shyr, C., Tan, G., Zhou, M., Lenhard, B., Sandelin, A., Wasserman, W.W.: JASPAR 2014: an extensively expanded and updated open-access database of transcription factor binding. Nucleic Acids Res. **42** (Database issue), D142–D147 (2014)
7. Huang, Y., Niu, B., Gao, Y., Fu, L., Li, W.: CD-HIT suite: a web server for clustering and comparing biological sequences. Bioinformatics **26**(5), 680–682 (2010)
8. Choi, S., Han, K.: Predicting protein-binding RNA nucleotides using the feature-based removal of data redundancy and the interaction propensity of nucleotide triplets. Comput. Biol. Med. **43**(11), 1687–1697 (2013)
9. You, Z.H., Chan, K.C., Hu, P.: Predicting protein-protein interactions from primary protein sequences using a novel multi-scale local feature representation scheme and the random forest. PLoS ONE **10**(5), e0125811 (2015)
10. Chang, C.C., Lin, C.J.: LIBSVM: a library for support vector machines. ACM Trans. Intell. Syst. Technol. **2**(3), 1–27 (2011)

Analysis and Comparison of Genomes of HIV-1 and HIV-2 Using Apriori Algorithm, Decision Tree, and Support Vector Machine

Yihyun Roh[1]([✉]), Seokhyun Yoon[1], Min Young Lee[1], Seongpil Jang[2],
and Taeseon Yoon[2]

[1] Hankuk Academy of Foreign Studies, Yongin, Republic of Korea
{yhr1501,tigerpaul09,mlee981001}@hafs.kr
[2] Department of Science, Hankuk Academy of Foreign Studies, Yongin, Republic of Korea
{heresphill,tsyoon}@hafs.hs.kr

Abstract. AIDS is caused by HIV, which can be divided into two strains: HIV-1 and HIV-2. Whereas HIV-1 is distributed around the world and is the major cause of global infections, HIV-2 is less infectious and transmissible and is therefore generally confined to West Africa. Thus this research aims to account for their difference by analyzing genome sequences of HIV-1 and HIV-2 using some methods: Apriori algorithm, Decision tree, and Support Vector Machine. Apriori demonstrates that HIV-1 has lysine, arginine, and serine as its typical amino acids, while HIV-2 has glycine, lysine, leucine, and arginine. Decision tree determines the significant positions of amino acids that can distinguish the two viruses: pos5 in 9 window, pos13 in 13 window, and pos16 in 19 window. SVM indicates that two viruses are seemingly similar but indeed different. The collective results provide a biologically verifiable background for making effective vaccines for HIV, especially for HIV-2.

Keywords: HIV-1 · HIV-2 · Amino acids · Bioinformatics · Data mining · Apriori algorithm · Decision tree · Support vector machine (SVM)

1   Introduction

Immunodeficiency is a state in which an immune system's capability of protecting oneself from infectious diseases is weakened or absent [1]. AIDS is acquired by an exposure to human immunodeficiency virus (HIV) [2]. This research aims to draw a comparison between the properties of HIV-1 and HIV-2 in Africa, in order to provide a biologically verifiable account for the difference in distribution of the two types [3, 4]. The research thus analyzes the genomic DNA sequences of two virus through Apriori, Decision Tree, and Support Vector Machine (SVM). The objects of analysis are chosen by few criteria: first, two strains from same area will demonstrate the different properties that have actually caused current difference in distribution, since both began their propagation in Africa; second, two strains from the same area will reduce the unnecessary variability that arises from regional difference.

© Springer International Publishing Switzerland 2016
D.-S. Huang et al. (Eds.): ICIC 2016, Part I, LNCS 9771, pp. 392–398, 2016.
DOI: 10.1007/978-3-319-42291-6_39

2   Materials and Methods

2.1   HIV

HIV-1 is the most common strain of HIV. It accounts for the 95 % of the global HIV infections. It originated from Common Chimpanzee [5]. HIV-1 is more virulent and infectious due to its short incubation period. HIV-2 is the other strain of HIV that is concentrated mostly in West African countries such as Senegal and Nigeria. It originated from Sooty Mangabey [4]. It is less pathogenic because the incubation period is longer than that of HIV-1. This characteristic accounts for the lower transmissibility and slower progression to AIDS [6, 7].

2.2   Window

In bioinformatics, a window is a region of fixed size over nucleotide sequence. Windowing sequences in adequate size is crucial in extracting genomic information from a given data. If the window size is too small, many windows will exhibit zero counts and almost no pattern can be observed [8]. Amino acids are designated as FASTA Format.

2.3   Apriori Algorithm

Apriori algorithm is an algorithm of extracting database by its frequency based on association rule. By fixed minimum support count, Apriori scans individual databases and finds out ones with high frequency. If an itemset is frequent, then all of its subsets or its supersets must be also frequent [9]. In this experiment, genomes of HIV-1 and 2 are analyzed through Apriori algorithm in 9, 13 and 19 window with 18 cycles. Itemsets are the amino acids, and an item is each amino acid in each virus. For each window, the minimum support is set as 0.1, meaning that rules that appear in over 10 % of whole instances are selected as best rules [10].

2.4   Decision Tree

Decision tree extracts rules with a data node and their possibilities. It is especially useful when target function is discrete valued, when it is describable by attribute-value pairs, or when the data sets are noisy trained. Each internal node represents a single trial which tests an attribute [11]. In this experiment, decision tree classifies the amino acids into class 1 and class 2, which refer to HIV-1 and HIV-2 in 9, 13, and 19 windows. This process finds out the major rules that govern each sequence [12].

2.5   SVM

Support Vector Machine is a method used to recognize patterns and classify data based on statistical learning theory. It is especially useful in two-group classifications. The main function idea of the machine is to map the input vectors to a very high dimension feature space. The machine elevates the dimensions until it achieves the right dimension to classify the sets properly. Characteristics of the decision surface allow high generalization ability [13]. In this experiment, SVM finds out that HIV-1 and HIV-2 are different, using different windows with 10-fold cross validation. In analysis, total four kernels are used: Normal, Sigmoid, Polynomial and RBF [14] (Fig. 1).

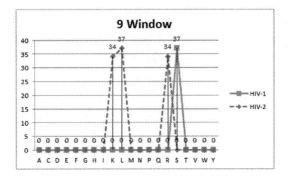

Fig. 1. 9 window Apriori algorithm result

3   Results

3.1   Apriori

Because minimum support is 0.1, rules that appear in at least 34 instances are extracted as the best rules. In HIV-1, serine with frequency of 37 is selected as the only rule. In HIV-2, K, L, and R are extracted as the amino acids that typify HIV-2 (Fig. 2).

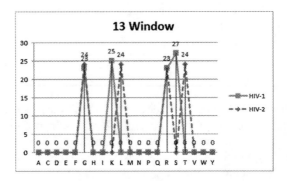

Fig. 2. 13 window Apriori algorithm result

The rules that appear in at least 23 instances are extracted as the best rules. In HIV-1, G, K, R, and S are selected as the typical amino acids. In HIV-2, G, L, R, and T are extracted as the amino acids that typify HIV-2 (Fig. 3).

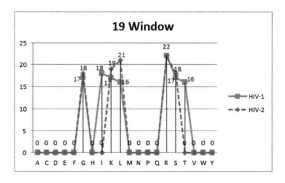

Fig. 3. 19 window Apriori algorithm result

The rules that appear in at least 16 instances are extracted as the best rules. In HIV-1, G, I, K, L, R, S, and T are selected as the typical amino acids. In HIV-2, G, K, L, R, and S are extracted as the amino acids that typify HIV-2. It is noticeable that all the typical amino acids in HIV-2 are included in those in HIV-1.

3.2 Decision Tree

Tables 1, 2, and 3 show the results of Decision Tree in 9, 13, and 19 window, respectively. 167, 105, and 151 rules are extracted from the entire amino acid data.

Table 1. Best rules found by 9 window Decision Tree

	Amino acid	N/A	Frequency	Coverage
Class 1	pos5 = _	10	0.811	0.0680
Class 2	pos3 = L	4	0.833	0.0272
	pos5 = D			

Table 2. Best rules found by 13 window Decision Tree

	Amino acid	N/A	Frequency	Coverage
Class 1	pos13 = _	6	0.885	0.0571
	pos13 = Y	4	0.783	0.0381
Class 2	pos5 = F	7	0.832	0.0667
	pos7 = P	4	0.832	0.0381
	pos11 = Q	3	0.833	0.0286
	pos13 = S			

In Table 1, one major rule is found in each HIV. That both rules involve pos5 indicates the significance of amino acid in this position. All rules show high frequency above 0.800. Furthermore, the rule found in HIV-1 has high number of N/A despite the smaller number of all extracted rules. It is also noticeable that the two rules in HIV-1 and HIV-2 have a big difference in coverage.

In Table 2, two major rules are found in HIV-1, while three major rules are found in HIV-2. Rules involving pos13 appear in almost all major rules. All rules show high frequency as well. The result shows overall stability and the high significance of pos13.

In Table 3, three major rules are found in HIV-1, two major rules are found in HIV-2. pos16 appears frequently in both class 1 and 2. This result shows rules with an average N/A of 9.2, a number that is significantly higher. One rule has a frequency of 0.900, a number that is highest of those of all rules in 9, 13, and 19 window. The coverage value is also higher in the rules involving pos16. Overall, the significance of pos16 is reinforced by high N/A, coverage, and frequency.

Table 3. Best rules found by 19 window Decision Tree

	Amino acid	N/A	Frequency	Coverage
Class 1	pos16 = M	10	0.900	0.066
	pos16 = W	10	0.785	0.066
	pos5 = F	7	0.762	0.046
Class 2	pos16 = D	10	0.837	0.066
	pos16 = I	9	0.792	0.060

3.3 Support Vector Machine

Among Normal, Polynomial, RBF, and Sigmoid, Normal has the highest number of SVs as 40,000, and other kernels have numbers all around 400. In regards to accuracy, while Normal, Polynomial, and Sigmoid have 50 % accuracy, RBF demonstrates perfect accuracy (Figs. 4 and 5).

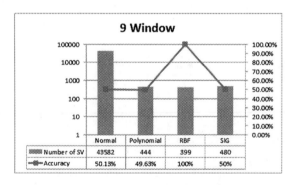

	Normal	Polynomial	RBF	SIG
Number of SV	43582	444	399	480
Accuracy	50.13%	49.63%	100%	50%

Fig. 4. 9 window SVM result

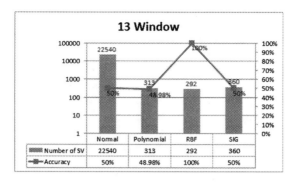

Fig. 5. 13 window SVM result

Among all four kernels, Normal has the highest number of SVs as 20,000, and other kernels have similar numbers all around 300. In regards to accuracy, while Normal, Polynomial, and Sigmoid have 50 % accuracy, RBF shows perfect accuracy (Fig. 6).

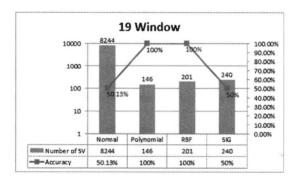

Fig. 6. 19 window SVM result

Among all four kernels, Normal has the highest number of Support Vectors, approximately 57 times higher than the lowest number in Polynomial, and other kernels have similar numbers around 150~250. In regards to accuracy, whereas Normal and Sigmoid, the linear kernels, have relatively low accuracy around 50 %, Polynomial and RBF, the nonlinear kernels, demonstrate 100 % accuracy.

4   Discussion and Conclusion

This research aims to discover the different characteristics of HIV-1 and HIV-2 that accounts for their global distributional difference through statistical analysis of amino acids. Apriori algorithm shows which amino acids play significant roles in each virus [15]. In collective results, HIV-1 has K, R, and S, while HIV-2 has G, K, L, and R. Decision Tree shows the exact rules that govern each virus and the positions of amino acids that distinguish HIV-1 and HIV-2: pos5, 13, and 16 in 9, 13, and 19 window, respectively. This result suggests that different amino acids in these positions are highly likely to account for the different characteristic of the two HIVs, resulting in their distinctive life cycles [16]. SVM classifies the two classes through four kernels. In the collective results, Normal kernel has exceptionally high number of SVs. Since the number of SVs is equal that of classification methods, Normal is ineffective. With regards to accuracy, Normal and Sigmoid have low accuracy around 50 %, whereas Polynomial and RBF demonstrate significantly higher accuracy, 100 % in several cases. These results suggest that nonlinear classification is reliable in classifying HIV-1 and HIV-2 [14].

References

1. Chinen, J., Shearer, W.T.: Secondary immunodeficiencies, including HIV infection. J. Allergy Clin. Immunol. **125**(2), S195–S203 (2010)
2. Sharp, P.M., Hahn, B.H.: Origins of HIV and the AIDS pandemic. Cold Spring Harb. Perspect. Med. **1**(1), a006841 (2011)
3. Hemelaar, J., et al.: Global and regional distribution of HIV-1 genetic subtypes and recombinants in 2004. Aids **20**(16), W13–W23 (2006)
4. Reeves, J.D., Doms, R.W.: Human immunodeficiency virus type 2. J. Gen. Virol. **83**(6), 1253–1265 (2002)
5. Keele, B.F., et al.: Chimpanzee reservoirs of pandemic and nonpandemic HIV-1. Science **313**(5786), 523–526 (2005)
6. Gilbert, P.B., et al.: Comparison of HIV-1 and HIV-2 infectivity from a prospective cohort study in Senegal. Stat. Med. **22**(4), 573–593 (2003)
7. Marlink, R., et al.: Reduced rate of disease development after HIV-2 infection as compared to HIV-1. Science **265**(5178), 1587–1590 (1994)
8. Kyte, J., Doolittle, R.F.: A simple method for displaying the hydropathic character of a protein. J. Mol. Biol. **157**(1), 105–132 (1982)
9. Creighton, C., Hanash, S.: Mining gene expression databases for association rules. Bioinformatics **19**(1), 79–86 (2003)
10. Go, E., Lee, S., Yoon, T.: Analysis of Ebolavirus with decision tree and Apriori algorithm. Int. J. Mach. Learn. Comput. **4**(6), 543 (2014)
11. Stiglic, G., et al.: Comprehensive decision tree models in bioinformatics. PLoS ONE **7**(3), e33812 (2012)
12. Kropp, S., Caulfield, V.I.C.: Data Mining and Bioinformatics. Faculty of Information Technology, Monash University, Caulfield (2004)
13. Hsu, C.-W., Chang, C.-C., Lin, C.-J.: A practical guide to support vector classification. 1–16 (2003)
14. Byvatov, E., Schneider, G.: Support vector machine applications in bioinformatics. Appl. Bioinform. **2**(2), 67–77 (2002)
15. Inokuchi, A., Washio, T., Motoda, H.: An Apriori-based algorithm for mining frequent substructures from graph data. In: Zighed, D.A., Komorowski, J., Żytkow, J.M. (eds.) PKDD 2000. LNCS (LNAI), vol. 1910, pp. 13–23. Springer, Heidelberg (2000)
16. Chen, X., Wang, M., Zhang, H.: The use of classification trees for bioinformatics. Wiley Interdiscip. Rev.: Data Min. Knowl. Disc. **1**(1), 55–63 (2011)

Haplotyping a Diploid Single Individual with a Fast and Accurate Enumeration Algorithm

Xixi Chen[2], Jingli Wu[1,2(✉)], and Longyu Li[2]

[1] Guangxi Key Lab of Multi-source Information Mining and Security,
Guangxi Normal University, Guilin 541004, China
wjlhappy@mailbox.gxnu.edu.cn
[2] College of Computer Science and Information Technology,
Guangxi Normal University, Guilin 541004, China

Abstract. The minimum error correction (MEC) model is one of the important computational models for determining haplotype information from sequencing data, i.e., single individual single nucleotide polymorphism (SNP) haplotyping, haplotype reconstruction or haplotype assembly. Due to the NP-hardness of the model, a fast and accurate enumeration algorithm is proposed for solving it. The presented algorithm reconstructs the SNP sites of a pair of haplotypes one after another. It enumerates two kinds of SNP values, i.e., $(0\ 1)^T$ and $(1\ 0)^T$, for the SNP site being reconstructed, and chooses the one with more support coming from the SNP fragments that are covering the corresponding SNP site. The experimental comparisons were conducted among the presented algorithm, the FAHR, the Fast Hare and the DGS algorithms. The results prove that our algorithm can get higher reconstruction rate than the other three algorithms.

Keywords: Sequence analysis · Haplotype · Minimum error correction (MEC) · Algorithm · Enumeration

1 Introduction

Genetic difference has received increasingly extensive attention with the development of high-throughput DNA sequencing technologies. It is well known that all human are almost identical at DNA level ($\sim 99.5\ \%$ identical). Genomic differences located in $\sim 0.5\ \%$ nucleotide bases answer for the observed diversities in our phenotypes. Among various genetic variations, single nucleotide polymorphisms (SNPs) are believed to be the most widespread form of variation, and their significance cannot be overvalued for medical, drug-design, diagnostic and forensic applications [1]. Due to linkage disequilibrium (LD) and the absence of recombination event, adjacent SNPs tend to inherit together from ancestors to descendants. A sequence of related SNPs on each copy of a pair of chromosomes is referred to as a haplotype. Studies indicate that haplotypes generally carry more genetic information than individual SNP, and they play a crucial role in gene expression and disease association

© Springer International Publishing Switzerland 2016
D.-S. Huang et al. (Eds.): ICIC 2016, Part I, LNCS 9771, pp. 399–411, 2016.
DOI: 10.1007/978-3-319-42291-6_40

studies [2, 3]. Since it is both time consuming and expensive to obtain haplotypes through biological experiments directly, it is common to determine haplotypes through computational methods. Three categories of computational methods exist for computing haplotype [4]: haplotyping a population based on population genotype data [5, 6]; haplotyping a population based on sequenced fragment data [7]; haplotyping a single individual based on sequenced fragment data [8]. The third category is studied in this paper.

The problem of haplotyping a single individual is also called haplotype reconstruction or haplotype assembly. In recent years, the diploid individual haplotype reconstruction problem has attracted considerable attention, and a considerable number of theories and algorithms have been presented. There are four popular formalizations for solving this problem, as follows: minimum fragment removal (MFR) [8], minimum SNP removal (MSR) [8], longest haplotype reconstruction (LHR) [8], and minimum error correction (MEC) [9]. A multitude of exact reconstruction algorithms were presented for solving these computational models. Lancia et al. [8], Lippert et al. [9] and Xie et al. [10, 11] proposed dynamic programming algorithms for the MFR and MSR models. Cilibrasi et al. [12] presented algorithm for the LHR model by using dynamic programming. Wang et al. [13] devised a branch and bound algorithm for solving the MEC model. He et al. [14] designed an exact algorithm for the MEC model by using dynamic programming. Due to NP-hard nature of these models, exact algorithms do not scale well for solving the general cases. In recent years, an amount of heuristic algorithms with good performances [15–20] have also been contrived. The Fast Hare algorithm proposed by Panconesi et al. [15] is a well-known heuristic one. Wang et al. [16] presented a clustering algorithm for solving the MEC model. Genovese et al. [17] devised a greedy heuristic algorithm SpeedHap for haplotyping a single individual. Levy et al. [18] designed a superior iterative algorithm for reconstructing haplotypes (we shall call it DGS as in [2]). Bansal et al. [19] reduced the MEC model to the MAX-CUT problem and proposed a greedy heuristic algorithm HapCUT for solving it. Chen et al. [20] designed a novel randomized algorithm for the problem. Geraci [2] has reviewed these algorithms and drawn the conclusion that the DGS algorithm has the best performance. Later in 2013, Wu et al. [4] described a fast and accurate heuristic algorithm FAHR, Aguiar and Istrail [21] proposed the minimum weighted edge removal (MWER) model and presented a graph theory-based algorithm for solving it. In 2014, Mazrouee and Wang [22] devised FastHap algorithm which takes advantage of fuzzy conflict graphs to reconstruct individual haplptypes.

In this paper, the problem of reconstructing individual haplotype is examined based on the MEC model. A fast and accurate algorithm Enumeration Haplotyping Diploid (EHD) is proposed for solving it. The experimental comparisons were performed among the EHD, the FAHR, the Fast Hare and the DGS algorithms. The results prove that algorithm EHD has better performance than the other three algorithms.

2 Definitions and Notations

A diploid individual enjoys a pair of haplotypes from a couple of homologous chromosomes for a given sequence of SNPs. Since only two kinds of possible nucleotides may appear at each SNP site, a SNP allele can be encoded as 0 or 1, where 0 indicates the major allele, and 1 indicates the minor allele. Therefore, a sequence over a 2-letter alphabet {0,1} can be applied to express a haplotype. Each SNP site can be either a homozygous one if it has the same allele on both haplotypes, or a heterozygous one if it has different alleles on both haplotypes.

Assume that the input to the single individual SNP haplotyping problem is a 2D array recording the aligned SNP fragments, called SNP matrix M of size $m \times n$, where m represents the number of SNP fragments generated by sequencing a pair of haplotype, and n denotes the number of SNPs that the union of all fragments cover [22], i.e., the length of the corresponding haplotypes. The entry M_{ij} ($i = 1,...,m$, $j = 1,..., n$) indicates the allele of the i-th fragments at the j-th SNP. $M_{ij} \in \{0,1, -\}$, where 0 and 1 encode two discovered alleles and $-$ represents an unknown allele, i.e., the i-th fragment does not cover the j-th SNP site, or the j-th SNP allele of the i-th fragment cannot be confidently ascertained.

Let $n_x^j (x \in \{0, 1\}, j = 1, ..., n)$ refer to the number of x entries in column j, and p_x^j records the ratio of x entries to the not null entries in the j-th column.

$$p_x^j = \frac{n_x^j}{n_x^j + n_{1-x}^j}, \quad x \in \{0, 1\}, \quad j = 1, ..., n \tag{1}$$

Given two strings $U = <u_1,...,u_n>$ and $V = <v_1,...,v_n>$, where u_j, $v_j \in \{0,1,-\}$ ($j = 1,...,n$), the Hamming Distance $D(U, V)$ of the two strings is defined by formula (2).

$$D(U,V) = \sum_{j=1}^{n} d(u_j, v_j) \tag{2}$$

where

$$d(u_j, v_j) = \begin{cases} 1 & \text{if } u_j \neq -, v_j \neq -, \text{ and } u_j \neq v_j, \\ 0 & \text{otherwise.} \end{cases} \tag{3}$$

$D(U, V)$ can be applied to calculate the distance between two fragments, a fragment and a haplotype, and two haplotypes. Specially, when $D(U, V)$ indicates the distance between two fragments, $D(U,V) > 0$ means fragment U *conflicts* with fragment V, and $D(U, V) = 0$ means they are *compatible* with each other. The conflict fragments generally come from different chromosome copies or contain sequencing errors. If the fragments in matrix M are all free of errors, they can be divided into two disjoint sets containing non-conflicting fragments. A pair of haplotypes can be assembled from the two sets. In this case, the matrix M is called *feasible* or *error-free*.

Based on the above mentioned definitions, the MEC model can be stated as follows: given a SNP matrix M, correct the minimum number of entries (0 into 1 and vice versa) so that the resulting matrix is *feasible*.

In this paper, reconstruction rate (RR) is adopted to calculate the accuracy of the pair of reconstructed haplotypes. Suppose that $h = (h_1, h_2)$ is the pair of initial haplotypes, and $h' = (h'_1, h'_2)$ is the pair of reconstructed ones. RR is defined as the proportion of SNPs that are reconstructed correctly [4, 13], as follows:

$$\text{RR}(h, h') = 1 - \frac{\min\{r_{11} + r_{22}, r_{12} + r_{21}\}}{2n} \tag{4}$$

where $r_{ij} = D(h_i, h'_j)$, $i, j = 1, 2$.

3   EHD Algorithm

In this section, a heuristics algorithm EHD is proposed. The input is an $m \times n$ matrix M and a parameter t, and the output is a pair of reconstructed haplotypes $h = (h_1, h_2)$ of length n. The first step of the algorithm is simplifying the input matrix by eliminating unnecessary information, which does not contribute to reconstructing haplotypes. The second step is building on an iteration of enumerating and choosing. For each SNP site being reconstructed, enumerate two kinds of SNP values and choose the one with more support coming from the SNP fragments. After this iteration process is completed, a pair of haplotypes $h' = (h'_1, h'_2)$ with only heterozygous SNP sites is built, for the SNP fragments in the simplified matrix having only heterozygous SNPs. The final step is augmenting the pair of haplotypes $h' = (h'_1, h'_2)$ by recovering the discarded sites. The details of each step will be introduced as follows.

3.1   Preprocessing

As indicated in [4, 15], in order to improve the reconstruction efficiency, the input matrix M is preprocessed. Delete column j ($j = 1, \ldots, n$) from matrix M where $p_0^j \leq t$ or $p_1^j \leq t$, and then delete all rows that have no 0 and no 1. Here the parameter t is set to 0.2. The dropped column is called as 1-field (resp. 0-field) when $p_0^j \leq t$ (resp. $p_1^j \leq t$). For convenience of description, the new SNP matrix is still represented by $M_{m \times n}$. The SNP sites in the new matrix are all heterozygous ones.

3.2 Reconstructing

As mentioned above, the EHD algorithm reconstructs the heterozygous SNP sites of a pair of haplotypes $h' = (h'_1, h'_2)$ one after another. Suppose the alleles at the first $j-1$ SNP sites have been inferred, and the j-th SNP site is under reconstruction. Since the retained SNP sites are all heterozygous, it is apparent that there are only two kinds of SNP values for the j-th SNP site, i.e.; $(h'_{1j} = 0, h'_{2j} = 1)$ or $(h'_{1j} = 1, h'_{2j} = 0)$. The EHD algorithm tries to choose the value with more support coming from the SNP fragments. The concrete method is described as follows.

Let S_{01} and S_{10} record the support degree of $(h'_{1j} = 0, h'_{2j} = 1)$ and $(h'_{1j} = 1, h'_{2j} = 0)$ respectively, and they are computed as (5) and (6).

$$S_{01} = \sum_{i=1}^{m} \max\left(\sum_{k=1}^{j-1} c\left(M_{ik}, M_{ij}, h'_{1k}, 0\right), \sum_{k=1}^{j-1} c\left(M_{ik}, M_{ij}, h'_{2k}, 1\right)\right), \quad j = 2, \ldots, n \tag{5}$$

$$S_{10} = \sum_{i=1}^{m} \max\left(\sum_{k=1}^{j-1} c\left(M_{ik}, M_{ij}, h'_{1k}, 1\right), \sum_{k=1}^{j-1} c\left(M_{ik}, M_{ij}, h'_{2k}, 0\right)\right), \quad j = 2, \ldots, n \tag{6}$$

where

$$c(x_1, x_2, y_1, y_2) = \begin{cases} 1 & \text{if } x_1 \neq -, x_2 \neq -, x_1 = y_1, \text{ and } x_2 = y_2 \\ 0 & \text{otherwise.} \end{cases} \tag{7}$$

Take formulation (5) for example, $\max\left(\sum_{k=1}^{j-1} c\left(M_{ik}, M_{ij}, h'_{1k}, 0\right), \sum_{k=1}^{j-1} c\left(M_{ik}, M_{ij}, h'_{2k}, 1\right)\right)$ counts the number of pairwise phase relationships, between the former $j-1$ SNP sites and the j-th SNP site, that exist in the i-th fragment and the solution with $(h'_{1j} = 0, h'_{2j} = 1)$ simultaneously. It measures the support degree of $(h'_{1j} = 0, h'_{2j} = 1)$ that comes from the i-th fragment, and S_{01} measures the support degree of $(h'_{1j} = 0, h'_{2j} = 1)$ that comes from all of the input fragments. Therefore, $(h'_{1j} = 0, h'_{2j} = 1)$ is chosen if S_{01} is greater than S_{10}, and $(h'_{1j} = 1, h'_{2j} = 0)$ is chosen otherwise.

3.3 Augmenting

The homozygous SNPs that are deleted during preprocessing must be inserted into haplotypes $h' = (h'_1, h'_2)$. Augment h'_1 and h'_2 by the bits of the columns discarded to produce h_1 and h_2. If the deleted column j is 0-field (resp. 1-field), both of the haplotypes are inserted with 0 (resp. 1) at position j. After this procedure, the final pair of haplotypes $h = (h_1, h_2)$ is obtained. The EHD algorithm is described as follows.

Algorithm. EHD

```
Input: an m×n SNP matrix M, a parameter t
Output: a pair of reconstructed haplotypes h=(h₁,h₂)
```

1. preprocess matrix M
2. $h'_{11}=0$, $h'_{21}=1$
3. for $j=2,...,n$ do
4. $S_{01}=0$; $S_{10}=0$;
5.   for $i=1,...m$ do
6. $c_0=0$; $c_1=0$;
7.   for $k=1,...,j-1$ do
8.   if $(M_{ik}!=- \&\& M_{ij}!=-)$ then
9.   if $(M_{ik}==h'_{1k} \&\& M_{ij}==0)$ then c_0++;
10.   else if $(M_{ik}==h'_{2k} \&\& M_{ij}==1)$ then c_1++;
11.   if $(c_0<=c_1)$ then $S_{01}=S_{01}+c_1$;
12.   else $S_{01}=S_{01}+c_0$;
13.   for $i=1,...m$ do
14. $c_0=0$; $c_1=0$;
15.   for $k=1,...,j-1$ do
16.   if $(M_{ik}!=- \&\& M_{ij}!=-)$ then
17.   if $(M_{ik}==h'_{1k} \&\& M_{ij}==1)$ then c_1++;
18.   else if $(M_{ik}==h'_{2k} \&\& M_{ij}==0)$ then c_0++;
19.   if $(c_0<=c_1)$ then $S_{10}=S_{10}+c_1$;
20.   else $S_{10}=S_{10}+c_0$;
21.   if $(S_{01}>=S_{10})$ then $h'_{1j}=0$, $h'_{2j}=1$;
22.   else $h'_{1j}=1$, $h'_{2j}=0$;
23. augment $h'=(h'_1,h'_2)$, and get the final result $h=(h_1,h_2)$
24. output $h=(h_1,h_2)$

Here the time complexity of algorithm EHD is discussed. Preprocessing SNP matrix and deleting redundant information take time $O(mn)$. Reconstructing a pair of haplotypes with only heterozygous SNP sites takes time $O(mn^2)$. Recovering the deleted columns takes time $O(n)$. In summary, the time complexity of the EHD algorithm is $O(mn^2)$.

4   Experimental Results

In this section, the performance of the EHD algorithm was assessed in terms of both accuracy and speed in comparison with three famous reconstruction algorithms, namely Fast Hare [15] and DGS [18], and the FAHR [4] algorithm which was proposed by us recently. In the experiments, the reconstruction rate (RR) and the running time were adopted to measure the accuracy and speed respectively. All the tests are conducted on a Windows 7 64bit Lenovo workstation with 3.0 GHz CPU and 2 GB RAM. The compiler was Microsoft Visual C++ 6.0.

4.1   Experimental Instances

The experiments were performed by employing the real haplotypes recorded in genotypes_chr1_CEU_r22_nr.b36_fwd.phase.gz[1], a publicly available file released by the International HapMap Project on 20 December, 2007. The file contains 120 haplotypes, which come from the chromosome 1 of 60 individuals in CEPH samples (Utah residents with ancestry from northern and western Europe). In the experiments, an individual was chosen at random from the 60 ones, and a pair of haplotypes with a specified length n can be got from the haplotypes of the chosen individual.

As to fragment data, two different sequencing simulators, CELSIM [23] and Metasim [24], were adopted to generate simulation fragment datasets for implementing testing. CELSIM was invoked to simulate shotgun sequencing platform. m_1 single SNP fragments and m_2 mate-pair SNP fragments were generated in order that each mate-pair fragment consisted of two single fragments of the same haplotype. A single fragment had a length ranging from f_{min} to f_{max}, and a mate-pair fragment had a length of $n/10$. The coverage was $c/2$ for both single SNP fragments and mate-pair ones, and the total coverage was c. Finally, reading errors were planted into the fragments with probability p_s. In realistic applications of shotgun sequencing, the values of f_{min} and f_{max} are 3 and 7 respectively, c ranges from 5 and 10, and p_s ranges from 2 % and 5 % [15]. Metasim was used to simulate 454 sequencing platform. m SNP fragments, including $m_1 = (1 - p_m) \times m$ single ones and $m_2 = p_m \times m$ mate-pair ones, were generated, where p_m denoted the probability of mate-pair fragments and was set to 0.25 in our experiments. A single fragment had an expected length of len, and a mate-pair fragment had a length of $3 \times len$. Since each mate-pair fragment consists of two single fragments of the same haplotype, the coverage c equals to $[(m_1 + 2m_2) \times len]/2n$. The test instances generated by CELSIM and Metasim simulators are called as CELSIM instances and Metasim instances respectively. We ran all the experiments with the coverage c, the reading error probability p_s (Metasim does not have this parameter), the length of haplotype n and the range of single fragment length $[f_{min}, f_{max}]$ (or the expected length len for Metasim) varied. The average over 100 repeated experiments at each parameter setting was calculated and presented.

4.2   Performance Evaluation

Tables 1, 2, 3 and 4 summarize the reconstruction rate results on CELSIM instances. In Table 1, the RRs of the above four algorithms are compared with different coverage c, where $f_{min} = 3$, $f_{max} = 7$, $n = 100$, and $p_s = 0.05$. It can be recognized that the increase of coverage contributes to reconstructing haplotypes, for more fragment information can be fed to the algorithm. From Table 1 we observe that with the increase of c, the RRs of the four algorithms all increase. The EHD algorithm generally achieves higher

---

[1] Download from http://hapmap.ncbi.nlm.nih.gov/downloads/phasing/2007-08_rel22/phased/.

Table 1. Reconstruction rates of different coverage (CELSIM instances).

c	RR			
	EHD	FAHR	DGS	Fast Hare
2	0.870	0.866	0.863	0.863
4	0.921	0.918	0.919	0.926
6	0.948	0.947	0.943	0.939
8	0.963	0.958	0.960	0.945
10	0.974	0.972	0.970	0.950

RR than the other three algorithms. The RR of the EHD algorithm changes from 0.870 to 0.974, the RR of the FAHR algorithm changes from 0.866 to .972, the RR of the DGS algorithm changes from 0.863 to 0.970, and the RR of the Fast Hare algorithm changes from 0.863 to 0.950.

In Table 2, six sets of parameters were set in terms of error rate p_s, where $f_{min} = 3$, $f_{max} = 7$, $c = 10$, $n = 100$. As can be seen from Table 2, algorithm EHD can obtain higher reconstruction rate than the other three algorithms under each p_s setting. Since error rate measures the discrepancy between sequencing fragments and original hap-

Table 2. Reconstruction rates of different error rate (CELSIM instances).

ps	RR			
	EHD	FAHR	DGS	Fast Hare
0.02	0.978	0.972	0.977	0.964
0.04	0.975	0.972	0.973	0.953
0.06	0.972	0.969	0.966	0.944
0.08	0.966	0.960	0.961	0.930
0.15	0.928	0.916	0.919	0.882
0.18	0.905	0.900	0.900	0.860

lotypes, the RRs of the four algorithms show general trend of decrease with the increase of error rate p_s. When p_s varies from 0.02 to 0.18, the RRs of the EHD, the FAHR, the DGS, and the Fast Hare algorithms decrease by about 7.5 %, 7.4 %, 7.9 % and 10.8 %, respectively.

In Table 3, when the haplotype length n varied from 100 to 1000, the four algorithms were tested with $f_{min} = 3$, $f_{max} = 7$, $c = 10$, and $p_s = 0.05$. The comparison results show that, in general, the RRs got by algorithm EHD are higher than the other

Table 3. Reconstruction rates of different haplotype length (CELSIM instances).

n	RR			
	EHD	FAHR	DGS	Fast Hare
100	0.974	0.972	0.970	0.950
300	0.954	0.949	0.953	0.953
500	0.941	0.942	0.940	0.939
1000	0.940	0.938	0.928	0.926

three algorithms. The increase of haplotype length plays negative role in reconstructing haplotypes. When n is 100, the RRs of the EHD, the FAHR, the DGS and the Fast Hare algorithms are 0.974, 0.972, 0.970, and 0.950, respectively. When n increases to 1000, the RRs of the four algorithms decrease to 0.940, 0.938, 0.928, and 0.926, respectively.

In Table 4, three sets of parameters were set in dealing with single SNP fragment length range $[f_{min}, f_{max}]$, where $c = 10$, $n = 10$, and $p_s = 0.05$. The comparison results show that the haplotypes reconstructed by the EHD algorithm have higher RRs than those obtained by the other three algorithms. The shortening of single fragment length plays a negative role in haplotype reconstruction, for it leads to decreasing the probability of fragments overlapping. In Table 4, with the decrease of the single SNP fragment length, the RRs of the four algorithms decrease by about 3.5 %, 3.3 %, 4.6 %, and 2.8 %, respectively.

Table 4. Reconstruction rates of different single fragment length range (CELSIM instances).

$[f_{min}, f_{max}]$	RR			
	EHD	FAHR	DGS	Fast Hare
(3, 7)	0.974	0.972	0.970	0.950
(2, 4)	0.953	0.955	0.954	0.948
(1, 2)	0.940	0.940	0 .925	0.923

Tables 5, 6, and 7 further show the reconstruction rate results on Metasim instances in dealing with coverage c, haplotype length n, and expected read length len, respectively. In Table 5, c varied from 10 to 50, where $len = 5$ and $n = 100$. In Table 6, n increased from 100 to 1000, where $c = 20$ and $len = 5$. In Table 7, three values of

Table 5. Reconstruction rates of different coverage (Metasim instances).

c	RR			
	EHD	FAHR	DGS	Fast Hare
10	0.982	0.970	0.943	0.937
20	0.989	0.973	0.946	0.945
30	0.992	0.981	0.944	0.942
40	0.994	0.986	0.950	0.950
50	0.995	0.988	0.951	0.947

Table 6. Reconstruction rates of different haplotype length (Metasim instances).

n	RR			
	EHD	FAHR	DGS	Fast Hare
100	0.989	0.973	0.946	0.945
300	0.957	0.949	0.941	0.931
500	0.939	0.932	0.923	0.915
1000	0.923	0.922	0.911	0.909

Table 7. Reconstruction rates of different expected read length (Metasim instances).

len	RR			
	EHD	FAHR	DGS	Fast Hare
3	0.937	0.925	0.915	0.913
5	0.989	0.973	0.946	0.945
10	0.985	0.974	0.951	0.941

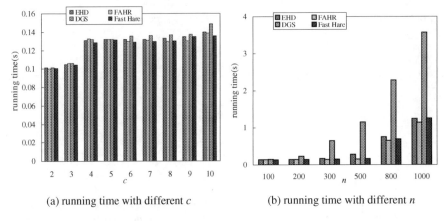

(a) running time with different c (b) running time with different n

Fig. 1. Running time comparisons on CELSIM instances ((a) different c with $f_{min} = 3, f_{max} = 7$, $n = 100$, and $p_s = 0.05$, (b) different n with $f_{min} = 3, f_{max} = 7$, $c = 10$, and $p_s = 0.05$).

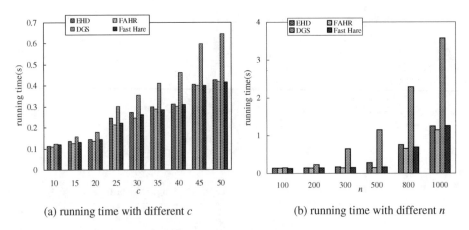

(a) running time with different c (b) running time with different n

Fig. 2. Running time comparisons on Metasim instances ((a) different c with $len = 5$, $n = 100$, (b) different n with $len = 5$, $c = 20$).

expected read length are tested, where $c = 20$ and $n = 100$. It can be observed from the three tables that under different parameter setting, algorithm EHD can get higher reconstruction accuracy than the other three algorithms. The above mentioned influence of each parameter on RR is also verified.

Figures 1 and 2 illustrate the running time comparisons on CELSIM and Metasim instances, respectively. It is apparent that the running time of the four algorithms essentially depends on the number of SNP fragments and the haplotype length. Since coverage mainly contributes to the number of fragments, test cases are generated in dealing with different coverage c and haplotype length n. As can be seen from these Figures, all of the four algorithms have high efficiency and run fast when solving problems with large c or n.

5   Conclusion

Haplotyping a single individual plays a significant role in better understanding of human complex genetic polymorphisms. In this paper, the problem is studied based on the minimum error correction model, and a practical algorithm EHD is proposed for solving this model by using the enumeration strategy. The EHD algorithm computes the support degrees for two kinds of enumerated values of a heterozygous SNP site, and chooses the one with higher support degree as the value of the reconstructed haplotypes on the SNP site. Based on the idea, the EHD algorithm can reconstruct a pair of haplotypes one SNP site after another. By utilizing simulation fragment datasets of both shotgun sequencing and 454 sequencing, extensive experiments were conducted to test and compare the performance of the EHD, the FAHR, the Fast Hare and the DGS algorithms. Experimental results have indicated that the EHD algorithm has rather good performance in terms of both accuracy and speed, and it can reconstruct haplotype pairs with higher RR than FAHR, Fast Hare and DGS in most instances.

Acknowledgments. The authors are grateful to anonymous referees for their helpful comments. This research is supported by the National Natural Science Foundation of China under Grant No.61363035 and No.61502111, Guangxi Natural Science Foundation under Grant No. 2015GXNSFAA139288, No. 2013GXNSFBA019263 and No. 2012GXNSFAA053219, Research Fund of Guangxi Key Lab of Multisource Information Mining & Security No. 14-A-03-02 and No. 15-A-03-02, "Bagui Scholar" Project Special Funds, Guangxi Collaborative Innovation Center of Multi-source Information Integration and Intelligent Processing.

References

1. Bafna, V., Istrail, S., Lancia, G., Rizzi, R.: Polynomial and APX-hard cases of the individual haplotyping problem. Theoret. Comput. Sci. **335**, 109–125 (2005)
2. Geraci, F.: A comparison of several algorithms for the single individual SNP haplotyping reconstruction problem. Bioinformatics **26**(18), 2217–2225 (2010)

3. Stephens, J.C., Schneider, J.A., Tanguay, D.A., Choi, J., Acharya, T., Stanley, S.E., Jiang, R., Messer, C.J., Chew, A., Han, J.H., Duan, J., Carr, J.L., Lee, M.S., Koshy, B., Kumar, A.M., Zhang, G., Newell, W.R., Windemuth, A., Xu, C., Kalbfleisch, T.S., Shaner, S.L., Arnold, K., Schulz, V., Drysdale, C.M., Nandabalan, K., Judson, R.S., Ruano, G., Vovis, G.F.: Haplotype variation and linkage disequilibrium in 313 human genes. Science **293**, 489–493 (2001)
4. Wu, J.L., Liang, B.B.: A fast and accurate algorithm for diploid individual haplotype reconstruction. J. Bioinform. Comput. Biol. **11**(4), 1350010 (2013)
5. Clark, A.G.: Inference of haplotypes from PCR-amplified samples of diploid populations. Mol. Biol. Evol. **7**(2), 111–122 (1990)
6. Gusfield, D.: Inference of haplotypes from samples of diploid populations: complexity and algorithms. J. Comput. Biol. **8**(3), 305–324 (2001)
7. O'Neil, S.T., Emrich, S.J.: Haplotype and minimum-chimerism consensus determination using short sequence data. BMC Genom. **13**(Suppl. 2), S4 (2012)
8. Lancia, G., Bafna, V., Istrail, S., Lippert, R., Schwartz, R.: SNPs problems, complexity, and algorithms. In: Meyer auf der Heide, F. (ed.) ESA 2001. LNCS, vol. 2161, pp. 182–193. Springer, Heidelberg (2001)
9. Lippert, R., Schwartza, R., Lancia, G., Istrail, S.: Algorithmic strategies for the SNPs haplotype assembly problem. Brief. Bioinform. **3**(1), 23–31 (2002)
10. Xie, M.Z., Chen, J.E., Wang, J.X.: Research on parameterized algorithms of the individual haplotyping problem. J. Bioinform. Comput. Biol. **5**(3), 795–816 (2007)
11. Xie, M.Z., Wang, J.X.: An improved (and practical) parameterized algorithm for the individual haplotyping problem MFR with mate-pairs. Algorithmica **52**, 250–266 (2008)
12. Cilibrasi, R., Iersel, L.V., Kelk, S., Tromp, J.: The complexity of the single individual SNP haplotyping problem. Algorithmica **49**(1), 13–36 (2007)
13. Wang, R.S., Wu, L.Y., Li, Z.P., Zhang, X.S.: Haplotype reconstruction from SNP fragments by minimum error correction. Bioinformatics **21**(10), 2456–2462 (2005)
14. He, D., Choi, A., Pipatsrisawat, K., Darwiche, A., Eskin, E.: Optimal algorithms for haplotype assembly from whole-genome sequence data. Bioinformatics **26**(12), i183 (2010)
15. Panconesi, A., Sozio, M.: Fast Hare: a fast heuristic for single individual SNP haplotype reconstruction. In: Jonassen, I., Kim, J. (eds.) WABI 2004. LNCS (LNBI), vol. 3240, pp. 266–277. Springer, Heidelberg (2004)
16. Wang, Y., Wang, E., Wang, R.S.: A clustering algorithm based on two distance functions for MEC model. Comput. Biol. Chem. **31**(2), 148–150 (2007)
17. Genovese, L.M., Geraci, F., Pellegrini, M.: SpeedHap: an accurate heuristic for the single individual SNP haplotyping problem with many gaps, high reading error rate and low coverage. IEEE/ACM Trans. Comput. Biol. Bioinform. **5**(4), 492–502 (2008)
18. Levy, S., Sutton, G., Ng, P.C., Feuk, L., Halpern, A.L., Walenz, B.P., Axelrod, N., Huang, J., Kirkness, E.F., Denisov, G., Lin, Y., MacDonald, J.R., Pang, A.W., Shago, M., Stockwell, T.B., Tsiamouri, A., Bafna, V., Bansal, V., Kravitz, S.A., Busam, D.A., Beeson, K.Y., McIntosh, T.C., Remington, K.A., Abril, J.F., Gill, J., Borman, J., Rogers, Y.H., Frazier, M.E., Scherer, S.W., Strausberg, R.L., Venter, J.C.: The diploid genome sequence of an individual human. PLoS Biol. **5**(10), 2113–2144 (2007)
19. Bansal, V., Bafna, V.: HapCUT: an efficient and accurate algorithm for the haplotype assembly problem. Bioinformatics **24**(16), i153–i159 (2008)
20. Chen, Z., Fu, B., Schweller, R., Yang, B., Zhao, Z., Zhu, B.: Linear time probabilistic algorithms for the singular haplotype reconstruction problem from SNP fragments. J. Comput. Biol. **15**(5), 535–546 (2008)
21. Aguiar, D., Istrail, S.: Haplotype assembly in polyploidy genomes and identical by descent shared tracts. Bioinformatics **29**(13), i352–i360 (2013)

22. Mazrouee, S., Wang, W.: FastHap: fast and accurate single individual haplotype reconstruction using fuzzy conflict graphs. Bioinformatics **30**(17), i371–i378 (2014)
23. Myers, G.: A dataset generator for whole genome shotgun sequencing. In: Lengauer, T., Schneider, R., Bork, P., et al. (eds.) ISMB 1999, pp. 202–210. AAAI Press, California (1999)
24. Richter, D.C., Ott, F., Auch, A.F., Schmid, R., Huson, D.H.: MetaSim—a sequencing simulator for genomics and metagenomics. PLoS ONE **3**(10), e3373 (2008)

Application of Machine Learning-Based Classification to Genomic Selection and Performance Improvement

Zhixu Qiu[1,2], Qian Cheng[1,2], Jie Song[1,2], Yunjia Tang[1,2], and Chuang Ma[1,2(✉)]

[1] State Kay Laboratory of Crop Stress Biology for Arid Areas, Northwest A&F University, Yangling 712100, Shaanxi, China
[2] Center of Bioinformatics, College of Life Sciences, Northwest A&F University, Yangling 712100, Shaanxi, China
cma@nwafu.edu.cn

Abstract. Genomic selection (GS) is a novel breeding strategy that selects individuals with high breeding value using computer programs. Although GS has long been practiced in the field of animal breeding, its application is still challenging in crops with high breeding efficiency, due to the limited training population size, the nature of genotype-environment interactions, and the complex interaction patterns between molecular markers. In this study, we developed a bioinformatics pipeline to perform machine learning (ML)-based classification for GS. We built a random forest-based ML classifier to produce an improved prediction performance, compared with four widely used GS prediction models on the maize GS dataset under study. We found that a reasonable ratio between positive and negative samples of training dataset is required in the ML-based GS classification system. Moreover, we recommended more careful selection of informative SNPs to build a ML-based GS model with high prediction performance.

Keywords: Genomic selection · Marker-assisted breeding · Relative efficiency · Machine learning · Random forest

1 Introduction

Genomic selection (GS) is a promising marker-assisted breeding paradigm that aims to improve breeding efficiency through computationally predicting the breeding value of individuals in a breeding population using information from genome-wide molecular markers (e.g., single nucleotide polymorphisms [SNPs]) [1]. During GS, a prediction model is firstly built with a training population for modeling relationships between high-throughput molecular markers and phenotype of individuals, and then employed to predict the breeding value of individuals in a testing (breeding) population, which are only genotyped but not phenotyped [2]. Individuals with higher prediction scores are finally selected for the breeding experiment. Although GS has been demonstrated to be effective in the breeding of dairy cattle [3], pig [4] and chicken [5], its application in crop breeding is still challenging, in term of high prediction performance, because of the deficiency of robust prediction models for the limited training population size [6],

© Springer International Publishing Switzerland 2016
D.-S. Huang et al. (Eds.): ICIC 2016, Part I, LNCS 9771, pp. 412–421, 2016.
DOI: 10.1007/978-3-319-42291-6_41

the nature of genotype-environment interactions [7], and the complex linkage disequilibrium and interaction patterns between molecular markers [8].

Numerical efforts have been made to develop GS prediction models with regression algorithms for predicting breeding values equal or close to real phenotypic values. Some of representative regression-based GS models are BayesA [1], BayesB [1], BayesC and ridge regression best linear unbiased prediction (rrBLUP) [10, 11]. However, in real breeding situation, it is not necessary to correctly predict phenotypic values of all individuals in a candidate population, because only individuals with high breeding value are selected for further breeding [12]. Therefore, GS has recently been regarded as a classification problem with two classes: individuals with higher phenotypic values and individuals with lower phenotypic values [13]. Some researchers even defined three classes: individuals with upper, middle and lower phenotypic values [14]. For this purpose, the classification-based GS started to be investigated with machine learning (ML) technologies, including random forest (RF) [13], support vector machine (SVM) [13] and probabilistic neural network (ANN) [14]. ML is a branch of artificial intelligence that employs various mathematical algorithms to allow computers "learn" from the experience and to perform prediction on new large datasets [15]. Instead of building a regression curve that fits all the training data, ML-based classification approaches estimate the probability of each individual that belongs to different classes. The superiority of ML-based classification over traditional regression-based approaches has been reported on several crop GS datasets [13]. Nevertheless, the application of ML in GS is still required to be explored, because very little is known about the direction toward the performance improvement of ML-based classification approaches.

Several factors may limit the performance of ML-based classification systems. One is the ratio between positive and negative samples (RPNS) in training dataset, which has been demonstrated in the ML-based prediction of mature miRNAs [16, 17], protein-protein interactions [18] and stress-related genes [19, 20]. For classification-based GS, the prediction model is required to be trained with positive and negative samples generated from the separation of training population according to phenotypes of individuals. However, the effect of RPNS on the prediction performance of ML-based GS classification approaches was rarely explored in the literature [14].

Another factor that influences the prediction accuracy is the number of informative features used to build ML-based prediction systems. In GS, thousands of molecular markers are usually used as the input features of ML-based prediction systems. Due to the limited training population size in many crop GS experiments, it is difficult to model the complex relationships between genome-wide molecular markers and phenotypic values [21]. Known that not all molecular markers are contributed to the trait phenotype [22], selecting a subset of molecular markers that is informative and small enough to deduce prediction models has become an important step toward effective GS [23]. Although hands of feature selection algorithms have been developed for the ML-based classification problems in the research area of bioinformatics and computational biology [24], it is still not clear whether these feature selection algorithms work well in the selection of informative molecular markers for improving the performance of ML-based classification systems in GS programs.

In this study, we developed a bioinformatics pipeline to perform ML-based classification for GS. We employed the random forest (RF) algorithm to build a ML-based classifier named rfGS, and explored the performance of rfGS affected by different factors on a maize GS dataset. We found that an optimized ratio between training positive and negative samples is required for ML-based GS models. Moreover, we confirmed that the selection of molecular markers is an important way of performance improvement, while the rrBLUP (ridge regression best linear unbiased prediction)-based SNP selecting yields better results than mean decrease accuracy (MDA) and mean decrease Gini (MDG), which are widely used in RF-based classification problems.

2 Methods and Materials

2.1 GS Data Set

The GS data set used in this study comprises individuals from 242 maize lines with each individual phenotyped for the grain yield under drought stress. These individuals were genotyped using 46374 single-nucleotide polymorphism (SNP) markers (Illumina MaizeSNP50 array). This data set can be publicly downloaded at the CIMMYT (International Maize and Wheat Improvement Center) website (http://repository.cimmyt.org/xmlui/handle/10883/2976).

2.2 GS Prediction Models

We built GS prediction models with four widely used regression algorithms (ridge regression best linear unbiased prediction [rrBLUP], BayesA, BayesB and BayesC) and one representative ML algorithm random forest (RF). For regression algorithms, the relationships between SNPs and phenotypic values can be generally expressed as $y = \eta + X\beta + ZA + e$, where y is the vector of phenotypic values, η is a common intercept, X is a full-rank design matrix for the fixed effects in β, which indicates the factor (e.g., population structure) influences phenotypes, $Z = \sum_k z_k$ is the allelic state at the locus k, $A = \sum_k a_k$ is marker effect at the locus k, and $e \sim N(0, \sigma_e^2)$ where e is the vector of random residual effects and σ_e^2 is the residual variance [9]. In Z, the allelic state of individuals can be encoded as a matrix of 0, 1 or 2 to a diploid genotype value of AA, AB, or BB, respectively [2].

For rrBLUP, $A \sim N(0, \lambda\sigma_a^2)$ is calculated as following formula:

$A = (Z^TZ + \lambda I_P)^{-1}Z^Ty$, where $\lambda = \dfrac{\sigma_e^2}{\sigma_a^2}$ is the ratio between the residual and marker variances. The rrBLUP algorithm was implemented using the "mixed.solve" function in R package rrBLUP (https://cran.r-project.org/web/packages/rrBLUP/index.html).

For the Bayesian regression analysis, the conditional distribution of A can be estimated using the user-given marker information and phenotypic values. The prior distribution can be estimated using different algorithms in the Bayesian framework. We selected BayesA (scaled-t prior), BayesB (two component mixture prior with a point of mass at zero and a scaled-t slab), BayesC (two component mixture prior with a point of mass at zero and a Gaussian slab), respectively. BayesA, BayesB and BayesC were implemented using the "BGLR" function in R package BGLR (https://cran.r-project.org/web/packages/BGLR/index.html).

Random forest, developed by Breiman [25], is a combination of random decision trees. Each tree in the forest is built using randomly selected samples and SNPs. RF outputs the probability of each sample to be the best class based on votes from all trees. RF is a powerful ML algorithm that has been widely applied in many classification problems [26, 27]. The RF algorithm was implemented using the R package random-forest (https://cran.r-project.org/web/packages/randomForest/index.html). The number of constructed decision trees (ntree) was set to be 500, other parameters were used default values.

2.3   SNP Selection

RF-Based SNP Selection. RF provides two built-in measurements for estimating the importance of each feature: MDA and MDG [28]. For a given feature, the MDA quantifies the mean decrease of the predictor when the value of this feature is randomly permuted in the out-of-bag samples, while the MDG calculates the quality of a split for every node of a tree by means of the Gini index. The higher MDA or MDG value indicates the more importance of the feature in the prediction. Both MDA and MDG were calculated by the R package randomforest.

rrBLUP-Based SNP Selection. The rrBLUP model estimates the marker effect for reflecting the importance of each SNP in the prediction of the correlation between genotype and phenotypic values. We selected informative SNPs according to the absolute values of marker effects.

2.4   Performance Evaluation

As previously described [13], the relative efficiency (RE) measurement was used to evaluate the prediction performance of each GS prediction model. Of note, other measurements, such as sensitivity, specificity and area under receiver operating characteristic (ROC), may be also interesting in the GS program. The RE was defined as below:

$$R(\alpha) = \frac{\mu'_\alpha - \mu}{\mu_\alpha - \mu},$$

where μ represents the mean phenotypic value of the whole GS dataset, μ_α denotes the mean of real phenotypic values of the top α individuals with extreme phenotypic values, μ_α is the mean of the real values of extreme individuals (ranked by the predicted values) that have the top α. RE ranges from -1 to 1. A higher RE value indicates a high degree that extreme individuals can be predicted by the classifier. The possible α value ranged from 10 % to 50 % was considered in this study.

Leave-one-out cross-validation (LOOCV) test was used to evaluate the prediction performance and robustness (Fig. 1). In the LOOCV, each individual was picked out in turn as an independent test sample, and all the remaining individuals were used as training samples for building the GS prediction model with rrBLUP, BayesA, BayesB, BayesC or RF algorithm (Fig. 1A–C). This process was repeated until each individual was used as test data one time (Fig. 1A–C). Because sampling strategy was used in the three Bayesian-related regression models and RF-based ML classification model, the LOOCV test was repeated 10 times for calculating the average performance of all tested GS algorithms at each possible percentile value (α).

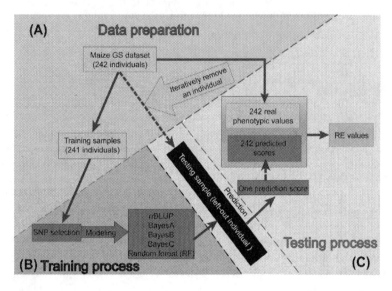

Fig. 1. Overview of LOO cross-validation test for performance evaluation of GS prediction models built with rrBLUP, BayesA, BayesB, BayesC and RF algorithms.

3 Results and Discussion

3.1 Performance Comparison Between rfGS and Four Representative GS Algorithms

The prediction performance of five algorithms (rrBLUP, BayesA, BayesB, BayesC and rfGS) was evaluated using the LOO cross validation test, which iteratively selected one individual as the testing sample and the other individuals as the training samples. The relative efficiency (RE) measurement was used to estimate the prediction accuracy of these algorithms for correctly selecting the best individuals at a given percentile value (α). As shown in Fig. 2, the RE of BayesA gradually decreases from 0.33 to 0.23, when α increases from 10 % to 50 %. Similar results are observed for BayesB and BayesC. Differently, the RE of rrBLUP remarkably decreases from 0.40 to 0.34 when α increases from 10 % to 15 %, but notably increases at higher percentile values (α = 18 %, 22 %, 27 %, 39 % and 47 %). rfGS shows a different pattern of RE compared to the other four algorithms, and reaches the highest RE value (0.53) when α is 14 %. These results indicate that the performance of all five algorithms is influenced by the percentile value.

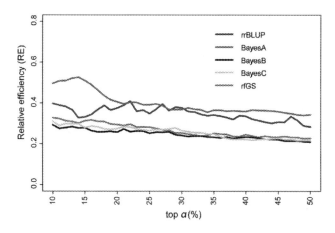

Fig. 2. The relative efficiency (RE) of five GS algorithms at different percentile values (α). (Color figure online)

Compared with BayesA, BayesB and BayesC, rrBLUP yields higher RE values at all tested percentile values. However, we found that the RE can be further improved by using rfGS for almost all tested percentile values. Our result suggests that compared with the widely used regression-based GS algorithms (BayesA, BayesB, BayesC and rrBLUP), RF-based ML classification system rfGS would be an alternative option for the GS program.

3.2 Performance of rfGS is Affected by the Ratio Between Training Positive and Negative Samples

We explored how the performance of rfGS changed with different ratios between positive and negative samples in the training dataset, by selecting the proportion of individuals in the best–worst classes to 20–80, 30–70, 40–60, 50–50 or 60–40.

In Fig. 3, it is shown that the RE of the setting 20–80 gradually decreases from 0.57 to 0.22, when α increases from 10 % to 50 %. Differently, the RE patterns under settings 30–70, 40–60, 50–50, have similar trend that the RE scores first increase when α increases from 10 % to 15 %, and then decrease when α increases from 15 % to 50 %. The RE under the setting 60–40, frequently fluctuates compared to other settings, and has a peak when α is 21 %. The different trends of these RE values under five proportion settings could be explained by the different ability of ML-based classifiers that identify the best individuals under the corresponding ratio. The setting 30–70 showed the best performance among the five different partitions evaluated. Overall, our findings show that the impacts of the ratio between training positive and negative samples on the performance of ML-based GS classifiers should not be neglected, and a reasonable proportion of best–worst classes in the training sets is important for GS program.

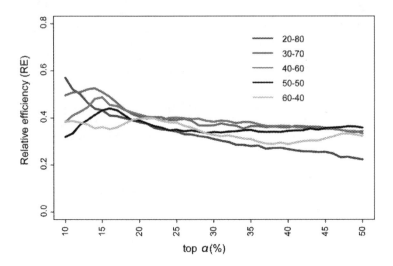

Fig. 3. The relative efficiency of rfGS is affected by the ratio between positive and negative samples in training dataset. (Color figure online)

3.3 Prediction Performance of rfGS Can Be Improved with SNP Selection Process

SNP selection is a process in which a subset of informative SNPs is selected for building GS prediction models. In ML-based classification, MDA and MDG are two powerful feature selection algorithms that are widely used in selected informative features from high-dimensional genomic data. In each round of LOOCV, we estimated the importance of each SNP using the MDA and MDG, respectively, and selected the top N (N = 50, 100, 500, 1000, 1500, 2000, 2500, 3000, 3500, 4000, 4500, 5000, 10000, 15000, 20000, 25000, 30000, 35000, 40000) to build GS prediction models (Fig. 4). We also performed the SNP selection based on the marker effects estimated by the rrBLUP algorithm (Fig. 4). Compared with using all 46374 SNPs, the proportion of predicted accuracy (RE) by selecting top 3000, 4500, 5000, 10000, 30000, 35000 SNPs using MDA increases from −4.48 % to 14.1 % (mean 5.59 % ± 3.47 %) when α increases from 28 % to 35 %. Meanwhile, the proportion is elevated from 0.46 % to 12.61 % (mean 6.29 % ± 2.68 %) by selecting top 4000, 5000, 15000, 20000, 35000 SNPs using MDG with α increasing from 24 % to 35 %. When top 100, 500, 3500, 15000 SNPs were selected, the proportion of prediction accuracy increases from −9.8 % to 34.79 % (mean 9.35 % ± 8.47 %) with a range of α from 20 % to 40 %. rfGS reaches the best performance when selecting the top 10000, 35000, 100 SNPs estimated with MDA, MDG, and rrBLUP algorithms, respectively. Compared to MDA and MDG, rrBLUP-based SNP selection requires the least SNPs to obtain the same prediction ability. It should be noted that, for the GS programs interested in the α ranged from 14 % to 16 % and from 40 % to 50 %, the predicted accuracy consistently decreases in all three SNP selection algorithms, suggesting that more powerful SNP algorithms are urgent to be developed.

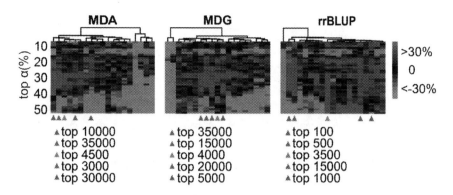

Fig. 4. The performance of mlDNA affected by different SNP selection algorithms. (Color figure online)

Overall, our result shows that the algorithms of selection important SNPs is effective for improving efficiency of GS, and rrBULP-based SNP selection is a promising approach.

4   Conclusions

In this study, we designed a bioinformatics pipeline to perform ML-based classification in GS, exemplified with the application of RF algorithm on a maize GS dataset. RF-based ML classification system rfGS outperforms the widely used regression-based GS algorithms (BayesA, BayesB, BayesC and rrBLUP) on the maize GS dataset under study. Some cautions are raised about the application of ML-based classification to GS. A reasonable proportion of training positive and negative samples is required to increase the prediction accuracy of ML-based GS model. Additionally, SNP selection is also viable to improve efficiency of GS, and rrBULP-based SNP selection is a promising algorithm. In the future, we will apply the graphics processing unit (GPU)-based acceleration technologies to perform the ML-based GS experiments with more complex ML algorithms (e.g., SVM, deep convolutional neural network) and more GS datasets.

Acknowledgement. This work was supported by the grants of the National Natural Science Foundation of China (No. 31570371), Agricultural Science and Technology Innovation and Research Project of Shaanxi Province, China (No. 2015NY011) and the Fund of Northwest A&F University (No. Z111021403 and Z109021514).

References

1. Meuwissen, T.H., Hayes, B.J., Goddard, M.E.: Prediction of total genetic value using genome-wide dense marker maps. Genetics **157**, 1819–1829 (2001)
2. Desta, Z.A., Ortiz, R.: Genomic selection: genome-wide prediction in plant improvement. Trends Plant Sci. **19**, 592–601 (2014)

3. Hayes, B.J., Bowman, P.J., Chamberlain, A.J., Goddard, M.E.: Invited review: genomic selection in dairy cattle: progress and challenges. J. Dairy Sci. **92**, 433–443 (2009)
4. Wellmann, R., Preuss, S., Tholen, E., Heinkel, J., Wimmers, K., Bennewitz, J.: Genomic selection using low density marker panels with application to a sire line in pigs. Genet. Sel. Evol. **45**, 28 (2013)
5. Wolc, A., Zhao, H.H., Arango, J., Settar, P., Fulton, J.E., O'Sullivan, N.P., Preisinger, R., Stricker, C., Habier, D., Fernando, R.L., Garrick, D.J., Lamont, S.J., Dekkers, J.C.: Response and inbreeding from a genomic selection experiment in layer chickens. Genet. Sel. Evol. **47**, 59 (2015)
6. Isidro, J., Jannink, J.L., Akdemir, D., Poland, J., Heslot, N., Sorrells, M.E.: Training set optimization under population structure in genomic selection. Theoret. Appl. Genet. **128**, 145–158 (2015)
7. Crossa, J., Perez, P., Hickey, J., Burgueno, J., Ornella, L., Ceron-Rojas, J., Zhang, X., Dreisigacker, S., Babu, R., Li, Y., Bonnett, D., Mathews, K.: Genomic prediction in CIMMYT maize and wheat breeding programs. Heredity **112**, 48–60 (2014)
8. Brito, F.V., Neto, J.B., Sargolzaei, M., Cobuci, J.A., Schenkel, F.S.: Accuracy of genomic selection in simulated populations mimicking the extent of linkage disequilibrium in beef cattle. BMC Genet. **12**, 80 (2011)
9. Habier, D., Fernando, R.L., Kizilkaya, K., Garrick, D.J.: Extension of the Bayesian alphabet for genomic selection. BMC Bioinform. **12**, 186 (2011)
10. Endelman, J.B.: Ridge regression and other kernels for genomic selection with R package rrBLUP. Plant Genome **4**, 250–255 (2011)
11. de Los Campos, G., Hickey, J.M., Pong-Wong, R., Daetwyler, H.D., Calus, M.P.: Whole-genome regression and prediction methods applied to plant and animal breeding. Genetics **193**, 327–345 (2013)
12. Blondel, M., Onogi, A., Iwata, H., Ueda, N.: A ranking approach to genomic selection. PLoS ONE **10**, 0128570 (2015)
13. Ornella, L., Perez, P., Tapia, E., Gonzalez-Camacho, J.M., Burgueno, J., Zhang, X., Singh, S., Vicente, F.S., Bonnett, D., Dreisigacker, S., Singh, R., Long, N., Crossa, J.: Genomic-enabled prediction with classification algorithms. Heredity **112**, 616–626 (2014)
14. Gonzalez-Camacho, J.M., Crossa, J., Perez-Rodriguez, P., Ornella, L., Gianola, D.: Genome-enabled prediction using probabilistic neural network classifiers. BMC Genom. **17**, 208 (2016)
15. Chen, X., Ishwaran, H.: Random forests for genomic data analysis. Genomics **99**, 323–329 (2012)
16. Sturm, M., Hackenberg, M., Langenberger, D., Frishman, D.: TargetSpy: a supervised machine learning approach for MicroRNA target prediction. BMC Bioinform. **11**, 292 (2010)
17. Cui, H., Zhai, J., Ma, C.: MiRLocator: machine learning-based prediction of mature MicroRNAs within plant pre-miRNA sequences. PLoS ONE **10**, e0142753 (2015)
18. Hamp, T., Rost, B.: More challenges for machine-learning protein interactions. Bioinformatics **31**, 1521–1525 (2015)
19. Shaik, R., Ramakrishna, W.: Machine learning approaches distinguish multiple stress conditions using stress-responsive genes and identify candidate genes for broad resistance in rice. Plant Physiol. **164**, 481–595 (2014)
20. Ma, C., Xin, M., Feldmann, K.A., Wang, X.: Machine learning-based differential network analysis: a study of stress-responsive transcriptomes in arabidopsis. Plant Cell **26**, 520–537 (2014)

21. Hickey, J.M., Dreisigacker, S., Crossa, J., Hearne, S., Babu, R., Prasanna, B.M., Grondona, M., Zambelli, A., Windhausen, V.S., Mathews, K., Gorjanc, G.: Evaluation of genomic selection training population designs and genotyping strategies in plant breeding programs using simulation. Crop Sci. **54**, 1476–1488 (2014)
22. Bermingham, M.L., Pong-Wong, R., Spiliopoulou, A., Hayward, C., Rudan, I., Campbell, H., Wright, A.F., Wilson, J.F., Agakov, F., Navarro, P., Haley, C.S.: Application of high-dimensional feature selection: evaluation for genomic prediction in man. Sci. Rep. **5**, 10312 (2015)
23. Long, N., Gianola, D., Rosa, G.J.M., Weigel, K.A., Avendano, S.: Machine learning classification procedure for selecting SNPs in genomic selection: application to early mortality in broilers. J. Anim. Breed. Genet. **124**, 377–389 (2007)
24. Adorjan, P., Distler, J., Lipscher, E., Model, F., Muller, J., Pelet, C., Braun, A., Florl, A.R., Gutig, D., Grabs, G., Howe, A., Kursar, M., Lesche, R., Leu, E., Lewin, A., Maier, S., Muller, V., Otto, T., Scholz, C., Schulz, W.A., Seifert, H.H., Schwope, I., Ziebarth, H., Berlin, K., Piepenbrock, C., Olek, A.: Tumour class prediction and discovery by microarray-based DNA methylation analysis. Nucleic Acids Res. **30**, e21 (2002)
25. Breiman, L.: Random forests. Mach. Learn. **45**, 5–32 (2001)
26. Lloyd, J.P., Seddon, A.E., Moghe, G.D., Simenc, M.C., Shiu, S.H.: Characteristics of plant essential genes allow for within- and between-species prediction of lethal mutant phenotypes. Plant Cell **27**, 2133–2147 (2015)
27. Panwar, B., Arora, A., Raghava, G.P.: Prediction and classification of NcRNAs using structural information. BMC Genom. **15**, 127 (2014)
28. Touw, W.G., Bayjanov, J.R., Overmars, L., Backus, L., Boekhorst, J., Wels, M., van Hijum, S.A.: data mining in the life sciences with random forest: a walk in the park or lost in the jungle? Brief. Bioinform. **14**, 315–326 (2013)

Prediction of Hot Spots Based on Physicochemical Features and Relative Accessible Surface Area of Amino Acid Sequence

ShanShan Hu[1], Peng Chen[1(✉)], Jun Zhang[2], and Bing Wang[3]

[1] Institute of Health Sciences, Anhui University, Hefei 230601, Anhui, China
pchen.ustc10@yahoo.com
[2] College of Electrical Engineering and Automation, Anhui University,
Hefei 230601, Anhui, China
[3] School of Electronics and Information Engineering, Tongji University,
Shanghai 804201, China

Abstract. Hot spot is dominant for understanding the mechanism of protein-protein interactions and can be applied as a target to drug design. Since experimental methods are costly and time-consuming, computational methods are prevalently applied as an useful tool in hot spot prediction through sequence or structure information. Here, we propose a new sequence-based model that combines physicochemical features with relative accessible surface area of amino acid sequence. The model consists of 83 classifiers involving IBk algorithm, where instances for one classifier are encoded by corresponding property extracted from 544 properties in AAindex1 database. Then several top performance classifiers with respect to F1 score are selected to be an ensemble by majority voting technique. The model outperforms other state-of-the-art computational methods, yields a F1 score of 0.80 on BID test set.

Keywords: Hot spots · Physicochemical features · Majority voting · IBk algorithm

1 Introduction

Protein-protein interactions play a fundamental role in many cellular processes such as signal transduction, cellular motion, and hormone-receptor interactions [1]. However, many principles of governing protein interactions are still not fully understood. It is revealed that hot spots is a small fraction of interfacial residues which contribute the majority of binding free energy ($\Delta\Delta G$). Actually, hot spots tend to be clustered tightly in the center of protein interfaces and surrounded by energetically less important residues [2]. It is turns out that hot spots have received extensive concerns not only as potential binding motifs for small molecule inhibitors of protein interactions, but also as the prime targets in rational drug design [3].

Alanine-scanning mutagenesis is a prevalent approach to identify hot spots by measuring the change in the free energy of binding when a certain residue is mutated to

© Springer International Publishing Switzerland 2016
D.-S. Huang et al. (Eds.): ICIC 2016, Part I, LNCS 9771, pp. 422–431, 2016.
DOI: 10.1007/978-3-319-42291-6_42

alanine [4]. However, experimental methods are costly and time-consuming [5]. A growing number of researchers prefer to develop various computational methods to predict hot spot residues in protein interfaces, such as energy function-based models, molecular dynamics (MD) simulations and machine learning approaches. Robetta [6] adopted a simple physical model to predict hot spot residues which is based on an energy function and an all-atom rotamer description of the side chains. Another method, FOLD-X Energy Function (FOLDEF) provided a fast and quantitative estimation of how significant the interactions contribute to the stability of protein complexes, with a full atomic description of proteins structures and a weighted different energy term [7]. Darnell et al. [8] developed two knowledge based models (K-FADE and K-CON) to predict hot spots. Recently, Shingate et al. [9] proposed a new ECMIS (Energetic Conserved Mass Index and Spatial Clustering) algorithm that uses the combination of diverse protein features and it achieved an accuracy of around 0.80. Wang et al. [10], developed an extreme learning machine (ELM) algorithm to predict hot spots which integrates hybrid features of the target residue and its spatial neighbor residues.

Despite marvelous progress has been achieved in hot spot prediction, most aforementioned methods required 3D structures of protein. But the structures of the most of proteins are still not resolved and the sequence information is provided only.

To address the issue of hot spot prediction from sequence only, we propose a model combining physicochemical characteristics of residues with relative accessible surface area (relASA) of amino acid sequences which are obtained from NetSurfP Website. Initially, 83 relatively independent physicochemical properties are extracted from AAindex1 data set [11]. Each property integrating relative accessible surface area (relASA) of residues are used to encode amino acid residues. The encoding vectors for residues are then input into one classifier with IBk algorithm [12]. As a result, 83 different classifiers are obtained by 83 physicochemical properties for predicting hot spots. According to F1 score, a combination of the top n individual classifiers are built to identify hot spots, which can further improve hot spot prediction. To demonstrate the power of our proposed method, benchmark datasets of ASEdb and BID are applied. Compared with state-of-the-art methods on BID test dataset, our method yields better predictions.

2 Methods

2.1 Datasets

In this study, training data set is derived from the Alanine Scanning Energetics Database (ASEdb) [13] and the proteins with sequence identity above 35 % are filtered out. In the database, the interface residue with the change of binding free energy $\Delta\Delta G \geq 2.0$ cal/mol is considered as hot spot and that with $\Delta\Delta G < 0.4$ cal/mol is considered as non-hot spot. The residue with the energy change between 0.4 and 2.0 are deleted. Thus, the training data set contains 149 interface residues with 58 hot spots and 91 non-hot spots. The binding interface database (BID) is as an independent test data to assess the predictive performance of our model. Instead of a single threshold, in BID alanine mutation data are divided into 'strong', 'intermediate', 'weak' and 'insignificant' interactions, and only 'strong' interaction strengths are considered as hot spots [14]. As a result, 112 interfacial residues are obtained where 54 residues are hot spots and 58 non-hot spots.

2.2   Feature Extraction

Encoding Protein Sequences. AAindex1 is a database containing 544 physico-chemical and biochemical properties for 20 types of amino acids. Since high related properties change the predictive results, relevant ones with a correlation coefficient more than 0.6 are removed like Chen's method [15], as did in AAindex1 itself. First, the correlation coefficients of a property that randomly selected from the 544 ones and the remaining ones are calculated. Then the number of relevant properties is counted for one property and the relevant numbers for 544 properties are ranked. From the top one property, all of the next properties related to the top one are removed. This process is not terminated until no pair of properties have correlation coefficient more than 0.6. At last, 83 properties are retained and then used to encode protein sequences, where the correlation coefficient between each two properties is less than 0.6. That is to say, the subset of AAindex1 database consists of above-mentioned 83 descriptors with 1×20 dimensions.

In this study, a sliding window with the length of 11 residues centered at a residue is adopted to encode for the residue. The window moves along whole protein sequence to encode for all sequence residues. The encoding schema for residues in the sliding window is shown in Fig. 1. The encoding method is inspired by Chou's method which proposed a novel concept of pseudo-amino acid composition, considering not merely amino acid composition but also the sequence-order information [16]. Moreover, pseudo-amino acid composition is widely applied in biological fields, such as identi-fication of DNA-binding proteins [17] and immunoglobulins [18].

$$
\begin{aligned}
\theta_1 &= \frac{1}{L-1} \sum_{i=1}^{L-1} [\Phi(R_{i+1}) - \Phi(R_i)]^2 \\
\theta_2 &= \frac{1}{L-2} \sum_{i=1}^{L-2} [\Phi(R_{i+2}) - \Phi(R_i)]^2 \\
\theta_3 &= \frac{1}{L-3} \sum_{i=1}^{L-3} [\Phi(R_{i+3}) - \Phi(R_i)]^2 \quad (k = 1, 2, \ldots, L-1) \\
&\vdots \\
\theta_k &= \frac{1}{L-k} \sum_{i=1}^{L-k} [\Phi(R_{i+k}) - \Phi(R_i)]^2
\end{aligned}
\tag{1}
$$

From the beginning to the end of the sequence, a central residue is encoded by a set of sequence order-correlated factors between neighbor residues. Equation (1) is cor-responding to Fig. 1. For a case when one of the beginning or the end of L/2 (L represents the length of sliding window) residues is regarded as central residue, no enough adjacent residues exist to encode the central residue. So the value of vacant

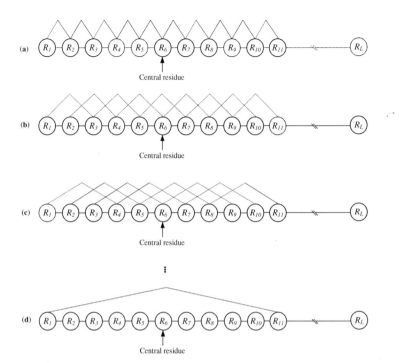

Fig. 1. Encoding schema for residues. Here R_1 represents the 1st residue, R_2 represents the 2nd residue, ..., and R_L represents the L-th residue, each of them belongs to 20 native amino acids. A schematic drawing to show (a) the first-tier, (b) the second-tier, and (3) the third-tier sequence order correlation mode along a protein sequence. Panel (a) reflects the correlation mode between all the nearest neighboring residues, panel (b) the sec-nearest neighboring residues, panel (c) the third-nearest neighboring residues, and panel (d) the 10th nearest neighboring residues that reflects the correlation model between the first and the last residues of the sliding window. In this graph, the sixth residue is the central residue and all the tier sequence order correlations are considered as encoding features representing the central residue.

place is set to zero. The central residue is encoded by a vector whose elements are calculated in Eq. (1). In Eq. (1), $\theta 1$ is called the first-tier correlation factor that reflects the sequence order correlation between all the nearest residues along a protein chain (Fig. 1a), $\theta 2$ is the second-tier correlation factor that reflects the sequence order correlation between all the second nearest residues (Fig. 1b), $\theta 3$ is the third-tier correlation factor that reflects the sequence order correlation between all the 3rd nearest residues (Fig. 1c), and so forth [19]. In Eq. (1), k reflects the relationship between neighbor residues; R represents one type of amino acids; $\Phi(R_i)$ is one of 83 descriptors for the i-th residue R_i.

Relative Accessible Surface Area (relASA). Accessible surface area (ASA) is the area of the surface determined by the center of a probe sphere, whose radius is the nominal radius of the solvent, as it rolls over the van der Waals surface of the molecule [20].

Relative accessible surface area (relASA) is the ratio of the calculated ASA over the referenced ASA [21]. NetSurfP web server (http://www.cbs.dtu.dk/services/NetSurfP/) used an ensemble of artificial neural networks to predict the surface accessibility and secondary structure of residues along an amino acid sequence and simultaneously provide a reliability score for each prediction, in the form of a Z-score [22]. The relative accessible surface area predicted by NetSurfP is added as the 11th feature to characterize the central residue. Therefore, every residue is represented by an input vector with 11 features. At last, all the input vectors are normalized to an initial value, say, 1.

2.3   Classifier Construction

For each one of the 83 descriptors from the AAindex1, a classifier is constructed based on IBk algorithm, a type of modified k-nearest neighbor algorithm. The conventional nearest neighbor algorithm is sensitive to the choice of the algorithm's similarity function and is computationally expensive. IBk algorithm can surmount these problems and parameter k can be selected by cross-validation [12]. After 83 classifiers are constructed, a combination of the n top classifiers is explored, where classifiers are ranked in terms of the F1 score in descend sort and thus the top 1 classifier performs the best among the 83 classifiers. For each combination, the residue is identified to be a hot spot if at least half of the n classifiers identified it to be a hot spot, otherwise it is predicted to be a non-hot spot.

2.4   Evaluation Criteria

Several widely used metrics are adopted to evaluate the prediction performance, including accuracy (Acc), precision (Pre), sensitivity (Sen), specificity (Spe), Matthews correlation coefficient (MCC) and F1 score (F1). F1 score is the primary evaluation metric in this study which gauges the balance between precision and sensitivity rates.

$$Acc = \frac{TP + TN}{TP + FP + TN + FN} \tag{2}$$

$$Spe = \frac{TN}{TN + FP} \tag{3}$$

$$Pre = \frac{TP}{TP + FP} \tag{4}$$

$$Sen = \frac{TP}{TP + FN} \tag{5}$$

$$F1 = \frac{2 \times Sen \times Pre}{Sen + Pre} \tag{6}$$

$$MCC = \frac{TP \times TN - FP \times FN}{\sqrt{(TP + FP)(TP + FN)(TN + FP)(TN + FN)}} \quad (7)$$

In which, TP, FP, TN and FN respectively represent the number of true positives (correctly predicted hot spots), the number of false positives (incorrectly predicted hot spots), the number of true negatives (correctly predicted non-hot spots) and the number of false negatives (incorrectly predicted non-hot spots).

3 Results

The 10-fold cross-validation technique is adopted which means that the dataset is randomly divided into 10 subsets with approximately the same number and one subset is used as the test set and the others the training set. The process is repeated until every 1-fold is as the test set. Figure 2 illustrates the performance of all 83 classifiers where five classifiers yield F1 scores greater than 0.7.

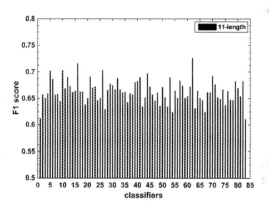

Fig. 2. The performance of 83 classifiers with 11-length of sliding window based on ASEdb cross-validation. The x-axis represents the 83 different classifiers that correspond to the 83 relatively independent descriptors from the AAindex1. The y-axis is the value of F1 score.

3.1 Prediction Performance on an Independent Test Set

BID database is adopted as an independent test dataset to further evaluate our proposed model. The 62th classifier is observed to be the best classifier among these five classifiers on training dataset with the highest Acc (0.71), Spe (0.97), Sen (0.58), Pre (0.97) and F1 score (0.73). While the 16th classifier is presented to be the optimal classifier on test dataset, which gets the highest Acc (0.79), Spe (0.89), Sen (0.72), Pre (0.91), MCC (0.60) and F1 score (0.81). Three of the top 5 classifiers, say (the 5th, 10th and 16th classifiers), on the test dataset perform better than the training dataset. The reason

Table 1. The prediction performance of the top 5 classifiers with 11-length sliding window on training dataset and test dataset, respectively.

Classifier		Acc	Spe	Sen	Pre	F1	MCC
5	Training	0.70	0.92	0.57	0.92	0.70	0.48
	Test	0.72	0.85	0.65	0.89	0.75	0.47
10	Training	0.70	0.91	0.57	0.92	0.70	0.48
	Test	0.65	0.82	0.59	0.91	0.71	0.36
16	Training	0.70	0.97	0.57	0.97	0.72	0.51
	Test	**0.79**	**0.89**	**0.72**	**0.91**	**0.81**	**0.60**
26	Training	0.69	0.94	0.56	0.95	0.70	0.48
	Test	0.64	0.74	0.59	0.82	0.69	0.31
62	Training	**0.71**	**0.97**	**0.58**	**0.97**	**0.73**	**0.53**
	Test	0.63	0.70	0.59	0.79	0.68	0.29

The first column represents the i-th classifier correspond to the i-th descriptor in the AAindex1.
The greatest values are highlighted in bold.

principally lies in that the ratio of hot spots and non-hot spots in the training dataset (the ratio is 0.64) is less than that in the test dataset (the ratio is 0.93). Thus, the test dataset is more balanced and it is more easily identified by our model, which suggests that our model creates less false positive and more false negative results (Table 1).

3.2   Construction of Classifier Ensemble

Since individual classifiers are instable and poorly robust, and may bring about considerable fluctuation and over-fitting, an ensemble of the top n classifiers is built as a new predictor. Figure 3 demonstrates F1 scores of all combinations of the top n classifiers, where n is in the range of 3 to 33. From Table 2, the ensemble of the top 23 classifiers performs the best and obtains an F1 score of 0.77 on ASEdb dataset by 10-fold cross-validation. Then the combination of the top 23 classifiers is validated on BID dataset and yields an F1 measure of 0.80. It should be stressed that despite the 16th classifier yields a slightly higher F1 score of 0.81 than the classifier ensemble on BID test dataset, individual classifier is sensitive to noise and not applicable to all datasets. Thus, the classifier ensemble technique can substantially improve prediction performance and offset disadvantages of individual classifier.

Table 2. Prediction performance of the combination of the top 23 classifiers on ASEdb and BID databases

Database	Acc	Spe	Sen	Pre	F1	Mcc	Ratio[&]
ASEdb	0.77	1.0	0.63	1.0	0.77	0.63	0.64
BID	0.76	1.0	0.67	1.0	0.80	0.60	0.93

[&]The last column represents the ratio of the number of hot spots and that of non-hot spots.

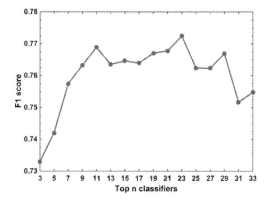

Fig. 3. Prediction performance of the combinations of the top n classifiers. Here n is in the range of 3 to 33 and n is odd for majority voting technique.

3.3 Comparison with Other Predictive Methods

To further validate the effectiveness of our model, we make comparison with other methods on the same BID test dataset such as Robetta [6], KFC [23], KFCA (KFC +Robetta) [8], ISIS [24], Tuncbag's method [25] and Chen's method [15]. It is shown in Table 3 that although Chen's method yielded the competitive highest sensitivity (Sen = 0.92), our model obtains a satisfactory precision (Pre = 1.0) and an improvement of F1 score (F1 = 0.80) by 4 % than Chen's method. Therefore, our model is superior to other existing methods for the prediction of hot spot residues.

Table 3. Comparison of hot spot prediction by different methods for the same BID test dataset

Method	Sen	Pre	F1
Robetta	0.57	0.63	0.60
KFC	0.36	0.51	0.42
KFCA[a]	0.48	0.53	0.51
ISIS	0.70	0.48	0.57
Tuncbag	0.59	0.73	0.65
Chen's method	**0.92**	0.65	0.76
Our method	0.67	**1.0**	**0.80**

[a]KFCA means the method combines KFC with Robetta
The greatest values are highlighted in bold.

4   Conclusion

This paper proposed a ensemble model based on IBk algorithm with the combination of physicochemical properties of amino acid sequences and the relative accessible surface area (relASA) of residues, to predict hot spot residues in protein-protein interfaces.

A number of 83 descriptors were filtered out from AAindex1 database, each of which was utilized to encode protein sequences by 11-length sliding window along protein sequences. The relASA of residues from NertsurfP website were adopted as another feature to encode residue. The ensemble of the n top classifiers was constructed by majority voting method, which largely improved prediction performance. As a result, when n equals to 23, the classifier ensemble yielded the highest F1 score of 0.77 on ASEdb database by 10-fold cross-validation. Further, BID database is used as an independent test data set to test the prediction results. It is observed that the prediction power of the proposed method surpass other state-of-the-art computational methods.

Acknowledgement. This work was supported by the National Natural Science Foundation of China (Nos. 61300058, 61472282, 61271098 and 61374181).

References

1. Chothia, C., Janin, J.: Principles of protein-protein recognition. Nature **256**(5520), 705–708 (1975)
2. Bogan, A.A., Thorn, K.S.: Anatomy of hot spots in protein interfaces. J. Mol. Biol. **280**(1), 1–9 (1998)
3. Brenke, R., Kozakov, D., Chuang, G.Y., Beglov, D., Hall, D., Landon, M.R., Mattos, C., Vajda, S.: Fragment-based identification of druggable 'hot spots' of proteins using Fourier domain correlation techniques. Bioinformatics **25**(5), 621–627 (2009)
4. Wells, J.A.: Systematic mutational analyses of protein-protein interfaces. Methods Enzymol. **202**, 390–411 (1991)
5. DeLano, W.L.: Unraveling hot spots in binding interfaces: progress and challenges. Curr. Opin. Struct. Biol. **12**(1), 14–20 (2002)
6. Kortemme, T., Baker, D.: A simple physical model for binding energy hot spots in protein-protein complexes. Proc. Nat. Acad. Sci. U.S.A. **99**(22), 14116–14121 (2002)
7. Guerois, R., Nielsen, J.E., Serrano, L.: Predicting changes in the stability of proteins and protein complexes: a study of more than 1000 mutations. J. Mol. Biol. **320**(2), 369–387 (2002)
8. Darnell, S.J., Page, D., Mitchell, J.C.: An automated decision-tree approach to predicting protein interaction hot spots. Proteins **68**(4), 813–823 (2007)
9. Shingate, P., Manoharan, M., Sukhwa, A., Sowdhamini, R.: ECMIS: computational approach for the identification of hotspots at protein-protein interfaces. BMC Bioinformatics **15**, 303 (2014)
10. Wang, L., Zhang, W., Gao, Q., Xiong, C.: Prediction of hot spots in protein interfaces using extreme learning machines with the information of spatial neighbour residues. IET Syst. Biol. **8**(4), 184–190 (2014)
11. Kawashima, S., Pokarowski, P., Pokarowska, M., Kolinski, A., Katayama, T., Kanehisa, M.: AAindex: amino acid index database, progress report 2008. Nucleic Acids Res. (Database Issue) **36**, D202–205 (2008)
12. Aha, D., Kibler, D., Albert, M.: Instance-based learning algorithms. Mach. Learn. **6**(1), 37–66 (1991)
13. Thorn, K.S., Bogan, A.A.: ASEdb: a database of alanine mutations and their effects on the free energy of binding in protein interactions. Bioinformatics **17**(3), 284–285 (2001)

14. Fischer, T.B., Arunachalam, K.V., Bailey, D., Mangual, V., Bakhru, S., Russo, R., Huang, D., Paczkowski, M., Lalchandani, V., Ramachandra, C.: The binding interface database (BID): a compilation of amino acid hot spots in protein interfaces. Bioinformatics **19**(11), 1453–1454 (2003)
15. Chen, P., Li, J., Wong, L., Kuwahara, H., Huang, J.Z., Gao, X.: Accurate prediction of hot spot residues through physicochemical characteristics of amino acid sequences. Proteins **81**(8), 1351–1362 (2013)
16. Chou, K.C.: Prediction of protein cellular attributes using pseudo-amino acid composition. Proteins **43**(3), 246–255 (2001)
17. Liu, B., Wang, S., Wang, X.: DNA binding protein identification by combining pseudo amino acid composition and profile-based protein representation. Sci. Rep. **5**, 15479 (2015)
18. Tang, H., Chen, W., Lin, H.: Identification of immunoglobulins using chou's pseudo amino acid composition with feature selection technique. Mol. BioSyst. **12**(4), 1269–1275 (2016)
19. Shen, H.B., Chou, K.C.: PseAAC: a flexible web server for generating various kinds of protein pseudo amino acid composition. Anal. Biochem. **373**(2), 386–388 (2008)
20. Martins, J.M., Ramos, R.M., Pimenta, A.C., Moreira, I.S.: Solvent-accessible surface area: how well can be applied to hot-spot detection? Proteins **82**(3), 479–490 (2014)
21. Chen, R., Chen, W., Yang, S., Wu, D., Wang, Y., Tian, Y., Shi, Y.: Rigorous assessment and integration of the sequence and structure based features to predict hot spots. BMC Bioinformatics **12**, 311 (2011)
22. Petersen, B., Petersen, T.N., Andersen, P., Nielsen, M., Lundegaard, C.: A generic method for assignment of reliability scores applied to solvent accessibility predictions. BMC Struct. Biol. **9**, 51 (2009)
23. Darnell, S.J., LeGault, L., Mitchell, J.C.: KFC server: interactive forecasting of protein interaction hot spots. Nucleic Acids Res. (Web Server Issue). **36**, W265–269 (2008)
24. Ofran, Y., Rost, B.: ISIS: interaction sites identified from sequence. Bioinformatics **23**(2), E13–E16 (2007)
25. Tuncbag, N., Gursoy, A., Keskin, O.: Identification of computational hot spots in protein interfaces: combining solvent accessibility and inter-residue potentials improves the accuracy. Bioinformatics **25**(12), 1513–1520 (2009)

Identification of Hot Regions in Protein-Protein Interactions Based on Detecting Local Community Structure

Xiaoli Lin[1,2] and Xiaolong Zhang[1(✉)]

[1] Hubei Key Laboratory of Intelligent Information Processing
and Real-time Industrial System, School of Computer Science and Technology,
Wuhan University of Science and Technology, Wuhan 430065, China
aneya@163.com, xiaolong.zhang@wust.edu.cn
[2] Information and Engineering Department of City College,
Wuhan University of Science and Technology, Wuhan 430083, China

Abstract. Hot regions can help proteins to exert their biological function and contribute to understand the molecular mechanism, which is the foundation of drug designs. In this paper, combining protein biological characteristics, a new method is proposed to predict protein hot regions. Firstly, we used support vector machine to predict the hot spots. Then, the local community structure detecting algorithm based on the identification of boundary nodes was proposed to predict the hot regions in protein-protein interactions. The experimental results demonstrate that the proposed method improves significantly the predictive accuracy and performance of protein hot regions.

Keywords: Hot regions · PPI · Local community structure · Classification

1 Introduction

Hot regions are the functionally important regions and play an important role in maintaining stability of protein-protein interaction. Prediction hot regions is one of the most challenging problems in computational molecular biology [1–3]. The structural information of an increasing number of protein sequences is stored, but a lot of information is useless, so it is necessary to predict protein hot regions by protein sequences [4]. The principles that govern the general properties of their interacting interfaces remain uncovered [5], resulting in the extremely difficult problem of hot regions identification directly from protein sequences.

Research experiment results demonstrate that binding energy of proteins is not uniformly distributed over their interaction surfaces. Only a small subset of contact residues contributes significantly to the binding free energy [5]. Beyond that, the researchers propose the hot regions can be constructed for comparison using hot spots with the experimental data from the alanine mutation energy database [6]. Then we will make some hot region predictions using the hot spots and non-hot spots.

The development of effective predictive models and methods is very necessary and helpful. Many research methods have been proposed for prediction of hot spots [7, 8]

© Springer International Publishing Switzerland 2016
D.-S. Huang et al. (Eds.): ICIC 2016, Part I, LNCS 9771, pp. 432–438, 2016.
DOI: 10.1007/978-3-319-42291-6_43

and hot regions [4, 5]. In the recent years, several computational approaches have been developed to identify hot spot residues at protein-protein interfaces [9, 10]. Hsu et al. [4] presented an approach to predict hot regions in protein-protein interactions. Keskin et al. [5] given the standard definition of hot regions and developed an algorithm to cluster hot spots into hot regions after studying the organization and contribution of structurally conserved hot spots residues. Tuncbag et al. [11], Cukuroglu et al. [12] and Carles et al. [13] did a lot of researches to identify and analyze protein-protein binding regions conformation. Nan and Zhang [14] used complex network and community method to predict hot regions. Hu et al. [15] predicted of hot regions in protein-protein interaction by combining density-based incremental clustering with feature-based classification. However, the prediction of hot regions is a very important but difficult problem. It is need to further improve the prediction accuracy.

In this paper, we propose the local community structure detecting algorithm to predict the hot regions at protein-protein interfaces based on the basic characteristics that contribute to hot spot interactions.

This paper consists of the follows sections. Section 2 describes the method of identification hot regions. Section 3 gives the experiment results and discusses the effectiveness of the method. Section 4 describes conclusion and future directions.

2   Method

2.1   Definition of Hot Regions

Reichman et al. [16] defined hot regions as the modules of hotspots which have at least two contacting hotspot neighbors in the interface. Also, Ahmad et al. [17] labeled hot regions as groups of at least three conserved residues. In this paper, we define hot region as clusters of residues with at least three residues from Keskin et al. [5]. Here two residues are defined as contacting if the distance between their atoms is smaller than 6.5 Å and each hot spot residue is assumed to be a perfect sphere with a specific volume, and the -atoms of the hot spot residues are the centers of these spheres [5].

2.2   Feature Selection

Feature selection process preserves the most original characteristics of protein residues, and improves the accuracy of the final results of the experiment. Feature selection technique is advantageous to avoid over fitting. It can also help to improve the performance of the model and to provide more efficient models.

Many physical and chemical characteristics of protein complexes are redundant and irrelevant, and these features must be removed in order to improve the accuracy and performance of the classifier. In our work, to obtain the best subset of features for predicting hot spots, a feature selection procedure is used to select optimal feature subset from the feature space. All features are evaluated by the minimum Redundancy Maximum Relevance (mRMR) algorithm [18] and then ranked according to their mRMR scores. Details of the results can be found in Table 1.

Table 1. The sorting results of top five features

Order	Name	Feature description	Score
1	RctASA	Relative change in total ASA upon complexation	0.023
2	RcsASA	Relative change in side-chain ASA upon complexation	0.021
3	BsASA	Bound side-chain ASA	0.018
4	UsASA	Unbound side-chain ASA	0.016
5	RctmPI	Relative change in total mean PI upon complexation	0.013

Then classification is performed based on the top features. Finally performance properties are analyzed, which can improved classification accuracy.

2.3   Classification

The hot regions is just composed of hot spots, thus the non-hot spots should be removed. In our work, a feature-based classification method is used to identify the hot spots and non-hot spots.

In this paper, a support vector machine (SVM) classifier is used, which is more and more widely used in the field of computational biology. Support vector machines, developed by Vapnik, are a set of related supervised learning methods used for classification and regression. It is a well-known classifier used to validate the application of the successful classification interface with hot region features [19].

Here SVM has been used to learn from a training set to classify residues as non-hot spots ($\triangle\triangle$G < 2.0 kcal/mol) and hot spots ($\triangle\triangle$G ≥ 2.0 kcal/mol).

2.4   Detecting Local Community Structure Based on the Identification of Boundary Nodes

It is generally known, the hot spots are closely combined in an area, rather than the uniform distribution of the protein interfaces homogeneously. Therefore, the hot spot residues can be clustered to a local regions (Here, we can call it community) using clustering methods. In this paper, a local community detecting algorithm is proposed to predict hot regions based on boundary nodes identification. Then, the similarity can be defined as

$$Similarity(i,j) = \frac{\sum w(i,t) * w(j,t)}{\sqrt{\sum w(i,t)^2}\sqrt{\sum w(j,t)^2}} \tag{1}$$

Where $w(i,t)$ is the weight of edge $E(i,t)$, and $w(j,t)$ is the weight of edge $E(j,t)$. The similarity of the two residues connection is larger, the more likely it belongs to the same hot region (Table 2).

Table 2. Local community detecting based on the identification of boundary nodes

---

Input: G (V, E), given residues
Output: Hot regions C of the residues

Begin
Initialization:
    N (C): adjacency sub graph;
    B (C): boundary node set;
For each do
    If Similarity(i,C)> Similarity(i,j) then
        Merge node i into the hot region C, ;
    else
        Set i to be a boundary node, ;
    End if
End for

End

2.5   Data Sets

In this paper, we use the same data set as in [11, 14, 15]. The data set is composed of protein complex, and the alanine mutation data of these complexes can be obtained. The alanine mutation data can be collected from Alanine Scanning Energetics Database, which is a searchable database of single alanine mutations in protein–protein, protein–nucleic acid, and protein–small molecule interactions [6]. In the case of a known structure, it includes the surface areas about mutated side chain and the PDB items.

This paper selects the alanine mutational data and gets rid of interface residues with binding energies between 0.4 kcal/mol and 2 kcal/mol, which results in the 255 residues. In the same way, the 65 residues with binding energies higher than 2.0 kal/mol are considered as hot spots and other 90 residues are considered as non-hot spots.

2.6   Measures of Prediction Performance

To evaluate the performances of hot regions prediction, three criterions are used in the hot regions prediction. The is the accuracy in the hot region prediction

$$R\_Precision = \frac{TP\_R}{TP\_R + FP\_R} \tag{2}$$

The true positive of the predicted hot region (*TP_R*) is the number of hot regions in the prediction results and also in real hot regions, and the false positive of the predicted hot region () is the number of the hot regions of in the prediction results but not in real hot regions. The is the coverage of hot regions prediction in the real hot regions

$$R\_Recall = \frac{TP\_R}{TP\_R + FP\_R} \tag{3}$$

The false negative () is the number of hot regions that are not in predicted results but in the real hot regions. The represents the balance performance between and

$$R\_F1 = \frac{2 * R\_Precision * R\_Recall}{R\_Precision + R\_Recall} \tag{4}$$

3   Experiment and Evaluation

To evaluate our method, we use the same data set as in [11, 14, 15], and compared the proposed method with the previous methods including those of Tuncbag et al. [11], Nan and Zhang [14] and Hu et al. [15].

Table 3 lists the accuracy, the coverage and the balance performance between accuracy and coverage about the hot regions prediction. It can be seen from the Table 3 that our method can predict hot region correctly with 0.786 (R_Precision) and 0.832 (R_Recall), that is to say, our method can correctly predict 78.6 % of the true hot regions, and 83.2 % of the predicted hot regions are true hot region. In addition, Tuncbag's results have very high accuracy (R_Precision = 1), but the coverage is too low (R_Recall = 0.200). Our R_F1 (0.809) is much better than those of Tuncbag (0.333) and Nan (0.500), and is slightly better than that of DICFC (0.737). From Table 3, it indicates that our method has the good performance for predicting hot regions in PPI.

Table 3. Comparison results with different methods to predict hot region

Method	Hot region		
	R-Precision	R_Recall	R_F1
Tuncbag	1	0.200	0.333
Nan	0.667	0.400	0.500
DICFC	0.778	0.700	0.737
Our method	0.786	**0.832**	**0.809**

In order to further evaluate the performance of our method, we provided three prediction examples of our research results, as shown in Table 4.

Table 4. Local community detecting based on the identification of boundary nodes

PDB ID	The initial hot regions		Hot regions based on local community detecting		Hot spot residues unpredicted
	Cluster no.	Hot spot residues	Cluster no.	Hot spot residues	
1DQJ	1	C20,C96, A32,A31	1	C20,C96, C97, A31,A32, A50, B33, B32	B50,C32
1EMV	1	B86,A52, A48,A49	1	B75,B86,A52,A48,A49, A53,A39	–
3HFM	1	H33,H50, H53	1	H33,H50,H53, Y96,Y97, H98	Y20
	2	–	2	L31,L32,L50	–

4 Conclusion

In this paper, the feature selection is performed using the mRMR method to identify the best subset of features for prediction hot spots. Based on the selected features, a prediction model for hot spots is created using support vector machine. Then, the local community structure detecting algorithm based on the identification of boundary nodes is used to locate hot regions with the hot spots predicted. Our experimental results show that our method has better prediction accuracy compared to previous methods. One of the future work is to consider the combination of different energy contributions to each other, which is still very much needed. In addition, we will use the better methods to enhance the prediction performance of the hot regions.

Acknowledgment. The authors thank the members of Machine Learning and Artificial Intelligence Laboratory, School of Computer Science and Technology, Wuhan University of Science and Technology, for their helpful discussion within seminars. This work was supported in part by National Natural Science Foundation of China (No. 61502356, 61273225, 61273303).

References

1. Hsu, C.M., Chen, C.Y., Liu, B.J.: MAGIIC-PRO: detecting functional signatures by efficient discovery of long patterns in protein sequences. Nucleic Acids Res. **34**, W356–W361 (2006)
2. Casari, G., Sander, C., Valencia, A.: A method to predict functional residues in proteins. Nat. Struct. Biol. **2**, 171–178 (1995)
3. Armon, A., Graur, D., Ben-Tal, N.: ConSurf: an algorithmic tool for the identification of functional regions in proteins by surface mapping of phylogenetic information. J. Mol. Biol. **307**, 447–463 (2001)
4. Hsu, C.M., Chen, C.Y., Liu, B.J., Huang, C.C.: Identification of hot regions in protein-protein interactions by sequential pattern mining. BMC Bioinform. **8**(Suppl. 5), S8 (2007)
5. Keskin, O., Ma, B.Y., Mol, R.J.: Hot regions in protein-protein interactions: the organization and contribution of structurally conserved hot spot residues. J. Mol. Biol. **345**, 1281–1294 (2005)

6. Thorn, K.S., Bogan, A.A.: ASEdb: a data base of alanine mutations and their effects on the free energy of binding in protein interactions. Bioinformatics **17**, 284–285 (2001)
7. Tuncbag, N., Gursoy, A., Keskin, O.: Identification of computational hot spots in protein interfaces: combining solvent accessibility and inter-residue potentials improves the accuracy. Bioinformatics **25**, 1513–1520 (2009)
8. Ezkurdia, I., Bartoli, L., Fariselli, P., Casadio, R., Valencia, A., et al.: Progress and challenges in predicting protein-protein interaction sites. Brief Bioinform. **10**, 233–246 (2009)
9. Lise, S., Buchan, D., Pontil, M., Jones, D.T.: Predictions of hot spot residues at protein-protein interfaces using support vector machines. PLoS ONE **6**, e16774 (2011)
10. Lise, S., Archambeau, C., Pontil, M., Jones, D.T.: Prediction of hot spot residues at protein-protein interfaces by combining machine learning and energy-based methods. BMC Bioinform. **10**, 365 (2009)
11. Tuncbag, N., Keskin, O., Gursoy, A.: HotPoint: hot spot prediction server for protein interfaces. Nucleic Acids Res. **38**, 402–406 (2010)
12. Engin, C., Gursoy, A., Keskin, O.: Analysis of hot region organization in hub proteins. Ann. Biomed. Eng. **38**, 2068–2078 (2010)
13. Carles, P., Fabian, G., Juan, F.: Prediction of protein-binding areas by small world residue networks and application to docking. BMC Bioinform. **12**, 378–388 (2011)
14. Nan, D.F., Zhang, X.L.: Prediction of hot regions in protein-protein interactions based on complex network and community detection. In: Bioinformatics and Biomedicine. pp. 17–23 (2013)
15. Hu, J., Zhang, X.L., Liu, X.M., Tang, J.S.: Prediction of hot regions in protein-protein interaction by combining density-based incremental clustering with feature-based classification. Comput. Biol. Med. **61**, 127–137 (2015)
16. Reichmann, D., Rahat, O., Albeck, S., Meged, R., Dym, O., Schreiber, G.: The modular architecture of protein-protein binding interfaces. Proc. Natl. Acad. Sci. **102**(1), 57–62 (2005)
17. Ahmad, S., Keskin, O., Sarai, A., Nussinov, R.: Protein–DNA interactions: structural, thermodynamic and clustering patterns of conserved residues in DNA-binding proteins. Nucleic Acids Res. **36**, 5922–5932 (2008)
18. Li, B.Q., Feng, K.Y., Li, C., Huang, T.: Prediction of protein-protein interaction sites by random forest algorithm with mRMR and IFS. PLoS ONE **7**(8), e43927 (2012)
19. Kyu-il, C., Dongsup, K., Doheon, L.: A feature-based approach to modeling protein-protein interaction hot spots. Nucleic Acids Res. **37**(8), 2672–2678 (2009)

Power Spectrum-Based Genomic Feature Extraction from High-Throughput ChIP-seq Sequences

Binhua Tang[1,2(✉)], Yufan Zhou[3], and Victor X. Jin[3]

[1] College of the Internet of Things,
Hohai University, Jiangsu 213022, China
bh.tang@outlook.com
[2] School of Public Health,
Shanghai Jiao Tong University, Shanghai 200025, China
[3] Department of Molecular Medicine and Biostatistics,
University of Texas Health Science Center,
San Antonio, TX 78229, USA

Abstract. Due to its enhanced accuracy and high-throughput capability in capturing genetic activities, recently Next Generation Sequencing technology is being applied prevalently in biomedical study for tackling diverse topics. Within the work, we propose a computational method for answering such questions as deciding optimal argument pairs (peak number, p-value threshold, selected bin size and false discovery rate) from estrogen receptor α ChIP-seq data, and detecting corresponding transcription factor binding sites. We employ a signal processing-based approach to extract inherent genomic features from the identified transcription factor binding sites, which illuminates novel evidence for further analysis and experimental validation. Thus eventually we attempt to exploit the potentiality of ChIP-seq for deep comprehension of inherent biological meanings from the high-throughput genomic sequences.

Keywords: ChIP-seq · Genomic feature · Optimal argument pair · Transcription factor binding site · Comprehensive analysis

1 Introduction

Next generation sequencing (NGS) combined with ChIP technology provides a genome-wide study perspective for current biomedical research and clinical diagnosis applications [1–3].

Data quality and inherent characteristics of those NGS sequencing profiles are directly related to the reliability and authenticity of analysis results. For example, ChIP-seq data characterize alteration evidence for transcription factor (TF) binding activities in response to chemical or environmental stimuli, but if the ChIP-seq data quality is below normal standard, any follow-up analysis may lead to inaccurate TF binding results, for example, inevitable loss of biological meaningful sites [4, 5].

And secondly, the mostly investigated items in ChIP-seq peak-calling procedures are peak number, false discovery rate (FDR) and corresponding bin size selected in

© Springer International Publishing Switzerland 2016
D.-S. Huang et al. (Eds.): ICIC 2016, Part I, LNCS 9771, pp. 439–447, 2016.
DOI: 10.1007/978-3-319-42291-6_44

each analysis. Without exception such arguments form impenetrable barriers for biologists and bioinformaticians to choose suitable conditions for analyzing experimental results. And to our knowledge, few literatures focus on such topics, thus herein we propose a flexible data feature detection algorithm for solving such an argument-optimization problem in peak-calling.

Within analysis of the estrogen receptor α (ERα) ChIP-seq data, we attempt to detect underlying genetic transcription factors from corresponding genome sequence, and also employ a frequency-based signal processing method for extracting inherent features from the ERα ChIP-seq data.

The following paragraphs are organized as: Sect. 2 raises the problem to solve, and briefly introduces related biological background and ChIP-seq data generated for follow-up analysis; Sect. 3 proposes computational methods and illustrates analysis results; together discussion and conclusion on the work are illustrated in Sect. 4.

2   Data and Methods

2.1   Biological Background and Data Source Description

In breast cancer cell content, a specific estrogen receptor α (ERα) is recognized as mediating genetic regulation through diverse nuclear signaling pathways and protein kinase cascades. Within those biological pathways, ER binds to corresponding estrogen response elements (ERE) at target genes' regulatory areas, and combines other components to control downstream transcriptional processes.

Thus elucidating inherent regulatory mechanisms of those specific ERs facilitates comprehensive understanding of ER-specific regulation in most breast cancer pathways and networks under investigation.

For our experiments of ChIP-seq data generation, the tissue samples were collected from patients diagnosed with breast cancer. Then the corresponding ChIP-seq data were generated with Illumina NGS platform. We collected the ChIP-seq data before-and-after estrogen (E2) treatment with replicate measurements, respectively.

The NGS sequencing platform provides short read length sequences of ∼36 base pairs capable for capturing ChIP-derived fragments. Then sequences are mapped back to a corresponding reference genome, where those frequently sequenced fragments will form peaks at specific regions.

Through global computational analysis of those identified peaks, it facilitates our further understanding of underlying genetic regulatory activities.

2.2   Analysis Methods for ChIP-Seq Data

Normally for examining data quality, one needs to analyze peak numbers under specific argument constraints. And we attempt to acquire optimal peak numbers by constraining specific arguments, which might be formalized as a class of optimal track analysis, illustrated as follows,

$$\arg \max_i P_i, i \in N$$

$$s.t. : f_i \leq \chi, \; b_i = \beta, \; p_i \leq \delta \tag{1}$$

where P_i denotes a set of optimal peak numbers under corresponding argument constraints, f_i stands for argument FDR, b_i for bin size and p_i for p-value threshold, χ, β and δ represent presupposed argument values, respectively.

Herein we define a track rate function (*TR*) to characterize underlying data features from diverse argument pair sets (peak number and FDR), depicted as follows,

$$TR_i = \frac{ATS_i}{STS_i} = \frac{\sum_{j=1}^{M} S(j)}{\sum_{k=1}^{N} S(k)}, \quad i \in N \tag{2}$$

where *ATS* represents actual track scoring function, *STS* represents the shortest track scoring function, and $S(\cdot)$ denotes corresponding score value for each track step, respectively.

Through the above optimal track estimation, we extract underlying genomic features in those ChIP-seq data, as shown in the below section. Moreover, we detect transcription factors binding sites; then through a frequency power spectrum method we attempt to acquire related genomic characteristics.

For a finite random variable sequence, its power spectrum is normally estimated from its autocorrelation sequence by use of discrete-time Fourier transform (DTFT), denoted as [6–8],

$$P(\omega) = \frac{1}{2\pi} \sum_{n=-\infty}^{\infty} C_{xx}(n) e^{-jn\omega} \tag{3}$$

where C_{xx} denotes autocorrelation sequence of a discrete signal x_n, defined as,

$$C_{xx}(i,j) = \frac{E[(X_i - \mu_i)(X_j - \mu_j)]}{\sigma_i \sigma_j} \tag{4}$$

where μ and σ stand for mean and variance, respectively. In our study, for consideration of the investigated data characteristics, we use 128 sampling points to calculate discrete Fourier transform, with the related sampling frequency 1 kHz.

3 Analysis and Results

3.1 Optimal Track Analysis with Arguments' Constraints

For the ChIP-seq dataset, we detect several optimal argument pairs for peak number and FDR with the corresponding argument constraints of bin size and p-value threshold [9–13]. Normally, we attempt to acquire the highest peak numbers under specific argument constraints following an optimal track set. The algorithmic pseudo-code for detecting an optimal track set is illustrated in Table 1.

Table 1. Algorithmic pseudo-code for detecting an optimal track set from ChIP-seq data.

Input:
   π_x: maximum FDR value;
   π_n: minimum FDR value;
   δ: incremental step.
Output:
   optimal track set: P.
begin:
index$= \pi_n$;
while index$\leq \pi_x$ **do**
   1. search a maximum peak number s.t. index
      and other arguments;
   2. index=index$+\delta$;
   3. save index's information to P.
end

With the algorithmic analysis on the real ERα ChIP-seq data, thus we get the optimal tracking result in Fig. 1, where black solid dot denotes a starting track point, arrows for intermediate points, and black solid square for the track end.

Totally there are seven intermediate points, one start and one end, respectively, with incremental step 0.0044. From Fig. 1, the maximum peak number is 15,597 with its FDR at 0.21498; and contrarily, the minimum peak number is 508 with its corresponding FDR at 0.

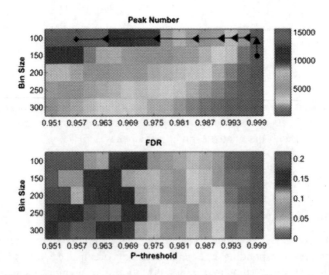

Fig. 1. The distributions of peak number and corresponding FDR with respect to two argument constraints (bin size and p-value threshold). The upper plot illustrates peak number distribution; the lower for FDR's distribution. (Color figure online)

Figure 1 also indicates that under condition of current arguments, the most suitable peak number exists when bin size is selected as 100, no matter which FDR constraint is chosen in the follow-up peak-calling procedures.

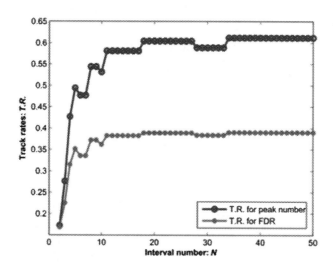

Fig. 2. The track rate (T.R.) distribution plot for peak number (dark blue) and FDR (green) with respect to interval number N. (Color figure online)

According to the predefined track rate in Eq. 2, the ERα ChIP-seq data's track rate values are 0.6117 for peak number and 0.39 for FDR, with its interval number $N = (\pi_x - \pi_n)/\delta = 50$. Figure 2 illustrates the track rate distribution with respect to interval number. As depicted, when interval number exceeds 40, both track rates will eventually stabilize to equilibria, respectively. The equilibria denotes the estimated optimal status for the peak number and FDR.

Furthermore, for further analyzing NGS sequences, we attempt to detect underlying sequence-based transcription factors, which actually facilitates understanding diverse biological regulatory mechanisms.

3.2   Frequency-Based Analysis of NGS Sequences

The basic idea for identifying transcription factor binding sites from genome-wide NGS sequences by use of the position weight matrix concept has been presented in the references [14, 15].

Using the position weight matrix, we have identified 487 transcription factor binding sites from the MCF-7 ChIP-seq data. Although all of those candidates need further experimental validation, from the computational aspect we can detect their corresponding frequencies of occurrence and calculate average scores to identify the inherent genomic data features. Figure 3 illustrates the percentage distribution of the occurrence.

444   B. Tang et al.

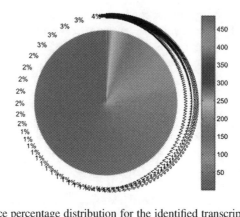

Fig. 3. The occurrence percentage distribution for the identified transcription factor candidates. Totally 487 transcription factor candidates are listed in occurrence percentage within pie chart. The overall distribution range is within 1% and 4%. (Color figure online)

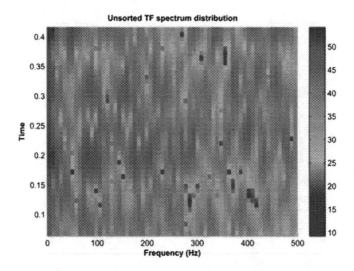

Fig. 4. The unsorted (randomized) transcription factor spectrum distribution based on their occurrence statistical information. The sampling frequency is 1 kHz in Fourier analysis. We may easily find that in randomized power spectrum distribution there exists regular high frequency spectrum regions at 100, 200, 300 and 400 Hz.

Figures 4 and 5 illustrate the corresponding spectrum distributions for the identified TF candidates' randomized and sorted occurrence information, respectively.

On Fig. 4, we may easily find that in randomized TF power spectrum distribution there exists regular high frequency spectrum regions at 100, 200, 300 and 400 Hz; and for the sorted TF spectrum on Fig. 5, we find the low frequency areas have the higher spectrum intensities around 20 to 40 Hz, and wave-like fluctuations exist at 120, 240 and 360 Hz, respectively. These features are also indicative of corresponding occurrence of transcription factors candidates in binding activities.

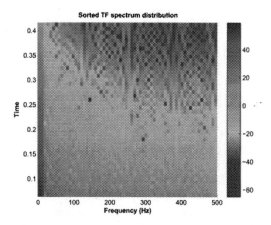

Fig. 5. The sorted TF spectrum distribution for their occurrence statistical information. The sampling frequency is 1 kHz. The upper panel exhibits periodic signs of wave-like fluctuations. (Color figure online)

Together we analyze the relationship between those TF candidates' calculated average scores and corresponding frequency of their occurrence, as shown in Fig. 6. We find that the calculated average scores for those TF candidates mostly remain around 0.965. And the most part of TF's frequency of occurrence is less than 0.01. Thus it provides clues for further study and validation for those candidates by use of other statistical and computational methods, for example, integrative analysis with other profiling, PCR information, histone modification, and other epigenetic level information.

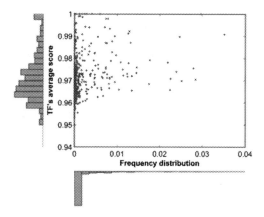

Fig. 6. The statistical distribution for the identified transcription factor candidates' average score information. The x-axis denotes frequency range distribution of their occurrence, with maximum 0.04; and the y-axis illustrates the TF candidates' average score distribution, with range between 0 and 1.

4 Discussion

Within the work, we sought to analyze ERα transcription factor binding and its relevant regulatory characteristics through extracting underlying genomic data features.

We utilized corresponding high-throughput ChIP-seq profiling platform to generate the NGS sequences from breast cancer MCF-7 cell lines. With globally determining an optimal peak number set under relative argument constraints, we discovered the inherent statistical features, thus we may quantitatively compare and determine their qualities within diverse data sources, thus which will guarantee the statistical confidence level of any further analysis for those genomic sequences, especially in the downstream interrogation of biologically meaningful results.

Furthermore, we extracted multiple features by use of a spectrum-based signal processing method. We identified statistical characteristics from transcription factor candidates by use of their frequency distribution of genome-wide occurrence, and we found there existed regular regions of high frequency spectrum, which is indicative of the existence of high occurrence of specific transcription factors candidates, both in randomized and sorted study cases.

Lastly, we analyzed calculated average score distribution and corresponding frequency of occurrence for those transcription factor candidates identified above, thus we may further combine those findings with other information in comprehensive analysis of the underlying transcriptional binding characteristics.

Besides its flexibility and enhanced accuracy in high-throughput genomic sequencing, NGS technology demands comprehensive analysis of multiple underlying features from those generated sequences to exploit its potentialities.

Our method proposes an applicable perspective for facilitating global comprehension of diverse biological mechanisms under investigation.

5 Availability

The MCF-7 ChIP-seq data interrogated in this work was deposited and is currently available at NCBI GEO with accession number: GSE35109.

Acknowledgments. This work was supported by the Fundamental Research Funds for China Central Universities [grant number 2016B08914 to BHT] and Changzhou Science & Technology Program [grant number CE20155050 to BHT]. This work made use of the resources supported by the NSFC-Guangdong Mutual Funds for Super Computing Program (China), and the Open Cloud Consortium (OCC)-sponsored project resource, which supported in part by grants from Gordon and Betty Moore Foundation and the National Science Foundation (USA) and major contributions from OCC members.

References

1. Mardis, E.R.: ChIP-seq: welcome to the new frontier. Nat. Methods **4**(8), 613–614 (2007)
2. Martinez, G.J., Rao, A.: Cooperative transcription factor complexes in control. Science **338** (6109), 891–892 (2012)

3. Kilpinen, H., Barrett, J.C.: How next-generation sequencing is transforming complex disease genetics. Trends Genetics (TIG) **29**(1), 23–30 (2013)
4. Chikina, D.M., Troyanskaya, O.G.: An effective statistical evaluation of ChIP-seq dataset similarity. Bioinformatics **28**(5), 607–613 (2012)
5. Furey, T.S.: ChIP-seq and beyond: new and improved methodologies to detect and characterize protein-DNA interactions. Nat. Rev. Genet. **13**, 840–852 (2012)
6. Oppenheim, A.V., Schafer, R.W.: Discrete-Time Signal Processing. Prentice Hall Signal Processing Series, 3rd edn. Prentice Hall, Upper Saddle River (2010). Ed. by A.V. Oppenheim
7. Tang, B., Hsu, H.K., Hsu, P.Y.: Hierarchical modularity in ERA transcriptional network is associated with distinct functions and implicates clinical outcomes. NPG Sci. Rep. **2**, 875 (2012)
8. Wang, S.L., Zhu, Y.H., Jia, W.: Robust classification method of tumor subtype by using correlation filters. IEEE/ACM Trans. Comput. Biol. Bioinform. **9**(2), 580–591 (2012)
9. Zhang, Y., Liu, T., Meyer, C.A., et al.: Model-based Analysis of ChIP-seq (MACS). Genome Biol. **9**(9), R137 (2008)
10. Lan, X., Bonneville, R., Apostolos, J., et al.: W-ChIPeaks: a comprehensive web application tool for processing chip-chip and ChIP-seq data. Bioinformatics **27**(3), 428–430 (2011)
11. Spyrou, C., Stark, R., Lynch, A.G., et al.: BayesPeak: Bayesian analysis of ChIP-seq data. BMC Bioinform. **10**(1), 299 (2009)
12. Fejes, A.P., Robertson, G., Bilenky, M., et al.: FindPeaks 3.1: a tool for identifying areas of enrichment from massively parallel short-read sequencing technology. Bioinformatics **24** (15), 1729–1730 (2008)
13. Zhu, L., Guo, W.L., Deng, S.P., et al.: ChIP-PIT: enhancing the analysis of ChIP-seq data using convex-relaxed pair-wise interaction tensor decomposition. IEEE/ACM Trans. Comput. Biol. Bioinform. **13**(1), 55–63 (2016)
14. Cheng, A.S.L., Jin, V.X., Fanet, M.: Combinatorial analysis of transcription factor partners reveals recruitment of C-MYC to estrogen Receptor-A responsive promoters. Mol. Cell **21** (3), 393–404 (2006)
15. Ou, Y.Y., Chen, S.-A., Gromiha, M.M.: Classification of transporters using efficient radial basis function networks with position-specific scoring matrices and biochemical properties. Proteins: Struct. Funct. Bioinform. **78**(7), 1789–1797 (2010)

Effective Protein Structure Prediction with the Improved LAPSO Algorithm in the *AB* Off-Lattice Model

Xiaoli Lin[1,2(✉)], Fengli Zhou[1], and Huayong Yang[1]

[1] Information and Engineering Department of City College,
Wuhan University of Science and Technology, Wuhan 430083, China
{thinkview,HuayongYang}@163.com
[2] Hubei Key Laboratory of Intelligent Information
Processing and Real-Time Industrial System,
School of Computer Science and Technology,
Wuhan University of Science and Technology, Wuhan 430065, China
aneya@163.com

Abstract. Protein structure prediction is defined as predicting the tertiary structure from the primary structure of the protein sequence. Because the real protein structure is very complex, it is necessary to adopt the simplified structure model for studying protein 3D space structure. In this paper, we introduce a kind of 3D *AB* off-lattice model for protein structure prediction, and the amino acids are labeled as two hydrophobic amino acids and hydrophilic amino acids. When the protein model is simplified, the optimization algorithm is also needed to use for searching the lowest energy conformation of the protein sequence based on the hypothesis theory. In this paper, a hybrid algorithm which combines PSO algorithm based on local adjust strategy (LAPSO) and genetic algorithm, was proposed to search the space structure of the protein with *AB* off-lattice model. Experimental results show that the minimal energy values obtained by the improved LAPSO are lower than those obtained by previous methods. The performance of our improved algorithm is better, and it can effectively solve the search problem of the protein space folding structure.

Keywords: Protein structure prediction · Local adjust · PSO · *AB* off-lattice model

1 Introduction

The biological functions of a protein are determined by its folding structure in space, and its spatial structure is completely determined by its primary structure [1]. Because the protein spatial folding structure is very complex and enormous, it is quite difficult to search for the lowest energy conformation of the protein from the amino acid sequence [2].

Over the years, many highly simplified and optimized models have been proposed. The *HP* model (Hydrophobic-polar model) [3] has become the most commonly tools and has been widely used in the study protein structure. The *HP*-lattice model is a

© Springer International Publishing Switzerland 2016
D.-S. Huang et al. (Eds.): ICIC 2016, Part I, LNCS 9771, pp. 448–454, 2016.
DOI: 10.1007/978-3-319-42291-6_45

simplified model which ignores local interactions, while local interactions might be very important for the local structure of the chains [4, 5]. Stillinger [6] studied a similar *AB* off-lattice protein model in two dimensions which uses only two types of residues, hydrophobic (*A*) and hydrophilic (*B*). Irbäck et al. [5] extended a two dimension (2D) to a three dimension (3D) in the *AB* off-lattice model, which takes account of the torsional energy implicitly. Many researchers have proposed other methods to studied the *AB* off-lattice models, such as the energy landscape paving minimizer (ELP) [7], the conformational space annealing (CSA) [8], the pruned-enriched-Rosenbluth method (PERM) [9] etc.

We also have done some researches about the *AB* off-lattice models, such as genetic-annealing algorithm (GAA) [10], the local adjust genetic algorithm (LAGA) [11], the local adjust tabu search algorithm (LATS) [12] f and the tabu search based on simulated annealing (SATS) [13].

This paper describes a protein structure prediction method that is based on the *AB* off-lattice model in three dimension. An improved hybrid algorithm (LAPSO), which combines the PSO algorithm based on local adjust strategy and genetic algorithm is applied to complete these tasks.

The rest of the paper is arranged as following. Section 2 briefly introduces the 3D *AB* off-lattice model. Section 3 describes the improved hybrid algorithm (LAPSO), which is used to search the conformation of the final state with a given protein sequence. Section 4 presents the experimental results and the evaluation of the algorithm. Section 5 concludes this paper and gives the possible future research directions.

2 The *AB* Off-Lattice Model

2.1 The Model in 3D

The 3D *AB* off-lattice model also consists of hydrophobic *A* monomers ($\sigma_i = +1$) and hydrophilic *B* monomers ($\sigma_i = -1$). The energy function is given by [5]

$$E = -k_1 \sum_{i=1}^{N-2} \hat{b}_i \cdot \hat{b}_{i+1} - k_2 \sum_{i=1}^{N-3} \hat{b}_i \cdot \hat{b}_{i+2} + \sum_{i=1}^{N-2} \sum_{j=i+2}^{N} E_{LJ}(r_{ij}; \sigma_i, \sigma_j) \quad (1)$$

The *N*-mer can be specified by the $N-1$ bond vectors $\hat{b}_i$ or by $N-2$ bond angles θ_i and $N-3$ torsional angles α_i.

The r_{ij} is just the usual Euclidean distance between sites *i* and *j*, and r_{ij} depends on the bond angles and torsional angles. These two angles are the degrees of freedom of the model. The species-dependent global interactions are given by the Lennard-Jones potential

$$E_{LJ}(r_{ij}; \sigma_i, \sigma_j) = 4C(\sigma_i, \sigma_j)\left(\frac{1}{r_{ij}^{12}} - \frac{1}{r_{ij}^{6}}\right) \quad (2)$$

The depth of the minimum of this potential $C(\sigma_i, \sigma_j)$ is chosen to favor the formation of a core of A residues and $\sigma_1, \ldots, \sigma_N$ is a binary string that specifies the primary sequence.

3   Methods Description

3.1   PSO Algorithm

Particle swarm optimization (PSO) algorithm is proposed by Kennedy and Eberhart in 1995, and it is a kind of swarm intelligence optimization algorithm [14, 15]. The algorithm has been very successful and widely used in complex combinatorial optimization problems [16]. The particle swarm optimization algorithm has many existing improved algorithms, such as the standard particle swarm optimization (SPSO) and the Gauss particle swarm optimization (GPSO). However, with the increase of the dimension, these methods are easy to fall into the local optimal solution.

3.2   Euclidean Interference Strategy

In the process of optimization, this paper sets up a Counter to count the unchanged times about the global best fitness. The position loci of each particle can be modified by

$$loc_i = loc_i + v_i \tag{3}$$

Where v_i is defined as

$$v_i = \begin{cases} wv_i + c_1 R_1 (pos_i - loc_i) + c_2 R_2 (Bestpos - loc_i) & Counter \leq k \\ wv_i + c_1 R_1 (pos_i - loc_i) + c_2 R_2 (Bestpos - loc_i) + \varepsilon_i & Counter > k \end{cases} \tag{4}$$

The Euclidean interference factor ε_i will decrease with the increase of distance d_i, where d_i is the distance from current particle i to the global best particle [17].

$$\varepsilon_i = 2v_{max}\left(\frac{1}{1 + \exp(-a/d_i)} - 0.5\right) \tag{5}$$

Besides, the several variables are defined as:

- loc_i is the current position.
- pos_i is the previous best position.
- $Bestpos$ is the global best position in whole swarm space.
- w is the weight, and decreases linearly from 0.9 to 0.4 [18].
- c_1 and c_2 are positive constants, usually setting c1 = c2 = 2.0.
- $R1$, $R2$ are random numbers in the range of (0,1).

3.3 Local Adjustment Strategy

In this paper, the local adjustment strategy can be used to improve the search ability and find the best solution in the off-lattice model.

Assuming that there is a good individual, it is a local optimal solution in the current search process. We use the adjustment strategy of our previous research [13] to guide the current individual to move toward the direction of the best global optimal solution. The local adjustment strategy greatly improves the speed and accuracy of the searching the minimum energy of protein sequence.

3.4 The Algorithm

The energy of each individual in population is calculated based on AB off-lattice model. Then population is rearranged from minimal energy to maximal energy. At the same time, the minimum energy value and the individual with minimal energy are stored. The steps of the LAPSO algorithm are described as Fig. 1.

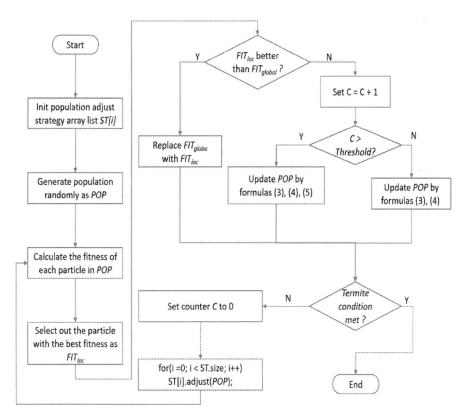

Fig. 1. The flow diagram of algorithm

4   Optimization Results and Discussion

This part uses the same Fibonacci sequences in Ref. [19] for comparison. Table 1 lists the Fibonacci sequences with $13 \leq n \leq 55$. The optimum lowest energies of these four Fibonacci sequences are given respectively. Table 2 shows that the results obtained with LAPSO are better than the results of the previous algorithms.

Table 1. Fibonacci sequences with $13 \leq n \leq 55$

Length	Sequence
13	ABBABBABABBAB
21	BABABBABABBABBABABBAB
34	ABBABBABABBABBABABBABABBABBABABBAB
55	BABABBABABBABBABABBAB ABBABBABABBABBABABBABABBABBABABBAB

In the Table 2, E_{ACMC} is the minimum energy obtained by the annealing contour Monte Carlo (ACMC) algorithm [20] while E_{ELP} and E_{CSA} are the minimum energy obtained by the energy landscape paving minimizer ELP [7] and the conformational space annealing (CSA) algorithm [8] respectively. E_{LAGA} is the minimum energy obtained by the local adjust genetic algorithm studied by us in [11], E_{SATS} is the Tabu search based on simulated annealing by us in [13].

It can be seen that the results obtained with LAPSO are smaller than those of the CSA for all the four sequences. For length 13, our result is equal to those of ACMC and SATS, and is smaller than other result. For length 21, our result is slightly smaller than those of ACMC and CSA, and equal to those of ELP, LAGA and SATS. For other cases, our results are smaller than the results obtained by other methods, which shows that LAPSO also has better performance for long sequence in 3D *AB* off-lattice model. Therefore, we can find that our method show much better performance with increasing sequence length.

Table 2. The minimum energies obtained by different algorithm for Fibonacci sequences with $13 \leq n \leq 55$ in 3D

Length	E_{ELP}	E_{ACMC}	E_{CSA}	E_{LAGA}	E_{SATS}	E_{LAPSO}
13	−26.498	**−26.507**	−26.471	−26.498	**−26.507**	**−26.507**
21	**−52.917**	−51.757	−52.787	**−52.917**	**−52.917**	**−52.917**
34	−92.746	−94.043	−97.732	−98.765	−99.876	**−100.483**
55	−172.696	−154.505	−173.980	−176.542	−178.986	**−179.132**

5   Conclusions

A hybrid algorithm that combines PSO algorithm based on local adjust strategy and genetic algorithms is developed to predict the protein structure with the AB off-lattice model in 3D space. The proposed algorithm can deal with the optimization problem and search for the minimum energy conformation of protein sequence. This paper proposes some new strategies to improve the prediction efficiency. Compared with the previous algorithms, this algorithm appears to be superior performance and has stronger capability of global optimization in the protein structure prediction. But, the model only considered two different interaction energy, so it cannot fully reflect the true nature of the protein sequence. We should use other models to study the interaction energy of more protein amino acids.

Acknowledgments. The authors thank the members of Machine Learning and Artificial Intelligence Laboratory, School of Computer Science and Technology, Wuhan University of Science and Technology, for their helpful discussion within seminars. This work was supported in part by National Natural Science Foundation of China (No. 61502356, 61273225, 61373109).

References

1. Anfinsen, C.B.: Principles that govern the folding of protein chains. Science **181**, 223–227 (1973)
2. Lopes, H.S.: Evolutionary algorithms for the protein folding problem: a review and current trends. In: Smolinski, T.G., Milanova, M.G., Hassanien, A.-E. (eds.) Computational Intelligence in Biomedicine and Bioinformatics. SCI, vol. 151, pp. 297–315. Springer, Heidelberg (2008)
3. Dill, K.A.: Theory for the folding and stability of globular proteins. Biochemistry **24**, 1501–1509 (1985)
4. Hart, W.E., Newman, A.: Protein structure prediction with lattice models. In: Aluru, S. (ed.) Handbook of Molecular Biology. Chapman & Hall/CRC Computer and Information Science Series, pp. 1–24. CRC Press, Boca Raton (2006)
5. Irbäck, A., Peterson, C., Potthast, F., Sommelius, O.: Local interactions and protein folding: a three-dimensional off-lattice approach. J. Chem. Phys. **107**, 273–282 (1997)
6. Stillinger, F.H., Head-Gordon, T., Hirshfel, C.L.: Toy model for protein folding. Phys. Rev. **E48**, 1469–1477 (1993)
7. Bachmann, M., Arkin, H., Janke, W.: Multicanonical study of coarse-grained off-lattice models for folding heteropolymers. Phys. Rev. **E71**, 031906 (2005)
8. Kim, S.-Y., Lee, S.B., Lee, J.: Structure optimization by conformational space annealing in an off-lattice protein model. Phys. Rev. **E72**, 011916 (2005)
9. Hsu, H.-P., Mehra, V., Grassberger, P.: Structure optimization in an off-lattice protein model. Phys. Rev. **E68**, 037703 (2003)
10. Zhang, X.L., Lin, X.L.: Effective protein folding prediction based on genetic-annealing algorithm in toy model. In: 2006 Workshop on Intelligent Computing Bioinformatics of CAS, pp. 21–26 (2006)
11. Zhang, X.L., Lin, X.L.: Effective 3D protein structure prediction with local adjustment genetic-annealing algorithm. Interdisc. Sci. Comput. Life Sci. **2**, 1–7 (2010)

12. Lin, X., Zhou, F.: 3D protein structure prediction with local adjust tabu search algorithm. In: Huang, D.-S., Gupta, P., Wang, L., Gromiha, M. (eds.) ICIC 2013. CCIS, vol. 375, pp. 106–111. Springer, Heidelberg (2013)
13. Lin, X.L., Zhang, X.L., Zhou, F.L.: Protein structure prediction with local adjust tabu search algorithm. BMC Bioinformatics **15**(Suppl. 15), S1 (2014)
14. Lecchini-Visintini, A., Lygeros, J., Maciejowski J.: Simulated annealing: rigorous finite-time guarantees for optimization on continuous domains. In: Advances in Neural Information Processing Systems 20, Proceedings of NIPS (2007)
15. Gatti, C.J., Hughes, R.E.: Optimization of muscle wrapping objects using simulated annealing. Ann. Biomed. Eng. **37**, 1342–1347 (2009)
16. Gao, S., Yang, J.Y.: Swarm Intelligence Algorithm and Applications, pp. 112–117. China Water Power Press, Beijing (2006)
17. Zhu, H.B., Pu, C.D., Lin, X.L.: Protein structure prediction with EPSO in toy model. In: 2009 Second International Conference on Intelligent Networks and Intelligent Systems, pp. 673–676 (2009)
18. Shi, Y., Eberhart, R.: Parameter selection in particle swarm optimization. In: Proceedings of the 1998 Annual Conference on Evolutionary Programming (1998)
19. Stillinger, F.H.: Collective aspects of protein folding illustrated by a toy model. Phys. Rev. **E52**, 2872–2877 (1995)
20. Liang, F.: Annealing contour Monte Carlo algorithm for structure optimization in an off-lattice protein model. J. Chem. Phys. **120**, 6756 (2004)

Prediction of Target Genes Based on Multiway Integration of High-Throughput Data

Wei-Li Guo[1], Kyungsook Han[2], and De-Shuang Huang[1(✉)]

[1] Institute of Machine Learning and Systems Biology, College of Electromics and Information Engineering, Tongji University, Shanghai 201804, China
guoweili_henu@126.com, dshuang@tongji.edu.cn
[2] Department of Computer Science and Engineering, Inha University, Incheon, South Korea
khan@inha.ac.kr

Abstract. In the past few years, ChIP-seq data have emerged as a powerful source to discover regulatory functional elements as well as disease related genetic reasons. However, owing to cost, or biological material availability, it is hard to do experiment for every transcription factor in every cell line, which restricts analysis requiring completed data to only those with existed experiments. The imputation of missing ChIP-seq data can help to solve the problem, while because of the massive scale of ChIP-seq data, traditional methods is unsuitable to treat such huge amount of data or it is time-consuming. In this paper, we proposed a tensor completion-based method for the imputation of ChIP-seq data by modeling the ChIP-seq dataset as a 3-way tensor pattern. The results show that the proposed method is better than state-of-the-art baseline methods with respect to imputation accuracy.

Keywords: ChIP-seq data · Transcription factor · Target gene · Latent factor · Trace norm

1 Introduction

Chromatin immunoprecipitation coupled with high-throughput sequencing (ChIP-seq) [1] enables the genome-wide identification of the binding sites of transcription factors (TFs) present in a given condition or cell types. With the ChIP-seq data which map the binding sites of transcription factors in some cell lines, we can construct transcriptional regulatory networks and identify specific signaling pathways for particular biological process and learn features of regulation that define cell specificity [2, 3].

Thanks to the ENCyclopedia Of DNA Elements (ENCODE) Projects [4, 5], genome-wide predictions of large scale of transcription factors have been generated. While owing to sample material availability, cost or time, there are only hundreds of ChIP-seq TF-cell line samples for about 200 transcription factors across about 100 cell lines. As a result, researches that explore the regulation of particular biological process have to focus on only several TFs with ChIP-seq data [2, 6, 7].

Recently, several researchers have proposed to address this problem. Wu et al. [8] enriched the analysis by integrating ChIP-seq data with large scale of publicly available

© Springer International Publishing Switzerland 2016
D.-S. Huang et al. (Eds.): ICIC 2016, Part I, LNCS 9771, pp. 455–460, 2016.
DOI: 10.1007/978-3-319-42291-6_46

gene expression data to discover potential TF regulatory activities in new biological contexts. This provide a cost effective way to expand knowledge from limited one ChIP-seq data to other research areas, however, the effective depends on particular collection of corresponding side information. Zhu et al. [9] proposed one method to enhancing the analysis of ChIP-seq data using convex-relaxed pair-wise tensor decomposition, which using the correlative relationship between observed and missing data, while the predictive performance remains to be improved.

In this paper, we use an efficient tensor completion algorithm to impute unobserved ChIP-seq sample values using correlated information of binding activity between TFs as well as cell lines, to enhance the available of ChIP-seq experiments. A ChIP-seq experimental sample data is converted into a binary number vector, representing the binding activities on the whole genes for one TF in a particular cell line. We adopt a 3-mode tensor pattern to model ChIP-seq data [10], in which the three modes represents TFs, cell lines and genes, separately. Then an high accuracy low rank tensor completion algorithm called HaLRTC is used to impute the unperformed ChIP-seq experiment samples [11, 12]. Our experiments show that our approach achieves higher recovery accuracy than the state-of-art methods and is particularly suitable for the imputation of massive scale ChIP-seq data.

2   Notations and Tensor Completion

2.1   Notations

Scalars are denoted by lower-case letters, e.g. x. And vectors by bold lower-case letters, e.g., **x**. Matrices are denoted by capital letters such as M, and their entries by corresponding lower-case letters, e.g., m_{ij}. The order of a tensor is the number of dimensions, also known as modes or ways. A N-way tensor is denoted by a calligraphic letter, for example, $\mathcal{X} \in \mathbb{R}^{I_1 \times I_2 \cdots \times I_N}$, and its entries are denoted by lower-case letters with subscripts, e.g., the $(i_1, i_2, \ldots, i_N)$ entry of an N-way tensor is written as $x_{i_1,i_2,\ldots,i_N}$.

2.2   Tensor and Tensor Completion

Tensors provide an effective and faithful representation of the structural properties of data, particularly when multidimensional data are involved. For instance, consider an online such as Amazon.com where users review various products over time. One can form a users × users × products tensor from the review text [13]. In real data, missing data can arise in various situations due to loss of information, problems in acquisition process or costly experiments [14]. Completion procedure estimates missing entries of array data by using available elements and correlation structural properties of data. In the recent years, techniques for tensor completion have attracted attention owing to their potential applications and flexibility, and has been successfully applied to a wide variety of problems, such as visual data prediction [11], link prediction [15], and social network analysis [16].

Nuclear norm based approaches are successfully used for solving the tensor completion problem. Nuclear norm minimization method based on low-rank property [11, 17, 18]. The convex optimization of nuclear norm has gain considerable attention in matrix completion, which essentially seeks the minimum rank under the condition of limited observed data.

3   Proposed Method for Target Gene Prediction

ChIP-seq dataset can be modeled as a 3-way tensor $\mathcal{X} \in \mathbb{R}^{n_c \times n_t \times n_g}$, in which the three modes represent cell lines, TFs and genes, separately. Each ChIP-seq experiment sample which corresponds to the binding activity of a particular TF in a cell line, it can be summarized into a $n_g \times 1$ binary vector, in which 1 represents the TF binds to the particular gene and n_g is the number of genes.

Based on the correlations of ChIP-seq data, the built tensor model can impute the missing data with keeping up the integrity of the original structure and exploiting multimode correlations simultaneously. A high accuracy low rank tensor completion algorithm (HaLRTC) [11] is used for imputing missing values. Let Ω be the set of values with observed data. With the correlation of ChIP-seq data, the tensor can be represented as a low-dimensional structure. Thus, the problem of imputing missing data can be solved by the optimization problem for low rank tensor completion:

$$\min_{\mathcal{X}} : rank(\mathcal{X}) \quad s.t. : \mathcal{X}_\Omega = \mathcal{T}_\Omega \tag{1}$$

where χ and K are 3-mode tensor and Ω is the set of values with observed data. While, the rank of tensor is NP hard and one common approach is to use trace norm to approximate the rank of tensor [11]. Using the definition of trace norm of a tensor in [11], the optimization in (1) can be written as:

$$\min_{\mathcal{X},} : \sum_{i=1}^{n} \alpha_i \left\| \mathcal{X}_{(i)} \right\|_* \quad s.t. : \mathcal{X}_\Omega = \mathcal{T}_\Omega, \tag{2}$$

where α_i satisfies $\alpha_i \geq 0$ and $\sum_{i=1}^{n} \alpha_i = 1$. To solve the optimization problem, additional tensors $\mathcal{M}_1, \ldots, \mathcal{M}_n$ are introduced and optimization in (2) is converted into:

$$\begin{aligned} \min_{\mathcal{X}, \mathcal{M}_i} : &\sum_{i=1}^{n} \alpha_i \left\| \mathcal{M}_i \right\|_* \\ s.t. : \quad &\mathcal{X}_{(i)} = \mathcal{M}_i \, for \quad i = 1, \ldots, n \\ &\mathcal{X}_\Omega = \mathcal{T}_\Omega, \end{aligned} \tag{3}$$

Then, a high accuracy algorithm ADMM [19] is used to solve the large-scale problem [11].

4   Experiments

In this section, we evaluated the predictive performance of our proposed method on ChIP-seq data by comparing to ChIP-PIT method, logistic regression method and one heuristic method, i.e. ChIP-Aveg used in paper [9]. All tests were performed using Matlab R2013a on a machine with 2 Xeon E5-2680v2 CPUs and 128 GB RAM.

4.1   Datasets and Settings

The dataset are the human transcription factor (TF) binding peaks based on ChIP-seq experiments generated by ENCODE Consortium [4], in which the peak represent the binding affinity of a TF to the genome. In particular, we got 539 ChIP-seq datasets from ENCODE UCSC website. These data represent 161 regulatory factors, the datasets span 91 human cell types and some are in various treatment conditions. As in paper [8], We consider a gene as target of a TF if binding peaks of the TF with high significance (FDR < 10 %) full into the window ([−10 kb, +5 kb]) around the transcription start site of the particular gene.

4.2   Evaluation Metrics and Results

In this paper, we use leave one out cross validation to evaluate the performance of proposed model. In each fold of the validation, we withhold 1 of the 593 observed data, and use the remaining data to train the model, then the withhold data is used to evaluate the predictive performance of the model. We use area under the receiver operating characteristic curve (AUC) quantitative metrics to assess the predictive ability of proposed method.

To provide perspective on the performance of our method in each metric, we compared it with three baselines, ChIP-PIT method [9], logistic regression method and one heuristic method, i.e. ChIP-Aveg, defined in [9].

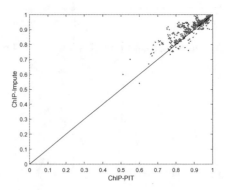

Fig. 1. The scatter plot shows the AUC predictive performance of ChIP-Impute (y axis) and ChIP-PIT (x axis) for each sample.

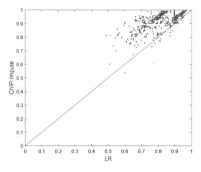 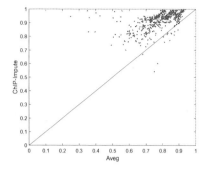

Fig. 2. (a) The scatter plot shows the AUC predictive performance of ChIP-Impute (y axis) and logistic regression method (x axis) for each sample. (b) The AUC predictive performance of ChIP-Impute (y axis) and ChhIP-Aveg (x axis) for each sample.

Table 1. The table shows the proportion of samples with higher AUC when comparing ChIP-Impute with three baselines.

	ChIP-PIT	LR	ChIP-Aveg
Students t-test (P-value)	2.39e-50	1.4e-72	6.15e-98
Higher AUC (proportion)	85.48 %	93.13 %	95.21 %
Higher AUC by 0.05	41.38 %	86.32 %	88.32 %

To assess the predictive performance of proposed method, we compare it with the three baselines on the metrics mentioned above. The results are shown in Figs. 1, 2, and Table 1. Results show that for more than 90 % samples, our proposed method gets significantly higher AUC compared with three baselines, which indicates that it outperforms ChIP-PIT for ChIP-seq imputation.

5 Conclusion

In this paper, a multiway tensor model is proposed to represent the ChIP-seq data, considering the fact that in TF binding activities across different cell lines are correlative and different TFs cooperate to regulate genes. Based the latent correlation, we employed a high accuracy low rank tensor completion (HaLRTC) algorithm to estimate the missing data without experiments by integrating a large collection of ChIP-seq data corresponding diverse TFs and cell lines. Experiments results show that the proposed method performs better than other state-of-art method. For further work, it is interesting to considering the potentially nonlinear correlation for imputation.

Acknowledgements. This work was supported by the grants of the National Science Foundation of China, Nos. 61133010, 61520106006, 31571364, 61532008, 61572364, 61373105, 61303111, 61411140249, 61402334, 61472282, 61472280, 61472173, 61572447, and 61373098, China Postdoctoral Science Foundation Grant, Nos. 2014M561513 and 2015M580352.

References

1. Furey, T.S.: ChIP–seq and beyond: new and improved methodologies to detect and characterize protein–DNA interactions. Nat. Rev. Genet. **13**(12), 840–852 (2012)
2. Chen, X., Xu, H., Yuan, P., Fang, F., Huss, M., Vega, V.B., Wong, E., Orlov, Y.L., Zhang, W., Jiang, J.: Integration of external signaling pathways with the core transcriptional network in embryonic stem cells. Cell **133**(6), 1106–1117 (2008)
3. Gerstein, M.B., Kundaje, A., Hariharan, M., Landt, S.G., Yan, K.-K., Cheng, C., Mu, X.J., Khurana, E., Rozowsky, J., Alexander, R.: Architecture of the human regulatory network derived from ENCODE data. Nature **489**(7414), 91–100 (2012)
4. Consortium, E.P.: An integrated encyclopedia of DNA elements in the human genome. Nature **489**(7414), 57–74 (2012)
5. Kundaje, A., Meuleman, W., Ernst, J., Bilenky, M., Yen, A., Heravi-Moussavi, A., Kheradpour, P., Zhang, Z., Wang, J., Ziller, M.J.: Integrative analysis of 111 reference human epigenomes. Nature **518**(7539), 317–330 (2015)
6. Malhotra, D., Portales-Casamar, E., Singh, A., Srivastava, S., Arenillas, D., Happel, C., Shyr, C., Wakabayashi, N., Kensler, T.W., Wasserman, W.W.: Global mapping of binding sites for Nrf2 identifies novel targets in cell survival response through ChIP-Seq profiling and network analysis. Nucleic Acids Res. **38**(17), 5718–5734 (2010)
7. Kunarso, G., Chia, N.-Y., Jeyakani, J., Hwang, C., Lu, X., Chan, Y.-S., Ng, H.-H., Bourque, G.: Transposable elements have rewired the core regulatory network of human embryonic stem cells. Nat. Genet. **42**(7), 631–634 (2010)
8. Wu, G., Yustein, J.T., McCall, M.N., Zilliox, M., Irizarry, R.A., Zeller, K., Dang, C.V., Ji, H.: ChIP-PED enhances the analysis of ChIP-seq and ChIP-chip data. Bioinformatics, btt108 (2013)
9. Zhu, L., Guo, W.-L., Deng, S.-P., Huang, D.-S.: ChIP-PIT: enhancing the analysis of ChIP-seq data using convex-relaxed pair-wise tensor decomposition
10. Kolda, T.G., Bader, B.W.: Tensor decompositions and applications. SIAM Rev. **51**(3), 455–500 (2009)
11. Liu, J., Musialski, P., Wonka, P., Ye, J.: Tensor completion for estimating missing values in visual data. IEEE Trans. Pattern Anal. Mach. Intell. **35**(1), 208–220 (2013)
12. Filipović, M., Jukić, A.: Tucker factorization with missing data with application to low-n-rank tensor completion. Multidimension. Syst. Signal Process. **26**(3), 677–692 (2015)
13. McAuley, J., Leskovec, J.: Hidden factors and hidden topics: understanding rating dimensions with review text. In: Proceedings of the 7th ACM Conference on Recommender systems 2013, pp. 165–172. ACM (2013)
14. Acar, E., Dunlavy, D.M., Kolda, T.G., Mørup, M.: Scalable tensor factorizations for incomplete data. Chemometr. Intell. Lab. Syst. **106**(1), 41–56 (2011)
15. Ermiş, B., Acar, E., Cemgil, A.T.: Link prediction via generalized coupled tensor factorisation (2012). arXiv preprint arXiv:1208.6231
16. Sun, J., Papadimitriou, S., Lin, C.-Y., Cao, N., Liu, S., Qian, W.: MultiVis: content-based social network exploration through multi-way visual analysis. In: SDM 2009, pp. 1063–1074. SIAM (2009)
17. Liu, Y., Shang, F., Cheng, H., Cheng, J., Tong, H.: Factor matrix trace norm minimization for low-rank tensor completion. In: SDM 2014, pp. 866–874. SIAM (2014)
18. Gandy, S., Recht, B., Yamada, I.: Tensor completion and low-n-rank tensor recovery via convex optimization. Inverse Prob. **27**(2), 025010 (2011)
19. Lin, Z., Chen, M., Ma, Y.: The augmented lagrange multiplier method for exact recovery of corrupted low-rank matrices (2010). arXiv preprint arXiv:1009.5055

Intelligent Computing in Scheduling

A Hybrid Genetic Algorithm for Dual-Resource Constrained Job Shop Scheduling Problem

Jingyao Li[1($\boxtimes$)] and Yuan Huang[2]

[1] The 365th Research Institution, Northwestern Polytechnical University,
Xi'an, Shaanxi, China
ljyao.6106@163.com
[2] School of Mechatronics, Northwestern Polytechnical University,
Xi'an, Shaanxi, China

Abstract. This paper presents a new scheduling approach based on Genetic Algorithm. This algorithm is developed to address the scheduling problem in manufacturing systems constrained by both machines and heterogeneous workers, which is called as Dual Resource Constrained Job Shop Scheduling Problem. In this algorithm, the evolutionary experience of parent chromosomes is inherited by branch population, which prevents premature and retain excellent gene. Some other optimization mechanisms, such as the elite evolutionary operator, the roulette selection operator with sector partition, the scheduling strategy based on compressed time window, are proposed to improve the algorithm. The performances of proposed approach are verified according to simulation experiments with random benchmark instances while related discussions are represented at last.

Keywords: Genetic algorithm · Dual resource constrained · Branch population · Sector partition · Compressed time window

1 Introduction

The type of system, where both machines and workers represent potential capacity constraints on the production capacity of system, is referred to as Dual Resource Constrained Job-shop Scheduling Problem [1] (DRCJSP), that is more complex than JSP and presents a number of additional technical challenges. One research highlight is the interaction between job dispatching and worker assignment due to un-fully staffed machines. It is necessary to study how these worker assignment, such as "when" rule, "where" rule [2], "Push/Pull" rule [3] and "who" rule [4, 5], affect and are affected by job dispatching rule. Another research highlight focuses on the improved utilization based on unique characteristics of workers, as this direction allows more accurate depictions of real manufacturing situations as the same as job dispatching. These worker relevant issues concern the effect of different characteristics levels among individual workers, such as cross-trained staffing levels [6], worker allocation [7], worker fatigue and recovery [8], learning and forgetting levels, [9] and so on. [10] The

© Springer International Publishing Switzerland 2016
D.-S. Huang et al. (Eds.): ICIC 2016, Part I, LNCS 9771, pp. 463–475, 2016.
DOI: 10.1007/978-3-319-42291-6_47

move of research in this direction allows better reflection of real manufacturing process and makes DRC research more valuable and relevant to industry.

Traditionally, DRCJSP has been addressed solving by analytical and simulation approaches [11]. However, it is difficult to model details of DRC system accurately due to above additional factors and unique worker characteristics when using analytical approach, while the efficiency of simulation modeling approach is undesirable as it may lead to high computational cost. Consequently, there is an observed move towards the use of meta-heuristic methods, which can quickly find a near optimal solution for DRCJSP. Following the pioneering work by Nelson [1], Maryam Hamedi [12] proposed a multi-objective Tabu Search (TS) algorithm based on Goal Programming (GP) approach. An effective variable neighborhood search (VNS) is presented by Lei [13], where two neighborhoods search procedures are sequentially executed to produce new solutions for two sub-problems respectively. Li [14] presented a hybrid scheduling algorithm, which is the combination of ACO and Simulated Annealing (SA) algorithm, for solving DRCJSP with heterogeneous workers. For the same problem, two meta-heuristic algorithms, namely SA and Vibration Damping Optimization (VDO), are proposed by Yazdani [15].

In this paper, a practical meta-heuristic algorithm, which is a hybrid of Genetic Algorithm (GA) and the Ant Colony Optimization (ACO) algorithm, is proposed to solve DRCJSP and referred to as Branch Population Genetic Algorithm (BPGA). BPGA introduces the idea of accumulating and transferring evolutionary experience via pheromone of ACO based on frame of GA, which ensures excellent global search performance depending on its parallel evolution mechanism of multi population. The branch chromosomes population, which is generated before each iteration, not only strengthens the population diversity, but also inherits the survival experience of parent population to accelerate convergence. On the other hands, some mechanisms are utilized to optimize performance of BPGA, such as the elite evolutionary operator, the roulette wheel selection operator based on sector segmentation, the scheduling strategy based on compressed time window, and so on.

2   Problem Description and Mathematical Model

2.1   Problem Description

A DRCJSP system may be described as follows: given a $n \times m \times w$ manufacturing system, where parts $\{P_1, \ldots, P_i, \ldots, P_n\}$ must be processed exactly on some machines $\{M_1, \ldots, M_k, \ldots, M_m\}$ operating by a set of workers $\{W_1, \ldots, W_l, \ldots, W_w\}$ during the plan cycle, each part has a certain delivery time and is composed of an aggregate of jobs which can be processed with several combinations of machines and workers. In the DRCJSP system, the number of workers is less than that of machines, at least one worker is capable of operating more than one machine, the number of skills and operating proficiency are both different among different workers. Therefore, the actual processing time and cost of each job are effected by both machine capacity and worker efficiency. The objective of DRCJSP is to find a scheme of job schedule and resources allocation with the optimal or near-optimal performance criteria.

2.2   Mathematical Model

The following are the symbols and variables used in the mathematical model of DRCJSP:

P_{ij} —The job j of part P_i, $i = \{1, 2, \ldots, n\}$; $t^P_{P_{ij} M_k}$ —The standard processing time of job P_{ij} processed by machine M_k, $k = \{1, 2, \ldots, m\}$; $e_{W_l M_k}$ —The work efficiency of worker W_l when operating machine M_k, $l = \{1, 2, \ldots, w\}$; $H_{P_{ij}M_kW_l} =$

$$\begin{cases} 1 & t^P_{P_{ij}M_k} > 0 \wedge e_{W_l M_k} > 0 \\ -1 & \text{else} \end{cases}$$ —The judge matrix of manufacturing resource combination; $T^S_{P_{ij} M_k W_l}$ —The start time of job P_{ij} processed by worker W_l and machine M_k; $T^P_{P_{ij} M_k W_l}$ —The real processing time of job processed by worker W_l and machine M_k; $T^E_{P_{ij} M_k W_l}$ —The finish time of job P_{ij} processed by worker W_l and machine M_k; $T^{PE}_{P_i}$ — The real finish time of part P_i; $T^E_{P_i}$—The delivery time of part P_i; R_{M_k}—Available time sets of worker M_k; R_{W_l} —Available time sets of machine W_l; C_{M_k} —Hourly operation cost of machine M_k; C_{W_l} —Hourly wage of worker W_l; C_{P_i} —Material cost of P_i; $C^{early}_{P_i}$ —Earliness penalties of P_i; $C^{late}_{P_i}$ —Tardiness penalties of P_i; C_{year} —Annual interest.

The mathematical model of DRCJSP for minimizing the *makespan* and *cost* is established as shown in Eqs. (1), (2) and (3).

$$\min F = [F_1 , F_2] \tag{1}$$

$$F_1 = \min(makespan) = \min(\max(T^E_{P_{ij} M_k W_l})) \tag{2}$$

$$F_2 = \min(cost) \tag{3}$$

According to the research of reference [16], the discount rate of fund can be set to zero and the affect after interest deduction can be ignored. Based on this hypothesis and the classical formula of production cost for JSP proposed by Shafei and Brunn [17], some factors, such as earliness/tardiness penalties, flexible resources and heterogeneous workers, have been introduced to the definition of production cost, as shown in Eq. (4).

$$\begin{aligned} cost = &\; C_{year} \times \sum_{i=1}^{n} (C_{P_i} \times (\max(T^{PE}_{P_i} , T^E_{P_i}) - T^{ST}_{P_{i1}})) + \\ &\; C_{year} \times \sum_{i=1}^{n} \sum_{j=1}^{n_i} ((C_{M_k} + C_{W_l}) \times T^P_{P_{ij} M_k W_l} \times (T^{ST}_{P_{i(j+1)} M_{k'} W_{l'}} - T^{ST}_{P_{ij} M_k W_l})) + \\ &\; \sum_{i=1}^{n} \sum_{j=1}^{n_i} ((C_{M_k} + C_{W_l}) \times T^P_{P_{ij} M_k W_l}) + \sum_{i=1}^{n} (C^{early}_{P_i} \times \max(0, T^E_{P_i} - T^{PE}_{P_i}) + C^{late}_{P_i} \times \max(0, T^{PE}_{P_i} - T^E_{P_i})) \end{aligned}$$

$$\tag{4}$$

The frontal two parts of this formula represents the inventory cost caused during the production process, the third part indicates the resource operation cost, while the last part is the penalties to early and tardy parts.

The DRCJSP system subjects to two kinds of constraint, known as the manufacturing process constraint and resource capability constraint. Each part can be processed as soon as the scheduling started while the order of each job was fixed, as shown in Eqs. (5) and (6).

$$T^S_{P_{ij}M_kW_l} \geq 0 \tag{5}$$

$$T^E_{P_{ij}M_kW_l} \leq T^S_{P_{i(j+1)}M_qW_r} \tag{6}$$

Delays during manufacturing process, caused by resources conflict, must be considered, as shown in Eq. (7).

$$T^E_{P_{ij}M_kW_l} \leq T^S_{P_{ij}M_kW_l} + T^P_{P_{ij}M_kW_l} \tag{7}$$

Compared to classical JSP, the actual processing time of each job is affected by both machine performance and worker efficiency, as shown in Eq. (8).

$$T^P_{P_{ij}M_kW_l} = t^P_{P_{ij}M_k} \Big/ e_{W_lM_k} \tag{8}$$

The job can be processed only if the correlative machine and worker were both available, as shown in Eqs. (9), (10) and (11)

$$H_{P_{ij}M_kW_l} H_{P_{xy}M_kW_r} (T^E_{P_{ij}M_kW_l} - T^S_{P_{xy}M_kW_r})(T^S_{P_{ij}M_kW_l} - T^E_{P_{xy}M_kW_r}) \geq 0 \tag{9}$$

$$H_{P_{ij}M_kW_l} H_{P_{xy}M_qW_r} (T^E_{P_{ij}M_kW_l} - T^S_{P_{xy}M_qW_r})(T^S_{P_{ij}M_kW_l} - T^E_{P_{xy}M_qW_r}) \geq 0 \tag{10}$$

$$R_{M_k} \cap R_{W_l} \cap [T^S_{P_{ij}M_kW_l}, T^E_{P_{ij}M_kW_l}] \neq \varPhi \tag{11}$$

3   BPGA Scheduling Algorithm

3.1   Step of BPGA Algorithm

Step 1: Initialize parameters of BPGA algorithm.
Step 2: The original chromosome population is generated based on the multi-dimensional coding method and the heuristic strategy library.
Step 3: The branch population of t_{th} iteration is generated as P^t_{branch} to form the evolution population $P^t_{evoluation}$ and P^{t-1}_{alive}.
Step 4: The value of *makespan* and *cost* of each chromosome are compared with their population mean value of $P^t_{evoluation}$ respectively. The elite population P^t_{alive} is then generated based on chromosomes that had at least one performance criteria better than its mean value. Other chromosomes form the normal population P^t_{normal}.

Step 5: The elite crossover operator and mutation operator are implemented based on P^t_{elite} and $P^{t-1}_{parento}$. After that, parent population, child population, and P^t_{normal}, are used to construct the selected child population P^t_{child}.

Step 6: The child population is decoded with the scheduling strategy based on compressed time window, and then the Pareto population $P^t_{parento}$, which was the set of the global Pareto solution of t_{th} iteration. Pareto solution is produced according to the Pareto dominance relationship comparisons among each chromosome.

Step 7: N evenly distributed chromosomes are selected from surplus child population via the roulette wheel selection operator to form the alive population P^t_{alive}.

Step 8: Execute local and global update of four kinds of pheromone based on $P^t_{parento}$ and P^t_{alive}.

Step 9: Go back to step3 if the iteration termination condition is not meet, else the program will terminate display the results.

3.2 Optimization Mechanisms of BPGA

Multi-dimensional Coding Method. Each gene G_i of the chromosome R_c is made up of four elements $G^p_i (p = 1,2,3,4)$, which indicate part, job, machine and worker respectively. These are shown in Fig. 1. With this coding method, each chromosome represents only one scheduling scheme and is convenient to decode.

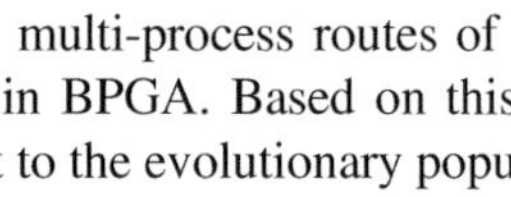

Fig. 1. Multi-dimensional coding method

Branch Population. In allusion to the multi-constraint and flexible process nature of double-objective DRCJSP, the multi-process routes of each job can be considered as multi-branch path of ant map in BPGA. Based on this hypothesis, the branch population, which was a supplement to the evolutionary population, was generated using the improved pseudo-random proportional state transition rule based on ant flow before each iteration, as shown in Eq. (12).

$$P_{P_{xy} P_{ij}} = \begin{cases} 1P \leq q \wedge \underset{(i,j,k,l)\in P_s}{\arg\max} \{ \left(\tau^{OO_1}_{P_{xy} P_{ij}} \times \tau^{OO_1}_{P_{xy} P_{ij}} \times \tau^{RO_1}_{P_{ij} M_k W_l} \times \tau^{RO_2}_{P_{ij} M_k W_l} \right) \times \eta^{\beta}_{P_{ij} M_k W_l} \times \left(n_{P_{xy} P_{ij}} \times n_{P_{ij} M_k W_l} \right)^{\gamma} \} \\ \dfrac{ \left(\tau^{OO_1}_{P_{xy} P_{ij}} \times \tau^{OO_1}_{P_{xy} P_{ij}} \times \tau^{RO_1}_{P_{ij} M_k W_l} \times \tau^{RO_2}_{P_{ij} M_k W_l} \right)^{\alpha} \times \eta^{\beta}_{P_{ij} M_k W_l} \times \left(n_{P_{xy} P_{ij}} \times n_{P_{ij} M_k W_l} \right)^{\gamma} }{ \sum\limits_{(i,j,k,l)\in P_s} \left(\left(\tau^{OO_1}_{P_{xy} P_{ij}} \times \tau^{OO_1}_{P_{xy} P_{ij}} \times \tau^{RO_1}_{P_{ij} M_k W_l} \times \tau^{RO_2}_{P_{ij} M_k W_l} \right)^{\alpha} \times \eta^{\beta}_{P_{ij} M_k W_l} \times \left(n_{P_{xy} P_{ij}} \times n_{P_{ij} M_k W_l} \right)^{\gamma} \right) } \quad P > q \end{cases}$$

(12)

The existence of the branch population strengthened the diversity of the evolutionary population and expanded the search space of BPGA effectively. On the other hand, with this branch population, the evolutionary experience of parent generation was inherited by the pheromone, which not only reflected the influence of each job on performance criteria but also avoided the loss of outstanding gene.

After each evolutionary iteration, the pheromone and ant flow were updated using the way of local update and global update based on $P_{evolution}^t$ and P_{Pareto}^t.

Elite Evolution Operator. BPGA proposed the elite evolutionary operator which encouraged evolution of elite chromosomes. The elite population was constructed with the chromosomes having at least one performance index that was better than the population mean value. Then implemented elite crossover operator and elite mutation based on P_{elite}^t and P_{pareto}^t, other chromosomes entered into the child population P_{child}^t and waited to be selected that strengthened the diversity of the population.

Elite crossover operation: firstly selected the king chromosomes, whose double performance criteria was better than the population mean value, out from P_{elite}^t and P_{pareto}^t, then executed one random crossover operator from the composite crossover operator library between each king chromosome and the other king ones, moreover, the crossover operator was executed between each king chromosome and two random elite chromosomes.

Elite mutation operation: In BPGA, a composite mutation operator library was constructed including common mutation operators and three resource mutation operators. Each king chromosome executed all the mutation operators in the library to increase the possibility of obtaining the Pareto solution; other elite chromosomes executed only one random mutation operator in the library respectively to avoid prematurity.

Roulette Wheel Selection Operator Based on Sector Segmentation. After the chromosomes decoding, a normalization operator was executed due to the major different of order of magnitudes between *makespan* and *cost*, consequently the objective space was compressed into an area of 1×1, where each solution of P_{child}^t was appeared as just one point. The solution angle θ, which was formed by the horizontal axes and the line connecting solution point and original point, was calculated according to Eq. (13) and expressed the distribution level of solution sets to some extent.

$$\theta = \arctan(f_N^{O_1}(A)/f_N^{O_2}(A)) \qquad (13)$$

After updated the Pareto population, some evenly distributed outstanding chromosomes were selected out from P_{child}^t using the roulette wheel selection operator based on sector segmentation to make up the evolutionary population $P_{evolution}^t$. The specific steps of the selection operator were as follows: ① Found two points A and B in the objective space which indicated the chromosomes with the optimal *makespan* or *cost* respectively. ② Used two lines to connect the original point and the two points above-mentioned respectively, thus an enclosed sector area that covered all the alternative chromosomes was shaped after prolonging such two lines to the edge of the objective space. ③ The big sector was divided into N small sectors equally according to the angle. ④ Selected one point out to accede to the $P^t_{evolution}$ from each sector area using the roulette wheel method based on the distance between the original point and the solution point. ⑤ Ended the selection iteration until the scale of $P^t_{evolution}$ was up to N, else returned to step ④.

Scheduling Strategy Based on Compressed Time Window. With the scheduling strategy based on compressed time window (SSBCTW), the global scheduling performance of existed scheduling scheme can be improved by means of compressing the occupied time window of a job and inserting it into the previous available time window of associated resource combination.

This paper proposed the judging criteria and specific steps of the SSBCTW, which was utilized as the decode strategy in the BPGA, for optimizing double objective as follows.

Step 1: In a multi-dimensional chromosome R_c generated by BPGA, the resource combination of each job was predefined as part of the gene. Therefore, all the jobs should be dispatched according to the gene sequence via the scheduling strategy based on the time window comparison [18] (SSBTWC) firstly, and confirmed the occupied time window $TWI[t^S_{TWI}, t^E_{TWI}]$ of each job, thus obtained the original scheduling scheme S_c and the available time window sets of resource combination $\mathbf{R}_{M_k W_l}$.

Step 2: Picked out the final job P_{ij} of S_c according to $\{P_{ij}|\ P_{ij} \subset S_c \wedge T^E_{P_{xy}} = \max(\{T^E_{P_{xy}}|\forall P_{xy} \subset S_c\})\}$.

When using the available time window $A[t^S_A, t^E_A]$ as the new occupied time window of P_{ij}, as shown in Fig. 2, the occupied quantity of $A[t^S_A, t^E_A]$ was $\Delta t_F = t^E_A - \max(t^S_A, T^E_{P_{i(j-1)}})$ while the maximum delay time of other jobs caused by the forward scheduling of P_{ij} was $\Delta t_A = T^P_{P_{ij}} - \Delta t_F = T^P_{P_{ij}} - t^E_A + \max(t^S_A, T^E_{P_{i(j-1)}})$, therefore the global *makespan* would be optimized only if $\Delta t_F > \Delta t_A$, which was summarized as shown in Eq. (14).

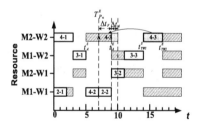

Fig. 2. Scheduling based on compressed time window

$$2 \times (t^E_A - \max(t^S_A, T^E_{P_{i(j-1)}})) > T^P_{P_{ij}} \tag{14}$$

Step 3: The available time windows sets $\mathbf{P}_A$ of the alternative resource combinations for job P_{ij} were listed as the alternative time windows set for SSBCTW, according to Eq. (15).

$$\mathbf{P}_A = \{A_x \,|\, \exists A_x \,[t^S_{A_x}, \; t^E_{A_x}] \subset \mathbf{R}_{M_k - W_l} \wedge t^E_{A_x} > T^E_{P_{i(j-1)}} \wedge t^S_{A_x} < t^S_{TWI}\} \qquad (15)$$

Step 4: The time window of $\mathbf{P}_A$ was judged case-by-case in ascending order of $t^S_{A_x}$, all the time windows, which could optimize the global *makespan* according to Eq. (18), were selected to form $\mathbf{P}_T = A_y[t^S_{A_y}, \; t^E_{A_y}]$, the process of SSBCTW would be ended if $\mathbf{P}_T = \emptyset$, else went to step 5 and further searched the time windows that optimized the global *cost*.

In this paper, *cost* was composed of four sections: material cost increment, resource cost increment, resource operation cost and earliness\tardiness penalties. During the rescheduling process of SSBCTW, the resource operation cost remained unchanged due to the fixed production resource combination and processing time. Meanwhile, the material cost increment and the earliness\tardiness penalties were both calculated only one time for one part, and would be unchanged or decreased if the part was finished before the deadline, therefore the variation of them caused by the movement of time window was ignored in this paper. In conclusion, the cost increment was calculated only based on resource cost increment, as shown in Eq. (16), in where the movement of the dispatched time window of P_{ij} was supposed as $\Delta t (\Delta t > 0$ when moving along the direction of time axis).

$$\Delta C_{p_{ij}} = C_{year} \times ((C_{M_k} + C_{W_l}) \times T^P_{P_{ij}M_kW_l} \times \Delta t) \qquad (16)$$

The global *cost*, which was different from *makespan*, was affected by the adjustment of each job rather than the final job only. However, increment of the global *cost* was too complex to calculate due to so many influenced jobs caused by the flexible route and the coupling relationship among each jobs during DRCJSP scheduling. On the other hands, only the jobs, which were processed by the same machine or worker with P_{ij} during $[T^S_{P_{i(j-1)}}, \; t^S_{TWI}]$, and their subsequent jobs would be delayed, therefore only the cost increment of these jobs needed to be considered. In conclusion, the *cost* increment was calculated as Eq. (17) supposed the movement of time window was $\Delta t = \max(t^S_A, \; T^E_{P_{ij}}) - t^S_{TWI}$ after scheduling of P_{ij} using time window $A[t^S_A, \; t^E_A]$ based on the SSBCTW.

$$\Delta C_F = C_{year} \times ((C_{M_k} + C_{W_l}) \times T^P_{P_{ij}M_kW_l} \times (\max(t^S_A, \; T^E_{P_{ij}}) - t^S_{TWI})) \qquad (17)$$

The upper bound of delay time of jobs affected by the SSBCTW was calculated as Eq. (18). The ΔC_A stood for the upper bound of *cost* increment sum of all delayed jobs, as shown in Eq. (19), in which $\mathbf{P}_A$ was the set of delayed jobs. Consequently, the global *cost* would be optimized if the time window, which satisfied Eq. (20), was selected for P_{ij} based on SSBCTW.

$$\Delta t_A = T_{P_{ij}}^P - \Delta t_A = T_{P_{ij}}^P - t_A^E + \max(t_A^S, T_{P_{i(j-1)}}^E) \tag{18}$$

$$\Delta C_A = \sum_{P_{ij} \subset \mathbf{P}_A} (C_{year} \times ((C_{M_k} + C_{W_l}) \times T_{P_{ij}M_kW_l}^P \times \Delta t_A)) \tag{19}$$

$$\Delta C_F + \Delta C_A < 0 \tag{20}$$

Step 5: The time window $A_y[t_{A_y}^S, t_{A_y}^E]$ of $\mathbf{P}_T$ was judged in ascending order of $t_{A_y}^S$ one by one, all the time windows which satisfied Eq. (20) was selected for scheduling of P_{ij} based on SSBCTW; else the A_y with the minimum $t_{A_y}^S$ was selected as the final scheduling time window for P_{ij} if there was no time window satisfied Eq. (20).

Step 6: Inserted the gene of chromosome R_c, which stood for P_{ij}, into the position after the gene which stood for $P_{i(j-1)}$ if $A_y[t_{A_y}^S, t_{A_y}^E]$ existed, and set up $T_{P_{ij}}^S = t_{A_y}^S$, then P_{ij} and its subsequent jobs in R_c were all rescheduled based on SSBTWC, the S_c and $\mathbf{R}_{M_k-W_l}$ were updated at last.

Step 7: Went to step2 for the next iteration until $\mathbf{P}_T = \emptyset$ or achieved a certain number of iteration NC_{max}.

4 Simulation Analysis

4.1 Construction of DRCJSP Benchmark Instance

This paper constructed the DRCJSP random benchmark instances based on the variables as shown in Table 1.

Table 1. Value of variables

Variables	Value	Variables	Value
n (Number of Parts)	U[3,5]	Processing Time	U[2,30]
n_i (Number of Jobs)	U[2,5]	Machine Cost	$10 \times \lceil (N_F^M)^2 / M_T \rceil$
m (Number of Machines)	U[3,5]	Worker Cost	$10 \sum_k e_{M_i W_l}$
w (Number of Workers)	$\lceil m \times 0.7 \rceil$	Material Cost	$10 \, n_i$
Number of Worker Skills	U[1,3]	Part Deadline	$T_{P_i}^E = \sum_j^{n_i} \sum_k^m t_{P_i M_i}^P \times Ran$
Worker Efficiency	U[0.5,1]	Earliness/ Tardiness Penalties	U[30,50]

Among them, N_F^M was the flexible value of machine and M_T represented the average processing time of all the jobs that could be processed by this machine.

4.2 Pareto Performance Evaluation Criteria

The feasible solution set was a series of points which were distributed in the solution space, with the *makespan* and *cost* as its x and y coordinates respectively. Therefore, this paper proposed four kinds of performance evaluation criteria based on the two-dimensional geometric characteristics of approximation the Pareto front as follows.

$\bar{S}_I$: The average of a product of x-coordinate and y-coordinate of all Pareto solutions, which was used for representing the degree of approaching to the Pareto front.

A_{range}: The angle of the sector area, which covered all the Pareto solutions, was used for reflecting the distribution range of the Pareto solution set.

$S_{\Delta A}$: The standard deviation of the angle difference between adjacent Pareto solutions, the smaller the $S_{\Delta A}$, the more distribution uniform of Pareto solutions.

$C_H = 100 \times (\bar{S}_I \times S_{\Delta A})/A_{range}$: The comprehensive evaluation criteria of Pareto performance.

4.3 Analysis on Algorithm Performance Comparison

The multi-objective evolutionary algorithm NSGAII [19] and DOIGAII [18] was used as the performance comparison algorithm of BPGA in this paper, the parameters were set as follows: maximum iteration times $NC_{max} = 200$, population scale $N = 200$, branch population scale $N_{ant} = 50$. Three algorithms were used for calculating 10 groups of DRCJSP random benchmark instances for 20 times respectively, four Pareto performance evaluation criteria of above mentioned algorithm were compared, as shown in Table 2.

For all benchmark instances, the $\bar{S}_I$ of BPGA and DOIGAII were better than NSGAII, which illustrated that the elite evolution operator can optimize the local search ability. The search space of BPGA was more close to the Pareto front than DOIGAII as BPGA's $\bar{S}_I$ was better than DOIGAII except instance 4 and 8. However, the A_{range} of NSGAII and BPGA was better obviously than DOIGAII, that was closely related to its characteristic of focusing on searching the area nearby double-objective extreme value, which presented bigger solution coverage area. BPGA not only optimized the population diversity by introducing branch population, but also strengthened solution uniform distribution based on sector segmentation selection operator, which caused better $S_{\Delta A}$ and C_H on most benchmark instances. In summary, although the distribution uniform degree of Pareto solution sets of BPGA and DOIGAII were roughly the same, the BPGA obtained Pareto solution set which was more close to the Pareto front, therefore BPGA had better scheduling performance and stronger robustness.

Table 2. Results of algorithm performance comparison

Benchmark	BPGA				DOIGAII				NSGAII			
	S_I	A_{range}	$S_{\Lambda\Lambda}$	C_H	S_I	A_{range}	$S_{\Lambda\Lambda}$	C_H	S_I	A_{range}	$S_{\Lambda\Lambda}$	C_H
Pro1	0.005	21.59	49.21	1.14	0.007	7.85	48.38	4.31	0.039	54.22	53.25	3.83
Pro2	0.003	36.74	47.69	0.39	0.015	9.43	50.85	8.09	0.047	61.50	51.00	3.90
Pro3	0.013	23.56	15.83	0.87	0.024	63.83	14.51	0.55	0.200	30.51	22.56	14.79
Pro4	0.012	44.58	36.75	0.99	0.003	34.42	62.63	0.55	0.044	52.59	53.89	4.51
Pro5	0.008	29.37	40.56	1.10	0.010	12.82	39.75	3.10	0.044	35.12	61.13	7.66
Pro6	0.004	38.30	85.32	0.89	0.009	13.11	79.88	5.48	0.053	16.22	76.65	25.05
Pro7	0.024	69.07	6.69	0.23	0.061	55.31	6.85	0.76	0.150	47.49	8.43	2.66
Pro8	0.016	20.91	59.61	4.56	0.004	15.70	69.68	1.78	0.122	36.73	42.53	14.13
Pro9	0.011	24.33	39.72	1.80	0.012	34.74	37.12	1.28	0.037	45.09	58.50	4.80
Pro10	0.008	20.93	43.84	1.68	0.008	16.50	65.25	3.16	0.041	42.74	56.25	5.40

5 Conclusion

In this paper, a branch population genetic algorithm was proposed after establishing of double-objective DRCJSP mathematical model. In BPGA, some performance optimization mechanisms were introduced. The branch population was generated to strengthen population diversity and accelerate the algorithm convergence; the elite evolutionary operator was utilized to optimize search ability; the roulette selection operator based on sector segmentation was proposed to decrease the computational complexity effectively and avoid algorithm prematurity; the scheduling strategy based on compressed time window was proposed to improve scheduling performance of BPGA. Finally, the effectiveness of BPGA algorithm for solving DRCJSP problem was verified by algorithm performance comparison experiment. In conclusion, the developed scheduling algorithm was proved to be useful for DRCJSP scheduling under the realistic conditions imposed by both machine and worker availability constraints.

References

1. Nelson, R.T.: Labor and machine limited production systems. Manag. Sci. **13**, 648–671 (1967)
2. Kher, H.V.: Examination of worker assignment and dispatching rules for managing vital customer priorities in dual resource constrained job shop environments. Comput. Oper. Res. **27**, 525–537 (2000)
3. Salum, L.A.: Using the when/where rules in dual resource constrained systems for a hybrid push-pull control. Int. J. Prod. Res. **47**, 1661–1677 (2009)
4. Berman, O., Larson, R.C.: A queueing control model for retail services having back room operations and cross-trained workers. Comput. Oper. Res. **31**, 201–222 (2004)
5. Bokhorst, J.A.C., Slomp, J., Gaalman, G.J.C.: On the who-rule in Dual Resource Constrained (DRC) manufacturing systems. Int. J. Prod. Res. **42**, 5049–5074 (2004)
6. Kim, S., Nembhard, D.: Cross-trained staffing levels with heterogeneous learning/forgetting. IEEE Trans. Eng. Manag. **57**, 560–574 (2010)
7. Lobo, B.J., Wilson, J.R., Thoney, K.A., Hodgson, T.J., King, R.E.: A practical method for evaluating worker allocations in large-scale dual resource constrained job shops. IIE Trans. **46**, 1209–1226 (2014)
8. Jaber, M.Y., Neumann, W.P.: Modelling worker fatigue and recovery in dual-resource constrained systems. Comput. Ind. Eng. **59**, 75–84 (2010)
9. Givi, Z.S., Jaber, M.Y., Neumann, W.P.: Production planning in DRC systems considering worker performance. Comput. Ind. Eng. **87**, 317–327 (2015)
10. Xu, J., Xu, X., Xie, S.: Recent developments in Dual Resource Constrained (DRC) system research. Eur. J. Oper. Res. **215**, 309–318 (2011)
11. Treleven, M.D., Elvers, D.A.: An investigation of labor assignment rules in a dual-constrained job shop. J. Oper. Manag. **6**, 51–68 (1985)
12. Hamedi, M., Esmaeilian, G., Ismail, N., Ariffin, M.: Capability-based virtual cellular manufacturing systemsformation in dual-resource constrained settings using Tabu Search. Comput. Ind. Eng. **62**, 953–971 (2012)
13. Lei, D., Guo, X.: Variable neighbourhood search for dual-resource constrained flexible job shop scheduling. Int. J. Prod. Res. **52**, 2519–2529 (2014)

14. Jingyao, L., Shudong, S., Yuan, H.: Adaptive hybrid ant colony optimization for solving Dual Resource Constrained job-shop scheduling problem. J. Softw. **6**, 584–594 (2011)
15. Yazdani, M., Zandieh, M., Tavakkoli-Moghaddam, R., Jolai, F.: Two meta-heuristic algorithms for the dual-resource constrained flexible job-shop scheduling problem. Scientia Iranica **22**, 1242–1257 (2015)
16. Rohleder, T.R., Scudder, G.D.: Comparing performance measures in dynamic job shops: economics vs. time. Int. J. Prod. Econ. **32**, 169–183 (1993)
17. Shafaei, B.P.R.: Workshop scheduling using practical (inaccurate) data part 1: the performance of heuristic scheduling rules in a dynamic job shop environment using a rolling time horizon approach. Int. J. Prod. Res. **37**, 3913–3925 (1999)
18. Jingyao, L., Shudong, S., Yuan, H.: Research on Dual Resource Constrained job shop scheduling based on time window. J. Mech. Eng. **46**, 150–159 (2011)
19. Kalyanmoy, D., Amrit, P., Sameer, A., Meyarivan, T.: A fast and elitist multi-objective genetic algorithm: NSGA-II. IEEE Trans. Evol. Comput. **6**, 182–197 (2002)

A Competitive Memetic Algorithm
for Carbon-Efficient Scheduling
of Distributed Flow-Shop

Jin Deng, Ling Wang$^{(\boxtimes)}$, Chuge Wu, Jingjing Wang,
and Xiaolong Zheng

Department of Automation, Tsinghua University, Beijing 100084, China
{dengjl3,wucgl5,wang-jjl5,
zhengxlll}@mails.tsinghua.edu.cn,
wangling@tsinghua.edu.cn

Abstract. Considering the energy conservation and emissions reduction, carbon-efficient scheduling becomes more and more important to the manufacturing industry. This paper addresses the multi-objective distributed permutation flow-shop scheduling problem (DPFSP) with makespan and total carbon emissions criteria (MODPFSP-Makespan-Carbon). Some properties to the problem are provided, and a competitive memetic algorithm (CMA) is proposed. In the CMA, some search operators compete with each other, and a local search procedure is embedded to enhance the exploitation. Meanwhile, the factory assignment adjustment is used for each job, and the speed adjustment is used to further improve the non-dominated solutions. To investigate the effect of parameter setting, full-factorial experiments are carried out. Moreover, numerical comparisons are given to demonstrate the effectiveness of the CMA.

Keywords: Carbon-efficient scheduling · Distributed shop scheduling · Multi-objective optimization

1 Introduction

Nowadays, global warming becomes a serious public problem. Carbon dioxide (CO_2) produced during the combustion process of fossil fuel is believed to be a critical reason that causes the global warming. As fossil fuel is the main source of energy, massive amounts of CO_2 are released to the atmosphere. Realizing the danger of climate, many policies and treaties are made to restrict the emissions of greenhouse gas. About half of the world's total energy consumption is contributed by industry sector [1]. Therefore, it is imperative for manufacturing industry to implement the energy-efficient scheduling in order to reduce carbon footprints.

Under the situation, much research has been carried out about energy-saving and low-carbon. In [2], several dispatching rules were proposed relying on the estimation of inter-arrival time between jobs to control the turn on/turn off time of machines, and a multi-objective mathematical programming model was developed to minimize the energy consumption and total completion time. In [3], the turn on/off framework was applied to the single machine scheduling, and a greedy randomized multi-objective

© Springer International Publishing Switzerland 2016
D.-S. Huang et al. (Eds.): ICIC 2016, Part I, LNCS 9771, pp. 476–488, 2016.
DOI: 10.1007/978-3-319-42291-6_48

adaptive search algorithm was proposed to optimize total energy consumption and total tardiness. The framework was extended to the flexible flow shop scheduling [4], and genetic algorithm (GA) and simulated annealing (SA) were hybridized to minimize the makespan and the total energy consumption. Another energy saving technique is speed scaling [5]. In such a case, machines can be run at varying speeds, and the lower speed results in lower energy consumption and longer processing time. Some researchers assumed that the speed range of machine is continuous adjustable. For example, the performance of several algorithms was studied in terms of the management of energy and temperature [6]. Others considered that there are a number of discrete speeds are available for machines. Two mixed integer programming (MIP) models were presented and their performances were investigated for the permutation flow shop scheduling problem (PFSP) with makespan and peak power consumption [7]. To reduce the carbon emissions and makespan for the PFSP, an NEH-Insertion procedure was developed based on the problem properties, and a multi-objective NEH and an iterative greedy (IG) algorithm were proposed [8]. To solve the multi-mode resource constrained project scheduling with makespan and carbon emissions criteria, a Pareto-based estimation of distribution algorithm (EDA) was proposed [9].

With market dispersing throughout the world, manufacturing is changing from the traditional pattern in one single factory into the co-production among multi-factories [10]. Distributed manufacturing enables companies to improve production efficiency and profit [11]. There exist two sub-problems: allocating jobs to suitable factories and scheduling jobs on machines in each factory. Since the coupled two sub-problems cannot be solved sequentially if high performance is required, distributed scheduling is more difficult to solve [12]. Currently, the distributed PFSP (DPFSP) has been a hot topic. In [12], six MIP models and two factory assignment rules as well as several heuristics and variable neighborhood descent methods were developed. Besides, tabu search [13], EDA [14], hybrid immune algorithm [15], IG [16], and scatter search [17] have been developed to solve the DPFSP. Moreover, the DPFSP with reentrant constraint [18] have also been studied. Most of the above research only considered the time-based objectives. In this paper, the multi-objective DPFSP with carbon emissions and makespan criteria is studied. Inspired from the good performance of memetic algorithms in solving the complex optimization problems [19], we extend the competitive memetic algorithm (CMA) for the DPFSP [20] to a multi-objective version. Especially, several specific search operators are designed according to the problem characteristics. Due to the complexity brought by the optimization of the carbon emission, the sharing phase is replaced by the adjustment phase in the CMA to effectively reduce the carbon emission. In addition, some properties will be analyzed, and the effectiveness will be demonstrated by numerical comparisons.

2   Problem Description

The following notions will be used to describe the MODPFSP-Makespan-Carbon.

f: total number of factories; n: total number of jobs;
m: number of machines in each factor; s: number of speeds alternative to machines;

S: discrete set of s different processing speeds, $S = \{v_1, v_2, \ldots, v_s\}$;
n_k: number of jobs in the factory k; $O_{i,j}$: operation of job i on machine j;
$p_{i,j}$: standard processing time of $O_{i,j}$; $V_{i,j}$: processing speed of $O_{i,j}$;
$C_{i,j}$: the completion time of $O_{i,j}$; $C(k)$: the completion time of factory k;
$PP_{j,v}$: energy consumption per unit time when machine j is run at speed v;
SP_j: energy consumption per unit time when machine j is run at standby mode;
π^k: the processing sequence in the factory k; Π: a schedule, $\Pi = (\pi^1, \pi^2, \ldots, \pi^f; V)$.

The problem is described as follows. There are f identical factories, each of which is a permutation flow shop with m machines. Each of n jobs can be assigned to any one of the f factories for processing. Each operation $O_{i,j}$ has a standard processing time $p_{i,j}$. Each machine can be run at s different speeds S, and it cannot change the speed during processing an operation. When operation $O_{i,j}$ is processing at speed $V_{i,j}$, the actual processing time becomes $p_{i,j}/V_{i,j}$. Machines will not be shut down before all jobs are completed. If there is no job processed on machine j, it will be run at a standby mode. The makespan $C_{\max}$ is calculated as follows.

$$C_{\pi^k(1),1} = p_{\pi^k(1),1}/V_{\pi^k(1),1} \tag{1}$$

$$C_{\pi^k(i),1} = C_{\pi^k(i-1),1} + p_{\pi^k(i),1}/V_{\pi^k(i),1} \tag{2}$$

$$C_{\pi^k(1),j} = C_{\pi^k(1),j-1} + p_{\pi^k(1),j}/V_{\pi^k(1),j} \tag{3}$$

$$C_{\pi^k(i),j} = \max\{C_{\pi^k(i-1),j}, C_{\pi^k(i),j-1}\} + p_{\pi^k(i),j}/V_{\pi^k(i),j} \tag{4}$$

$$C(k) = C_{\pi^k(n_k),m} \tag{5}$$

$$C_{\max} = \max\{C(1), C(2), \ldots, C(f)\} \tag{6}$$

Let $x_{kjv}(t)$ and $y_{kj}(t)$ be the following binary variables:

$$x_{kjv}(t) = \begin{cases} 1, & \text{if machine } j \text{ in factory } k \text{ is run at speed } v \text{ at time } t \\ 0, & \text{otherwise} \end{cases} \tag{7}$$

$$y_{kj}(t) = \begin{cases} 1, & \text{if machine } j \text{ in factory } k \text{ is run at standby mode at time } t \\ 0, & \text{otherwise} \end{cases} \tag{8}$$

The total carbon emissions (TCE) can be calculated as follows:

$$TCE = \varepsilon \cdot TEC$$
$$= \varepsilon \cdot \sum_{k=1}^{f} \int_{0}^{C(k)} \left(\sum_{v=1}^{s} \sum_{j=1}^{m} PP_{jv} \cdot x_{kjv}(t) + \sum_{j=1}^{m} SP_j \cdot y_{kj}(t) \right) dt \tag{9}$$

where TEC is the total energy consumption and ε refers to the emissions per unit of consumed energy.

A Gantt chart of problem with 2 factories is shown in Fig. 1. Since $C(1) > C(2)$, $C_{max} = C(1)$. Take the situation at time t_1 to explain how *TCE* is calculated. In the factory F_1, machine M_1 is in the standby mode, while M_2 and M_3 are in processing mode. Assuming that M_2 and M_3 are run at speed v and u, the energy consumption of factory F_1 at time t_1 is $P(t_1) = SP_1 + PP_{2v} + PP_{3u}$. Similarly, the energy consumption of each factory at different time points can be calculated as Fig. 2. The area between the curve and time axis is energy consumption of the corresponding factory. Then, *TCE* can be obtained by accumulating the energy consumption of each factory.

For the MODPFSP-Makespan-Carbon, it assumes that when $V_{i,j}$ is increased from v to u, the energy consumption increases while processing time decreases.

$$p_{i,j}/u < p_{i,j}/v \tag{10}$$

$$PP_{j,u} \cdot p_{i,j}/u > PP_{j,v} \cdot p_{i,j}/v \tag{11}$$

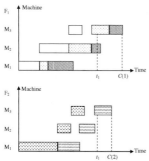

Fig. 1. Gantt chart of the DPFSP

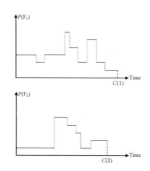

Fig. 2. Energy consumption

Based on the assumption, two properties for the PFSP were given in [8]. Here, one more property is given and extended to those of the MODPFSP-Makespan-Carbon.

Property 1 [8]. If two schedules π_1 and π_2 satisfy (1) $\forall i \in \{1, 2, \ldots, n\}$, $j \in \{1, 2, \ldots, m\}$, $V_{i,j}(\pi_1) = V_{i,j}(\pi_2)$, (2) $C_{max}(\pi_1) < C_{max}(\pi_2)$, then, $TCE(\pi_1) < TCE(\pi_2)$. That is $\pi_1 \succ \pi_2$.

Property 2 [8]. If two schedules π_1 and π_2 satisfy (1) $C_{max}(\pi_1) = C_{max}(\pi_2)$, (2) $\forall i \in \{1, 2, \ldots, n\}, j \in \{1, 2, \ldots, m\}$, $V_{i,j}(\pi_1) \leq V_{i,j}(\pi_2)$, (3) $\exists i \in \{1, 2, \ldots, n\}$, $j \in \{1, 2, \ldots, m\}$, $V_{i,j}(\pi_1) < V_{i,j}(\pi_2)$, then, $TCE(\pi_1) < TCE(\pi_2)$. That is $\pi_1 \succ \pi_2$.

Property 3. If two schedules π_1 and π_2 satisfy (1) $\forall i \in \{1, 2, \ldots, n\}, j \in \{1, 2, \ldots, m\}$, $V_{i,j}(\pi_1) = V_{i,j}(\pi_2)$, (2) $TCE(\pi_1) < TCE(\pi_2)$, then, $C_{max}(\pi_1) < C_{max}(\pi_2)$. That is $\pi_1 \succ \pi_2$.

Proof. The *TCE* of π_1 and π_2 can be calculated in the following ways:

$$TCE(\pi_1) = \varepsilon \cdot \left(\sum\nolimits_{j=1}^{m} SP_j \times t_j^{\text{idle}}(\pi_1) + \sum\nolimits_{i=1}^{n} \sum\nolimits_{j=1}^{m} PP_{j,V_{i,j}} \times p_{i,j}/V_{i,j} \right) \qquad (12)$$

$$TCE(\pi_2) = \varepsilon \cdot \left(\sum\nolimits_{j=1}^{m} SP_j \times t_j^{\text{idle}}(\pi_2) + \sum\nolimits_{i=1}^{n} \sum\nolimits_{j=1}^{m} PP_{j,V_{i,j}} \times p_{i,j}/V_{i,j} \right) \qquad (13)$$

where t_j^{idle} represents the total idle time of machine M_j.

Since $TCE(\pi_1) < TCE(\pi_2)$, it has $\sum_{j=1}^{m} t_j^{\text{idle}}(\pi_1) < \sum_{j=1}^{m} t_j^{\text{idle}}(\pi_2)$. Thus, $\exists j' = 1, 2, \ldots, m$, $t_{j'}^{\text{idle}}(\pi_1) < t_{j'}^{\text{idle}}(\pi_2)$. So, $t_{j'}^{\text{idle}}(\pi_1) + \sum_{i=1}^{n} p_{i,j'}/V_{i,j'} < t_{j'}^{\text{idle}}(\pi_2) + \sum_{i=1}^{n} p_{i,j'}/V_{i,j'}$. According to the definition of $C_{\max}$ in a PFSP, it has $C_{\max}(\pi_1) = t_j^{\text{idle}}(\pi_1) + \sum_{i=1}^{n} p_{i,j}/V_{i,j}$, $\forall j = 1, 2, \ldots, m$. Therefore, $C_{\max}(\pi_1) < C_{\max}(\pi_2)$.

Property 4. For a schedule Π, keep the speeds of all operations unchanged and change the job processing sequence in the factory with maximum completion time (denoted as F_m). If the completion time of F_m is decreased, then the carbon emissions of F_m are also decreased. That is, if $C_{\max}(\Pi)$ is decreased, then $TCE(\Pi)$ is decreased.

Property 5. For a schedule Π, keep the speeds of all operations unchanged and change the job processing sequence in F_m. If the carbon emissions of F_m are decreased, then the completion time of F_m is also decreased. That is, if $TCE(\Pi)$ is decreased, then $C_{\max}(\Pi)$ is decreased.

Property 6. For a schedule Π, if it keeps the completion time of each factory unchanged and slows down the speeds of some operations, the $TCE(\Pi)$ will be decreased while $C_{\max}(\Pi)$ will remain the same.

3   CMA for MODPFSP-Makespan-Carbon

3.1   Encoding Scheme

In the CMA, an individual X_l is represented by a job-factory matrix J-F and a velocity matrix A. J-F is a 2-by-n matrix, where the first row is job permutation sequence and the second row is factory assignment sequence. A is a n-by-m matrix, where element $A_{i,j} \in \{1, 2, \ldots, s\}$ represents the processing speed of $O_{i,j}$. An instance with $f = 2$, $n = 5$, $m = 4$, $s = 3$ is shown in Fig. 3. J-F implies that jobs J_1 and J_2 are assigned to F_1 with the processing sequence $\pi^1 = \{1, 2\}$, and jobs J_3, J_5, J_4 are assigned to F_2 with the sequence $\pi^2 = \{3, 5, 4\}$. In matrix A, for example, $A_{2,3} = 3$ means that the operation $O_{2,3}$ is processed at speed v_3.

$$J - F = \begin{bmatrix} 3 & 1 & 5 & 4 & 2 \\ 2 & 1 & 2 & 2 & 1 \end{bmatrix}, \quad A = \begin{bmatrix} 2 & 3 & 1 & 2 \\ 1 & 3 & 2 & 1 \\ 1 & 1 & 3 & 2 \\ 3 & 2 & 1 & 3 \\ 2 & 1 & 3 & 1 \end{bmatrix}$$

Fig. 3. An example for encoding scheme

3.2 Solution Updating Mechanism, Initialization and Archive

In the CMA, once a new individual X_l' is generated, it will be compared with its original one X_l, and the acceptance rule is based on the dominance relationship between the two individuals: (1) If $X_l' \succ X_l$, $X_l = X_l'$; (2) If $X_l \succ X_l'$, remain X_l unchanged; (3) If X_l and X_l' are non-dominated, then randomly choose one as new X_l.

In the initialization phase, all population size (PS) individuals are generated randomly to achieve enough diversity. Besides, a Pareto archive (PA) is used to record the explored non-dominated solutions and a temporal archive (TA) is used to store the newly found non-dominated solutions in each generation.

3.3 Competition

Adjusting the job processing sequence or the processing speed will impact the objective value, so three operators are designed, including SU, SD and CS.

SU: the operator is to increase the speed of an operation for optimizing makespan. Since the $C_{\max}$ of the DPFSP will be decreased only by improving the schedule in F_m, SU is designed to randomly choose an operation $O_{i,j}$ from the factory F_m; if $A_{i,j} = a < s$, then increase $A_{i,j}$ to b ($a < b \leq s$).

SD: the operator is to decrease the speed of an operation for optimizing TCE. Since the reduction of carbon emissions of any factory contributes to the reduction of TCE, factory F_k is randomly selected in SD. Then, randomly choose an operator $O_{i,j}$ from F_k; if $A_{i,j} = a > 1$, then decrease $A_{i,j}$ to b ($1 \leq b < a$).

CS: based on the Properties 4 and 5, the operator is to change job processing sequence in a factory to decrease $C_{\max}$ and TCE simultaneously. CS is designed to randomly select a job J^* from F_m, then insert J^* into a new position in F_m.

In each generation, the objective of each individual is normalized as follows:

$$g_p(X_l) = (f_p(X_l) - f_p^{\min})/(f_p^{\max} - f_p^{\min}) \tag{14}$$

where $g_p(X_l)$ denotes the normalized value of the p-th objective, $f_p(X_l)$ denotes the value of the p-th objective, $f_p^{\max}$ and $f_p^{\min}$ denote the maximum and minimum value of the p-th objective in the current population.

It can be seen from Fig. 4 that the normalized objective space is divided into three areas by α_1, α_2 and α_3. Obviously, individuals belonging to Ω_1 has better performance on the TCE while is relatively weaker on the $C_{\max}$. Individuals in Ω_2 are just the opposite. Individuals in Ω_3 have better balance between the two objectives. Since individuals in Ω_1, Ω_2 and Ω_3 have different features, we make a distinction among them by choosing different operators for individuals in different areas. To be specific, individuals in Ω_1 execute operator SU to make an emphasis on optimizing $C_{\max}$, and those in Ω_2 execute operator SD to focus on the reduction of TCE, and those in Ω_3 carry out operator CS to optimizing both of the two objectives.

The size of the corresponding area is decided by the angles α_1, α_2 and α_3, namely, the use ranges of SU, SD and CS is controlled by the three angles. In the beginning of the CMA, we set $\alpha_1 = \alpha_2 = \alpha_3 = \pi/6$. Later, the performance of the operators may be different at different evolution phase. Besides, to adaptively adjust the use ranges of SU, SD and CS, a competition is performed among α_1, α_2 and α_3 based on the performance of operators in each generation.

To evaluate the three operators, the score of operator K denoted will be calculated after every execution. Because SU and SD are designed to optimize single objective, the score of SU or SD is calculated as follows:

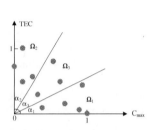

```
1: for i = 1 → LS do
2:     if F_m = F_c then
3:         CS(X*)
4:     else
5:         SU(X*)
6:         CS(X*)
7:         SD_2(X*)
8:         CS_2(X*)
9:     end if
10: end for
```

Fig. 4. Normalized objective space   **Fig. 5.** Pseudocode of local search procedure

$$S_r(K) = \max\{0, f*(X_l) - f*(X'_l)\}/f*(X_l) \tag{15}$$

where $f^*(X_l) = f_1(X_l)$ when $K = SU$, and $f^*(X_l) = f_2(X_l)$ when $K = SD$.

CS is to optimize both C_{max} and TCE, so its score is calculated as follows:

$$S_r(CS) = \max\{0, f_1(X_l) - f_1(X'_l)\}/f_1(X_l) + \max\{0, f_2(X_l) - f_2(X'_l)\}/f_2(X_l) \tag{16}$$

Let IN_q be the number of individuals in the area Ω_q. The average score of operator K is calculated as $AVS(K) = \sum_{r=1}^{IN*} S_r(K)/IN*$, where $IN* = IN_1$ when $K = SU$, $IN* = IN_2$ when $K = SD$, and $IN* = IN_3$ when $K = CS$.

Then, the values of α_1, α_2 and α_3 are redefined as follows:

$$\alpha_q = (\pi/2 - 3\beta) * AVS(K)/\sum_K AVS(K) + \beta \tag{17}$$

where β is a small angle that guarantees $\alpha_q \neq 0$. Here, it sets $\beta = \pi/60$.

3.4   Local Intensification

It is widely recognized that local search is helpful to intensify the exploitation ability of memetic algorithms [19]. Based on the SD and CS, two more local search operators SD_2 and CS_2 are presented.

SD_2: randomly select an operation $O_{i,j}$ from the factory with maximum energy consumption (denoted as F_c), if $A_{i,j} = a > 1$, then decrease $A_{i,j}$ to b ($1 \leq b < a$).

CS_2: randomly select a job from F_c, and insert it into a new position in F_c.

In the local intensification phase, a non-dominated solution in the current population is selected to perform local search for *LS* times. The local search procedure which includes SU, CS, SD_2 and CS_2 is illustrated in Fig. 5. When $F_m = F_c$, only CS is performed to avoid the repeated modification of the processing speeds.

3.5 Adjustment

There are two steps in the adjustment phase. The first step is factory assignment adjustment, and the second step is speed adjustment.

In the factory adjustment, four adjusting schemes are designed. (1) Randomly select a job from factory F_m, and insert it into all possible positions of another factory. (2) Randomly select a job from factory F_m, and exchange its position with all jobs in another factory. (3) Randomly select a job from factory F_c, and insert it into all possible positions of another factory. (4) Randomly select a job from factory F_c, and exchange its position with all jobs in another factory.

Each individual chooses one of the above schemes to search better factory assignments. Let P^* be the set of non-dominated solutions that newly generated when X_l execute the factory assignment adjustment. Then, randomly select a non-dominated solution X^* from $X_l \cup P*$, and set $X_l = X^*$.

In the speed adjustment, according to the Property 5, a solution can be improved by adjusting the processing speeds without deteriorating the completion time of each factory. Therefore, the speed adjustment is performed on each solution in the TA. After the adjustment, solutions in the TA are used to update the PA. Since the completion time of a factory will not be longer if the critical path [21] remains the same, the speed adjustment is implemented on the operations that are not on the critical paths (called non-critical operations). An example of speed adjustment is shown in Fig. 6. Firstly, the critical path of each factory is found, as the arrowed line. Secondly, for each factory, the non-critical operations are selected to execute the speed adjustment from the final job back to the first. For example, the operations in Fig. 6(a) are to be adjusted in the order $\{O_{3,2} \rightarrow O_{3,1} \rightarrow O_{2,1} \rightarrow O_{1,3}\}$.

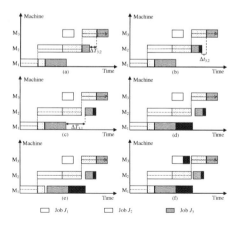

Fig. 6. Illustration of speed adjustment

Step 1: Put all the non-critical operations into the poor Pr^* in order.
Step 2: Select the first operator $O_{i,j}$ from Pr^*.
Step 3: Calculate the maximum extension time (MET) of $O_{i,j}$ as $\Delta T_{i,j} = \min\{ST_{i+1,j}, ST_{i,j+1}\} - C_{i,j}$,
where $ST_{i,j}$ denotes the starting time of $O_{i,j}$, and set $ST_{n+1,j} = \infty$, $ST_{i,m+1} = \infty$. As shown in Fig.
6(a), the MET of $O_{3,2}$ is $\Delta T_{3,2} = ST_{3,3} - C_{3,2}$.
Step 4: Set $V_{i,j} = 1, 2, \ldots, A_{i,j}$ step by step until the condition $p_{i,j} / V_{i,j} - p_{i,j} / A_{i,j} \leq \Delta T_{i,j}$ is satisfied.
Then, set $A_{i,j} = V_{i,j}$, $C_{i,j} = ST_{i,j} + p_{i,j} / A_{i,j}$.
Step 5: Calculate the maximum backward time (MBT) of $O_{i,j}$ as $\Delta t_{i,j} = \min\{ST_{i+1,j}, ST_{i,j+1}\} - C_{i,j}$.
As shown in Fig. 6(b), the MBT of $O_{3,2}$ is $\Delta t_{3,2} = ST_{3,3} - C_{3,2}$.
Step 6 Set $C_{i,j} = C_{i,j} + \Delta t_{3,2}$, $ST_{i,j} = ST_{i,j} + \Delta t_{3,2}$.
Step 7: Remove $O_{i,j}$ from Pr^*.
Step8: If $Pr^* \neq \emptyset$, go to **Step 2**.

Fig. 7. The procedure of speed adjustment

The procedure of speed adjustment for one factory is described as Fig. 7. And the flowchart of the CMA is illustrated in Fig. 8.

4   Computational Results

The CMA is coded in C language, and all the tests are run on the same PC with an Intel (R) core(TM) i5-3470 CPU @ 3.2 GHz/ 8 GB RAM under Microsoft Windows 7. The stopping criterion is set as $0.5 \times n$ seconds CPU time.

Since there is no benchmark for the MODOFSP-Makespan-Carbon, we generate test instances based on the test data as [8]. To be specific, $f = \{2, 3, 4, 5\}$, $n = \{20, 40, 60, 80, 100\}$, $m = \{4, 8, 16\}$, $v = \{1, 1.3, 1.55, 1.75, 2.10\}$, $p_{i,j}$ is uniformly distributed

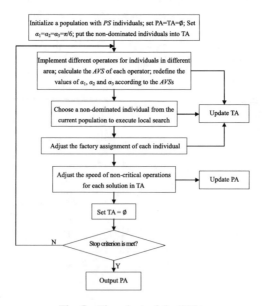

Fig. 8. Flowchart of the CMA

within $5 \sim 50$, $PP_{j,v} = 4 \times v^2$, $SP_j = 1$. Clearly, there are 15 combinations of $n \times m$. For each combination, 10 instances are randomly generated, and each instance is extended to $f = \{2, 3, 4, 5\}$. Thus, it has $15 \times 10 \times 4 = 600$ instances in total for evaluation.

The CMA contains two parameters: PS and LS. To investigate the influences of PS and LS on the performance of the CMA, we set PS with four levels $\{10, 20, 30, 40\}$ and LS with four levels $\{0, 100, 200, 300\}$, and then 4^2 full-factorial experiments are employed. To carry out the experiments, 60 instances are generated randomly, where each corresponds to a combination of $f \times n \times m$. For each instance, 16 combinations of $PS \times LS$ are tested. For each combination of $PS \times LS$, the CMA is run 10 times independently and the obtained non-dominated solutions E_{c_i} ($c\_i = 1, 2, ..., 16$) are stored. The final non-dominated solutions FE are obtained by integrating E_1, E_2, ..., E_{16}. Then, the contribution of a certain combination (CON) is calculated as $CON(c\_i) = |E'_{c\_i}|/|FE|$, where $E'_{c\_i} = \{X_l \in E_{c\_i} | \exists X_{l'} \in FE, X_l = X_{l'}\}$.

After all the instances are tested, the average CON of each combination is calculated as the response variable (RV) value. The results are listed in Tables 1 and 2, and the interval plots of PS and LS are shown in Fig. 9.

Table 1. RVs of full-factorial experiments.

Experiment Number	Factors PS	Factors LS	RV(%)	Experiment Number	Factors PS	Factors LS	RV(%)	Experiment Number	Factors PS	Factors LS	RV(%)	Experiment Number	Factors PS	Factors LS	RV(%)
1	1	1	0.0338991	5	2	1	0.0512754	9	3	1	0.0277986	13	4	1	0.0231631
2	1	2	0.0564796	6	2	2	0.0873351	10	3	2	0.0637920	14	4	2	0.0586763
3	1	3	0.0800603	7	2	3	0.1041840	11	3	3	0.0551406	15	4	3	0.0650945
4	1	4	0.0786400	8	2	4	0.0824610	12	3	4	0.0610678	16	4	4	0.0726370

Table 2. Result of analysis of variance.

Source	DF	Seq SS	Adj SS	Adj MS	F	p
PS	3	0.0020926	0.0020926	0.0006975	10.61	0.003
LS	3	0.0045512	0.0045512	0.0015171	23.08	0.000
Error	9	0.0005915	0.0005915	0.0000657		
Total	15	0.0072353				

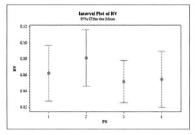
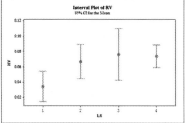

Fig. 9. Interval plot

From the Table 2, it can be seen that the influences of *PS* and *LS* are both significant with 95 % confidence interval. From Fig. 9, we know that the value of *PS* should neither be too small nor too large. A large *PS* may lead to an insufficient evolution, while a small *PS* is harmful to the diversity of the population. Similarly, a large *LS* is benefit to the exploitation, but a too large *LS* costs much of computation time on the local minima. According to the results of experiments, an appropriate combination of parameters is suggested as $PS = 20$ and $LS = 200$.

Since there is no published paper for solving the MODPFSP-Makespan-Carbon, the CMA is compared with the NSGA-II [22] and random algorithm (RA). In the NSGA-II, the population size is equal to *PS* in the CMA, and the crossover rate and mutation rate are set as 0.9 and $1/n$ as suggested in [22]. The stopping criteria of NSGA-II and RA are also set as $0.5 \times n$ seconds CPU time. There are several performance metrics for multi-objective problems [23]. In this paper, we focus on the quality of the obtained non-dominated solutions. Thus, the coverage metric (CM) is used for evaluation. The CM is defined as follows:

$$C(E_1, E_2) = |\{X_2 \in E_2 | \exists X_1 \in E_1, X_2 \prec X_1 \, or \, X_2 = X_1\}|/|E_2| \qquad (21)$$

where $C(E_1, E_2)$ denotes the percentage of the solutions in E_2 that are dominated by or the same as the solutions in E_1.

For each instance, the CMA, NSGA-II and RA are run 10 times independently within $0.5 \times n$ seconds CPU time. The CM is applied to pairwise comparison between the CMA and NSGA-II as well as the CMA and RA. For the same combination of $f \times n \times m$, the average CM of 10 instances is calculated. The comparison results are listed in Tables 3, 4, 5 and 6 grouped by different number of f. From Tables 3, 4, 5 and 6, it can be seen that the proposed CMA is superior to NSGA-II and RA at all sets of instances. Besides, hypothesis testing is carried out on C(CMA,NSGA-II) and C (NSGA-II,CMA) as well as C(CMA,RA) and C(RA,CMA), and all the resulted p-values are equal to 0. So, it is demonstrated that the difference between C(CMA, NSGA-II) and C(NSGA-II,CMA) as well as the difference between C(CMA,RA) and C

Table 3. Comparisons of algorithms ($f = 2$).

$n \times m$	CM			
	C(CMA,NSGA-II)	C(NSGA-II,CMA)	C(CMA,RA)	C(RA,CMA)
20×4	0.99	0.00	1.00	0.00
20×8	1.00	0.00	1.00	0.00
20×16	1.00	0.00	1.00	0.00
40×4	0.96	0.01	1.00	0.00
40×8	1.00	0.00	1.00	0.00
40×16	1.00	0.00	1.00	0.00
60×4	0.75	0.03	1.00	0.00
60×8	1.00	0.00	1.00	0.00
60×16	1.00	0.00	1.00	0.00
80×4	0.87	0.01	1.00	0.00
80×8	0.97	0.00	1.00	0.00
80×16	1.00	0.00	1.00	0.00
100×4	0.93	0.01	1.00	0.00
100×8	0.83	0.00	1.00	0.00
100×16	1.00	0.00	1.00	0.00

Table 4. Comparisons of algorithms ($f = 3$).

$n \times m$	CM			
	C(CMA,NSGA-II)	C(NSGA-II,CMA)	C(CMA,RA)	C(RA,CMA)
20×4	1.00	0.00	1.00	0.00
20×8	1.00	0.00	1.00	0.00
20×16	0.85	0.00	1.00	0.00
40×4	1.00	0.00	1.00	0.00
40×8	1.00	0.00	1.00	0.00
40×16	1.00	0.00	1.00	0.00
60×4	1.00	0.00	1.00	0.00
60×8	1.00	0.00	1.00	0.00
60×16	1.00	0.00	1.00	0.00
80×4	0.99	0.00	1.00	0.00
80×8	1.00	0.00	1.00	0.00
80×16	1.00	0.00	1.00	0.00
100×4	0.93	0.01	1.00	0.00
100×8	1.00	0.00	1.00	0.00
100×16	1.00	0.00	1.00	0.00

Table 5. Comparisons of algorithms ($f = 4$).

$n \times m$	CM			
	C(CMA,NSGA-II)	C(NSGA-II,CMA)	C(CMA,RA)	C(RA,CMA)
20×4	1.00	0.00	1.00	0.00
20×8	1.00	0.00	1.00	0.00
20×16	0.44	0.03	1.00	0.00
40×4	1.00	0.00	1.00	0.00
40×8	1.00	0.00	1.00	0.00
40×16	0.96	0.00	1.00	0.00
60×4	1.00	0.00	1.00	0.00
60×8	1.00	0.00	1.00	0.00
60×16	1.00	0.00	1.00	0.00
80×4	1.00	0.00	1.00	0.00
80×8	1.00	0.00	1.00	0.00
80×16	1.00	0.00	1.00	0.00
100×4	0.99	0.00	1.00	0.00
100×8	1.00	0.00	1.00	0.00
100×16	1.00	0.00	1.00	0.00

Table 6. Comparisons of algorithms ($f = 5$).

$n \times m$	CM			
	C(CMA,NSGA-II)	C(NSGA-II,CMA)	C(CMA,RA)	C(RA,CMA)
20×4	1.00	0.00	1.00	0.00
20×8	0.95	0.00	1.00	0.00
20×16	0.25	0.16	0.97	0.00
40×4	1.00	0.00	1.00	0.00
40×8	1.00	0.00	1.00	0.00
40×16	0.97	0.00	1.00	0.00
60×4	1.00	0.00	1.00	0.00
60×8	1.00	0.00	1.00	0.00
60×16	1.00	0.00	1.00	0.00
80×4	1.00	0.00	1.00	0.00
80×8	1.00	0.00	1.00	0.00
80×16	1.00	0.00	1.00	0.00
100×4	1.00	0.00	1.00	0.00
100×8	1.00	0.00	1.00	0.00
100×16	1.00	0.00	1.00	0.00

(RA,CMA) are significant with 95 % confidence interval. Thus, it is concluded that the CMA is more effective than the NSGA-II and RA in terms of the quality of the obtained solutions.

5 Conclusions

This is the first work to consider the carbon-efficient scheduling for the distributed permutation flow shop scheduling problem with makespan and total carbon emissions criteria. Some properties were analyzed, a competitive memetic algorithm was proposed, the effect of parameter setting was investigated, and the effectiveness of the designed CMA was demonstrated. Future work could focus on the design of the new search operators and new mechanisms to perform competition. It is also interesting to studying the carbon-efficient scheduling for other distributed scheduling problems.

Acknowledgement. This research is supported by the National Key Basic Research and Development Program of China (No. 2013CB329503) and the National Science Fund for Distinguished Young Scholars of China (No. 61525304).

References

1. Fang, K., Uhan, N., Zhao, F., Sutherland, J.W.: A new approach to scheduling in manufacturing for power consumption and carbon footprint reduction. J. Manuf. Syst. **30**, 234–240 (2011)
2. Mouzon, G., Yildirim, M.B., Twomey, J.: Operational methods for minimization of energy consumption of manufacturing equipment. Int. J. Prod. Res. **45**, 4247–4271 (2007)
3. Mouzon, G., Yildirim, M.B.: A framework to minimise total energy consumption and total tardiness on a single machine. Int. J. Sustain. Eng. **1**, 105–116 (2008)
4. Dai, M., Tang, D., Giret, A., Salido, M.A., Li, W.D.: Energy-efficient scheduling for a flexible flow shop using an improved genetic-simulated annealing algorithm. Robotics Comput.-Integr. Manuf. **29**, 418–429 (2013)

5. Yao F., Demers A., Shenker S.: A scheduling model for reduced CPU energy. In: 36th Annual Symposium on Foundations of Computer Science, pp. 374–382 (1995)
6. Bansal, N., Kimbrel, T., Pruhs, K.: Speed scaling to manage energy and temperature. J. ACM **54**, 3 (2007)
7. Fang, K., Uhan, N.A., Zhao, F., Sutherland, J.W.: Flow shop scheduling with peak power consumption constraints. Ann. Oper. Res. **206**, 115–145 (2013)
8. Ding, J.Y., Song, S., Wu, C.: Carbon-efficient scheduling of flow shops by multi-objective optimization. Eur. J. Oper. Res. **248**, 758–771 (2016)
9. Zheng, H., Wang, L.: Reduction of carbon emissions and project makespan by a Pareto-based estimation of distribution algorithm. Int. J. Prod. Econ. **164**, 421–432 (2015)
10. Pinedo, M.L.: Scheduling: Theory, Algorithms, and Systems. Springer, Berlin (2012)
11. Wang, B.: Integrated Product, Process and Enterprise Design. Chapman & Hall, London (1997)
12. Naderi, B., Ruiz, R.: The distributed permutation flowshop scheduling problem. Comput. Oper. Res. **37**, 754–768 (2010)
13. Gao, J., Chen, R., Deng, W.: An efficient tabu search algorithm for the distributed permutation flowshop scheduling problem. Int. J. Prod. Res. **51**, 641–651 (2013)
14. Wang, S., Wang, L., Liu, M., Xu, Y.: An effective estimation of distribution algorithm for solving the distributed permutation flow-shop scheduling problem. Int. J. Prod. Econ. **145**, 387–396 (2013)
15. Xu, Y., Wang, L., Wang, S., Liu, M.: An effective hybrid immune algorithm for solving the distributed permutation flow-shop scheduling problem. Eng. Optim. **46**, 1269–1283 (2014)
16. Fernandez-Viagas, V., Framinan, J.: A bounded-search iterated greedy algorithm for the distributed permutation flowshop scheduling problem. Int. J. Prod. Res. **53**, 1111–1123 (2015)
17. Naderi, B., Ruiz, R.: A scatter search algorithm for the distributed permutation flowshop scheduling problem. Eur. J. Oper. Res. **239**, 323–334 (2014)
18. Rifai, A.P., Nguyen, H.T., Dawal, S.Z.M.: Multi-objective adaptive large neighborhood search for distributed reentrant permutation flow shop scheduling. Appl. Soft Comput. **40**, 42–57 (2016)
19. Ong, Y.S., Lim, M., Chen, X.: Research frontier-memetic computation-past, present and future. IEEE Comput. Intell. Mag. **5**, 24–31 (2010)
20. Deng J., Wang L., Wang S.: A competitive memetic algorithm for the distributed flow shop scheduling problem. In: 2014 IEEE International Conference on Automation Science and Engineering, pp. 107–112. IEEE Press, New York (2014)
21. Nowicki, E., Smutnicki, C.: A fast tabu search algorithm for the permutation flow-shop problem. Eur. J. Oper. Res. **91**, 160–175 (1996)
22. Deb, K., Pratap, A., Agarwal, S., Meyarivan, T.: A fast and elitist multiobjective genetic algorithm: NSGA-II. IEEE Trans. on Evol. Comput. **6**, 182–197 (2002)
23. Li, B., Wang, L., Liu, B.: An effective PSO-based hybrid algorithm for multiobjective permutation flow shop scheduling. IEEE Trans. Syst. Man Cybern. Part A: Syst. Hum. **38**, 818–831 (2008)

A Developed NSGA-II Algorithm for Multi-objective Chiller Loading Optimization Problems

Pei-yong Duan[1], Yong Wang[1], Hong-yan Sang[1], Cun-gang Wang[1],
Min-yong Qi[1], and Jun-qing Li[1,2(✉)]

[1] School of Computer, Liaocheng University, Liaocheng 252059, China
Lijunqing.cn@gmail.com
[2] State Key Laboratory of Synthetical Automation for Process Industries,
Northeastern University, Shenyang 110819, China

Abstract. During recent years, for its simplicity and efficiency, the non-dominated-sorting algorithm (NSGA-II) has been widely applied to solve multi-objective optimization problems. However, in the canonical NSGA-II, the resulted population may have multiple individuals with the same fitness values, and which makes the resulted population lack of diversity. To solve this kind of problem, in this study, we propose a developed NSGA-II algorithm (hereafter called NSGA-II-D). In NSGA-II-D, a novel duplicate individuals cleaning procedure is embedded to delete the individuals the same fitness values with other ones. Then, the proposed algorithm is tested on the well-known ZDT1 instance to verify the efficiency and performance. Finally, to solve the realisitc optimization problem in intelligent building system, we select a well-known optimal chiller loading (OCL) problem to test the ability to maintain population diversity. Experimental results on the benchmarks show the efficiency and effectiveness of the proposed algorithm.

Keywords: Non-dominated-sorting algorithm · Optimal chiller loading (OCL) · Multi-objective optimization · Population diversity

1 Introduction

During recent years, multi-objective evolutionary algorithms (MOEA) have enjoyed more and more attention both in academic and industrial areas. Instead of providing a single solution for the decision maker, the multi-objective algorithms can present a set of optimal solutions or Pareto-optimal solutions in a single run. All of these Pareto-optimal solutions generally non-dominate with each other, and therefore, the decision maker can select one or several solutions in a prefer way. Therefore, compared with the weighted sum method which can only provide one solution at a run time, multi-objective evolutionary algorithms always have following advantages: (1) population diversity, which means that the algorithm can provide many solutions with different optimal results in different objectives; (2) the user will have different prefer selection for different scenario.

© Springer International Publishing Switzerland 2016
D.-S. Huang et al. (Eds.): ICIC 2016, Part I, LNCS 9771, pp. 489–497, 2016.
DOI: 10.1007/978-3-319-42291-6_49

Because MOEA has so many advantages, during recent decades, MOEA enjoyed more attentions. However, MOEA also have many shortcomings which make it receive more and more research attentions. The main issues in considering the designing the multi-objective optimization algorithms are as follows: (1) lack of pressure to converge to the Pareto front, in most cases, because of the problem structure, the algorithm lacks the convergence pressure to the Pareto front; (2) it is hard to maintain the population diversity. The main reason is that it is a tradeoff to maintain the population diversity while increase the algorithm performance. If the algorithm has the ability to maintain the population diversity, it is common lack of performance convergence and vice versa. Therefore, how to make a balance between the performance and diversity is the main issue of the many-objective evolutionary algorithm.

The rest of this paper is organized as follows: Sect. 2 briefly review the literature about the multi-objective optimization algorithms. Next, the optimal chiller loading (OCL) problem is presented in Sect. 3. In Sect. 4, the duplicated individuals cleaning procedure is described. Section 5 illustrates the experimental results and makes comparisons to the performances of algorithms from the literature to demonstrate the superiority of the proposed algorithm. Finally, the last section presents the concluding remarks and future research directions.

2   Literature Review

During recent years, multi-objective optimization algorithms have enjoyed more attention. The main development of the multi-objective algorithms can be divided into three stages, i.e., before 2000, between 2000 and 2007, and after 2007.

Before 2000, most of the multi-objective evolutionary algorithms used non-dominated sorting and sharing approach. Normally, most of these MOEA algorithms have following drawbacks [1]: (i) The computational complexity is $O(MN^3)$, where M is the objective number and N is the number of variables. (ii) Most of the algorithms have not used elitism approach, that is, the optimal solutions found so far will be lost in the following evolutional stages, which decrease the performance and convergence ability of the algorithms. (iii) Most of the algorithms have the need for specifying a sharing parameter, which makes the algorithms parameter related and hard to apply to different realistic problems.

After 2000, Deb's NSGA-II and MOEA/D have been applied to solve many kinds of optimization problems. The main challenges of the multi-objective optimization algorithms can be found in Table 1.

Table 1. Challenges of the MOEA algorithms

Challenges	Methods	References
Non-dominated population	Pareto partial dominance	Sato et al. [2]
		Aguirre and Tanaka [3]

(Continued)

Table 1. (Continued)

Challenges	Methods	References
	Subset of the objectives	
	ε-domination principle	ε-MOEA [4], ε-NSGA-II [5], Borg-MOEA [6] and AGE-II [7].
	Approximate Pareto set	Papadimitriou and Yannakakis [8] and Erlebach et al. [9]
	User preference	K. Deb. [10]. Reference points (e.g., [11, 12]) Weights in the objective space (e.g., [13, 14])
Computational efficiency	Hyper-volume	Hyper-volume [15–17].
	Well-spread reference points	Deb and Jain [18]
Recombination	Recombination scheme	Deb and Jain [18]
Visualization	Objective reduction methods	K. Deb and D. K [18]
	Parallel coordinates plot	A. Inselberg and B. Dimsdale [19]
	Heat maps	A. Pryke, S. Mostaghim, and A. Nazemi. [20]
	Sophisticated methods	Decision maps [21];Geodesic maps [22]

3 The OCL Problem

In an HVAC system, a multi-chiller system always contains multiple chillers that are connected to each other by parallel or series piping. The parallel structure is used to provide operational flexibility, standby capacity and less disruption maintenance.

The part load ratio (PLR) is commonly used for each chiller, which is the ratio of the chiller load to the design capacity for the chiller. Considering a multi-chiller system, the power consumption for each chiller is a function of its PLR value, which is computed as follows:

$$P_i = a_i + b_i\text{PLR}_i + c_i\text{PLR}_i^2 + d_i\text{PLR}_i^3 \tag{1}$$

where a_i, b_i, c_i, and d_i are coefficients of interpolation for consumed power versus PLR of the ith chiller in the multi-chiller system.

Based on the PLR computed above-mentioned, the objective function of the system is to minimize the total sum of the PLR of all chillers, i.e., $f_1 = \sum_{i=1}^{N} P_i$, where N is the number of chillers in the system, and P_i is the consumed power energy by i^{th} chiller.

The constraint of the system is that each chiller should satisfy its assigned capacity limitation, which is computed as follows:

$$f_2 = \left| \sum_{i=1}^{N} PLR_i \times \overline{RT_i} - CL \right| \tag{2}$$

where $\overline{RT_i}$ is the capacity of ith chiller and CL is demanded cooling load of the system.

4 The Proposed Algorithm

In the canonical NSGA-II algorithm, the parent population (population size n) is used to generate the neighboring population or child population (population size n). After that, both the parent population and the neighboring population are utilized to combine into a mating population with population size $2n$. Then, the mating population is used to apply the non-dominate-sorting procedure and the selection mechanism to generate the population for the next generation. However, the canonical NSGA-II algorithm doesn't consider whether there exist duplicate individuals in the mating population. Therefore, in the resulted population, there may be some individuals with the same objectives, which make the population lack of diversity. To release this kind of disadvantage, we propose a *clean_duplicate* procedure, which is described in Fig. 1.

```
Procedure clean_duplicate
Input: parent population pa, neighboring population chd
Output: mating population mp
Begin
   For i: 1 to n
      Insert the i^th individual of chd into pa
   End for
   Initialize a flag vector fv in which each element is assigned to 0
   For i: 1 to 2n-1
      For j: i+1 to 2n
         If individual i has the same fitness values with individual j
            Set the flag fv_j with -1
         End for
      End for
   For i: 1 to 2n-1
      If individual fv_i =0 then
         Insert the individual i into the population mp
      End if
   End for
   Let sel=2n-|mp| and i=0
   Do while i<sel
      Randomly generate a solution k
      If there is none individual in mp has the same fitness with k
         Insert the individual k into the population mp
      End if
   End do
End
```

Fig. 1. The pseudo code of the *clean_duplicate* procedure

5 Experimental Results

5.1 Experiments Parameters

According to our detailed experiments, the parameters for the systems are set as follows: (i) the crossover probability is set to 0.3; (ii) the mutation probability is set to 0.02; (iii) the population size is set to 300; (iv) the maximum iteration times is set to 5000.

5.2 Efficiency for ZDT1

For solving the well-known ZDT1 instance, we present the experimental results of the canonical NSGA-II and the proposed NSGA-II-D algorithms. Figure 2 gives the results of the NSGA-II, with the population size and iteration times equal to 200 and 500, respectively. The experimental results due to the proposed NSGA-II-D algorithm are presented in Fig. 3 with the same parameters.

It can be concluded from Figs. 2 and 3 that: (1) the proposed NSGA-II-D algorithm achieves a well-distributed population performance, which is obviously better than the results of the canonical NSGA-II algorithm; (2) considering the performance, the proposed NSGA-II-D can achieve nearly all of the Pareto solutions in the Pareto front while the canonical NSGA-II algorithm ignore lots of the solutions after 500 iterations; (3) the comparison for the ZDT1 instance shows the efficiency of the proposed NSGA-II-D algorithm.

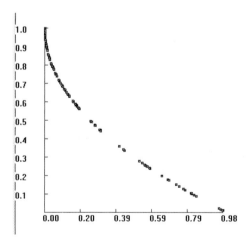

Fig. 2. population size, iteration time = (200,500), NSGA-II

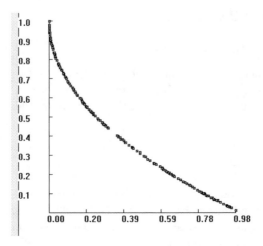

Fig. 3. population size, iteration time = (200,500), NSGA-II-D

5.3 Efficiency for OCL-1

One of the well-known OCL instance is shown in Table 2. The problem is to consider an area with five different types of chillers. The detailed parameters for these chillers are list in the following four columns. The last column describes the power capacity for each chiller.

For solving the well-known OCL instance, we present the experimental results of the canonical NSGA-II and the proposed NSGA-II-D algorithms. Figure 4 gives the results of the NSGA-II, with the population size and iteration times equal to 300 and 5000, respectively. The experimental results due to the proposed NSGA-II-D algorithm are presented in Fig. 5 with the same parameters.

It can be concluded from Figs. 4 and 5 that: (1) the proposed NSGA-II-D algorithm achieves a well-distributed population performance, which is obviously better than the results of the canonical NSGA-II algorithm; (2) considering the performance, the proposed NSGA-II-D can achieve nearly all of the Pareto solutions in the Pareto front while the canonical NSGA-II algorithm ignore lots of the solutions after 5000 iterations; (3) the comparison for the OCL instance shows the efficiency of the proposed NSGA-II-D algorithm.

Table 2. Chiller data

System	Chiller	a_i	b_i	c_i	d_i	Capacity (RT)
1	1	100.95	818.61	−973.4	788.55	800
	2	66.598	606.34	−380.58	275.95	800
	3	130.09	304.50	14.377	99.8	800
	4	104.09	166.57	−430.13	512.53	450
	5	−67.15	1177.79	−2174.53	1456.53	450

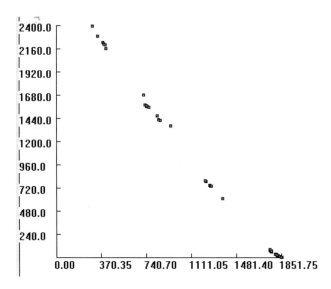

Fig. 4. population size, iteration time = (300,5000), NSGA-II

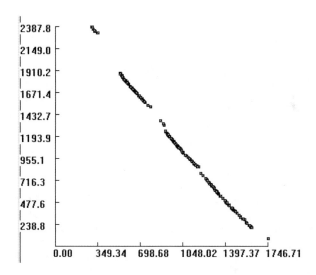

Fig. 5. population size, iteration time = (300,5000), NSGA-II-D

6 Conclusion

In this paper, we first review the literature about the multi-objective algorithms. Then, the three challenges of the multi-objective algorithms are discussed. Next, we proposed a simple procedure for cleaning the duplicated individuals in the canonical

NSGA-II algorithm. Experimental on several well-known benchmark instances show that the proposed algorithm is superior to several existing approaches in both solution quality and population diversity abilities.

Acknowledgments. This research is partially supported by National Science Foundation of China under Grant 61573178, 61374187 and 61503170, basic scientific research foundation of Northeastern University under Grant N110208001, starting foundation of Northeastern University under Grant 29321006, Science Foundation of Liaoning Province in China (2013020016), Key Laboratory Basic Research Foundation of Education Department of Liaoning Province (LZ2014014), Science Research and Development of Provincial Department of Public Education of Shandong under Grant J12LN39, Shandong Province Higher Educational Science and Technology Program (J14LN28), and Postdoctoral Science Foundation of China (2015T80798, 2014M552040).

References

1. Deb, K., Pratap, A., Agarwal, S., et al.: A fast and elitist multiobjective genetic algorithm: NSGA-II. IEEE Trans. Evol. Comput. **6**(2), 182–197 (2002)
2. Sato, H., Aguirre, H.E., Tanaka, K.: Pareto partial dominance MOEA and hybrid archiving strategy included CDAS in many-objective optimization. IEEE Congr. Evol. Comput. (CEC) **2010**, 1–8 (2010)
3. Aguirre, H., Tanaka, K.: Many-objective optimization by space partitioning and adaptive ε-ranking on MNK-landscapes. In: Ehrgott, M., Fonseca, C.M., Gandibleux, X., Hao, J.-K., Sevaux, M. (eds.) EMO 2009. LNCS, vol. 5467, pp. 407–422. Springer, Heidelberg (2009)
4. Deb, K., Mohan, M., Mishra, S.: A fast multi-objective evolutionary algorithm for finding well-spread pareto-optimal solutions. KanGAL report 2003002, Indian Institute of Technology, Kanpur, India (2003)
5. Kollat, J.B., Reed, P.: Comparison of multi-objective evolutionary algorithms for long-term monitoring design. Adv. Water Resour. **29**(6), 792–807 (2006)
6. Hadka, D., Reed, P.: Borg: an auto-adaptive many-objective evolutionary computing framework. Evol. Comput. **21**(2), 231–259 (2013)
7. Wagner, M., Bringmann, K., Friedrich, T., Neumann, F.: Efficient optimization of many objectives by approximation-guided evolution. Eur. J. Oper. Res. (2014). (ISSN 0377-2217)
8. Papadimitriou, C.H., Yannakakis, M.: On the approximability of trade- offs and optimal access of web sources. In 41st Annual Symposium on Foundations of Computer Science (FOCS) 2000, pp. 86–92. IEEE Press (2000)
9. Erlebach, T., Kellerer, H., Pferschy, U.: Approximating multi-objective knapsack problems. In: Dehne, F., Sack, J.-R., Tamassia, R. (eds.) WADS 2001. LNCS, vol. 2125, pp. 210–221. Springer, Heidelberg (2001)
10. Deb, K.: Recent developments in evolutionary multi-objective optimization. Trends in Multiple Criteria Decision Analysis. volume 142 of International Series in Operations Research & Management Science, pp. 339–368. Springer, Heidelberg (2010). ISBN 978-1-4419-5903-4
11. Deb, K., Sundar, J.: Reference point based multi-objective optimization using evolutionary algorithms. In: Proceedings of the 8th Annual Conference on Genetic and Evolutionary Computation, pp. 635–642. ACM (2006)

12. Nguyen, A.Q., Wagner, M., Neumann, F.: User preferences for approximation-guided multi-objective evolution. In: Dick, G., Browne, W.N., Whigham, P., Zhang, M., Bui, L.T., Ishibuchi, H., Jin, Y., Li, X., Shi, Y., Singh, P., Tan, K.C., Tang, K. (eds.) SEAL 2014. LNCS, vol. 8886, pp. 251–262. Springer, Heidelberg (2014)
13. Friedrich, T., Kroeger, T., Neumann, F.: Weighted preferences in evolutionary multi-objective optimization. In: Wang, D., Reynolds, M. (eds.) AI 2011. LNCS, vol. 7106, pp. 291–300. Springer, Heidelberg (2011)
14. Bader, J., Zitzler, E.: Hype: an algorithm for fast hypervolume- based many-objective optimization. Evol. Comput. 19(1), 45–76 (2011). ISSN 1063-6560
15. While, L.P., Barone, H.L., Huband, S.: A faster algorithm for calculating hypervolume. IEEE Trans. Evol. Comput. 10(1), 29–38 (2006). ISSN 1089-778X
16. Bringmann, K., Friedrich, T.: Approximating the volume of unions and intersections of high-dimensional geometric objects. In: Hong, S.-H., Nagamochi, H., Fukunaga, T. (eds.) ISAAC 2008. LNCS, vol. 5369, pp. 436–447. Springer, Heidelberg (2008)
17. Deb, K., Jain, H.: An evolutionary many-objective optimization algorithm using reference-point-based non-dominated sorting approach, part I: solving problems with box constraints. IEEE Trans. Evol. Comput. 18(4), 577–601 (2014). ISSN 1089-778X
18. Deb, K., Saxena, D.K.: On finding pareto-optimal solutions through dimensionality reduction for certain large-dimensional multi-objective optimization problems. KanGAL report 2005011, Indian Institute of Technology, Kanpur, India (2005)
19. Inselberg, A., Dimsdale, B.: Parallel coordinates: a tool for visualizing multi-dimensional geometry. IEEE Conf. Vis. 1990, 361–378 (1990)
20. Pryke, A., Mostaghim, S., Nazemi, A.: Heatmap visualization of population based multi objective algorithms. In: Obayashi, S., Deb, K., Poloni, C., Hiroyasu, T., Murata, T. (eds.) EMO 2007. LNCS, vol. 4403, pp. 361–375. Springer, Heidelberg (2007)
21. Mohammadi, A., Omidvar, M., Li, X., Deb, K.: Integrating user preferences and decomposition methods for many-objective optimization. IEEE Congr. Evol. Comput. (CEC) 2014, 421–428 (2014)
22. Tenenbaum, J.B., de Silva, V., Langford, J.C.: A global geometric framework for nonlinear dimensionality reduction. Science 290(5500), 2319–2323 (2000)

Crane Scheduling in Coordination
of Production and Transportation Process

Xie Xie[✉]

Key Laboratory of Manufacturing Industrial and Integrated Automation,
Shenyang University, Shenyang 110044, Liaoning, China
xiexie8118@gmail.com

Abstract. The paper takes iron and steel making of the iron and steel enterprise for background, researches a crane scheduling problem in coordination of production and transportation process with transport vehicle capacity limitation consideration and transportation time consideration to minimize makespan. Two special cases are analyzed polynomially solvable and the general case is further demonstrated NP-hard. We propose a heuristic algorithm based on crane coordinated. Computational results are given to show the efficiency of the proposed algorithm compared with that of the staged algorithm from practical production.

Keywords: Crane scheduling · Complexity analysis · Heuristic algorithm

1 Introduction

Research on traditional crane scheduling problem divided crane operation and production scheduling separately. It is commonly taking production in the first place and crane operations in a subordinate position, that is, arranging production scheduling first and then scheduling crane. However, in the practical of a steelmaking shop, crane often connects with other transportation tools. Since the limitations of number and capacity for the transportation tools, the transfer of materials between productions is limited. It is difficult to effectively implement even an optimal scheduling without considering transportation. Moreover, most of the transported objectives in the iron and steel enterprises have high temperature and large unit price, the transporting time of each process is demanding due to the continuous operation, effectively carrying out crane scheduling between production and transportation helps to reduce energy consumption, improves the efficiency of production equipment and transportation tools, and ensures real-time performance requirements and production.

Our problems arising from steelmaking which is a subsystem in a steelmaking shop. We propose a crane scheduling problem in coordination of production and transportation process. As illustrated in Fig. 1, the melted steel in ladles that are transported firstly by trolleys after blast furnace and then by cranes mounted on the tracks over the trolley railway to converter. If trolley arrives and crane is busy, then trolley must wait in its designated position and thus idle times in converter appear. Usually, waiting times should not be too long for the molten steel temperature drop. Once hot consumption generates and the steelmaking time in converter increases.

D.-S. Huang et al. (Eds.): ICIC 2016, Part I, LNCS 9771, pp. 498–504, 2016.
DOI: 10.1007/978-3-319-42291-6_50

In order to avoid this waste, improve the utilization rate of converter, and reduce the transportation cost before steel-making production, the objectives to be scheduled concern about not only trolley (transport vehicle) waiting time is not more than a certain value but also the capacity of the transport vehicle is limited.

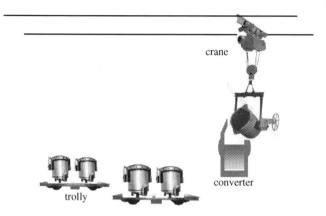

Fig. 1. Figure for coordination between transportation and one machine production with crane scheduling

To the best of our knowledge, most existing literatures on crane scheduling problem deal with two settings: one is hoist scheduling in electroplating lines and the other is quay or yard crane scheduling in container terminals. Philips and Unger [1] first proposed hoist scheduling problem in electroplating line. They developed a mixed integer linear programming (MILP) model to determine the optimal cycle time. However, in the last thirty years, people have been concerned about the scheduling of single crane and the simultaneous scheduling of multiple cranes is ignored. One important reason is that multi-crane scheduling problem is more complex than single crane scheduling problem. In practical, as a result of a cross on the existence of at least two cranes, it is necessary to take into account crane allocation for avoiding the collision between cranes. For double-crane scheduling problem, Lei and Wang [2] used a named 'zoned' method to propose a heuristic algorithm using a partitioning approach by which a system is partitioned into two sets of contiguous workstations and each hoist is assigned to a set. A sequence of alternative partitions is evaluated. Leung et al. [3] presented a MILP formulation for finding the minimum-time cycle for lines with multiple hoists and present valid inequalities for this problem. Kats and Levner [4] proposed an algorithm that provides an exact solution for more complicated case of multiple cranes and solves the problem in O(n2logn). Moreover, Daganzo [5] first proposed quay crane scheduling and presented exact and approximate solution methods. The approximation methods are based on optimality principles and are easy to implement. The exact methods can only be used for a few ships.

Guan et al. [6] developed both exact and heuristic solution approaches for the problem. For small-sized instances, they proposed a time-space network flow formulation and applied an exact solution approach to obtain an optimal solution. For medium-sized and large-sized instances, they proposed a Lagrangian relaxation approach and two heuristics respectively. Tang et al. [7] addressed the joint quay crane and truck scheduling problem at a container terminal, considering the coordination of the two types of equipment to reduce their idle time between performing two successive tasks. They proposed a MILP and an improved particle swarm optimization algorithm. Kaveshgar and Huynh [8] developed a mixed integer programming model for scheduling quay cranes and yard trucks jointly. The integrated model explicitly considered real-world operational constraints such as precedence relationships between containers, blocking, quay crane interference, and quay crane safety margin. To solve the integrated optimization model, a genetic algorithm (GA) combined with a greedy algorithm was developed.

While a number of problems discussed in the literature above are similar, none appear to address the particular crane scheduling problem in coordination of production and transportation process we present in this paper. We regard steel ladle as a job, trolley as a transport vehicle and converter as a machine, that is, job in batches which carried by transport vehicle for process on a single machine, once a job arrives, crane lifts the job and put it into the machine to start processing. The paper considers a single crane scheduling problem in coordination of production and transportation process with the number of transport vehicle, transport vehicle capacity limitation, and transportation time consideration to minimize makespan. If the heat job is transported to the pre processed buffer and can not be started immediately, the actual processing time increases linearly with the increase of waiting time. Since actual processing time and the waiting time is related that it will lead to the increase of the actual processing time of a job and the maximum completion time of the job delayed greatly.

2 Problem Description

The problem can be described as follows, a set of n jobs $N = \{J_1, J_2, ..., J_n\}$ are first transported to buffer area and then lifted by crane to machine. There is infinite buffer space in front of the machine. The job transportation time depends (1) the transportation time in trolley t_i^1, (2) the time from crane lifting to converter t_i^2, once a steel ladle is lifted by crane to machine, the empty steel ladle is carried to trolley. We denote $t_i^{2'}$ as the transportation time from crane to trolley while $t_i^{1'}$ as the returning time by trolley to its original position. The number of transport vehicle is $v = 1$ and transport vehicle capacity is C.

If a vehicle can transport more than one job once, the transportation time is defined as the longest transportation time of a job. The processing time of a job is defined as $P_i = b_i + a_i w_i$, where $b_i \geq 0$ is the basic time, without loss of generality, we have $b_i \geq t_i^{2'}$, w_i is the waiting time for job j before processing, $a_i \geq 0$ is the increased processing time in the unit waiting time, where $a_i = 0$ means that the cold job which the actual processing time is not increased with the increase of the waiting time. The waiting time for job j w_j is defined as the time window $[t_i, t_i']$, where t_i is the arrival time for job j. In the paper, $t_i = t_i^1 + t_i^2$, t_i' is the starting processing time and the objective is minimize makespan. Using the conventional notation, the problem can be described as by Granham et al. [9] $T \rightarrow 1 \rightarrow 1 \mid c = C, P_i = b_i + a_i w_i \mid C_{\max}$.

3 Complexity Analysis

This section studies scheduling models with only one transport vehicle problem for a variety of situations. The problem is expressed as $T \rightarrow 1 \rightarrow 1 \mid c = C, P_i = b_i + a_i w_i \mid C_{\max}$. Each job is first transported by a vehicle and then lifted by crane to machine.

If we do not consider the return time of vehicle and crane, the problem with no temperature drop for job, a single vehicle with the given capacity, at this time, we take trolley and crane transports as processing machines respectively, $t_i = t_i^1 + t_i^2$ is considered as the processing time on the first machine. The problem is equivalent to a traditional problem $F_2 \| C_{\max}$ solved by Johnson rule [10] optimally.

Lemma 1. If we do not consider the return time of vehicle and crane, problem $T \rightarrow 1 \rightarrow 1 \mid P_i = b_i + 0 w_i \mid C_{\max}$ is optimally solved by Johnson rule.

Now we research the problem $T \rightarrow 1 \rightarrow 1 \mid P_i = b_i + 0 w_i \mid C_{\max}$ with return time of vehicle and crane $t_i^{1'}$ and $t_i^{2'}$ are constant where $t > 0$.

Since all jobs are available at time zero and $b_i \geq t_i^{2'}$, problem $T \rightarrow 1 \rightarrow 1 \mid P_i = b_i + 0 w_i \mid C_{\max}$ is considered as the problem with an extra transportation time. The time $t_i^1 + t_i^2 + t_i^{1'}$ is also considered as the processing time on the first machine (dummy). Define P_i the processing time on the second machine, the case can be also optimally solved by Johnson rule. Therefore, the return time of vehicle and crane are constant, our problem can be solved optimally in polynomial time.

Considering the case that job processing time is a linear function of the waiting time, the problem $T \rightarrow 1 \mid P_i = b_i + a_i w_i \mid C_{\max}$ is reduced to a strongly NP-hard problem studied by Sriskandarajah and Goyal [11] where the processing time on the second machine is $P_i = b_i + a_i w_i$. Therefore, even we do not consider the return time of vehicle and crane, problem $T \rightarrow 1 \rightarrow 1 \mid P_i = b_i + a_i w_i \mid C_{\max}$ is still strongly NP-hard.

Lemma 2. Even without considering the return time of vehicle and crane, problem $T \rightarrow 1 \rightarrow 1 \mid P_i = b_i + a_i w_i \mid C_{\max}$ is still strongly NP-hard.

For solving the problem, we propose a heuristic algorithm.

4 Heuristic Algorithm

We propose a heuristic algorithm for the problem $T \rightarrow 1 \rightarrow 1 \mid c = k, P_i = b_i + a_i w_i \mid C_{\max}$ (where $t_i^2 = 0$) with the vehicle capacity given as a constant. Without loss of generality, we assume the number of job is the integer multiple that of vehicle capacity $n = l \, k$ and maximum capacity limit for all transport vehicles. If not, we can add some jobs with basic processing time, deterioration factor and transport time are zero for the total job number $n = l \, k$.

The optimal algorithm commonly used in actual production stage is: dividing the order of the transportation, processing production into several stages, optimal solution solved in the first stage as the input values in the second stage, in turn, until the optimal solution obtained in the last stage. In this part, we regard the transportation problem as the first stage, and machine processing as the second stage. First consider the case with the shortest transportation time, then we obtain the arrival time for all jobs. The production in the second stage arranges according to the time of arrival of the job to obtain minimum the maximum completion time. We use the following algorithm 1 to obtain the lower bound.

Define T_{ik} is the time for reaching the machine in the ith time when vehicle capacity k, and C'_{ik} is the total processing time for transporting the number of $(i-1)\,k$ jobs. The definition of C'_{ik} for the vehicle before the I transport to all parts before k jobs in the processing machine and the initial processing time, C_{ik} is the maximum completion time for transporting job in the ith time.

Algorithm

Step 1: All jobs in accordance with the transport time of the order $t^1_1+t^2_1\leq t^1_2+t^2_2\leq$

$\ldots\leq t^1_{lk}+t^2_{lk}$ here we have $T_{ik}=\displaystyle\sum_{j=1}^{i} t^1_{j\times k} + t^2_{j\times k}$, $i=1,\ldots,l$.

Step 2: according to the sequence in Step 1, take k for a number of jobs in a batch, where each batch of internal job in non-decreasing b_i/a_i.

Step 3: The initial processing time of all jobs is regarded as the actual processing time, that is without taking into account the deterioration of the job for $i=1,\ldots,l$,

$C_{(i+1)k}=\max\{t_k+\displaystyle\sum_{j=1}^{i} C_{jk}{}', T_{(i+1)k}\}+C_{(i+1)k}{}'$.

Step 4: If a number of the last two jobs $i-1, i$ satisfies 1) $k_i\geq t_i$ and $t_i+a_i\,k_i\geq t_{i-1}$ or 2) $k_i<t_i$ and $(a_i+1)\,k_i\geq t_{i-1}$, then job i is split as a single batch.

According to Step 4, from the front split out some small batch, and recombining the immediately behind the number may bring smaller objective function value. Then we give the following step for merging those mall batches.

Step 5: According to Step 4 the split batch and if two adjacent batches are combined (the number of jobs is more than k), then merger batches until there is no batch can continue to merge.

5   Computational Results

The algorithm was programmed in C language and run on a PC with Pentium-IV (2.40 GHz) CPU using the windows XP operating system. In the numerical experiment, for each combination of $n\in\{20,30,40,50,60,70,80,90,100\}$ and k $\{2,4,6,8,10,12,14,16\}$, 20 examples are generated $C_{\max}$ by algorithm and the total number is 1440 where the transportation time t^1 and t^2_i, the basis job processing time b_i, coefficient of deterioration a_i obey the uniform distribution [1,100], [1,100] and [0,1] respectively. Define the objective function of the algorithm is C^A and the lower bound is C^S. $(C^S - C^A)/C^S$ represents the improvement of our algorithm (Table 1).

Table 1. Average improvement from separate algorithm to our Algorithm

k	2	4	6	8	10	12	14	16
n = 20	3.530	20.034	31.322	41.581	62.446	2.675	83.112	83.936
n = 30	5.112	21.952	40.901	47.183	69.155	70.523	89.678	86.319
n = 40	5.103	21.396	39.622	52.118	66.518	77.045	85.188	88.704
n = 50	4.582	20.748	40.923	59.521	69.323	80.985	80.895	87.985
n = 60	4.699	22.647	38.014	56.043	70.829	75.344	82.396	89.978
n = 70	4.195	24.152	40.438	52.637	70.239	76.771	84.744	90.235
n = 80	4.439	22.143	38.394	52.031	67.427	74.871	85.813	90.150
n = 90	4.801	21.466	37.756	56.212	71.101	78.038	85.494	89.484
n = 100	4.825	22.434	38.522	56.703	70.322	77.541	86.130	90.010
Average	4.587	21.886	38.432	52.670	68.596	75.977	84.828	88.533

(1) The calculation results show that when the vehicle capacity is greater than 1, the coordination algorithm has obvious improvement with respect to the discrete algorithm. The proposed heuristic algorithm is not less than the minimum average improvement 4 % of the phased algorithm, and the maximum average improvement can be achieved 88 %.
(2) With the capacity of the vehicle increasing, our proposed approximate algorithm is improved obviously more and more. This is because of the number of more jobs more than less jobs in the same batch for the heuristic improvement, thereby decreasing the maximum completion time is more obvious.
(3) For fixed vehicle capacity, when the number of the job is increased, the improvement of the coordination algorithm with respect to the discrete optimal algorithm is not obvious.

6 Conclusions

The paper takes the iron and steel-making process in the iron and steel enterprises as background, considers crane scheduling problem in coordination of a single machine production and trolley transportation with the number of transport vehicles and capacity limitation, the objective function is to minimize the maximum completion time. We analyze the property of the problem, even if without considering the return time of trolley and crane, the problem is also a strong NP-hard. A heuristic algorithm is constructed, and further the numerical calculation is carried out. The experimental results show that the proposed heuristic algorithm is not less than the minimum average improvement 4 % of the phased algorithm, and the maximum average improvement can be achieved 88 %.

Acknowledgements. This research is supported by National Natural Science Foundation of China (Grant No. 71201104). The colleges and universities of Liaoning Province outstanding talent support plan (LJQ2014133).

References

1. Phillips, L.W., Unger, P.S.: Mathematical programming solution of a hoist scheduling problem. AIIE Trans. **28**, 219–225 (1976)
2. Lei, L., Wang, T.J.: The minimum common-cycle algorithm for cyclic scheduling of two material handling hoists with time window constraints. Manag. Sci. **37**, 1629–1639 (1991)
3. Leung, J.M.Y., Zhang, G.Q., Yang, X.G., Mak, R., Lam, K.: Optimal cyclic multi-hoist scheduling: a mixed integer programming approach. Oper. Res. **52**(6), 965–976 (2004)
4. Kats, V., Levner, E.: Cyclic scheduling in a robotic production line. J. Sched. **47**, 23–41 (2002)
5. Daganzo, C.F.: The crane scheduling problem. Transp. Res. B **23**(3), 159–175 (1989)
6. Guan, Y.P., Yang, K.-H., Zhou, Z.L.: The crane scheduling problem: models and solution approaches. Ann. Oper. Res. **203**, 119–139 (2013)
7. Tang, L.X., Zhao, J., Liu, J.Y.: Modeling and solution of the joint quay crane and truck scheduling problem. Eur. J. Oper. Res. **236**, 978–990 (2014)
8. Kaveshgar, N., Huynh, N.: Integrated quay crane and yard truck scheduling for unloading inbound containers. Int. J. Prod. Econ. **159**, 168–177 (2015)
9. Graham, R.L., Lawler, E.L., Lenstra, J.K., Rinnooy Kan, A.H.G.: Optimization and approximation in deterministic sequencing and scheduling theory: a Survey. Ann. Discrete Math. **5**, 287–326 (1979)
10. Johnson, S.M.: Optimal two- and three-stage production schedules with setup times included. Naval Res. Logistics Q. **1**(1), 61–68 (1954)
11. Sriskandarajah, C., Goyal, S.K.: Scheduling of a two-machine flowshop with processing time linearly dependent on job waiting-time. J. Oper. Res. Soc. **40**(10), 907–921 (1989)

Hybrid Estimation of Distribution Algorithm for No-Wait Flow-Shop Scheduling Problem with Sequence-Dependent Setup Times and Release Dates

Zi-Qi Zhang, Bin Qian[✉], Rong Hu, Chang-Sheng Zhang,
and Zi-Hui Li

Department of Automation, Kunming University of Science and Technology,
Kunming 650500, China
bin.qian@vip.163.com

Abstract. This paper proposes an innovative hybrid estimation of distribution algorithm (HEDA) for the no-wait flow-shop scheduling problem (NFSSP) with sequence dependent setup times (SDSTs) and release dates (RDs) to minimize the total completion time (TCT), which has been proved to be typically NP-hard combinatorial optimization problem with strong engineering background. Firstly, a speed-up evaluation method is developed according to the property of NFSSP with SDSTs and RDs. Secondly, the genetic information both order of jobs and the promising blocks of jobs are concerned to generate the guided probabilistic model. Thirdly, after the HEDA based global exploration, a problem dependent local search is developed to emphasize exploitation. Due to the reasonable balance between HEDA based global search and problem-dependent local search as well as the comprehensive utilization of the speed-up evaluation, TCT-NFSSP with SDSTs and RDs can be solved effectively and efficiently. Computational results and comparisons demonstrate the superiority of HEDA in terms of searching quality, robustness, and efficiency.

Keywords: Estimation of distribution algorithm · No-wait flow-shop scheduling problem · Sequence-dependent setup times · Release dates · Local search · Speed-up evaluation

1   Introduction

Flow shop scheduling problems (FSSPs) have attracted much attention in the manufacturing systems of contemporary enterprises and wide research in both computer science and operation research fields. The no-wait flow shop problem (NFSSP) is a special case of FSSPs, in which there should be no waiting time between successive operations of jobs. That is, under the no-wait circumstances, each job is required to be processed continuously from start to end either on or between machines without any interruption. In addition, each machine can handle no more than one job at a time and each job has to visit each machine exactly once. In NFSSP, it is usually assuming the release time of all jobs is zero and the setup time on each machine is included in the job

© Springer International Publishing Switzerland 2016
D.-S. Huang et al. (Eds.): ICIC 2016, Part I, LNCS 9771, pp. 505–516, 2016.
DOI: 10.1007/978-3-319-42291-6_51

processing time. However, the setup times are very important in some practical applications as noted in Allahverdi [1], which need to be explicitly treated and the release dates are usually nonzero. For the total completion time criterion, the NFSSP with SDSTs and RDs is classified as $Fm/no - wait, ST_{sd}, r_j/ \sum C_j$, which can also be identified as TCT-NFSSP with SDSTs and RDs. For the computational complexity of the NFSSP, Garey and Johnson proved that it was NP-hard. Because it turns out that the $1//\sum C_j$ is NP-hard and it also reduces to $Fm/no - wait, ST_{sd}, r_j/\sum C_j$ (i.e., $1//\sum C_j \propto Fm/no - wait, ST_{sd}, r_j/\sum C_j)$, thus it can be undoubtedly concluded that $Fm/no - wait, ST_{sd}, r_j/\sum C_j$ is also NP-hard [2]. Bianco et al. [3] further indicated that the NFSSP with SDSTs and RDs was equivalent to the asymmetric travelling salesman problem with additional visiting time constraints, which is proved to be strongly NP-hard. Because of the NFSSP with SDSTs is proven to be NP-hard, let alone with RDs, many heuristics and meta-heuristics evolutionary algorithms (EAs) have been developed recent years. Some of the developing algorithms assumed that the setup times can be ignored by considering setup times either negligible or as part of the processing times. A corresponding importance survey of setup times in NFSSP is provided by Allahverdi et al. [2]. Ruiz and Stützle [4] proposed the iterated greedy algorithms incorporated local search (IG_LS). Experimental results showed that IG_LS is the state-of-the-art methods. Qian et al. [5] proposed hybrid differential evolution (HDE) to minimize the makespan of the NFSSP with SDSTs and RDs. Simulation results and comparisons demonstrate the superiority of HDE in terms of searching quality, robustness, and efficiency. Very recently, Ding et al. [6] developed a tabu-mechanism improved iterated greedy (TMIIG) algorithm to enhance the exploration ability, which combined with insert, swap, and double-insert neighborhood search to obtain better solution. The results showed that the TMIIG is more effective than other existing well-performing heuristic algorithms. Allahverdi and Aydilek [7] investigated the improved simulated annealing algorithm (ISA-2) to address NFSSP with setup times and computational analysis indicated that the proposed ISA-2 performed significantly better than the other algorithms. Samarghandi and ElMekkawy [8] proposed particle swarm optimization (PPSO) with matrix coding for NFSSP with SDSTs. Computational results show that the PPSO have a better performance than other algorithms. Moreover, as we can see from the literature reviews, the researches on this concerned problem with both SDSTs and RDs are considerably scarce. Therefore, it is meaningful to develop effective algorithm more lucubrate for addressing this problem.

As a novel probabilistic model, population-based incremental learning (PBIL) introduced by Baluja [9] for addressing TSP and FSSP. PBIL generally uses an independence probability model to guide search direction, which is the early model of the EDA developed by Mühlenbein and Paass. The evolution of EDAs is regarded as a statistical learning, whose model is updated by the current superior solutions, contrary to crossover and mutation operator in GAs. Owing to its outstanding global exploration ability and inherent parallelism, EDA has attracted much attention in the field of EAs and has already been extensively applied to deal with the production scheduling. Particularly, Ceberio et al. [10] reviewed the EDAs in combinatorial optimization problems, which is of great importance given excellent efforts in the area of EDAs for researchers. Pan and Ruiz [11] presented the probabilistic model by taking into account both job permutation and promising similar blocks of jobs, and also employed an efficient NEH initialization and local search to improve the performance of the EDA. Furthermore, Wang et al. [12] proposed an effective bi-population EDA with local search scheme based on the critical path to address the flexible job-shop scheduling problem. Recently, Wang et al. [13] proposed an order-based estimation of distribution algorithm which adopted optimal computing budget allocation technique to provide a reliable identification to the promising solutions. It appears that the tradeoff between a proper model with updating mechanism are crucial to EDAs.

Consequently, to the best of our knowledge and review the development of EDAs, there is no any published work on EDAs for the NFSSP with SDSTs and RDs. In the proposed HEDA, EDA is adopted to find the superior solutions and guide the search direction to the promising regions, and then the local search using speed-up evaluation is designed to emphasize the exploitation from those promising regions.

The remainder of this paper are organized as follows. Section 2 introduces the TCT-NFSSP with SDSTs and RDs. Section 3 presents HEDA in details, such as speed-up evaluation, HEDA global search, and problem-dependent local search, respectively. Computational comparisons are presented and analyzed in Sect. 4. Finally, Sect. 5 gives some concluding remarks and suggestions of future research.

2 TCT-NFSSP with SDSTs and RDs

The NFSSP with SDSTs and RDs can be described as follows. There are n jobs and m machines. Each of n jobs will be continuously processed on all machines $1, 2, \ldots, m$, respectively. To satisfy the no-wait restriction, each job must be processed without interruptions between consecutive machines and each machine just can process no more than one job. In a FSSP with SDSTs, setup times dependent upon both the current and the immediately preceding jobs at each machine. Nevertheless, if a machine is ready to process a job but the job has not been completely released yet, it must stay idle until the release date of the job meets requirement.

2.1 NFSSP with SDSTs

Denote $\pi = [j_1, j_2, \ldots, j_n]$ the permutation of jobs, $p_{j_i,l}$ the processing time of job j_i on machine l, sp_{j_i} the total processing time of job j_i on all machines, $ML_{j_i,l}$ the minimum delay on the machine l between the completion of job j_{i-1} and j_i, $L_{j_{i-1}j_i}$ the minimum delay on the first machine between the start of job j_{i-1} and j_i, $s_{j_{i-1}j_i,l}$ the completely determinate SDSTs between job j_{i-1} and j_i on machine l. Let $p_{j_0,l} = 0$, $l = 1, \ldots, m$ for initial job. Then $ML_{j_i,l}$ can be calculated as follows:

$$ML_{j_i,l} = \begin{cases} \max\{s_{j_{i-1}j_i,1} + p_{j_i,1} - p_{j_{i-1},2}, s_{j_{i-1}j_i,2}\} + p_{j_i,2}, l = 2 \\ \max\{ML_{j_i,l-1} - p_{j_{i-1},l}, s_{j_{i-1}j_i,l}\} + p_{j_i,l}, l = 3, \ldots, m \end{cases}. \tag{1}$$

Therefore, the total completion time $C_T(\pi)$ (i.e., $C_T(\pi) = \sum C_j$) is as follows:

$$C_T(\pi) = \sum_{i=1}^{n}(n+1-i)ML_{j_i,m}. \tag{2}$$

Accordingly, L_{j_{i-1},j_i} can be calculated by utilizing the following formula (Fig. 1):

$$L_{j_{i-1},j_i} = ML_{j_i,m} + sp_{j_{i-1}} - sp_{j_i}. \tag{3}$$

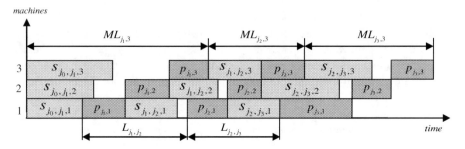

Fig. 1. Gantt chart for NFSSP with SDSTs when $n = 3$ and $m = 3$.

2.2   TCT-NFSSP with SDSTs and RDs

Denote r_{j_i} the arrival time of job j_i, St_{j_i} the process starts time of job j_i on machine 1, C_{j_i} the completion time of job j_i on machine j_i. Then St_{j_i} can be written as follows:

$$St_{j_i} = \begin{cases} \max\{ML_{j_i,m} - sp_{j_i}, r_{j_i}\}, \ i = 1 \\ St_{j_{i-1}} + \max\{L_{j_{i-1},j_i}, r_{j_i} - St_{j_{i-1}}\}, \ i = 2,\ldots,n \end{cases}. \tag{4}$$

Hence, C_{j_i} and $\mathrm{TCT}(\pi)$ can be calculated as follows:

$$C_{j_i} = St_{j_i} + sp_{j_i}, i = 1,\ldots,n, \tag{5}$$

$$C_T(\pi) = \sum_{i=1}^{n} C_{j_i}. \tag{6}$$

The aim of this paper is to find a schedule π^* in the set of all schedules Π such that

$$\mathrm{TCT}(\pi^*) = \min_{\pi \in \Pi} \mathrm{TCT}(\pi) = \min_{\pi \in \Pi} C_T(\pi). \tag{7}$$

3   HEDA for TCT-NFSSP with SDSTs and RDs

In this section, we will propose the HEDA for TCT-NFSSP with SDSTs and RDs after explaining the solution representation, speedup evaluation method, probabilistic model, updating mechanism and new population generation method of global search, and problem dependent local search, respectively.

3.1   Solution Representation and Population Initialization

Because of the properties of TCT-NFSSP with SDSTs and RDs, we adopt the permutation coding based solution representation, that is, every individual is a feasible solution. For example, $[\pi_1, \pi_2, \ldots, \pi_5] = [5, 2, 1, 4, 3]$ is an individual when the scale of problem n is set to 5, which convey that the first processing job is job 5, and then followed by job 2. Finally, the last processing job in the sequence is job 3. Similarly, the decoding for TCT-NFSSP-SDSTs-RDs is in strict accordance with specific order of the sorting of jobs for processing successively. In this paper, HEDA adopt the random initialization method to produce individuals as the initial population, which not only guarantee the diversification in population, but also can realize the fair comparison for HEDA with other algorithms, for the sake of verify the effectiveness for HEDA.

3.2   Speed-Up Evaluation Method

Based on the expressions of the TCT-NFSSPs with SDSTs and RDs in Sect. 2, L_{j_{i-1},j_i} is only decided by the job j_{i-1} and j_i. Accordingly, by utilizing this property, speedup evaluation can be adopted to reduce the computing complexity (CC) of TCT(π). That is, L_{j_{i-1},j_i}, sp_{j_i} and $\sum_{i=1}^{n} sp_{j_i}$ can be calculated and saved in the initial phase, and then can be used as constant values in the global exploration and local exploration of HEDA, which can reduce the CC of TCT(π) from $O(nm)$ to $O(n)$.

3.3   EDA-Based Global Search

3.3.1   Probabilistic Model

Probabilistic model is adopted to investigate the genetic information among the high quality individuals, which guides the promising direction. Hence, the model has a key influence on the performances for the HEDA, and obviously, a good model can enhance the algorithm's efficiency and effectiveness for optimizing the problem considered. In this study, the probabilistic model of HEDA is denoted by $\mathbf{P(gen)}$ as follows:

$$\mathbf{P(gen)} = \begin{bmatrix} \mathbf{P_1(gen)} \\ \cdot \\ \cdot \\ \cdot \\ \mathbf{P_n(gen)} \end{bmatrix} = \begin{bmatrix} P_{11}(gen) & \cdot & \cdot & \cdot & P_{1n}(gen) \\ \cdot & \cdot & & & \cdot \\ \cdot & & \cdot & & \cdot \\ \cdot & & & \cdot & \cdot \\ P_{n1}(gen) & \cdot & \cdot & \cdot & P_{nn}(gen) \end{bmatrix}_{n \times n} \tag{8}$$

where $\mathbf{P_i(gen)} = [P_{i1}(gen), P_{i2}(gen), \cdots, P_{in}(gen)]$ denote the ith vector of $\mathbf{P(gen)}$ in gen generation among the HEDA's population. Specifically, $P_{ij}(gen)$ is the corresponding probability of job j appearing in the ith position of the sequence π at the generation gen. What is more noteworthy point being that the probability matrix is a random matrix, in which each element $P_{ij}(gen)(i, j = 1, \cdots, n)$ present a probability for the certain event. Additionally, for a certain position, sum of the probability for all jobs in this specific position i for corresponding solution sequence inevitable to 1, noted as

$\sum_{j=1}^{n} P_{ij}(gen) = 1$, $gen \geq 0$. The value of the probability matrix $\mathbf{P(gen)}$ reflect the relationship of priority processing for the promising jobs in the considering sequence. For the element of probability matrix under the influence of the selected superior individuals, the value of the $P_{ij}(gen)$ $(i,j = 1, \cdots, n)$ is higher, the job j is selected in the position i more likely to have higher possibility.

The HEDA generate offspring individuals by sampling from the probability matrix, which means each offspring individual generate from $\mathbf{P(gen)}$ through the roulette wheel selection. Therefore, the specific value of internal elements in $\mathbf{P(gen)}$ determine the composition of the generated individuals in the population. Additionally, we set $P_{ij}(0) = 1/(n \times n)$ $(i,j = 1, \cdots, n)$ when the initialization for HEDA $(gen = 0)$. which make the $\mathbf{P(gen)}$ comply with uniform distribution to guarantee the diversification of the initial population, and also can accumulate more quality individual's information appropriately in the initial phase during the first time update $\mathbf{P(gen)}$.

3.3.2 Updating Mechanism

Probability model determines the search direction and the update mechanism have great influence on the performance. To make sure the $\mathbf{P(gen)}$ accurately estimate and effectively update the probability distribution of the superior sub-population (the elite individuals) and guide HEDA to the promising search direction, the two probability matrix $\mathbf{\eta(gen)}$ and $\mathbf{\xi(gen)}$ are proposed in this Section, which record the information of the order of jobs in sequence and the similar blocks of jobs in the selected individuals, respectively. Therefore, $\mathbf{\eta(gen)}$ and $\mathbf{\xi(gen)}$ are described as follows:

$$\mathbf{\eta(gen)} = \begin{bmatrix} \mathbf{\eta_1(gen)} \\ \cdot \\ \cdot \\ \cdot \\ \mathbf{\eta_n(gen)} \end{bmatrix} = \begin{bmatrix} \eta_{11}(gen) & \cdot & \cdot & \cdot & \eta_{1n}(gen) \\ \cdot & & \cdot & & \cdot \\ \cdot & & & \cdot & \cdot \\ \cdot & & & \cdot & \cdot \\ \eta_{n1}(gen) & \cdot & \cdot & \cdot & \eta_{nn}(gen) \end{bmatrix}_{n \times n} \tag{9}$$

$$\mathbf{\xi(gen)} = \begin{bmatrix} \mathbf{\xi_1(gen)} \\ \cdot \\ \cdot \\ \cdot \\ \mathbf{\xi_n(gen)} \end{bmatrix} = \begin{bmatrix} \xi_{11}(gen) & \cdot & \cdot & \cdot & \xi_{1n}(gen) \\ \cdot & & \cdot & & \cdot \\ \cdot & & & \cdot & \cdot \\ \cdot & & & \cdot & \cdot \\ \xi_{n1}(gen) & \cdot & \cdot & \cdot & \xi_{nn}(gen) \end{bmatrix}_{n \times n} \tag{10}$$

where each element $\eta_{ij}(gen)$ in $\mathbf{\eta(gen)}$ records the probability information for the number of times that job j appears before or in position i in the selected individuals. Additionally, $\mathbf{\eta(gen)}$ can efficiently slow down the convergent rate and effectively prevent premature convergence for HEDA. $\mathbf{\eta(gen)}$ can be calculated as follows:

$$\eta_{ij}(gen) = \sum_{s=1}^{Sbest} X_{ij}^s(gen) , \quad i,j = 1, \ldots, n \tag{11}$$

To make the HEDA not fall into local optimum easily, if the job j appears in position i, then set the row j from i to n lines as 1, otherwise 0. X_{ij}^s is a binary variable and the expressed using formula as follows:

$$X_{ij}^s(gen) = \begin{cases} 1, & \text{if job } j \text{ appears before or in position } i \\ 0, & \text{else} \end{cases} \qquad (12)$$

It can effectively avoid the probability of just only one position in each column of the probability matrix become larger and also efficiently increase diversity of the population. Using this update method can improve the sampling efficiency of the algorithm at the same time. In addition, to avoid falling into local optimum, the normalized operation should be performed to $\boldsymbol{\eta}(\mathbf{gen})$ after each update operation.

The calculation of $\boldsymbol{\xi}(\mathbf{gen})$ inspired by the schema theorem and the hypothesis of building block, and each element $\xi_{ij}(gen)$ in $\boldsymbol{\xi}(\mathbf{gen})$ records probability of the number of times that job i appears immediately after job j when job j is in the previous position of the job i placed. Then the $\boldsymbol{\xi}(\mathbf{gen})$ can be calculated as follows:

$$\xi_{ij}(gen) = \sum_{s=1}^{Sbest} Y_{ij}^s(gen) , \quad i,j = 1,\ldots,n \qquad (13)$$

$$Y_{ij}^s(gen) = \begin{cases} 1, & \text{if job } j \text{ appears immediately after job } i \\ 0, & \text{else} \end{cases} \qquad (14)$$

It is also need to do the same normalized operation for the $\boldsymbol{\xi}(\mathbf{gen})$ after update operation, so it is avoided to fall into local optimum as well.

Then, $\boldsymbol{\eta}(\mathbf{gen})$ and $\boldsymbol{\xi}(\mathbf{gen})$ indicate the importance of the order of jobs and the similar blocks of jobs in the selected promising solutions, respectively. Therefore, the probabilistic model can be updated by utilizing the information of the $\boldsymbol{\eta}(\mathbf{gen})$ and $\boldsymbol{\xi}(\mathbf{gen})$. Similar to the Hebb learning principle, the update mechanism as follows:

$$\mathbf{P_i}(\mathbf{gen}+\mathbf{1}) = \begin{cases} r \times \mathbf{P_i}(\mathbf{gen}) + (1-r) \times \boldsymbol{\eta_i}(\mathbf{gen}+\mathbf{1}), i = 1 \\ r \times \mathbf{P_i}(\mathbf{gen}) + (1-r) \times (\delta_1\boldsymbol{\eta_i}(\mathbf{gen}+\mathbf{1}) + \delta_2\boldsymbol{\xi_{[i-1]}}(\mathbf{gen}+\mathbf{1}))/\omega, i = 2,3,\ldots,n \end{cases} \qquad (15)$$

where $[i-1]$ present the selected job placed in position $i-1$. $r(0 < r \le 1)$ is learning rate. Specifically, if $r = 0$ which means $\mathbf{P}(\mathbf{gen})$ update directly without considering inertia. Therefore, $\omega = \sum_{j=1}^{n} (\delta_1\eta_{ij}(gen+1) + \delta_2\xi_{i-1,j}(gen+1))$ $(i = 2,3,\ldots,n)$, δ_1 and δ_2 are two parameters used for the diversification of the population which also indicate the importance of $\boldsymbol{\eta}(\mathbf{gen})$ and $\boldsymbol{\xi}(\mathbf{gen})$. Apparently, it is noteworthy that the updating mechanism is more advantageous to the enhancement of the promising genetic information, which efficiently improve the guidance of proposed HEDA global search.

3.3.3   New Population Generation Method
Sampling to generate new population that means all jobs for each position are selected dependent on the probabilistic matrix $\mathbf{P}(\mathbf{gen})$ which mentioned in Sect. 3.3.2. Denote *popsize* the size of the population and $\mathbf{pop}(\mathbf{gen})$ the generated population for HEDA. The procedure to generate a new population is described in detail as follows:

Step 1: Set the control parameter $p = 1$ for generating population.
Step 2: Randomly generate a probability r where $r \in [0,1)$.

Step 3: Get a candidate job j_c by the roulette wheel selection scheme.
   Step 3.1: If $r \in [0, P_{i1}(gen))(i \in \{1,\ldots,n\})$, then set $j_c = 1$, and go to Step 4.
   Step 3.2: If $r \in [\sum_{l=1}^{h-1} P_{il}(gen), \sum_{l=1}^{h} P_{il}(gen))$ and $(i \in \{1,\ldots,n\}, h = 2,\ldots,n)$, then select candidate job $j_c = h$, and go to Step 4.
Step 4: If candidate job j_c do not repeat with the selected jobs, then return j_c.
Step 5: Put the selected j_c into the corresponding position and repeat the sampling method based on $\mathbf{P}(gen)$ until generate a feasible solution.
Step 6: Set $p = p + 1$.
Step 7: If $p \leq popsize$, then go to Step 2. Otherwise output a new $\mathbf{pop}(gen)$.

Accordingly, HEDA sampling method adequately present the distribution of the correlated relationship between the variables in the solution sequence, constructing and capturing the promising blocks in the field of the optimal solution.

3.4   Problem-Dependent Local Search

3.4.1   Insert-based Neighborhoods

Because *Insert-based* neighborhood is more effective than other neighborhoods, we adopt *Insert* here for the local search to carry out intensification. Let $Insert(\pi, i, l)$ denote the insertion of j_i before j_l in the ith dimension of π when $i > l$ and after j_l when $i < l$. The *Insert-based* neighborhood of π can be presented as:

$$N_{Insert}(\pi) = \{\pi^{n,i,l} = Insert(\pi, i, l) | l \neq i, i-1; i, l = 1, 2, \ldots, n\}. \quad (16)$$

where $l \neq i - 1$ due to $\pi^{n,i-1,i} = \pi^{n,i,i-1}$. Apparently, the neighbor size is $(n-1)^2$.

3.4.2   Speedup Scanning Method for Insert-based Neighborhoods

Denote $FindBestN_{Insert}(\pi)$ the scanning procedure of finding the best neighbor $\pi^{n,i,l}_{best}$ in $N_{Insert}(\pi)$, where $s = min(i,l)$, and $\pi^{n,i,l} = [j'_1, j'_2, \ldots, j'_s, \ldots, j'_n]$. Then, for $s = 2, \ldots, n-1$ and $k = 1, \ldots, s-1$, it has $j_k = j'_k$, $C_{j_{s-1}} = C_{j'_{s-1}}$, and $\sum_{k=1}^{s-1} C_{j'_k} = \sum_{k=1}^{s-1} C_{jk}$. Thus, according to (6), it has

$$TCT(\pi^{n,i,l}) = \sum_{k=1}^{s-1} C_{j'_k} + \sum_{k=s}^{n} C_{j'_k} = \sum_{k=1}^{s-1} C_{jk} + \sum_{k=s}^{n} C_{j'_k}. \quad (17)$$

Therefore, in $FindBestN_{Insert}(\pi)$, $St_{j_{s-1}}$ and $\sum_{k=1}^{s-1} C_{jk}$ $(s-1 = 1, \ldots, n-2)$ can be calculated and saved simultaneously in advance before scanning the neighbors in $N_{Insert}(\pi)$, and they can be treated as constant values for calculating $TCT(\pi^{n,i,l})$, which can reduce the computing complexity (CC) of $FindBestN_{Insert}(\pi)$ to some extent. In other words, if $s > 1$, $St_{j'_{s-1}}$ and $\sum_{k=1}^{s-1} C_{j'_k}$ need not be computed and can be replaced with $St_{j_{s-1}}$ and $\sum_{k=1}^{s-1} C_{jk}$, respectively, and $St_{j'_s}$ can be computed directly from $St_{j_{s-1}}$. In HEDA's problem dependent local search, the above speed-up scanning method is utilized in $FindBestN_{Insert}(\pi)$, which is denoted as $SP\_FindBestN_{Insert}(\pi)$.

3.5   Procedure of HEDA

According to the proposed HEDA above, innovative EDA-based global search and problem dependent local search, the procedure of HEDA is as follows:

Step 0: Parameter initialization. Denote π_{best} the global best individual, and set the critical parameters of HEDA: r, *popsize*, $Sbest = \varphi \times popsize$, respectively.
Step 1: Calculate and save sp_{j_i} and L_{j_{i-1},j_i} $(j_{i-1},j_i \in 1,\ldots,n)$ in advance.
Step 2: Population initialization. Generate HEDA's initial individuals randomly.
Step 3: Calculate $TCT(\pi)$ of initial HEDA's population using speed-up evaluation method, then evaluate to get π_{best}, and set $gen = 1$.
Step 4: Initialization **P(0)** and select *Sbest* individuals to generate **P(gen)**.
Step 5: Evolution phase. Sample to generate offspring from **P(gen)** and calculate $TCT(\pi)$ to update π_{best}.
Step 6: Local search. Apply $SP\_FindBestN_{Insert}(\pi)$ to π_{best} and renew π_{best}.
Step 7: Set $gen = gen + 1$. If $gen \leq MaxGen$, then go to Step 5.
Step 8: Output π_{best}.

It can be seen that not only does HEDA adopt probabilistic model to execute global exploration, but it also applies problem-dependent local search to perform exploitation for the best individual. Because both exploration and exploitation are reasonable balanced, it is expected to achieve good results for TCT-NFSSP with SDSTs and RDs.

4   Simulation Results and Comparisons

4.1   Experimental Setup

Some randomly generated instances are adopted to test the performance of HEDA. That is, the $n \times m$ combinations are carried out with $\{20, 30, 50, 70, 100\} \times \{5, 10, 20\}$. The processing time $p_{j_i,l}$ and the setup time $s_{j_{i-1},j_i,l}$ are generated from two different uniform distribution [1, 100] and [0, 100], respectively. The job arrival time r_{j_i} is randomly generated from a uniform distribution $[0, 150n\alpha]$, where the parameter α is used to control the speed of arrival times. The level of α are set to 0, 0.2, 0.4, 0.6, 0.8, 1 and 1.5, respectively. Therefore, we have a total of 105 different instances.

Additionally, denote $\pi_{ini}(\alpha)$ the initial permutation in which all jobs are ranked by ascending order of job's release date at level α. Accordingly, $TCT(\pi(\alpha))$ the total completion time of the permutation $\pi(\alpha)$ at the level α, $TCT_{avg}(\pi(\alpha))$ the average value of $TCT(\pi(\alpha))$, $ARI(\alpha) = (TCT_{avg}(\pi(\alpha)) - TCT(\pi_{ini}(\alpha)))/TCT(\pi_{ini}(\alpha)) \times 100\%$ the average percentage improvement over $TCT(\pi_{ini}(\alpha))$, $SD(\alpha)$ the standard deviation of $TCT(\pi(\alpha))$ at the level α, where S_α the set of all values of α, and $|S_\alpha|$ the number of different values in S_α. Whereupon we define two metrics to evaluate the performances of the compared with state-of-the-art evolutionary algorithm optimizers, i.e., $ARI = \sum_{\alpha \in S\alpha} ARI(\alpha)/|S_\alpha|$ and $SD = \sum_{\alpha \in S\alpha} SD(\alpha)/|S_\alpha|$, respectively. Based on previous experiments, parameters are set as follows: the population size *popsize* = 80, the learning rate $r = 0.7$, the *Sbest* $= 0.1 \times popsize$, $\delta_1 = 0.7$ and $\delta_2 = 0.3$. All algorithms have been re-implemented and coded in Embarcadero Studio XE8 platform and all tests independently run 30 times on Inspur Servers with a 2.6 GHz processor/64 GB RAM. Therefore, the results are impartial and completely comparable.

4.2 Computational Results and Comparisons

To verify the effectiveness of HEDA, we have re-implemented the existing state-of-the-art algorithms IG_LS [4], TMIIG [6], ISA_2 [7] and PPSO [8] for comprehensive comparisons. Since the above algorithms are not designed for the problem considered here, we have adapted them by utilizing the TCT criterion presented in Sect. 2. The statistical results for *ARI* and *SD* with the same maximum elapsed CPU time limit of $t = 10 \times n \times m$ milliseconds as a termination criterion are reported in Table 1, where Average is a statistical average and the optimal result in bold.

Apparently, from Table 1, it is shown that the HEDA can achieve better performance with respect to solution quality than IG_LS, TMIIG, ISA_2 and PPSO in the different scale problems with various α levels considering all the metrics. Besides, TMIIG is also a potential competitive algorithm better than IG. Meanwhile, ISA_2 and PPSO have high dependence on the parameter for addressing the different problems. Moreover, standard deviation (SD) values by the HEDA are comparatively smallest than other algorithms for all the instance, which safely concluded that the proposed HEDA is a new state-of-the-art algorithm with excellent quality and robustness for TCT-NFSSP with SDSTs and RDs. The superiority of the HEDA owes to some aspects as follow: (1) With the well-designed probability model and the suitable updating mechanism, it is helpful to explore the search space effectively. (2) With the insert-based local search, it is helpful to exploit the promising regions and improve the

Table 1. Comparison of *ARI* and *SD* of IG_LS, TMIIG, ISA_2 and PPSO

Instance	IG_LS [4]		TMIIG [6]		ISA_2 [7]		PPSO [8]		HEDA	
n, m	*ARI*	*SD*	*ARI*	*SD*	*ARI*	*SD*	*ARI*	*SD*	*ARI*	*SD*
20, 5	16.810	1.204	17.389	0.946	11.666	1.447	13.530	1.066	**17.849**	**0.778**
20, 10	16.937	1.121	17.379	1.105	11.701	1.539	13.156	1.244	**17.735**	**0.668**
20, 20	13.233	0.452	**13.655**	0.412	9.358	1.371	11.211	1.109	13.653	**0.326**
30, 5	17.959	0.878	18.341	0.696	10.008	1.228	12.503	1.172	**18.585**	**0.565**
30, 10	17.013	1.065	18.508	0.718	9.707	1.434	12.188	1.022	**18.854**	**0.675**
30, 20	13.946	0.597	14.635	0.613	11.808	1.082	11.635	0.813	**14.869**	**0.567**
50, 5	21.943	0.792	22.541	0.651	16.158	0.971	18.118	0.812	**22.841**	**0.445**
50, 10	19.015	0.653	19.250	0.548	13.919	1.226	16.544	0.940	**19.823**	**0.398**
50, 20	17.877	0.641	17.928	0.501	9.867	0.938	12.278	0.796	**18.040**	**0.391**
70, 5	24.379	0.835	25.279	0.615	18.118	1.188	21.703	0.806	**25.796**	**0.402**
70, 10	24.007	0.732	24.878	0.681	17.805	1.138	21.431	0.979	**25.316**	**0.352**
70, 20	21.405	0.775	21.853	0.527	15.897	0.951	18.290	0.818	**22.380**	**0.330**
100, 10	22.153	0.562	22.715	0.518	16.246	1.107	19.836	0.870	**23.637**	**0.406**
100, 20	21.327	0.483	21.844	0.442	15.658	0.972	18.387	0.534	**22.578**	**0.373**
100, 40	24.239	0.538	24.969	0.504	18.027	0.861	21.622	0.701	**25.516**	**0.486**
Average	19.483	0.755	20.071	0.636	13.730	1.150	16.162	0.965	**20.498**	**0.477**

good schedule by enhancing the exploitation capability. (3) With the suitable parameter setting, it is helpful to obtain satisfactory schedules with TCT criterion. With the above merits, the HEDA is more effective than the existing algorithms.

5 Conclusions and Future Research

To the best of the current authors' knowledge, this is the first report on the application of the HEDA for the no-wait flow-shop scheduling problem (NFSSP) with sequence-dependent setup times (SDSTs) and release dates (RDs) to minimize the total completion time (TCT). The superiority of the proposed HEDA is mainly due to the fact that EDA-based global search adequately applies the genetic information of the order of jobs and the promising blocks of jobs to execute exploration for promising regions, while the problem dependent local search adopts speed-up methods based on problem's properties to further exploit these regions. Due to the reasonable hybridization of exploration and exploitation, HEDA's search behavior can be enriched and its search ability can be greatly enhanced. Moreover, the speedup evaluation method is designed to reduce the computing complexity of evaluation solutions successfully. Therefore, both the effectiveness of searching solutions and the efficiency of evaluating solutions were considered, and the influence of parameter setting was investigated as well. According to the computational results and statistical analyses, the HEDA outperforms all other considered algorithms with the effectiveness and robustness for the TCT-NFSSP with SDSTs and RDs. Our future work is to develop some effective algorithms to deal with the flexible distributed re-entrant scheduling problems.

Acknowledgments. This research was partially supported by the National Science Foundation of China under Grant 60904081, the Applied Basic Research Foundation of Yunnan Province under Grant 2015FB136, the 2012 Academic and Technical Leader Candidate Project for Young and Middle-Aged Persons of Yunnan Province under Grant 2012HB011, and the Discipline Construction Team Project of Kunming University of Science and Technology under Grant 14078212.

References

1. Allahverdi, A.: The third comprehensive survey on scheduling problems with setup times/costs. Eur. J. Oper. Res. **246**, 345–378 (2015)
2. Allahverdi, A., Ng, C.T., Cheng, T.C.E., Kovalyov, M.Y.: A survey of scheduling problems with setup times or costs. Eur. J. Oper. Res. **187**(3), 985–1032 (2008)
3. Bianco, L., Dell'Olmo, P., Giordani, S.: Flow shop no-wait scheduling with sequence dependent setup times and release dates. INFOR **37**(1), 3–19 (1999)
4. Ruiz, R., Stützle, T.: An iterated greedy heuristic for the sequence dependent setup times flowshop problem with makespan and weighted tardiness objectives. Eur. J. Oper. Res. **187**(3), 1143–1159 (2008)
5. Qian, B., Zhou, H.-B., Hu, R., Xiang, F.-H.: Hybrid differential evolution optimization for no-wait flow-shop scheduling with sequence-dependent setup times and release dates. In: Huang, D.-S., Gan, Y., Bevilacqua, V., Figueroa, J.C. (eds.) ICIC 2011. LNCS, vol. 6838, pp. 600–611. Springer, Heidelberg (2011)

6. Ding, J.Y., Song, S.J., Gupta, J.N.D., et al.: An improved iterated greedy algorithm with a Tabu-based reconstruction strategy for the no-wait flowshop scheduling problem. Appl. Soft Comput. **30**, 604–613 (2014)
7. Allahverdi, A., Aydilek, H.: Total completion time with makespan constraint in no-wait flowshops with setup times. Eur. J. Oper. Res. **238**(1), 724–734 (2014)
8. Samarghandi, H., ElMekkawy, T.Y.: Solving the no-wait flow-shop problem with sequence dependent set-up times. Int. J. Comput. Integr. Manuf. **27**(3), 213–228 (2014)
9. Baluja, S.: Population-based incremental learning: a method for integrating genetic search based function optimization and competitive learning. Technical report CMU-CS-94-193 (1994)
10. Ceberio, J., Irurozki, E., Mendiburu, A., Lozano, J.A.: A review on estimation of distribution algorithms in permutation-based combinatorial optimization problems. Prog. Artif. Intell. **1** (1), 103–117 (2012)
11. Pan, Q.K., Ruiz, R.: An estimation of distribution algorithm for lot-streaming flow shop problems with setup times. Omega—Int. J. Manag. Sci. **40**(2), 166–180 (2012)
12. Wang, L., Wang, S.Y., Xu, Y., Zhou, G., Liu, M.: A bi-population based estimation of distribution algorithm for the flexible job-shop scheduling problem. Comput. Ind. Eng. **62**, 917–926 (2012)
13. Wang, S.Y., Wang, L., Liu, M., Xu, Y.: An effective estimation of distribution algorithm for solving the distributed permutation flow-shop scheduling problem. Int. J. Prod. Econ. **145**, 387–396 (2013)

A Discrete Invasive Weed Optimization Algorithm for the No-Wait Lot-Streaming Flow Shop Scheduling Problems

Hong-Yan Sang, Pei-Yong Duan$^{(\boxtimes)}$, and Jun-Qing Li

School of Computer Science,
Liaocheng University, Liaocheng 252059, People's Republic of China
sanghongyanlcu@163.com, duanpeiyong@lcu.edu.cn,
lijunqing@lcu-cs.com

Abstract. The no-wait lot-streaming flow shop scheduling has important applications in modern industry. This paper deals with the makespan for the problems with equal-size sublots. A fast calculation method is designed to reduce the time complexity. A discrete invasive weed optimization (DIWO) algorithm is proposed. In the proposed DIWO algorithm, job permutation representation is utilized, Nawaz–Enscore–Ham heuristic is used to generate initial solutions with high quality. A reference local search procedure is employed to perform local exploitation. Extensive computational simulations and comparisons are provided, which demonstrate the effectiveness of the proposed DIWO algorithm.

Keywords: Invasive weed optimization · Lot-streaming · No-wait · Scheduling

1 Introduction

The lot-streaming flowshop scheduling problem (LFSP) has important applications in practical situations where a job is divided into many identical items. Since the late 1980s, lot- streaming flowshop scheduling technique has been extensively studied in academic as well as industrial fields [1]. Yoon and Ventura [2] presented sixteen pairwise interchange methods to search for the best sequence for an m-machine lot-streaming flowshop problem, where a linear programming (LP) formulation was designed to obtain optimal sublot completion times. Later, the same authors [3] proposed a hybrid genetic algorithm (HGA) by incorporating the LP and a pairwise interchange method into the traditional genetic algorithm. Some effective methods including: genetic algorithm (GA), hybrid genetic algorithm (HGA), ant colony optimization (ACO) algorithm, tabu search (TS) and threshold accepting (TA) algorithm were proposed to the lot-streaming with the objective of minimizing makespan and total flow time by Marimuthu et al. [4–6]. Then, Tseng and Liao [7] developed a discrete particle swarm optimization (DPSO). In the DPSO, a net benefit of movement (NBM) algorithm instead of the LP method was utilized to produce the optimal allocation of the sublots for a given sequence.

© Springer International Publishing Switzerland 2016
D.-S. Huang et al. (Eds.): ICIC 2016, Part I, LNCS 9771, pp. 517–526, 2016.
DOI: 10.1007/978-3-319-42291-6_52

The no-wait lot-streaming flow shop scheduling problems (nwLFSP) can be found in metallurgy, chemical industry, pharmaceutical industry, and so on. In the scheduling, each sublot must be processed continuously from the first machine to the last machine. There is no interruption in this process and the sublot is no-wait. Obviously, the problem is very difficult to solve. Many researchers studied no-wait lot streaming flowshop scheduling problems. Sriskandarajah and Wagneur [8] considered lot-streaming and scheduling multiple products in two-machine no-wait flowshops. They devised an efficient heuristic for the problem of simultaneous lot streaming and scheduling of multiple products.

For the m-machine no-wait flowshop scheduling problems, Kumar, Bag and Sriskandarajah [9] obtained optimal continuous-sized sublots in single product case and multi-product case, and then used GA to optimize the number of sublots for each product with the objective of minimizing makespan. Chen and Steiner [10] presented a new linear programming formulation for the problem. A dynamic programming algorithm was developed to generate the profiles for each product on discrete lot-streaming [11]. An adaptive genetic algorithm was proposed to solve flow shop scheduling with no-wait flexible lot streaming [12, 13]. More recently, Zhang and Wang [14] designed a grouped fruit-fly optimization algorithm for the no-wait lot streaming flow shop scheduling with intermingling consistent-size sublots. This paper deals with nwLFSP with equal-size sublots. The number of sublots for each job is fixed. It needs to find the optimal sequence of jobs on machines.

Invasive Weed Optimization (IWO) is a new swarm algorithm. The algorithm is inspired by a common phenomenon in agriculture that is colonization of invasive weeds. It was first proposed by Mehrabian and Lucas in 2006 [15]. Recently, it has been successfully applied to solve traveling salesman problem [16], multi-objective portfolio optimization problem [17], the inverse Stefan problem [18], lot-streaming flowshop scheduling problems [19], and many others [20, 21]. For IWO is robust and efficient, we will propose a discrete invasive weed optimization algorithm (DIWO) to solve the problem considered in the paper.

The rest of the paper is organized as follows. In Sect. 2, description of the problem is introduced and a fast calculation method is presented. In Sect. 3, the detail of the presented DIWO algorithm is given. The computational results and comparisons are described in Sect. 4. Finally, Sect. 5 concludes the paper.

2   Description of the Problems

2.1   Problem Description

The no-wait lot-streaming flowshop scheduling problem consists of m stages, one machine per stage, where n jobs must be processed. All the jobs are processed in an identical processing order on all the machines. In the lot-streaming environment, each job consists of many identical items. Each job can be split into l_j number of sublots with equal size. Sublots are treated as separate entities in production. All the sublots of any job should be processed continuously on any machine, and each sublot should be processed continuously from its start on the first machine to its completion on the last one without any waiting time between machines [14].

The following assumptions are made: (1) all the sublots of jobs are independent and available for processing at time zero; (2) no preemption is allowed; (3) at any time, no sublot can be processed on more than one machine; (4) no machine can process more than one sublot simultaneously.

The notation used is presented below.

$j \in J = \{1, 2, \ldots, n\}$ set of jobs

$m \in M = \{1, 2, \ldots, m\}$ set of machines

l_j sublots number of job j

$p_{i,j}$ sublot processing time of job j on machine i

$C_{i,k,q}$ completion time of qth sublot of the kth job processed on machine i

$C_{\max}$ makespan or maximum completion time

Note that for a given job j:

$$X_{j,k} = \begin{cases} 1 & \textit{if job j is assigned to the kth job processed} \\ 0 & \textit{otherwise,} \end{cases} \quad j, k \in \{1, 2, 3, \ldots, n\};$$

Then, the sublots number of the kth job processed $\partial_k = \sum_{j=1}^{n} (X_{j,k} \cdot l_j)$.

The objective is to minimize makespan:

$$\min(C_{\max}) = \min \left(\max_{k=1}^{n} \left(C_{m,k,\partial_k} \right) \right) \tag{1}$$

There is no waiting time between machines for each sublot, so adjacent sublots have lag on the first stage. $\pi = \{\pi_1, \pi_2, \ldots, \pi_n\}$ is one sequence of jobs. For this sequence, $E_{\pi_j}(\pi)$ is the lag between adjacent sublots of job π_j, and $E_{\pi_j,\pi_{j+1}}(\pi)$ is the lag between the last sublot of job π_j and the first sublot of π_{j+1}. The calculate formula is below.

$$E_{\pi_j}(\pi) = \max_{1 \leq i \leq m} \left\{ p_{i,\pi_j} \right\} \tag{2}$$

$$E_{\pi_j,\pi_{j+1}}(\pi) = \max \left\{ \max_{2 \leq k \leq m} \left\{ \sum_{i=1}^{k} p_{i,\pi_j} - \sum_{i=1}^{k-1} p_{i,\pi_{j+1}} \right\}, p_{1,\pi_j} \right\} \tag{3}$$

Then, the makespan of the sequence π is:

$$C_{\max}(\pi) = \sum_{j=1}^{n-1} (l_{\pi_j} - 1) \times E_{\pi_j}(\pi) + E_{\pi_j,\pi_{j+1}}(\pi)) + (l_{\pi_n} - 1) \times E_{\pi_n}(\pi) + \sum_{i=1}^{m} p_{i,\pi_n} \tag{4}$$

A scheduling Gantt chart of sequence $\{1, 2, 3\}$ is shown in Fig. 1.

machine m

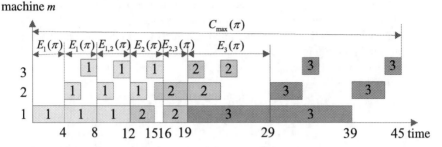

Fig. 1. The no-wait lot-streaming flowshop scheduling Gantt chart

2.2   A Fast Calculation Method

The time complexity of $E_{\pi_j,\pi_{j+1}}(\pi)$ and $C_{\max}(\pi)$ are $O(m^2)$, $O(nm^2)$ separately. A fast calculation method is designed to reduce the time complexity. $ST_{i,j,q}$ and $CT_{i,j,q}$ are the starting time and finishing time of the qth sublot of job j on machine i. The procedure of the fast calculation method can be summarized as in Fig. 2.

Using the method, the complexity of $E_{\pi_j,\pi_{j+1}}(\pi)$ is reduced to $O(m)$. The makespan can be calculated by the method and (4). Then, the time complexity of $C_{\max}(\pi)$ is $O(nm)$.

3   The Proposed Discrete Invasive Weed Optimization Algorithm

The basic IWO algorithm was designed for continuous optimization, which can't be used for the nwLFSP. To apply the IWO algorithm for solving the nwLFSP with the makespan criterion, a discrete IWO (DIWO) algorithm is proposed. We details it as follows:

3.1   Solution Representation

A population of initial solution $\pi = \{\pi_1, \pi_2, \ldots, \pi_n\}$ is being dispread over the n dimensional searches pace with the job permutation representation, as in Table 1.

Table 1. Solution representation

Individual dimensional	1	2	3	4	5	6
π_j		3	1	2	4	6	5
Sequence		3	1	2	4	6	5

3.2   Population Initialization

The Nawaz-Enscore-Ham (NEH) heuristic [22] is based on the idea that the job with larger processing times should be scheduled earlier. We generate one initial solution

Procedure fast calculation method

begin

$ST_{1,\pi_j,l_{\pi_j}} = 0$

$CT_{1,\pi_j,l_{\pi_j}} = p_{1,\pi_j}$

for $(i = 2; i \le m; i++)$

$ST_{i,\pi_j,l_{\pi_j}} = CT_{i-1,\pi_j,l_{\pi_j}}$

$CT_{i,\pi_j,l_{\pi_j}} = ST_{i,\pi_j,l_{\pi_j}} + p_{i,\pi_j}$

endfor

$ST_{1,\pi_{j+1},1} = CT_{1,\pi_j,l_{\pi_j}}$

$CT_{1,l_{\pi_{j+1}},1} = ST_{1,\pi_{j+1},1} + p_{1,\pi_{j+1}}$

for $(i = 2; i \le m; i++)$

$ST_{i,\pi_{j+1},1} = \max\{CT_{i,\pi_j,l_{\pi_j}}, CT_{i-1,\pi_{j+1},1}\}$

$CT_{i,\pi_{j+1},1} = ST_{i,\pi_{j+1},1} + p_{i,\pi_{j+1}}$

endfor

for $(i = m-1; i \ge 1; i--)$

$CT_{i,\pi_{j+1},1} = ST_{i+1,\pi_{j+1},1}$

$ST_{i,\pi_{j+1},1} = CT_{i,\pi_{j+1},1} - p_{i,\pi_{j+1}}$

endfor

$E_{\pi_j,\pi_{j+1}}(\pi) = ST_{1,\pi_{j+1},1} - ST_{1,\pi_j,l_{\pi_j}}$

end

Fig. 2. The fast calculation method

using NEH. The remaining solutions are produced randomly in the solution space. For keeping the diversity of the population, each individual in the initial population is different from others, that is $\pi_i \ne \pi_j, i, j \in \{1, 2, \ldots, PS\}$.

3.3 Reproduction

The lower the weed's fitness is, the more seeds it produces. The formula of a number of s_i seeds for a weed Π_i is

$$s_i = floor\left(s_{\max} - \frac{C_{\max}(\Pi_i) - C_{\max}(\Pi_{best})}{C_{\max}(\Pi_{worst}) - C_{\max}(\Pi_{best})} \times (s_{\max} - s_{\min})\right) \qquad (5)$$

Where Π_{worst} and Π_{best} respectively denote the worst and best solution in the population, $floor()$ is a function which rounds the elements to the nearest integers towards minus infinity,$C_{max}()$ is the objective function which returns the makespan value of a job permutation. s_{max} and s_{min} are the maximum and minimum number of seeds for the weeds, separately.

3.4   Spatial Dispersal

The generated seeds are being randomly scattered over the search space according to a normally distribution $N(0, \sigma^2)$ with mean equal to zero. Then the standard deviation for the normal distribution is calculated in an alternative way as follows.

$$\sigma_{iter} = \tan(0.875 \times \frac{iter_{max} - iter}{iter_{max}}) \times (\sigma_0 - \sigma_f) + \sigma_f \tag{6}$$

Where $\tan()$ is a tangent function. It can better balance global exploration and the local exploitation by using tangent function.

Follow the steps; a normally distributed random number N can be received. For the current permutation, we execute $floor(abs(N))$ insertion moves. Among the formula, $abs()$ is absolute value function. And, an insertion move to $\pi = (\pi_1, \pi_2, \ldots, \pi_n)$, denoted as $v(\pi, j, k)$, $j, k \in \{1, 2, \ldots, n\}$ and $k \neq j$, generates a permutation π' by removing the job from π in position j and reinserting it into another position k.

3.5   Competitive Exclusion

After all the weeds generate their seeds, the weeds are ranked together with their parents with respect to their makespan values. And weeds with lower fitness are eliminated; *PS* best solutions without repetition are selected as the weeds for the next iteration. That is, there are no identical job permutations in the population.

3.6   Referenced Local Search

The referenced local search is embedded in the proposed DIWO algorithm to enhance the exploitation capability of the algorithm. The method is applied to the best solution of the population after the initialization and applied to each generated seed with a probability P_{LS}. Let $\pi = \{\pi_1, \pi_2, \ldots, \pi_n\}$ be the sequence needed to be improved, the best sequence $\pi^b = \{\pi_1^b, \pi_2^b, \ldots, \pi_n^b\}$ is the referenced sequence. The pseudo code of the referenced local search is given as follow (Fig. 3).

Procedure referenced local search
begin

    count = 0

  while (*count* < *n*)

  for $\left(i = 1; i \leq n; i++\right)$

    remove π_i^b *from sequence* π

    insert π_i^b *to all the other different possible positions of* π

    denote the best obtained sequence as π'

    if $(C_{\max}(\pi') < C_{\max}(\pi))$

        $\pi = \pi'$

        count = 0

    else

        count = *count* + 1

    endif

  endfor

  endwhile

end

Fig. 3. the referenced local search

4 Computational Results and Comparisons

In this section, we test the performance of the proposed DIWO algorithm for solving nwLFSP with makespan criterion. Some algorithms published in the literature are compared, including: tabu search (TS) [4], simulated annealing with insertion neighborhood (Sai) [4], simulated annealing with swap neighborhood (SAs) [4], hybrid genetic algorithm (HGA) [3], ant colony optimization (ACO) [6], threshold accepting with insertion neighborhood (Tai) [6], threshold accepting with swap neighborhood (TAs) [6], discrete particle swarm optimization algorithm (DPSO) [7].

For this problem, 24 problem sizes each with the following sizes ($m \in \{10, 30, 50, 70, 90, 110\}, n \in \{5, 10, 15, 20\}$) are randomly generated, and the related data is given by the discrete uniform distributions as follows: $p_{i,j} \in U(1, 31)$, $l_j \in U(1, 6)$. The parameters of DIWO are: $PS = 10$, $s_{\max} = 3$, $s_{\min} = 1$, $P_{LS} = 0.15$, $\sigma_0 = 5$, $\sigma_f = 0.5$. The parameters of other algorithms are in accordance with the original paper. We coded all the algorithms in Visual C++ and conducted experiments on a computer of P4 CPU, 3.0 GHz with 2 GB memory. To make a fair comparison, all algorithms adopted the same maximum CPU time limit of $0.01 \times m \times n$ seconds as a termination criterion. For each instance, we carried out 5 independent replications. For each replication, we calculated the relative percentage increase (RPI) as follows:

$$RPI(C_i) = \frac{C_i - C^*}{C^*} \times 100 \qquad (7)$$

Where C_i is the makespan generated in the *ith* replication by an algorithm, and C^* is the smallest makespan found by all the algorithms in 5 replications. We calculate the average relative percentage increase (ARPI) over the 5 replications as statistics for the solution quality.

ARPI of all problem size is in Table 2. As seen from the table, we can draw the conclusion that the DIWO algorithm has the optimal performance compared with the rest of algorithms. It has the smallest ARPI value 0.07, which is far better than the rest of algorithms. The DIWO algorithm can find the best solution for each problem size. Hence, it is concluded that the DIWO algorithm is clearly superior to the HGA, DPSO, ACO, SA, TA and TS algorithms for solving the no-wait lot-streaming flow shop scheduling problem.

Table 2. Comparison of the presented DIWO algorithm with other algorithms

n × m	DIWO	ACO	DPSO	HGA	SAi	SAs	TAi	TAs	TS
10 × 5	**0.00**	0.09	0.01	0.05	0.22	0.55	0.52	1.01	0.15
10 × 10	**0.00**	0.02	**0.00**	0.01	0.12	0.81	0.48	1.54	**0.00**
10 × 15	**0.00**	0.18	0.01	0.18	0.28	0.72	0.52	2.00	**0.00**
10 × 20	**0.00**	0.09	0.03	0.05	0.19	0.90	1.07	2.95	**0.00**
30 × 5	**0.06**	0.61	0.34	0.68	0.78	1.55	1.13	1.81	0.35
30 × 10	**0.02**	0.62	0.33	0.81	0.96	1.85	1.35	2.53	0.34
30 × 15	**0.00**	0.98	0.34	0.74	1.31	2.44	1.59	3.18	0.28
30 × 20	**0.03**	0.71	0.27	0.82	1.14	2.33	1.53	3.12	0.33
50 × 5	**0.10**	0.66	1.13	0.76	0.74	1.18	1.07	1.46	0.39
50 × 10	**0.10**	0.81	1.64	1.05	1.06	2.07	1.50	2.69	0.70
50 × 15	**0.07**	0.88	1.66	1.06	1.21	2.38	1.53	2.79	0.61
50 × 20	**0.07**	0.92	1.81	1.29	1.37	2.65	1.81	3.36	0.65
70 × 5	**0.11**	0.68	2.35	0.81	0.65	0.98	0.98	1.42	0.39
70 × 10	**0.09**	0.74	2.70	1.14	1.03	1.91	1.42	2.59	0.58
70 × 15	**0.10**	0.72	2.88	1.24	1.11	2.34	1.58	2.82	0.69
70 × 20	**0.10**	0.88	3.13	1.47	1.35	2.67	1.86	3.37	0.66
90 × 5	**0.11**	0.78	3.85	0.74	0.58	0.83	0.93	1.32	0.27
90 × 10	**0.10**	0.87	4.04	1.01	0.76	1.59	1.10	1.95	0.52
90 × 15	**0.13**	0.86	4.58	1.50	1.14	2.21	1.50	2.79	0.73
90 × 20	**0.05**	0.96	4.09	1.31	1.21	2.40	1.60	3.05	0.84
110 × 5	**0.13**	0.87	4.86	0.79	0.45	0.62	0.86	1.06	0.21
110 × 10	**0.08**	1.06	5.73	1.04	0.68	1.53	1.12	2.07	0.62
110 × 15	**0.10**	0.91	5.54	1.10	0.85	1.95	1.14	2.57	0.78
110 × 20	**0.13**	1.28	6.35	1.50	1.11	2.58	1.55	3.08	0.82
Average	**0.07**	0.72	2.40	0.88	0.85	1.71	1.24	2.36	0.45

Statistical analysis is carried out to verify the effectiveness and validity of the DIWO algorithm. The ANOVA results are shown in Fig. 4. Figure 4 reveals the interaction and 95 % HSD confidence intervals between the types of algorithms at the same maximum elapsed CPU time. It can be easily observed that the DIWO algorithm has the best performance, which operates significantly better than the other algorithms.

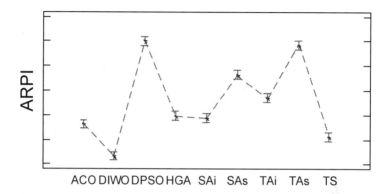

Fig. 4. Means plot and 95 % Tukey HSD confidence intervals for the interaction between algorithms

5 Conclusions

This paper addressed the nwLFSP with the makespan criterion, which has important applications in metallurgy, chemical, pharmaceutical, and many other industries. We first analyzed the problem and designed a fast calculation method to reduce the time complexity. Then a discrete invasive weed optimization algorithm was presented. In the algorithm, job permutation representation was utilized. NEH heuristic was used to generate initial solutions with high quality. A reference local search procedure was employed to perform local exploitation ability. The results showed that the DIWO outperforms other algorithms when solving the nwLFSP.

Acknowledgements. This research is partially supported by National Foundation of China (515775212, 61503170, 61573178, and 61374187), Shandong Province Higher Educational Science and Technology Program (J14LN28).

References

1. Chang, J.H., Chiu, H.N.: A comprehensive review of lot streaming. Int. J. Prod. Res. **4**(8), 1515–1536 (2005)
2. Yoon, S.H., Ventura, J.A.: Minimizing the mean weighted absolute deviation from due dates in lot-streaming flow shop scheduling. Comput. Oper. Res. **29**, 1301–1315 (2002)

3. Yoon, S.H., Ventura, J.A.: An application of genetic algorithms to lot-streaming flow shop scheduling. IIE Trans. **34**, 779–787 (2002)
4. Marimuthu, S., Ponnambalam, S.G., Jawahar, N.: Tabu search and simulated annealing algorithms for scheduling in flow shops with lot streaming. Proc. Inst. Mech. Eng. Part B J. Eng. Manuf. **221**, 317–331 (2007)
5. Marimulthu, S., Ponnambalam, S.G., Jawahar, N.: Evolutionary algorithms for scheduling m-machine flow shop with lot streaming. Robot. Comput.-Integr. Manuf. **24**, 125–139 (2008)
6. Marimulthu, S., Ponnambalam, S.G., Jawaha, N.: Threshold accepting and ant-colony optimization algorithm for scheduling m-machine flow shop with lot streaming. J. Mater. Process. Technol. **209**, 1026–1041 (2009)
7. Tseng, C.T., Liao, C.J.: A discrete particle swarm optimization for lot-streaming flow shop scheduling problem. Eur. J. Oper. Res. **191**, 360–373 (2008)
8. Sriskandarajah, C., Wagneur, E.: Lot streaming and scheduling multiple products in two-machine no-wait flowshops. IIE Trans. **31**(8), 695–707 (1999)
9. Kumar, S., Bagchi, T.P., Sriskandarajah, C.: Lot streaming and scheduling heuristics for m-machine no-wait flowshops. Comput. Ind. Eng. **38**(1), 149–172 (2000)
10. Chen, J., Steiner, G.: On discrete lot streaming in no-wait flow shops. IIE Trans. **35**(2), 91–101 (2003)
11. Hall, N.G., Laporte, G., Selvarajah, E.: Scheduling and lot streaming in flowshops with no-wait in process. J. Sched. **6**(4), 339–354 (2003)
12. Kim, K., Jeong, I.J.: Flow shop scheduling with no-wait flexible lot streaming using an adaptive genetic algorithm. Int. J. Adv. Manuf. Technol. **44**(11–12), 1181–1190 (2009)
13. Kim, K.W., Jeong, I.J.: Flow shop scheduling with no-wait flexible lot streaming using adaptive genetic algorithm. In: International Conference on Computational Science and its Applications, ICCSA 2007. IEEE, 474–479 (2007)
14. Zhang, P., Wang, L.: Grouped fruit-fly optimization algorithm for the no-wait lot streaming flow shop scheduling. In: Huang, D.-S., Jo, K.-H., Wang, L. (eds.) ICIC 2014. LNCS, vol. 8589, pp. 664–674. Springer, Heidelberg (2014)
15. Mehrabian, A.R., Lucas, C.: A novel numerical optimization algorithm inspired from weed colonization. Ecol. Inform. **1**, 355–366 (2006)
16. Zhou, Y., Luo, Q., Chen, H.: A discrete invasive weed optimization algorithm for solving traveling salesman problem. Neurocomputing **151**, 1227–1236 (2015)
17. Pouya, A.R., Solimanpur, M., Rezaee, M.J.: Solving multi-objective portfolio optimization problem using invasive weed optimization. Swarm Evol. Comput. **28**, 42–57 (2016)
18. Hetmaniok, E., Slota, D., Zielonka, A.: Experimental verification of approximate solution of the inverse Stefan problem obtained by applying the invasive weed optimization algorithm. Therm. Sci. **19**(Suppl. 1), 205–212 (2015)
19. Sang, H.Y., Pan, Q.K., Duan, P.Y.: An effective discrete invasive weed optimization algorithm for lot-streaming flowshop scheduling problems. J. Intell. Manuf. 1–13 (2015). doi:10.1007/s10845-015-1182-x
20. Zhou, Y., Chen, H., Zhou, G.: Invasive weed optimization algorithm for optimization no-idle flow shop scheduling problem. Neurocomputing **137**(5), 285–292 (2014)
21. Mehrabian, A.R., Koma, A.Y.: A novel technique for optimal placement of piezoelectric actuators on smart structures. J. Franklin Inst. **348**, 12–23 (2011)
22. Nawaz, M., Enscore, E.J.R., Ham, I.: A heuristic algorithm for the m machine, n job flowshop sequencing problem. Omega-Int. J. Manag. Sci. **11**(1), 91–95 (1983)

Two-Stage Flow-Open Shop Scheduling Problem to Minimize Makespan

Tao Ren[1], Bingqian Liu[1], Peng Zhao[1], Huawei Yuan[2], Haiyan Li[3],
and Danyu Bai[4(✉)]

[1] Software College, Northeastern University,
Shenyang 110819, People's Republic of China
[2] School of Computer Science and Engineering, Northeastern University,
Shenyang 110819, People's Republic of China
[3] School of Information Science and Engineering, Northeastern University,
Shenyang 110819, People's Republic of China
[4] School of Economics and Management,
Shenyang University of Chemical Technology,
Shenyang 110142, People's Republic of China
mikebdy@163.com

Abstract. This paper investigates a novel shop scheduling model, in which each job is processed first in a flow shop and then in an open shop. The objective is to optimize the maximum completion time, i.e. makespan. For large-scale problems, the asymptotic optimality of the dense scheduling (DS) algorithm is proven in the sense of probability limit. Furthermore, a DS-based heuristic algorithm is presented to obtain approximate solution. For moderate-scale problems, a discrete differential evolution algorithm is provided to achieve high-quality solutions. A series of random simulations are executed to demonstrate the effectiveness of the proposed algorithms.

Keywords: Flow-open shop · Makespan · Asymptotic analysis · Discrete differential evolution

1 Introduction

The flow-open shop model is a combination of two processing stages. The first stage is a flow shop, in which each job passes through the machines in identical sequence without pre-emption. The second stage is an open shop, in which the jobs that finished processing in the first stage can arbitrarily choose the route that is taken in passing through the machines. The objective is to determine a schedule that minimizes the maximum completion time, i.e. makespan.

To better understand the new model, the process that a patient seeks medical care in the emergency department is considered as an instance. The patient is first evaluated by the triage nurse, and then, is initially examined by the specialist. The two steps cannot be reversed, and can be treated as a flow shop. After that, the patient must undertakes a checklist including blood test, X-ray examination, ultrasound scan, etc., prescribed by the doctor. The examination order is arbitrary, and can be treated as an open

© Springer International Publishing Switzerland 2016
D.-S. Huang et al. (Eds.): ICIC 2016, Part I, LNCS 9771, pp. 527–535, 2016.
DOI: 10.1007/978-3-319-42291-6_53

shop. Therefore, the whole process of seeking medical care is obviously a flow-open shop scheduling problem.

The first stage of the model is a flow shop total completion time (FSTCT) problem, in which the completion time of each job is served as the release date in the next stage. Garey et al. [1] reported the strong NP-hardness of the two-machine FSTCT problem, which implies that the flow-open shop makespan problem is strongly NP-hard even if the second stage is omitted. An overview of the FSTCT problem can be found in Bai and Ren [2]. The second stage is an open shop makespan problem with release dates whose strong NP-hardness is presented by Lawer et al. [3] even if only two machines are involved in the system. A survey of the open shop makespan problem with release dates is presented in Bai and Tang [4].

This paper handles the flow shop and open shop as a whole system. To the best of the authors' knowledge, no researcher has yet addressed the flow-open shop scheduling problem. For this NP-hard problem, the asymptotic optimality of the dense scheduling (DS) algorithm is proven in the sense of probability limit. This result reveals that the DS algorithm can serve as an optimal schedule for sufficiently large-scale problems. Such property is more beneficial for an industrial scheduling environment, in which thousands of jobs are usually processed on one or more machines. Furthermore, the SPT-LPT-DS heuristic, i.e. execute shortest processing time (SPT)-DS rule at the first stage and longest processing time (LPT)-DS rule at the second stage, is designed to obtain approximate solution. As intelligent optimizer is more efficient in achieving a high-quality solution within a specified time, a discrete differential evolution (DDE) algorithm is provided for moderate-scale problems. A set of random experiments are executed at the end of the paper to demonstrate the convergence of the SPT-LPT-DS heuristic and the performance of the DDE algorithm.

The remainder of the paper is organized as follows: The formulating expression of the flow-open shop problem is given in Sect. 2. The asymptotic analysis of the DS algorithm and the SPT-LPT-DS heuristic are introduced in Sect. 3. The DDE algorithm and the numerical experiments are provided in Sects. 4 and 5, respectively. Section 6 concludes the paper.

2 Problem Statement

The flow-open shop is a combination of the classical flow and open shops. In the first stage (flow shop), each job $j, j = 1, 2, \ldots, n$, has to be processed on machine $i, i = 1, 2$, follow the same route, i.e., they are first processed on machine 1, then on machine 2. The machine processes the jobs on a first-come, first-served manner. In the second stage (open shop), the finished jobs at the first stage have to be executed on machine i, $i = 3, 4$, where the job route passing through the two machines is arbitrary chosen. The operation of job $j, j = 1, 2, \ldots, n$, on machine $i, i = 1, 2, 3$ and 4, is denoted by $O(i, j)$, having an associated processing time $p(i, j)$. Pre-emption and delay of the operations are not allowed, and the jobs are independent. At a time, a feasible schedule constrains that a machine cannot process more than one operation and that no job can be assigned to more than one machine. In the succeeding sections, the completion time of job j, $j = 1, 2, \ldots, n$, on machine $i, i = 1, 2, 3$ and 4, is denoted by $C(i, j)$. For convenience, let C_j denote the completion time of job j on the final machine. The objective is to determine a schedule, with given processing times of the n jobs, to minimize makespan, i.e., $C_{max} = min\{C_1, C_2, ..., C_n\}$.

3   Asymptotic Analysis of DS Algorithm and DS-Based Heuristic

The DS algorithm provides a feasible solution for the open shop scheduling problem, in which any machine is idle if and only if no job is currently available on the machine. Bai and Tang [4] proved the asymptotic optimality of the DS algorithm for the open shop makespan problem with release dates. This section generalizes the result to the flow-open shop makespan problem.

As no idle time appears on the first machine, the maximum value of the completion time on this machine is independent of the job sequence. Therefore, an intuitional lower bound, LB1, is presented

$$C_{LB1} = \sum_{j=1}^{n} p(1,j)$$

where C_{LB1} denotes the objective value of LB1. For any feasible schedule S of the problem, there at least exist one critical path that satisfies

$$C_{max}(S) = \max_{1 \leq j_1 \leq j_2 \leq j_3 \leq n} \left\{ \sum_{k=1}^{j_1} p(1,k) + \sum_{k=j_1}^{j_2} p(2,k) + \sum_{k=j_2}^{j_3} p(i_3,k) + \sum_{k=j_3}^{n} p(i_4,k) \right\}$$

where i_3 and i_4 denote the sequence in which the critical jobs pass through the machines at the second stage.

Theorem 1. Let the processing times of job j, $p(i,j)$, $j = 1, 2, ..., n$, $i = 1, 2, 3$ and 4, be independent random variables having the same continuous distribution defined on $(0, 1]$. With probability one,

$$\lim_{n \to \infty} \frac{C_{LB1}}{n} = \lim_{n \to \infty} \frac{C_{max}(OPT)}{n} = \lim_{n \to \infty} \frac{C_{max}(DS)}{n}$$

where $C_{max}(OPT)$ and $C_{max}(DS)$ denote the objective values of the optimal solution and DS algorithm, respectively.

Proof. Let G denote the set including the operations forming the critical path in a dense schedule. Therefore, $|G| = O(n)$. Without loss of generality, it is assumed that $|G| = n + a$, where a is a positive constant, and $\lim_{n \to \infty} a/n = 0$. With the law of Large Numbers,

$$0 \leq \lim_{n \to \infty} \frac{1}{n} (C_{max}(DS) - C_{max}(OPT))$$

$$\leq \lim_{n \to \infty} \frac{1}{n} (C_{max}(DS) - C_{LB1})$$

$$= \lim_{n \to \infty} \frac{1}{n} \max_{1 \leq j_1 \leq j_2 \leq j_3 \leq n} \left\{ \sum_{k=1}^{j_1} p(1,k) + \sum_{k=j_1}^{j_2} p(2,k) \right. \tag{1}$$

$$\left. + \sum_{k=j_2}^{j_3} p(i_3,k) + \sum_{k=j_3}^{n} p(i_4,k) \right\} - \lim_{n \to \infty} \frac{1}{n} \sum_{j=1}^{n} p(1,j)$$

$$\leq E(p) + \lim_{n \to \infty} \frac{a}{n} - E(p) = 0$$

where $E(p)$ denotes the expectation value of the processing times. Rearranging inequality (1) yields the result of the theorem.

In the DS algorithm, the available operations are randomly scheduled for each time when a machine is idle. However, in the actual scheduling process, giving preference to the job with short processing times on the two machines at the flow shop stage can decrease the time when the job waits at the open shop stage. Bai et al. [5] pointed out that the LPT-DS algorithm can effectively minimize the makespan for the open shop scheduling problem. On the basis of these ideas, the SPT-LPT-DS heuristic is formally expressed as follows.

SPT-LPT-DS Heuristic

Step1: At the flow shop stage, calculate $FP_j = p(1, j) + p(2, j)$, and schedule the jobs according to the non-decreasing order of value FP_j. If some jobs simultaneously satisfy the condition, give preference to the job with the smallest processing time on machine 1.

Step2: At the open shop stage, calculate $OP_j = p(3, j) + p(4, j)$, and schedule the job with largest value OP_j among the available jobs once a machine is idle. If some jobs simultaneously satisfy the condition, give preference to the job with the largest processing time on this machine.

Step3: All the jobs are scheduled. Calculate the makespan value.

Although Theorem 1 provides the asymptotic optimality, the worst case performance of LB1 is unbounded, as shown in the following example.

Example 1. Consider an instance that includes two jobs, J_1 and J_2. The processing time of job J_2 on machine 2 is $p(2, 2) = 1$. The remaining processing times are all equal to ε, where ε is an arbitrarily small number on $(0, 1)$. The optimal schedule is $\{J_1, J_2\}$. Therefore,

$$\frac{C_{max}(OPT)}{C_{LB1}} = \frac{1 + 4\varepsilon}{2\varepsilon} \to +\infty$$

as $\varepsilon \to 0$. Referring to the structure of the problem, another lower, LB2, is presented

$$C_{LB2} = \max\left\{\max_{1\leq i\leq 2}\left\{\min_{1\leq j\leq n}\{p(i-1,j)\} + \sum_{j'=1}^{n}p(i,j') + \min_{1\leq j''\leq n}\sum_{i'=i+1}^{4}p(i',j'')\right\},\right.$$

$$\left.\max_{3\leq i\leq 4}\left\{\min_{1\leq j\leq n}\{p(1,j) + p(2,j)\} + \sum_{j'=1}^{n}p(i,j')\right\}\right\}$$

where $p(0, j) = 0$.

Theorem 2. Under the assumption of Theorem 1, with probability one,

$$\lim_{n\to\infty}\frac{C_{LB1}}{n} = \lim_{n\to\infty}\frac{C_{LB2}}{n}$$

Proof. Obviously,

$$C_{LB1} \leq C_{LB2} \leq C_{\max}(OPT) \tag{2}$$

Dividing inequality (2) by n and taking limit yield the result of the theorem.

A numerical example is presented to show the SPT-LPT-DS schedule and the calculation of LB2.

Example 2. Consider a flow-open shop makespan problem with five jobs. The processing times of job j, $j = 1, 2, 3, 4, 5$, on machine i, $i = 1, 2, 3, 4$, are given in the following form.

	J_1	J_2	J_3	J_4	J_5
M_1	2	7	3	5	4
M_2	2	4	4	3	6
M_3	3	6	7	1	3
M_4	4	3	6	5	8

According to the SPT-LPT-DS heuristic, the final schedule is presented in Fig. 1, and the associated makespan value is 34. Calculating the lower bound value with lower bound LB2 yields $C_{LB2} = \max\{30, 27, 24, 30\} = 30$. Thus, the gap between the heuristic and LB2 is $(34 - 30)/30 \approx 13.33\%$.

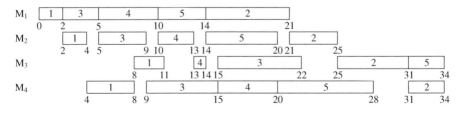

Fig. 1. The SPT-LPT-DS schedule in Example 2.

4   The DDE Algorithm

Within a given time, intelligent optimization algorithm is an effective way to obtain a near-optimal solution for moderate-scale problems. Differential evolution (DE) is a one of the latest evolutionary optimizer [6] for global optimization over continuous search spaces.

Given that each chromosome is a vector of floating point numbers, which is invalid for mapping a discrete search space, the canonical DE algorithm cannot be directly applied to solve scheduling problems. This section introduces a DDE algorithm by transforming the individuals into job-based permutations to solve the flow-open shop makespan problem. In the encoding process, a feasible schedule is represented as a permutation according to the job index on machine 1. In the decoding process, a

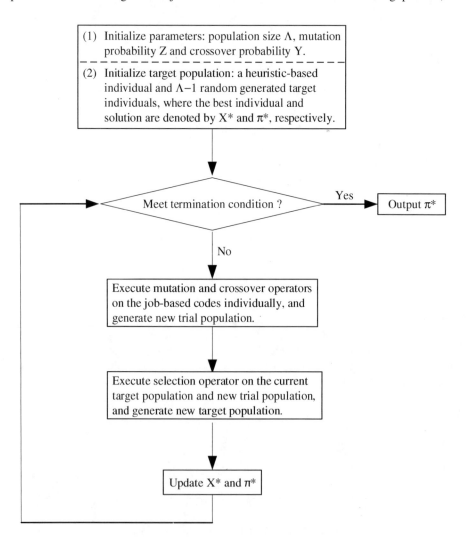

Fig. 2. Flowchart of the DDE algorithm

job-based permutation is recovered to a feasible schedule, in which each job is assigned at the open shop stage according to the LPT-DS rule. A bi-point insertion is used in the crossover operation. A general procedure of the DDE algorithm is presented in Fig. 2. The reader can refer to the concrete steps of the algorithm in Bai et al. [5].

5   Computational Results

A series of computational experiments are designed to demonstrate the convergence of the DS-based heuristic and the performance of the DDE algorithm in different size problems. The algorithm is coded in C++ language and implemented on a PC with an Intel Core i7-2600 CPU (3.4 GHz × 4) and 8 GB RAM. The processing times of the jobs are randomly generated from a discrete uniform distribution on [1, 100] and a discrete normal distribution with expectation 200 and variance 100^2. Ten random tests for each scale of the problem are independently executed, and the averages are reported in the tables.

5.1   Convergence of the DS-Based Heuristic

The tests are proposed to reveal the asymptotic optimality of the SPT-LPT-DS heuristics for large-scale problems. Given the strong NP-hardness of the problem, LB2 is therefore adopted as substitute to the optimal schedule. Mean gap percentage $MGP = \frac{C_{max}(H) - C_{LB2}}{C_{LB2}} \times 100\%$ is employed, where $C_{max}(H)$ denotes the makespan values of the SPT-LPT-DS heuristic. The machine-job combinations are 4 machines with 100, 200, 500 and 1000 jobs.

The data in Table 1 imply that the asymptotic optimality of SPT-LPT-DS heuristic is independent of the processing time distributions. It is obvious that the ratios approach zero as the number of jobs increases (Fig. 3). These results indicate a convergence trend in the tested schedules, confirming the asymptotic optimality of the heuristic.

Table 1. Results of the SPT-LPT-DS heuristic (%)

	Uniform				Normal			
Jobs	100	200	500	1000	100	200	500	1000
MGPs	3.267	1.433	0.965	0.793	3.426	1.757	0.874	0.595

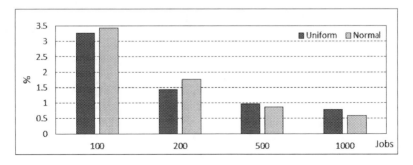

Fig. 3. Convergence of the DS-based heuristic

5.2   Performance of the DDE Algorithm

Mean improvement percentage $MIP = \frac{C_{max}(OS) - C_{max}(FS)}{C_{max}(FS)} \times 100\%$ is employed, where $C_{max}(OS)$ denotes the makespan value of the minimum initial solution; and $C_{max}(FS)$ denotes the makespan value of the DDE algorithm. The problem scale is 4 machines with 20, 40, 60 and 80 jobs. To reveal the effect of the DDE algorithm on the improvement, random feasible schedule is employed to generate the initial population. With a set of trials, the parameters of the algorithm are determined as: population size $\Lambda = 10$, mutation probability $Z = 0.3$, crossover probability $Y = 0.8$, and maximum iteration $T = 100$. For saving computation time, the procedure is terminated if the current optimal solution remains unimproved after 10 successive iterations.

The data in Table 2 show that the performance of the DDE algorithm is also independent of the processing time distributions. The improvement of the algorithm gradually weakens as the number of jobs increases (Fig. 4). A possible explanation is that more available operations provide more processing chances for each machine, which shortens idle time of machines and enhances the quality of the initial solution. When the number of jobs is larger than 100, the improved effect of the algorithm can be disregarded.

Table 2. Results of the DDE algorithm (%)

	Uniform				Normal			
Jobs	20	40	60	80	20	40	60	80
MIPs	7.662	7.015	5.658	3.461	8.228	5.392	4.435	4.673

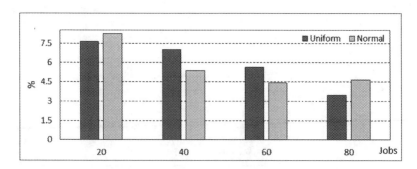

Fig. 4. Performance of the DDE algorithm

6   Conclusions

This paper proves the asymptotic optimality of the DS algorithm in the sense of probabilistic limit, and designs a DS-based heuristic for the flow-open shop makespan problem. The DDE algorithm is presented to further promote the quality of solution for moderate-scale problems, in which a bi-point insertion method is introduced in the crossover operation to obtain high-quality final solution. Numerical simulations demonstrate the effectiveness of the proposed algorithms.

Acknowledgements. This work is partially supported by National Natural Science Foundation of China (51305073, 61473073, 61104074, and 71201107), Fundamental Research Funds for the Central Universities (L1517004), and Program for Liaoning Excellent Talents in University (LJQ2014028).

References

1. Garey, M.R., Johnson, D.S., Sethi, R.: The complexity of flow shop and job shop scheduling. Math. Oper. Res. **1**, 117–129 (1976)
2. Bai, D., Ren, T.: New approximation algorithms for flow shop total completion time problem. Eng. Optim. **45**, 1091–1105 (2013)
3. Lawer, E.L., Lenstra, J.K., Rinnooy Kan, A.H.G.: Minimizing maximum lateness in a two-machine open shop. Math. Oper. Res. **6**, 153–158 (1981)
4. Bai, D., Tang, L.: Open shop scheduling problem to minimize makespan with release dates. Appl. Math. Model. **37**, 2008–2015 (2013)
5. Bai, D., Zhang, Z., Zhang, Q.: Flexible open shop scheduling problem to minimize makespan. Comput. Oper. Res. **67**, 207–215 (2016)
6. Storn, R., Price, K.: Differential evolution – a simple and efficient heuristic for global optimization over continuous spaces. J. Glob. Optim. **11**, 341–354 (1997)

An Improved Quantum-Inspired Evolution Algorithm for No-Wait Flow Shop Scheduling Problem to Minimize Makespan

Jin-Xi Zhao, Bin Qian$^{(\boxtimes)}$, Rong Hu, Chang-Sheng Zhang, and Zi-Hui Li

Department of Automation, Kunming University of Science and Technology, Kunming 650500, China
bin.qian@vip.163.com

Abstract. In this paper, an improved quantum-inspired evolution algorithm (IQEA_M) with a special designed local search is proposed to deal with the no-wait flow shop scheduling problem (NFSSP) with sequence-independent setup times (SISTs) and release dates (RDs), which has been proved to be strongly NP-hard. The criterion is to minimize makespan. The method was tested with other literature methods. Experimental results show that IQEA_M presented the best performance regarding other algorithm.

Keywords: Minimize makespan · No-wait flow shop scheduling problem · Quantum-inspired evolution · Sequence-independent setup times and release dates

1 Introduction

No-wait flow shop scheduling problem is one of the common models in the modern manufacturing industry [1]. This paper considers the no-wait flow shop scheduling problem (NFSSP) with sequence-independent setup times (SISTs) and release dates (RDs), and the criterion is to minimize makespan. That means, once job j start processing, it cannot be stopped until the end of the last machine processing. The job must be processed from the first machine to the last one. Each job has a release dates, before the release dates the job cannot be processed. Every machine has setup time, and the setup time are sequence-independent, which means with different job sequence there have same setup time. The NFSSP was proven to be NP-hard [2]. Through literature research, the research of QEA on NFSSP with SISTs and RDs is very limited. It's meaning full to improve the algorithm for the considered problem.

QEA was born to solve combinatorial optimization problems. Since Narayanan's QGA [3] and Han's QEA [4], quantum evolution algorithm has been widely concerned as a new field of research. Quantum evolutionary algorithm was used to solve the knapsack problem [4], traveling salesman problem [5, 11], FSSP [6–8] and other combinatorial optimization problem. In recent years, the research of quantum evolution algorithm is divided into two directions, one is to improve the QEA itself [9, 10], the other is mixed with other algorithms [11, 12].

© Springer International Publishing Switzerland 2016
D.-S. Huang et al. (Eds.): ICIC 2016, Part I, LNCS 9771, pp. 536–547, 2016.
DOI: 10.1007/978-3-319-42291-6_54

In the current paper, an improved quantum-inspired evolution algorithm (IQEA_M) is presented to deal with the NFSSP with SISTs and RDs. Firstly, we improve the traditional QEA algorithm, a new observation method is proposed. Secondly, a set of specially designed local search algorithm is proposed. In our IQEA_M, the IQEA search is used for global search and guide the whole search to the promising regions, while an improved interchange and insert-based local search is developed to emphasize exploitation from those regions. By comparing with quantum algorithms and other algorithms in recent years, the results show the effectiveness of the presented IQEA_M.

The remainder of this paper is organized as follows. Section 2 introduces the mathematical model of NFSSP with SISTs and RDs. Section 3 presents IQEA_M in details. Section 4 provides simulation results and comparisons. Finally, Sect. 5 gives some conclusion and future research.

2 NFSSP with SISTs and RDs

The NFSSP with SISTs and RDs problem can be describe as follows. There are n jobs and m machines. Each of n jobs will be sequentially processed on machine. The processing time of each job on each machine is deterministic. At any time, preemption is forbidden and each machine can process at most one job. To satisfy the no-wait restriction, each job must be processed without interruptions between consecutive machines. Thus, all jobs are processed in the same sequence on all machines. In a flow-shop with SISTs, setup must be performed between the completion time of one job and the start time of another job on each machine, and setup time only depends on the current jobs at each machine. In a flow-shop with RDs, if a machine is ready to process a job but the job has not been released yet, it stays idle until the release date of the job.

2.1 NFSSP with SISTs

Let $\pi = [j_1, j_2, \ldots, j_n]$ denote the schedule or permutation of jobs to be processed, $p_{j_i,l}$ the processing time of job j_i on machine l, sp_{j_i} the total processing time of job j_i on all machines, $ML_{j_i,l}$ the minimum delay on the machine l between the completion of job j_{i-1} and j_i, $L_{j_{i-1}j_i}$ the minimum delay on the first machine between the start of job j_{i-1} and j_i, $s_{j_{i-1},l}$ the sequence-independent setup time of job j_{i-1} on machine l. Let $p_{j_0,l} = 0$ for $l = 1, \ldots, m$. Then $ML_{j_i,l}$ can be calculated follows:

$$ML_{j_i,l} = \begin{cases} max\{s_{j_{i-1}j_i,1} + p_{j_i,1} - p_{j_{i-1},2}, s_{j_{i-1},2}\} + p_{j_i,2}, l = 2 \\ max\{ML_{j_i,l-1} - p_{j_{i-1},l}, s_{j_{i-1},l}\} + p_{j_i,l}, l = 3, \ldots, m \end{cases} \quad (1)$$

Accordingly, $L_{j_{i-1}j_i}$ can be calculated by using the following formula:

$$L_{j_{i-1}j_i} = ML_{j_i,m} + sp_{j_{i-1}} - sp_{j_i} \quad (2)$$

2.2   Makespan of NFSSP with SISTs

Denote r_{j_i} the arrival time of job j_i, St_{j_i} the process starting time of job j_i on machine l, C_{j_i} the completion time of job j_i on machine m, d_{j_i} the due date of job j_i. Then St_{j_i} can be written as follows:

$$St_{j_i} = \begin{cases} max\{ML_{j_i,m} - sp_{j_i}, r_{j_i}\}, i = 1 \\ St_{j_{i-1}} + max\{L_{j_{i-1},j_i}, r_{j_i} - St_{j_{i-1}}\}, i = 2, \ldots, n \end{cases} \tag{3}$$

So C_{j_i} can be calculated as follows:

$$C_{j_i} = St_{j_i} + sp_{j_i}, i = 1, \ldots, n \tag{4}$$

The aim of this paper is to find a permutation π^* in the set of all permutations Π such that

$$C_{j_n}(\pi^*) = \min_{\pi \in \Pi} C_{j_n}(\pi) \tag{5}$$

3   IQEA_M

The whole process of IQEA_M is described in this chapter. IQEA perform global search, improved interchange and insert search perform local search. In order to jump out of a local optimum, we try other two kinds of local search.

3.1   Global Search

Qubit encoding is crucial for quantum evolutionary algorithm. In order to maintain the characteristics of probability and improve the efficiency of the algorithm, we propose a new encoding method. This section provides a detailed description of IQEA design. Qubit observation model are summarized below:

In quantum computing, the use of $|0>$ and $|1>$ represents a single qubit ground states. Any state of a single quantum bit $|\varphi>$ represents a linear combination of these two ground states, as follows:

$$|\varphi| = \alpha|0> + \beta|1> \tag{6}$$

Where α and β are a pair of complex numbers, called quantum state probability amplitude. That is according to the probability $|\alpha|^2$ to making a quantum bit $|\varphi>$ collapsed to the ground state $|0>$, and according to the probability $|\beta|^2$ to making qubit collapsed to the ground state $|1>$ of each observation. And satisfied:

$$|\alpha|^2 + |\beta|^2 = 1 \tag{7}$$

Therefore, Eq. (9) can also be expressed as means of trigonometric functions:

$$|\varphi> = cos\theta|0> + sin\theta|1> \tag{8}$$

θ is phase of $|\varphi>$, $\alpha = cos\theta$, $\beta = sin\theta$.

Qubit Encoding. Qubit observation model defined herein as a 2 dimensional qubit probability matrix β.

$$\beta = \begin{bmatrix} |\beta_{1,1}|^2 & |\beta_{1,2}|^2 & \cdots & |\beta_{1,n}|^2 \\ |\beta_{1,2}|^2 & |\beta_{2,2}|^2 & \cdots & |\beta_{2,n}|^2 \\ \vdots & \vdots & |\beta_{i,j}|^2 & \vdots \\ |\beta_{n,1}|^2 & |\beta_{n,2}|^2 & \cdots & |\beta_{n,n}|^2 \end{bmatrix} \tag{9}$$

Let the job's permutation $\pi = [\pi_1, \pi_2, \ldots, \pi_i, \ldots, \pi_n]$. $|\beta_{i,j}|^2$ represents the probability of job j at π_i, $|\beta_{i,j}|^2$ initial value of $\frac{1}{n}$. The way to generate the permutation is as follows:

Step 1: $c = 1$; set $s1 = [1,2,\ldots,n]$; set $s2 = [1,2,\ldots,n]$; $t = 1$;

Step 2: For $i = 1$ to n

Step 2.1: Randomly generate an integer i, $i \in s1$; $t = \Sigma_{p \in s2}|\beta_{i,p}|^2$;

　　　　　　Randomly generate a decimal k; $w = 0$; $j = 1$;

Step 2.2: While c

　　　　　$w = w + |\beta_{s1[i],s2[j]}|^2$;

　　　　　If $k * t < w$

　　　　　　$\pi_i = s2[j]$;

　　　　　　Delete i from s1;

　　　　　　Delete $s2[j]$ from s2;

　　　　　　Break;

　　　　　End

Step 2.3:　　$j = j + 1$;

　　　　　End

　　End

This is a new way of observing the qubits. It not only avoids the time consuming of observation steps, but also retains the advantage of probability selection.

Update Probability Amplitude. At previous qubit update mechanism, it is generally used to increase or decrease a fixed angle (such as $0.05°$), or based on the current adaptation value to change probability amplitude with dynamic. Such update operations often required n^2 times. This paper uses the index function to update the qubits, avoids the complicated calculation steps, and only need $2n$ times to complete the update operations. In the initial stage of the algorithm, the convergence speed is fast. With the increase of the number of iterations, the convergence speed is more and more slow. Update speed depends on variable s to control, $0<s<1$. s closer to 0, the faster the convergence rate. s closer to 1, the slower the convergence rate.

From Fig. 1, the smaller the s, the faster the evolution.

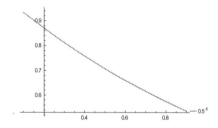

Fig. 1. When $|\beta_{i,\pi b_i}|^2 = 0.5$, the $|\beta_{i,\pi b_i}|^{2*}$ curve.

From Fig. 2, with $|\beta_{i,\pi b_i}|^2$ close to 1, the evolution rate is getting slower and slower until close to 0.

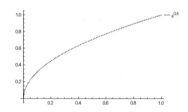

Fig. 2. When s $= 0.5$, the $|\beta_{i,\pi b_i}|^{2*}$ curve.

Update operation steps are as follows:

Step 1: Perform this update operation for all individuals;

Let $\pi b = [\pi b_1, \pi b_2, \dots, \pi b_n]$ is the permutation of the current optimal solution, and β is the qubit matrix to perform the update operation;

Let β^* represent the updated elements;

Step 2: For $i = 1$ to n

Step 2.1: $|\beta_{i,\pi b_i}|^{2^*} = (|\beta_{i,\pi b_i}|^2)^s$;

Step 2.2: $a = |\beta_{i,\pi b_i}|^{2^*} - (|\beta_{i,\pi b_i}|^2)^s$; $b = \frac{1-a}{1-|\beta_{i,\pi b_i}|^{2^*}}$;

Step 2.3: For $j = 1$ to n

If $j = \pi b_i$

Continue;

End

$|\beta_{i,j}|^{2^*} = b * |\beta_{i,j}|^2$;

End

End

Mutation Method. For probability evolutionary algorithms such as QEA and EDA, excessive crossover and mutation are unnecessary. To try to escape from the local optimum, we use a mutation method after reaching a certain variation of conditions. When searching fall into local optimum, we randomly selected part of individuals and change their probability amplitude to the initial value.

3.2 Local Search

Because of the problems to solve are NP-hard problem, solution space is very complicated, so a fast and efficient local search algorithm is very important. We use specially designed local search algorithm to quickly improve global solutions. It will keep IQEA_M iterative in fast and maintaining the diversity of the population, to avoid falling into a local optimum.

Improved Interchange Search. To accelerate the local search, we only need to search part of the permutation of interchange search. We suppose the global search's permutation is π, and the evaluation of π is kr. And *evaluation p* means using evaluation function to calculate the value of permutation p.

Step 1: Let *iisp = empty array; iisr = infinite number;*

Step 2: For $i = 1 \ to \ n$

Step 2.1:   For $j = 1 \ to \ n$

$p = \pi$;

exchange $p[i]$ and $p[j]$;

$r = evaluation \ p$;

If $r < iisr$

$iisr = r$;

$iisp = p$;

If $iisr < kr$

$\pi = p$;

End

End

End

End

That is, this algorithm in the process of each cycle, with a new permutation to replace the old permutation.

Improved Insert Search. Similar to the improved interchange search, improved insert search needn't search whole permutation of insert search. Let the result of improved interchange search's permutation is π, and the evaluation of π is kr.

Step 1: Let *iisp = empty array; iisr = infinite number;*

tp = empty array;

Step 2: For $i = 1 \ to \ n$

Step 2.1:   For $j = 1 \ to \ n$

$p = k$; $tp = p[i]$; $p = p \ drop \ p[i]$;

Evaluation every permutation of tp insert to p's every position;

$r = evaluation \ p$;

If r $< iisr$

iisr = r;

iisp = p;

If iisr $< kr$

π = p;

End

End

End

End

Parity Exchange Search. We try to use a new local search here. When the algorithm falling into a local optimum, then we will use this algorithm. We suppose the local optimum permutation is π and the evaluation of π is kr.

Step 1: q = take integer of $\sqrt{n}$;
Step 2: If q is odd-numbered
$$q = q - 1;$$
    End
Step 3: For $i = 1\ to\ n - q$
Step 3.1: $p = \pi$;
Step 3.2: p = exchange adjacent odd numbered and even
    numbered job of p from column i to $i + q$;
Step 3.3: Implement improved interchange search and improved insert search on p;
    End

Block Search. We also try block search here to jump out local optimum. We suppose the input permutation is π and the evaluation of π is kr.

Step 1: q = random integers from 2 to $\sqrt{n}$;
Step 2: For $i = 1\ to\ n - q$
Step 2.1: $a = \pi[i\ to\ i + q]$;
Step 2.2: $p = \pi$ without column i to i + q;
Step 2.3: Insert a to every position of p and use improved interchange search and
    improved insert search to it; Let pr and pl is the optimal fitness value
    and permutation founded in this step;
Step 2.4: If $pr < kr$
$$\pi = pr;$$
    End
    End

3.3   Overall Description of IQEA_M Algorithm

According to the above IQEA and special designed local search, the procedure of IQEA_M is presented as follows:

Step 1: Initialize the qubit population; $c = 1$; bp is initialized to an empty set; $bpr = infinite\ number$;

Step 2: While c

Step 2.1: Observation the qubit matrix; evaluation every permutation; select the best permutation $p1$;

Step 2.2: Implementation the improved interchange search and improved insert search to the $p1$ and get the output permutation $p2$ and $p2r$;

Step 2.3: If $p2r < bpr$

$\qquad\qquad bp = p2$; $bpr = p2r$; $d = 0$;

$\qquad$ else

$\qquad\qquad d = 1$;

$\qquad$ End

Step 2.4: If $d = 1$

$\qquad\qquad\qquad$ Implementation parity exchange search and block search to $p2$ and the output is permutation $p3$ and evaluation result $p3r$;

Step 2.5: $\qquad$ If $p3r < bpr$

$\qquad\qquad\qquad e = 0$;

$\qquad\qquad$ else

$\qquad\qquad\qquad e = 1$;

$\qquad\qquad$ End

Step 2.6: $\qquad$ If $e = 1$

$\qquad\qquad\qquad$ Implementation qubit mutation; $e = 0$;

$\qquad\qquad$ End

Step 2.7: $\qquad$ Use bp to update probability amplitude;

Step 2.8: $\qquad$ If exit criteria not met

$\qquad\qquad\qquad c = 1$;

$\qquad\qquad$ else

$\qquad\qquad\qquad c = 0$;

$\qquad\qquad$ End

$\qquad$ End

The exit criteria can be set like maximum iterate times or running time.

4 Test Results and Compare

4.1 Experiment Setup

In order to evaluate the performance of IQEA_M, we compare it with QEA and other algorithms in recent years. Some random generated instances with different scales are used to test the performance of IQEA_M. That is, the $n \times m$ combinations are: $\{20, 30, 50, 70, 100\} \times \{10, 20\}$. The processing time $p_{j_i,l}$ and the setup time $s_{j_{i-1},l}$ are generated from a uniform distribution $[1, 100]$. The job arrival time r_{j_i} is an integer that is randomly generated in $[0, 150n\alpha]$, where the parameter α is used to control the jobs' arrival speeds. The values of α are set to 0, 0.2, 0.4, 0.6, respectively. For each instance at each α, every algorithm independently run 20 replications for comparison. Thus, it has a total of 40 different instances.

In IQEA_M, we use the following parameters: the population size popize $= 10$, the convergence rate $s = 0.9$, the maximum non-improved times are 5, the mutation probability are 0.5. Here $s = 0.9$, so the IQEA_M will evolve at a slower rate. To make a fair comparison, all the compared algorithms use the same runtime limit of $n * m/100$ s as a termination criterion. We code all procedures in Matlab 2016a and run all tests on an Intel Core i7 2630QM with 16 GB memory.

4.2 Performance Metrics

Denote $\pi_{ini}(\alpha)$ the permutation in which jobs are ranked by ascending value of job's release date at α, $M(\pi(\alpha))$ the makespan of the permutation $\pi(\alpha)$ at α, $avg\_M(\pi(\alpha))$ the average value of $M(\pi(\alpha))$, $best\_M(\pi(\alpha))$ the best value of $M(\pi(\alpha))$, $worst\_M(\pi(\alpha))$ the worst value of $M(\pi(\alpha))$, $ARI(\alpha) = (M(\pi_{ini}(\alpha)) - avg\_M(\pi(\alpha))) / M(\pi_{ini}(\alpha)) \times 100\%$ the average percentage improvement over $M(\pi(\alpha))$, $BEI(\alpha) = (M(\pi_{ini}(\alpha)) - best\_M(\pi(\alpha))) / M(\pi_{ini}(\alpha)) \times 100\%$ the best percentage improvement over $M(\pi(\alpha))$, $WRI(\alpha) = (M(\pi_{ini}(\alpha)) - worst\_M(\pi(\alpha))) / M(\pi_{ini}(\alpha)) \times 100\%$ the worst percentage improvement over $M(\pi(\alpha))$, $SD(\alpha)$ the standard deviation of $M(\pi(\alpha))$ at α, S_α the set of all values of α, and $|S_\alpha|$ the number of different values in S_α. Then, we define four metrics to evaluate the performances of the compared algorithms, i.e., $ARI = \sum_{\alpha \in S_\alpha} ARI(\alpha)/|S_\alpha|$, $BEI = \sum_{\alpha \in S_\alpha} BEI(\alpha)/|S_\alpha|$, $WRI = \sum_{\alpha \in S_\alpha} WRI(\alpha)/|S_\alpha|$, and $SD = \sum_{\alpha \in S_\alpha} SD(\alpha)/|S_\alpha|$.

4.3 Comparisons IQEA_M, GA, IQEA, QUARTS

To show the effectiveness of IQEA_M, we carry out some comparisons with GA [13], IQEA [14] and QUTARS [15]. Here we only show the comparison results of ARI and BEI.

Table 1. Comparisons of ARI and BEI of GA, IQEA and QUARTS and IQEA_M

Instances	GA		IQEA		QUARTS		IQEA_M	
n, m	ARI	BEI	ARI	BEI	ARI	BEI	ARI	BEI
20,10	11.44	16.14	9.36	15.13	16.86	19.02	**17.97**	**20.63**
20,20	9.06	12.13	7.24	11.30	13.12	13.79	**14.67**	**16.27**
30,10	8.46	11.45	7.04	10.67	13.56	14.14	**14.29**	**15.64**
30,20	13.83	16.99	11.61	16.81	19.75	21.49	**20.31**	**22.23**
50,10	11.17	12.81	9.96	12.63	14.78	15.48	**15.86**	**17.03**
50,20	11.34	16.24	9.25	12.16	15.37	16.26	**17.62**	**19.58**
70,10	9.81	11.65	8.25	10.75	14.69	15.25	**16.09**	**17.11**
70,20	11.55	14.65	10.18	14.46	18.21	19.71	**18.76**	**21.27**
100,10	10.53	12.51	9.22	13.13	15.00	15.97	**15.62**	**17.03**
100,20	9.44	12.12	8.26	11.29	15.04	15.82	**16.18**	**17.87**

The test results of the four algorithms are shown in Table 1. From Table 1, it can be concluded that the searching quality of IQEA_M is better than that of QUTARS and is superior or comparable to that of GA and IQEA. Therefore, it is concluded that IQEA_M is an effective and robust algorithm for dealing with M-NFSSP with SISTs and RDs.

5   Conclusion and Future Work

In our presented algorithm, IQEA-based global search was used to perform exploration for promising regions within the entire solution space, while a special local search based on problem's properties was developed to stress exploitation in these regions. Due to the hybridization of IQEA and local search, IQEA_M's search behavior can be enriched and its search ability can be enhanced. Simulations and comparisons showed the effectiveness and robustness of IQEA_M. The future work is to develop some effective IQEA-based algorithms for dynamical scheduling and reentrant scheduling.

Acknowledgements. This research was partially supported by the National Science Foundation of China under Grant 60904081, the Applied Basic Research Foundation of Yunnan Province under Grant 2015FB136, the 2012 Academic and Technical Leader Candidate Project for Young and Middle-Aged Persons of Yunnan Province under Grant 2012HB011, and the Discipline Construction Team Project of Kunming University of Science and Technology under Grant 14078212.

References

1. Raaymakers, W., Hoogeveen, J.: Scheduling multipurpose batch process industries with no-wait restrictions by simulated annealing. Eur. J. Oper. Res. **126**, 131–151 (2000)
2. Lenstra, J.K., Rinnooy Kan, A.H.G.: Computational complexity of discrete optimization problems. Ann. Discret. Math. **4**(1), 21–40 (1979)
3. Narayanan, A., Moore, M.: Quantum-inspired genetic algorithms. In: Proceedings of the 1996 IEEE International Conference on Evolutionary Computation (ICEC 1996), pp. 61–66 (2010)
4. Han, K.H., Kim, J.H.: Genetic quantum algorithm and its application to combinatorial optimization problem. In: Proceedings of the 2000 Congress on Evolutionary Computation, vol. 2, pp. 1354–1360. IEEE (2000)
5. Talbi, H., Draa, A., Batouche, M.: A new quantum-inspired genetic algorithm for solving the travelling salesman problem. In: IEEE International Conference on Industrial Technology, IEEE ICIT 2004, vol. 3, pp. 1192–1197 (2004)
6. Wang, L., Wu, H., Tang, F., Zheng, D.-z.: A hybrid quantum-inspired genetic algorithm for flow shop scheduling. In: Huang, D.-S., Zhang, X.-P., Huang, G.-B. (eds.) ICIC 2005. LNCS, vol. 3645, pp. 636–644. Springer, Heidelberg (2005)
7. Bin-Bin, L., Ling, W.: A hybrid quantum-inspired genetic algorithm for multiobjective flow shop scheduling. IEEE Trans. Syst. Man Cybern. Part B Cybern. **37**(3), 576–591 (2007). A Publication of the IEEE Systems Man and Cybernetics Society
8. Niu, Q., Zhou, T., Ma, S.: A quantum-inspired immune algorithm for hybrid flow shop with makespan criterion. J. Univ. Comput. Sci. **15**(4), 765–785 (2009)
9. Deng, G., Wei, M., Su, Q., Zhao, M.: An effective co-evolutionary quantum genetic algorithm for the no-wait flow shop scheduling problem. Adv. Mech. Eng. **7** (2015)

10. Gu, J., Gu, M., Gu, X.: A mutualism quantum genetic algorithm to optimize the flow shop scheduling with pickup and delivery considerations. Math. Probl. Eng. **2015**, 1–17 (2015)
11. Ma, Y., Tian, W.J., Fan, Y.Y.: Improved quantum ant colony algorithm for solving TSP problem. In: IEEE Workshop on Electronics, Computer and Applications, pp. 453–456 (2014)
12. Latif, M.S., Zhou, H., Amir, M.: A hybrid quantum estimation of distribution algorithm (Q-EDA) for flow-shop scheduling. In: Ninth International Conference on Natural Computation, vol. 23, pp. 654–658. IEEE (2013)
13. Samarghandi, H.: Studying the effect of server side-constraints on the makespan of the no-wait flow-shop problem with sequence-dependent set-up times. Int. J. Prod. Res. **53**(9), 2652–2673 (2015)
14. Cui, L., Wang, L., Deng, J., Zhang, J.: A new improved quantum evolution algorithm with local search procedure for capacitated vehicle routing problem. Math. Probl. Eng. **2013**(3), 1–17 (2013)
15. Nagano, M.S., Miyata, H.H., Araújo, D.C.: A constructive heuristic for total flowtime minimization in a no-wait flowshop with sequence-dependent setup times. J. Manuf. Syst. **36**, 224–230 (2014)

An Enhanced Memetic Algorithm
for Combinational Disruption Management
in Sequence-Dependent Permutation Flowshop

Xiao-pan Liu[1], Feng Liu[2(✉)], and Jian-jun Wang[1]

[1] Institute of Systems Engineering, Dalian University of Technology,
Dalian 116023, China
[2] School of Management Science and Engineering,
Dongbei University of Finance and Economics, Dalian 116025, China
liufengapollo@163.com

Abstract. In this paper, a combinational disruption management problem in permutation flowshop with sequence dependent setup time is studied. Six types of disruption events, namely machine breakdown, processing time variation, setup times variation, new job arrival, job priority upgrading, and job cancellation, are considered simultaneously. Thus, a bi-objective rescheduling model is built, considering both the schedule makespan and the sequence deviation from the baseline schedule. After analysis, the problem is found to be NP-hard. From the perspectives of improving initial solution qualities and balancing the global search and local search, a two-stage memetic algorithm is proposed. Finally, computational experiments were carried out to validate the effectiveness of initial solution improvement strategy and search balance strategy of the proposed algorithm in solving different combinational disruption problems.

Keywords: Disruption management · Setup times · Combinational disruption · Memetic algorithm · Pareto front

1 Introduction

As a kind of classic scheduling problem and an important combinational optimization problem, Permutation Flowshop Scheduling (PFS) has wide application background in engineering, industrials and services such as metallurgical industry, food processing and healthcare. At the same time, many uncertain factors are often encountered in the process of actual production. With disruption events as machine breakdown, processing time variation, the initial optimal schedule performs poorly, and is even no longer feasible. Disruption management in PFS, therefore, has attracted enormous attention in academic communities.

The literature of disruption management in PFS are reviewed as follows. The makespan minimization problem in flexible flow shop after machine breakdown was studied by Wang and Choi [1]. The original problem was decomposed into multiple clustered scheduling problems that were easier to solve and computational experiments showed that the method was more efficient than separate SPT rule and GA. A similar flexible flow shop scheduling problem after machine breakdown was studied by Wang

© Springer International Publishing Switzerland 2016
D.-S. Huang et al. (Eds.): ICIC 2016, Part I, LNCS 9771, pp. 548–559, 2016.
DOI: 10.1007/978-3-319-42291-6_55

et al. [2] and a FL-HEDA based on fuzzy logic was put forward to efficiently solve the problem. For new job arrival, Weng *et al.* [3] designed the routing strategies for distributed flowshop scheduling, and combined the strategies with assignment policy for new job arrival in JIT production. Liu *et al.* [4] used probability information to describe the distribution of job processing time in the flow shops, and designed an improved GA to maximize the probability of on-time completion.

The above studies in flow shops address disruption events separately, but it is crucial to deal with combinational disruption management problem, for a variety of disruption events may influence the current scheduling scheme in the manufacturing process simultaneously [5]. As the situation becomes more complex, the combinatorial disruption management problem is more complicated than that with single disruption event. Please refer to Katragjini *et al.* [5], Rahmani *et al.* [6], Li *et al.* [7], and Rahmani and Heydari [8] for researches related to flowshop environment.

Besides single disruption event was extended into combinational disruption, basic processing environment also become more complex. At present, most researches on disruption management in PFS didn't take setup time into consideration, or treated setup time as a part of sequence-independent job processing time. However, when the setup time is associated with the processing sequence, its influence on the scheduling results cannot be ignored [9]. The optimization of expectation of completion time with machine breakdown in the environment that setup time is related to the processing sequence was studied by Gholami *et al.* [9]. The random key genetic algorithm was applied to obtain the near-optimal result, and the expectation objective was evaluated with the simulator constructed by combining the event-driven strategy with right shift heuristic method. The same problem was also solved with anneal simulation and immune algorithms by Zandieh and Gholami [10]. Considering a different type of disruption, Kianfar *et al.* [11] studied average delay time minimization problem with job dynamic arrival in flexible flowshop, and proposed search-based assignment policy giving priorities to the collaborative scheduling. Then a mixed method based on genetic algorithm was designed to solve the problem effectively.

In this paper, a combinational disruption management problem in PFS with sequence dependent setup times is studied. Six types of disruption events, namely machine breakdown, processing time variation, setup times variation, new job arrival, job priority upgrading, and job cancellation, are considered simultaneously. Both schedule makespan and sequence deviation from the baseline schedule are taken into consideration when rescheduling. The rest of this paper is organized as follows. In Sect. 2 the problem is formally stated. In Sect. 3 a memetic algorithm combining local search and global search is proposed. The effectiveness of the algorithm is verified through numerical experiments in Sect. 5. Finally in Sect. 6 conclusion is given and some future directions are pointed out.

2   Problem Statement

The job processing is limited by the processing environment to meet the following basic constraint conditions: (1) Each job is processed by m operations, and all pass through the machines $M_1, M_2, \ldots, M_m$ in a given order to complete the processing operations. (2) Each operation of each job can only be carried out on one machine, and each machine is dedicated to one operation. (3) One machine can process at most one job at a time. (4) Pre-emption is not allowed during machining. The baseline schedule of minimizing C_{max}, is obtained by using the Hybrid Genetic Algorithm (HGA) proposed by Mirabi [12].

In the actual production, there are often various disruption events, which affect the smooth implementation of the baseline schedule. According to the source, we divide them into two major categories: *external* and *internal* events. The former refers to the events caused by changes in the external market demand, such as the change of the job priority, job cancellation, and new job arrival, etc. The latter refers to the events caused by the obsolete production equipment in the workshop or change in technology, including machine breakdown, change of processing time and setup time, etc. Considering the circumstances that the six disruption events happen in a combinatorial manner simultaneously, we make the following assumptions.

A1. For machine breakdown, its timing and duration are unknown in advance. After the machine recovers, the jobs which are interrupted from processing by the machine breakdown are continued to process the unfinished operations, i.e., the *pre-empt-resume* mode is adopted.

A2. Jobs affected by four types of disruption events including the change of processing time, the change of setup time, the upgrading of the priority, and job cancellation are randomly generated in the unfinished job set.

A3. Jobs that are already completed on M_1 when disruptions occur are not included in the unfinished job set.

A4. When a job's priority is upgraded, it is arranged after the already completed job set for processing in a timely manner. When a new job arrives in the production system, it is arranged to the end of the initial processing sequence. When a job is cancelled, it will be removed from the unfinished job set.

A5. All of the jobs affected by disruption events, after being processed according to assumptions A1 $\sim$ A4, should be postponed keeping their order. The processing schedule will also be adjusted accordingly.

With the above assumptions, we can minimize the deviation objective from the baseline schedule under the circumstances that the six disruption events happen in a combinatorial way. But it may lead to a certain degree of performance deterioration with respect to C_{max}. Therefore, in order to meet the preferences of different decision-makers, we have to reschedule the set of unfinished jobs except the jobs whose priority are upgraded, to obtain a high quality Pareto Front between C_{max} and deviation.

If the disruption events don't include the job priority upgrading, denote the job being processed on the first machine as J_0. Otherwise, denote the upgraded job as J_0. The completion time of job J_0 on the machine $M_i(i = 1, 2, \cdots, m)$ is $C_{i,0}$, which is regarded as the new 0 time. For the n jobs arranged to be processed after J_0, re-index them according to the processing sequence in the initial processing schedule as $J_1, J_2, \cdots, J_n$.

Consider the situation that the setup time is dependent on the processing sequence, if the job J_k on the machine M_i is arranged after the processing of J_j, denote the setup time as s_{ijk}. If the workpiece J_k is arranged after the processing of J_0, which also means to be the first to process after the new time 0, then we need to consider the time the machine has been occupied after the 0 time; we can regard it as part of the setup time in order to facilitate the handling, that also means the setup time $s_{i,0,k} = (C_{i,0} - C_{1,0}) + s_{i,0,k}$. Denote the processing time that workpiece J_j on machine M_i as $p_{i,j}$. For any feasible processing schedule $\pi = (\sigma_1, \sigma_2, \cdots, \sigma_n)$, define the following two objectives.

For the schedule performance, the maximum completion time of the production system is adopted. C_{i,σ_k} is the completion time for the job σ_k on the machine M_i, as a result, there is a relationship as follows:

$$C_{1,\sigma_1} = p_{1,\sigma_1} + s_{1,0,\sigma_1}$$

$$C_{i,\sigma_1} = \max\{C_{i-1,\sigma_1}, s_{i,0,\sigma_1}\} + p_{i,\sigma_1} \quad i = 2, \cdots, m$$

$$C_{1,\sigma_k} = C_{1,\sigma_{k-1}} + s_{1,\sigma_{k-1}\sigma_k} + p_{1,\sigma_k} \quad k = 2, \ldots, n$$

$$C_{i,\sigma_k} = \max\{C_{i,\sigma_{k-1}} + s_{i,\sigma_{k-1},\sigma_k}, C_{i-1,\sigma_k}\} + p_{i,\sigma_k} \quad i = 2, \ldots, m; k = 2, \ldots, n$$

Let $f_1(\pi)$ represent the *makespan*, then $f_1(\pi) = \min C_{m,\sigma_n}$.

For the deviation objective, the impact of disruption events is measured by the relative order of the π with respect to the initial processing order, whose meaning is the impact of interference events on reallocating resources throughout the entire production cycle of the processing system. D_{ij} indicates that compared with "J_i is arranged directly in front of J_{i+1}" in the initial processing sequence, the change in processing order when J_i is arranged to be processed directly in front of J_j. Therefore, we can conclude that:

$$D_{i,j} = \begin{cases} j - i - 1 & i < j \\ 0 & i = j \\ j - i + n & i > j \end{cases}$$

Let $f_2(\pi)$ be the sequence disturbance, then $f_2(\pi) = \min(D_{0,\sigma_1} + \sum_{i=1}^{n-1} D_{\sigma_i,\sigma_{i+1}} + D_{\sigma_n,0})$.

To summarize, by using the classical triples $\alpha|\beta|\gamma$, the problem can be expressed as $Fm|prmu, s_{ijk}, post - mgt|(f_1(\pi), f_2(\pi))$, the "$Fm$" means that there are m machines. "$prmu$" means that in the flow shop, the jobs access to machines with a fixed order. "s_{ijk}" means that the setup time associated with the processing order is considered. The "$post - mgt$" indicates the disruption events can only be dealt with after they ends. The "$(f_1(\pi), f_2(\pi))$" indicates a bi-objective scheduling problem.

3   An Enhanced Memetic Algorithm

We proposed an enhanced memetic algorithm (EMA) to solve the problem. In fact, a memetic algorithm (MA) is rather an idea that can be described as cooperation between global population evolution and local individual learning than a specific algorithm.

Hence, different combinations global search strategy with local search strategy will result different MAs. Meanwhile, the construction strategy of initial population and the balance of global search and local search are two key points to affect quality of Pareto front, which are also our focus of improvement.

3.1 Initial Population Construction Strategy

For multi-objective intelligent optimization algorithm, the quality of initial population will affect Pareto front's quality and algorithm convergence speed. In order to improve the quality of initial population, we apply three parts to construct initial population. They are respectively some good solutions for initial objective f_1, some good solutions for deviation objective f_2, and some random solutions for maintaining diversity of initial population.

For initial objective f_1, the problem can be solved by using HGA algorithm [12]. We set the final population as pop_1. For deviation objective f_2, there isn't existing algorithm to solve. However, we can transform it into initial scheduling problem. First, we construct an initial scheduling problem whose scale is one machine and n jobs, whose processing time and setup times are as follows:

$$p_{1,j} = 0, \quad j = 1, 2, \ldots, n; \ s_{1,j,k} = D_{j,k} \quad j, k = 1, 2, \ldots, n$$

Thus, the result of the initial schedule problem is equal to the deviation objective. If job sequence remain unchanged, that is to say $s_{1,j,k} = D_{j,j+1} = 0$, we can get the deviation objective $f_2 = 0$. And the bigger the sequence deviation is, the larger the result. Finally we set the final population as pop_2.

When two single-objective good populations pop_1 and pop_2 are achieved, we can't combine two populations directly because of the limit on population size and the high maturity. Therefore, we combine both of the two populations and perform non-dominated sorting. Then we take a certain proportion of better individuals. Finally, the rest are made up by solutions generated randomly.

3.2 Global Search Strategy

NSGA-II, as one of the most effective multi-objective evolutionary algorithms, is used as the global search engine [16]. The genetic operators mainly include selection, crossover and mutation. Because this part isn't the research focus and there have already been many classic operators, we apply existing operators.

For selection operator, we refer to the literature [13]. We first replace the worst r individuals with the best r individuals, then we adopt the classical tournament selection operation. We set $P_r = r/popsize$ as replacement rate. Two-point crossover and insertion mutation operators are used respectively, as Ishibuchi et al. [14] shows the result that the combination of two-point crossover operator and insertion mutation operator is best to solve the sequence-dependent PFS problem. We set the crossover probability and the mutation probability as p_c and p_m respectively.

3.3   Local Search Strategy

Local search is the process of the individuals' continuing improvement after genetic operators, which need to identify target individual for local search. The key factors of local search, including target individual, neighbourhood structure, acceptance criteria, search step, and search intensity are elaborated as follows.

(1) *Target individual*: In order to balance the effectiveness and diversity of the solutions, we generate random number $0 < w < 1$ after each genetic operation. Then compare $w * f_1 + (1 - w) * f_2$ among parents and its offsprings, and select the individual with minimum value as target individual.

(2) *Neighborhood structure*: In the research of multi-objective scheduling problem of permutation flow shop, Ishibuchi *et al.* [14] used the operation that insert neighborhood to get better results. Subramanian *et al.* [15] also showed that inserting two or three consecutive jobs from the current position to another location perform well. Therefore, we apply insertion operation for consecutive multi-jobs, the number of jobs genes is controlled within three, as shown in Fig. 1.

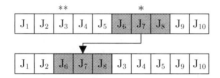

Fig. 1. Neighborhood structure

(3) *Search step*: namely the number of iterations of local search, recorded as N_{LS}.

(4) *Acceptance criteria*: common acceptance criteria is that when a new individual dominates target one, accept it as a new solution. However it's easy to miss the solution which is better overall. So we take a criteria that as long as a new individual is not dominated by the target individual, accept it as a new solution.

(5) *Search intensity*: it includes two aspects: search frequency and the search probability, which are set as T_{LS} and P_{LS}. The intensity of local search has a direct impact on the trade-off between global search and local search, which has a critical impact on the performance of the algorithm. Which is explained in details in the Sect. 3.4.

The local search will be executed every T_{LS} generations. Before it starts, first generate random weight coefficient $0 < w < 1$, when the genetic operation is completed. Then select the target individuals and set the number of them as *NUM*. The procedure of local search for this generation can described as follows.

Step 1. Initialize the number of individuals that can been found by local search $n_{ls} = 0$, the number of individuals to be accepted $n_{accept} = 0$, and the number of searched target individual to be local searched $i = 0$.

Step 2. If $i < NUN$, then $i = i + 1$, the depth of search $j = 0$; else, EMA ends.

Step 3. Generate a random number of $[0, 1]$ as $rand$, if $rand < P_{LS}$, then perform local search; else, go to Step 2.

Step 4. If $j < N_{LS}$, then carry out local search, and $j = j + 1$; else, $n_{ls} = n_{ls} + N_{LS}$ and go to Step 2.

Step 5. Judging whether the new individual is dominated by target individual, if not, accept it as a new solution, let $n_{accept} = n_{accept} + 1$, $n_{ls} = n_{ls} + j$ and go to Step 2; else, go to Step 4.

3.4 The Balance Strategy of Global Search and Local Search

Global search focuses on exploring broader solution space, while local search focuses on further improvement of individuals. So it is very important to balance the global search and local search under the limited computing resource. We introduce the concept of the local search's efficiency and the calculating formula of local search's efficiency η_t in the generation t is as $\eta_t = n_{accept}/n_{ls}$.

The situation $\eta_t = 0$ can be seen as algorithm convergence. We take it as a boundary to divide the memetic algorithm into two stages. Because the population maturity is different, we adopt different to adjust the trade-off between global and local search. In the first stage, the population is not convergent. Local search should be conducted every generation. Global and local search are balanced by the local search probability, and make P_{LS} smaller when η_t get smaller. Therefore, we set two parameter settings as: $T_{LS} = 1, P_{LS} = 1 - \eta_1 + \eta_t$. In the second stage: the population has already been convergent, global search should get further strengthened in order to enhance the diversity of solutions. Therefore, the two parameter settings are set as: $T_{LS} \in N^+, P_{LS} = 1$. The algorithm termination condition is controlled by applying functional evaluation 100 thousand times after a pilot test.

4 Numerical Experiments

To verify the effectiveness of the EMA in addressing six types of disruption events, we use NSGA-II, one of the most popular multi-objective algorithm, as a benchmark method for comparison [16]. The genetic operations of NSGA-II are in conformance with the EMA in this process. In this section, we firstly present five performance metrics for measuring Pareto front's quality from different perspectives, then design and explain the experimental data and algorithm parameters, finally compare and analyze the experimental results. All algorithms are coded in MATLAB language of version 7.10.0. The computer's CPU is Intel(R) Core(TM) 2 Duo CPU, the memory is 2G, and the main frequency is 3.00 GHz.

4.1 Performance Metrics

We introduce five performance indexes which can evaluate the performance of the multi-objective algorithm in terms of proximity and diversity.

HV (hyper-volume): the performance index which can evaluate the Pareto Front's convergence and diversity in a comprehensive manner [17]. The larger the value of *HV*, the better the comprehensive performance of the non-dominated solution set would be.

GD (Generation distance): the performance index which can evaluate the proximity of the non-dominated solution set, the smaller the value of *GD*, the better the proximity would be.

NDS_DIS (Non-dominated distance): the performance index which can evaluate the average distance between the non-dominated solution set and the ideal point [18].

ONVG (Overall Non-dominated Vector Generation): number of non-dominated solutions, the most direct index which can compare the diversity of different algorithms [13].

NDS_NUM (Non-dominated set number): the performance index which can evaluate the quality of diversity of the non-dominated solution set [18].

4.2   Numerical Experiment Design

The parameters of the initial scheduling problem are set according to [12]. The job processing time is uniformly sampled from the interval $[1, 99]$, the setup time is uniform sampled from the interval $[1, 9]$. In order to verify the general applicability on different scale of problem, consider the situations that the number of jobs $n = 25$, 50, 100 and the number of machines $m = 3$, 5, 10 respectively. 10 instances are randomly generated for each combination of n and m. The parameters regarding to six types of disruption events are as follows.

(1) *Machine breakdown*. The probability of random failure on each machine is equal. In order to make the problems fit the size of the scale of the initial problem, set the occurrence time of breakdown from a uniform distribution on the interval $[0, C_{1,5}]$ in the baseline schedule. Let the breakdown duration follow a uniform distribution $[1, 99]$.

(2) *Change of processing time*. The job processing time is randomly regenerated following the uniform distribution $[1, 99]$.

(3) *Change of setup time*. The job setup time is randomly regenerated following the uniform distribution $[1, 9]$.

(4) *New job arrival*. The processing and setup time of new jobs are consistent with the settings of the initial scheduling problem.

In order to verify the universality of the combination of different disruption events, each numerical example consider three types of disruption combinations.

(1) The first kind is the combination of disruption events occurring due to the change of external demand, including the arrival of the new workpieces, the change of job priority and the job cancellation.

(2) The second kind is those occurring due to the obsolete production equipment in the workshop or change in technology, including machine breakdown, change of processing time and setup time.

(3) The third kind is the complete combined interference events that the external and internal interference events mentioned above occur simultaneously.

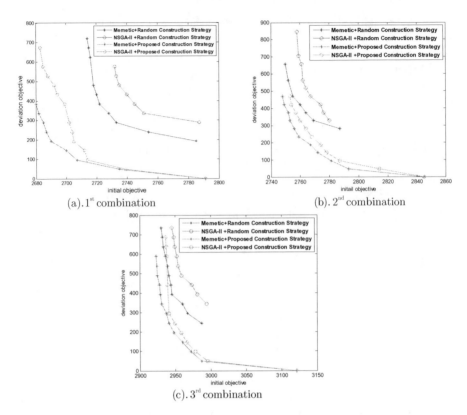

(a). 1st combination

(b). 2nd combination

(c). 3rd combination

Fig. 2. The comparison of Pareto Fronts under different event combination

In terms of tuning of algorithm parameters, by the random numerical experiment under the circumstance of six disruption events occur in combination and the scale of 100 jobs and 10 machines, we can adjust and determine the parameters of NSGA-II algorithm and EMA. Firstly, NSGA-II includes three parameters named population.

size (pop_size), crossover probability (P_c) and mutation probability (P_m), taking values from sets $\{100, 150, 200\}$, $\{0.7, 0.8, 0.9\}$, and $\{0.01, 0.05, 0.1\}$ respectively. We choose the combination of parameters with the minimum average HV of five repetitions. Then we tune the parameters of EMA, and share the values of common parameters with NSGA-II. The three distinctive parameters namely proportion of copy elimination (P_r), cycle (T_{LS}) and step length (N_{LS}) of local search are respectively taken from sets $\{0.1, 0.2, 0.3\}$, $\{3, 4, 5\}$, and $\{1, 2, 3\}$. The determination of these three parameters is the same as the previous one. At last, the parameter values involved in the algorithm are shown. $pop\_size = 150$, $P_c = 0.8$, $P_m = 0.1$, $P_r = 0.2$, $T_{LS} = 3$, and $N_{LS} = 2$.

4.3   Results Analysis

For problems with 5 machines and 50 jobs, we perform 10 repetitions of experiments following 3 different combinations respectively. As shown in Fig. 2, by combining EMA and NSGA-II with the random construction strategy of initial population and the proposed construction strategy, we can obtain four strategies. When the different combination of events happen, the Pareto front obtained by the enhanced memetic algorithm will completely dominate those obtained by NSGA-II when they have same construction strategy of initial population. The Pareto front obtained by using the proposed construction strategy will completely dominate those obtained by using the random construction strategy of initial population under the same algorithm. Hence the effectiveness of the proposed construction strategy of initial population and the enhanced memetic algorithm is strait forward verified.

In order to verify the general applicability of the proposed algorithm with respect to different problem size, we consider all the cases of $n = 25, 50, 100$ and $m = 3, 5, 10$ respectively. The result of error bar on HV is shown in Fig. 3. With the same number of machines and the increasing of number of jobs, the advantages of the proposed algorithm compared with classical NSGA-II become more evident. The reason can be attributed to two points: first, with the increase of n, the solution space is also increasing. The enhanced memetic algorithm can use local search to concentrate the limited computing resources to the search direction of the optimal solution. Second, to a certain extent, the proposed construction of initial population can indicate the direction for later search, which avoids blind search. In the process of n staying same and m increasing continuously, the increase of HV is only due to the increase of the corresponding makespan, so we don't need to perform too much analysis here.

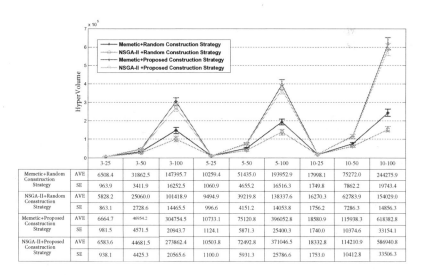

		3-25	3-50	3-100	5-25	5-50	5-100	10-25	10-50	10-100
Memetic+Random Consstruction Strategy	AVE	6508.4	31862.5	147395.7	10259.4	51435.0	193952.9	17998.1	75272.0	244275.9
	SE	963.9	3411.9	16252.5	1060.9	4655.2	16516.3	1749.8	7862.2	19743.4
NSGA-II+Random Consstruction Strategy	AVE	5828.2	25060.0	101418.9	9494.9	39219.8	138337.6	16270.3	62783.9	154029.0
	SE	863.1	2728.6	14465.5	996.6	4151.2	14053.8	1756.2	7286.3	14856.3
Memetic+Proposed Consstruction Strategy	AVE	6664.7	46954.2	304754.5	10733.1	75120.8	396052.8	18580.9	115938.3	618382.8
	SE	981.5	4571.5	20943.7	1124.1	5871.3	25400.3	1740.0	10374.6	33154.1
NSGA-II+Proposed Conssttruction Strategy	AVE	6583.6	44681.5	273862.4	10503.8	72492.8	371046.5	18332.8	114210.9	586940.8
	SE	938.1	4425.3	20565.6	1100.0	5931.3	25786.6	1753.0	10412.8	33506.3

Fig. 3. The error bar chart of HV under various problem sizes

5   Conclusion

In the permutation flow shop environment with sequence-dependent setup time, a disruption management problem is studied with six events including machine breakdown, change of processing time and setup time, new job arrival, change of job priority and job cancellation.

After the occurrence of combinatorial disruption events, rescheduling is carried out in order to maintain a good scheduling performance of makespan, at the meantime, taking the disturbance to the production workshop into consideration. In order to get a better Pareto front, this paper proposes an enhanced bi-objective memetic algorithm with a construction strategy of generating initial population, and a strategy of balancing the global and local search.

Numerical experiments show that, (1) through the construction strategy of the initial population we can effectively improve the quality of the Pareto front. (2) The effective Pareto front obtained by the improved memetic algorithm performs better than those obtained by the simple NSGA-II algorithm. (3) The proposed algorithm has general applicability in different combination of disruption events and different scale of problem.

Acknowledgement. This research was supported by the National Natural Science Foundation of China (71502023, 71271039), the New Century Excellent Talents in University (NCET-13-0082), and the Fundamental Research Funds for the Central Universities (DUT14YQ211).

References

1. Wang, K., Choi, S.H.: A decomposition-based approach to flexible flow shop scheduling under machine breakdown. Int. J. Prod. Res. **50**(1), 215–234 (2012)
2. Wang, K., Huang, Y., Qin, H.: A fuzzy logic-based hybrid estimation of distribution algorithm for distributed permutation flowshop scheduling problems under machine breakdown. J. Oper. Res. Soc. **67**(1), 68–82 (2016)
3. Weng, W., Wei, X., Fujimura, S.: Dynamic routing strategies for JIT production in hybrid flow shops. Comput. Oper. Res. **39**(12), 3316–3324 (2012)
4. Liu, Q., Ullah, S., Zhang, C.: An improved genetic algorithm for robust permutation flowshop scheduling. Int. J. Adv. Manuf. Technol. **56**(1–4), 345–354 (2011)
5. Katragjini, K., Vallada, E., Ruiz, R.: Flow shop rescheduling under different types of disruption. Int. J. Prod. Res. **51**(3), 780–797 (2013)
6. Rahmani, D., Heydari, M., Makui, A., et al.: A new approach to reducing the effects of stochastic disruptions in flexible flow shop problems with stability and nervousness. Int. J. Manag. Sci. Eng. Manag. **8**(3), 173–178 (2013)
7. Li, J., Pan, Q., Mao, K.: A discrete teaching-learning-based optimisation algorithm for realistic flowshop rescheduling problems. Eng. Appl. Artif. Intell. **37**, 279–292 (2015)
8. Rahmani, D., Heydari, M.: Robust and stable flow shop scheduling with unexpected arrivals of new jobs and uncertain processing times. J. Manuf. Syst. **33**(1), 84–92 (2014)
9. Gholami, M., Zandieh, M., Alem-Tabriz, A.: Scheduling hybrid flow shop with sequence-dependent setup times and machines with random breakdowns. Int. J. Adv. Manuf. Technol. **42**(1–2), 189–201 (2009)

10. Zandieh, M., Gholami, M.: An immune algorithm for scheduling a hybrid flow shop with sequence-dependent setup times and machines with random breakdowns. Int. J. Prod. Res. **47**(24), 6999–7027 (2009)
11. Kianfar, K., Ghomi, S.M.T.F., Jadid, A.O.: Study of stochastic sequence-dependent flexible flow shop via developing a dispatching rule and a hybrid GA. Eng. Appl. Artif. Intell. **25**(3), 494–506 (2012)
12. Mirabi, M.: A novel hybrid genetic algorithm to solve the sequence-dependent permutation flow-shop scheduling problem. Int. J. Adv. Manuf. Technol. **71**(1–4), 429–437 (2014)
13. Li, B.B., Wang, L.: A hybrid quantum-inspired genetic algorithm for multiobjective flow shop scheduling. IEEE Trans. Syst. Man Cybern. B Cybern. **37**(3), 576–591 (2007)
14. Ishibuchi, H., Yoshida, T., Murata, T.: Balance between genetic search and local search in memetic algorithms for multiobjective permutation flowshop scheduling. IEEE Trans. Evol. Comput. **7**(2), 204–223 (2003)
15. Subramanian, A., Battarra, M., Potts, C.N.: An iterated local search heuristic for the single machine total weighted tardiness scheduling problem with sequence-dependent setup times. Int. J. Prod. Res. **52**(9), 2729–2742 (2014)
16. Deb, K., Pratap, A., Agarwal, S., et al.: A fast and elitist multiobjective genetic algorithm: NSGA-II. IEEE Trans. Evol. Comput. **6**(2), 182–197 (2002)
17. Bhuvana, J., Aravindan, C.: Memetic algorithm with preferential local search using adaptive weights for multi-objective optimization problems. Soft Comput. **20**(4), 1365–1388 (2015)
18. Qian, B., Wang, L., Huang, D.X., et al.: Scheduling multi-objective job shops using a memetic algorithm based on differential evolution. Int. J. Adv. Manuf. Technol. **35**(9–10), 1014–1027 (2008)

Information Security

Event Space-Correlation Analysis Algorithm Based on Ant Colony Optimization

Si Liu, Guo-Ning Lv[(✉)], and Cong Feng

College of Information Science and Technology,
Zhengzhou Normal University, Zhengzhou, China
ukyosama@163.com

Abstract. Historical disaster events are taken as a case for space-correlation analysis, three-dimensional disasters space-time complex network are modeled and chain relationship of disaster nodes are mined by looking for similar space vector in network. Then transformed the vector discover problem into a path optimization problem and solved by using ant colony algorithm, where the pheromone parameter in the process of optimal-path finding is concerned as the algorithm result, in order to solve the problem of path competition which existed when only to solve the optimal path. Experimental results of MATLAB show that this method has high accuracy and practicality.

Keywords: Complex network · Space correlation · Disaster chain · Ant Colony Optimization · Pheromone

1 Introduction

More and more facts have shown that drought, floods, earthquake and other disaster events often do not exist in isolation, but have some connections. Such as, after 8.5-magnitude earthquake in Indonesia Sumatra in March 2005, major flood struck China's Pearl River Basin in July, and after the 8.3-magnitude earthquake in Sumatra in September 2007, snow and ice storms struck the south of China in early 2008, because of this similar chain-like cognate phenomenon of disasters, the disaster chain and disaster evolution mechanism [1–4] have been got more and more attention, many researchers of our country believe that the occurrence of the Sumatra earthquake is directly related to the above two disasters of our country, "(earthquake causes) the evaporation of sea water causes clouds and rain, they become flood in summer, become winter rain or snow in winter" [5–7], and proposed the disaster's homology, chain nature and rhythmicity which reflected by the relationship between disasters, has a high value for the prediction of disasters. However, the relationship between disasters is not often instantaneous and self-evident, but is a large span of geographical and time, so the analysis and excavation of the relationship between disasters is a very complex problem.

Now the disaster correlation studies are mainly statistical analysis of disaster records, however, due to the space migration and time delayed etc. complex characteristics of the disaster correlation, it is difficult to effectively extract the association rules, thus the efficiency and credibility are low. Literature [8, 9] was based on time series similarity matching, achieved the correlation analysis for the earthquake areas,

© Springer International Publishing Switzerland 2016
D.-S. Huang et al. (Eds.): ICIC 2016, Part I, LNCS 9771, pp. 563–570, 2016.
DOI: 10.1007/978-3-319-42291-6_56

and achieved very good results, but the model can only obtain generally relevance degree, but cannot determine the causal relationship between the regions, namely, it shows the law of the region A cause the earthquake of region B, or shows the law of the region B cause the earthquake of region A, or both A and B are the areas of frequent earthquake, they shows a fake matching.

In view of this, this paper attempts to describe the relationship between disasters with a historical three-dimensional space-time network model, this model reflects the causal relationship between the disasters by the vector space, thus transform the problem of disaster correlation analysis into the problem of excavating the similar vector of network, and by the ant colony algorithm [10–15] to improve velocity and precision of excavation, the experiments show that this method has a good application value.

2 Space-Time Network Model of the Historical Disasters

China has a long history, the account of natural disaster events in history books are characterized by huge number, rich type, long sequence, strong continuity etc., and these records provide a valuable source of data for analyzing and summing the law of the disaster. So this paper takes the historical disasters as the analysis object of the disaster correlation, and maps various disasters in the history records to the map of China by GIS technology, then arranges the multi-layer disaster map according to the fixed time span (year or season) in order to build a three-dimensional network model of historical disasters as shown in Fig. 1.

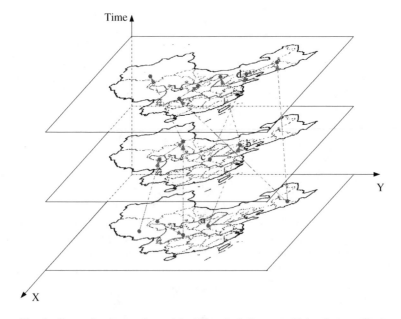

Fig. 1. Space-time network model of historical disasters (Color figure online)

The blue spots represent disaster node, the red directed lines represent correlation vector (disaster chain, shows the previous disaster trigger the following disaster) in Fig. 1, the three-dimensional disaster vector can conveniently describe the evolving rule of disasters on time and space, so how to determine the disaster correlation vector needs to be addressed in this paper. Multiple disaster nodes distributed each historical time layer, each node can choose a large number of nodes which can be linked with it, and these nodes into a chain in the same layer or cross-layer, so only rely on analyzing the distribution rules of the entire three-dimensional network nodes to make sure the linked objects. Set the disaster node in the network is n, n can be defined as $n = (x, y, z, type)$, where x and y represent coordinates of disaster node on the map, z represents the layer of the disaster node, type represents the category of disaster. For a node pair (n_i, n_j), if can find a number of node pairs which have the similar location and the same time span in the other layer of the network, then space vector which connected by these nodes is similar, the position of node a and node c are similar and they are separated by a layer in Fig. 1, the position of node b and node d are similar and they are separated by a layer, so the vector ab is similar with vector cd, if there are many vectors which are similar with the vector ab in the three-dimensional network, we can consider this type of vector space is disaster correlation vector, so can determine the chaining rules of the corresponding node location to the vector. Through the method of disaster correlation analysis which based on vector, not only can provide probability etc. basis of being a chain for the further study of disaster network evolution model, but also can provide more specific advices for preventing disaster for the local governments.

In order to make the object of study more clearly, this paper simplify the above model, only study the law of a single type and adjacent layers in single region of chaining disaster, such as Henan Province, study which areas of the earthquake in the underlying map will lead to flood in the adjacent upper map of Henan Province (earthquake-water disasters chain), then the problem becomes what vector can be found a number of similar vectors between the other adjacent layers in the all connected vectors that lower earthquake nodes point to the upper Henan Province floods nodes. If takes the interlamination vectors between these layers as a connection path between the connected layers and interlamination is tantamount to solving all possible paths from the bottom level to the top level, how to make the sub-paths in the total paths as much as possible similar, can be transformed into a path optimization problem.

Obviously, the above path optimization problem belongs to np-hard problem, the ant colony algorithm which as a classic metaheuristic has better computational efficiency on the problem of path optimization. In addition, obtaining an optimal path is not a perfect scheme for solving the above problem, because optimal path can only select a sub-path between every two levels, but it is likely to have many types of similar sub-paths between the two layers, that is, there is the phenomenon of "competition". However, what we are looking for is not the path which has the most parallels in the similar sub-paths, but as long as has a certain amount parallels in the sub-path is valuable path, ant colony algorithm builds a new solution by releasing pheromone on the path, this way provides a guideline for us to solve the "competition", in the algorithm, the more pheromone owned by a sub-path, the greater probability is selected by the ants, the algorithm converges can be considered if a path has accumulated a number of considerable pheromones, then this path is excellent, therefore, in the using of ant colony algorithm, is let the ant colony algorithm to find the optimal path as the media, and the ultimate aim is to extract the sub-path which with high pheromones at the end of the algorithm, in a sense, it is a reverse using of the ant colony algorithm.

3   Description of Algorithm Process

The description of the algorithm is as follows; set the number of ants as m.

1. Data Initialization

Step 1. Import network nodes topology and generate the node path matrix, then initialize the pheromone matrix $pheromone[n][n]$, where the value of array element $pheromone [i][j]$ (i is an arbitrary node in lower layer, j is an arbitrary node in the selected region of adjacent upper layer) which represents the pheromone content of the path from node i to j is set a uniform initial value as shown in formula (1).

$$pheromone[i][j] = v_{init\_pheromone} \tag{1}$$

Step 2. Initialize the parameters of ants and algorithm, such as the maximum iterations $epoch\_max$, the length of global optimal path $path\_best$, the lower limit of pheromone content $pheromone\_min$ and so on.

2. Iteration

When the iteration number is equal to $epoch\_max$, the iteration end; otherwise circulating the following acts:

Step 1. Clear the way-finding memory of all ants on the previous iteration;

Step 2. All ants were randomly assigned to an initial node in the lowest layer;

Step 3. Every ant builds a full path: in each step of the path construction, ant k ($k \in [1,m]$) in accordance with the roulette wheel selection method to decide the next node to go. The characteristic of this selection method is that the value of $pheromone [i][j]$ corresponding to a path is greater, the probability of the path chosen by ant k is greater. So on the one hand ants tend to choose the paths which is considered better currently, on the other hand the other paths are also have the chance of being selected that can avoid algorithm falling into local convergence prematurely. The probability of an ant in node i will choose the node j as its next visit node is shown in formula (2), where N represents the set of optional nodes in next layer.

$$p_{ij} = \frac{pheromone[i][j]}{\sum\limits_{v \in N} pheromone[i][v]}, \quad j \in N \tag{2}$$

Step 4. Calculate the fitness value of each path constructed by ants, then reduce the pheromone content of each sub-path has been visited (set the reduction as V_{sub}), so that the probability of other paths chosen by ants in next iteration is increased which could prevent the algorithm falling into a local optimum. If the pheromone content of a path is less than $pheromone\_min$, then make it equal to $pheromone\_min$, which can avoid the pheromone content of a path is too low so the path can hardly be selected. Algorithm need to find a full path contains the most similar sub-paths, so the fitness value is obtained by comparing the sub-path similarity, set the coordinate of starting node i of sub-path $p_{i,j}$ as (x_i, y_i, z_i), and the coordinate of tai node j as (x_j, y_j, z_j), the coordinate of starting node u of sub-path p_{uv} as (x_u, y_u, z_u), and the coordinate of tai node v as (x_v, y_v, z_v), then the similarity judging method for these two sub-paths is shown in formula (3).

$$sim = \sum_{\substack{a = i, b = u; \\ a = j, b = v}} \sqrt{(x_a - x_b)^2 + (y_a - y_b)^2} \leq threshold \qquad (3)$$

where *threshold* represents the similarity threshold, two sub-paths are similar if the value of *sim* below the predefined threshold, then the similarity determination value (represent by h) of these two sub-paths is equal to 1. However, due to the error increases phenomenon in the comparing process, for example, there are three sub-paths set as a, b and c, a and b are similar, a and c are also similar, but b and c are not necessarily still similar, so in this paper the fitness value is determined by comparing the sub-paths one by one, as shown in formula (4).

$$fitness = \alpha * \sum_{sim_{ij} \leq threshold} h_{ij} \qquad (4)$$

where, α is the adjustment factor, i, j represent for any two sub-paths in constructed full path.

Step 5. Select the path which has the highest fitness value in this iteration and increase the pheromone content (set the increment as V_{add}) of this path, which is to reward the best path in this iteration so the probability of this path chosen by ants in next iteration is increased, and V_{add} is significantly greater than V_{sub}. If the fitness value of this path is more than *path_best*, then use it to replace *path_best*, and record the node order of this path;

Step 6. Go to the next iteration.

3. Iteration Complete

Extract the sub-paths whose pheromone content is higher than the set threshold, and combine the similar sub-paths with same kind, then return to network and conduct accurate data processing, so as to figure out the actual similar number of these sub-paths.

4   Case Study

In order to verify the performance of the proposed algorithm, we programmed the algorithm under MATLAB 7.0, the experimental data are the nationwide drought records and locust plague records of Henan in Ming and Qing Dynasty (1368–1912), so as to study which region of China has a drought, will more likely to cause the occurrence of locusts in Henan, the data derived from historical disaster databases which constructed by relying Fund, where the record years of Henan locust are 216 years, and in the 216 years with drought records at the same time in China there are 211 years, this paper used the corresponding historical disaster records of the 211 years to establish the disaster network.

In specific experiment, the number of ants in the algorithm is set to 10, used time layer as year to conduct the network construction, the disaster geographical positions are sampling point marked by China map under the range of 1280 * 1024, the initial pheromone content of each path is 1, the pheromone reward for best path is 0.1, pheromone volatilization is 0.01, the similarity determine threshold is set to 15.0, the number of iterations is set to 2000. After observed repeated tests, set the sub-path whose pheromone content is greater than 7 is the similar sub-path after the end of iteration, and finally sorted out a total of seven groups of similar sub-paths, and the vector information represented by these similar sub-paths is shown in Table 1, due to disasters node is determined according to the affected center which in fact has a certain affected range, and when combine the similar vectors, the affected disaster range may further expand. The vector information in Table 1 has been translated into specific locations, from the table we can found the end points of vector are mostly distributed in the northern part of Henan, it has great relevance with the historical locust records in Henan are also concentrated in the northern part. According to the experimental results, it can be seen in history after the drought occurred in Nanjing, Taiyuan, Shijiazhuang, there are many locust records appeared in Henan (in particular, Nanjing-northern Henan vector has a very prominent number of similar records), and therefore these vectors have a high disaster relevance, worthwhile for the Government to pay early attention when the similar incident occurs.

Table 1. Path information matched with the actual history

Vector starting point	Vector end point	Average pheromone content	Number of similar records in history
Nanjing, Yangzhou	Northern Henan	30.7	94
Taiyuan	Jiaozuo, Zhengzhou	23.2	65
Baoding, Shijiazhua-ng	Kaifeng	17.4	50
Northern Shandong	Northern Henan	15.2	44
Linfen	Henan Province	13.4	39
Wuhan	Kaifeng	8.1	33
Quzhou	Shangqiu	7.5	21

Figure 2 is the parameter monitoring chart for the global highest value when the program is running, parameters, you can see the fitness value increased steeply in algorithm running prophase, and improved and converged steadily in the later. In addition, because the algorithm is not purposed to obtain the optimal path, the set number of cycles can be further reduced, the distribution of sub-path pheromone content can be considered mature when the change of fitness value began to slow, so the computational efficiency of the proposed algorithm is actually higher than the normal ant colony algorithm.

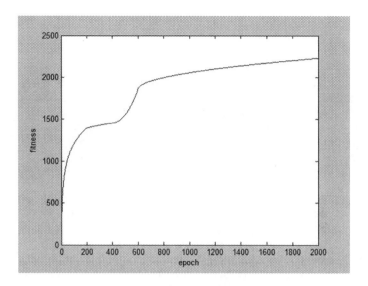

Fig. 2. Maximum fitness value in iteration

5 Conclusion

This paper through the establishment of three-dimensional disaster network to make the analysis of disaster correlation more clarity on the geographical and time scales, and converted the purpose of using ant colony algorithm from solving the optimal path to extract high pheromone sub-paths, so that to solve the problem of path competition. However, the "single" disaster chain solved in this paper is only a basic relational form of disasters evolution, and for further disaster evolutionary modeling, it still need to do more in-depth discussion on the circumstances of composite hazard, cross-layer.

Acknowledgment. This paper is sponsored by the National Natural Science Foundation of China (NSFC, Grant 61572447).

References

1. Girvan, M., Newman, M.: Community structure in social and biological networks. Proc. Natl. Acad. Sci. **9**(12), 7821–7826 (2002)
2. Newman, J.: Fast algorithm for detecting community structure in networks. Phys. Rev. E **69** (6), 066133 (2004)

3. Guimerà, R., Amaral, L.: Functional cartography of complex metabolic networks. Nature **433**(7028), 895–900 (2005)
4. Newman, J.: Detecting community structure in networks. Eur. Phys. J. B **38**(2), 321–330 (2004)
5. Duch, J., Arenas, A.: Community detection in complex networks using extremal optimization. Phys. Rev. E **72**(2), 027104 (2005)
6. Blondel, V.D., Guillaume, J.L., Lambiotte, R., Lefebvre, E.: Fast unfolding of communities in large networks. J. Stat. Mech. Theor. Exp. **10**, 10008 (2010)
7. Lü, Z., Huang, W.: Iterated tabu search for identifying community structure in complex networks. Phys. Rev. E **80**(2), 026130 (2009)
8. Yang, B., Cheung, W., Liu, J.: Community mining from signed social networks. IEEE Trans. Knowl. Data Eng. **19**(10), 1333–1348 (2007)
9. Palla, G., Derényi, I., Farkas, I., Vicsek, T.: Uncovering the overlapping community structure of complex networks in nature and society. Nature **435**(7043), 814–818 (2005)
10. Raghavan, U., Albert, R., Kumara, S.: Near linear-time algorithm to detect community structures in large-scale networks. Phys. Rev. E **76**(3), 036106 (2007)
11. Rosvall, M., Bergstrom, C.T.: An information-theoretic framework for resolving community structure in complex networks. Proc. Natl. Acad. Sci U.S.A. **104**(18), 7327–7331 (2007)
12. Jin, D., Yang, B., Liu, J., Liu, D.: Ant colony optimization based on random walk for community detection in complex networks. J. Softw. **23**(3), 451–464 (2012)
13. He, Z., Wang, J., Liu, S.: TSP-Chord: an improved chord model with physical topology awareness. In: 2012 International Conference on Information and Computer Networks, vol. 27, pp. 176–180 (2012)
14. Zachary, W.: An information flow model for conflict and fission in small groups. J. Anthropol. Res. **33**(4), 452–473 (1977)
15. Lusseau, D.: The emergent properties of a dolphin social network. In: Proceedings of The Royal Society B. Biological Sciences, vol. 270, pp. 186–188 (2003)

Research on Information Security of College Ideology

Shan-shan Gu[1,2(✉)] and Yu Shi[2]

[1] Wuhan University of Technology, Wuhan, China
manbanpaiya@163.com
[2] Zhengzhou Normal University, Zhengzhou, China

Abstract. Although technical and procedural measures helping to improve the information security of college ideology, there is an increased need to accommodate teachers-student, social and organizational factors. The purpose of this paper is to investigate the role of ideology information security in college. In that regard, a construction model is developed and tested. Overall, analyses indicate strong support for the validation of the proposed theoretical model.

Keywords: Information security · College ideology · Informatization · Construction model

1 Introduction

Ideology is the sum of the concepts of political and economic direct contact with a certain society, which including politics, law, ideology, morality, literature, art, religion, philosophy and other social sciences and other forms of consciousness, it systematically, consciously, directly reflect economic formation of society and political system and ideology, is composed of the superstructure of concept ideology system in various forms of social consciousness. Ideology is the forerunner of people thought and the behavior of, it as an important part of national political system is very sensitive areas, to countries with fundamental significance of an objective existence, it between political power and often exist a kind of affinity, provide a kind of legal support for the political power. The core is political idea and political belief. Marx doctrine is the ideology of the proletariat, which represents the interests of the proletariat, and plays a huge role in promoting the construction of socialist material and spiritual civilization. And with the process of globalization, with the change of the situation at home and abroad, the field of ideology of our country is quietly and carefully and deeply. Among them, in the field of mainstream ideology occupy the dominant position of the Marxism to withstand the various erosion, confusion and struggle, is particularly worthy of attention, so the information security of ideology arises.

Ideological security is an important part of national political security and cultural security, and it is an important part of the national security system. "Ideology is not only the ruling class of faith and the rule of the rationality of theoretical basis; but also more important to maintain the function of the political system of the state". Therefore, Western hostile forces leave no stone unturned to ideology of our country intensify

© Springer International Publishing Switzerland 2016
D.-S. Huang et al. (Eds.): ICIC 2016, Part I, LNCS 9771, pp. 571–579, 2016.
DOI: 10.1007/978-3-319-42291-6_57

penetration, safeguard the security of ideology become social nationalism is facing major problems [1].

The college is in the forefront of the ideological field, and is an important position to cultivate qualified builders and reliable successors of socialism with Chinese characteristics. For a long time, the Western hostile forces taking colleges and universities as the target of Westernization, the young students and intellectuals as a key object, and we vying for positions, for youth, the battle for hearts and minds [2]. Various ideological and cultural exchanges and integration, all kinds of social ideology clash contest, a variety of network information interchange diffusion, teachers and students in universities and colleges of ideas, value orientation and behavior produced profound effect. And this undoubtedly on safety construction of university ideology proposed very urgent task.

College information security and data protection have become important concerns and challenges facing organizations and users. Despite the effort and money that organizations spend to secure their assets, many incidents of data breaches and information loss continue to happen every year. Today, organizations realize that securing information is a continuous and complex task. The burden of keeping information secure rests on the shoulders of all organizational functions and members [3]. On this view, users must be aware of their roles and responsibilities in protecting information assets and of how to respond to any potential threat [4]. As a result, security awareness programs should focus on enlightening users on how to effectively protect information assets.

Werlinger, et al. [5] compared the levels of information security and ethics awareness of students in diverse university environments, and found that technology universities' students were more aware of information systems security and ethics than those who attended a liberal arts university. Ng and Xu [6] examined the effect of code communication on ethical behavior through its effects on code awareness and understanding and found that code understanding by IS professionals was a significant determinant of their ethical behavior. Siponen, et al. [7] investigated why ethics and/or punishment may or may not serve to deter people from breaching information security measures. The study found that punishment and ethics training can be effective in reducing threats to information security. Other studies proposed frameworks for teaching information security ethics [8]. The few studies found in security research investigated the role of ethics with other constructs. This study investigates the role of ethics in explaining compliance behavior.

Colleges play an important role in the development of a country and society, and play an increasingly important role. The university is the production of social ideology and the intersection of the frontier, is the students ideological and theoretical study of views amalgamate, hit the window, and the college students are the main force in the future development of the country. Therefore, to grasp the college ideological education initiative and responsibility. Colleges and universities in the new period ideology and existence question also has its particularity: (1) globalization and the concept of "Westernization" of the mainstream ideology pose a threat; (2) the safety awareness of College Ideology desalination; (3) traditional ideological education in the university has gradually cannot adapt to the era of rapid development of network.

Throughout the domestic scholars in the study, at present the overall domestic relates to the ideology of the content is more, mainly from the ideology of the encountered challenges, the ideological work and implementation of network point of view of ideology in Colleges and universities were analyzed. This shows that with the continuous development of the information age, many scholars have been aware of the importance and urgency of strengthening the construction of university ideology. However, these studies is research on the traditional ways of ideological construction, and clearly put forward "the ideology of information security" and as the research object is few, the information security of university ideology control should have a variety of ways, from the point of view of information to control the high school ideology is the problem of this paper is to study.

2   Challenges of Ideological Information Security in Colleges

At present, our country is in an important historical period of deepening the reform, various social contradictions and problems appear realistic social problems and problems of ideological understanding of mutual influence, various forces and the trend of competing sound, the struggle in the ideological field unusually sharp. To strengthen the construction of Ideological Security in Colleges and universities can effectively deal with the new changes in the field of ideology in recent years, in view of the actual situation of college students and the ideological situation of college students, to grasp the fundamental theory of Ideological and theoretical construction. To strengthen and improve the ideological and political education in Colleges and universities, the content of Ideological and political education in Colleges and Universities.

The University at the forefront of the development of the Internet, teachers and students in Colleges and universities is the largest network group. Network development enhances the socialist ideology of radiation and influence, enhance the degree of college teachers and students agree to the socialist ideology, provide technical support and guarantee for the innovation and development of socialist ideology. Overall, the dominant position of Chinese ideology construction, always adhere to the Marx doctrine, the mainstream is healthy and positive with positive energy. But network throughout the world in all kinds of thoughts and cultures are surging, the cultural soft power of our country is still relatively weak, the Western hostile forces deliberately to me were ideological and cultural penetration, spare no effort to sell western political philosophy, system and value concept, put forward the new challenge to the College Ideology Construction, which is mainly manifested in the following aspects.

Our country is now in the golden period of development of socialist reform and opening up. At the same time, various social contradictions prominent, intensified; the rapid growth of social wealth is also no restrained to make people between the gap between the rich and the poor will continue to widen; in the social reality of touch, can easily lead to social mentality of instability, intensification of the various social focus and hot issues, the eruption of social contradictions often touches vulnerable sensitive nerve. The development of the network makes up for the past people to express their ideas and interests of a single channel and not smooth. Due to the wide range of network communication, fast speed and has openness, autonomy, hidden features, desire checks on network speech operation difficulty is great, and the become people to express personal desire and reflect the preferred place of the will of the subject. On the Internet, on behalf of the "social conscience" intellectuals can not only free access to information, and the ability to independently review, which was originally a thing that is worth to advocate, but virtual network and occult characteristics, making the identity of members of the network is also symbolic of the, reducing the binding the moral self-discipline of Internet users and online behavior of their own, resulting in Network Behavior Anomie. This kind of thing once, through the network spread and the rapid spread, which quickly became the focus of public opinion, caused the wide attention of the society at the same time, social contradictions intensifies, and affect social stability. In addition, ideological work under the network environment to the traditional ideological work of the object of the reality, certainty and can be mastered, such as the characteristics of a thorough subversion. The new characteristics of network communication, as well as a series of new situations, which have brought the challenge to the dominant power of the socialist ideology in our country. Colleges and universities ideological and is not fully mature, to network information and statements lack due examination and the ability to identify, so the networking environment to the ideological work in Colleges and universities in our country bring strong impact.

"Construction of ideology is an important function of mass communication; ideology and the media always have a very close relationship" [9]. Before the rise of the Internet technology, the main carrier of ideology of our country spread is printed media (news-papers, magazines, books, etc.) and the power of media (broadcast, television, radio, etc.), has obvious mandatory, from top to bottom of the theory of indoctrination, this mode of transmission effectively highlighting the authority of the mainstream ideology and greatly limits to prevent the negative information dissemination and diffusion. It is in this long period of publicity and practice that our country has formed the centralized and unified control mode of socialist ideology. "The spread of ideology has always been the media; the nature is also an important part of the new media dissemination" [10]. Today, with the development of new technology revolution and knowledge economy, with information technology as the carrier of new media to its mass, free, democratic, interactive characteristics and fast means of communication, so that people can through the full expression of the will, and let people feel oneself is a historical "dramatis personae", at the same time, it is the "author" of historical drama. But some phenomena which are produced by this trend also constitute the existence of "heterogeneity" with the mainstream ideology.

The field of Internet and network culture has become the western developed capitalist countries in foreign culture and expansion of the new strategic and ideological struggle, the main battlefield. At present, domestic and foreign hostile forces in support of the western developed countries led by the United States has been built thousands of dedi-cated website, use the most fashionable, the most effective network means of commu-nication, to our country implementation of uninterrupted all-round, public opinion warfare, spare no effort to all of China's political, economic, cultural and other fields within the social affairs of wanton distortion and slander. Thus, the ideological concept, value orientation and behavior mode of college teachers and students have a profound impact on the traditional beliefs and ideas have been shaken or changed, and even appeared to discredit the Marx doctrine; The field of social ideology has changed from the stable development period in the past to the "extremely active period" and "issue multiple period", which has realistic problems, such as the belief dilution, the weakening of the mainstream ideology and the difficulty of public opinion guidance. All this shows that in the era of new media, the United States and other Western capitalist countries to socialist countries in the development of the implementation of the "network culture hegemony", by virtue of the Internet to vigorously implement "culture colonialism". Therefore, under the background of network of our country of capitalist culture and ideology of the resistance and defense work will become very difficult.

3   Construction of Ideological Information Security in Colleges

The media is the main channel of the mechanism of ideology construction. Network characteristics of new media, and it displays the global, social, subjectivity and inter-activity of communication attributes and modes of transmission, the traditional social ideology construction ways to produce a disruptive influence, this also makes the network under the background of the ideology security building design showing a different from the traditional new features.

Media and ideology between very close relationships between the performance of: ideology construction is often dependent on mass communication, mass media is an important carrier of the spread of ideology. In this sense, the new media should become contemporary ideological construction indispensable elements. As Thompson said: "the analysis of the ideology in the modern society, must take the nature and influence of mass media on the core position, although the only place of mass communication is not ideology operation." In fact, instead of the traditional print media and new media audio-visual media carrier, has become the main position of communication theory of ideology and values.

"Education and media has increasingly become capitalist society the main ideolog-ical machine", Western hostile forces in our country are cultural and ideological infil-tration, is bound to take all means to occupy the position of this front, through the network of teachers and students of colleges and universities in our country of confusion, for audiences at the same time, interference for colleges and universities to achieve the strategic functions of China's mainstream ideology propaganda, can be said to kill two birds with one stone. Therefore, we have attached great importance to the college as we are the strategic value of the main position of socialist ideology and propaganda, and consciousness to strengthen college ideology construction to strengthen our country ideology security construction importance.

The information asymmetry between educators and students, propaganda and the audience increasingly eliminated, formation and adapt to the network's openness, equality, interaction characteristics of openness, participation, responsiveness and mode of education. The characteristic of new media dissemination of ideology is a two-way interaction or multi-directional interaction. The ideology of the era of network construction often depending on behalf of the mainstream ideology of the official, engaged in the ideological work of intellectuals and experts and the majority of the public the three levels of cohesion and interaction. To make ideology persuasive and maintain the vitality, official will not blindly arrogant, underestimate the public cognitive ability; intellectuals and experts cannot due to grasp the "Discourse", despise the public understanding; they must face up to the network condition under, "education" and "education" and "propaganda" and "audience" symmetry of information between the reality, and deeply aware of, in the network era, the public and civil society is ideology is able to exist and maintain the sustainable development of an important basis, once lost, such groups and platform, the consciousness of the form will become lonely "monologue", and lose the meaning of its existence. Therefore, compliance with the development of network times, to establish an open, participatory, responsive and inclusive education mode of college ideological construction is necessary and timely.

Network is to destroy the student is interactive, instant image and characteristics of the ideological information transfer new mode, changed the "centralized" and "top-down" traditional mode of transmission. Because there is no center network structure, regardless of rank, coupled with public network, and the diversity, random, self and scattered characteristics, making communication paradigm has fundamentally changed, and gradually showed a "center" and "to power wafers" and "level" of the trend. This is undoubtedly the subversion of the authority of the previous structure, while the structure of social relations, social interaction and even people's ideas have had a profound impact. This impact on the specific performance, it not only expanded and enhanced the audience to accept the information of autonomy, but also to enhance the audience's release of information, communication and diffusion capacity. Network communication "whispers" tendency is also reflected in, people don't just listen to one sided report and explain, are no longer easily accept, mainstream ideology identity point of view, and began to dare to challenge authority, to subvert the traditional. The individual expression makes the individual in the ideological construction mechanism in an unprecedented upgrade, which greatly mobilize the enthusiasm and initiative of the Internet users to participate in the construction of ideology. They verify the channels greatly increased, greatly improve the ability to question, the general will not be easy to accept top-down propaganda.

Network information is the extension of people's social life in virtual space. Compared with the traditional media and information dissemination way, the development of information network has a series of unique advantages, the large amount of information, interactivity, timeliness strong, making bodies in the social network can be in a very short period of time through a variety of ways, such as massive accumulation of micro blogging, micro channel, post, reprint and trackers and other views, opinions and voices, so that a social focus and a hot issue with the fastest rising. The modern network information media gradually become in today's China society in the various views, ideas, thoughts and ideas of the collision between the land and become the broad masses of the people to exercise its right to express, the right to participate in, the important position of the right and the supervision right to know, and timely release, transfer the birthplace of all of the news media and the incidents of corruption. But at the same time due to some reports, freedom of speech, freedom of information serious misrepresentation and deception, fraud is prevalent, immoral, such as network spread of Mei Mei "" dry Lulu indecent photos and other typical network events. Its essence is the values of socialist moral bottom line and the mainstream value impact and challenge. By means of establishing the network bottom line of thinking, for network information space to draw a line in the sand and is non quality standard, so that people clearly recognize that what is the bottom line of the network information space, and break the bottom line will bring what kind of consequence, providing a direction guide for the public network behavior, strictly of the network information resources, examination and supervision, network information released to maximize transfer positive energy, cohesion "is exciting and publicize and promote the socialist main value view, create a positive, warm heart, fairness and justice of the cultural and moral atmosphere around.

Table 1. Descriptive statistics of respondents

	Item	Freq.	Percent
Gender	Male	338	52.2
	Female	312	47.8
Age	17–29 years	452	69.5
	30–39 years	112	17.2
	40–49 years	53	8.3
	>50 years	33	5.0
Education	High school	17	2.6
	Collage	251	38.6
	Bachelor's degree	182	28.0
	Master	113	17.3
	PhD	87	13.5
Experience	1–5	233	33.7
	6–10	191	29.4
	11–15	103	16.9
	16–20	70	11.7
	>20	53	8.3
Computer use at work (hrs./day)	Mean	6.35	
	Std. deviation	4.12	
Using the computer (years)	Mean	7.25	
	Std. deviation	5.42	

4   Case Study

The research study participants were students, teachers and employees working at seven colleges in Henan. The survey questionnaire was distributed on paper to 1000 randomly-selected people at different age levels in all college departments participating in the study. The participants were given one week to complete and return the questionnaire. The identities of participants were kept confidential. In total, 801 questionnaires were returned. 99 of them indicated that they were unaware of their colleges' information security, so they were excluded. Another 41 questionnaires were later eliminated because of incomplete answers, and 11 were eliminated because of unreliable responses. For the purpose of data analysis, 650 usable responses were received, and the effective response rate was 65.0 %.

Table 1 summarizes respondents' descriptive statistics. Participants reported using different computer software such as spreadsheets, e-mail, programming languages, and database applications. The sample was quite evenly distributed in terms of the responsibilities of the respondents and in terms of their managerial levels.

Next, discriminant validity was assessed. The square root of the AVE for each construct was noted to be higher than the inter-construct correlations (refer Table 2). Also, from the cross-loading matrix, it was found that, as recommended, all measurement items loaded higher than 0.789 on their underlying construct, and loaded very low, less than 0.40, on other constructs. All constructs in the model.

Table 2. Discriminant validity, AVE, and CR

	CR	AVE	1	2	3	4	5	6	7	8
ATT	0.96	0.94	0.95							
EEG	0.95	0.75	0.12	0.90						
FOR	0.92	0.82	0.15	0.16	0.89					
IC	0.93	0.78	0.22	0.08	0.13	0.91				
MO	0.94	0.79	0.31	0.22	0.23	0.18	0.88			
SE	0.93	0.82	0.24	0.15	0.14	0.32	0.25	0.85		
SN	0.93	0.78	0.16	0.35	0.32	0.28	0.15	0.31	0.86	
UTI	0.94	0.83	0.19	0.21	0.18	0.32	0.13	0.27	0.34	0.85

5   Conclusion

Strengthening the construction of Ideological Security in Colleges and universities is an important strategic task of the party building work in Colleges and universities. Network of our country socialism ideology dominance and ideological construction work has brought a strong impact, network break the myth of discourse hegemony of the traditional media, the control of ideology is facing challenges, the network to make Chinese and Western culture and ideology conflict more directly, defenses of ideology of our country put forward a severe test. Under the background of the network of college ideology construction has many characteristics: network broke the monopoly of traditional media discourse, has become a new and most dynamic, the most influential mass media; ideological education establishments and the environment is open, invisible, virtual and infinite; the spread of ideology has an interactive, immediacy and image; ideological education mode with openness, participation, responsiveness and strengthening the construction of Ideological Security in Colleges and Universities under the background of network should strengthen the work of organizational guarantee, ideological guarantee, system guarantee and carrier guarantee.

Acknowledgment. This paper is sponsored by the National Natural Science Foundation of China (NSFC, Grant U1204703, U1304614).

References

1. Chen, C.C., Shaw, R., Yang, S.C.: Mitigating information security risks by increasing user security awareness. a case study of an information security awareness system. Inf. Technol. Learn. Perform. J. **24**, 1–15 (2006)
2. Schlienger, T., Teufel, S.: Analyzing information security culture. increased trust by an appropriate information security culture. In: Proceedings of the 14th International Workshop on Database and Expert Systems Applications (DEXA 2003), pp. 405–409 (2003)
3. Bulgurcu, B., Cavusoglu, H., Benbasat, I.: Information security policy compliance: an empirical study of rationality based beliefs and information security awareness. MIS Q. **34**, 523–548 (2010)
4. Zhang, J., Reithel, B.J., Li, H.: Impact of perceived technical protection on security behaviors. Inf. Manag. Comput. Secur. **17**, 330–340 (2009)
5. Werlinger, R., Hawkey, K., Beznosov, K.: Human, organizational and technological challenges of implementing IT security in organizations. In: Proceedings of the Second International Symposium on Human Aspects of Information Security and Assurnace (HAISA 2008), Plymouth, UK, pp. 35–47 (2008)
6. Ng, B.Y., Xu, Y.: Studying users' computer security behavior using the health belief model. In: 11th Pacific-Asia Conference on Information Systems, pp. 423–437 (2007)
7. Siponen, M., Pahnila, S., Mahmood, M.A.: Compliance with information security policies. Empirical Invest. Comput. **43**, 64–71 (2010)
8. Ruighaver, A.B., Maynard, S.B., Warren, M.: Ethical decision making: improving the quality of acceptable use policies. Comput. Secur. **29**, 731–736 (2010)
9. Alder, G., Schminke, M., Noel, T., Kuenz, M.: Employee reactions to internet monitoring: the moderating role of ethical orientation. J. Bus. Ethics **80**, 481–498 (2008)
10. Stahl, B.C.: The ethical nature of critical research in information systems. Inf. Syst. J. **18**, 137–163 (2008)

A Network Protocol Reverse Engineering Method Based on Dynamic Taint Propagation Similarity

Weiming Li[1], Meirong Ai[1], and Bo Jin[2(✉)]

[1] Huazhong University of Science and Technology, Wuhan, China
lwm@hust.edu.cn, emily.amr@qq.com
[2] The Third Research Institute of Ministry of Public Security, Shanghai, China
jinbo@stars.org.cn

Abstract. Automatic network protocol reverse engineering is very important for many network applications such as fuzz testing and intrusion detection. Since sequences alignment on network traces is limited by the lack of semantic information, recent researches focus on dynamic taint analysis. But current dynamic taint based methods need heuristics rules to handle different network protocols which make them too complex to run automatically and efficiently. Our approach is inspired by the observation that different fields of network protocol message are processed in different execution path of the binary application, while the bytes of same message field are processed by highly similar instructions sequence. After analyzing the similarity of dynamic taint propagation and adjusting boundaries according to keywords and separators, we can identify the field boundaries not only accurately but also fully automatically. Evaluated by real-world protocol implementations (FTP, HTTP, DNS, etc.), the result shows our method is more accurate and simpler than exist methods.

Keywords: Network protocol reverse engineering · Dynamic taint propagation · Similarity

1 Introduction

Network protocol is the foundation of Internet, its security has a significant impact on the stable operation of the Internet. Understanding and analyzing network protocols is an important research area. According to their openness, current network protocols can be divided into two groups: open and closed network protocols. The former has public documents on their specifications, such as HTTP and FTP, while the latter does not for commercial interests or security considerations, such as Oracle's TNS and Microsoft's Skype protocol.

This work was supported by the National Science Foundation of China No. 61370230 and Opening Project of Key Lab of Information Network Security of Ministry of Public Security (C14603).

D.-S. Huang et al. (Eds.): ICIC 2016, Part I, LNCS 9771, pp. 580–592, 2016.
DOI: 10.1007/978-3-319-42291-6_58

Automatic network protocol reverse engineering is the automatic process of identifying message field boundaries, searching semantic information and extracting the protocol message formats from the closed protocols. It is very useful for network security, including fuzz testing [1], semantic analysis [2], fingerprint generation, replay protocol sessions [3], etc.

Current automatic network protocol reverse engineering methods more or less rely on some heuristics rules to handle different protocols. In this paper we define dynamic taint propagation similarity, and propose an automatic network protocol reverse method based on the concept. We also present the method of field boundaries adjustment according to keywords and separators to overcome under tainting. The prototype system is built to automatically obtain network protocol message formats without any prior knowledge.

We organize the rest of the paper as follows. We discuss common network protocol reverse engineering methods in Sect. 2 and the main idea of our method in Sect. 3. We describe message field boundaries extracting based on dynamic taint propagation similarity and how to handle under tainting in Sect. 4. We present how to get the semantic info of one message field in Sect. 5. We discuss the experiments in Sect. 6 and future work in Sect. 7.

2 Related Work

According to approaches of network protocol reverse engineering, Narayan et al. [4] divides them into three main kinds: manual analysis, sequence alignment of network trace and dynamic taint analysis.

2.1 Method Based on Sequences Alignment

This method analyzes network protocols, using the sequence alignment algorithm in bioinformatics to align messages with similar byte patterns. It is used in PI [5], Discoverer [6], GAPA [7], etc. It can handle text protocol based on network traces. However, this method can only find message field boundaries at a coarser level and can't reverse the precise semantic information of message fields.

2.2 Method Based on Dynamic Taint Analysis

The method based on dynamic taint analysis is the most popular method in automatic network protocol reverse engineering. By monitoring how the application process network messages at binary instruction level, this method can get detailed semantic information and accurate reverse results. There are some research projects based on this method, such as Polyglot [8], AutoFormat [9], "Automatic Network Protocol Analysis" [10], Prospex [11], Dispatcher [12], Tupni [13], Rosetta [14], and [15], etc.

Polyglot extracts protocol message fields, such as separator, keyword, length field and direction field, based on features of applications processing. While using separators and direction fields to identify field boundaries is too coarse to find many specific fields. And, it performs better on text protocols than binary protocols.

AutoFormat identifies field boundaries and infers hierarchical relationship between each field through analyzing function call stack and taint related instructions, and then finds useful semantic information, such as parallel fields (their positions are convertible). The disadvantage of AutoFormat is that an application function always contains different process flows of various fields, so identifying field boundaries at function level is too coarse to distinguish some successive fields. It also perform poorly on binary protocols.

The method of "Automatic Network Protocol Analysis" [10] is improved on the basis of Polyglot, and it can identify a message field is variable or constant. In addition, it uses multiple sequence alignment to determine whether a field is optional. However, it has the same drawbacks of Polyglot and is more complex than Polyglot.

Both Prospex and Dispatcher use the technique of Polyglot to analyze the format of messages as first step. Prospex infers network protocol state machine during the whole client and server interaction process by classifying the messages in the same session according to their format information. While Dispatcher propose novel techniques to extract the format of the protocol messages sent by an application, and enable extracting the complete protocol format even when only one endpoint's implementation of the protocol is available. Due to the limits of Polyglot, their field boundaries identifying method is not accurate enough. And, the combination of heuristic rules and static analysis in identifying semantic information makes their methods complicated.

During taint data propagation, "under tainting" and "over tainting" have g reat impacts on the completeness of taint analysis. To cope with these problems, DTA++ [14] presents a technique combined static analysis with symbolic execution to analyze the possible locations of under tainting.

3   Our Method

Identifying field boundaries of previous work is not accurate enough. Most of the methods need heuristic rules or static analysis, which is not automatic enough and not conducive to be embedded into security applications. Therefore, in this paper we present a new network protocol reverse method based on dynamic taint propagation similarity. Our main idea can be summarized as follows:

Firstly, we use dynamic taint tracking to trace the execution at binary instruction level when an application processing network protocol messages. Secondly, we identify message field boundaries based on taint propagation similarity calculated from multiple sequences alignment result. We identify the message field semantic information according to features of how the message field is processed, then adjust field boundaries and enrich their attributes to increase the accuracy of protocol format reverse engineering. Finally, we use semantic information to solve under-tainting problem. The entire process is simple and complete automatic without manually intervention or any prior knowledge. The message field boundaries identifying results are highly consistent with the protocol syntax. Our system architecture is shown as Fig. 1.

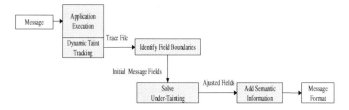

Fig. 1. System architecture

4 Identify Field Boundaries Based on Similarity

In the current automatic methods, field boundaries identifying is based on network trace multiple sequences alignment, or semantic information such as separators, keywords and direction fields. These two methods are coarse-grained. This paper presents a more accurate method based on similarity of instructions sequence, which is not only for text protocols, but also for binary protocols.

To implement the method, we design the shadow memory as Fig. 2, and a hash table record all tainted memory bytes. Every tainted byte is linked to tags which point to the original network message bytes, and it means that the byte is affected by these bytes in the network message. We also log all the binary instructions processing every tainted byte, then we can reconstruct an array of instructions processing a specific network message byte and its tainted bytes.

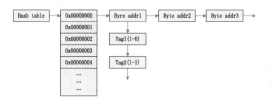

Fig. 2. Shadow memory

4.1 Principle

As we know, a binary application processes different semantic parts of a network message using different instructions. For example, the header and data segment of HTTP messages are processed by different functions of Apache server. Various semantic parts of the header such as keywords and command parameters are processed by different instructions. On the other hand, the different bytes at the same semantic part are analogously processed. The instructions of searching every bytes of a keyword, such as "HTTP", are almost the same in Apache server. According to this principle, we proposed a field boundaries identifying method based on similarity of instructions sequences, and the definition and calculation of similarity are elaborated as follows.

A network protocol message can be considered as a sequence of bytes and expressed as (1):

$$packet = [byte_1, byte_2 \ldots \ldots, byte_n] \tag{1}$$

We tag every bytes of a network message as taint data. Every bytes of the message is processed by various instructions, and then taint is propagated to other memory bytes. We express the sequence of instructions operating on a specific byte and its tainted bytes as follow:

$$insSeq = [i_1, i_2 \ldots \ldots, i_n] \tag{2}$$

The propagation similarity of two different bytes of a message can be defined by the similarity of their instructions sequences as (3). The simins denotes the similarity between two instructions sequences.

$$similarity(byte_1, byte_2) = sim_{ins}(insSeq_1, insSeq_2) \tag{3}$$

We use Needleman-Wunsch multiple sequences alignment algorithm to calculate the similarity of instructions sequences, and the details are as follows:

1. insSeq 1 represents the instructions sequence of byte1 and the sequence length is len1, insSeq 2 represents the instructions sequence of byte2 and the length is len2;
2. the relative distance between insSeq 1 and insSeq 2 is defined as (4);
3. the similarity between insSeq 1 and insSeq 2 is calculated as (5);

For a given byte $byte_y$ and a field, $field_a[byte_x, byte_{x+1} \ldots \ldots, byte_{y-1}]$ is the average similarity between $byte_y$ and all bytes in $filed_a$, which is defined as (6).

$$distance(insSeq_1, insSeq_1) = Needleman(insSeq_1, insSeq_1) \tag{4}$$

$$sim_{ins}(insSeq_1, insSeq_2) = 1 - distance(insSeq_1, insSeq_1)/(len_1 + len_2) \tag{5}$$

$$sim_{average}(byte_y, field_a) = (sim_{ins}(byte_y, byte_x) + \ldots \ldots + sim_{ins}(byte_y, byte_{y-1}))/(y-x) \tag{6}$$

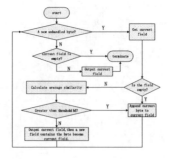

Fig. 3. Flow diagram of field boundaries identifying

If $sim_{average}(byte_y, field_a)$ is greater than a threshold value = M,then we append $byte_y$ to $field_a$. Otherwise, we end $field_a$ as an integrated field and start a new field with $byte_y$.

For a given message and its instructions sequences of all bytes, the flow diagram of field boundaries identifying based on taint propagation similarity is shown as Fig. 3.

After a large number of experiments, we find setting the threshold value M to 0.7 can identify field boundaries precisely.

4.2 Needleman-Wunsch and POA Algorithm

We merely use Needleman-Wunsch algorithm to calculate the relative distance of two sequences for similarity and do not care the complete matching results. Therefore, we use two rolling vectors to process Needleman-Wunsch algorithm for the final distance. Such process can significantly reduce memory consumption and accelerate sequences alignment.

After plentiful experiments, we find that Needleman-Wunsch algorithm and POA can take advantage of their own strengths in different situation:

1. Needleman-Wunsch algorithm is easy to implement and its space complexity is O(N), and N is the maximal length of two sequences. It is applicable to short protocol message.
2. POA is complex to implement but applicable to longer message, and its space complexity is O(N) as well.
3. Needleman-Wunsch algorithm can calculate the average similarity more accurately than POA, but need more Comparisons.
4. POA merges the instrument sequences of bytes in the same field, which reduces the comparisons but increases length of instructions sequence. And it usually leads to a smaller similarity value compared with Needleman-Wunsch algorithm.

In view of execution time and accuracy, we eventually chose Needleman-Wunsch algorithm to calculate the taint propagation similarity.

4.3 Solution of Under-Tainting

Under-tainting, also known as implicit flow or control-flow taint, is a problem that tainted data impacts control flow, then the control flow affects other clean data. Such indirect taint propagation can't be tracked by traditional dynamic taint tracking methods. It breaches the completeness of dynamic taint analysis and is very harmful to taint propagation similarity.

The most common under-tainting occurs in keywords processing of network applications. To accelerate searching speed, applications use several beginning characters instead of the whole string to compare with keywords table. This leads to incorrect taint propagation similarity, and a keyword field may be divided into two message fields. For example, FTP server identifies keywords, such as USER and PASS, only through comparing the first byte of a command. If the beginning several bytes is unique of all the keywords, then FTP server jump over the rest characters to process the follow-up command's arguments. In order to cope with this under-tainting problem, we propose a method to identify keyword and separator fields.

As location of a keyword or a separator in protocol message is relatively fixed, network applications usually use string comparison to identify keywords. While several bytes in message are processed by same successive comparison instructions and all comparison result is true, these bytes maybe keywords or separators. The details are shown as follows:

1. Separators
   (a) Extract all compare instructions that comparing taint byte with constant character.
   (b) For every constant character, build a taint tag set which include all the tags of taint data compared with this constant character.
   (c) Process the taint tag set of every constant character. A given constant character can become a candidate character when its taint tag set has several contiguous taints tag. "Contiguous" means the bytes that taint tags represented are adjacent in the message.
   (d) Use candidate character to extract real separators from the message.
   (e) Finally, join the contiguous separators in the message to be a full separator as long as possible.
2. Keywords
   (a) Extract successful comparison instructions that comparing tainted bytes with constant character. "Successful" means the comparison result is true.
   (b) Append all taint tags in these comparison instructions to a set.
   (c) Remove taint tags that have been identified as separators from the set.
   (d) Extract the contiguous taint tags as blocks in the rest of the set. If the block size is greater than two bytes, we can identify the corresponding contiguous bytes in a message as a keyword.

In order to solve the under-tainting problem, we use keywords and separators to adjust the field boundaries. We merge the fields that belong to the same keyword and divide the field containing separators. This approach significantly enhances the accuracy of field boundaries identifying. The accuracy of before-and-after adjustment is shown in Fig. 4., based on testing three messages of servU, Apache, Savant and Miniweb.

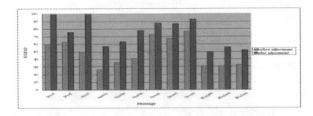

Fig. 4. Accuracy before and after adjustment for solving under-tainting (Color figure online)

5   Semantic Information

Our field boundaries identifying method can't provide enough semantic information of fields. We present some techniques to make up the shortage based on the processing instructions characteristics of different fields. Our approach can identify length fields, target fields and some value type fields such as string and byte string.

5.1   Length Field

Length fields can be divided into two types: fixed-length fields and variable-length fields. Applications typically use a loop to process variable-length field and terminate the loop according to length field. Therefore, only the comparison instructions that terminate the loop frequently access the length field. We define these comparison instructions as length field related instructions and the others as length field unrelated instructions, and present the method to identify length fields as follows.

1. Extract all comparison instructions with taint data.
2. For every comparison instruction, append all taint tags of taint data to its taint tag set.
3. Process all the comparison instructions. If a comparison instruction's taint tag set contains multiple bytes in the protocol message, categorizing it as length field unrelated instruction.
4. For every byte in the message, build a list including all comparison instructions that accessed to this byte. Then, remove the length field unrelated instructions from all the lists.
5. If a given list contains multiple instructions and all the instructions is length field related, the corresponding byte is a length field.
6. Merge the continuous length fields whose list contains the same length field related instructions.

5.2   Value Type Identification

We identify the message fields as string fields or binary fields according to if the fields are printable. If a byte is printable, we identify it belong to string field, otherwise binary field. However, it is impossible to identify field type in some cases. For example, we can find field "ab*\r" has printable characters "ab*" and a binary byte "\r", but cannot identify "ab*\r" as a string field or a binary field. Considering the fact that bytes in a field are same type, we divide the field "ab*\r" into a string field "ab*" and a binary field of "\r". This approach enhances the accuracy of field identifying and makes the results more close to the real protocol format.

6 Experiment and Evaluation

We implement our approach and build a network protocol reverse engineering system based on the dynamic binary instrumentation tool PIN. We define the field identifying accuracy to measure accuracy of field boundaries identifying. If the count of fields identified by our approach is M, the count of accurate fields by manually identifying is S, so we define accuracy as (7).

$$W = S/M \tag{7}$$

We evaluate it through testing several common network protocols, such as FTP, HTTP, DNS, etc. Moreover, we compare our field identifying results with Wireshark. Finally, we compare our results with Polyglot [8].

6.1 FTP Message

We test two FTP servers: ServU 4.0 and Warftp 1.65, and then compare FTP message reverse results with Wireshark. The comparisons are shown as Table 1.

Table 1. Comparisons of FTP reverse results

Field count	SERVU4.0 commands					Warftpd 1.65 commands				
	USER	PASS	PORT	LIST	QUIT	USER	PASS	PORT	LIST	QUIT
Wireshark	4	4	5	2	2	4	4	5	2	2
Our approach	5	5	17	3	3	4	2	15	2	2
Common	3	3	2	1	1	1	1	2	2	2
Accurate	5	5	13	3	2	2	2	11	2	2
Accuracy	100 %	100 %	76 %	100 %	100 %	50 %	100 %	73 %	100 %	100 %

Table 1 shows that our approach works well in field boundaries identifying. Our approach can identify more fields than Wireshark according to real protocol message processing in application. For example, Wireshark identifies "\r\n" as one field but our approach identifies it as two fields: '\r' and '\n', because ServU 4.0 only use '\r' as end of a FTP command. Wireshark identifies the IP address in PORT message as one field, but our approach identifies it as several fields because of separator '.'.

6.2 Compare with Polyglot

We compare the results of semantic information identification with Polyglot in HTTP protocol test. The comparisons are shown as Tables 2 and 3.

In Table 2, the results are similar except the differences that we successfully identify extra separator '/' and 'H', but fail in '\r\n' and '.'. As Miniweb locates 'HTTP/1.1' in the message through searching 'H' at a big loop, our method identifies 'H' as a separator.

Table 2. Comparisons of separators

Separator	Apache 2.2.4		Savant 3.1		Miniweb 0.8.1	
	Polyglot	Our method	Polyglot	Our method	Polyglot	Our method
0x0d0a('\r\n')	Yes	-	Yes	Yes	Yes	-
0x0d('\r')	-	-	-	-	-	Yes
0x2f('/')	Yes	Yes	Yes	Yes	-	Yes
0x2e('.')	Yes	Yes	Yes	-	Yes	Yes
0x20(' ')	-	-	Yes	Yes	Yes	Yes
0x3a20(': ')	Yes	-	-	-	-	-
0x3a(':')	-	Yes	-	-	-	-
0x48('H')	-	-	—	-	-	Yes

The comparisons in Table 3 show that our results are basically same as Polyglot, and our method can identify 'Keep-Alive' of Savant in addition. Due to Miniweb's imperfect implementation of HTTP protocol, its taint propagation is so incomplete that both Polyglot and our approach underperform.

Table 3. Comparisons of keywords

Keyword	Apache 2.2.4		Savant 3.1		Miniweb 0.8.1	
	Polyglot	Our method	Polyglot	Our method	Polyglot	Our method
GET	Yes	Yes	Yes	Yes	Yes	Yes
Host	Yes	Yes	-	-	Yes	-
User-agent	-	-	Yes	Yes	-	-
Accept	Yes	Yes	Yes	Yes	-	-
Accept-language	Accept	Accept-	Yes	Yes	-	-
Accept-encoding	Accept-	Accept-	Accept-	Accept-	-	-
Accept-charset	Accept-	Accept-	Accept-	Accept-	-	-
Keep-alive	Yes	Yes	-	Yes	-	-
Connection	Yes	Yes	Yes	Yes-	Yes	-
HTTP/	Yes	Yes	-	-	-	-
HTTP/1.	-	-	Yes	Yes	-	-
.html	-	-	-	Yes	-	-

6.3   DNS Binary Message

We reverse engineer the DNS reply message of nslookup in Windows, and compare the results with Wireshark. The comparisons are shown as Table 4. Our method is also suitable for binary protocols and gain good results.

Table 4. Comparisons of DNS reverse results

Protocol format of DNS		Wireshark		Our method		
		Fields division	Semantic information	Fields division	Semantic information	Value type
Header	Identification	0x00 0x02	Fixed	0x00 0x02 0x81 0x80	Data	Binary
	Flags	0x81 0x80	Fixed			
	Total questions	0x00 0x01	Fixed	0x00 0x01	Data	Binary
	Total answers	0x00 0x01	Fixed	0x00 0x01	Data	Binary
	Total authority	0x00 0x03	Fixed	0x00 0x03	Data	Binary
	Total additional	0x00 0x02	Fixed	0x00 0x02	Data	Binary
Query name	Label length	0x03	Fixed	0x03	Length	-
	label	'www'	Variable	'www'	Data	String
	Label length	0x04	Fixed	0x04	Length	-
	Label	'hust'	Variable	'hust'	Data	String
	Label length	0x03	Fixed	0x03	Length	-
	Label	'edu'	Variable	'edu'	Data	String
	Label length	0x02	Fixed	0x02	Length	-
	Label	'cn'	Variable	'cn'	Data	String
	Type	0x00 0x01	Fixed	0x00 0x01 0x00 0x01	Data	Binary

6.4   Closed Zombie Protocol Message

We use SDbot (server) and mIRC (client) to set up a zombie network, and reverse engineer the messages of the interactive process. The results are shown as Table 5.

The results of fields identifying and semantic information identification are accurate, which means our approach works well even on closed protocols.

Table 5. Results of reversing unpublished zombie network protocol

Range of field division	Value	Semantic information	Value type
[1, 24]	':ace! ~ ace@xxxxxxxxxxxxxx'	Keyword	String
[25, 26]	'x'	Separator	String
[27, 33]	'PRIVMSG'	Keyword	String
[34, 34]	' '	Separator	String
[35, 43]	'#channel1'	Data	String
[44, 44]	' '	Separator	String
[45, 51]	':.login'	Keyword	String
[52, 52]	' '	Data	String
[53, 59]	'hust452'	Keyword	String
[60, 60]	'\r'	Data	Binary
[61, 61]	'\n'	Data	Binary
[1, 6]	':uyjge'	Keyword	String
[7, 29]	'! ~ uyjge@DP-201208301102'	Data	String
[30, 30]	' '	Separator	String
[31, 34]	'NICK'	Keyword	String
[35, 35]	' '	Separator	String
[36, 36]	':'	Data	String
[37, 40]	'dtva'	Data	String
[41, 41]	'\r'	Data	Binary
[42, 42]	'\n'	Data	Binary

7 Conclusion

In this paper we present a network protocol reverse engineering method based on dynamic taint propagation similarity. We implement a prototype system and evaluate it using FTP, HTTP and DNS protocols. The results demonstrate that the method is simple, universal and accurate. As far as under-tainting and over-tainting are concerned, our method can only solve a special case, and it needs more future work to solve them completely.

References

1. Cui, B., Wang, F., Hao, Y., et al.: WhirlingFuzzwork: a taint-analysis-based API in-memory fuzzing framework. Soft Comput. 1–14 (2016). http://dx.doi.org/10.1007/s00500-015-2017-6
2. Bossert, G., Guihéry, F., Hiet, G.: Towards automated protocol reverse engineering using semantic information. In: Proceedings of the 9th ACM Symposium on Information, Computer and Communications Security, pp. 51–62. ACM (2014)
3. Newsome, J., Brumley, D., Franklin, J., Song, D.: Replayer: automatic protocol replay by binary analysis. In: 13th ACM Conference on Computer and Communications Security (CCS 2006), pp. 311–321 (2006)

4. Narayan, J., Shukla, S.K., Clancy, T.C.: A survey of automatic protocol reverse engineering tools. ACM Comput. Surv. (CSUR) **48**(3), 40 (2015)
5. Beddoe, M.A.: Network protocol analysis using bioinformatics algorithms. http://www.4tphi.net/awalters/PI/PI.html
6. Cui, W., Kannan, J., Wang, H.J.: Discoverer, automatic protocol description generation from network traces. In: Proceedings of the USENIX Security Symposium, pp. 143–157. USENIX Association, Berkeley, USA (2007)
7. Borisov, N., Brumley, D., Wang, H.J., et al.: A generic application-level protocol analyzer and its language. In: Proceedings of the 14th Symposium on Network and Distributed System Security (NDSS) (2007)
8. Caballero, J., Song, D.: Polyglot: automatic extraction of protocol message format using dynamic binary analysis. In: ACM Conference on Computer and Communications Security (CCS 2007), Alexandria, Virginia, USA, pp. 317–329 (2007)
9. Lin, Z., Jiang, X., Xu, D., et al.: Automatic protocol format reverse engineering through context-aware monitored execution. In: Proceedings of the Network and Distributed System Security Symposium, San Diego, CA, pp. 37–53 (2008)
10. Wondracek, G., Comparetti, P.M., Kruegel, C., et al.: Automatic network protocol analysis. In: Proceedings of the Network and Distributed System Security Symposium, San Diego, CA, pp. 125–133 (2008)
11. Cui, W., Peinado, M., Chen, K., et al.: Tupni: automatic reverse engineering of input formats. In: Proceedings of the ACM Conference on Computer and Communications Security, pp. 391–402. ACM, New York (2008)
12. Comparetti, P.M., Wondracek, G., Kruegel, C., et al.: Prospex: Protocol specification extraction. In: Proceedings of the IEEE Symposium on Security and Privacy, pp. 110–125. IEEE Computer Society, Oakland (2009)
13. Caballero, J., Poosankam, P., Kreibich, C., et al.: Dispatcher: enabling active botnet infiltration using automatic protocol reverse-engineering. In: Proceedings of the ACM Conference on Computer and Communications Security, Chicago, IL, pp. 77–89 (2009)
14. Caballero, J., Song, D.: Rosetta: extracting protocol semantics using binary analysis with applications to protocol replay and NAT rewriting. Technical report, 69–84. Carnegie Mellon University (2008)
15. Kang, M.G., Camant, S.M., Poosankam, P., et al.: DTA++: dynamic taint analysis with targeted control-flow propagation. In: Proceedings of the 18th Annual Network and Distributed System Security Symposium, February 2011

The Music Channel Access System Based on Internet

Wen Li[✉]

Zhengzhou Normal University,
Zhengzhou, China
1837956@qq.com

Abstract. Mobile Internet technology development, driven by the intelligent terminal equipment fast update, the application scenarios and achieve the diversity will become more endless. Change the traditional computer based terminal system, terminal value-added application value creation, with the operators more open access capability, which requires communication operators should not only to open the Internet learning thoughts also need to combine their own advantages, integration platform for the ability of their own, can drive the channels of the interconnection and interaction to create mutually beneficial and win-win.

Keywords: Mobile Internet · Channel access technology · Open platform technology

1 Introduction

With the development of science and technology, the development of technology of the computer information industry experienced the mainframe era, minicomputer era, the era of personal computers and desktop Internet era. In the first 10 years of this century, the mobile Internet technology which has an enormous potential for development has formed through the integration of Internet technology and mobile communication network. The development of mobile Internet has changed the pattern of the development of the communication industry. Many Internet companies and mobile operators have joined the innovation research of the mobile Internet technology.

In China, with the promotion of 3G technology and the evolution of the three operators and network fusion, many operators have the ability that support a full service operation in mobile Internet. The focus on the competition among the telecom operators has changed from voice service to the value-added business competition. Such as, mobile reading business, mobile video services, the business of mobile phone game, mobile music and some other value-added services. In addition, the mobile music service has led the innovation of the traditional record company business model, because it has fused with the music industry fully.

© Springer International Publishing Switzerland 2016
D.-S. Huang et al. (Eds.): ICIC 2016, Part I, LNCS 9771, pp. 593–599, 2016.
DOI: 10.1007/978-3-319-42291-6_59

2 Definition and Evolution of the Mobile Internet

Mobile Internet is the use of mobile wireless modems to access the traditional Internet. And the mobile wireless modem has two kinds, one is the mobile phone terminal in the integration of Modem function modules, and the other is an independent device, such as the USB Modem.

Mobile Internet has two aspects of meaning. The first aspect is the Internet mobile access, that is, users can experience Internet business through the access mode of the mobile; The other meaning is that the traditional mobile service has mixed with the Internet, and traditional telecom operators are able to make various value-added application services access to the Internet platform use by using the channel of the mobile Internet. The mobile Internet includes three factors, which are mobile terminal, mobile network and application services.

The mobile Internet development process of mobile communications operators can be divided into four periods. The first period is analog standard for mobile communication system. And the second generation is the digital cellular mobile communication system with GSM and CDMA IS-95 represented. The third generation is the broadband multimedia communication, that is, the popular 3G network currently. The fourth generation is the integrated multifunction broadband mobile communication system, which can not only tell transmission technology outside but also included the told wireless information access systems, mobile platform technology, password security technology such as: LTE (long term evolution) follows the quasi 4G standard. The evolution of mobile communication network has been shown in the below figure.

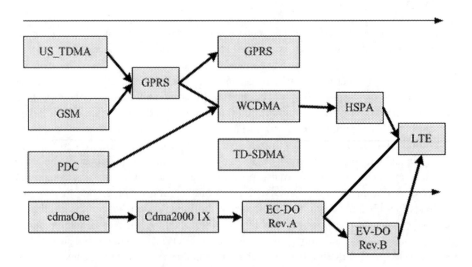

3 Wireless Music Business Technology

With the innovation of mobile technology and the increase of the speed of the network broadband, a variety of new mobile value-added services emerge as the times require. Among them, mobile phones and other wireless communication terminal music download business become with the fastest developing speed in 3G value-added services. Japanese and South Korean mobile operator KDDI and SKT started the earliest in the development of the wireless music business and their development has made great success. At present, with the development of mobile Internet technology and the evolve of the terminal technology gradually, the development of the global wireless music is very quick because of the participation and efforts of all aspects of the industry chain Wireless music download service has become the key business for the development of the mobile Internet operators in the development of the mobile Internet.

Wireless music business refers to the users of mobile phones or other wireless terminal who can experience music content related business by using the Internet, SMS (short message service) message, WAP (Wireless Application Protocol), APP application and some other ways. The development goals of wireless music business is to make users can experience the music business anywhere through good copyright and channel control, through diverse access way, through diversification of product form.

4 Introduction of Channel Management Theory

Channel is one sort of the marketing 4P theory (marketing strategy), which is the abbreviation of produce, price, place and promotion, proposed by scholars, Professor McCarthy in the sixties of the 20th century. And channel is the most important marketing strategy and means to improve the market competitiveness of enterprise as a portfolio theory.

Channel construction is an important strategy of marketing and dependence of production enterprises, and an important factor of competition of enterprises. The promotional capability of channel determines the purchasing power of production in peculiar market, so channel construction is an important mean for production enterprises to provide service value and hold customers in palm. Channel management became an important task for the mobile operators as the focus of the competition of telecom operators in the era of 3G network. Mobile operators improve channel management capabilities through information management. At present telecom operators have the following problems in the business channel access management: channel management system chaos, no Unified channel management mechanism. There is a security risk in channel access, and there is no uniform secure access mechanism and post processing mechanism, causing channel malicious consumption, misallocation of resources, and less channel memory hierarchical management and channel management differences.

5   Security Design of Access System

Internet access to mobile network increased technical difficulty of access security due to the rapid growth of wireless mobile users and the fusion of traditional internet and mobile network business. It will lead to illegal use of illegal access providers, which will cause adverse negative social impact and economic loss to the operator, as a result of the complexity of network application and expansion, the value-added business profits in a large space, and the vulnerabilities of business process design.

Taking into account the characteristics of channel access system, the project in the system design process, there are several ways to solve the safety of the interface as follow:

(1)   Access security design from access source control: such as through the distribution of identity ID authentication; through the dynamic password encryption interface encryption; through the application of the system to access IP restrictions.

(2)   Access security design from the security mechanism of the business system: such as interface authorization management and access channel capacity subdivision; increasing the complexity of interface flow through constraining the business process specification, such as using SMS to verify the acknowledgment mode, which is often taken by the mobile operators and so on.

(3)   Other auxiliary safety measures, such as strengthening the link security through HTTPS, widely using security communications with the internet is the technology used by bank payment interfaces at present.

(4)   The monitoring design of abnormal data: taking hierarchical control means to the channel access, establishing risk control measures, setting up monitoring threshold, alarming the abnormal situation or interrupting the channel service request directly.

6   Conceptual Model Design

Through the analysis of demand and function, we need an access system to establish the conceptual model to complete the tasks of conceptual model, such as: defining the boundaries of the system, determining the system subject domain, and completing usage scenarios according to the analysis of the demand function. There are eight themes including below contents.

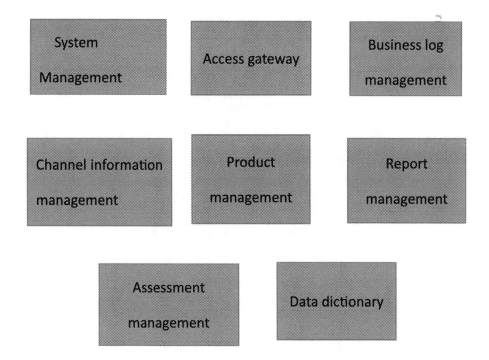

7   Source Code Examples Related Channel Product Distribution Management

The product information related channel access system come from the underlying platform and include the wireless music product information which contains the product name, product price, product type and other information.

After the introduction of content, the underlying platform synchronizes the product information timing of channel access system periodically. In product management module, channel access system synchronizes data with the product information launched by starting the thread to monitor the underlying platform, and writes this part of the content data to the database table.

It is the main implementation for the program to receive content synchronization request message from the underlying platform, to return the results message and to synchronize the data to channel access providers through the callback function.

The system processing flow chart is shown below:

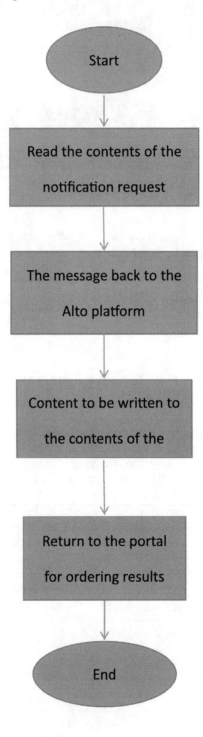

Through the acceptance test of channel access system, the system can achieve the application and approval process of the channel access, product configuration process of the channel access, product channel distribution process, the WWW and WAP service flows, IVR service flows, SMS access service flows, in order to meet the business needs of the Department.

8 Conclusion

The development of mobile internet technology drives fast update of the intelligent terminal equipment makes the diversity of application scenarios and implementation model become more endless the. As the access capabilities of operators become more open, it requires communication operators not only open the internet learning thoughts, but also have the ability to combine their own advantages through integrating their own platform, what can drive the interconnection and interaction of channels to create mutually beneficial and win-win.

References

1. Yu, X.: The development and thinking of mobile internet. Telecommun. Netw. Technol. **12**, 21–27 (2008)
2. Zhang, C., Liu, L.: Mobile Internet Technology and Service, pp. 56–100. Publishing House of Electronics Industry, Beijing (2012)
3. China Communication Association: Dialogue Mobile Internet, pp. 221–225. People's Posts and Telecommunications Publishing House, Beijing (2010)
4. Li, Y.: 3G Business and Related Technology. People's Posts and Telecommunications Publishing House, Beijing (2007)
5. Zhou, L.: Mobile internet business innovation analysis. Mod. Telecommun. Technol. **7**, 36–40 (2009)
6. Zhang, Z.: Open Technology of Next Generation Network, pp. 7–45. Publishing House of Electronics Industry, Beijing (2004)
7. Zhang, H.: Introduction to Software Engineering, pp. 102–145. Tsinghua University Press, Beijing (2005)
8. Liu, J.: Network Information Security, pp. 22–37. Mechanical Industry Press, Beijing (2003)
9. Zhang, L., Yao, S.: The achievement of identity management by cross system. Telecommun. Eng. Technol. Stand. **7**, 19–22 (2009)
10. Liao, J.: Research on the application mechanism of mobile internet. Mob. Commun. **19**, 51–54 (2009)

Research on Information Security of Electronic Commerce Logistics System

Guoning Lv[✉], Min Gao, and Xiaoyu Ji

College of Information Science and Technology, Zhengzhou Normal University,
Zhengzhou, China
lvguoning@126.com

Abstract. Nowadays the logistics system has become an indispensable element for economic commerce activities, while information security is an important foundation in ensuring the development of electronic commerce logistics system. The purpose of this paper is to investigate the role of information security in electronic commerce logistics system. In that regard, an information security model is developed to protect the electronic commerce logistics system.

Keywords: Information security · Electronic commerce · Logistics system · Security model

1   Introduction

With the development and popularization of the Internet and smart phones, electronic commerce in the commodity trading market occupied an indispensable position and gradually become more commodity trading preferred further affects the social economic development and people's life. In the growing environment for the development of electricity suppliers, electronic commerce logistics information system security directly affects the success or failure of the electricity business transactions [1].

With the advent of the Internet era, electronic commerce relying on network technology, the rapid development of communication technology. The so-called e-commerce, it is the business model that combines business activities and electronic information technology, the relative traditional business model, the whole business process of e-commerce is basically in the electronic and networking. E-commerce on the traditional business model is not only reflected in the pattern of change, but also to the economic and industrial structure upgrade has brought great changes. Due to the high efficiency and low cost, it can provide more business opportunities for small and medium sized companies. However, based on the characteristics of the Internet itself, openness, sharing and seamless connectivity, there are many security risks and these risks will threaten the security of e-commerce information to a certain extent? E-commerce information security involves not only the system security, also relates to the application and database security, to avoid potential safety problems brought by the loss and destruction, e-commerce and Internet Security Technology in the field of has been committed to the electronic commerce information security protection technology research. And before the analysis of the

© Springer International Publishing Switzerland 2016
D.-S. Huang et al. (Eds.): ICIC 2016, Part I, LNCS 9771, pp. 600–611, 2016.
DOI: 10.1007/978-3-319-42291-6_60

current e-commerce information security technology, we must first deal with e-commerce information security risks have a certain understanding [2].

In recent years, the new concept of security began to lead the trend of information security management, to enhance the community's attention to the electricity business logistics information system security management. In the electricity business logistics often appear product and customer information leakage and other security threats. At present, trading business in the people's life more and more frequently, business logistics industry has been an unprecedented development, under the new security concept of electronic business logistics information system security management research is not only the research focus of electrical business, is of concern to society. In such a large environment and need to explore new security concept business logistics information system security management index and electronic business logistics information system security management influence factors, and put forward the corresponding management measures, provide the basis for the further improvement of the domestic electricity supplier logistics information system security management system, and to provide theory and practice guidance for business logistics information system security management.

2 Related Study

In recent years, the new concept of security began to lead the trend of information security management, to enhance the community's attention to the electricity business logistics information system security management. In the electricity business logistics often appear product and customer information leakage and other security threats. At present, trading business in the people's life more and more frequently, business logistics industry has been an unprecedented development, under the new security concept of electronic business logistics information system security management research is not only the research focus of electrical business, is of concern to society. In such a large environment and need to explore new security concept business logistics information system security management index and electronic business logistics information system security management influence factors, and put forward the corresponding management measures, provide the basis for the further improvement of the domestic electricity supplier logistics information system security management system, and to provide theory and practice guidance for business logistics information system security management.

The most common security problem in the electronic commerce is the theft and tampering of the electronic commerce information [3]. Because the process of e-commerce in modem transmitted between the plaintext information did not use encryption measures, the intruder can be use this loophole these information interception down and carries on the analysis, by looking for the e-commerce information rules and format, so as to grasp the electronic commerce in the process of transmission of information content. After mastering the rules and format of the information, the intruder can add two same types of demodulator to the network modem. Through this measure, the information data of the original network modem can be transmitted to the demodulator at the other end under the control of an intruder. In the crime of e-commerce information, tampering with information is a very common crime, any router or gateway can be applied. After stealing the intruder, the e-commerce information data master of information, of data information has been passed by the implementation of arbitrarily change, confidential information on the Internet, the file of the master, sneaked into each other inside the network, to legitimate Internet users impostor, of counterfeit electronic commerce information transmitted or received, resulting in more serious consequences. Actually, these endanger the security of e-commerce information all down the behavior of speaking all belong to data attacks, the attacker through intercepted electronic commerce basic information, by forging or forced to interrupt the transmission of information in electronic commerce, the e-commerce information for correcting errors and so on, to the information safety of electronic commerce brought great risks.

Based on the hidden dangers existing in the electronic commerce information security, the present electronic commerce has the urgent demand to the information security protection service. E-commerce requirements of e-commerce information is confidential, to ensure that enterprises some files or commercial information confidentiality, ensure that it will not be illegally intercepted or crack, to make illegal even intercepted confidence was unable to read the meaning, thus for the electronic commerce enterprise user privacy provides great protection. In addition, data integrity and non-repudiation of the data is also required for the protection of e-commerce information security. In the process of transmitting data, we should ensure the authenticity of data transmission, ensure that it will not be tampered with in the middle, and keep the consistency of the information sent by the original users. Data transmission process in the transmission and reception of the two sides also need to ensure transparency and non-repudiation of information, to avoid a variety of disputes in the process of electronic commerce transactions. In addition, the electronic commerce information security protection also requires to determine the correctness of the trader, that is, to determine the other side is not to cheat the information of the third party, the loss and leakage of information to effectively prevent.

In order to meet the requirements of the security protection of the electronic commerce, the appropriate electronic commerce information security protection technology should be selected. At present, the technology of electronic commerce information security protection is mainly used in encryption technology, authentication technology and firewall technology, SSL, SET and so on.

2.1 Encryption Technology

Encryption technology [4] is mainly for data encryption, through the sensitive clear text data through the identification of the password into a more difficult to identify the encrypted data to protect the electronic commerce information security technology. The secret key of the use of different, through the same encryption algorithm on the same plaintext is encrypted, making it a different cipher text, when the need to open the cipher text, can use the key implementation of reduction, let the cipher text in the form of plaintext data now reading before, this process i.e. decrypt. Symmetric key encryption and asymmetric key encryption is the most typical two types of encryption technology.

Symmetric key technology is used to control the process of using the same key in the process of encryption and decryption, the secret of the key is the key factor of the density of the technology. When sending information, the two sides must exchange the keys to each other, DES, 3DES, etc. is the common block cipher algorithm. The symmetric key system not only has fast encryption and decryption speed, but also has a relatively small amount of digital computing and high security. However, the symmetric key in management is more difficult, need to save more keys, easy to leak secret key in the process of transfer, the security of the data has a direct impact. Asymmetric key technology is to use different keys to encrypt and decrypt control, in which the encryption key is open. In the calculation, the public encryption key or the cipher text is not possible to calculate the decryption key, and the non-symmetric key technology can be used to realize the protection of the electronic commerce information security. In the application of the encryption technology, each user is composed of a public key and a private key, and the two keys cannot be released from one another. Non symmetric key distribution technology is simple, easy to manage, and can realize digital signature and authentication, but the processing speed is relatively slow symmetric key technology.

Encryption is based on a certain password algorithm to convert the text data into a difficult to identify the encrypted data. In the same encryption algorithm, using different keys, the same plaintext will be encrypted for different cipher texts, decryption also need to correct key can will be restored to the correct plaintext cipher text, so as to ensure the confidentiality of all users. Therefore, the data encryption technology to meet the requirements of the integrity of information is an active security strategy. Based on the key using the different methods of encryption technology can be divided into symmetric encryption and asymmetric encryption two. In the encryption and decryption process if using the same key called "symmetric encryption" and its characteristics is a small amount of computation, fast, but the insurance density is highly dependent on the secrecy of the key, once the secret key is compromised, is data security will be lost. Therefore, the secrecy of the key and management become weak links. "No symmetric key encryption technology, encryption and decryption keys are not identical, user generated a pair of keys for encryption and decryption keys, to preserve their decryption keys, known for the private key; public key encryption to send to each other, known as a public key. Its characteristic is unable to calculate the private key from the public key [2], good secrecy, but the large amount of computation.

2.2 Authentication Technology

Security transactions [5] in electronic commerce rely solely on the basic encryption technology, although it can be a certain degree of protection, but this is not completely protected. Therefore, the information authentication and identity authentication technology is also indispensable in the process of electronic commerce information security protection. The content of the authentication technology is more extensive, in addition to digital abstract, digital signature, digital envelope, digital certificate, etc. Authentication technology in the application of electronic commerce information security, mainly in order to identify the true identity of the trader, to avoid the occurrence of e-commerce information is tampered with, forgery, delete and other potential risks. The digital technology is mainly with the help of one-way hash function promote e-commerce information file conversion operations, and of a certain fixed length of the code acquisition and the add files by information transmission to the recipient, the recipient needs through agreed function conversion, to ensure that the results and send the same code can be that file is complete has not been tampered with. Digital signature is the sender's private key used to encrypt the document summary, attached in the original common transmission to the recipient, encrypted abstract only by the sender's public key to unlock, similar to the autograph, of information integrity, the authenticity is a certain protection role. Digital envelope sender information encryption using symmetric keys, to be encrypted with the public key of the receiver, the receiver will be "envelope" and file, which need to use their own private key to make "envelope" open, in the process in the available symmetric keys to open the file, to ensure the confidentiality of the document. Digital certificate is a digital identity card issued by the PKI user CA actuator, for online transactions of non-repudiation guarantees, and to provide protection for the integrity and reliability of information security and electronic signature.

(1) Digital signature. Digital signature is used to ensure the integrity of the information in the information transmission process and provide information sender authentication and non-repudiation of. Using public key algorithm is the main technology to realize digital signature [6]. In daily life, usually the file signature to ensure authenticity, prevent repudiation. Under the electronic commerce environment, resulting in a digital signature and document summary with the sender's private key encryption, and the original transmitted to a receiver, receiver only with the sender's public key can decrypt the encrypted summary, because only the sender has its own private key. The documents that are encrypted by the sender's public key can only be sent after the sender's private key is encrypted, as if it is an autograph. Digital signature can realize the identification and verification of the original file, to ensure that the file integrity, authenticity and the sender to send a document cannot deny.

(2) Digital envelope. The digital envelope is used to ensure that only the specified recipient can read the contents of the document. In digital envelope, the sender uses a symmetric key to encrypt the information, then the symmetric key using the receiver's public key to encrypt (this part of said digital envelope), followed by the "envelope" with the document sent to a receiver, receiver to use his private key to open the "envelope", get the symmetric key, and then use the symmetric key to unlock the file. The security of this technology is more excellent.

(3) Digital digests. The digital technology is using one-way hash function operation conversion of important elements in the file after a fixed length of the code, and will distribute the join files together to give the recipient at the time of information transmission, the receiver receives the documents by prior agreement with the same function into line the same transform operation, if the results obtained and sent to the same code, that the file has not been tampered with, on the other hand, it can think file has been tampered with. The digital technology can ensure the integrity of documents.

(4) Digital time stamp. Document signing date and signature are key contents of the written document to prevent forgery. Similarly, in electronic commerce transactions also need the time information of file protection, digital time stamp service can provide safety protection of electronic documents published, signed time, become documents conform to the transaction time requirements of effective certificate.

(5) Digital certificate. Digital certificate is the user's digital ID card, digital certificate is based on PKI, it is the core elements of the PKI implementation agency CA issued, in compliance with the X.509 standard. Its role is to ensure that the online banking, e-commerce and e-government, online identity authentication; ensure that online transactions of undeniable recognition; ensure the integrity and security of online transactions and data transmission; ensure reliability of electronic signature [7].

2.3 Firewall Technology

In the solution of enterprise network security problem, the firewall technology [8] is compared with the traditional technology, which is to protect the security of the electronic commerce information in the firewall by restricting the public data and service into the firewall. Firewall technology mainly includes packet filter and application proxy. In the whole process of the development of firewall technology, through the way of packet filtering, the current e-commerce information security protection technology has developed two different versions of the static and dynamic. Static version of the router technology with a period, for each packet is in accord with the rules defined for review; through to dynamically set the packet filtering rules of the dynamic version, in the filtering rules to filter items, and automatically add or update. Application agents mainly include the first generation application gateway firewall and the second generation adaptive proxy firewall. The former can be used to hide the data from the original, which is generally accepted. The latter can protect all data communication in the network, and the most prominent advantage is security. This kind of firewall network security technology, in addition to other than the security protection technology is simple and practical and, also has relatively high transparency, corresponding security goal can also be achieved without changing the premise of the original network application system. However, with many customers for communication on business activities carried out, electronic business outstanding demand, firewall technology on the grade of electronic commerce information security requirements may alone be able to assume, is suitable to be used as a basis of information security.

Firewall is the traditional solution of enterprise network security, that is, the public data and services placed outside the firewall, so that the external users of the firewall's internal resources will be restricted access. Firewall uses filtering technology, filtering usually make the network performance is reduced by more than 50 %. As a kind of network security technology, firewall has the characteristics of simple and practical, and high transparency, cannot modify the original network application system to achieve a certain security requirements. However, the business needs of business to business electronic commerce is communication to carry out business activities with a large number of customers, firewall can only be as one of the foundation of information security means, can't bear the grade of electronic commerce information security needs.

2.4 SSL and SET Technology

In the technology of electronic commerce information security, network communication protocol is a kind of technology which is often used. This technology mainly includes two kinds, namely SSL (Secure Socket Layer Protocol) [9] and SET (secure electronic transaction announcement) [10]. These two protocols have become the standard of network security in the electronic commerce information. SSL is aimed at the whole process of the dialogue between the computers, through the overall encryption protocol to ensure the security of e-commerce information. The protocol provides security services including three, first to maintain the confidentiality of data, and the second to ensure the legitimacy of client and server, the third is the integrity of the data maintenance. In the method, the SSL security protocol usually choose to hash function and secret sharing, through these methods between the client and server secure channel was created to ensure the electronic commerce information reaches the destination of complete accuracy. SET is an electronic payment system protocol based on credit card, which is based on credit card in Internet environment. Under the function of the digital signature, the SET protocol can guarantee the integrity of the information of the electronic commerce, and can give the message source to the authentication. SET protocol for dual signature technology to be used, can provide more stringent protection of customer privacy to prevent the occurrence of user information is violated. Under the protection of the double signature technology, the consistency between the order information and the payment information, the merchant and the bank need to be verified together to prevent the information from being modified. And SET compatibility is strong, can run on different hardware and operating system platform.

SET is based on the credit card electronic payment system based on the protocol, mainly used in B to protect the security of payment information in C mode. The SET protocol to ensure the integrity of the message by digital signature and message source authentication, digital signature and message encryption using the same principles. The SET protocol uses the double signature technology can ensure customer privacy will not be violated. The cardholder's order information and payment information correspond to each other. The dual signature technology ensures that businesses and banks can verify the consistency of ordering information and payment information, in order to determine whether the ordering information and the payment information are modified in the process of transmission. At the same time, businesses can only see the order information, and the bank can only see the payment information, so as to ensure user privacy. Set protocol to reach 5 goals: ensure all the participants information on electronic commerce in the corresponding isolation; guarantee the secure transmission of information on the Internet, to prevent being stolen or tampered with; to solve the problem of multi certification; transactions to ensure the real-time and all the payment process can be online, follow the form of BDZ trade, standard protocols and message formats, prompting different manufacturers to develop the software has functions of compatibility and interoperability, and can run on different hardware and operating system platform [11].

Facing the ubiquitous Internet information security, electronic commerce and related units or enterprises in the expansion of e-commerce activities in, information security hidden trouble to deal with a comprehensive understanding of, pay more attention to and when the e-commerce information security protection technology be selection and application, electronic business information on the units or enterprises try to protect.

3   Information Security Model of E-Commerce Logistics System

Electronic commerce logistics system, despite the introduction of the modern Internet technology, but it is mainly the operation or to trade mainly, the process mainly involves the cash flow, network and logistics and logistics as the terminal part of the business, which relates to the amount of information is most, under the environment of new security concept, importance of electronic business logistics and information security management is more and more attention to. Foreign scholars of logistics safety management system were a lot of research, and more mature theoretical research results are made, according to the organizational structure of the logistics information system within the organization presents available information security management method and the corresponding management system is designed.

On the basis of drawing lessons from foreign theoretical research, the national logistics information system security management system is established, and the corresponding safety management rules of logistics information system are introduced. Domestic scholars Wu pointed out that the logistics information security management system construction is an important part of the rapid development of the electricity business enterprise, based on this proposed logistics information risk management model [12]. Lin analyzed the security environment of the Internet of things logistics system information system, at the same time based on the Internet of things environment to build a logistics information system security management sensing model [13]. Zhang analysis of the importance of logistics under e-commerce environment, also of electronic business logistics information management were classified stressed the importance of logistics information system security management, and points out that only the security management of electronic business logistics information system and business enterprise to the healthy and orderly development of [14]. Yang put forward logistics information leakage will cause cannot be estimated economic losses to the enterprise and the customer, and even related to the survival and development of enterprises. Therefore, the safety of logistics information system is an essential information security management work [15]. Gao analyzed the security requirements of logistics information system under the multi-level security information system [16]. Rao through the analysis of the business enterprise logistics information security management between the gap and evaluate the risk of business logistics and information security management process in the presence of, and finally construct the Jingdong electronic business information system security management system [17].

In our business logistics field of information security management, technical means of prevention is still one of the main measures of the safety management, insufficient understanding of the importance of electronic business logistics information system security management, the technology did not play its due role. Although some researchers by relevant standards of domestic and international information security management system to understand and learn gradually in the electronic business logistics information system security field implement management system, and achieved certain results. But electricity supplier logistics information security management of the state still exist some shortcomings, did not develop a unified standard and complete business logistics information system security management system that does not form a business logistics information security management system. And some researchers to build the business logistics information system safety management mode, from the perspective of management and technology emphasizes the fundamental theory and methods of logistics information system of safety management, but the most of the research is still in system management level, but also did not rise to the information system security management level. Therefore, this paper studies the security management mode of logistics information system in the perspective of the new security concept.

In this paper, in view of the new security concept of e-commerce logistics information system security management analysis for business logistics process, in the logistics and information security management based on, combined with the new security view of management concept, proposed the new concept of security business logistics information system security management model:

First of all, in the electricity business logistics information system security management model on the first level, the establishment of a new security concept under the electricity business logistics system of the overall information security management model. The whole logistics information security management mode mainly by society, science and technology enterprise with three aspects, in the whole logistics information security management level mainly is to ensure that the logistics and information security management of the target company to achieve real-time feedback. The entire logistics information security management level mainly includes five processes and two service support processes.

Secondly, in the electronic business logistics information system security management model of the second levels, the establishment of a new security concept of e-commerce logistics information system security management technology dimensions. Logistics information system security management needs to be based on the new security concept, the use of electronic business logistics information system security management technology, the establishment of e-commerce logistics information system security management model.

Finally, in business logistics information system security management pattern on the surface of the third level, the establishment of national electricity supplier logistics information system security management level protection system, choose the right logistics information security level of protection, to ensure the realization of scientific and reasonable safety risk control and management method.

E-commerce logistics information system security management model construction is the core for business logistics information system of the level of safety management, combined with the electricity supplier logistics operation process and information system security management level together. E-commerce logistics information system security management model construction mainly includes the information and communication technology, Internet technology, security technology and security management technology. Electronic business logistics information system security management is mainly on data related to the logistics information system to protect and prevent the information is destroyed, altered and leakage, and electricity supplier logistics information integrity, reliability and authenticity protection.

(1) Assessment and prediction of the security risks of the logistics information system. The electricity business logistics information system security risk is divided into 1~5 level, the greater the number, the higher the risk accordingly. E-commerce logistics information system security risk assessment by referring to the design of the risk assessment method, to predict the E-commerce logistics information system security risk level, each risk level indicates the degree of damage to the corresponding risk.

(2) Analysis of the results of the assessment and prediction of the security risks of the business logistics information system. Through the analysis of the security risk assessment and prediction of the results found that the electricity suppliers logistics information system security management problems and deficiencies, and make timely and effective feedback.

(3) According to the analysis of data to develop electricity suppliers logistics information system security measures. Building information security protection measures is a major move to help the electricity suppliers logistics to solve security risks. Through on the results of electronic business logistics information system risk assessment system and abstraction of analysis, from the complex data identify business logistics information system risk points, further warning, provide scientific data for the business enterprise to business enterprise can quickly find the logistics information system, and with reference to the corresponding security risk measures to resist giving appropriate feedback.

(4) The establishment of the logistics information system security feedback system, real-time detection of electricity suppliers in the business process of logistics information systems in the existing problems.

4   Conclusion

In this paper a new concept of security under the premise of safety management of electricity supplier logistics information system is discussed, through summarizing the theoretical research achievements of the domestic and abroad, the electricity supplier logistics information system security management are analyzed, then put forward three aspects included safety management mode of electricity supplier logistics information system, to further explore the specific content includes electricity supplier logistics information system, according to the new security concept four steps under the electricity supplier logistics information system security management, construct the new security concept of logistics information system security management structure, and provides a theoretical basis for domestic enterprises to establish a safety management strategy of logistics information, and solve the problem response evaluation logistics information system security risk can be better, the formation of safe operation of electronic commerce logistics information system strong Management system further help e-commerce enterprises to solve bottleneck problems existing in the development process of logistics information is not safe, to cater to the development trend of the new security concept, in order to provide a reference for the security of e-commerce logistics information system security.

Acknowledgments. This paper is sponsored by the National Natural Science Foundation of China (NSFC, Grant U1204703, U1304614).

References

1. Fienberg, S.E.: Privacy and confidentiality in an e-commerce world: data mining, data warehousing, matching and disclosure limitation. Stat. Sci. **21**, 143–154 (2006)
2. Friedman, J.H., Popescu, B.E.: Predictive learning via rule ensembles (2005). stat.stanford.edu/~jhf/#selected
3. Ghose, A., Sundararajan, A.: Evaluating pricing strategy using e-commerce data: evidence and estimation challenges. Stat. Sci. **21**, 131–142 (2006)
4. Hastie, T., Tibshirani, R., Friedman, J.: The Elements of Statistical Learning. Springer, New York (2001). MR1851606
5. Hui, K.-L, Png, I.P.L.: The economics of privacy. In: Handbook on Economics and Information Systems (2006, to appear)
6. Karr, A.F., Lin, X., Sanil, A.P., Reiter, J.P.: Secure regression on distributed databases. J. Comput. Graph. Stat. **14**, 263–279 (2005). MR2160813
7. Karr, A.F., Sanil, A.P., Banks, D.L.: Data quality: a statistical perspective. Stat. Methodol. **3**, 137–173 (2006)
8. Kohavi, R., Mason, L., Parekh, R., Zheng, Z.: Lessons and challenges from mining retail e-commerce data. Mach. Learn. **57**, 83–113 (2004)
9. Liggett, W., Buckley, C.: System performance and natural language expression of information needs. Inf. Retrieval **8**, 101–128 (2005)
10. Madigan, D.: Statistics and the war on spam. In: Peck, R., Casella, G., Cobb, G., Hoerl, R., Nolan, D., Starbuck, R., Stern, H. (eds.) Statistics: A Guide to the Unknown, 4th edn, pp. 135–147. Thomson Brooks/Cole, Belmont (2005)
11. Moe, W., Fader, P.S.: Dynamic conversion behavior at e-commerce sites. Manag. Sci. **50**, 326–335 (2004)
12. Shmueli, G., Jank, W.: Visualizing online auctions. J. Comput. Graph. Stat. **14**, 299–319 (2005). MR2160815
13. Shmueli, G., Jank, W.: Modeling the dynamics of online auctions: a modern statistical approach. In: Kauffman, R., Tallon, P. (eds.) Economics, Information Systems and E-commerce Research II: Advanced Empirical Methods. Sharpe, Armonk, NY (2006, to appear)
14. Sismeiro, C., Bucklin, R.E.: Modeling purchase behavior at an e-commerce web site: a task completion approach. J. Mark. Res. **41**, 306–323 (2004)

15. Bulgurcu, B., Cavusoglu, H., Benbasat, I.: Information security policy compliance: an empirical study of rationality based beliefs and information security awareness. MIS Q. **34**, 523–548 (2010)
16. Chen, C.C., Shaw, R., Yang, S.C.: Mitigating information security risks by increasing user security awareness: a case study of an information security awareness system. Inf. Technol. Learn. Perform. J. **24**, 1–15 (2006)
17. Schlienger, T., Teufel, S.: Analyzing information security culture: increased trust by an appropriate information security culture. In: Proceedings of the 14th International Workshop on Database and Expert Systems Applications (DEXA 2003), pp. 405–409 (2003)

Advances in Swarm Intelligence:
Algorithms and Applications

Discrete Interior Search Algorithm
for Multi-resource Fair Allocation
in Heterogeneous Cloud Computing Systems

Xi Liu, Xiaolu Zhang, Weidong Li, and Xuejie Zhang$^{(\boxtimes)}$

School of Information Science and Engineering,
Yunnan University, No. 2, North Cuihu Road, Wuhua District,
Kunmming 650091, Yunnan, People's Republic of China
lxghost@126.com, {zxl,weidong,xjzhang}@ynu.edu.cn

Abstract. The mechanism of resource allocation for cloud computing not only affects the users' fairness, but also has a significant impact on resource utilization. Most current resource allocation models did not take into account the indivisible demands, the heterogeneity servers, and the situations multi-server. Dominant resource fairness allocation in heterogeneous systems (DRFH) is a fair and efficient resource allocation mechanism. But solving the DRFH problem is NP-hard. There are significant gaps between solutions obtained by existing heuristic algorithms and optimal solutions. They cannot effectively use server resources, resulting in a waste of resources of servers. In this paper, we propose a novel discrete interior search algorithm (DISA) to solve indivisible demands in heterogeneous servers, with a specific repair operator and *task-fit* value. Experimental results demonstrate that DISA can well adapt to dynamic changes in user resource request type, obtain the near-optimal solutions, maximize the value of minimum global dominant share and resource utilization.

Keywords: Dominant resource fairness · Interior search algorithm · Multi-resource fair allocation · Heterogeneous cloud

1 Introduction

Cloud computing resource allocation problem is that the server resources allocated to each user according to the needs of different requests. And users can efficient use of resources on the server, in order to save energy and reduce the cost. However, there are a large of heterogeneous servers in cloud computing environment, so how to efficiently allocate server resources is a key issue and the current hot issue need to be resolved. Therefore, Wang et al. [1] proposed dominant resource fairness in heterogeneous environment (DRFH) where resource are pooled by a large number of heterogeneous servers. DRFH is to maximize the user's minimum global dominant share in heterogeneous cloud computing systems. It well solved the resource allocation in heterogeneous system.

Towards addressing the indivisible and divisible demands, it is easy to find optimization solution by solving linear programming for divisible demands. But the tasks of users are indivisible in real world system. For example, user who gets 1/2 of

© Springer International Publishing Switzerland 2016
D.-S. Huang et al. (Eds.): ICIC 2016, Part I, LNCS 9771, pp. 615–626, 2016.
DOI: 10.1007/978-3-319-42291-6_61

resource request can complete 1/2 its task. Some cloud computing platform cannot save any compute results, so user gets final result after must complete its whole task [2]. So this is unrealistic, and it's meaningless that the resources which are available to the user cannot run a task completely. Finding the optimal integer solution of DRFH problem for indivisible demands is often impossible. Wang et al. [1] designed a simple heuristic algorithm, called Best-Fit algorithm, to find a feasible allocation. Best-Fit cannot get optimal solutions of *global dominant resource*, leading to lower resource utilization. Zhu and Oh [3] extended to the distributed environment to solve resource allocation problem. They designed a heuristic algorithm to find a feasible allocation with different owners, called Distribute-DRFH. Although the Best-Fit algorithm [1] and its variant [3] are simple and can find a better solution than the traditional scheduling algorithm, we think that there is still room for improvement.

Thus, it is necessary to design a new algorithm, which satisfies the goal and also be efficient. Intelligent optimization algorithm has many advantages, such as self-organizing, adaptive, robust, not dependent on the specific environment and other fine features, to become one of efficient methods for solving complex optimization problems. Interior search algorithm (ISA) [4] is a new swarm intelligence algorithm. In addition to the advantages of intelligent optimization algorithm, ISA has some advantages, such as simple, fast convergence, easy to implement, fewer parameters and powerful for global optimization problems. It can find the near-optimization allocations in NP-hard problem. But ISA is used to solve continuous optimization problem and cannot be used to discrete optimization problem. Therefore, based on the ISA idea, a new discrete interior search algorithm (DISA) is designed and applied to resource allocation problem.

In this paper, we focus on the multi-resource fair allocation in heterogeneous cloud computing systems. This paper can be divided into three parts: (1) an integer representation to encode the solutions are used to search for a discrete problem space; (2) repair operator algorithm allow the generation of infeasible discrete solutions and repair to the infeasible discrete solutions; (3) the DISA is proposed to solve the discrete DRFH problem. We evaluate the performance of DISA, Best-Fit, Dis-tribute-DRFH and optimization solution (OPT). The *global dominant resource* and resource utilization by employing DISA are closer to the OPT and much better than Best-Fit and Distribute-DRFH. The experimental results show that DISA can avoid falling into local optimal solutions, adapted to different resource request, efficiently utilize of server resource, strong environment adapt-ability and higher robustness.

2 Related Work

Max-min fairness has been proposed [5] to solve fair resource allocation problem. The basic idea is to maximize minimum dominant resource allocation. Ghodsi et al. [6] first proposed the dominant resource fairness (DRF) mechanism for multi-resource fair allocation in cloud computing systems. DRF expand to multiple types of resources are allocated and to ensure the fairness of distribution based on the max min model. Parkes et al. [7] extended DRF to zero demands and indivisible demand. Friedman et al. [8] proposed a model for fair strategy-proof in a realistic model of cloud computing centers. Li et al. [9] designed a generalized dynamic dominant resource fairness mechanism, and developed a combinatorial optimal algorithm to find a fair allocation. Zarchy et al. [10] not only developed a framework for fair resource allocation, but also analyzed Lexicographically-Max-Min-Fair (LMMF) mechanism and the Nash-Bargaining (NB) mechanism for fairly allocating resources. Also, some unifying multi-resource allocation frameworks are developed [11, 12], developed for generalizing the DRF measure. Li et al. [13] extended DRF to the finite case, where the number of tasks for each user is bounded. Our previous research work also promoted a dynamic DRF mechanism to bounded number of tasks [14]. However, the assumptions of DRF and other mechanism were unreasonable. First, they assumed divisible demands, meaning, maybe some users get resource are not able to fulfill a task. Second, this mechanism without taking into account the heterogeneity of servers.

Traditional search methods may not find the optimal solution in a reasonable time, so need to find an algorithm to quickly obtain the optimal or near-optimal solutions. In order to deal with the engineering problems, existing researches use the swarm optimization algorithm to solve and achieve good results. Resource allocation in heterogeneous cloud systems is a challenge, while swarm optimization algorithm successfully in a wide range of current cloud computing applications [15, 16]. However, the current study were not considered heterogeneous cloud computing diversity and user resource request type. The multi-resource fair allocation in heterogeneous cloud computing systems is very complicated. Traditional swarm optimization algorithms have some shortcomings, so need improve the algorithms to find the optimal solution.

3 Problem Statement

The resource pool is composed of many heterogeneous servers, denoted by $S = \{1, \ldots, L\}$. Each sever has m resource type denoted by $R = \{1, \ldots, m\}$. For each server $l \in S$, let $\mathbf{c}_l = (c_{l1}, \ldots, c_{lm})^T$ be the total amount of each resource for server $l \in S$, and c_{lr} be the total amount of resource r available in server l. Let $\mathbf{c}_l = (c_{l1}, \ldots, c_{lm})^T$ be the resource matrix for all resource. Without loss of generality, assume that

$$\sum_{l \in S} c_{lr} = 1, \forall r \in R \tag{1}$$

Let $U = \{1, \ldots, n\}$ be the set of cloud users sharing the cloud system. For each user i, let $\mathbf{D}_i = (D_{i1}, \ldots, D_{im})$ be its resource demand vector, where D_{ir} is the ratio of requirements r for user each task to the total amount of resources in systems, assuming that $D_{ir} > 0$ for all i, r. Define the dominant resource demand of each user r_i^*:

$$r_i^* \in \arg\max_{r \in R} D_{ir} \tag{2}$$

For each user i and server l, let $\mathbf{A}_{il} = (A_{il1}, \ldots, A_{ilm})^T$ be the resource allocation vector, where A_{ilr} is the share of resource r allocated to user i in server l. Let $\mathbf{A}_i = (\mathbf{A}_{i1}, \ldots, \mathbf{A}_{iL})$ be the allocation matrix of user i, and $\mathbf{A} = (\mathbf{A}_1, \ldots, \mathbf{A}_n)$ represents the overall allocation for all users. Let $N_{il}(\mathbf{A}_{il}) = min_{r \in R} \lfloor A_{ilr}/D_{ir} \rfloor$ be the number of tasks of user i can run on server l under allocation $\mathbf{A}_i$. The total number of tasks of user i can run under allocation $\mathbf{A}_i$ is hence

$$N_i(\mathbf{A}) = \sum_{l \in S} \min_{r \in R} \left\lceil \frac{A_{ilr}}{Dir} \right\rceil \tag{3}$$

Given the overall allocation $\mathbf{A}$, the *global dominant share* of user i is defined as

$$G_i(\mathbf{A}_i) = N_i(\mathbf{A}_i)D_{ir_i^*} \tag{4}$$

The goal of DRFH mechanism proposed in [1] is to maximize the minimum *global dominant resource* share for all users under capacity constraints. It can be represented as the following mathematical formula:

$$\begin{aligned} &\max_{i \in U} \min G_i(\mathbf{A}_i) \\ &s.t. \sum_{i \in U} A_{ilr} \leq c_{lr}, \forall l \in S, r \in R \end{aligned} \tag{5}$$

When the tasks are indivisible, we only consider the non-waste allocation, i.e., there is a non-negative integer z such that $A_{ilr} = zD_{ir}$ for all i, l, r, Wang et al. [1] proposed the Best-Fit DRFH algorithm, which is described as follows. Let $\bar{\mathbf{c}}_l = (\bar{c}_{l1}, \ldots, \bar{c}_{lm})$ is the available resource vector, where $\bar{c}_{lr}$ is the share of resource r remaining available in server l. The *fitness* of the task of user i for server l is defined as

$$H(i, l) = \|\mathbf{D}_i/D_{i1} - \bar{\mathbf{c}}_l/c_{l1}\| = \sum_{r \in R} |D_{ir}/D_{i1} - \bar{c}_{lr}/c_{l1}| \tag{6}$$

Best-Fit DRFH chooses user i with the lowest *global dominant share* and schedules user i's tasks to server l with the least $H(i, l)$ at each iteration. To the best of knowledge, Best-Fit is the best current algorithm which can lead to a DRFH allocation with highest resource utilization.

Distribute-DRFH [3] is proposed to solve resource allocation in distribute environment. Let P be resource providers set. For each resource provider $p \in P$, let $\mathbf{c}_{ip} = (c_{p1}, \ldots, c_{pm})^T$ be its normalized resource capacity. Let $\mathbf{A}_{ip} = (c_{ip1}, \ldots, c_{ipm})^T$ be the resource allocation vector for user i on resource provider p. The Distribute-DRFH can be defined as:

$$\begin{cases} Maximize\ g \\ \min_{i \in U} \sum_{p \in P} (\min(A_{ipr}/D_{ir})D_{ir_i^*}) \geq g, \forall p \in P, r \in R \\ \sum_{i \in U} A_{ipr} \leq c_{pr}, \forall p \in P, r \in R \end{cases} \tag{7}$$

4   Solution Construction

4.1   Solution Representation

For convenience, let $x_{il} = N_{il}(\mathbf{A}_{il})$ be the number of tasks user i can schedule on server l under a feasible allocation, for $i = 1, \ldots, n$ and $l = 1, \ldots, L$, where L is the number of servers. An equivalent representation $\mathbf{X}$ of the DRFH allocation with indivisible solution for (5) is definition as following $n \times L$ matrix.

$$\mathbf{X} = \begin{pmatrix} x_{11} & x_{12} & \cdots & x_{1L} \\ x_{21} & x_{22} & \cdots & x_{2L} \\ \cdots & \cdots & \cdots & \cdots \\ x_{n1} & x_{n2} & \cdots & x_{nL} \end{pmatrix} \tag{8}$$

Here, row i represents the number of tasks of user running on each server in the solution, and the sum of the i-th row $\sum_{l=1}^{L} x_{il}$ represents the total number of tasks of each user. The DRFH mechanism (5) can be stated as the following integer linear program (ILP):

$$\begin{cases} Maximize\ g \\ \sum_{l \in S} x_{il} D_{ir_i^*} \geq g, \forall i \in U \\ \sum_{i \in U} x_{il} D_{ir} \leq c_{lr}, \forall l \in S, r \in R \\ x_{il} \in \mathbb{Z}^+ \cup \{0\}, \forall i \in U, l \in S \end{cases} \tag{9}$$

Clearly, by the capacity constraints, for each user i and server l, x_{il} is not more than

$$UB_{il} = \min_{r \in R} \left\lceil \frac{c_{ir}}{D_{ir}} \right\rceil, \forall i \in U, l \in S \tag{10}$$

in any feasible solution.

4.2   Repair Operator

After DISA has completed search, the solution may not be feasible, such as exceeding resource limits. So we need to design an algorithm that provide the feasible solution to DRFH allocation. The algorithm, called repair operator, is based on greedy heuristic approach, and it is given in Algorithm 1. The repair operator algorithm has two input parameters, the solution of DRFH allocation X and the matrix of resource capacity $\mathbf{C}$. The algorithm has the output parameter $\mathbf{X}$, the feasible solution.

Algorithm 1.   Pseudo-code of the repair operator

Input: $\mathbf{X}$;   matrix of DRFH allocation
Input: $\mathbf{C}$;   matrix of resource capacities
for $i=1$ to n

if $\sum_{i=1}^{n} x_{il} D_{ir} > c_{lr}$ for some $r \in R$ then

  $k \leftarrow$ who has the maximum global dominant   share satisfying $x_{kl} > 0$

   set $x_{kl} = x_{kl} - 1$;

if $\sum_{i=1}^{n} x_{il} D_{ir} < c_{lr}, \forall r$ then

  $k \leftarrow$ who has the minimum task-fit value

   if $\sum_{i=1}^{n} x_{il} D_{ir} + D_{kr} < c_{lr}, \forall r$ then

    set $x_{kl} = x_{kl} + 1$;

Output: $\mathbf{X}$;

The repair operator algorithm consists of two phases. The first phase examines the capacity constraints, and decreases the number of tasks of the user who has the maximum global dominant share (lines 4–6). The second phase increases the number of tasks of the user who has the minimum *task-fit* value without violating the capacity constraints (lines 7–10). The *task-fit* $B(i,l)$ of user i for server l is defined as

$$B(i,l) = \sum_{r \in R} \left| \bar{c}_{ir} / D_{il} - \bar{c}_{ir_i^*} / D_{ir_i^*} \right| \tag{11}$$

5   Discrete Interior Search Algorithm

It is obvious that standard ISA cannot be used to search discrete solutions, so we propose a Discrete Interior Search Algorithm (DISA) to solve multi-resource fair allocation problem. To design the DISA, we use integrates local search. For a given iteration, DISA search better solutions in range of small integers. The individuals of solutions are randomly divided two groups. The individuals of composition group are changed randomly to find the better fitness within a large space, and the mirror group are find some other better fitness near the global best solution. The detailed algorithm is described below.

Algorithm 2.   Pseudo-code of DISA

Input: **C** ;matrix of resource capacities

for $k=1$ to number of interiors

   $\mathbf{X}^k = \mathbf{0}$;

$g = 0$;

$\mathbf{X}^* = \mathbf{0}$;

Initialize of the population;

for $cycle = 1$ to   maxiteration

   if $\mathbf{X}^k$ is global best   then

   $x_{il}^j = \max(x_{il}^{j-1} + rand(-\gamma, \gamma), 0)$;

   else **if** $r_1 \leq a$ then

   $w_{il}^j = rand(\min(x_{il}^{j-1}, GB_{il}^{j-1}), \max(x_{il}^{j-1}, GB_{il}^{j-1}))$;

   $x_{il}^j = \max(2w_{il}^j - x_{il}^{j-1}, 0)$;

   else

   $x_{il}^j = rand(0, UB_{il})$;

for $i = 1$ to number of interiors

   evaluate the $f(\mathbf{X}^i)$;

   if $g < f(\mathbf{X}^i)$ then

   $g = f(\mathbf{X}^i)$;

   $\mathbf{X}^* = \mathbf{X}^i$;

Output: $\mathbf{X}^*, g$;

---

1. We define the UB_{il} (upper bounds) which are generated. The number of element is K. Let $\mathbf{X} = (\mathbf{X}_1, \ldots, \mathbf{X}_K)$ is the set of solutions for all elements. The function *rand* is randomly generated integer solution within a limited search space.
2. Find the fittest individual. The problem of this paper is maximum objective function.
3. We define a parameter a. All individuals are randomly divided into two groups: composition group and mirror group by a. If $r_1 \leq a$, individual into the mirror group or else into the composition group, we set $a = 0.4$. DISA use r_1 to control the size of mirror group or composition group.
4. The global optimization solution perhaps around in the current optimization solution, so the individual which is current global best searches around its position by using the random search. We define a parameter γ which is the maximum length of random exploration. We set GB^j is the best solution at the *j-th* iteration. It can be formulated as

$$GB_{il}^j = \max(GB_{il}^{j-1} + rand(-\gamma, \gamma), 0), for\, i = 1, \ldots, n, l = 1, \ldots, L \qquad (12)$$

5. To avoid falling into local optimum and maintain the diversity of population, the individuals of composition randomly search solution in all search space

$$x_{il}^j = rand(0, UB_{il}), for\, i = 1, \ldots, n, l = 1, \ldots, L \qquad (13)$$

6. The individuals of mirror group are searching solutions space depend on the mirror position. First, the position of mirror are randomly placed between its location and the position of global best:

$$w_{il}^j = rand(\min(x_{il}^{j-1}, GB_{il}^{j-1}), \max(x_{il}^{j-1}, GB_{il}^{j-1})) \qquad (14)$$

Second, the positions of individuals are placed on the virtual mirror:

$$x_{il}^j = \max(2w_{il}^j - x_{il}^{j-1}, 0), for\, i = 1, \ldots, n, l = 1, \ldots, L \qquad (15)$$

7. Evaluate the *fitness* of each individual. We define a parameter β which is control the convergence speed. We define a parameter r_2 which is a random value between 0 and 1. Then update each location if its *fitness* is improved for revival design or $r_2 \leq \beta$.

$$X = \begin{cases} X^j, & f(X^j) > f(X)\ \ or\ \ r_2 \leq \beta \\ X, & else \end{cases} \qquad (16)$$

We can set β is a small positive number in order to accept the suboptimal solutions and prevent too fast convergence. For element k, $f(X^k)$ is her fitness, which is *global dominant resource*.

8. If do not reach the termination condition, repeat from step 2.
The pseudo code of the proposed DISA is presented in Algorithm 2.

6 Simulations

In this section, we compare the performance of the DISA, Best-Fit, Distribute-DRFH and OPT. As our data we use traces of real workloads on a Google compute cell, from a 7 h period in [8]. This workload have tasks of each user and server configuration of each server. We assume each user has infinite tasks. The server configurations are summarized in Table 1 from the Google's cluster [17]. Experimental platform environment is using C# in Visual studio 2013.

Figure 1 shows that the *global dominant resource* and resource utilization, respectively. We compared the values of the *global dominant resource* and resources utilization of DISA, Best-Fit and Distribute-DRFH. The experimentally evaluated the DISA with the numbers of individuals is 20, and the number of servers is 100. The configurations of 100 servers are randomly selected. The number of iterations is 1500.

Table 1. Configuration of servers in one of Google's clusters [17].

Number of servers	CPUs	Memory
510	0.25	0.25
6	0.50	0.03
3	0.50	0.06
97	0.50	0.12
10188	0.50	0.25
21731	0.50	0.50
2983	0.50	0.74
7	0.50	0.97
5	1.00	0.50
2281	1.00	1.00

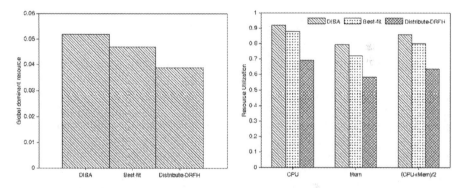

Fig. 1. Global dominant resource and resource utilization.

As for the DISA implementations, we see that DISA leads to higher resource utilization and *global dominant resource* than Best-Fit and Distribute-DRFH at all times. DISA is best, because it uses the interactive individuals, a global exchange of information and searching optimal solutions within a global space.

Figure 2 shows that global dominant resource and execution time, when users request types are compute intensive and memory intensive, respectively. The number of users is 3 and the number of servers is 5, which randomly selected from the Google cluster servers in Table 1. We randomly choose users which satisfy the CPU requirement of each selected user are not less than the memory requirement when compute intensive, and the memory requirement are higher than the compute intensive. We noticed that the *global dominant resource* obtained by Distribute-DRFH is better than Best-Fit when memory intensive, but when compute intensive, the situation is reversed. The *global dominant resource* obtained by Best-Fit and Distribute-DRFH are far from the optimal solution. We can see that *global dominant resource* of DISA are close to the optimal solution, regardless of the task's requirement types, marking it suitable for maximizing minimum *global dominant share* in heterogeneous systems.

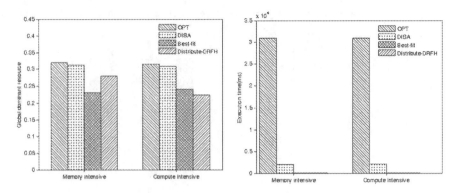

Fig. 2. Global dominant resource and execution time.

We can see that OPT has the highest execution time among four algorithms, and DISA obtains reasonable solutions very fast.

Figure 3 shows the resource utilization for the above benchmarks. The resource utilization of DISA are closer to the optimal solutions than Best-Fit and Distribute-DRFH. For example, the CPU utilization of OPT, DISA, Best-Fit, and Distribute-DRFH for compute intensive are 80 %, 77.2 %, 70 %, and 67.6 %, the memory utilization of OPT, DISA, Best-fit, and Distribute-DRFH are 95.2 %, 93 %, 75.5 %, and 71 %, respectively. As shown in the figures, Best-Fit and Distribute-DRFH obtain the resource utilization are less than that of DISA. This is due to the fact that for compute intensive or memory intensive, DISA can reach better resource utilization.

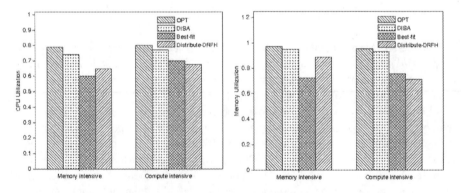

Fig. 3. CPU and memory resource utilization.

Figure 4 shows the *global dominant resource* and resource utilization of DISA, Best-Fit, and Distribute-DRFH, respectively. We randomly selected servers and users data and simulate it on a cloud computing system consists of 1,500 servers. We set to $\{20, 21, \ldots, 40\}$ of the number of users and assume the number of tasks for each user is

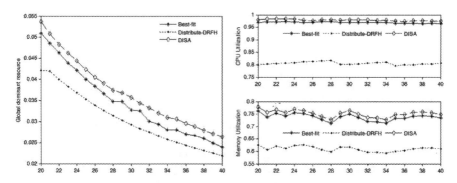

Fig. 4. Global dominant resource and resource utilization (L = 1500)

infinite. The experimentally evaluated the DISA with the numbers of individuals is 20, and the number of iterations is 1000. We can see that the DISA significantly outperform the Best-Fit, and Distribute-DRFH with much higher *global dominant resource*, mainly because the DISA can reach solution from (9). As for the DISA implementations, we can see that DISA leads to uniformly higher resource utilization than Best-Fit and Distribute-DRFH at all times.

Table 2 shows the P-value of *t* test with different sample sizes. We use paired *t* test to compare DISA, Best-Fit and Distribute-DRFH for statistic analysis. Let $H_0 : u_d = 0$, $H_1 : u_d \neq 0$, $\alpha = 0.05$. We observe that p-values are less than 0.05, so reject H_0 and accept H_1. This is due to the fact that DISA leads to higher *global dominant resource* than Best-Fit and Distribute-DRFH.

Table 2. P-values of t test.

Sample sizes	P-value (DISA: Best-Fit)	P-value (DISA: Distribute-DRFH)
$n = 5, L = 10$	0.042	0.000012
$n = 20, L = 100$	0.046	0.000002
$n = 50, L = 200$	0.049	0.000001
$n = 60, L = 300$	0.016	0.0000004

7 Conclusion and Future Work

In this paper we have proposed a novel algorithm with new operators for maximizing minimum *global dominant share* in heterogeneous servers. We have experimentally evaluated the DISA algorithm. Experimental results demonstrate that the proposed algorithm is effective and competitive in composing good results, and reveal that, in a smaller execution time, better solutions with better fitness values for large problems. Further validate and improve performance and parameter optimization are the next step.

Acknowledgement. The work is supported in part by the National Natural Science Foundation of China [No. 61170222, 11301466, 11361048], and the Natural Science Foundation of Yunnan Province of China [No. 2014FB114].

References

1. Wang, W., Liang, B., Li, B.: Multi-resource fair allocation in heterogeneous cloud computing systems. IEEE Trans. Parallel Distrib. Syst. **26**(10), 2822–2835 (2015)
2. Psomas, C., Schwartz, J.: Beyond beyond dominant resource fairness: indivisible resource allocation in clusters. Technical report, Berkeley (2013)
3. Zhu, Q., Oh, J.C.: An approach to dominant resource fairness in distributed environment. In: Ali, M., Kwon, Y.S., Lee, C.-H., Kim, J., Kim, Y. (eds.) IEA/AIE 2015. LNCS, vol. 9101, pp. 141–150. Springer, Heidelberg (2015)
4. Gandomi, A.H.: Interior search algorithm (ISA): a novel approach for global optimization. ISA Trans. **53**(4), 1168–1183 (2014)
5. Max-Min Fairness [EB/OL]. http://en.wikipedia.org/wiki/Max-min_fairness. Accessed 10 June 2015
6. Ghodsi, A., Zaharia, M., Hindman, B., et al.: Dominant resource fairness: fair allocation of multiple resource types. In: NSDI 2011: 8th USENIX Symposium on Networked Systems Design and Implementation, pp. 323–336 (2011)
7. Parkes, D.C., Procaccia, A.D., Shah, N.: Beyond dominant resource fairness: extensions, limitations, and indivisibilities. Proc. Sixteenth ACM Conf. Econ. Comput. **3**(1), 808–825 (2015)
8. Friedman, E., Ghodsi, A., Psomas, CA.: Strategyproof allocation of discrete jobs on multiple machines. In: Proceedings of the Fifteenth ACM Conference on Economics and Computation, pp. 529–546 (2014)
9. Li, W., Liu, X., Zhang, X., Zhang, X.: Dynamic fair allocation of multiple resources with bounded number of tasks in cloud computing systems. Multiagent Grid Syst. Int. J. **11**, 245–247 (2016)
10. Zarchy, D., Hay, D., Schapira, M.: Capturing resource tradeoffs in fair multi-resource allocation. In: 2015 IEEE Conference on Computer Communications (INFOCOM), pp. 1062–1070 (2015)
11. Gutman, A., Nisan, N.: Fair allocation without trade. In: International Conference on Autonomous Agents and Multiagent Systems. International Foundation for Autonomous Agents and Multiagent Systems, pp. 719–728 (2012)
12. Joe, W.C., Sen, S., Lan, T., Chiang, M.: Multi-resource allocation: fairness-efficiency tradeoffs in a unifying framework. IEEE/ACM Trans. Netw. **21**(6), 1785–1798 (2013)
13. Li, W., Liu, X., Zhang, X., Zhang, X.: Multi-resource fair allocation with bounded number of tasks in cloud computing systems. Eprint Arxiv, pp. 1410–1255 (2014)
14. Liu, X., Zhang, X., Zhang, X., Li, W.: Dynamic fair division of multiple resources with satiable agents in cloud computing systems. In: Big Data and Cloud Computing (BDCloud), pp. 131–136 (2015)
15. Pacini, E., Mateos, C., Garino, CG.: Multi-objective swarm intelligence schedulers for online scientific clouds. Computing 1–28 (2014)
16. Shen, H., Liu, G.P., Chandler, H.: Swarm intelligence based file replication and consistency maintenance in structured P2P file sharing systems. IEEE Trans. Comput. **64**(1), 2953–2967 (2015)
17. Wilkes, J., Reiss, C.: Google ClusterData2011_2. https://code.google.com/p/googleclusterdata/

A Modified Bacterial Foraging Optimization Algorithm for Global Optimization

Xiaohui Yan$^{(\boxtimes)}$, Zhicong Zhang, Jianwen Guo,
Shuai Li, and Shaoyong Zhao

Department of Mechanical Engineering,
Dongguan University of Technology, Dongguan 523808, China
yxhsunshine@gmail.com

Abstract. To improve the optimization ability of Bacterial Foraging Optimization (BFO), A Modified Bacterial Foraging Optimization algorithm is proposed, which we named MBFO. In MBFO, tumble directions of bacteria are guided by the global best of the population to make bacteria search the optimization area more effectively. Then, chemotactic step size of each bacterium will change dynamically to adapt with the environment. Meanwhile, in reproduction loop, all individuals will be chosen with a probability. To test the global optimization ability of MBFO, we tested it on ten classic benchmark functions. Original BFO, PSO and GA are used for comparison. Experiment results show that MBFO algorithm has significant improvements compared with original BFO and it performs best on most functions among the compared algorithms.

Keywords: Modified bacterial foraging optimization · Chemotaxis · Adaptive strategies

1 Introduction

Swarm intelligence algorithms are a kind of powerful optimization algorithms which inspired by the social behaviors of animal swarms in nature. By interaction and cooperation of the individuals who have only simple behaviors, complex collective intelligence could emerge on the level of swarm. Recent years, many swarm intelligence algorithms have been proposed, such as Ant Colony Optimization (ACO) [1], Particle Swarm Optimization (PSO) [2] and Bacterial Foraging Optimization (BFO) [3] et al. BFO is proposed by Passino in 2002. It is inspired by the foraging and chemotactic behaviors of E. coli bacteria. Recently, BFO algorithm and its variants have been used for many numerical optimization [4] and engineering optimization problems [5, 6].

However, original BFO algorithm has some defects. First, the tumble angles are generated randomly. Useful information has not been fully utilized. Second, the chemotactic step size in BFO is a constant. It will make the population hard to converge to the optimal point at the end stage. In this paper, we proposed a Modified Bacterial Foraging Optimization (MBFO) algorithm. Several adaptive strategies are used in MBFO to improve its optimization ability. First, the tumble angles are generated towards the direction of global best, which enhances its optimization ability and convergence speed. Second, adaptive step size is employed. Chemotactic step size of

© Springer International Publishing Switzerland 2016
D.-S. Huang et al. (Eds.): ICIC 2016, Part I, LNCS 9771, pp. 627–635, 2016.
DOI: 10.1007/978-3-319-42291-6_62

each bacterium will increase or decrease depending on whether its position becomes better or not. This could make the bacteria use different search strategies in different stages. Third, in reproduction loop, all individuals will be chosen with a probability related with their fitness.

The rest of the paper is organized as followed. In Sect. 2, the original BFO algorithm is introduced. In Sect. 3, the MBFO algorithm is proposed and its strategies are described in detail. In Sect. 4, we test the proposed MBFO and other three algorithms on ten benchmark functions. Results are presented and discussed. Finally, conclusions are drawn in Sect. 5.

2 Bacterial Foraging Optimization

BFO algorithm is inspired by the foraging and chemotaxis behavior of E. coli bacterium. The E. coli bacterium is one of the earliest bacterium which has been researched. It has a plasma membrane, cell wall, and capsule that contains the cytoplasm and nucleoid. Besides, it has several flagella used for locomotion.

If the flagella rotate counterclockwise, they will push the cell and so the bacterium runs towards. And if the flagella rotate clockwise, they will pull the cell and the bacterium tumbles [3]. With running and tumbling, bacteria try to find food and avoid harmful phenomena, and the bacteria seem to intentionally move as a group. By simulating the foraging process of bacteria, Passino proposed the BFO algorithm. The main steps of BFO are explained as followed.

2.1 Chemotaxis

As it described above, a chemotactic step is define to be a tumble followed by a tumble or a tumble followed by a run. The formula of position change of a bacterium is given as Eq. (1). θ_i^t is the position of the ith bacterium in the tth chemotaxis step. $C(i)$ presents the chemotactic step size. $\Phi(i)$ is a randomly produced unit vector which stands for the tumble angle. In each chemotactic step, the bacterium generated a random tumble direction firstly. Then the bacterium moves in the direction. If the nutrient concentration in the new position is better than the last position, the bacterium will run one more step in the same direction. If the nutrient get worse or the maximum run step is reached, it will finish the run and wait for the next tumble. The maximum run times in the algorithm is controlled by a parameter called N_s.

$$\theta_i^{t+1} = \theta_i^t + C(i)\phi(i) \tag{1}$$

2.2 Reproduction

For every N_c times of chemotactic steps, a reproduction step is taken in the bacteria population. The bacteria population is sorted in descending order by their nutrient values. Then the S_r least healthy bacteria die and the other S_r healthiest bacteria each split into two bacteria, which are placed at the same location. Usually, S_r equals half of the population size to keep the population size as a constant, which is convenient in coding the algorithm. By reproduction operator, individuals with higher nutrient are survived and reproduced, which guarantees the potential optimal areas are searched more carefully.

2.3   Eliminate and Dispersal

After every several times of reproduction steps, an eliminate-dispersal event happens. For each bacterium, it will be dispersed to a random location according to a certain probability. This operator could enhance the diversity of the algorithm.

3   Modified Bacterial Foraging Optimization

3.1   Global Best Guided Tumble Direction

Chemotaxis is the core mechanism of BFO algorithm. It controls the bacteria move to the nutrient area and escape from the bad area [7]. However, in the chemotactic steps of BFO, the tumble directions are generated randomly. Information carried by the bacteria in rich nutrient positions is not utilized. As a result, the convergence speed of BFO is not fast. Many computation resources are wasted as they search randomly and lake of cooperation.

In MBFO, we use global best of the population to guide the tumble direction of the bacteria individuals. The tumble directions are generated as Eq. (2). θ_{gbest} is the global best of the whole population found so far. *Dim* is the dimension of the problem and *rand*() is a random vector. The tumble direction is then normalized as unit vector as Eq. (3). And the positions are still produced according Eq. (1). By this way, bacteria can find the better optimum direction as useful information of the population is fully utilized.

$$\Delta(i) = \text{rand}(1, \text{dim}) * (\theta_{\text{gbest}} - \theta_i) \tag{2}$$

$$\phi(i) = \Delta(i) \Big/ \Delta(i)^{\text{T}} \cdot \Delta(i) \tag{3}$$

3.2   Adaptive Chemotactic Step Size

As it mentioned above, the chemotactic step size is a constant in original BFO. It will make the algorithm hard to converge to the optimal point [8]. For an optimization algorithm, it is important to balance its exploration ability and exploitation ability. Generally, we should enhance the exploration ability to search larger areas in the early stage of an algorithm. And in the end stage of the algorithm, we should enhance the exploitation ability of the algorithm, and make it search more carefully to converge to the optimal point.

In MBFO, we use a simple adaptive strategy to adjust the chemotactic step size. In chemotactic part, if the nutrient concentration in the new position is higher than the last position (fitness of the individual improved), the chemotactic step size C will decrease. Otherwise, chemotactic step size C will increase, just like in Eq. (4). A new parameter λ is introduced to control the amplitude of variation. λ is larger than one. In general, the times of each individual been improved are more than the times of individual hasn't been improved, so the step size C will reduce in the overall trend. In the early stage of MBFO algorithm, larger step size is used to guarantee the exploration ability. And at the end stage, we use smaller step size to make sure the algorithm can converge to the optimal point.

$$C(i) = \begin{cases} C(i)/\lambda, & \text{if } fitness(i) \text{ impoved} \\ C(i) * \lambda, & \text{if } fitness(i) \text{ didn't impoved} \end{cases} \tag{4}$$

3.3   Reproduction

In the reproduction part of original BFO, half of bacterium individuals with higher fitness split into two and the rest half die. This strategy is in accord with the law of survival of the fittest. However, it may reduce the diversity of the population. It may be a better way if we give less chance to those individual with lower fitness. All the individuals will be chose with a probability related with their fitness.

4   Experiments

In this section, we will test the optimization ability of MBFO algorithm on several classic benchmark functions. Original BFO, PSO and GA are used for comparison [9].

4.1   Benchmark Functions

The benchmark functions for our tests are listed in Table 1. They are widely adopted by many researchers to test the global optimization ability of their algorithms [10, 11]. Among them, f_1 and f_2 are unimodal functions with independent variables. f_3–f_7 are unimodal functions with dependent variables. f_8 is a multimodal function with independent variables. f_9 and f_{10} are multimodal functions with dependent variables. In our tests, all ten functions are with dimension of 20.

To compare these algorithms fairly, we use number of function evaluations (FEs) instead of number of iterations as termination criterion in our tests. It is also used in many other works [12, 13]. All algorithms were terminated after 50,000 times of function evaluations.

Table 1. Benchmark functions.

	Function	Formulation	Variable ranges	f (x*)
f_1	Powers	$f(x) = \sum_{i=1}^{n} \lvert x_i \rvert^{i+1}$	$[-1, 1]$	0
f_2	Sumsquares	$f(x) = \sum_{i=1}^{n} i x_i^2$	$[-10, 10]$	0
f_3	Quadric	$f(x) = \sum_{i=1}^{D} \left(\sum_{j=1}^{i} x_j \right)^2$	$[-10, 10]$	0
f_4	Zakharov	$f(x) = \sum_{i=1}^{n} x_i^2 + (\sum_{i=1}^{n} 0.5 i x_i)^2 + (\sum_{i=1}^{n} 0.5 i x_i)^4$	$[-5, 10]$	0
f_5	Dixon_Price	$f(x) = (x_i - 1)^2 + \sum_{i=2}^{n} i(2x_i^2 - x_{i-1})^2$	$[-10, 10]$	0
f_6	Schwefel2.22	$f(x) = \sum_{i=1}^{n} \lvert x_i \rvert + \prod_{i=1}^{n} \lvert x_i \rvert$	$[-10, 10]$	0
f_7	Rosenbrock	$f(x) = \sum_{i=1}^{D-1} \left(100(x_i^2 - x_{i+1})^2 + (1 - x_i)^2 \right)$	$[-15, 15]$	0
f_8	Rastrigin	$f(x) = \sum_{i=1}^{D} \left(x_i^2 - 10\cos(2\pi x_i) + 10 \right)$	$[-10, 10]$	0
f_9	Ackley	$f(x) = 20 + e - 20e^{\left(-0.2\sqrt{\frac{1}{D}\sum_{i=1}^{D} x_i^2} \right)} - e^{\left(\frac{1}{D}\sum_{i=1}^{D} \cos(2\pi x_i) \right)}$	$[-32.768, 32.768]$	0
f_{10}	Griewank	$f(x) = \frac{1}{4000} \left(\sum_{i=1}^{D} x_i^2 \right) - \left(\prod_{i=1}^{D} \cos(\frac{x_i}{\sqrt{i}}) \right) + 1$	$[-600, 600]$	0

4.2 Parameter Settings

The population sizes S of all algorithms are 50. In MBFO and BFO, the parameters are set as followed: $N_c = 50$, $N_s = 4$, $N_{re} = 4$, $N_{ed} = 10$, $P_e = 0.25$. Most of these parameters are the same with they are in reference [3]. N_{ed} is lager as the termination criterion has changed. Actually, the algorithm will terminate after 50,000 times of function evaluations and the dispersal loop won't happen for ten times. For MBFO, $\lambda = 1.04$, the initial step size $C_s = 0.1(Ub - Lb)$, where Lb and Ub refer the lower bound and upper bound of the variables of the problems. λ is an important parameter. We found the MBFO performed best on most functions when it equals 1.04. For BFO, $C = 0.1$, $S_r = S/2 = 25$. In PSO algorithm, ω decreased from 0.9 to 0.7. $C1 = C2 = 2.0$. $V_{min} = 0.1 \times Lb$, $V_{max} = 0.1 \times Ub$. In GA, P_c is 0.95 and P_m is 0.1 [12].

4.3 Experiment Results and Statistical Analysis

The results of MBFO, BFO, PSO and GA on the benchmark functions are listed in Table 2. Best values of the four algorithms on each function are marked as bold. Convergence plots of the algorithms are shown in Fig. 1.

It is obvious that MBFO performed best on most benchmark functions, which could be seen in Fig. 1 and Table 2. It obtained the best values on eight of all ten functions. PSO obtained the best mean values on Sumsquares and Griewank. However, MBFO obtained the best minimum values on these two functions and the mean values are only a little worse than that of PSO. BFO and GA performed much worse than MBFO and

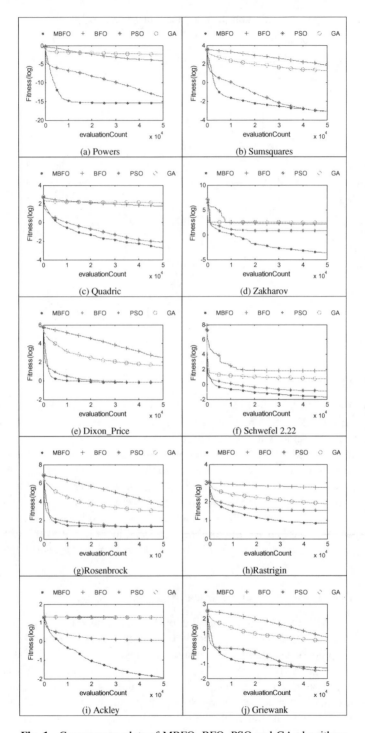

Fig. 1. Convergence plots of MBFO, BFO, PSO and GA algorithms

Table 2. Results obtained by MBFO, BFO, PSO and GA algorithms

Function		MBFO	BFO	PSO	GA
f_1	Mean	**4.27226e−016**	8.27042e−05	1.47070e−14	4.29984e−03
	Std	**1.31618e−15**	3.87442e−05	3.52524e−14	3.05427e−03
	Min	**5.09380e−27**	2.34883e−05	2.90549e−16	5.52865e−04
	Max	**4.75325e−15**	1.82960e−04	1.55753e−13	1.25895e−02
f_2	Mean	8.74679e−04	7.37280e+01	**8.37737e−04**	1.91917e+01
	Std	**7.94262e−04**	3.38091e+01	8.37166e−04	9.18486e+00
	Min	**4.35049e−06**	2.63560e+01	7.28788e−05	5.96596e+00
	Max	3.18269e−03	1.66385e+02	**3.11113e−03**	3.94321e+01
f_3	Mean	**1.72393e−03**	5.63869e+01	7.40234e−03	1.34907e+02
	Std	**1.77246e−−03**	8.82473e+00	3.17884e−03	3.62091e+01
	Min	**5.17414e−05**	4.23874e+01	2.09904e−03	6.18860e+01
	Max	**6.56173e−03**	8.08187e+01	1.29267e−02	1.92699e+02
f_4	Mean	**2.27702e−04**	1.24447e+02	6.43050e+00	2.74215e+02
	Std	**2.85556e−04**	2.76073e+01	2.29774e+01	7.43172e+01
	Min	**2.52866e−05**	7.82614e+01	1.03203e−03	1.66097e+02
	Max	**1.29145e−03**	1.72520e+02	1.00321e+02	4.60655e+02
f_5	Mean	**6.66772e−01**	3.02408e+02	6.69970e−01	4.04645e+01
	Std	**1.48317e−04**	1.58603e+02	4.84982e−03	2.71803e+01
	Min	**6.66671e−01**	6.72138e+01	6.66728e−01	1.07043e+01
	Max	**6.67253e−01**	6.97590e+02	6.88166e−01	1.02536e+02
f_6	Mean	**1.94715e−02**	6.01819e+01	1.34640e−01	5.14183e+00
	Std	**1.46431e−02**	1.24675e+01	1.62577e−01	9.88989e−01
	Min	**6.47249e−05**	2.34808e+01	3.01375e−02	3.47260e+00
	Max	**6.24126e−02**	8.27823e+01	6.53869e−01	6.75526e+00
f_7	Mean	**2.15232e+01**	4.23148e+03	2.35672e+01	9.85453e+02
	Std	**1.79283e+01**	1.91979e+03	2.17772e+01	7.99281e+02
	Min	9.19354e+00	1.02792e+03	**6.33827e+00**	2.51013e+02
	Max	**7.68225e+01**	7.13025e+03	9.43940e+01	3.27989e+03
f_8	Mean	**6.93457e+00**	5.55598e+02	3.29143e+01	7.50825e+01
	Std	**1.76076e+00**	7.23093e+01	6.84820e+00	1.55240e+01
	Min	**3.35444e+00**	4.01815e+02	1.99807e+01	4.86647e+01
	Max	**9.97962e+00**	6.69667e+02	4.77728e+01	1.04591e+02
f_9	Mean	**1.12142e−02**	1.96431e+01	1.12613e+00	1.88870e+01
	Std	**9.02245e−03**	3.44747e−01	7.82778e−01	5.47478e−01
	Min	**3.94373e−04**	1.85911e+01	9.21620e−03	1.78632e+01
	Max	**3.20858e−02**	2.00634e+01	2.31717e+00	1.96080e+01
f_{10}	Mean	5.10526e−02	5.76614e+00	**3.32566e−02**	3.24207e+00
	Std	3.71396e−02	1.20456e+00	**2.83369e−02**	7.69213e−01
	Min	**2.20532e−04**	3.40059e+00	3.13035e−03	1.49529e−01
	Max	1.41206e−01	7.21522e+00	**1.12975e−01**	5.41748e−02

X. Yan et al.

PSO on all functions. On Powers, Zakharov, Schwefel 2.22, Rastrigin and Ackley functions, MBFO performed better than PSO algorithm obviously. Especially on Zakharov and Ackley, MBFO obtained remarkable results while the other three algorithms all performed badly. On Dixon_Price and Rosenbrock functions, both MBFO and PSO were trapped into local optimum as the global minimum these two functions are inside a long, narrow, flat valley. It is difficult to converge to the global minimum. MBFO performed a little better than PSO on these two functions. The convergence curves are almost overlapped, as seen in Fig. 1(e) and (g). In general, MBFO shows significant improvement over the original BFO algorithm. And its optimization ability is the best among the four algorithms on most functions.

As a new parameter λ is added in MBFO to control its step size, in the experiments, we also test the impact of different values of parameter λ on the optimization ability of MBFO. In our tests, we find the results are sensitive to the value of λ. If λ is larger than 1.06, the step sizes will reduce to very small numbers close to zero in the early stage and the population barely evolves anymore. If λ is too small and close to one, the step sizes change frequently and may not reduce in the last stage of the algorithm. And the finally results are unsatisfactory too. In our tests, the algorithm obtains best optimization results on most functions when parameter λ is set as 1.04 or close to 1.04. For lake of space, we will not list the experiment results in this paper. However, we will keep attention to the new parameter and discuss it in future work.

5 Conclusions

This paper analyzes the shortages of original BFO algorithm. To overcome these shortages, we proposed a modified bacterial foraging optimization algorithm, which we called MBFO. In MBFO, three strategies are used to enhance its optimization ability. First, the tumble angles in chemotactic steps are no longer generated randomly. Instead, population information is used as guide and the bacteria individuals will swim towards the direction of the best position found so far. Second, the chemotactic step size will adjust adaptively according to whether the bacteria's position has been improved, which could balance the exploration ability and exploitation ability of the algorithm in different stages. Third, in reproduction loop, all individuals have a probability to stay in the population related with their fitness, which make the diversity of the population better.

To verify the optimization ability of the proposed MBFO algorithm, we tested it on ten classic benchmark functions. BFO, PSO and GA algorithm are employed for comparison. The results show that MBFO performed best on eight functions of all ten. On the rest two, it is only a little worse than PSO. In general, MBFO algorithm shows significant improvements over original BFO, and is a competitive algorithm among these classical comparison algorithms.

Acknowledgement. This work was supported by the Project of National Natural Science Foundation of China (Grant No. 71201026, 61503373), Project of Natural Science Foundation of Guangdong (Grant No. 2015A030310274, 2015A030313649), Project of Dongguan Social Science and Technology Development (Grant No. 2013108101011) and Project of Dongguan Industrial Science and Technology Development (Grant No. 2015222119).

References

1. Dorigo, M., Gambardella, L.M.: Ant colony system: a cooperating learning approach to the travelling salesman problem. IEEE Trans. Evol. Comput. **1**(1), 53–66 (1997)
2. Kennedy, J., Eberhart, R.C.: Particle swarm optimization. In: Proceedings of the 1995 IEEE International Conference on Neural Networks, vol. 4, pp. 1942–1948 (1995)
3. Passino, K.M.: Biomimicry of bacterial foraging for distributed optimization and control. IEEE Control Syst. Mag. **22**, 52–67 (2002)
4. Yıldız, Y.E., Altun, O.: Hybrid achievement oriented computational chemotaxis in bacterial foraging optimization: a comparative study on numerical benchmark. Soft. Comput. **19**(12), 3647–3663 (2015)
5. Niu, B., Wang, C., Liu, J., Gan, J., Yuan, L.: Improved bacterial foraging optimization algorithm with information communication mechanism for nurse scheduling. In: Huang, D.-S., Jo, K.-H., Hussain, A. (eds.) ICIC 2015. LNCS, vol. 9226, pp. 701–707. Springer, Heidelberg (2015)
6. Bhushan, B., Singh, M.: Adaptive control of DC motor using bacterial foraging algorithm. Appl. Soft Comput. **11**(8), 4913–4920 (2011)
7. Xu, X., Chen, H.: Adaptive computational chemotaxis based on field in bacterial foraging optimization. Soft. Comput. **18**(4), 797–807 (2014)
8. Yan, X., Zhu, Y., Zhang, H., Chen, H., Niu, B.: An adaptive bacterial foraging optimization algorithm with lifecycle and social learning. Discrete Dyn. Nat. Soc. Article ID 409478, 20 p (2012)
9. Kennedy, J.: Particle Swarm Optimization. In: Sammut, C., Webb, G.I. (eds.) Encyclopedia of Machine Learning, pp. 760–766. Springer, New York (2010)
10. Karaboga, Dervis, Akay, Bahriye: A comparative study of artificial bee colony algorithm. Appl. Math. Comput. **214**(1), 108–132 (2009)
11. van den Bergh, F., Engelbrecht, A.P.: A cooperative approach to particle swarm optimization. IEEE Trans. Evol. Comput. **8**(3), 225–239 (2004)
12. Liang, J., Qin, A.K., Suganthan, P.N., Baskar, S.: Comprehensive learning particle swarm optimizer for global optimization of multimodal functions. IEEE Trans. Evol. Comput. **10**, 281–295 (2006)
13. Nickabadi, A., Ebadzadeh, M.M., Safabakhsh, R.: A novel particle swarm optimization algorithm with adaptive inertia weight. Appl. Soft Comput. **11**(4), 3658–3670 (2011)

An Augmented Artificial Bee Colony with Hybrid Learning for Traveling Salesman Problem

Guozheng Hu[1], Xianghua Chu[1(✉)], Ben Niu[1(✉)], Li Li[1(✉)],
Dechang Lin[2], and Yao Liu[1]

[1] College of Management, Shenzhen University, Shenzhen, China
x.chu@szu.edu.cn,
{drniuben,llii318}@163.com
[2] Medical Business School, Guangdong Pharmaceutical University,
Guangzhou, China

Abstract. Traveling salesman problem (TSP) is a renowned NP-hard combinatorial optimization model which widely studied in the operation research community, such as transportation, logistics and industries areas. To address the problem effectively and efficiently, in this paper, a new meta-heuristic method, named hybrid learning artificial bee colony, is proposed based on the simply yet powerful swarm intelligence method, artificial bee colony algorithm. In HLABC, two different learning strategies are adopted in the employed bee phase and the onlooker bee phase. The updating mechanism for food source position is enhanced by employing global best food source. Experimental results on TSP problems with various city sizes indicate the effectiveness of the proposed algorithm.

Keywords: Traveling salesman problem · Artificial bee colony algorithm · Hybrid learning

1 Introduction

Traveling salesman problem (TSP) is a well-known NP-hard combinatorial optimization problem [1, 2] since there is no accurate or efficient methods have been found. The salesman need to find a shortest closed route which visits all the cities in a given set [3], and the complexity of TSP increases exponentially with scale growth. Many other NP complete problems such as the postal road, Hamiltonian cycle and production scheduling can be summed up in the TSP model, which make solving TSP problems not only theoretically necessary, but also practically meaningful in reality.

With the inspiration of swarm intelligence, a number of intelligent algorithms have been proposed to address TSP problems. Different types of TSP problems including TSP with time window [4, 5], stochastic TSP [6, 7], asymmetric TSP [8, 9] and TSP with precedence constraints [10, 11], were handled by genetic algorithm (GA) [12, 13], ant colony optimization (ACO) [14–16] and particle swarm optimization (PSO) [17, 18] etc.

Artificial bee colony is recently proposed method mimicking the information exchange behavior, i.e. dancing among the bees [19]. Initially, ABC was proposed for numerical optimization problems and many studies indicate that ABC outperformed to

© Springer International Publishing Switzerland 2016
D.-S. Huang et al. (Eds.): ICIC 2016, Part I, LNCS 9771, pp. 636–643, 2016.
DOI: 10.1007/978-3-319-42291-6_63

other methods. Later on it was extended to the discrete and combinatorial type problems [20]. In this paper, we present a new hybrid learning mechanism artificial bee colony algorithm for solving versions of the classical traveling salesman problem, and compared the performance with other algorithms. The remainder of the paper is structured as follows: In Sect. 2, a brief introduction of ABC, and a demonstrating of the ABC with a new hybrid learning mechanism followed in Sect. 3. In Sect. 4, experiments are set and results are shown. Finally, drawn the conclusions in Sect. 5.

2 Background of ABC

In ABC, the position of a food source is responded to a possible solution of optimization problems. The quality of a food source is related to the value of nectar (fitness). Employed bee is associated with a potential food source where they are currently exploiting. The food source information is then shared with onlookers nearby hive [20]. There are two types of unemployed foragers: onlookers that select and refine food sources and scout bees that search for new food sources.

The procedure of ABC comprises of four phases [21]: (1) initialization phase, where the control parameters and the possible food sources (solutions) are initially randomly. (2) employed bees phase, in which each employed bee is assigned to an individual food source and the corresponding fitness is evaluated. The promising information is shared in next step after returning to the hive. (3) onlooker bees phase, onlooker bees probabilistically choose the food sources depending on the quality provided by the employed bees. The qualities are related to the fitness of the food sources, and selection technique such as roulette wheel selection is employed. The chosen food source would be used to generate a new food source by exploiting the surrounds. (4) scout bees phase, if the solutions correspond with employed bees are unable to improve in a predetermined number of iterations, known as the "limit", they would be abandoned. Then the scouts start to search for new solutions randomly. As a result, the inferior food sources would be replaced by the promising ones in this phase. The last three phases iterate to approach the global optima.

The general algorithmic structure of the ABC optimization approach is given as Table 1:

Table 1. The general structure of ABC

Initialization
DO while stop criteria not met
Employed bees assigned to an individual food source and evaluate their fitness
Onlookers choose their food sources depending upon the fitness shared by employed bees
Send the scouts for exploring new food sources
Update and memorize food sources and global best information
Terminate and output results

3   Hybrid Learning Mechanism of ABC

As discussed in the previous section, the algorithm structure of ABC basically includes two operators: (1) the exploration operator based on the employed phase, and (2) the exploitation operator based on the onlooker phase. It is worth noticing that the same learning operator is adopted for both of the two phases. Since exploration and exploitation have different search interests during the optimization, the same learning mechanism may not be able to meet the needs of both of them at the same time. More specifically, for a current learning equation, the guiding exemplar is randomly selected from the neighbor which is good for global exploration, whilst it is less convergent for the exploitation. This may lead to low exploitation accuracy [21]. Therefore, we developed an augmented ABC by conducting different learning strategies in the employed bee phase and the onlooker bee phase, which is named as hybrid learning artificial bee colony (HLABC).

In the proposed technique, the position of food source is updated using the following function in onlooker phase:

$$v_{i,d} = x_{i,d} + \left(x_{i,d} - x_{k,d} \right) \times r + \left(x_{i,d} - x_{best} \right) \times r \qquad (1)$$

where v_{id} refers to the new potential food source, r is a random number between [0, 1], k denotes another random number different from i, x_{best} is the current best food source found by the population. In Eq. (1), the individual not only randomly learns from its neighbor, but also move towards the best food source which was found so far. This accelerates the convergence of the individuals which benefits the exploitation capability. Thus, in HLABC, the employed bee phase updates the bees not changed, while Eq. (1) is adopted for the onlooker bee phase.

The main procedures of HLABC used to address TSP are as follows: (1) initialize a set of vectors as potential solutions and evaluate their fitness. (2) employed bee is assigned to explore the area around its neighbors. (3) onlooker bees probabilistically choose the food sources provided by the employed bees. The chosen vector would be used to generate a new vector by learning from a random individual and the global optimal vector. (4) if a solution correspond with employed bees are unable to improve in certain iterations, it would be replaced by new food source that found by a scout. The performance of HLABC is analyzed in the following section.

4   Experimental Comparison

4.1   General Experimental Settings

To comprehensively evaluate the performance of HLABC on solving TSP, 13 TSP problems with various city sizes are employed in the experiment [22, 23]. For fair comparison, the maximum numbers of iterations (MNI) are different on the problems with various city sizes. For, MNI is set to be100; For MNI is set to be 200 and For MNI is set to be 500. The dimensions of TSP are equal to the city sizes. Each of the experiments is independently run 30 times. The performance of HLABC is compared with that of three existing algorithms including artificial bee colony [24] (ABC), genetic algorithm (GA) [13] and particle swarm optimization algorithm (PSO) [17], in order to justify the effectiveness of the proposed method. The population size is 100 for the involved algorithms. The mean results of the best solution found by the algorithms have been recorded for analysis.

The particular settings for each algorithm are given as follows:

The GA has four mechanisms when generate new chromosomes in each iteration: selection mechanism, the better the fitness values of chromosome, the greater the probability of be selected; crossover mechanism with the rate of 0.95; mutation mechanism with the rate of 0.05; and the reverse mechanism. A child chromosome with better fitness value was added to the population in child production scheme, and linear ranking fitness function was employed.

In PSO, a particular particle learns from the best solution of itself and the global best solution. When proposed PSO to TSP problems, vectors are generated randomly as swarms population. A particular vector corresponding to an individual particle and learn a period of vector from the best solution of itself and the global best solution, two rand numbers which less than the vector's length are used to define the length of period. Thus, there are no other parameters need to set.

In the original ABC algorithm, only one dimension is selected to update in each iteration, we also update one dimension in each iteration. The original ABC algorithm has the same parameter setting with the HLABC algorithm. For HLABC, the percentage of the employed bees and the onlooker bees are equally 50 % of the population size. The number of limits is set to 10.

4.2 Experimental Results and Discussions

Table 2 presents the result of the 30 runs of the four algorithms on the TSP problems. The best results are underlined in the table

From the Table 2, it is observed HLABC performance better than the contrastive algorithms. PSO also performance well when the city scale not very large, that testifies learning form the global best solution is an effective strategies and better than GA update mechanisms in dealing with TSP problems. The result of HLABC is significantly improved than that of the original ABC on most of the TSP problems in terms of mean results. This demonstrates the effectiveness of the proposed technique compared with original ABC. The new update mechanism which learns from the global best solution is able to employ existing best information to guide the search. It also indicates the capability of HLABC in getting out of the local optima while searching. However, HLABC cannot obtain the best solutions, more future works need to do to improve that algorithm for solving TSP problem.

Figures 1, 2, 3 and 4 show the convergence characteristics of the algorithms on four representative cities scales TSP problems: Ulysses16, Eil51, Pro136 and FL417. In our experiences, it can be observed that HLABC converges faster than the origin ABC, GA and PSO algorithms, and improves global solution until the late stage of the optimization. The result indicates that the proposed strategy significantly enhances ABC's convergence speed and accuracy for solving TSP problems.

Table 2. Results obtained by GA, PSO, ABC and HLABC algorithms

TSPs	MNI	The best solution	GA		PSO		ABC		HLABC	
			Mean	SD	Mean	SD	Mean	SD	Mean	SD
Ulysses16	100	7.20E+01	7.42E+01	3.48E-01	7.41E+01	2.36E-01	7.63E+01	1.59E+00	7.42E+01	2.64E-01
Ulysses22		7.40E+01	7.62E+01	5.48E-01	7.84E+01	6.55E-01	9.58E+01	1.54E+00	7.69E+01	1.42E+00
Chn31		1.54E+01	1.79E+01	6.71E-01	1.83E+01	4.75E-01	2.44E+01	5.63E-01	1.66E+01	5.97E-01
Att48		3.35E+04	5.54E+04	2.20E+03	5.58E+04	1.83E+03	8.55E+04	3.86E+03	4.58E+04	4.85E+03
Eil51		4.26E+02	8.39E+02	2.72E+01	7.18E+02	1.20E+01	1.02E+03	5.79E+01	6.39E+02	2.50E+01
Eil76	200	5.38E+02	1.15E+03	1.36E+01	1.07E+03	3.16E+01	1.50E+03	3.29E+01	8.62E+02	1.79E+01
Eil101		6.29E+02	1.72E+03	7.13E+01	1.49E+03	6.78E+01	2.20E+03	2.93E+01	1.28E+03	3.45E+01
Pro107		4.43E+04	2.08E+05	1.89E+04	1.61E+05	5.77E+03	3.10E+05	1.05E+04	1.07E+05	1.08E+04
Pro136		9.68E+04	4.36E+05	1.64E+04	3.30E+05	5.37E+03	5.26E+05	8.27E+03	2.82E+05	4.99E+03
Pro152		7.37E+04	5.56E+05	1.70E+04	3.92E+05	1.50E+04	6.85E+05	1.69E+04	3.23E+05	1.74E+04
KroA200	500	2.94E+04	1.41E+05	5.67E+03	1.34E+05	1.72E+03	1.90E+05	1.92E+03	9.46E+04	3.38E+03
Lin318		4.20E+04	3.47E+05	4.84E+03	2.30E+05	7.01E+03	3.51E+05	8.14E+04	2.14E+05	5.04E+04
Fl417		1.19E+04	2.57E+05	3.58E+03	1.57E+05	3.50E+05	2.90E+05	5.09E+04	1.32E+05	1.40E+04

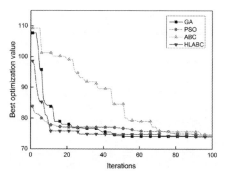

Fig. 1. Convergence curves of Ulysses16

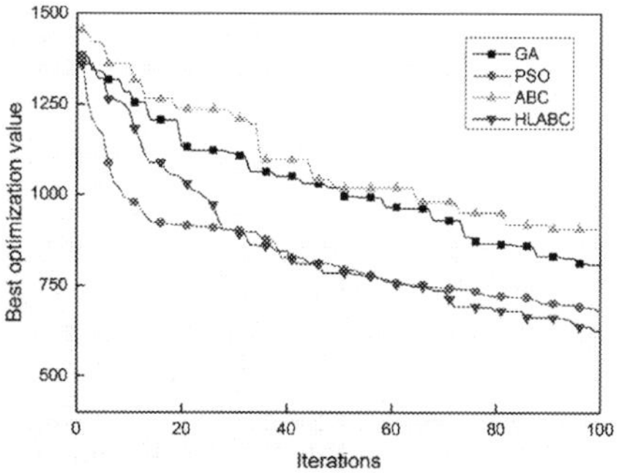

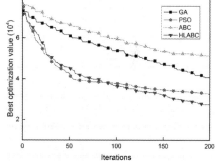

Fig. 2. Convergence curves of Eil51

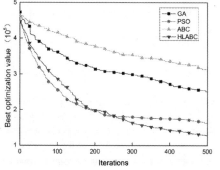

Fig. 4. Convergence curves of Fl417

Fig. 3. Convergence curves of Pro136

5 Conclusion

Traveling salesman problem is a well-known NP-hard combinatorial problem in operational research. To effectively address TSP, this paper proposes a novel algorithm based on artificial bee colony, termed as hybrid learning artificial bee colony (HLABC) for traveling salesman problem. In HLABC, a novel hybrid learning strategy is developed to enhance the exploration capability and exploitation capability of ABC. The performance of the proposed method is compared with GA, PSO and ABC in terms of mean error. Experimental results demonstrate that HLABC outperforms the original ABC and GA algorithms in terms of solution accuracy and convergence speed. Also, it is seen from the experiment that the exploitation capability is improved on most of the TSPs.

While promising, there is a large margin of improvement for future research. First, HLABC still somewhat converge slow on large-scale TSPs. Second, the solution accuracy on more high dimensional TSPs should be verified and enhanced. Last but not least, the application of HLABC to address more real world complex combinatorial optimization problems will also be studied in the future work.

Acknowledgment. This work was supported by the national natural science foundation of china (71501132, 71571120 and 71371127), and the Natural Science Foundation of Guangdong Province (2016A030310067 and 2015A030313556).

References

1. Laporte, G.: The traveling salesman problem: an overview of exact and approximate algorithms. Eur. J. Oper. Res. **59**(2), 231–247 (1992)
2. Lawler, E.L.: The traveling salesman problem; a guided tour of combinatorial optimization. Math. Gaz. **58**, 535–536 (1985)
3. Reinelt, G.: A traveling salesman problem library. J. Oper. Res. Soc. **11**, 19–21 (1992)
4. Focacci, F., Lodi, A., Milano, M.: A hybrid exact algorithm for the TSPTW. Inf. J. Comput. **14**, 403–417 (2002)
5. Ascheuer, N., Fischetti, M., Grötschel, M.: Solving the asymmetric travelling salesman problem with time windows by branch-and-cut. Math. Program. **90**, 475–506 (2001)
6. Chang, T.S., Wan, Y., Ooi, W.T.: A stochastic dynamic traveling salesman problem with hard time windows. Eur. J. Oper. Res. **198**, 748–759 (2009)
7. Bellman, R., Roosta, M.: A stochastic travelling salesman problem. Stoch. Anal. Appl. **1**, 159–161 (1983)
8. Majumdar, J., Bhunia, A.K.: Genetic algorithm for asymmetric traveling salesman problem with imprecise travel times. J. Comput. Appl. Math. **235**, 3063–3078 (2011)
9. Carpeneto, G., Toth, P.: Some new branching and bounding criteria for the asymmetric travelling salesman problem. Manag. Sci. **26**, 736–743 (1980)
10. Moon, C., Kim, J., Choi, G., Seo, Y.: An efficient genetic algorithm for the traveling salesman problem with precedence constraints. Eur. J. Oper. Res. **140**, 606–617 (2002)
11. Bianco, L., Mingozzi, A., Ricciardelli, S.: The traveling salesman problem with cumulative costs. Networks **1990**, 81–91 (1992)
12. Choi, I.C., Kim, S.I., Kim, H.S.: A genetic algorithm with a mixed region search for the asymmetric traveling salesman problem. Comput. Oper. Res. **30**, 773–786 (2003)
13. Jana, N., Rameshbabu, T.K., Kar, S.: Genetic algorithm for the travelling salesman problem using new crossover and mutation operators. In: Information and Management Sciences–Processings of the Ninth International Conference on Information and Management Sciences (2010)
14. Brezina Jr., I., Čičková, Z.: Solving the travelling salesman problem using the ant colony optimization. Int. Sci. J. Manag. Inf. Syst. **6**, 10–14 (2011)
15. Manfrin, M., Birattari, M., Stützle, T., Dorigo, M.: Parallel ant colony optimization for the traveling salesman problem. In: Dorigo, M., Gambardella, L.M., Birattari, M., Martinoli, A., Poli, R., Stützle, T. (eds.) ANTS 2006. LNCS, vol. 4150, pp. 224–234. Springer, Heidelberg (2006)
16. Mavrovouniotis, M., Yang, S.: Ant colony optimization with immigrants schemes for the dynamic travelling salesman problem with traffic factors. Appl. Soft Comput. **13**, 4023–4037 (2013)
17. Bouzidi, M., Riffi, M.E.: Discrete novel hybrid particle swarm optimization to solve travelling salesman problem. In: The Workshop on Codes, Cryptography and Communication Systems, pp. 17–20 (2014)
18. Zhong, W.L., Zhang, J., Chen, W.N.: A novel discrete particle swarm optimization to solve traveling salesman problem. In: IEEE Congress on Evolutionary Computation, 2007, CEC 2007, pp. 3283–3287 (2007)

19. Karaboga, D., Basturk, B.: A powerful and efficient algorithm for numerical function optimization: artificial bee colony (Abc) algorithm. J. Global Optim. **39**, 459–471 (2007)
20. Karaboga, D., Gorkemli, B., Ozturk, C., Karaboga, N.: A comprehensive survey: artificial bee colony (Abc) algorithm and applications. Artif. Intell. Rev. **42**, 21–57 (2014)
21. Alqattan, Z.N., Abdullah, R.: A hybrid artificial bee colony algorithm for numerical function optimization. Int. J. Mod. Phys. C **26**, 127–132 (2015)
22. Taş, D., Gendreau, M., Jabali, O., Laporte, G.: The traveling salesman problem with time-dependent service times. Eur. J. Oper. Res. **248**, 372–383 (2015)
23. Maity, S., Roy, A., Maiti, M.: An imprecise multi-objective genetic algorithm for uncertain constrained multi-objective solid travelling salesman problem. Expert Syst. Appl. **46**, 196–223 (2015)
24. Zhu, G., Kwong, S.: Gbest-guided artificial bee colony algorithm for numerical function optimization. Appl. Math. Comput. **217**, 3166–3173 (2010)

Particle Swarm Optimizer
with Full Information

Yanmin Liu[1]([⊠]), Chengqi Li[2], Xiangbiao Wu[1], Qingyu Zeng[1],
Rui Liu[1], and Tao Huang[1]

[1] School of Mathematics and Computer Science,
Zunyi Normal College, Zunyi 563002, China
yanmin7813@gmail.com
[2] School of Mathematics and Science, Wuyi University,
Jiangmen 529020, Guangzhou, China

Abstract. In order to improve the particle swarm optimizer (PSO) for solving complex multimodal problems, an improved PSO with full information and mutation operator (PSOFIM) is proposed base basic PSO and mutation thought. In PSOFIM, a novel mutation is adopted to improve the history optimal position of particle (*pbest*) by disturbance in operation of each dimension. Additionally, a full information strategy for each particle is introduced to make best use of each dimension of each particle to ensure the information utility for swarm topology where each particle learns from its neighborhood information for his optimal position to improve itself study ability, whose strategies improve the swarm fly to the probability of the optimal solution. The simulation experiment results of benchmark function tests show PSOFIM has better performance than the basic PSO algorithm.

Keywords: Particle swarm optimizer · Mutation operator · Simulation

1 Introduction

Optimization problem has been an active research field in the domain of the evolutionary computation. As the optimization problems of the real world become more and more complex, the optimization methods of the optimization problem are put forward higher request. Usually, an unconstrained optimization problem with n minimization problem can be formulated as follows:

$$\min f(X) \ subject \ to \ X = (x_1, x_1, \ldots, x_1, x_n) \in \Omega \subseteq R^n \tag{1}$$

Particle swarm optimization (PSO for short) [1] is a relatively new optimization technology, which is similar to genetic algorithms (GA) on calculation method, but the difference is that PSO does not use of crossing and mutation operations but conducts performance by imitating herd, birds, fish swarm behavior in the search space. Additionally, the concept of PSO is simple, less control parameters, easy to implement and when searching the optimization result there is nothing with the initial value. Therefore, since the PSO is proposed, it has received the extensive attention of the academia

© Springer International Publishing Switzerland 2016
D.-S. Huang et al. (Eds.): ICIC 2016, Part I, LNCS 9771, pp. 644–650, 2016.
DOI: 10.1007/978-3-319-42291-6_64

[2–6], and the experiences show that PSO algorithm has great advantage in solving the problem of most of the actual performance. But similar to other evolutionary, PSO is easy to trap in local optimal solution in solving complex multimodal problem. Therefore, in order to improve the population's ability to jump out of local optimal solution, in this paper, we proposed an improved PSO with full information and mutation operator (PSOFIM) based on the literature [3].

2   Particle Swarm Optimization (PSO)

Particle swarm optimization is a kind of random optimization algorithm based on swarm intelligence. In PSO, each particle flies toward the optimal solution through mutual learning. In the learning process, the learning strategies of each particle are determined by the topology of the swarm. Depending on the neighborhood topology structure, PSO can be divided into Local version particle swarm optimization (Local PSO, LPSO) and the global version particle swarm optimization (Global PSO, GPSO). Among them, the neighborhood topology structure of the GPSO algorithm is shown in Fig. 1(a), and that of the LPSO algorithm as shown in Fig. 1(b).

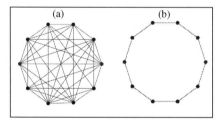

Fig. 1. Neigborhood topology

The evolution equation of GPSO and LPSO can be described as:

$$\overline{v_i}(t) = \overline{v_i}(t-1) + \varphi_1 r_1 (\overline{p_i}(t) - \overline{x_i}(t-1)) + \varphi_2 r_2 (\overline{p_g}(t) - \overline{x_i}(t-1))$$
$$\overline{x_i}(t) = \overline{v_i}(t) + \overline{x_i}(t-1)$$
(2)

$$\overline{v_i}(t) = \overline{v_i}(t-1) + \varphi_1 r_1 (\overline{p_i}(t) - \overline{x_i}(t-1)) + \varphi_2 r_2 (\overline{p_{neighbor_i}}(t) - \overline{x_i}(t-1))$$
$$\overline{x_i}(t) = \overline{v_i}(t) + \overline{x_i}(t-1)$$
(3)

where, $\overline{p_i}$ is the experienced best position of particle i, called "cognitive part" of the particles; $\overline{p_g}$ is the best position for the whole swarm, also known as the "social part" of the particles; $\overline{p_{neighbor_i}}$ is the experienced best position of neighborhood of particle i, also called "cognitive part" of the particles; $\overline{v_i}(t)$ and $\overline{x_i}(t)$ denote velocity and position of particle at iteration t; φ_1 and φ_2 is the particle's speed constant to balance the global search and local search for the whole swarm, whose value usually is within 0 and 2; r_1 and r_2 is uniformly distributed random vector between 0 and 1. These two kinds of algorithm are the same in essence, just when updating the particle velocity the parts of "social learning" have different choices.

Algorithm 1
1:**Begin**
2:Initialize position, velocity, *pbest$_i$* and *gbest.*
3:**For** each particle
4: **For** each dimension
5:      Updating particles' position and velocity.
6:    **If** fitness(x_i) < fitness(*pbest$_i$*)
7:         *pbest$_i$*=x_i , *pbestval$_i$*= fitness(x_i).
8:    **End**
9: **Endfor**
10:**Endfor**
11:Index=find(*pbestval$_i$*)
12:**If** fitness(*gbest*) > min(*pbestval$_i$*)
13:   *gbestval*= *pbestval$_i$* , *gbest*=x_{index}.
14:**End**
15:**End begin**

To enhance convergence probability of PSO algorithm to the optimal solution, in the literature [2], the author analyze PSO convergence mode and put forward PSO variants with the contraction factor to guarantee the convergence of PSO. The velocity update formula can be described as:

$$\overline{v}_i(t) = \chi\{\overline{v}_i(t-1) + \varphi_1 r_1(\overline{p}_i(t) - \overline{x}_i(t-1)) + \varphi_2 r_2(\overline{p}_g(t) - \overline{x}_i(t-1))\}$$
$$\overline{x}_i(t) = \overline{v}_i(t) + \overline{x}_i(t-1)$$
$$\chi = 2/\left|2 - \phi - \sqrt{\phi^2 - 4\phi}\right|, \phi = \phi_1 + \phi_2, \phi > 4$$

(4)

Algorithm 1 gives the basic PSO pseudo code.

3 Particle Swarm Optimizer with Full Information

3.1 Mutation Operator

From the population evolution equation (the Eqs. (2) and (3)), it can show that when each particle itself gets into a local optimal solution, this phenomenon will cause the population into a local optimal solution. In order to change this situation, the mutation operation is introduced to generate new the optimal location of the particles themselves (*pbest*), which will lead particles in flight. In order to make full use of the optimum position of each particle, the rules that conducts mutation operation for pbest is its performance of no improvement for five generations. Equation (5) gives Mutation operation process.

$$Npbest_i^k = pbest_i^k(1 + r)$$

(5)

where, r is a random number with the mean value 0 and the variance 1; $pbest_i^k$ is the k dimension of *pbest* of particle i; $Npbest_i^k$ is the new produced particle. Equation (6) gives the selection rule whether the new *pbest* is accepted after mutation operation.

$$pbest_i^k = \begin{cases} Npbest_i^k & f(pbest) \geq f(Npbest) \\ pbest_i^k & f(pbest) < f(Npbest) \end{cases} \tag{6}$$

3.2 Particle Swarm Optimizer with Full Information

In literature [3, 6], the author concluded that if learning exemplars of each dimension of a particle is the corresponding dimension of the same particle, it can cause "two steps forward and one step back" phenomenon, namely premature convergence phenomenon. Therefore, based on literature [3] this paper puts forward the complete information PSO algorithm (PSOFIM), which greatly improves the population's ability to jump out of local optimal solution. Equation (7) gives velocity and position updating formula of PSOFIM.

$$\overline{v_i}(t) = \chi\{\overline{v_i}(t-1) + \varphi(\overline{p_m}(t) - \overline{x_i}(t-1)) + r \cdot (\overline{p_i}(t) - \overline{x_i}(t-1))\}$$
$$\overline{x_i}(t) = \overline{v_i}(t) + \overline{x_i}(t-1) \tag{7}$$

$$\overline{p_m} = \frac{\sum_{k \in N} W(k)\overline{\varphi_k} \cdot p_k}{\sum_{k \in N} W(k)\overline{\varphi_k}}$$
$$\overline{\varphi_k} = \overline{U}[0, \frac{\varphi_{\max}}{N}], \forall k \in N \tag{8}$$

where, N is the number of neighborhood of particle i; p_k is *pbest* of neighborhood of particle i so far; $\overline{U}[\min, \max]$ is a random number between the minimum and maximum values; $W(k)$ balance weight, it is 1 here; $\varphi_{\max}$ is the sum of acceleration constant, here it is 4.1, and $\varphi = \varphi_1 + \varphi_1 + \cdots + \varphi_n$. r is uniform random number in [0, 0.5].

Algorithm steps are described as follows:

Step 1: Randomly initializing position and velocity in the population
Step 2: Setting *pbest* as the current position, and *gbest* and *pbest* as optimal particle.
Step 3: Determining whether satisfy the standard of algorithm stopping, if met, turned to step 7, else if Step 4.
Step 4: According to the Eq. (7), updating the particle velocity, position, *pbest* and *gbest*.
Step 5: If meet the conditions, performing mutation *pbest*, else if Step 6.
Step 6: If algorithm meets stopping criteria, stopping; else if turn to step 4;
Step 7: Outputting the result.

4 Simulation Experiments and Analysis

To test the performance of the algorithm, we select the Sphere, and Rastrigin Schwefel functions to test improved PSO performance. Benchmark function expressions are shown in [2], respectively, and their features and the curve of the function shape are shown in Table 1 and in Fig. 2. From function shape, we can conclude that Rastrigin and Schwefel functions are complex multimodal functions with multiple local optimal solutions. Additionally, we compare the proposed PSO (PSOFIM) with LPSO and GPSO algorithm. The experiment of various parameter settings are as follows: each algorithm contains 30 particles, each test function independently runs 30 times, and each time runs 3×10^4 function evaluation. In LPSO and GPSO algorithm, $\varphi_1 = \varphi_2 = 2$. Figure 3 shows the convergence of the algorithm features graph. Table 2 shows the performance results for the different algorithms in 30 run independently, where B is the best operation result; W is the worst run results; M is the average run results.

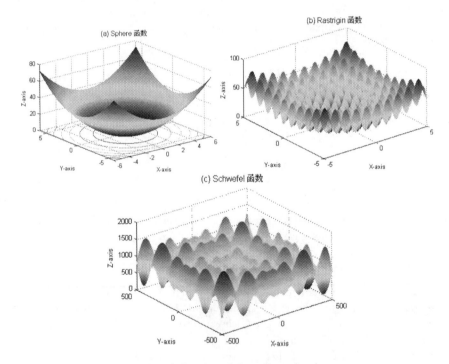

Fig. 2. Benchmark function of three-dimensional stereogram

Table 1. Benchmark function characteristics

f	n	Search space	Global optimum (x*)	f(x*)
Sphere	30	$[-100, 100]^{30}$	[0, 0, ..., 0]	0
Rastrigin	30	$[-5.12, 5.12]^{30}$	[0, 0, ..., 0]	0
Schwefel	30	$[-500, 500]^{30}$	[420.96, ..., 420.96]	0

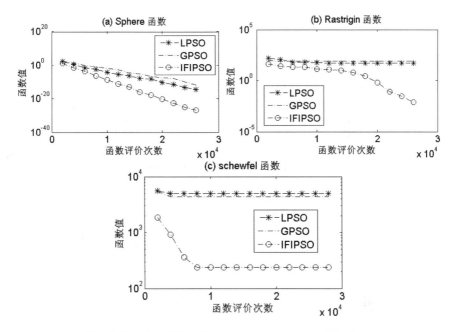

Fig. 3. Benchmark function convergence characteristic figure

From the algorithm convergence characteristics in Fig. 3, we can see that on the multimodal problems Rastrigin and Schwefel function, LPSO and GPSO algorithm has been into the local optimal solution at the beginning of iteration, i.e., appearing premature convergence. However, the proposed PSOFIM algorithm can effectively avoid premature convergence, which mainly adopted complete information with novel the updating process of the particle velocity, not only by themselves, but also by the most information of other particles in the particle neighborhood. From Table 2 we can also see that the proposed algorithm is obviously superior to the basic particle swarm optimization. Therefore, it can conclude that PSOFIM is an effective improvement based on the basic PSO algorithm.

Table 2. Mean and confidence interval

Algorithm		Sphere	Rastrigin	Schwefel
LPSO	B	4.5502e−017	5.1234e+001	4.9945e+003
	W	1.8432e−014	9.3621e+001	5.2032e+003
	M	4.9367e−016	6.9326e+001	5.1306e+003
GPSO	B	4.7546e−014	8.2542e+001	4.3474e+003
	W	2.8431e−011	9.9327e+001	6.0836e+003
	M	6.1132e−013	8.7926e+001	5.1134e+003
PSOFIM	B	2.9897e−032	4.5354e−007	2.3687e+002
	W	1.4574e−031	3.1943e−006	2.9104e+002
	M	7.9213e−032	9.1732e−007	2.5731e+002

5   Conclusions

In order to improve the particle swarm optimization for solving complex multimodal problems, this paper proposes an improved particle swarm optimizer with full information and mutation operator (PSOFIM) where the complete information and mutation strategies are introduced to make the particles information of their neighborhood and its own history study. Additionally, to make full use of the history information of each particle, the mutation operator is conducted in terms of several generations to enhance the diversity of population. The results of simulation experiment show that PSOFIM algorithm is an effective improvement of the basic PSO algorithm.

Acknowledgments. This work is supported by the National Natural Science Foundation of China (Grants nos. 71461027, 71471158). Science and technology talent training object of Guizhou province outstanding youth (Qian ke he ren zi [2015] 06); Guizhou province natural science foundation in China (Qian Jiao He KY [2014] 295); 2013, 2014 and 2015 Zunyi 15851 talents elite project funding; Zunyi innovative talent team (Zunyi KH (2015) 38); Project of teaching quality and teaching reform of higher education in Guizhou Province (Qian Jiao gaofa [2013] 446, [2015] 337), College students' innovative entrepreneurial training plan (201410664004, 201510664016); Guizhou science and technology cooperation plan (Qian Ke He LH zi [2015] 7050, Qian Ke He J zi LKZS [2014] 30, Qian Ke He LH zi [2016] 7028); Zunyi Normal College Research Funded Project (2012 BSJJ19).

References

1. Kennedy, J., Eberhart, R.: Particle swarm optimization. In: Proceedings of IEEE International Conference on Neural Networks, Piscataway, NJ, pp. 1942–1948 (1995)
2. Clerc, M., Kennedy, J.: The particle swarm explosion, stability, and convergence in a multi-dimensional complex space. IEEE Trans. Evol. Comput. **6**(1), 58–73 (2002)
3. Mendes, R., Kennedy, J.: The fully informed particle swarm: simpler, maybe better. IEEE Trans. Evol. Comput. **8**(3), 204–210 (2004)
4. Hao, R., Wang, Y.J.: Escape an improved adaptive particle swarm optimization and experimental analysis. J. Softw. **13**(12), 2036–2044 (2005)
5. Liu, Y.M., Zhao, Q.Z.: A kind of particle swarm algorithm based on dynamic neighbor and mutation factor. Control Decis. **25**(7), 968–974 (2010)
6. Liang, J.J., Qin, A.K., Suganthan, P.N.: Comprehensive learning particle swarm optimizer for global optimization of multimodal functions. IEEE Trans. Evol. Comput. **10**(3), 281–295 (2006)

Location Selection of Multiple Logistics Distribution Center Based on Particle Swarm Optimization

Qingyu Zeng[1(✉)], Chengqi Li[2], Xiangbiao Wu[1], Shengjie Long[1],
Zhuanzhou Zhang[1], Rui Liu[1], Tao Huang[1], and Yanmin Liu[1]

[1] School of Mathematics and Science, Zunyi Normal College,
Zunyi 563002, Guizhou, China
963738522@qq.com
[2] School of Mathematics and Science,
Wuyi University, Jiangmen 529020, Guangzhou, China

Abstract. Distribution center is an important pivot position at the logistics system. This paper presents the site selection's model of logistics distribution center based on decision matrix C, which adopts particle swarm optimization (PSO) to find the best combination of site selection by structuring the iteration of decision matrix. At the same time, simulation experiments are conducting for the site selection of logistics distribution center with 4 candidate centers and 10 distribution points, and the results show that PSO can get the best solution of distribution center in 90 % success rate of the best solution and the average research time of the approximate 3.5 s. From simulation experiment, PSO is efficient, accurate and suitable for the model optimization of distribution center, and therefore, it can be regarded as an effective method for the site selection's model of logistics distribution.

Keywords: Logistics · Distribution center · Particle swarm optimization

1 Introduction

Based on the importance of distribution center's position, many researchers carry out the corresponding research and build a series of algorithm and model, and the research hotspots mainly include gravity method, P median method, mathematical programming method, heuristic approaches that can solve the NP-hard problem simulation method and some mixed method [1, 2]. But every coin has two sides, and there are some disadvantages in those methods, for example, the gravity and numerical analysis method only can solve the single location of logistics distribution center's model (actually, the linear programming method is usually used in the multiple location of logistics distribution center) and so on. In recent years, many researchers develop new intelligent methods to optimize the logistics model such as genetic algorithm (GA) and ant colony algorithm (ACO) [3, 4] to solve some complex optimization problem. PSO, GA and ACO belong to a same stochastic searching algorithm by imitating life behavior between the elements of relevance of the solution, and they also achieved

© Springer International Publishing Switzerland 2016
D.-S. Huang et al. (Eds.): ICIC 2016, Part I, LNCS 9771, pp. 651–658, 2016.
DOI: 10.1007/978-3-319-42291-6_65

good results in dispatch of logistics distribution vehicle routing [5, 6]. This paper presents the location distribution center's model algorithm based on decision matrix C, and gives solving method of multiple logistics distribution center.

2 Logistics Distribution Center Model

For a long time, the logistic is on the dispersive diversification pattern with low facilities utilization rate, unreasonable layout, redundant construction and so on. In view of this phenomenon, we take the distribution center location of multiple distributions as research object, and the total cost of logistics distribution mainly include distribution operation cost and built distribution center cost. On the model of the location of distribution center, p positions are selected from m positions (the building points of candidate distribution center) and the distribution centers with the reasonable scale are conducted by the n distribution points with some goods at the same time. Altogether, the choose point that makes the operation cost and building cost lowest can satisfy the distribution requirement as the corresponding scale to build the distribution center. Some symbols are given: n is distribution points, (x_i, y_i) is the coordinate of i point with demand a_i; (X_j, Y_j) is the j distribution center with factory point (Z_x, Z_y); $s_{ij} = \sqrt{(x_i - X_j)^2 + (y_i - Y_j)^2}$ is the distance from distribution center j to distribution point i, and $d_j = \sqrt{(Z_x - X_j)^2 + (Z_y - Y_j)^2}$ is the distance from factory to distribution center j.

So the operation cost of distribution center is:

$$f = a \times diag(s \times C^T) \times k + a \times C \times d^T \times k \tag{1}$$

where C is the decision matrix b_{ij} with value 0 and 1. If the distribution center j transports some goods for distribution point i, then $b_{ij} = 1$, otherwise $b_{ij} = 0$. C^T is the transport matrix of matrix C, $s = s_{ij} (i = 1, 2, \ldots, m, j = 1, 2, \ldots, n)$, $diag(s \times C^T)$ is the main diagonal vector with the elements of extraction matrix $s \times C^T$, k stands the rate of transportation expenses.

Supposing the land area of distribution center built is in proportion to the number logistics distribution and the location center of distribution is different, as a result the land price and cost of building are different. Here, h is the required construction cost of distribution center, D_j is the land price of distribution center's for every distribution volume. So the distribution cost of is:

$$g = a \times C \times D^T \times h \tag{2}$$

where, the requirement of distribution center's location is finding the suitable decision matrix C to make the system's cost $M = f + g$ lowest.

3 Particle Swarm Optimization

PSO was a random search algorithm based on population in 1995, and the evolution process of PSO is as follows:

$$v_{id}^{k+1} = w \times v_{id}^{k} + c_1 \times rand_1^k \times (Pbest_{id}^k - x_{id}^k) + c_2 \times rand_2^k \times (Gbest_d^k - x_{id}^k) \quad (3)$$

$$x_{id}^{k+1} = x_{id}^k + v_{id}^{k+1} \quad (4)$$

where, v_{id}^k is the d dimension of particle i speed in the k iteration; the v_{id}^k is current location of the d dimension of particle i speed in the k iteration; $i = 1, 2, \ldots, M$ is the size of population; the c_1 and c_2 are the acceleration factor, and the suitable c_1 and c_2 can accelerate constringency; w is called momentum term, where the superior w suits the large scale exploration, and the minor w suits local mining; the $rand_1$ and $rand_2$ are the random number between $[0, 1]$; $Pbest_{id}^k$ is the position that particle i that had experienced the best position; $Gbest_d^k$ is the best position that in entire population.

4 Design of Multi Logistics Distribution Center

4.1 Expression of Particle Structure

The structure space matrix of $N \times n$ with N particle and n distribution point is initially constructed, and every point in the matrix is equivalent to the center of distribution point. When all distribution points are finished in terms of problems, it means that the solution of building distribution center has affirmed. In order to express and calculate expediently, firstly, numbering the distribution center in sequence, each element represents a particle of the elective center. So a particle that is in the search space of position matrix lives in the i line, called x_i (position of particle). The corresponding relations of the point and position of particle are the following,

$$\begin{array}{cccccccc} \text{Distribution point}: & 1 & 2 & \ldots & i & \ldots & n \\ x_i: & x_{i1}, & x_{i2}, & \ldots, & x_{ii}, & \ldots, & x_{in} \end{array}$$

where $x_{i1}, x_{i2}, \ldots, x_{ii}, \ldots, x_{in} \in \{1, 2, \ldots, m\}$ is the position of particle, and for the arbitrary two distribution point $j, k \in \{1, 2, \ldots, n\}$, there is $x_{ij} = x_{ik}$. At the same time, the distribution point j, k are chose in the same candidate distribution center.

The biggest advantage of the representation is that every distribution can get the service of building and limit every particle to choose of distribution center for all distribution points at the same time, which makes the feasibility degree of the solution achieve great improvement.

4.2   The Structure of Decision Matrix and Preferences of Particle Swarm

(1) The Structure of Decision Matrix

The other key factor how to realize solution of model with PSO is the structure of decision matrix C, and here, the building of decision matrix is confirmed by the position of particle effectively. If there are m distribution point and m candidate distribution center, then the vector of particle i in searching space is $x_i = (x_{i1}, x_{i2}, \ldots, x_{ij}, \ldots, x_{in})$, and $x_{i1}, x_{i2}, \ldots, x_{ij}, \ldots, \in \{1, 2, \ldots, m\}$. Additionally $x_{ij} = k$ stands that distribution center k transports some goods for distribution j, so we can structure the decision matrix:

$$C = b_{jk} = \begin{cases} 1 \\ 0 \end{cases}.$$ If the distribution center k transports some goods for distribution

point j, then $b_{jk} = 1$; otherwise, $b_{jk} = 0$.

(2) Parameter Selection

Because every dimension number in the particle position is the integer between 1 and m, the number of candidate center m usually is the minor integer within 10, and the parameter w shouldn't choose too large, here $w = 0.25$. The individual evolution is learning process from swarm and its own experience, so for learning factor c_1 and c_2 should choose the large number. In view of the simulation experiment with 10 distribution point and 4 distribution center's situation, c_1 and c_2 are 4 and 2 respectively.

4.3   The Process of Algorithm Implementation

As described before, the algorithm of PSO algorithm is continuous optimization algorithm, so we need some strategies to satisfy PSO to implement the multi logistics distribution center location that belongs to the problem of integer programming.

Step 1. Initializing the particle swarm

(1) Each dimension of a particle randomly chooses a integer between $1 \sim m$ (number of distribution center) to construct a matrix with $N \times n$ dimension, where N is the number of particle, and n is the number of distribution point.

(2) The velocity matrix of each position is the random number of normal distribution with $N \times n$ elements.

(3) Using fitting function value (the function is described by the requirement of distribution center system) evaluates the sufficiency of all particles.

(4) Taking the particle position of initializing particle as the individual historical optimal solution, and the initial fitting function value as the individual historical optimum p. Next, finding the globally optimal solution pg and the global optimum $gbest$ in the swarm.

Step 2. Carrying out the following steps repeatedly until the termination condition is satisfied or the maximum iteration is reached.

(1) Calculating the speed for each particle with formula (3);
(2) Updating the position with formula (4), and rounding up the integer of element of particle position. If the position exceeds the search scope, the boundary is as the corresponding value;
(3) Using fitting function value to evaluate the sufficiency of all particles;
(4) If the current fitting function value of a particle is better than that of its historical value, the current particle position will be the individual optimal historical solution;
(5) Finding the globally optimal solution and the globally optimum.

5 Simulation Analyses of Experimental Results

5.1 The Example of Simulation Analysis

In order to prove the effectiveness of the proposed method, the simulation experiments were designed in the platforms of MATLAB software platform. The corresponding parameter setting is the following: the logistics distribution's rectangular area is between (0, 0) to (100, 100), and the distribution points is 10.

Table 1 gives the corresponding coordinate, and Table 2 gives the coordinate and land price for four candidate distribution centers in the rectangular area. In order to calculate convenience, and don't cause mistake situation, the rate of transportation cast is 1 and the coordinate of factory is setting (48, 55). In Fig. 1, we can see that the distribution point with the candidate distribution center 3 is empty which means that there is no any supply distribution point. Additionally, it also can conclude that the candidate distribution center 3 is redundant. From Fig. 2, we can see that the speed of PSO convergence is fast, which prove the rationality of the model. Additionally, the proposed thought not only can confirm the position of distribution but also compute the quantity and the scale of distribution.

Table 1. The situation of coordinate of distribution point and requirement

Distribution point number	1	2	3	4	5	6	7	8	9	10
Abscissa	24	21	71	93	43	88	40	30	56	15
Ordinate	81	22	61	52	65	46	8	69	21	57
Distribution quantity	1	0.5	0.6	2.0	1.2	1.4	1.1	2.0	2.5	1.1

Table 2. The coordinate of candidate distribution center and land price

Distribution point number	1	2	3	4
Abscissa	22	89	73	44
Ordinate	73	45	89	17
Land price	1.5	1.8	0.3	1

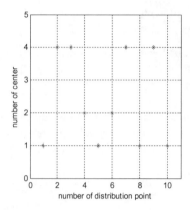

Fig. 1. Relationship of particle choice distribution center

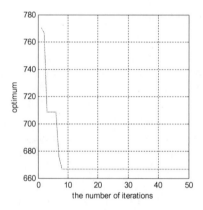

Fig. 2. Relationship of iterations and the global optimum

In Table 3, we see that the distribution quantity mainly decide the building scale of distribution, so the building scale of every building distribution center only needs calculate the sum of distribution point's quantity according to the optimization result. If the distribution center selection is in (1, 5, 8, 10), the corresponding building should satisfy the scale of demand quantity in distribution point 1, 5, 8, 10, else if there is no offered distribution point in the distribution center 3, the distribution center doesn't be built. In a word, PSO not only overcomes the limitation of the traditional algorithm but also can solve the problem of the large-scale multiple logistics distribution center location.

Table 3. Scheme of distribution center location

Distribution number	1	2	3	4
Distribution offered	1,5,8,10	3,4,6		2,7,9
Distribution scale	5.3	4.0	0.0	4.1

5.2 Effect of Particle Number on Algorithm

In this section, we investigate the effect the change of particle number based on the problem of distribution point randomly. Table 4 gives the results of the simulation problem with 4 candidate distribution centers and 10 distribution points, here the different particle population is adopted in 50 times. The experiment results show that the algorithm can avoid local optimum when the particle population is $15 \sim 20$ with the 90 % rate of search success.

Table 4. The correlation between the particle number and optimization results

Distribution point number	Optimum	Particle population	Times of reaching optimal path	Average time of reaching optimal path	Average iteration of reaching optimal path	Average optimum of result of 50 experiment
10	6666.606	200	24	0.936	9.25	695.959
		400	35	1.558	8.80	678.526
		600	43	2.425	8.05	672.925
		800	46	3.057	8.02	671.315
		1000	47	3.543	8.00	668.041

6 Conclusion

The distribution center lives on an important pivot position in the logistics system, and the choice of logistics distribution center is the key for the logistics system.

In order to effectively deal with the model of multi logistics distribution center location, this paper proposed the method based on PSO. To test the effectiveness of the proposed algorithm, the simulation experiments are conducting for the site selection of logistics distribution center with 4 candidate centers and 10 distribution points. Simulation results show that PSO can get the best solution of distribution center in 90 % success rate of the best solution and the average research time of the approximate 3.5 s.

Acknowledgments. This work is supported by the National Natural Science Foundation of China (Grants nos. 71461027, 71471158). Guizhou science and technology cooperation plan (Qian Ke He LH zi [2016]7028, Qian Ke He LH zi [2015]7050, [2015]7005, Qian Ke He J zi LKZS [2014]30). Science and technology talent training object of Guizhou province outstanding youth (Qian ke he ren zi [2015]06). Guizhou province natural science foundation in China (Qian Jiao He KY [2014]295); 2013, 2014 and 2015 Zunyi 15851 talents elite project funding; Zhunyi innovative talent team (Zunyi KH (2015) 38); Project of teaching quality and teaching reform of higher education in Guizhou Province (Qian Jiao gaofa [2013]446, [2015]337), College students' innovative entrepreneurial training plan (201410664004, 201510664016).

References

1. Yuan, Y.X.: A scaled central path for linear programming. J. Comput. Math. **19**(1), 35–40 (2001)
2. Konstantinos, G.Z., Konstantinos, N.A.: A heuristic algorithm for solving hazardous materials distribution problems. Eur. J. Oper. Res. **152**, 507–519 (2004)
3. Qian, J., Pang, X.H., Wu, Z.M.: An improved genetic algorithm for allocation optimization of distribution centers. J. Shanghai Jiaotong Uni. (Sci.) **E9**(4), 73–76 (2004)
4. Dorigo, M., Gambardella, L.M.: Ant colony system: a cooperative learning approach to the traveling salesman problem. IEEE Trans. Evol. Comput. **1**(1), 53–66 (1997)
5. Gao, Z.H.: Application of particle swarm optimization to continuous location of distribution center. Comput. Appl. **28**(9), 2401–2403 (2008)
6. Wang, T.J., Wu, Y.C.: Study on optimization of logistics distribution route based on chaotic PSO. Comput. Eng. Appl. **47**(29), 218–221 (2011)

Effects of Simulated Annealing Strategy on Swarm Intelligence Algorithm

Yanmin Liu[1(✉)], Chengqi Li[2], Qingyu Zeng[1], Zhuanzou Zhang[1], Rui Liu[1], and Tao Huang[1]

[1] School of Mathematics and Computer Science,
Zunyi Normal College, Zunyi 563002, China
yanmin7813@gmail.com
[2] School of Mathematics and Science, Wuyi University,
Jiangmen 529020, Guangzhou, China

Abstract. Swarm intelligence algorithm (SI) is a kind of stochastic search algorithm based on swarm. Similar to other evolutionary algorithm, when solving the complicated multimodal problem using SI, it is easy to have premature convergence. So, to promote the optimization of swarm intelligence algorithm, the typical algorithm (Particle swarm optimizer) of swarm intelligence algorithm is selected to explore some strategies how to improve the performance. In this paper, we explore the follow research: firstly, the mutation operation is introduced to produce new learn example for each individual in itself evolution process; secondly, in the view of the idea of simulated annealing, the range strategy of fitness of each individual is proposed; finally, to make best use of each individual information, the comprehensive learning strategy is adopted to improve each individual evolution mechanism.

Keywords: Particle swarm optimizer · Simulated annealing · Strategy

1 Introduction

Swarm intelligence algorithm is a kind of stochastic search algorithm based on swarm, but similar to other optimization algorithms, it is prone to premature phenomenon. So, to promote the optimization of swarm intelligence algorithm, we chose the typical algorithm PSO of swarm intelligence algorithm to explore some strategies how to improve the performance. Firstly, the mutation operation is introduced to produce new learn example for each individual in itself evolution process. Secondly, in the view of the idea of simulated annealing, the range strategy of fitness of each individual is proposed. Finally, to make best use of individual information, the comprehensive learning strategy is adopted to improve each individual evolution mechanism.

Particle swarm optimizer (PSO) [1] is a kind of evolutionary algorithm based on population, and its evolution is a cooperation and competition among the individuals by producing new individual. Because of simple operation of PSO algorithm and the advantages of fast convergence speed, it has been widely applied to engineering, science and technology in the process of optimization. Thought of PSO is similar to other evolution algorithm, like genetic algorithm (GA), ant colony optimization (ACO) and

D.-S. Huang et al. (Eds.): ICIC 2016, Part I, LNCS 9771, pp. 659–666, 2016.
DOI: 10.1007/978-3-319-42291-6_66

so on, so PSO algorithm is one of the biggest flaw is easy to fall into local optimal solution. In order to conquer this defect, scholars at home and abroad put forward many improved algorithms [2–5] by various strategies. On the basis of previous studies, this paper puts forward a kind of improved Particle swarm algorithm with simulated annealing strategy (PSOSA) where the mutation based on simulated annealing thought is introduced and a comprehensive learning strategy is adopted to improve the efficiency of the algorithm.

In PSO, each particle velocity and position updating equation of GPSO and LPSO can be described as follows,

$$\overline{v_i}(t+1) = \overline{v_i}(t) + \varphi_1 r_1 (\overline{p_i}(t) - \overline{x_i}(t)) + \varphi_2 r_2 (\overline{p_g}(t) - \overline{x_i}(t))$$
$$\overline{x_i}(t+1) = \overline{v_i}(t+1) + \overline{x_i}(t) \tag{1}$$

$$\overline{v_i}(t+1) = \overline{v_i}(t) + \varphi_1 r_1 (\overline{p_i}(t) - \overline{x_i}(t)) + \varphi_2 r_2 (\overline{p_{neighbor_i}}(t) - \overline{x_i}(t))$$
$$\overline{x_i}(t+1) = \overline{v_i}(t+1) + \overline{x_i}(t) \tag{2}$$

where at the iteration time t, $\overline{p_i}(t)$ is the experienced best position of particle i, called "cognitive part" of the particles. $\overline{p_g}(t)$ is the best position in the whole swarm, called " social part" of the particles. The nature of these two kinds of algorithm is the same, just when the particle velocity is updated, and the choice of social learning samples for each individual has difference.

2 Simulated Annealing Strategy

Simulated annealing algorithm [6] is based on simulated annealing process of high temperature physical object to solve optimization problems. The basic idea of the algorithm is that along with the temperature falling, the objection kicks at a certain probability characteristics randomly in the solution space to find the global optimal solution of the objective function at a given initial temperature. The physics thought that is applied in optimization field can effectively solve local search default of evolution algorithm. The specific steps of the algorithm described as follows:

Step 1 Initializing the annealing temperature T_k ($k = 0$), and producing random initial solution x_0.
Step 2 Under the temperature T_k, producing new feasible solution x' in the field of x.
Step 3 Calculating the difference ∇f of $f(x')$ and $f(x)$ of the new fitness function value and.
Step 4 If $\nabla f < 0$, accepting x', namely $x = x'$; otherwise accepting x' in probability $exp(-\nabla f / T_K)$.
Step 5 Annealing process. This paper uses the index of cooling strategy, namely $T_{k+1} = \alpha \cdot T_k$, where α is the temperature decrement that is value range in [0.8, 0.98] commonly. α is closer to 1, the slower the temperature reduction.
Step 6 If the convergence criterion is satisfied, the annealing process is over; Otherwise, turn to step 2.
Step 7 Outputting result.

Annealing temperature T controls the solving process and accepts inferior solution by its probability $\exp(-\Delta f/T_k)$. Based on this thought, the simulation process can ensure the algorithm to jump out of local extremism points.

3 Particle Swarm Optimizer with Simulated Annealing Strategy

3.1 Mutation Operation Based Simulated Annealing

According to the particle velocity updating formulas (1) and (2), if the history optimal position of particle (*pbest*) and the global optimal position (*gbest*), long stay at a particular location, the optimal position can be regarded as trapping in local optimal solution. Under such conditions, if the swarm has no corresponding measures to change the status quo, the algorithm will appear premature convergence [7]. In view of this, scholars have proposed many improvement strategies, in the literature [8], the author discusses the mutation operator and simulated its role, the experiments shows that the performances have certain effect to improve algorithm. Fortunately inspired by this thought, this paper uses the mutation operation to improve the population's ability to jump out of local optimal solution.

But different from the existing mutation operation method, we proposed a mutation operator with simulated annealing thought which combine physical object merit with PSO characteristic to improve evolutionary algorithm performance. In the mutation operator of the swarm performance, the history optimal position of particle (*pbest*) and the global optimal position (*gbest*) can produce new learning examples for swarm based on simulated annealing process. Here, in order to make full use of the most information of the particles, we rule that *gbest* and *pbest* of the swarm and each particle have no improvement for the continuous three generations, where the Eqs. (3) and (4) give the mutation process.

It is worth noting that the new particles produced by mutation operator are likely to be inferior to the original particles, so we borrow the idea of simulated annealing that allows new objective function change within permissible range to ensure the effective utilization of the particle information.

$$pbest_i = pbest_i \times (1+r), i = 1, 2, \ldots, N \tag{3}$$

$$gbest = gbest \times (1+r) \tag{4}$$

where, r is the normal distribution random number with average 0 and the variance 1; N is the swarm size; $pbest_i$ corresponds to the history optimal position of particle i.

3.2 Particle Swarm Optimizer with Simulated Annealing Strategy (PSOSA)

In updating process of particle velocity and position, if each dimension of particle learns from the corresponding dimension of the same particle, it will cause the reduction of the swarm diversity and make population into local optimal solution. Therefore, here the literature [4] proposes generalized learning strategies, to update the particle's position and speed. Equation (5) gives velocity and position updating formula of the particle.

$$\overline{v}_i(t+1) = w \cdot \overline{v}_i(t) + \varphi_1 r_1 \left(\overline{p_{f_i(d)}}(t) - \overline{x}_i(t) \right) + \varphi_2 r_2 \left(\overline{p_g}(t) - \overline{x}_i(t) \right)$$
$$\overline{x}_i(t+1) = \overline{v}_i(t+1) + \overline{x}_i(t) \tag{5}$$

$$w = w_{\max} - \frac{w_{\max} - w_{\min}}{T} \times t \tag{6}$$

where, w is called inertia weight, and here we adopted dynamical iterative changes according to the Eq. (6); w_{max} and w_{max} represent the maximum and minimum value of inertia weight respectively, and $w_{max} = 0.7$, $w_{min} = 0.2$ in this paper; $\overline{p_{f_i(d)}}(t)$ is the history optimal position of the corresponding any particle at the iteration t, and due to the limitation of space, please refer to the literature [4] The other symbols is consistent with the Eq. (1). Algorithm 1 gives PSOSA pseudo code.

Algorithm 1

Begin

Initialize particles' position, velocity, $pbest_i$, $gbest$, N=0,

$V_{max}=0.25(X_{max}\text{-} X_{min})$

Computer the diversity

while (fitcount <Max_FES) && (k<iteration)

    For each particle (i=1:ps)

        Updating particles' velocity and position in terms of Eq.(5).

        Update particles' *pbest* using particle comparison criteria.

        If *N==3*

        Mutation operation in terms of SA idea.

        N=0.

        End

    End for

    Update *gbest* using particle comparison criteria.

    Output results

End Begin

---

4 Role Analysis of Simulated Annealing Strategy

In order to test the effects of the role of simulated annealing strategy on Swarm intelligence algorithm, PSO is adopted to conduct search mission. In simulation part, Sphere, Rosenbrock, Ackley, Griekwanks, Griewank functions are selected to conduct the simulation experiments, whose characteristics are shown in [3] respectively. Here, we want to test the population diversity to analyze the simulated annealing, and usually if the population diversity is higher than other strategies, it indicates that the method is effective. As three benchmark functions are selected, the normalized value is computed by Eq. (7).

Figure 1 gives the simulation figure of the population diversity and normalized value alone with fitness evaluation, where normalized value is the synthetical value by the normalized function value of Rosenbrock, Ackley and Griewank. Here we can see that the simulated annealing strategy is more effective than LPSO and GPSO, so it can improve the swarm diversity and increase the ability to escape from local optimal for the whole swarm. In order to test the strategy of simulated annealing strategy, Fig. 2 give the box statistical figure on different test function, where 1 is the algorithm without simulated annealing strategy, and 2 is the algorithm with the strategy.

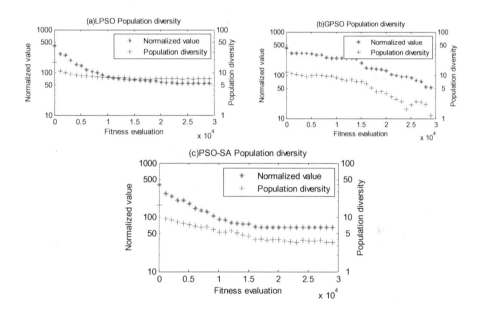

Fig. 1. Simulation figure of the population diversity and normalized value

$$X = \frac{x_{ij} - \mu_{ij}}{\sigma_{ij}} \tag{7}$$

Where, X is the normalized value; x_{ij} is the i benchmark function value of the j algorithm; μ_{ij} and σ_{ij} are the mean and standard deviation respectively.

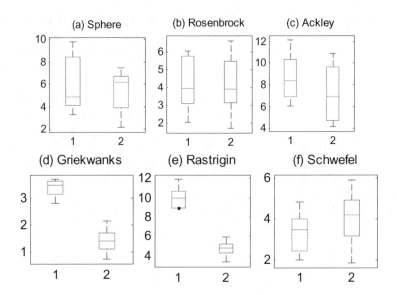

Fig. 2. The box statistical figure on different test function

Additionally LPSO and GPSO algorithm are selected to compare with the proposed algorithm, where LPSO is local version population topology and LPSO is global version population topology [4]. The parameters of the algorithm experimental setting are as follows: each algorithm contains 30 particles, 30 independent running and 3×10^4 function evaluation for each benchmark function. Table 1 shows the different algorithms average and 95 % confidence interval in 30 independent running where the best operation results are marked by boldface, and we can be seen that the best operation results of PSOSA algorithm for 30 independent running on three benchmark functions are better than other algorithms. Therefore the mutation based on simulated annealing thought and comprehensive learning strategies are effective for improving PSO performance.

Table 1. The mean and the confidence interval after 3×10^4 function evaluations

Algorithm	Rosenbrock	Ackley	Griewank
LPSO	2.42e+001 ± 1.12e+001	1.89e+000 ± 1.13e+000	4.66e−002 ± 3.76e−002
GPSO	2.41e+001 ± 1.74e+001	1.38e+001 ± 2.8168e+001	3.28e−001 ± 1.45e−001
PSOSA	**1.43e−001 ± 2.18e−002**	**1.11e−006 ± 2.07e−006**	**6.61e−004 ± 2.73e−004**

In order to improve the performance of swarm intelligence algorithm, the mutation operator and comprehensive learning strategies are also introduced in the proposed algorithm, and in order to verify whether the strategy increases the computational complexity, we use the simulation platform of ThinkPad SL400 Matlab7.0 (tic and toc command) to test the complexity. Table 2 shows the running time of various algorithms, where B is the best perform time in independent running 30 times running time, W represents the worst run time and M is the average running time. From the above results, we can be seen that the PSOSA computational time is in the same order of magnitude with other algorithms, therefore the introduction of strategy did not increase the computational complexity.

Table 2. Running time of the algorithm (Unit: second)

Algorithm		Rosenbrock	Ackley	Griewank
LPSO	B	13.4	21.8	22.1
	W	17.8	25.4	29.4
	M	15.8	23.7	25.7
GPSO	B	13.5	19.4	20.5
	W	16.3	22.6	27.8
	M	14.9	21.7	23.6
PSOSA	B	28.3	39.9	42.1
	W	32.4	44.2	48.8
	M	31.6	43.1	47.2

5   Conclusion

In order to conquer the premature convergence phenomenon for swarm intelligence algorithm, we explore the effects of the mutation operation, simulated annealing and the comprehensive learning strategy on swarm intelligence algorithm. Through the simulation experiments of the benchmark function, the results show that the proposed strategies is the effective method that improve the population diversity, and the corresponding PSOSA algorithm is an effective improvement for the basic PSO algorithm.

Acknowledgments. This work is supported by the National Natural Science Foundation of China (Grants nos. 71461027, 71471158). Science and technology talent training object of Guizhou province outstanding youth (Qian ke he ren zi [2015] 06). Guizhou province natural science foundation in China (Qian Jiao He KY [2014] 295); 2013, 2014 and 2015 Zunyi 15851 talents elite project funding; Zhunyi innovative talent team (Zunyi KH (2015) 38); Project of teaching quality and teaching reform of higher education in Guizhou province (Qian Jiao gaofa [2013] 446, [2015] 337), College students' innovative entrepreneurial training plan (201410664004, 201510664016); Guizhou science and technology cooperation plan (Qian Ke He LH zi [2015] 7050, [2015] 7005, [2016] 7028, Qian Ke He J zi LKZS [2014] 30); Zunyi Normal College Research Funded Project (2012 BSJJ19).

References

1. Kennedy, J., Eberhart, R.C.: Particle swarm optimization. In: Proceedings of IEEE International Conference on Neural Networks. Piscataway, USA, pp. 1942–1948 (1995)
2. Luo, C.Y., Chen, M.Y.: Adaptive particle swarm optimization algorithm II. Control Decis. **24**(6), 1135–1144 (2009)
3. Parsopoulos, E., Vrahatis, M.N.: Parameter selection and adaptation in unified particle swarm optimization. Math. Comput. Model. **46**(2), 198–213 (2007)
4. Liang, J.J., Qin, A.K.: Comprehensive learning particle swarm optimizer for global optimization of multimodal functions. IEEE Trans. Evol. Comput. **10**(3), 281–295 (2006)
5. Chen, M.R., Li, X.: A novel particle swarm optimizer hybridized with extremal optimization. Appl. Soft Comput. **10**(2), 367–373 (2010)
6. Meteopolis, N., Rosenbluth, A.W.: Equation of state calculations by fast computing machines. J. Chem. Phys. **21**(6), 1087–1092 (1953)
7. Ratnaweera, A., Halgamuge, S.: Self-organizing hierarchical particle swarm optimizer with time-varying acceleration coefficients. IEEE Trans. Evol. Comput. **8**(3), 240–255 (2004)
8. Andrews, P.S.: An investigation into mutation operators for particle swarm optimization. In: IEEE Congress on Evolutionary Computation, Vancouver, Canada, pp. 1044–1051 (2006)

A Complex Encoding Flower Pollination Algorithm for Global Numerical Optimization

Chengyan Zhao[1] and Yongquan Zhou[1,2(✉)]

[1] College of Information Science and Engineering,
Guangxi University for Nationalities, Nanning 530006, China
`yongquanzhou@126.com`
[2] Key Laboratory of Guangxi High Schools Complex System
and Computational Intelligence, Nanning 530006, China

Abstract. Flower pollination algorithm (FPA) is proposed to cause the attention of researchers. And this paper presents a new flower pollination algorithm with complex-valued encoding (CFPA) in which the update of populations will be divided into two parts, the real part and the imaginary part. This approach can expand the amount of information contained in the individual gene and enhances the diversity of individual population. Numerical experiments have been carried out based on the comparison with particle swarm optimization (PSO) and original flower pollination algorithm (FPA).

Keywords: Complex-valued encoding · Flower pollination · Diversity · Numerical experiments

1 Introduction

The flower pollination algorithm is a new metaheuristic algorithm which is proposed by Xin-She Yang in 2012 [1, 24]. It inspired by the cross pollination and self pollination of the flowers. According to bionics principle, a pollen gamete represents a point of search space N. Self-pollination and cross-pollination of a pollen gamete can be considered as a process of search and optimization. The corresponding position of a pollen gamete can be understood as the objective function of the solution problem. And the evolution process of the individual is regarded as the update of optimal solution. In addition, in the algorithm, the mutual transformation process of the global search and local search are dynamic controlled by a switch probability P. So the balance problem of global search and local search is solved in well. In the other hand, the algorithm use a Levy fight, it makes the algorithm with better capability of global optimization. Since FPA was published by Yang, there are many scholars are engaged in the research of the algorithm on international, and it has been applied in various fields [14–23].

But in this paper, the proposed method we present is the flower pollination algorithm based on complex diploid encoding (CFPA) [3, 4] which the variable can be represented by two parameters (i.e., the real part and the imaginary part) and the two parts can be updated in parallel. The modules and angles of their corresponding complex number determined the independent variables of their objective function [12, 13]. Thus, by this

© Springer International Publishing Switzerland 2016
D.-S. Huang et al. (Eds.): ICIC 2016, Part I, LNCS 9771, pp. 667–678, 2016.
DOI: 10.1007/978-3-319-42291-6_67

means, the new algorithm can enriched the diversity of the population and improve the performance of the algorithm, the dimensions for denoting can be expanded too.

2   Flower Pollination Algorithm

Flower pollination algorithm is an optimization algorithm which simulates the flower pollination behavior, we can idealize the characteristics of pollination process as the following rules [11]:

(1) Biotic and cross-pollination can be considered as global pollination process with pollen-carrying pollinators move in a way that obeys Levy flights.
(2) Abiotic and self-pollination can be regarded as local pollination.
(3) Flower constancy (develops from the pollination tend to visit exclusive certain flower species while by passing other flower species) is considered to be a reproduction probability that is proportional to the similarity of two flower involved.
(4) A switch probability P controls the transformation between the global pollination and the local pollination.

For simplicity, we assume that each plant only has one flower, and each flower only produce one pollen gamete. And we supposed the dimension of search space is N, the position of the pollen i at the time t is X_i^t. So the update of the position X_i^{t+1} at the time $t+1$ can be represented mathematically as following formulas (rule1, global pollination):

$$X_i^{t+1} = X_i^t + L\left(X_i^t - g_*\right) \tag{1}$$

where L is the strength of the pollination, which essentially can be considered as the step size. Since the characteristic of insects may move over a long distance with various distance steps, it can be mimicked by the Lěvy flight [2, 10].

$$L \sim \frac{\lambda \Gamma(\lambda) \sin(\pi\lambda/2)}{\pi} \frac{1}{s^{1+\lambda}} \quad s > > s_0 > 0 \tag{2}$$

where $\Gamma(\lambda)$ is the standard gamma function.
And the rule2 and rule3 can be represented as (local pollination).

$$X_i^{t+1} = X_i^t + \varepsilon\left(X_j^t - X_k^t\right) \quad \varepsilon \in [0, 1] \tag{3}$$

where X_j^t and X_k^t are the pollens from the different flowers of the same species, which essentially mimic the flower constancy [7] in a limited neighborhood.

Finally there is a switch probability P (according to rule 4) to control the switch between global pollination and local pollination. And on the basis of our simulations when $P = 0.8$ works more better.

The implementation steps of flower pollination algorithm as follows:

1. Initialize the basic parameters: population size N, best solution g_*, a switch probability $P \in [0, 1]$;
2. While (t < MaxGeneration)
   For $i = 1 : N$.

   If $rand < P$, update the current position (global pollination) according to (1);

3. Else
4. Draw from a uniform distribution in $[0, 1]$. Randomly choose j and k among all the solutions. The current position of pollens (local pollination) can be updated via (2);
5. Evaluate new solutions, if new solutions are better, update them in the population;
6. Find the current best solution.

3   Complex-Valued Flower Pollination Algorithm (CFPA)

3.1   The Method of the Complex-Valued Encoding Method

In nature, the chromosome of complex biological-tissue use double-chain structure and multiple-chain structure. For amphiploid, in the process of reproduction, the parents body provide respectively a daughter chromosome with one chromosome. And because of the 2D features of complex-valued encoding, the amphiploid can be represent atived by it in this picture. Specifically, we can use a complex-valued to express a pair of alleles of the chromosome which is that the real part and the imaginary part of the complex-valued can be considered the real gene and the imaginary gene. And in CFPA, each individual has two equal length gene chains and each gene has two alleles, the two alleles correspond to the real part and the imaginary part. For example, a question with M variables:

$$z_k = x_k + iy_k, k = 1, 2, \ldots M$$

recorded as:

$$((x_1, x_2, \ldots, x_M), (y_1, y_2, \ldots, y_M))$$

and the double-chain structure of the chromosome can be noted as:

$$r = (x_1, x_2, \ldots, x_M), \ d = (y_1, y_2, \ldots, y_M)$$

the r is real gene cluster, d is the imaginary gene cluster.

3.1.1   Initialize Population
Base on the definition interval of the problem $[A_k, B_k], k = 1, 2, \ldots, 2M'$, generate $2M$. Complex modulus and $2M$ phase angle randomly.

$$\rho_k \in \left[0, \frac{B_k - A_k}{2}\right], k = 1, 2, \ldots, 2M \tag{4}$$

$$\theta \in [-2\pi, 2\pi], k = 1, 2, \ldots, 2M \tag{5}$$

$$X_{Rk} + iX_{Ik} = \rho_k(\cos\theta_k + i\sin\theta_k), k = 1, 2, \ldots, 2M \tag{6}$$

So the real gene cluster and the imaginary gene cluster is:

$$r = (x_1, x_2, \ldots, x_{2M}), \ d = (y_1, y_2, \ldots, y_{2M})$$

The population can be shown in:

$$((x_1, x_2, \ldots, x_{2M}), (y_1, y_2, \ldots, y_{2M}))$$

Thus we get $2M$ real parts and $2M$ imaginary parts, and the update as following way.

3.1.2   The Updating Method of CFPA

Update the real parts:

$$X_{Rk}^{t+1} = X_{Rk}^t + L\left(X_{Rk}^t - g_{*1}\right) \tag{7}$$

$$X_{Rk}^{t+1} = X_{Rk}^t + \varepsilon\left(X_{Rj}^t - X_{Rp}^t\right) \tag{8}$$

Update the imaginary parts:

$$X_{Ik}^{t+1} = X_{Ik}^t + L\left(X_{Rk}^t - g_{*2}\right) \tag{9}$$

$$X_{Ik}^{t+1} = X_{Ik}^t + \varepsilon\left(X_{Ij}^t - X_{Ip}^t\right) \tag{10}$$

where the X_R^t, X_I^t are the pollen current position of real and imaginary, g_{*1}, g_{*2} are the current best solution of the real parts and the imaginary parts.

3.1.3   The Calculation Method of Fitness Value
To calculation the fitness value, the complex number need to converted into the real number first. So the value of the real number can be determined by complex modulus and the sign id determined by the angle:

$$\rho_k = \sqrt{X_{Rk}^2 + X_{Ik}^2}, k = 1, 2, \ldots, M \tag{11}$$

$$X_k = \rho_k \mathrm{sgn}\left(\sin\left(\frac{X_{Ik}}{\rho_k}\right)\right) + \frac{B_k + A_k}{2}, k = 1, 2, \ldots, M \tag{12}$$

where X_k represents the converted real variables.

3.2 CFPA Algorithm

Based on the analysis, the steps of the CFPA can be described as following:

Begin:

Step1: Initialize the flower pollination, $\rho_k \in \left[0, \dfrac{B_k - A_k}{2}\right], \theta_k \in [-2\pi, 2\pi]$;

Step 2: According to Eq.(6), get $2M$ complex numbers;

Step 3: Calculate the fitness value from Eqs. (11),(12);

Step 4: Find the global best solution;

Step 5: For $t = 1 : N\_iter$

      For $i = 1 : n$

        If $rand > p$;

Step 6: (global pollination) $L = Levy(d)$

      update the real parts according (7) ,update the imaginary parts according (9);

Step 7: Else

      (local pollination) $epsilon = rand, JK = randperm(n)$

      update the real parts according (8) ,update the imaginary parts according (10);

Step 8: If fitness improves (better solutions found), update then;

Step 9: Convert to real variables according to (11),(12) ;

Step 10: Calculate fitness X_k;

Step 11: Get the best solution;

Step 12: End While

Step 13: End .

4 Simulation Experiments and Results Analysis

4.1 Test Functions

In this paper, we select ten standard benchmark [8, 9] function to verify the effectiveness of the proposed algorithm. Firstly, there is a point that should be understood is for all of the test function, each of them can have varied dimensions. However, in this paper, the ten test functions which we use tend to in higher dimensions. Because higher-dimensions problems are more challenging, and a new algorithm should be tested against a wide range of functions in terms of function properties and dimensions. Like Table 1.

Table 1. Test functions

No.	Name	Functions	Dim	Scope
F1	Sphere	$f(x) = \sum_{i=1}^{D} x_i^2$	128	$[-100, 100]$
F2	Rosenbrock	$f(x) = \sum_{i=1}^{D} \left[100(x_{i+1} - x_i^2)^2 + (1 - x_i^2)^2 \right]$	30	$[-2.048, 2.048]$
F3	Drop Wave	$f_x = -\dfrac{1 + \cos\left(12\sqrt{x_1^2 + x_2^2}\right)}{\frac{1}{2}(x_1^2 + x_2^2) + 2}$	128	$[-5.12, 5.12]$
F4	Quartic	$f(x) = \sum_{i=1}^{D} i \cdot x^4 + rand(0, 1)$	30	$[-1.28, 1.28]$
F5	Yang	$f(x) = \sum_{i=1}^{D} \varepsilon_i |x_i - 1/i|, \varepsilon \in U[0,1]$	30	$[-5, 5]$
F6	Rastrigin	$f(x) = \sum_{i=1}^{D} \left[x_i^2 - 10\cos 2\pi x_i + 10 \right]$	30	$[-5.12, 5.12]$
F7	Ackley	$f(x) = -20\exp\left(-0.2\sqrt{\frac{1}{D}\sum_{i=1}^{D} x_i^2}\right) - \exp\left(\frac{1}{D}\sum_{i=1}^{D}\cos 2\pi x_i\right) + 20 + e$	30	$[-32.768, 32.768]$
F8	Zach-arov	$f(x) = \sum_{i=1}^{D} x_i^2 + \left(\frac{1}{2}\sum_{i=1}^{D} i x_i\right)^2 + \left(\frac{1}{2}\sum_{i=1}^{D} i x_i\right)^4$	30	$[-10, 10]$
F9	DJongs_3	$f(x) = \sum_{i=1}^{D} |x_i|^{i+1}$	30	$[-1, 1]$
F10	Scha-ffer_f6	$f(x) = \dfrac{\sin^2\sqrt{(x_1^2 + x_2^2)} - 0.5}{\left[1 + 0.001(x_1^2 + x_2^2)\right]^2} - 0.5$	30	$[-100, 100]$

4.2   Comparison of Experiment Results

For the above simulation results, there are three comparison algorithms. They are particle swarm optimization (PSO) [5, 6], flower pollination algorithm(FPA) [1], and the new algorithm complex-valued flower pollination algorithm (CFPA). For each of them, we have carried out 30 independent runs with a population size N = 30. In PSO the learning factor is $c_1 = c_2 = 1.4962$ with the linear decreasing inertia weight $\omega_{max} = 0.9$, $\omega_{min} = 0.4$, for FPA and CFPA, probability switch $P = 0.8$. In addition, for CFPA, The range of complex modulus and phase angle is $[A_k, B_k]$, where $\rho_k \in \left[0, \frac{B_k - A_k}{2}\right]$, $\theta_k \in [-2\pi, 2\pi]$ is the range of variables.

Table 2. Simulation results

Benchmark functions	Method	Result			
		Best	Mean	Worst	Std
Sphere	PSO	12.1378	52.2827	122.3189	26.2592
	FPA	128.3763	202.3881	331.1395	48.8051
	CFPA	*0*	*0*	*0*	*0*
Rosenbrock	PSO	34.0991	133.7277	1.2224e + 03	215.2269
	FPA	204.7071	381.7045	620.3883	109.0036
	CFPA	*28.8583*	*28.9173*	*28.9967*	*0.0310*
Quartic	PSO	0.0124	0.0638	0.1401	0.0382
	FPA	0.1483	0.4464	0.9638	0.2059
	CFPA	*2.5665e−07*	*1.4696e−05*	*6.3402e−05*	*1.3566e−05*
Yang	PSO	3.4901	7.6313	16.3483	3.1982
	FPA	6.8320	11.2754	16.4241	2.2259
	CFPA	*0.6558*	*0.7566*	*0.8494*	*0.0563*
Rastrigin	PSO	67.8764	126.5305	214.0366	33.1873
	FPA	234.4679	311.2577	414.1017	34.4972
	CFPA	*0*	*0*	*0*	*0*
Ackley	PSO	4.8767	7.0103	9.4318	1.4414
	FPA	11.6444	13.8136	16.8938	1.4190
	CFPA	*8.8818e−16*	*8.8818e−16*	*8.8818e−16*	*0*
DeJongs function_3	PSO	34.0844	1.3140e + 11	1.9496e + 12	4.2447e + 11
	FPA	3.1337e + 10	1.7015e + 17	3.0554e + 18	6.3366e + 17
	CFPA	*0*	*0*	*0*	*0*
Schaffer_f6	PSO	−1	−0.9948	−0.9903	0.0049
	FPA	−1.0000	−0.9909	−0.9902	0.0022
	CFPA	*−1*	*−1*	*−1*	*0*
DropWave	PSO	−0.0921	−0.0455	−0.0196	0.0175
	FPA	−0.0180	−0.9909	−0.9902	0.0022
	CFPA	*−1*	*−1*	*−1*	*0*
Zacharov	PSO	18.8287	230.7708	815.3811	221.9632
	FPA	81.3344	281.1685	1.1182e + 03	258.3430
	CFPA	*0*	*0*	*0*	*0*

Besides, the maximum number of iteration is 1000 for each algorithm.

In Table 2, the Best, Mean, Worst and Std. represent the optimal fitness value, mean fitness value, worst fitness value and standard deviation. Bold and italicized results mean that the algorithm is better.

Seen from Table 2, we get that CFPA can find the optimal solution for F_1,F_5,F_7 and F_{10} has a very strong robustness. For other functions, the optimal fitness value and mean fitness value is better than those of PSO and FPA. In the other hand, standard deviation of CFPA is less than that of PSO and FPA. It means that CPFA has better stability in the optimization of high-dimensional unimodal function.

For the three categories test functions, Figs. 1, 2, 3, 4, 5 and 6 are the fitness evolution curve, Figs. 7, 8, 9, 10, 11 and 12 are the anova test of the global minimum and Figs. 13, 14, 15, 16, 17 and 18 are the comparisons of optimal fitness value.

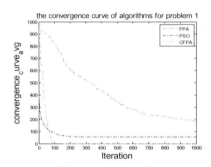

Fig. 1. Evolution curve of fitness for F_1 (Color figure online)

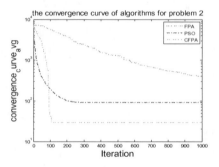

Fig. 2. Evolution curve of fitness for F_2 (Color figure online)

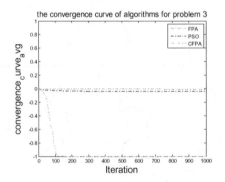

Fig. 3. Evolution curve of fitness for F_3 (Color figure online)

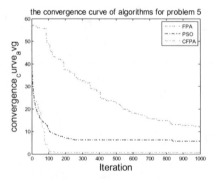

Fig. 4. Evolution curve of fitness for F_5 (Color figure online)

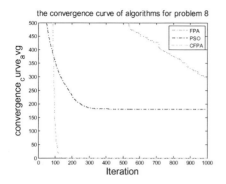

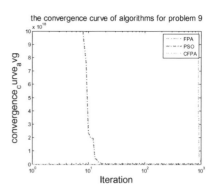

Fig. 5. Evolution curve of fitness for F_8 (Color figure online)

Fig. 6. Evolution curve of fitness for F_9 (Color figure online)

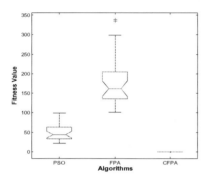

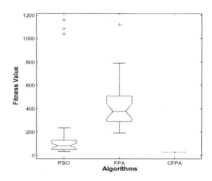

Fig. 7. ANOVA test of minimum for F_1

Fig. 8. ANOVA test of minimum for F_2

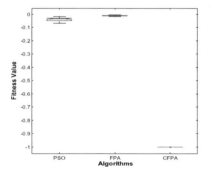

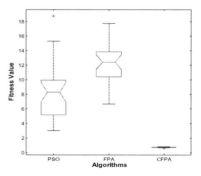

Fig. 9. ANOVA test of minimum for F_3

Fig. 10. ANOVA test of minimum for F_5

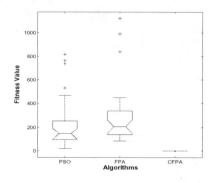

Fig. 11. ANOVA test of minimum for F_8

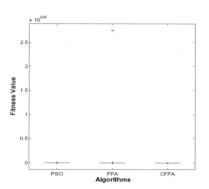

Fig. 12. ANOVA test of minimum for F_9

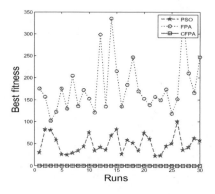

Fig. 13. Comparison of optimal fitness for F_1

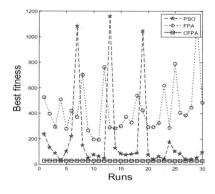

Fig. 14. Comparison of optimal fitness for F_2

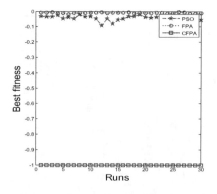

Fig. 15. Comparison of optimal fitness for F_3

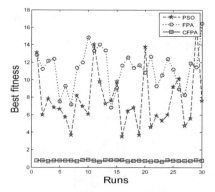

Fig. 16. Comparison of optimal fitness for F_5

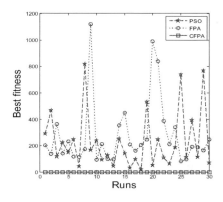

Fig. 17. Comparison of optimal fitness for F_8

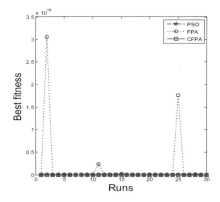

Fig. 18. Comparison of optimal fitness for F_9

5 Conclusions

In this paper, we tested ten benchmark functions. The results of comparison apparently show that convergence speed, precision of optimization and robustness of the new proposed algorithm are all better than FPA and PSO. The new flower pollination algorithm based on complex-valued encoding with the two-dimensional features of complex-valued encoding increases the diversity of population and improves the optimization performance of the algorithm.

Acknowledgments. This work is supported by National Science Foundation of China under Grants No. 61463007, 61563008.

References

1. Yang, X.-S.: Flower pollination algorithm for global optimization. In: Durand-Lose, J., Jonoska, N. (eds.) UCNC 2012. LNCS, vol. 7445, pp. 240–249. Springer, Heidelberg (2012)
2. Pavlyukevich, I.: Levy flights, non-local search and simulated annealing. J. Comput. Phys. **226**, 1830–1844 (2007)
3. Chen, D.-b., Li, H.-j., Li, Z.: Particle swarm optimization based on complex-valued encoding and application in function optimization. Comput. Eng. Appl. **45**(10), 59–61 (2009). (in Chinese)
4. Casasent, D., Natarajan, S.: A classifier neural network with complex-valued weights and square-law nonlinearities. Neural. Netw. **8**(6), 989–998 (1995)
5. Kennedy, J., Eberhart, R.: Particle swarm optimization. In: Proceedings of IEEE International Conference on Neural Networks, pp. 1942–1948. IEEE Press, Piscataway (1995)
6. Kennedy, J., Eberhart, R., Shi, Y.: Swarm intelligence. Academic Press, Cambridge (2001)
7. Chittka, L., Thomson, J.D., Waser, N.M.: Flower constancy, insect psychology, and plant evolution. Naturwissenschaften **86**, 361–377 (1999)

8. Yang, X.S.: Appendix A: test problems in optimization. In: Yang, X.S. (ed.) Engineering optimization, pp. 261–266. John Wiley & Sons, Hoboken (2010)
9. Tang, K., Yao, X., Suganthan, P.N., et al.: Benchmark functions for the CEC 2008 special session and competition on large scale global optimization. University of Science and Technology of China, Hefei (2007)
10. Reynolds, A.M., Frye, M.A.: Free-flight odor tracking in Drosophila is consistent with an optimal intermittent scale-free search. PLoS ONE **2**, e354 (2007)
11. Yang, X.S.: Engineering Optimization: An Introduction with Metaheuristic Applications. Wiley, USA (2010)
12. Yang, X.-S.: A new metaheuristic Bat-Inspired Algorithm. In: González, J.R., Pelta, D.A., Cruz, C., Terrazas, G., Krasnogor, N. (eds.) NICSO 2010. SCI, vol. 284, pp. 65–74. Springer, Heidelberg (2010)
13. Li, L., Zhou, Y.: A novel complex-valued bat algorithm. Neural Comput. Appl. **25**, 1369–1381 (2014)
14. Abdel-Raouf, O., Abdel-Baset, M., El-henawy, I.: A new hybrid flower pollination algorithm for solving constrained global optimization problems. Int. J. Appl. Oper. Res. **4**(2), 1–13 (2014). Spring
15. Kaur Johal, N., Singh, S., Kundra, H.: A hybrid FPAB/BBO algorithm for satellite image classification. Int. J. Comput. Appl. **6**(5), 0975–8887 (2010)
16. Sharawi, M., Emary, E., AlySaroit, I., El-Mahdy, H.: Flower pollination optimization algorithm for wireless sensor network lifetime global optimization. Int. J. Soft Comput. Eng. (IJSCE) **4**(3), 54–59 (2014)
17. El-henawy, I., Ismail, M.: An improved chaotic flower pollination algorithm for solving large integer programming problems. Int. J. Digit. Content Technol. Appl. (JDCTA) **8**(3), 72–81 (2014)
18. Yang, X.-S., Karamanoglu, M., He, X.: Multi-objective Flower Algorithm for Optimization. Procedia Comput. Sci. **18**, 861–868 (2013)
19. Harikrishnan, R., Jawahar Senthil Kumar, V., Sridevi Ponmalar, P.: Nature inspired flower pollen algorithm for WSN localization problem. ARPN J. Eng. Appl. Sci. **10**(5), 2122–2125 (2015)
20. Singh, P., Kaur, N., Kaur, L.: Satellite image classification by hybridization of FPAB algorithm and bacterial chemotaxis. Int. J. Comput. Technol. Electron. Eng. (IJCTEE) **1**(3), 21–27 (2011)
21. Kaur, G., Singh, D.: Pollination based optimization for color image segmentation. Int. J. Comput. Eng. Technol. (IJCET) **3**(2), 407–414 (2012)
22. ZeinEldin, R.A.: A hybrid SS-SA approach for solving multi-objective optimization problems. Eur. J. Sci. Res. **121**(3), 310–320 (2014)
23. Balasubramani, K., Marcus, K.: A study on flower pollination algorithm and its applications. Int. J. Appl. Innov. Eng. Manag. (IJAIEM) **3**(11), 230–235 (2014)
24. Fister Jr., I., Yang, X.-S., Fister, I., Brest, J., Fister, D.: A brief review of nature-inspired algorithms for optimization. Elektrotehniski Vestnik **80**(3), 116–122 (2013)

An Fruit Fly Optimization Algorithm with Dimension by Dimension Improvement

Haiyun Li, Haifeng Li$^{(\boxtimes)}$, and Kaibin Wei

School of Electronic Information and Electrical Engineering,
Tianshui Normal University, Tianshui 741001, China
lihaifeng8848@gmail.com

Abstract. To overcome the shortages of interference phenomena among dimensions, slow convergence rate and low accuracy, a new fruit fly optimization algorithm with dimension by dimension improvement is proposed. In addition, in order to speed up the algorithm convergence rate and avoid algorithm falling into local optimums, a Lévy flight mechanism is introduced to speed up the algorithm convergence rate and enhance the ability to jump out of the local optimum. The simulation experiments show that the proposed algorithm greatly speeds up the convergence rate and significantly improves the qualities of the solutions. Meanwhile, the results also reveal that the proposed algorithm is competitive for continuous function optimization compared with the basic fruit fly optimization algorithm and other algorithms.

Keywords: Fruit fly optimization · Multi-dimension function optimization · Dimension by dimension improvement

1 Introduction

Fruit fly optimization algorithm (FOA) [1, 2] was firstly proposed by Professor Pan in June 2011 and is a new global optimization algorithm inspired by the simulation of the foraging behavior of fruit flies. The fruit flies, by virtue of their better ability to smell and sight, feel various smell in the air to find food, and fly along the direction to food. FOA has been successfully applied to solve the function extreme value, the fine tuning Z-SCORE model coefficient, the generalized regression neural network parameter optimization and the support vector machine parameter optimization. Compared with other swarm intelligence algorithms, FOA has the advantages of simple algorithm, easy implementation of program code and less running time.

Many improved versions of the FOA algorithm have been proposed. The application of combining the reverse learning strategy to the basic FOA algorithm [3], to a certain extent, improved the convergence rate and accuracy of the algorithm. Cheng and Liu proposed a new FOA algorithm based on chaotic map [4], the diversity of the population was improved by using chaos mapping. All mutated versions of FOA in iterative producing new solutions have one thing in common, that is, for a fruit fly, the solution is evaluated after the completion of the update for all dimensions. For multi-dimensional optimization problems, however, there exists a mutual interference phenomenon, the quality of solutions and the convergence rate of the algorithm will be

© Springer International Publishing Switzerland 2016
D.-S. Huang et al. (Eds.): ICIC 2016, Part I, LNCS 9771, pp. 679–690, 2016.
DOI: 10.1007/978-3-319-42291-6_68

affected using the evaluation strategy after the update to the values of all dimensions. Therefore, Zhong et al. proposed an iterative improvement strategy which can effectively avoid the mutual interference between dimensions [5]. In addition, dimension by dimension improvement strategy been used in some swarm intelligence algorithms, such as flower pollination algorithm [6] and cuckoo search algorithm [7]. To reduce the dependence of the algorithm to the evaluation of the objective function, the iterative improvement strategy was used to construct a new evaluation operator for evaluating the updated value. In this paper, we adopt the idea of iterative improvement strategy, and put forward the fruit fly optimization algorithm with dimension by dimension improvement (DDIFOA). In the process of updating the coordinate position, the DDIFOA algorithm introduces the Lévy flight mechanism, together with the dimension by dimension improvement strategy to avoid the mutual interference between dimensions. The proposed algorithm can effectively utilize the local search ability of the single dimension information to enhance the quality of solutions. It would be specially mentioned that DDIFOA algorithm still retains the original objective function instead of reconstructing a new evaluation operator for a new evaluation, which is conducive to the practical application and promotion for DDIFOA algorithm. In order to verify the effectiveness and performance of the algorithm, the DDIFOA algorithm is tested on some benchmark functions. The simulation results show that the DDIFOA algorithm can effectively accelerate the convergence rate and enhance the convergence accuracy. The comparison experiment results of 9 benchmark functions show that the proposed algorithm has better performance.

The rest of this paper is organized as follows. The basics of fruit fly optimization algorithm is presented in the next section, and the fruit fly optimization algorithm with dimension by dimension improvement is presented in Sect. 3, and the concept of dimension update and evaluation based on the greed strategy and Lévy flight trajectory are explained, followed by the detailed discussion of proposed DDIFOA algorithm. The experiment results are illustrated in detail in Sect. 4. Finally, some remarks and conclusions are provided in Sect. 5.

2   Fruit Fly Optimization Algorithm

The basic FOA [1] is a global search optimization algorithm based on the simulation of the foraging behavior of the fruit fly. In nature, fruit flies mainly dependent on the rotten fruit as the staple food which can be seen everywhere, such as the human habitat orchard, vegetable market. Furies flies can feel the various smell in the air to find food depending on their superior olfactory and visual ability, then fly toward the food. Firstly, we set the population size *pop_size*. For a fruit fly i, the corresponding position is $X_i = (x_{i1}, x_{i2}, \cdots, x_{in})$, where n represents the scale of the problem, so x_i is a feasible solution of the problem mathematical model describing the fruit fly optimization algorithm which can be divided into the following steps:

Step 1. Initialize the population size *sizepop*, the maximum number of iterations *Maxgen*, randomly initialize the fruit flies' positions.

Step 2. Each individual search food with smell, each dimension of an individual will search randomly, *RandomValue* is a search distance generated by randomly, *RandomValue* $\in (0, 1)$, the random search formula is as follow:

$$x_{ij} = x_{ij} + RandomValue \tag{1}$$

Step 3. Since the food location cannot be known in advance, the distance to the origin is thus estimated first $Dist_i$, then the smell concentration judgment value S_i is calculated, and this value is the reciprocal of distance:

$$Dist_i = \sqrt{\sum_{j=1}^{n} x_{ij}^2} \tag{2}$$

$$S_i = {}^1\!/Dist_i \tag{3}$$

Step 4. Substitute smell concentration judgment value (S_i) into smell concentration judgments function (or called fitness function) so as to find the smell concentration (*Smell_i*) of the individual location of the fruit fly:

$$Smell_i = Function(S_i) \tag{4}$$

Step 5. Find out the fruit fly with maximal smell concentration (the optimal individual) among the fruit fly swarm:

$$[bestSmell\ bestIndex] = \min(Smell) \tag{5}$$

Step 6. Record the best smell concentration value *bestSmell* and the corresponding position *Pbest*, and at this moment, the fruit fly swarm will fly towards that location within vision:

$$Smellbest = bestSmell \tag{6}$$

$$Pbest = X_{bestIndex} \tag{7}$$

Step 7. Enter iterative optimization to repeat the implementation of **step 2** to **step 6**, terminate the algorithm until reaching the maximum number of iterations.

3 Fruit Fly Optimization Algorithm with Dimension by Dimension Improvement

3.1 Dimension Update and Evaluation Based on the Greed Strategy

In the basic FOA algorithm, we implement the random method for the initialization of the population, and generate the best individual through the evaluation after updating all dimensional values of each individual at each iteration. This strategy can strengthen the local search ability of the algorithm to some extent. For multi-objective function optimization problems, however, it may adversely affect the convergence rate and the quality of solutions. Assuming the optimization function is $f(x) = x_1^2 + x_2^2 + x_3^2$, the objective function value is 0.75 whose position is $X_i^t = (0.5, 0.5, 0.5)$. If the position X_i^g is updated to $X_i^{t+1} = (0, 1, 1)$ (t is the current generation) in accordance with the formula (1), the corresponding objective function value is 2. Because the objective function value of the individual is not improved, the current solution is still retained. However, as seen from the two solutions, the first dimension has evolved into an optimal value 0 actually. The objective function value becomes poor just due to the degeneration of other dimensional values, eventually leading to the updated solution is discarded due to the low quality.

Update evaluation strategy based on dimension by dimension improvement can avoid the problem of interference phenomena among dimensions. Each dimensional value is considered separately, that is, only one dimensional value is updated and other dimensional values are kept unchanged to produce a new solution, then using the greedy strategy determine the current solution. This strategy makes the evolution of the solution not be ignored because of some dimensional degradation. The algorithm can effectively guide the local search by taking advantage of the single dimension information of the evolution, and obtain a higher quality solution.

3.2 Lévy Flight Trajectory

Lévy flight trajectory is a stable distribution proposed by Paul and described by Benoist. Studies show that many of the typical features of Lévy flight are verified by the flight behavior of many animals and insects. In early 1996, Viswanathan proved that albatross forage by Lévy flight mode. They use satellite positioning system to study the wandering albatross foraging behavior, finding that the flight interval obeys power-law distribution. They explain their findings with the spatial distribution of food scales on the surface of the sea [8–10]. Mercadier et al. [11] also found Lévy flight in hot atomic vapor photon. Reynolds et al. studied the foraging locus of fruit fly, finding that the frequency of the linear part of its flight path is consistent with negative quadratic of scale-free.

Lévy flight is a random walk process based on short distance exploring with the occasional long distance walking, which is good for individuals to jump out the local optimum. Its step size obeys continuous heavy tailed distribution. In this paper, Lévy flight is introduced into basic fruit fly algorithm. It can expand the scope of the search, increase the diversity of population, and help FOA to jump out of the local optimum. The location update formula is as follow:

$$x_{ij} = x_{ij} + \alpha\, sign(rand - 0.5) \oplus Le'vy \tag{8}$$

where $j = 1, 2, \wedge, n$, α is a random parameter, we generally set $\alpha = 1$, $rand \in [0, 1]$, the symbolic $\oplus$ represents point multiplication. Random step size Lévy is derived from the Lévy distribution:

$$Le'vy \sim \mu = t^{-\lambda},\ 1 < \lambda \leq 3 \tag{9}$$

FOA is simple to plus a random number to each dimension and is easy to trap in the local optimums. DDIFOA introduces the update and evaluation strategy based on dimension by dimension improvement to avoid inter-dimensional interference while combining the Lévy flight mechanism during the population renew process.

4   Numerical Simulation Experiment

In order to test the performance of the DDIFOA algorithm for solving multi-dimensional function optimization problems, 9 benchmark functions are selected and are shown in Table 1. The benchmark functions contain single-peak functions and

Table 1. Benchmark functions

Benchmark	Function	D	Scope	f_{min}
Sphere	$f_1 = \sum_{i=1}^{n} x_i^2$	30	$[-100,100]$	0
Schwefel-2.22	$f_2 = \sum_{i=1}^{n} |x_i| + \prod_{i=1}^{n} |x_i|$	30	$[-10,10]$	0
Schwefel-1.2	$f_3 = \sum_{i=1}^{n} \left(\sum_{j=1}^{i} x_j \right)^2$	30	$[-100,100]$	0
Schwefel-2.21	$f_4 = \max\{|x_i|, 1 \leq i \leq n\}$	30	$[-100,100]$	0
Quartic	$f_5 = \sum_{i=1}^{n} i x_i^4 + random[0, 1)$	30	$[-1.28,1.28]$	0
Powell sum	$f_6(x) = \sum_{i=1}^{n} |x_i|^{i+1}$.	30	$[-1,1]$	0
Rastrigin	$f_7 = \sum_{i=1}^{n} \left[x_i^2 - 10\cos(2\pi x_i) + 10 \right]$	30	$[-5.12,5.12]$	0
Ackley	$f_8(x) = 20 + e - 20e^{-\frac{1}{5}\sqrt{\frac{1}{n}\sum_{i=1}^{n} x_i^2}} - e^{\frac{1}{n}\sum_{i=1}^{n} \cos(2\pi x_i)}$	30	$[-32,32]$	0
Griewank	$f_9(x) = \sum_{i=1}^{n} \frac{x_i^2}{4000} - \prod_{i=1}^{n} \cos(x_i/\sqrt{i}) + 1$	30	$[-600,600]$	0

multi-peak functions. We carry out the contrast experiments with dimensions between 30 and 100 for 9 benchmark functions. We compare DDIFOA with PSO [12], CS [13], GSO [14], GSA [15], ABC [16] and AMO [17].

4.1   Simulation Platform

The proposed algorithm is implemented in MATLAB R2012 (a). Operating system: Windows XP. CPU: AMD Athlon (tm) II X4 640 Processor with 3.01 GHz. RAM: 3 GB.

4.2   Parameter Setting

The parameter settings of FOA and DDIFOA are consistent: the population size are 50. For PSO [12], we use inertia weight from $w = 0.6$, learning factors $c_1 = c_2 = 2$, the population size is 50. For CS [13], the probability of abandoned $Pa = 0.25$, the population size is 50. For GSO [14], the initial head angle φ^0 is set to $\pi/4$, the constant a is given by $round(\sqrt{n+1})$ where n is the dimension of the search space, the maximum pursuit angle θ_{max} is π/a^2, the maximum turning angle α_{max} is set to $\theta_{max}/2$, the population size is 50. For GSA [15], $G_0=100$, $\alpha = 20$, the population size is 50. For ABC [16] and AMO [17], the population size is 50. The maximum number of iterations of each algorithm is set to be 1000.

4.3   Application in Function Optimization Problems

(1)   Compared with Basic FOA Algorithm

The comparison results of FOA and DDIFOA in 30 and 100 dimensions are shown in Table 2, the results obtained by the 20 time independent runs. The best, mean, worst and std represent the optimal value, mean value, worst value and standard deviation, respectively. As can be seen from Table 2, in the 30 dimensions, for the functions f_1 to f_9, the DDIFOA algorithm is better than the basic FOA in terms of the best, worst, mean and standard deviation. For the single-peak functions f_1 to f_6, compared with the mean value, the quality of DDIFOA algorithm is far better than that of the basic FOA algorithm, at least higher 189, 95, 177, 92, 5 and 200 orders of magnitude, respectively. Comparison with standard deviation, DDIFOA algorithm is also better than the basic FOA algorithm, we can see that the DDIFOA algorithm has high robustness. For the multi-peak functions f_7 to f_9, the advantage of DDIFOA algorithm is more obvious, and the optimal value of the function is obtained. And the basic FOA algorithm does not reach the optimal value of the function, especially for function f_7, the basic FOA shows a great deal of instability, the difference between the best and the worst is 5 orders of magnitude. From Table 2, DDIFOA algorithm is better than the basic FOA algorithm in accuracy and stability.

Table 2. Experiment results of benchmark functions for FOA and DDIFOA (D = 30)

Function	Algorithm	D = 30				D = 100			
		Best	Mean	Worst	Std	Best	Mean	Worst	Std
f_1	FOA	2.415e−05	0.191	0.846	0.341	8.0310e−05	0.4486	8.031e−05	0.933
	DDIFOA	1.817e−225	2.868e−189	5.736e−188	0	8.609e−217	2.12e−178	4.230e−177	0
f_2	FOA	2.691e−02	1.0983	4.8482	1.9114	8.965e−2	14.910	14.910	4.3780
	DDIFOA	9.715e−113	2.519e−95	2.497e−94	7.666e−95	2.948e−107	1.286e−91	2.096e−90	1.286e−91
f_3	FOA	1.1466	4224.4	27658.3	8862.7	423.027	1794419.9	10154921.7	1794419.9
	DDIFOA	1.061e−225	5.279e−177	1.056e−175	0	6.248e−219	6.070e−182	6.248e−219	0
f_4	FOA	8.969e−4	9.078e−4	9.231e−4	5.320e−06	8.979e−4	9.089e−4	9.194e−4	6.050e−06
	DDIFOA	1.483e−110	7.932e−92	1.486e−90	3.319e−91	5.708e−112	4.485e−92	8.189e−91	1.830e−91
f_5	FOA	1.859e−3	6.987e−2	1.0518	0.2403	3.59835e−3	0.2331	4.5479	1.0156
	DDIFOA	9.869e−08	4.607e−05	1.737e−4	4.721e−05	2.936e−06	3.578e−05	1.244e−4	3.371e−05
f_6	FOA	8.065e−07	8.236e−07	8.431e−07	8.496e−09	8.038e−07	8.210e−07	8.449e−07	1.147e−08
	DDIFOA	1.230e−256	7.398e−207	1.436e−205	0	4.718e−246	9.470e−208	1.894e−206	0
f_7	FOA	4.835e−3	20.388	117.847	42.211	1.612e−2	427.740	427.740	158.143
	DDIFOA	0	0	0	0	0	0	0	0
f_8	FOA	3.647e−3	7.602e−2	1.451	0.324	3.638e−3	0.302	1.579	0.613
	DDIFOA	0	0	0	0	0	0	0	0
f_9	FOA	1.613e−06	5.879e−3	6.488e−2	1.820e−2	2.122e−06	2.536e−2	0.267	6.907e−2
	DDIFOA	0	0	0	0	0	0	0	0

In the 100 dimensions, for functions f_1 to f_9, the DDIFOA algorithm shows a better performance, no matter in the best, worst, mean and standard deviation are significantly better than those of the basic FOA algorithm. For the single-peak functions f_1 to f_6, compared with the mean value, the accuracy of the DDIFOA algorithm is better than that of the basic FOA. Comparison with standard deviation, DDIFOA algorithm is also better than the basic FOA algorithm, and has a strong stability. For multi-peak functions f_7 to f_9, the performance of DDIFOA algorithm is not reduced when increase dimension, and still can solve the optimal value of the function, and the basic FOA algorithm cannot find the optimal value. In particular, for function f_7, the basic FOA shows a great deal of instability, the standard deviation reached 158. Seen from Table 2, improved DDIFOA algorithm still shows good performance in high dimension, the quality and stability of DDIFOA are much better than those of the basic FOA algorithm. Figures 1, 2, 3 and 4 are the objective value evolution curves of FOA and DDIFOA in the 100 dimensions for some part of the functions (f_2, f_5, f_7 and f_8).

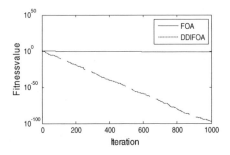

Fig. 1. Evolution curves for $f_2(D = 100)$ (Color figure online)

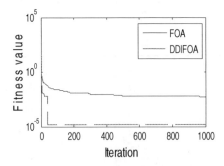

Fig. 2. Evolution curves for f_5 ($D = 100$) (Color figure online)

(2) **Comparison with Other Swarm Intelligence Algorithms**

In order to test the performance of the DDIFOA algorithm, the improved algorithm is compared with PSO, CS, GSO, GSA, ABC and AMO respectively. Table 3 shows

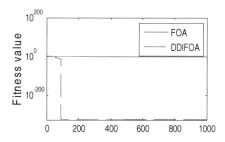

Fig. 3. Evolution curves for f_7 (D = 100) (Color figure online)

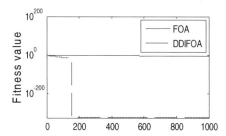

Fig. 4. Evolution curves for f_8 (D = 100) (Color figure online)

the comparison results of the mean and the standard deviation in 20 times independent runs with 30 dimensions.

Comparison of each function in Table 3, we can see that the DDIFOA algorithm is better than the other contrast algorithms (PSO, CS, GSO, GSA, ABC and AMO). For the single-peak functions f_1 to f_6, from the mean value, DDIFOA algorithm is significantly better than other comparative algorithms. For functions f_1 to f_6, the mean values are higher at least 155, 65, 186, 95, 2 and 60 orders of magnitude. Seen from the standard deviation, the orders of magnitude of functions f_1, f_2, f_3, f_4 and f_6 are also significantly higher than the other comparison algorithms, function f_5 is also at least 1 orders of magnitude higher. For multi-peak functions f_7 to f_9, seen from the mean value, DDIFOA has reached the optimal value. For function f_7, only DDIFOA achieves the optimal value. For other functions, GSA reaches the optimal value on function f_9, AMO achieves optimal values on f_7 and f_9. For the standard deviation, DDIFOA algorithm is better than other algorithms, and the algorithm is more stable. By means of the analysis of the mean and standard deviation, DDIFOA algorithm is superior to the other algorithms in the convergence rate and convergence accuracy.

Figures 5, 6, 7 and 8 are the fitness evolution curves for functions f_1, f_4, f_6 and f_9 in the 30 dimensions. For single-peak functions f_1, f_4 and f_6, DDIFOA not only convergence speed, but also the convergence accuracy is significantly better than other contrast algorithms. For f_9, GSA also reaches the optimal value, but from Fig. 8 we can see that GSA only converges to less than 20 order of magnitude in the 1000 generation. And compared with GSA, DDIFOA has a faster convergence rate in the case of reaching the optimal value.

Table 3. Experiment results of benchmark functions for different algorithms (D = 30)

Functions		PSO	CS	GSO	GSA	ABC	AMO	DDIFOA
f_1	Mean	3.33E−10	5.66E−06	1.95E−08	3.37E−18	7.61E−16	8.65E−40	**1.55E−195**
	Std	7.04E−10	2.86E−06	1.16E−08	8.09E−19	1.73E−16	1.04E−39	**1.32E−194**
f_2	Mean	6.66E−11	0.002	3.70E−05	8.92E−09	1.42E−15	8.23E−32	**3.30E−97**
	Std	9.26E−11	8.10E−04	8.62E−05	1.33E−09	5.53E−16	3.41E−32	**8.01E−94**
f_3	Mean	2.9847	0.0014	5.7829	0.1126	2.40E+03	8.89E−04	**3.85E−190**
	Std	2.2778	6.10E−04	3.6813	0.1266	656.96	8.73E−04	**3.64E−185**
f_4	Mean	7.9997	3.2388	0.1078	9.93E−10	18.5227	2.86E−05	**5.94E−105**
	Std	2.5351	0.6644	3.99E−02	1.19E−10	4.2477	2.35E−05	**2.59E−115**
f_5	Mean	0.0135	0.0096	7.37E−02	0.0039	0.0324	0.0017	**6.47E−05**
	Std	0.0041	0.0028	9.25E−02	0.0021	0.0059	4.71E−04	**5.79E−05**
f_6	Mean	5.79E−10	1.28E−06	1.85E−23	9.42E−18	1.01E−13	1.68E−175	**1.25E−235**
	Std	3.89E−09	7.55E−07	4.13E−23	1.40E−17	1.57E−13	2.56E−175	**3.53E−220**
f_7	Mean	18.2675	51.2202	1.0179	7.2831	1.64E−07	0	**0**
	Std	4.7965	8.1069	0.9509	1.8991	3.61E−07	0	**0**
f_8	Mean	3.87E−06	2.375	2.65E−05	1.47E−09	1.19E−09	4.44E−15	**0**
	Std	2.86E−06	1.1238	3.08E−05	1.44E−10	5.01E−10	0	**0**
f_9	Mean	0.0168	4.49E−05	3.07E−02	0	6.92E−13	0	**0**
	Std	0.0205	8.96E−05	3.08E−02	0	1.05E−12	0	**0**

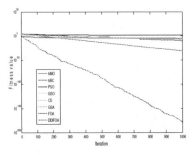

Fig. 5. Evolution curves for f_1 (D = 30) (Color figure online)

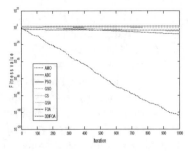

Fig. 6. Evolution curves for f_4 (D = 30) (Color figure online)

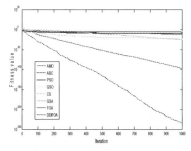

Fig. 7. Evolution curves for f_6 (D = 30) (Color figure online)

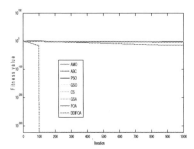

Fig. 8. Evolution curves for f_9 (D = 30) (Color figure online)

5 Conclusions

Fruit fly optimization algorithm is a novel swarm intelligence algorithm, which has the advantages of simple structure and easy implementation. To overcome the shortages of interference phenomena among dimensions, slow convergence rate and low accuracy, a new fruit fly optimization algorithm with dimension by dimension improvement is proposed. Using the updating strategy based on dimension by dimension improvement, the algorithm updates the value of each dimension and generates a new solution with the values of other dimensions. Then the current position of a fruit fly is updated to the better one, and the process is repeatedly performed until the completion of the update for all dimensions at each iteration. In addition, in order to speed up the algorithm convergence rate and avoid algorithm falling into local optimums, a Lévy flight mechanism is introduced to speed up the algorithm convergence rate and enhance the ability to jump out of the local optimum. The simulation experiments show that the proposed strategy greatly speeds up the convergence rate and significantly improves the quality of the solutions. Meanwhile, the results also reveal the proposed algorithm is competitive for continuous function optimization problems compared with the basic fruit fly optimization algorithm and other intelligent algorithms. In future, we will use some new evolution strategy to improve the optimization performance of fruit fly optimization algorithm and apply it to solve practical problems.

Acknowledgement. This work is supported by Educational Commission of Gansu Province of China (No. 2013B-078)

References

1. Pan, W.T.: A new fruit fly optimization algorithm: taking the financial distress model as an example. Knowl.-Based Syst. **26**, 69–74 (2012)
2. Pan, W.T.: Fruit fly optimization algorithm. Tsang Hai Book Publishing Co., Taipei, pp. 10–12 (2011)
3. Han, J.Y., Liu, C.Z.: Fruit fly optimization algorithm with opposition learning policy. Comput Appl. Softw. **4**, 157–160 (2014)
4. Cheng, H., Liu, C.Z.: Mixed fruit fly optimization algorithm based on chaotic mapping. Comput. Eng. **39**(5), 218-221 (2013)
5. Zhong, Y.W., Liu, X., Wang, L.J.: Particle swarm optimisation algorithm with iterative improvement strategy for multi-dimensional function optimisation problems. Int. J. Innov. Comput. Appl. **4**(3), 223–232 (2012)
6. Wang, R., Zhou, Y.Q.: Flower pollination algorithm with dimension by dimension improvement. Math. Probl. Eng. **32**(4), 666–674 (2014)
7. Wang, L.J., Yin, Y.L., Zhong, Y.W.: A cuckoo search algorithm with dimension by dimension improvement. J. Softw. **24**(11), 2687–2698 (2013)
8. Viswanathan, G.M., et al.: Lévy flights search patterns of wandering albatrosses. Nature **381**, 13–15 (1996)
9. Viswanathan, G.M., et al.: Lévy flights in random searchs. Phys. A **282**, 1–12 (2000)
10. Viswanathan, G.M., et al.: Lévy flights search patterns of biological organisms. Phys. A **295**, 85–88 (2001)
11. Mercadier, N., Guerin, W., Chevrollier, M., et al.: Lévy flights of photons in hot atomic vapours. Nat. Phys. **5**(8), 602–605 (2009)
12. Kennedy, J.: Particle swarm optimization. In: Encyclopedia of Machine Learning, pp. 760–766. Springer US (2011)
13. Yang, X.S., Deb, S.: Cuckoo search via Lévy flights. World Congress on Nature & Biologically Inspired Computing, 2009, NaBIC 2009, pp. 210–214. IEEE (2009)
14. He, S., Wu, Q.H., Saunders, J.R.: Group search optimizer: an optimization algorithm inspired by animal searching behavior. IEEE Trans. Evol. Comput. **13**(5), 973–990 (2009)
15. Rashedi, E., Nezamabadi-Pour, H., Saryazdi, S.: GSA: a gravitational search algorithm. Inf. Sci. **179**(13), 2232–2248 (2009)
16. Karaboga, D., Basturk, B.: A powerful and efficient algorithm for numerical function optimization: artificial bee colony (ABC) algorithm. J. Global Optim. **39**(3), 459–471 (2007)
17. Li, X., Zhang, J., Yin, M.: Animal migration optimization: an optimization algorithm inspired by animal migration behavior. Neural Comput. Appl. **24**(7–8), 1867–1877 (2014)

Median-Oriented Bat Algorithm for Function Optimization

Limin Zhao and Haifeng Li[✉]

School of Electronic Information and Electrical Engineering,
Tianshui Normal University, Tianshui 741001, China
lihaifeng8848@gmail.com

Abstract. The bat algorithm is easily trapped into local optima, the population diversity is poor, and the optimizing precision is bad. In order to overcome these disadvantages, this paper presents a median bat algorithm (MBA) to avoid local optima and carry out a global search over entire search space. The proposed algorithm adopts the median position of the bats. And the median and worst bats are combined to the basic bat algorithm to achieve a better balance between the global search ability and local search ability. The simulation results of 10 standard benchmark functions show that the proposed algorithm is effective and feasible in both low-dimensional and high-dimensional case. Compared to the basic bat algorithm, particle swarm optimization and CLSPSO, the proposed algorithm can get high precision and can almost reach the theoretical value.

Keywords: Bat algorithm · Median-oriented · Function optimization

1 Introduction

The nature-inspired optimization algorithms derived from the simulation of biological group behaviors in natural world. With a simple and parallel implement, strong robustness, good optimization results and so on, the nature-inspired optimization algorithms have become the focus of study. In recent years, some novel swarm intelligence algorithms have been proposed, such as Artificial Bee Colony Optimization (ABC) [1, 2], Shuffled Frog Leaping Algorithm (SFLA) [3], Artificial Fish Swarm Algorithm (AFSA) [4], Cuckoo Search (CS) [5], Monkey Algorithm (MA) [6], Firefly Algorithm (FA) [7], Glowworm Swarm Optimization algorithm (GSO) [8], Flower Pollination Algorithm (FPA) [9], Wind Driven Optimization (WDO) [10], Charged System Search (CSS) [11] and so on.

First proposed by Yang [12] in 2010, Bat Algorithm (BA) was originated from the simulation of echolocation behavior in bats. Bats use a type of sonar called echolocation to detect prey, and avoid obstacles in the dark. When searching their prey, the bats emit ultrasonic pulses. During flight to the prey, loudness will decrease while the pulse emission will gradually increase, which can make the bat locate the prey more accurately. Applications of BA algorithm span the areas of constrained optimization tasks [13], global engineering optimization [14], multi-objective optimization [15], structural optimization [16], and discrete size optimization of steel frames [17].

© Springer International Publishing Switzerland 2016
D.-S. Huang et al. (Eds.): ICIC 2016, Part I, LNCS 9771, pp. 691–702, 2016.
DOI: 10.1007/978-3-319-42291-6_69

In this paper, we propose a novel median-oriented bat algorithm (MBA) for the function optimization problem. The proposed algorithm adopts the median and worst bat individuals [18] to avoid premature convergence. As a result, the global search ability of MBA is improved and the proposed algorithm can avoid trapping in the local optimum. Simulation results demonstrate the effectiveness and robustness of the proposed algorithm. MBA can get a more accurate solution for the optimization problems. In MBA, the mutation operation in DE is added to the bat algorithm to accelerate the global convergence speed.

The remainder of this paper is organized as follows. Section 2 introduces the basic bat algorithm. In Sect. 3, median-oriented bat algorithm is introduced. The experimental results and comparison results are given in Sect. 4. Finally, some relevant conclusions are presented in Sect. 5.

2   The Basic Bat Algorithm

2.1   The Update of Velocity and Position

Initialize the bat population randomly. Supposed the dimension of search space is n, the position of the bat i at time t is x_i^t and the velocity is v_i^t. Therefore, the position x_i^{t+1} and velocity v_i^{t+1} at time $t+1$ are updated by the following formula:

$$f_i^t = f_{\min} + (f_{\max} - f_{\min})\beta \tag{1}$$

$$v_i^{t+1} = v_i^t + (x_i^t - Gbest)f_i^t \tag{2}$$

$$x_i^{t+1} = x_i^t + v_i^{t+1} \tag{3}$$

Where, f_i represents the pulse frequency emitted by bat i at the current moment. $f_{\max}$ and $f_{\min}$ represent the maximum and minimum values of pulse frequency respectively. β is a random number in $[0, 1]$ and $Gbest$ represents the current global optimal solution.

Select a bat from the bat population randomly, and update the corresponding position of the bat according to Eq. (1). This random walk can be understood as a process of local search, which produces a new solution by the chosen solution.

$$x_{new}(i) = x_{old} + \varepsilon A^t \tag{4}$$

Where, x_{old} represents a random solution selected from the current optimal solutions, A^t is the loudness, and ε is a random vector, and its arrays are random values in $[-1, 1]$.

2.2   Loudness and Pulse Emission

Usually, at the beginning of the search, loudness is strong and pulse emission is small. When a bat has found its prey, the loudness decreases while pulse emission gradually increases. Loudness $A(i)$ and pulse emission $r(i)$ are updated according to Eq. (2) and Eq. (3):

$$r^{t+1}(i) = r^0(i) \times [1 - \exp(-\gamma t)] \tag{5}$$

$$A^{t+1}(i) = \alpha A^t(i) \tag{6}$$

Where, both $0 < \alpha < 1$ and $\gamma > 0$ are constants. $A(i)=0$ means that a bat has just found its prey and temporarily stop emitting any sound. It is not hard to find that when $t \to \infty$, we can get $A^t(i) \to 0$ and $r^t(i) = r^0(i)$.

2.3 The Implementation Steps of Bat Algorithm

Step 1: Initialize the basic parameters: attenuation coefficient of loudness α, increasing coefficient of pulse emission γ, the maximum loudness A^0 and maximum pulse emission r^0 and the maximum number of iterations *Maxgen*;

Step 2: Define pulse frequency $f_i \in [f_{min}, f_{max}]$;

Step 3: Initialize the bat population x and v;

Step 4: Enter the main loop. If $rand < r_i$, update the velocity and current position of the bat according to Eqs. (2) and (3). Otherwise, make a random disturbance for position of the bat, and go to *Step 5*;

Step 5: If $rand < A_i$ and $f(x_i) < f(x^*)$, accept the new solutions, and fly to the new position;

Step 6: If $f(x_i) < f_{min}$, replace the best bat, and adjust the loudness and pulse emission according to Eqs. (5) and (6);

Step 7: Evaluate the bat population, and find out the best bat and its position;

Step 8: If termination condition is met (i.e., reach maximum number of iterations or satisfy the search accuracy), go to step 9; Otherwise, go to step 4, and execute the next search.

Step 9: Output the best fitness values and global optimal solution.

Where, *rand* is a uniform distribution in $[0, 1]$.

3 Median-Oriented Bat Algorithm

In this section, a novel median-oriented bat algorithm (MBA) is presented to enhance the performance of the basic bat algorithm [19–22]. In BA, each bat moves toward good solutions based on the best solution. MBA is a global search algorithm.

$$stepnow = (iterMax - iter)^3 * (step_{ini} - step_{final})/(iterMax)^3 + step_{final} \tag{7}$$

$$v_i^{t+1} = v_i^t + f_i^t * (x_i^t - Gbest) + stepnow * rand * (x_i^t - Gmedian - Gworst) \tag{8}$$

$$x_i^{t+1} = x_i^t + v_i^{t+1} \tag{9}$$

Due to the proposed algorithm considering the best bat individual, the median bat individual and the worst bat individual, this is equivalent to adopt a compromise solution. The coordination of the bat population of individuals is conducive to cover a wider range of bat population and increase the diversity of bat population.

Pseudo code of MBA

1: BEGIN
2:　　　　*Initialize the bat population*
3:　　　*For all X_i do*
4:　　　　　*Calculate fitness $f(X_i)$*
5:　　*End for*
6:　　　*Get the best solution Gbest , median solution Gmedian and worst*
solution Gworst
7:　　　*iter ←1*
8:　　*While iter <= iterMax*
9:　　　　*Calculate stepnow*
10:　　　　*Update the position and velocity as follows:*
11:　　　　$v_i^{t+1} = v_i^t + f_i^t * (x_i^t - Gbest) + stepnow * rand * (x_i^t - Gmedian - Gworst)$
12:　　　　*If rand > r_i then*
13:　　　　　*Generate a local solution around the best solution*
14:　　　　*End If*
15:　　　　*Generate a new solution randomly*
16:　　　　*If rand < p_m then*
17:　　　　　*Implement the differential evolution mutate operation*
18:　　　　*End if*
19:　　　　*Calculate fitness $f(X_i)$*
20:　　　　*If (rand < A_i and $f(X_i) < f(X_*))$ then*
21:　　　　　*Accept the new solutions*
22:　　　　　*Reduce A_i and increase r_i*
23:　　　　*End if*
24:　　　　*Get the best solution*
25:　　　　*iter ← iter +1*
26:　　*End While*
27:　　　*Memorize the best solution Gbest*
28: END

4   Simulation Experiments and Discussion

4.1   Simulation Platform

All the algorithms are implemented in Matlab R2012 (a). The test environment is set up on a computer with AMD Athlon (tm) II X4 640 Processor, 3.00 GHz, 4 GB RAM, running on Windows 7.

4.2   Benchmark Functions

In order to verify the effectiveness of the proposed algorithm, we select 10 standard benchmark functions [23] in Table 1 to detect the searching capability of the proposed

Table 1. Benchmark functions

No	D	Name	Benchmark function	Scope
F_1	30/100	Sphere	$f(x) = \sum_{i=1}^{n} x_i^2$	$[-100, 100]$
F_2	30/100	Step	$f(x) = \sum_{i=1}^{n-1} (\lfloor xi + 0.5 \rfloor)^2$	$[-100, 100]$
F_3	30/100	Quartic	$f(x) = \sum_{i=1}^{n} x_i^4 + random[0, 1)$	$[-1.28, 1.28]$
F_4	30/100	Rastrigin	$f(x) = \sum_{i=1}^{n} [x_i^2 - 10\cos(2\pi x_i) + 10]$	$[-5.12, 5.12]$
F_5	30/100	Rosenbrock	$f(x) = \sum_{i=1}^{n-1} [(x_i - 1)^2 + 100(x_i^2 - x_{i+1})^2]$	$[-2.048, 2.048]$
F_6	30/100	Ackley	$f(x) = -20\exp(-0.2\sqrt{\frac{1}{n}\sum_{i=1}^{n} x_i^2})$ $- \exp(\frac{1}{n}\sum_{i=1}^{n}\cos 2\pi x_i)) + 20 + e$	$[-32.768, 32.768]$
F_7	30/100	Griewank	$f(x) = \frac{1}{4000}\sum_{i=1}^{n}(x_i^2) - \prod_{i=1}^{n}\cos(\frac{x_i}{\sqrt{i}}) + 1$	$[-600, 600]$
F_8	30/100	Weierstrass	$f(x) = \sum_{i=1}^{n}(\sum_{k=0}^{kmax}[a^k\cos(2\pi b^k(x_i + 0.5))])$ $-n\sum_{k=0}^{kmax}[a^k\cos(2\pi b^k \times 0.5)]$ $a = 0.5, b = 3, kmax = 20$	$[-0.5, 0.5]$
F_9	30/100	Cosine mixture	$f(x) = \sum_{i=1}^{n} x_i^2 - 0.1\sum_{i=1}^{n}\cos(5\pi x_i)$	$[-1, 1]$
F_{10}	30/100	Alpine	$f(x) = \sum_{i=1}^{n} |x_i\sin(x_i) + 0.1x_i|$	$[-10, 10]$

algorithm. The proposed algorithm in this paper (i.e., MBA) is compared with PSO, CLSPSO and BA.

4.3   Parameter Setting

In PSO, we use linear decreasing inertia weight is $\omega_{max} = 0.9$, $\omega_{min} = 0.4$, and learning factor is $C_1 = C_2 = 1.4962$. In CLSPSO, inertia weight and learning factor are the same as in PSO. The times of chaotic search is $MaxC = 10$. In BA, the parameters are generally set as follows: pulse frequency range is $f_i \in [-1, 1]$, the maximum loudness is $A^0 = 0.9$, minimum pulse emission is $r^0 = 0.5$, attenuation coefficient of loudness is $\alpha = 0.95$, increasing coefficient of pulse emission is $\gamma = 0.05$. In MBA, the basic parameters are the same as in BA, and $stepLen\_ini = 5 step\ Len\_final = 0$.

4.4   Experimental Results

In order to evaluate the performance of MBA, sixteen benchmark functions are adopted in this paper. In this section, the population size $popsize = 50$ and maximum number of iterations $iterMax = 2000$, and MBA is compared to DE and BA. The test results are get from 50 independent run times.

4.4.1   Experimental Results of Low-Dimension Case

The comparison results of all the algorithms on all the functions are recorded in Table 2. The best, mean, worst and std represent the optimal value, mean value, worst value and standard deviation, respectively. We can see that the performance of MBA exhibits significantly better than that of other algorithms. For the functions F_1, F_2, F_4, F_7, F_8, F_9, F_{10}, the MBA algorithm can obtain the theoretical optimal values in all runs. For the function F_3, compared with the mean value, the quality of MBA algorithm is far better than PSO, CLSPSO, BA with at least higher 5, 5, and 3 orders of magnitude, respectively. For function F_5, the best value and the mean value of MBA are both better than those of other algorithms. For function F_6, compared with the mean value, the quality of MBA algorithm is far better than PSO, CLSPSO, BA with at least higher 16, 15, and 17 orders of magnitude, respectively. By Comparison with standard deviation, MBA is also better than PSO, CLSPSO, BA, we can see that the MBA algorithm has strong robustness. And for functions F_1, F_2, F_4, F_6, F_7, F_8, F_9, F_{10}, the standard deviations are 0. That is, the MBA algorithm obtains the same global optimal value in all runs. For the basic BA algorithm, the best values are inferior to those of MBA, even to the worse values of MBA on all functions. The results of CLSPSO are a little better than those of PSO, but obvious worse than those of MBA. Figure 1 shows the mean fitness of four algorithms on the function F_1 to F_{10}, when the mean value is not 0, MBA has a longer and downward column. When the mean value is 0, we cannot find the bar for MBA algorithm.

Table 2. Experimental results for function from F_1 to F_{10} (D = 30)

Fun	Algorithm	Best	Worst	Mean	Std
F_1	PSO	0.0115	0.2843	0.0979	0.0702
	CLSPSO	2.5669e−12	8.2722e−04	5.7519e−05	1.3171e−04
	BA	8.4173e−04	0.00133	0.0011	1.1398e−04
	MBA	*0*	*0*	*0*	*0*
F_2	PSO	24	226	96.48	41.1172
	CLSPSO	0	5	0.1	0.7071
	BA	34605	61144	48338	6398.5881
	MBA	*0*	*0*	*0*	*0*
F_3	PSO	0.0099	2.7316	0.2126	0.5970
	CLSPSO	0.0012	2.7361	0.1303	0.5374
	BA	0.0021	0.0108	0.0050	0.0019
	MBA	*8.0885e−07*	*5.6234e−05*	*1.3090e−05*	*1.1655e−05*
F_4	PSO	33.3947	1.4977e+02	79.0019	28.1500
	CLSPSO	4.2064e−12	1.61300e+02	27.5633	41.6538
	BA	1.3447e+02	2.5789e+02	1.9631e+02	30.3170
	MBA	*0*	*0*	*0*	*0*
F_5	PSO	28.1084	8.8179e+02	56.3213	1.2249e+02
	CLSPSO	3.2750e−06	1.2833e+02	6.5676	20.1953
	BA	21.4382	83.3127	27.2583	8.3434
	MBA	*1.0689e−05*	*3.9982*	*0.2400*	*0.9589*
F_6	PSO	2.0649	6.2396	3.8940	0.8686
	CLSPSO	1.9754e−06	0.2757	0.0481	0.0562
	BA	18.3337	19.4259	19.0089	0.2248
	MBA	*8.8818e−16*	*8.8818e−16*	*8.8818e−16*	*0*
F_7	PSO	0.5668	2.0102	1.2231	0.2560
	CLSPSO	5.9094e−07	1.0303	0.2590	0.2940
	BA	1.8765e+02	4.1058e+02	3.2020e+02	45.7655
	MBA	*0*	*0*	*0*	*0*
F_8	PSO	10.7982	26.1507	18.2713	3.8755
	CLSPSO	0.0023	25.9101	3.2922	5.9277
	BA	22.5364	41.3711	31.5940	3.4232
	MBA	*0*	*0*	*0*	*0*
F_9	PSO	−2.6592	−0.3120	−1.7452	0.5675
	CLSPSO	−2.999999	−0.1170	−2.7675	0.7129
	BA	0.1125	3.7779	1.6354	0.7540
	MBA	*−3*	*−3*	*−3*	*0*
F_{10}	PSO	0.4080	8.0998	3.9332	1.9486
	CLSPSO	3.8542e−06	7.8113	0.5980	1.6200
	BA	1.9301	16.4262	6.4830	3.1697
	MBA	*0*	*0*	*0*	*0*

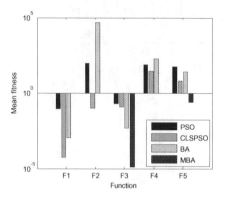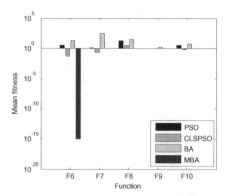

Fig. 1. Mean fitness of four algorithms for F_1 to $F_{10}(D = 30)$. (Color figure online)

4.4.2 Experimental Results of High-Dimension Case

In order validate the performance of the proposed algorithm further, we implement the experiment on 100 dimensions for the all algorithms and keep the parameters unchanged.

The comparison results for high-dimension case are shown in Table 3. Seen from the results, the optimization performance of MBA is the best. For the functions F_1, F_2, F_4, F_7, F_8, F_9, F_{10}, the MBA algorithm still obtains the theoretical optimal values in all runs without a doubt. Only the precision of the best value descends for solving the F5

Table 3. Experimental results for function from F_1 to F_{10} (D = 100)

Fun	Algorithm	Best	Worst	Mean	Std
F_1	PSO	0.8546	45.5550	10.4782	11.0198
	CLSPSO	1.9934e−16	30.1745	0.6042	4.2672
	BA	8.1807e−04	0.0013	0.0011	1.2315e−04
	MBA	*0*	*0*	*0*	*0*
F_2	PSO	99	3369	1257.86	7.5513e+02
	CLSPSO	0	0	0	0
	BA	25644	62332	47257	6.6565e+03
	MBA	*0*	*0*	*0*	*0*
F_3	PSO	0.1522	1.4610e+02	23.3675	38.9468
	CLSPSO	0.0017	1.4303e+02	10.1692	31.2951
	BA	0.0012	0.00992	0.0046	0.0016
	MBA	*3.6753e−06*	*4.3172e−04*	*9.7659e−05*	*9.3351e−05*
F_4	PSO	4.2493e + 02	7.4633e + 02	5.9333e + 02	73.6489
	CLSPSO	1.5224e−04	6.9392e+02	2.7107e+02	2.7534e+02
	BA	1.3754e+02	2.6587e+02	2.0306e+02	25.5434

(Continued)

Table 3. (*Continued*)

Fun	Algorithm	Best	Worst	Mean	Std
	MBA	*0*	*0*	*0*	*0*
F_5	PSO	1.1518e+02	2.7031e+03	5.9723e+02	6.1093e+02
	CLSPSO	1.0889e−05	2.7735e+03	2.6267e+02	6.9857e+02
	BA	11.0558	87.9935	29.0136	14.1507
	MBA	*2.3174*	*2.5904*	*2.5121*	*0.0664*
F_6	PSO	2.1735	8.1533	5.6369	1.1818
	CLSPSO	2.0846e−06	0.3539	0.0763	0.0912
	BA	18.4867	19.4188	19.0488	0.2185
	MBA	*8.8818e−16*	*8.8818e−16*	*8.8818e−16*	*0*
F_7	PSO	3.2679	25.5472	9.9551	4.7114
	CLSPSO	2.2958e−06	0.9778	0.2083	0.2596
	BA	1.8777e+02	3.7630e+02	3.1151e+02	42.2589
	MBA	*0*	*0*	*0*	*0*
F_8	PSO	31.5448	95.0606	67.5116	12.1533
	CLSPSO	0.0458	88.4343	44.8663	33.5603
	BA	23.6779	39.8344	31.3530	3.9133
	MBA	*0*	*0*	*0*	*0*
F_9	PSO	−7.2753	1.3841	−3.8135	1.6853
	CLSPSO	−9.999999999	0.8495	−7.4278	3.4429
	BA	−0.0349	3.7838	1.5973	0.9812
	MBA	*−10*	*−10*	*−10*	*0*
F_{10}	PSO	12.8953	53.5153	25.7114	7.7965
	CLSPSO	5.3529e−06	48.2857	3.5625	9.8779
	BA	1.2815	13.0499	5.8033	2.3913
	MBA	*0*	*0*	*0*	*0*

function. The standard deviations of MBA are better than those of other algorithms on all functions. It demonstrates that MBA has strong robustness and good global search ability, and MBA does not reduce the accuracy of the solutions as the dimension increases. However, the precision of solutions of other three algorithms, including the best value, the mean value, the worse value, the median value and the standard deviation, will decrease with the increase of the dimension. Figures 2 and 3 show the results of the ANOVA tests for all algorithms on F_1 and F_2. PSO and BA have long tail on the functions F_1, F_2, and more singular points. That is, the methods are not robust and are not acceptable for the function optimization. The precision of BA is inferior to those of PSO and CLSPSO, but better stability and robustness.

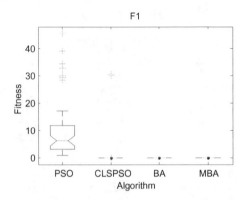

Fig. 2. ANOVA tests for $F_1(D = 100)$. (Color figure online)

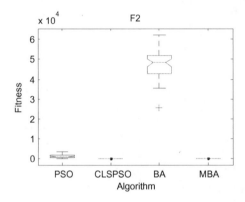

Fig. 3. ANOVA tests for $F_2(D = 100)$. (Color figure online)

5   Conclusions

This paper presented a novel Median-oriented bat algorithm (MBA) for function optimization problem. The proposed algorithm adopts the median and worst bat individuals to avoid premature convergence. MBA has an excellent ability of global search owing to its diversity caused by the probabilistic representation. The simulation experiments show that the proposed algorithm is a feasible and effective way for function optimization. The optimization ability of MBA does not show a significant decline with increasing the dimension. The proposed algorithm is suitable for low - dimensional and high-dimensional case.

References

1. Xu, C.F., Duan, H.B., Liu, F.: Chaotic artificial bee colony approach to Uninhabited Combat Air Vehicle (UCAV) path planning. Aerosp. Sci. Technol. **14**(8), 535–541 (2010)
2. Karaboga, D., Basturk, B.: A powerful and efficient algorithm for numerical function optimization: artificial bee colony (ABC) algorithm. J. Global Optim. **39**(3), 459–471 (2007)
3. Li, X., Luo, J.P., et al.: An improved shuffled frog-leaping algorithm with extremal optimisation for continuous optimization. Inf. Sci. **192**, 143–151 (2012)
4. Neshat, M., Sepidnam, G., et al.: Artificial fish swarm algorithm: a survey of the state-of-the-art, hybridization, combinatorial and indicative applications. Artif. Intell. Rev. **42**, 965–997 (2012). doi:10.1007/s10462-012-9342-2
5. Gandomi, A.H., Yang, X.S., Alavi, A.H.: Cuckoo search algorithm: a meta-heuristic approach to solve structural optimization problems. Eng. Comput. **29**, 17–35 (2013)
6. Zhao, R.Q., Tang, W.S.: Monkey algorithm for global numerical optimization. J. Uncertain Syst. **2**(3), 164–175 (2008)
7. Gandomi, A.H., Yang, X.S., Talatahari, S., et al.: Firefly algorithm with chaos. Commun. Nonlinear Sci. Numer. Simul. **18**(1), 89–98 (2013)
8. Wu, B., Qian, C.H., et al.: The improvement of glowworm swarm optimization for continuous optimization problems. Expert Syst. Appl. **39**(7), 6335–6342 (2012)
9. Yang, X.-S.: Flower pollination algorithm for global optimization. In: Durand-Lose, J., Jonoska, N. (eds.) UCNC 2012. LNCS, vol. 7445, pp. 240–249. Springer, Heidelberg (2012)
10. Bayraktar, Z., Komurcu M., Werner D.H.: Wind driven optimization (WDO): a novel nature-inspired optimization algorithm and its application to electromagnetics. In: 2010 IEEE International Symposium on Antennas and Propagation Society International Symposium (APSURSI) (2010)
11. Kaveh, A., Talatahari, S.: A novel heuristic optimization method: charged system search. Acta Mech. **213**, 267–289 (2010)
12. Yang, X.-S.: A New metaheuristic bat-inspired algorithm. In: González, J.R., Pelta, D.A., Cruz, C., Terrazas, G., Krasnogor, N. (eds.) NICSO 2010. SCI, vol. 284, pp. 65–74. Springer, Heidelberg (2010)
13. Gandomi, A.H., Yang, X.S., Alavi, A.H., et al.: Bat algorithm for constrained optimization tasks. Neural Comput. Appl. **22**, 1239–1255 (2013)
14. Yang, X.S., Gandomi, A.H.: Bat algorithm: a novel approach for global engineering optimization. Eng. Comput. **29**(5), 464–483 (2012)
15. Yang, X.S.: Bat algorithm for multi-objective optimization. Int. J. Bio-Inspir. Comput. **3**(5), 267–274 (2011)
16. Hasançebi, O., Teke, T., Pekcan, O.: A bat-inspired algorithm for structural optimization. Comput. Struct. **128**, 77–90 (2013)
17. Hasançebi, O., Carbas, S.: Bat inspired algorithm for discrete size optimization of steel frames. Adv. Eng. Softw. **67**, 173–185 (2014)
18. Beheshti, Z., Hj, S.M., Hasan, S.S.: MPSO: median-oriented particle swarm optimization. Appl. Math. Comput. **219**, 5817–5836 (2013)
19. Zhou, Y.Q., Xie, J., Li, L.L., Ma, M.Z.: Cloud model bat algorithm. Sci. World J. **2014**, 11 (2014). doi:10.1155/2014/237102. Article ID 237102
20. Li, L.L., Zhou, Y.Q.: A novel complex-valued bat algorithm. Neural Comput. Appl. **25**, 1369–1381 (2014). doi:10.1007/s00521-014-1624-y

21. Gandomi, A.H., Yang, X.S.: Chaotic bat algorithm. J. Comput. Sci. **5**, 224–232 (2014)
22. Zhou, Y. Q., Li, L.L., Ma, M.Z.: A complex-valued encoding bat algorithm for solving 0–1 knapsack problem. Neural Process. Lett. 1–24 (2015) DOI: 10.1007/s11063-015-9465-y
23. Karaboga, D., Akay, B.: A comparative study of Artificial Bee Colony algorithm. Appl. Math. Comput. **214**, 108–132 (2009)

An Improved Artificial Bee Colony Algorithm for Solving Extremal Optimization of Function Problem

Yunfei Yi[1,3], Gang Fang[2], Yangqian Su[1],
Jian Miao[1(✉)], and Zhi Yin[4]

[1] College of Computer and Information Engineering,
Hechi University, Yizhou 546300, China
gxyiyf@163.com
[2] Computer School, Wuhan University, Wuhan 43007, China
[3] Key Laboratory of Guangxi High Schools Complex System
and Computational Intelligence, Nanning 530006, China
[4] Hechi Municipal People's Government Office, Hechi 547000, China

Abstract. Some defects of artificial bee colony algorithm such as low efficiency, slow convergence rate, may lead to a fall into local optimum. In order to deal with these problems, some thoughts in genetic algorithm are introduced to improve bee colony algorithm in this paper. Specifically, a factor, which used to memorize the current global optimal position, is added to the follower bee operator to improve the global convergence speed and accuracy of the bee colony algorithm. Additionally, inertia factor and search factor are also adopted to change the proportion between the factors that affect the global convergence speed and local convergence speed, in order to accelerate the speed of bee colony algorithm applied in function extremum optimization. The experimental results show that the improved algorithm leads to fast convergence, high efficiency and robust performance.

Keywords: Bee colony algorithm · Leader bee · Follower bee · Function extremum optimization of · Follower bee searching operator

1 Introduction

Bee colony algorithm (BC algorithm) was proposed in 1995 by Seeley [1], it is a model of intelligent optimization algorithm established based on biological behavior of bees. The cooperation of each individual bee is to achieve the group searching results, and each bee searches for the relatively optimal solution in the entire solution space of the local optimization process. Every individual bee is responsible for one single task at each level during the global searching process. With the help of wagging dance, smell and other, bees are able to exchange information among the colony; the entire group is able to fulfill various tasks such as feeding, collecting honey, building nests and many other types of work. The advantage of BC algorithm is that the optimal solution can be obtained without knowing the problem details. Specifically, the proposed method is a method that can obtain the optimal value of current iteration by comparing the fitness

© Springer International Publishing Switzerland 2016
D.-S. Huang et al. (Eds.): ICIC 2016, Part I, LNCS 9771, pp. 703–713, 2016.
DOI: 10.1007/978-3-319-42291-6_70

function in the iterative process of the algorithm, then the values in all iterative process are compared to select the optimal one, and the global optimal solution is obtained when the algorithm ends. BC algorithm is widely applied in function optimization problems. Therefore, relevant researches on this topic make sense.

Current research on BC algorithm is mainly focused on the imitation of two mechanisms, gathering honey and reproducing offspring. Artificial bee colony algorithm [2–4] (ABC algorithm) imitates intelligent behavior of bees, and used to solve the problem of numerical function optimization. It has been used in the field of Artificial neural network training, bound constrained problem and the realization of the fuzzy clustering [5], as it's good performance and highly robustness [6]. Nowadays has been stepped into various areas of researches, such as Least secondary spanning tree problem [7], flow shop schedule [8, 9].

This paper it is studied the behavior of honey bees to imitate and make exhaustive study on the advantages and disadvantages of artificial bee colony algorithm. In addition, an optimization method is proposed to improve follower bee operator, and the improved algorithm is applied to solve function extreme problems.

The remainder of this paper is organized as follows: Sect. 2, the artificial bee colony algorithm is stated mathematically. Section 3, the Improved ABC algorithm is stated mathematically. The simulation results and analysis are provided in Sect. 4. Finally, we end the paper with some conclusions and future work in Sect. 5.

2 Artificial Bee Colony Algorithm

Like single particle swarm optimization algorithm (PSO algorithm) and differential evolution algorithm (DE algorithm) [10], ABC algorithm utilizes only common controlling parameters, such as population size and the maximum loop count. Served as an optimization tool, ABC model provides a colony-based search process to search for the global optimal solution [5, 11, 12]. In this search process, the nectar source called food location is constantly modified with the searching of the algorithm, in order to find the nectar source with more honey, until the source with the richest honey is found. The mutual information exchange between bees is of great importance in the form of swarm intelligence. Dance is a very important area for information exchanging where bees dance the "wagtail dance" in the area to exchange information with each other. In this way, food source information is shared with other bees, and the degree of exploitation efficiency of the food source is represented by the time that leader bee last dancing, so follower bees are able to watch a large number of wagtail dance and according to the exploitation degree of the food source to choose which food source to gather honey. Therefore, the higher yield of food source, the more likely the food source be chosen by follower bee, which means the probability food source attract follow bees is proportional to the yield of the source. BC algorithm mainly consists of four components:

2.1   Nectar Source

BC algorithms use nectar source to represent the feasible solutions, and the quality of solutions are represented by the fitness of the nectar source. It should be noted that all individual bee is treated as leader bee at the beginning of the algorithm.

2.2   Recruiting Follower Bee Operator

Refers to the relevant information of nectar source provided by leader bees in the dance region, recruiting follower bee operator is namely by fitness, to select which honey source to gather honey. This selection criterion is calculating according to certain probability, generally the roulette method, to determine which nectar source to choose. When use the roulette method, the probability of each bee being chosen is according to the proportion that the nectar fitness accounts for the total sum of alternative nectar fitness, namely:

$$P_i = fitness_i / \sum_{j=1}^{N} fit \tag{1}$$

where P_i represents the probability of the i-th nectar source being selected; *fitness* denotes the fitness of the i-th nectar; N is the total number of nectar source.

When follower bee completes the honey collection in the selected nectar source, it is supposed to return to the dance area, then possesses three options as following:

(1)   Dance the wagtail dance in the dance area to share information (become leader bee). The purpose of dancing wagtail dance is attracts more bees to the nectar source then collect more honey. After that back to the food source and gather honey.

(2)   Not to recruit bees but return to the previous place in which collected honey to go on gathering honey.

(3)   Abandon the previous nectar source, and then become a non-employment bee.

However, for non-employment bees, they will either be scouts bee searching for nearby food source, or watching the wagtail dance then become follower bee recruited by leader bee. Then follower bees follow corresponding leader bee arriving at appropriate nectar source to make honey.

2.3   Follower Bee Searching for Nectar Operator

In BC algorithm, the position of leader bee is updated by the follower bee searching for nectar operator (*fs* operator for short). Similar to mutation operator in GA, the operator performs like this: get new location by random searching in individual field, then execute mutation with certain probability, add individual location information afterwards to generate new location. The formula is as follows:

$$X_i(t+1) = X_i(t) + r \times (X_i(t) - X_k(t)) \tag{2}$$

where $X_i(t+1)$ represents the location of the i-th nectar in the $t+1$ generation; $X_i(t)$ denotes the original location of the i-th nectar; r is random number in the range $[-1,1]$; k is a random number in $\{1,2,3, \ldots, N\}$ and not equal to i, $X_k(t)$ represents the k-th individual in the t-th generation. It can be seen from this equation that with the decrease of the difference between X_i and X_k, the location change of X_i is also getting smaller, and with gradual close to the optimal solution, the step size is decrease adaptable.

2.4   Abandon Nectar Source Operator

With the process of gathering honey, the nectar can be collected less and less, and could eventually dry up. In this case, the fitness of this nectar source is very low which makes the solution quality becomes bad relatively. In order to deal with the desperate case, an upper limit of collecting honey in the nectar source has been set. When the exploiting counts one nectar source reaches the upper limit but there's no one with higher fitness found, it is necessary to search for a new random nectar source in the neighbor area and abandon the original source. This will not only improve the quality but also increase the diversity of solution, and it's more likely to get the global optimal solution. The operation of randomly generating solution is as follows:

$$X_i(t+1) = r \times X_i(t) \tag{3}$$

3   Improved ABC Algorithm

3.1   Follower Bee Operator Optimization

According to the above analysis, when utilizing the fs operator, formula (2) only consists of previous location of nectar source and the ratio that evolve towards a random individual, which leads the evolution process of nectar source to possess certain contingency. Each evolution is executed towards a randomly selected individual in the mechanism, the chance remains too strong which will result in that the convergence rate is not fast enough and the accuracy is not high, though it can be more effective to avoid falling into local optimum. The reason is nectar source location updated its evolutionary direction by one merely random individual location, and this location may be far away from the global optimum location, which makes the algorithm will degenerate when implemented. As a result, we propose an optimization method of the follower bee operator.

In this method, a new nectar source location to the fs operator is added; the location is generated by modifying some parts or the whole genes of the original individual nectar source using the current global optimal position. The algorithm will make it possible that with certain probability the operator evolve towards the location of the current global optimum during each search process. As a result, the algorithm will not fall into the local optimal location and able to converge to the global optimum quickly.

However, the added current optimal location can influence the *fs* operator by the situation as follow: as the current location of the global optimum may not be the ultimate global optimal value, if the current location of the global optimum is just an local optimal value before this search process and the operator is completely under the impact of the global optimal position, it is likely that the operator would fall into local optimum prematurely. Therefore, this added current global optimal location should be judged with a random probability r to decide the proportion, and then affect the *fs* operator. The improved *fs* operator is performed as the following formula:

$$X_i(t+1) = X_i(t) + r_1 \times (X_i(t) - X_k(t)) + r_2 \times (X_i(t) - X_b(t)) \tag{4}$$

where r_1 and r_2 are random numbers in the range $[-1,1]$ respectively; $X_b(t)$ represents the current global optimal position; $(X_i(t) - X_k(t))$ denotes random change to part of or the whole genes of an individual; $r_2 \times (X_i(t) - X_b(t))$ represents changing part of or all of the individual genes with the current global optimal position.

3.2 Objective Function and Parameter Settings

In this experiment, Goldstein-Price function is used as the evaluation function, which is defined as follows:

$$\begin{aligned} f(x_1,x_2) =&[1 + (x_1 + x_2 + 1)^2 \times (19 - 14x_1 + 3x_1^2 - 14x_2 + 6x_1x_2 + 3x_2^2)] \times \\ &[30 + (2x_1 - 3x_2)^2 \times (18 - 32x_1 + 12x_1^2 \times 48x_2 - 36x_1x_2 + 27x_2^2)], -2 \leq x_i \leq 2(i=1,2) \end{aligned}$$

$$(5)$$

The reason why Goldstein-Price function is adopted is that the function has only one global minimum point and 3 current global minimum points in the definition domain $[-2, 2]$. The global minimum point value is 3 at point $(0, -1)$.

Parameter setting is an important part in the improved *fs* operator method, method to determine the selected parameters previously is often empirical or carried out by virtue of large amounts of tests. In order to get a better parameter setting, we carry out a new orthogonal experiment method to decide the principal parameters in this paper,. There are three main parameters in the method, they are: the number of nectar source, the upper limit of exploiting the nectar source, the number of follower bees. Experiments are implemented on the three parameters with three levels respectively. Three representative values are selected, the details of values set shown in Table 1. The corresponding orthogonal design table and results are shown in Table 2.

Table 1. Level and factor table

Factors	Nectar source	Exploitation upper limit	Follower bee
1	20	60	60
2	40	80	80
3	60	100	100

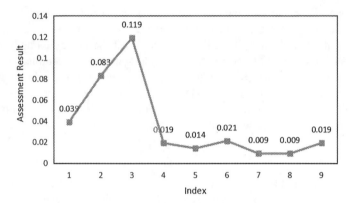

Fig. 1. Compression of assessment result

Table 2. The table of orthogonal experimental design and result

Index	Nectar source number	Exploitation upper limit	Follower bee number	Mean iteration number	Convergence number	Assessment result
1	20	60	60	461	18	0.039
2	20	80	80	421	35	0.083
3	20	100	100	336	40	0.119
4	40	60	80	474	9	0.019
5	40	80	100	508	7	0.014
6	40	100	60	566	12	0.021
7	60	60	100	665	6	0.009
8	60	80	60	588	6	0.009
9	60	100	80	423	8	0.019

Based on the evaluation results in orthogonal experimental design part of Table 2 the appropriate combination of parameters is being define. It can be seen from Table 2 and Fig. 1, No. 3 combination (20, 100, 100) achieves the best results, so this combination is adopted in this paper for setting parameters.

3.3 Algorithm Flow

Step 1: initialize parameters before the algorithm starts. The settings include: set the random function, the number of nectar source, the upper exploitation limit, the number of follower bees, dimension and the solution space range. Then, initialize the information of each nectar, included the global optimum position and the fitness of initial position of all nectar sources calculated according to Eq. (5).

Step 2: utilize the roulette method of Eq. (1) to lead each follower bee to select corresponding nectar source and gather honey.

Step 3: after selecting appropriate nectar source, follower bees will collect honey in the nectar source, and add one exploitation count after completing the collecting. Then the follower bees return to the dance area, sharing the information and continue to choose a nectar source to collecting honey using the roulette method.

Step 4: If the degree of exploitation has reached the upper limit, discard the nectar and find new nectar according to Eq. (3). Otherwise, search for a new nectar source nearby according to Eq. (4), and judge if the fitness of the new nectar source is higher than the original one. If true, discard the original nectar and use the new one, and set the exploitation count of new nectar source to 1; otherwise, continue to use the original nectar source for implement exploitation.

Step 5: after one search, calculate the fitness value of all nectar source and compare the global optimum, set the better one as the new global optimal nectar, according to Eq. (5).

Step 6: judge if the end condition is satisfied. If the global optimum position is found or the maximum number of iteration has reached the limits, the algorithm halts; otherwise jump to Step 2 and continue to loop.

4   Experimental Results and Analysis

The simulation environment: Intel (R) Core (TM) i3 M 370 @2.40 GHz, 2 GB RAM, Win7 operation system, simulation software: C++ language under the VC platform, Visual Studio 2010 platform. Due to the domain of Goldstein-Price function is in the range of [−2, 2], within which only one global minimum point exists that the value at the point (0, −1) is 3 (Table 3).

Table 3. Setting of parameter

Parameter	Nectar source number	Solution space range	Exploitation upper limit	Maximum evolution count	Follower bee number	Dimension
Value	20	[−2,2]	100	1000	100	2

It can be seen from Fig. 2 and Table 5 that when using the original ABC algorithm for solving the extreme value of Goldstein-Price function, 40 out of 50 times the original algorithm gets the results which meet the accuracy condition within the iteration limit. In those 40 times, the total iteration count is 13457, average number of iterations per time is 336. While the improved ABC algorithm gets the global optimal value 48 times in 50 experiments, and the total iterations count is 12148, mean iterations count is 253 (Table 4).

It can be easily seen from Fig. 2 that when solving the global optimal solution, the improved ABC algorithm outperforms the original ABC algorithm for 27 times in terms of the solution accuracy, and there are three times the two algorithms get the

Table 4. The experimental result

Number	ABC algo.		Improved ABC algo.		Number	ABC algo.		Improved ABC algo.	
	Iteration count	Optimal solution	Iteration count	Optimal solution		Iteration count	Optimal solution	Iteration count	Optimal solution
1	1000	3.00015	634	3.00008	26	493	3.00006	163	3.00001
2	351	3.00003	201	3.00004	27	406	3.00006	150	3.00001
3	399	3.00003	795	3.00008	28	628	3.00007	24	3.00004
4	24	3.00006	436	3	29	28	3.00001	1000	3.00013
5	1000	3.00049	364	3.00002	30	1000	3.00011	193	3.00003
6	143	3.00001	136	3.00002	31	1000	3.00016	253	3.00001
7	296	3.00002	58	3.00009	32	548	3.00005	40	3.00002
8	1000	3.00033	805	3.00006	33	313	3	58	3.00004
9	261	3.00002	78	3.00004	34	208	3.00007	154	3.00008
10	1000	3.00013	79	3.00009	35	91	3.00001	92	3.00007
11	621	3.00008	143	3	36	1000	3.00013	656	3.00003
12	378	3.00005	266	3.00001	37	1000	3.00017	267	3.00004
13	58	3.00001	176	3.00004	38	644	3.00009	1000	3.00013
14	451	3.00009	504	3	39	287	3.00009	655	3.00006
15	89	3.00005	36	3.00005	40	783	3.00003	83	3
16	87	3.00003	645	3.00007	41	171	3.00005	508	3.00006
17	480	3	155	3.00001	42	214	3.00004	101	3.00003
18	711	3.00007	71	3.00001	43	176	3.00002	284	3.00009
19	592	3.00009	232	3.00008	44	124	3.00002	239	3.00003
20	709	3.00002	321	3.00007	45	50	3.00005	55	3.00003
21	1000	3.00011	12	3.00004	46	381	3	70	3
22	640	3.00005	175	3.00007	47	471	3.00004	80	3.00009
23	87	3.00008	344	3.00002	48	380	3.00001	513	3.00001
24	43	3.00006	161	3.00001	49	377	3.00005	76	3.00009
25	1000	3.00017	266	3.0000	50	264	3.00001	341	3

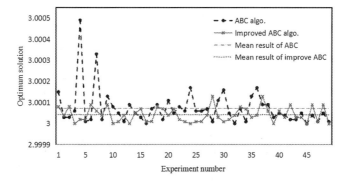

Fig. 2. Comparison of global optimum

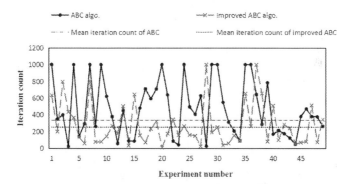

Fig. 3. Comparison of convergence rate

Table 5. The experimental result contrast

	Mean optimal result	Standard deviation of optimal result	Convergence count	Total iteration count	Mean iteration count
ABC algo.	3.0000726	$8.37331*10^{-5}$	40	13457	336
IABC algo.	3.0000428	$3.46877*10^{-5}$	48	12148	253

same results. In Fig. 3 on the comparison of convergence, the proposed algorithm is better than the original ABC algorithm in 33 times with regard to the convergence speed. It can be concluded that the improved ABC algorithm outperforms the original ABC algorithm either in terms of the count of solution that meets the condition in 50 experiments, or the mean iterations count in those experiments which get the desired solutions. The mechanism that adding the factor which saves the information of current global position to the original *fs* operator and leads the algorithm to search for the local optimum in the neighborhood area of nectar source and simultaneously evolves towards the global optimal position ever is the main reason of the better performance. In this way, not only the probability of obtaining global optimal location but also the convergence speed gets increased. As a result, it is demonstrated that the improved algorithm converges faster, gets solutions with higher accuracy and is more robust when solving function extreme value problems.

5   Conclusion and Future Work

Currently, ABC algorithm is being investigated widely [13–16]. It will certainly be the hot topic for a certain period of time in the future, because of the big challenge on the continuous improvement and application of ABC algorithm. More importantly, there is a high academic value in studying the algorithm. The algorithm in this paper is based on the basic ABC algorithm and makes improvements on the *fs* operator to apply the algorithm in solving function extreme value problems. By means of adding factor that saves current global optimal position to the *fs* operator, the mutation is implemented. Experiment results show that the improved ABC algorithm achieves significant improvement in terms of the average number of iterations and the times count of finding out optimal solution. Although the performance of ABC algorithm is greatly improved in this paper, there still remain many aspects for further studied. For instance, it's still unclear how to set the ratio that the additive memory factor affects the *fs* operator in the improved *fs* operator. Additionally, in the parameter settings, no specific relationship between the parameters and the algorithm efficiency has been found. All of these are worthy of being investigated and it is necessary that more scholars make continuous efforts to study the BC algorithm and come up with better ideas.

Acknowledgments. This work was supported by the Scientific Research Project of Hechi University (No. XJ2016KQ01), National Undergraduate Training Programs for Innovation and Entrepreneurship (No. 201610605029), Open Foundation of Guangxi Key Laboratory of Hybrid Computation and IC Design Analysis (No. HCIC201411), and the Education Scientific Research Foundation of Guangxi Province (No. KY2015YB254).

References

1. Seeley, T.D.: The Wisdom of the Hive: The Social Physiology of Honey Bee Colonies. Harvard University Press, Cambridge (1995)
2. Teodorović, D., Orco, M.D.: Bee colony optimization a cooperative learning approach to complex transportation problems. In: Proceedings of the 10th EWGT Meeting, Poznan, 13–16 September 2005
3. Karaboga, D., Basturk, B.: A powerful and efficient algorithm for numerical function optimization: artificial bee colony (ABC) algorithm. J. Glob. Optim. **39**(3), 459–471 (2007)
4. Karaboga, D.: Artificial bee colony algorithm. Scholarpedia **5**(3), 24–32 (2010)

5. Akay, B., Karaboga, D.: A modified Artificial Bee Colony algorithm for real-parameter optimization. Inf. Sci. **192**(6), 120–142 (2012)
6. AdiSrikanth, Kulkarni, N.J., Naveen, K.V.: Test case optimization using artificial bee colony algorithm. Commun. Comput. Inf. Sci. **192**, 570–579 (2011)
7. Sundar, S., Singh, A.: A swarm intelligence approach to the quadratic minimum spanning tree problem. Inf. Sci. **180**(17), 3182–3191 (2010)
8. Sang, H., Pan, Q.: Artificial bee colony algorithm for lot-streaming flow shop scheduling problem. Inf. Sci. **181**(12), 2455–2468 (2011)
9. Li, X., Yin, M.: A discrete artificial bee colony algorithm with composite mutation strategies for permutation flow shop scheduling problem. Scientia Iranica **19**(6), 1921–1935 (2012)
10. Karaboga, D., Akay, B.: A comparative study of Artificial Bee Colony algorithm. Appl. Math. Comput. **214**(1), 108–132 (2009)
11. Zhu, G., Kwong, S.: Gbest-guided artificial bee colony algorithm for numerical function optimization. Appl. Math. Comput. **217**(7), 3166–3173 (2010)
12. Guo, P., Cheng, W., Liang, J.: Global artificial bee colony search algorithm for numerical function optimization. In: Seventh International Conference on Natural Computation, ICNC 2011, Shanghai, China, vol. 08, pp. 1280–1283 (2011)
13. Abu-Mouti, F.S., El-Hawary, M.E.: Overview of artificial bee colony (ABC) algorithm and its applications. In: 2012 IEEE International Systems Conference (SysCon), pp. 1–6 (2012)
14. Zhang, C., Zheng, J., Zhou, Y.: Two modified Artificial Bee Colony algorithms inspired by grenade explosion method. Neurocomputing **151**, 1198–1207 (2015)
15. Ozturk, C., Hancer, E., Karaboga, D.: Improved clustering criterion for image clustering with artificial bee colony algorithm. Formal Pattern Anal. Appl. **18**(3), 587–599 (2015)
16. Wang, B.: A novel Artificial Bee Colony Algorithm based on modified search strategy and generalized opposition-based learning. J. Intell. Fuzzy Syst. Appl. Eng. Technol. **28**(3), 1023–1037 (2015)

Modified Cuckoo Search Algorithm for Solving Permutation Flow Shop Problem

Hong-Qing Zheng[1(✉)], Yong-Quan Zhou[2], and Cong Xie[1]

[1] College of Information Engineering,
Guangxi University of Foreign Languages, Nanning 530222, China
zhq7972@sina.com
[2] College of Information Science and Engineering,
Guangxi University for Nationalities, Nanning 530006, China

Abstract. In this paper, a modified cuckoo search (MCS) algorithm is proposed for solving the permutation flow shop scheduling problem (PFSP). Firstly, to make CS suitable for solving PFSPs, the largest position value (LPV) rule is presented to convert the continuous values of individuals in CS to job permutations. Secondly, after the CS-based exploration, a simple but efficient local search, which is designed according to the PFSPs' landscape, is applied to emphasize exploitation. In addition, the proposed algorithm is combined with the path relinking. Simulation results and comparisons based on benchmarks demonstrate the MCS is an effective approach for flow shop scheduling problems.

Keywords: Flow-shop scheduling · Hybrid algorithm · Local search · Modified cuckoo search algorithm

1 Introduce

Permutation flow shop scheduling problem is a well-known combinatorial optimization problem, it has been proved to be a typical NP problems [1]. Because of its important theoretical significance and engineering value, the PFSP has become a hot research topic in the current scheduling problem. The method of solving this problem usually includes 3 kinds: accurate algorithm [2, 3] and heuristic algorithm [4, 5] and meta-heuristic algorithm [6–8]. Exact algorithms generally only suitable for small scale problems, the heuristic algorithm is constructed, although it can solve large-scale problems, but the quality of the solution is often not high; meta-heuristic algorithm is optimized by imitating nature of certain phenomena and processes, which is generally from a solution as the starting point, it will continue to search the search space until reaches certain conditions and obtain the problem of the optimal solution or approximate optimal solution. Therefore, the meta-heuristic algorithm is a good method to solve the PFSP.

© Springer International Publishing Switzerland 2016.
D.-S. Huang et al. (Eds.): ICIC 2016, Part I, LNCS 9771, pp. 714–721, 2016.
DOI: 10.1007/978-3-319-42291-6_71

Recently, a new heuristic search algorithm, called Cuckoo Search (CS) [9], has been developed by Yang and Deb. The CS is a search algorithm based on the interesting breeding behavior such as brood parasitism of certain species of cuckoos. Each nest within the swarm is represented by a vector in multidimensional search space, the CS algorithm also determines how to update the position of cuckoo lay egg. Each cuckoo updates it position of lay egg based on current step size via Lévy flights. It has been shown that this simple model has been applied successfully to continuous nonlinear function, engineering optimization problem, etc. The CS was originally developed for continuous valued spaces, but many problems are, however, defined for discrete valued spaces where the domain of the variables is finite. We proposed an improved CS algorithm for solving PFSPs.

The remainder of this paper is organized as follows: Sect. 2, the PFSP is stated mathematically. Section 3 discusses the modified cuckoo search algorithm. The simulation results and comparisons are provided in Sect. 4. Finally, we end the paper with some conclusions in Sect. 5.

2 Formulation of PFSP

The permutation flow shop scheduling with n jobs and m machines is commonly defined as follows: each of job is to be sequentially processed on machine $k = 1, 2, \cdots, m$. The processing time of each job on machine k is given. At any time, each machine can process at most one job and each job can be processed on at most one machine. The sequence in which the jobs are to be processed is the same for each machine. The aim is to find a sequence for processing all jobs on all machines to optimize one or more objectives. In this paper, we want to optimize the objectives: the makespan C_{max} stands for the objective of minimizing the max completion time. Here, we suppose $\pi = (\pi_1, \pi_2, \cdots, \pi_n)$ to be any a processing sequence of all jobs. Suppose $C(\pi_j, k)$ is the completion time of job π_j on machine k and $t_{\pi_j,k}$ is the processing time of job t_{π_j} on machine k, then the mathematical formulation of the permutation flow shop problem can be described as follows:

$$C(\pi_1, 1) = t_{\pi_1,1}$$
$$C(\pi_j, 1) = C(\pi_j - 1, 1) + t_{\pi_j,1} \qquad j = 2, \ldots, n$$
$$C(\pi_1, k) = C(\pi_1, k - 1) + t_{\pi_1,k} \quad k = 2, \ldots, m$$
$$C(\pi_j, k) = \max(C(\pi_j - 1, k), C(\pi_j, k - 1)) + t_{\pi_j,k}$$
$$j = 2, \ldots, n \qquad k = 2, \ldots, m$$

So the maximum completion time can be described as:

$$C_{max}(\pi) = C(\pi_n, m)$$

The goal of flow scheduling is to find a best solution π^*, To satisfy the following conditions:

$$C_{\max}(\pi^*) \le (\pi_n, m), \quad \forall \pi \in \Pi$$

3   Modified Cuckoo Search

In this part, two field search structure in MCS are discussed: one is to establish in the successive values of the bird's nest, usually in CS Eq. (1) is used. However, another is uses the path relinking in the scheduling, and will be the best solution to keep the next generation. Then use local search. Two field search structure as shown in Figs. 1 and 2.

The updating formula of path and position that Cuckoo searches nest is as follows:

$$x_i^{(t+1)} = x_i^{(t)} + \partial^{\oplus} L(\lambda), \quad i = 1, 2, \cdots, n \tag{1}$$

3.1   Representation of Solution

A simple example of the rule is shown in Table 1. In Table 1, we have six jobs, so the job is from 1 to 6, we suppose the nest is: $nest_i = [3.89, 2.93, 3.07, -0.87, -0.21, 3.14]$, by sorting the $nest_i$ in descending order. Finally, the order of job is the processing order $\pi = [143652]$ which stands for the solution of the PFSP.

Table 1. A simple example of the rule for representation of solution

Job	1	2	3	4	5	6
$nest_i$	3.89	2.93	3.07	-0.87	-0.21	3.14
π	1	4	3	6	5	2

3.2   Swap and Inversion Strategy

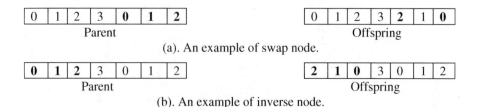

(a). An example of swap node.

(b). An example of inverse node.

Fig. 1. Local search

3.3   Path Relinking

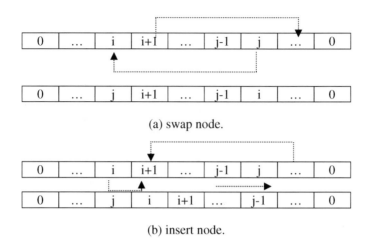

(a) swap node.

(b) insert node.

Fig. 2. Path relinking.

3.4   General Description of MCS

As the CS has good global search ability, so we propose a MCS based on the CS, the procedure of MCS is as follows:

Step 1: Initialize control parameters
Set the value of control parameters for MCS, the population size of MCS is 40, the rate of abandoned nest $p_a = 0.25$.

Step 2: Initialize the population
Determine initial population $nest^0 = [nest_1^0; nest_2^0; \cdots ; nest_n^0]$, where $nest_i^0 = [x_1, x_2, \cdots x_j]$, n is the population scale, j is the number of jobs, then by sorting the $nest_i^0$ in descending order. In this step, the initial processing sequence of each job on all machines is generated randomly.

Step 3: Evaluate the population
Obtain objective values (C_{max}) by evaluating $nest_i^0$, store the best one into $Best_0$, store the best individual into $Best_{0\_1}$ and best job sequence into $Best_{0\_2}$. In this step, we perform the evaluation operation based on job sequence and get the objectives value by calculating C_{max} and $Best_{0\_2}$ for all jobs, respectively.

Step 4: Perform the evolution
One is update the $nest^0$ by using Eq. (1), on the other hand is update $Best_{0\_2}$ by using path relinking, and then use the local search algorithm.

Step 5: Stopping condition check
If the stopping condition $t < itermax$ (represents the maximum number of iterations) is met or the optimum is found, output the optimum, else $t = t + 1$ and go to Step 4.

4   Experimental Results

4.1   Testing Benchmarks

To test the performance of the proposed MCS, computational simulation is carried out with some well-studied benchmarks.

In this paper, a total of 36 test examples, 29 problems that were contributed to the OR-Library are selected. The first eight problems were called car1, car2 through car8 by Carlier [10]. The other 21 problems were called rec01, rec03 through rec41 by Reeves [11]. In addition, the algorithm is applied to seven benchmark problems from http://mistic.heig-vd.ch/taillard/. Who used them to compare the performances of simulated annealing, genetic algorithm and neighborhood search and found these problems to be particularly difficulty. Thus far these problems have been used as benchmarks for study with different methods by many researchers.

4.2   Simulation Results and Comparisons

All computational experiments are conducted with Matlab R2010a, and run on Celeron (R) Dual-core CPU T3100, 1.90 GHz with 2 GB memory capacity. The essential parameters of MCS model for the PFSP are set as follows: we set population size as 40, maximum generation as $n \times m$, local search in consecutive $n \times m/10$ generations.

We run each algorithm 20 times for every problem, and the statistical results are summarized in Table 1, where BRE, ARE denote the best, average relative errors with C* (lower bound or optimal makespan) respectively.

From Table 1, it can be seen that the results obtained by MCS are much better than PGA. But Compared with HQGA [12], we found that the results of Rec17, Rec23, Rec25, Rec33, Rec35 are better than it, Other similar or slightly worse. Convergence curve of some test problems have been shown in Figs. 3 and 4 (Table 3).

Table 2. Comparison of the experimental result

P	n, m	C	MCS		HQGA		PGA	
			BRE	ARE	BRE	ARE	BRE	ARE
Car1	11,5	7038	0	0	0	0	0	0
Car2	13,4	7166	0	0	0	0	0	0.21
Car3	12,5	7312	0	0	0	0	0	0.86
Car4	14,4	8003	0	0	0	0	0	0.08
Car5	10,6	7720	0	0	0	0	0	0.09
Car6	8,9	8505	0	0	0	0	0	0.26
Car7	7,7	6590	0	0	0	0	0	0
Car8	8,8	8366	0	0	0	0	0	0.18
Rec01	20,5	1247	0	0.15	0	0.14	1.28	4.45
Rec03	20,5	1109	0	0.06	0	0.17	0.72	2.18
Rec05	20,5	1242	0.24	0.35	0.24	0.34	0.24	1.50

(*Continued*)

Table 2. *(Continued)*

P	n, m	C	MCS		HQGA		PGA	
			BRE	ARE	BRE	ARE	BRE	ARE
Rec07	20,10	1566	**0**	**1.21**	0	1.02	1.85	2.98
Rec09	20,10	1537	**0**	**0.54**	0	0.64	2.60	4.12
Rec11	20,10	1431	**0**	**0.78**	0	0.67	4.05	6.17
Rec13	20,15	1930	**0.25**	**1.01**	0.16	1.07	3.01	4.66
Rec15	20,15	1950	**0.05**	**0.86**	0.05	0.97	2.21	3.96
Rec17	20,15	1902	**0**	**0.68**	0.63	1.68	3.26	5.86
Rec19	30,10	2093	**1.05**	**2.07**	0.29	1.43	5.88	7.39
Rec21	30,10	2017	**1.43**	**1.63**	1.44	1.63	5.16	7.05
Rec23	30,10	2011	**0.59**	**1.90**	0.50	1.20	5.92	7.56
Rec25	30,15	2513	**0.55**	**2.86**	0.77	1.87	6.29	7.85
Rec27	30,15	2373	**1.22**	**1.73**	0.97	1.83	6.41	8.14
Rec29	30,15	2287	**0.91**	**2.96**	0.35	1.97	8.53	10.53
Rec31	50,10	3045	**2.03**	**3.24**	1.05	2.50	8.83	10.16
Rec33	50,10	3114	**0.51**	**0.98**	0.83	0.91	5.84	7.07
Rec35	50,10	3277	**0**	**0**	0	0.15	2.72	3.88
Rec37	75,20	4951	**3.02**	**5.63**	2.52	4.33	13.78	15.27
Rec39	75,20	5087	**2.90**	**4.79**	1.65	2.71	12.04	13.06
Rec41	75,20	4960	**4.71**	**5.99**	3.13	4.15	13.48	15.86

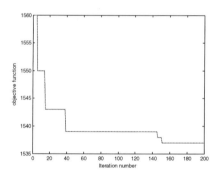

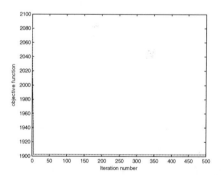

Fig. 3. The graph of rec09 function convergence

Fig. 4. The graph of rec17 function convergence

To further show the effectiveness of MCS, we carry on some comparisons with popular hybrid algorithms. We make the comparisons between NPSO [13] and the gravitational search algorithm (IGSA) proposed by Lian et al. [7], and the statistical results are summarized in Table 2.

From Table 2, it can be seen that the results of only Ta011 is better than NPSO and IGSA, Ta001, Ta031 and Ta061 are almost, Others results slightly worse. Convergence curve of some test problems have been shown in Figs. 5 and 6.

Table 3. Comparison of the experimental result

PFSP	Scale n,m	OPT	NPSO			IGSA			MCS		
			Min	Max	Avg	Min	Max	Avg	Min	Max	Avg
Ta001	20,5	1278	1278	1297	1297.9	1278	1297	1278	**1278**	**1278**	**1278**
Ta011	20,10	1582	1582	1639	1605.8	1583	1614	1600.7	**1582**	**1603**	**1592**
Ta021	20,20	2297	2297	2367	2334.9	2297	2356	2331.4	**2310**	**2328**	**2318**
Ta031	50,5	2724	2724	2729	2425	2724	2724	2427	**2724**	**2724**	**2427**
Ta041	50,10	2991	3034	3129	3086.9	3025	3046	3032.2	**3047**	**3126**	**3086**
Ta051	50,20	3771 ~ 3847	3938	3989	3964.3	3933	3952	3940.7	**3936**	**3978**	**3969**
Ta061	100,5	5493	5493	5495	5493.0	5493	5493	5493.0	**5493**	**5493**	**5493**

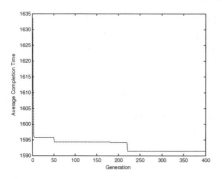 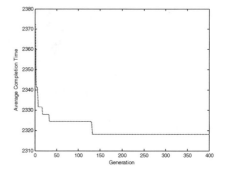

Fig. 5. The graph of Ta011 function convergence

Fig. 6. The graph of Ta021 function convergence

5 Conclusion

In terms of the discrete space optimization problem such as PFSP, we proposed a modified CS algorithm. Path relinking and local search algorithm are added to a cuckoo search algorithm for improving CS algorithm's performance. The experimental results show that this algorithm is considerably effective to the PFSP with moderate size, however, with the increasing scale of the problem, the nest's iterations need increase and they are more and more difficult to jump out of the approximate best solution. Generally speaking, the larger scale of the nest swarms needs more evolutionary time so the more quantities of the best solutions are required, and the computational complexity will subsequently increase greatly. In our future work, we will develop effective CS for kings of scheduling problems, and incorporate a suitable way into CS to propose powerful algorithms for stochastic scheduling problems.

Acknowledgments. This work is supported by the Project of Guangxi High School Science Foundation under Grant no. KY2015YB539.

References

1. Garey, E.L., Johnson, D.S., Sethi, R.: The complexity of flow-shop and job-shop scheduling. Math. Oper. Res. **1**, 117–129 (1976)
2. Croce, F.D., Narayan, V., Tadei, R.: The two-machine total completion time flow shop problem. Eur. J. Oper. Res. **90**, 227–237 (1996)
3. Croce, F.D., Ghirardim, M., Tadei, R.: An improved branch-and-bound algorithm for the two machine total completion time flow-shop problem. Eur. J. Oper. Res. **139**, 293–301 (2002)
4. Palmer, D.S.: Sequencing jobs through a multistage process in the minimum total time: a quick method of obtaining a near-optimum. Oper. Res. Q. **16**, 101–107 (1965)
5. Gupta, J.N.D.: Heuristic algorithms for multistage flowshop scheduling problem. AIIE Trans. **4**, 11–18 (1972)
6. Zhou, J.-H., Ye, C.-M., Sheng, X.-H.: Research on permutation flow-shop scheduling problem by intelligent water drop algorithm. Comput. Sci. **40**(9), 250–253 (2013)
7. Shen, J., Wang, L., Wang, S.: Chemical reaction optimization algorithm for the distributed permutation flowshop scheduling problem. J. Tsinghua Univ. (Sci. Technol.) **55**(11), 1184–1189 (2015)
8. Zheng, X., Wang, L., Wang, S.: A hybrid discrete fruit fly optimization algorithm for solving permutation flow-shop scheduling problem. Control Theor. Appl. **31**(2), 159–164 (2014)
9. Yang, X.S., Deb, S.: Cuckoo search via Levy flights. In: Proceedings of World Congress on Nature and Biologically Inspired Computing, pp. 210–214. IEEE Publications, India (2009)
10. Carlier, J.: Ordonnancements a contraintes disjonctives. R.A.I.R.O. Recherche Operationelle/Oper. Res. **12**, 333–351 (1978)
11. Reeves, C.R.: A genetic algorithm for flow-shop sequencing. Comput. Oper. Res. **22**, 5–13 (1995)
12. Wang, L., Wu, H., Tang, F., Zheng, D.-Z.: A hybrid quantum-inspired genetic algorithm for flow shop scheduling. In: Huang, D.-S., Zhang, X.-P., Huang, G.-B. (eds.) ICIC 2005. LNCS, vol. 3645, pp. 636–644. Springer, Heidelberg (2005)
13. Lian, Z.G., Gu, X.S., Jiao, B.: A novel particle swarm optimization algorithm for permutation flow-shop scheduling to minimize makespan. Chaos, Solitons Fractals **35**, 851–861 (2008)

Application of Improved Cuckoo Search Algorithm to Path Planning Unmanned Aerial Vehicle

Cong Xie and Hongqing Zheng[(✉)]

College of Information Engineering,
Guangxi University of Foreign Languages, Nanning 530222, China
zhq7972@sina.com

Abstract. Path planning of an Unmanned Aerial Vehicle (UAV) is NP complete problem. It is a hard problem to solve, especially when the number of control points is high and the number of radar is more, even more so. At present, the intelligent algorithm becomes the mainstream method of UAV route planning problem. For this question, this paper proposed an improved hybrid cuckoo search algorithm, combined with the crossover and mutation operator of genetic algorithm. Simulation results show that when the number of control points is high and the number of radar is more, this method can offer a safe and effective path planning for unmanned aerial vehicle.

Keywords: UAV · Cuckoo search algorithm · Genetic algorithm · Route planning

1 Introduction

UAV is a kind of Unmanned and reusable aircraft, which flight along the route of the default or autonomously. In order to complete a variety of military and civilian missions, unmanned aerial vehicle control system must handle many problems, such as task collaboration, path planning and path tracking. Especially, the path planning is the key technology of unmanned aerial vehicle navigation and control.

Study of Route Planning began in 1960 [1], when many approaches have been proposed to solve this problem, such as the A* algorithm [2], Voronoi graph search algorithm [3] and the dynamic programming algorithm [4], and so on. Since the route planning is NP complete problem, the computation time is increasing rapidly when the scale of the problem becomes larger and the environment is more complex. The above methods cannot solve the constraints of tasks well. In recent years, many new heuristic algorithms have been successfully applied to the route planning issues, such as genetic algorithms [5, 6], ant colony optimization (ACO) [7] and the bee colony algorithm [8]. In the literature [5], 6 radar threat areas are simulated by introducing the immigration operator and artificial selection operator to improve the standard genetic algorithm. In the literature [6], 7 radar threat areas are simulated by using the strategy of selection for ensuring quality and the scheme of improved code. literature [7], 4 radar threat areas are simulated by adopting the strategy of keeping optimal solution and self-adaptive

© Springer International Publishing Switzerland 2016
D.-S. Huang et al. (Eds.): ICIC 2016, Part I, LNCS 9771, pp. 722–729, 2016.
DOI: 10.1007/978-3-319-42291-6_72

route selection in the earlier stage, and improving the updating rules of pheromone and Volatile factor in the period. In the literature [8], 5 radar threat areas are simulated by designing the route coding scheme and the randomized initial route generation algorithm. The route use tournament selection.

In the literature [9–11], proposed some new method. The algorithm above improved algorithm in the application of the UAV route planning from different angles, but when the number of threats increases and the environment is more complex, whether the above algorithms can be applied correctly remain to be studied. Aiming at the above problems, the article designs a hybrid cuckoo search algorithm which simulate under multiple control points and complex environment. The simulation experiment results show that this method can find the optimal solution of the problem within an acceptable time and in a more complex environment.

Cuckoo search algorithm is a new global search algorithm, which complete the process of searching optimization through the behavior that Cuckoo randomly selecting the nest to lay eggs in Levy flight. It has been widely used in practical problems, such as engineering optimization, but it rarely is used in route planning. In order to solve the above problems, this paper proposes an improved cuckoo search algorithm which combination genetic operators.

This article is organized as follows: Sect. 2 describes the route coding and evaluation of performance. Section 3 discusses an improved search algorithm cuckoo combination genetic operators. The validity of the algorithm is verified in Sect. 4. Finally, we end this paper with some conclusions and future work in Sect. 5.

2 Route Planning

2.1 Flight Environment and Task Descriptions

Unmanned aerial vehicle need to plan route before flying and automatically generate the best route according to the terrain and the enemy information. Because of unmanned aerial vehicle flight level is higher and the terrain can't realize the shelter, only consider the level of track, converting 3 d route planning to two dimensional problems. So the main task of path planning is how to avoid risks and choose the shortest route.

2.2 Routing Coding

The coding of solution which y coordinates indicates in literature [7] determines the feasibility and efficiency of the algorithm. Suppose navigation electronic map is 50 km × 50 km, circles represent a threat area, the radius is the effective distance of threat area, planning space is shown in Fig. 1. With S as the origin of coordinates, T for the target point. Rectangular coordinate system regards the straight line that ST is located in as the x axis and regards the line that ST is vertical to as y axis. Then the ST is divided into n parts whose length is equal, that is $L_1, L_2, \ldots, L_n$. The unmanned aerial vehicle route will pass the point where $L_1, L_2, \ldots, L_n$ are located. Now, the abscissa is fixed and the ordinate changes between -25 to 25, so the coding of solution can be expressed as formula (1) and route coding can be expressed as formula (2). With this kind of the coding method, when the number of control points is more, the route will be smoother. Certainly, time and the difficulty of calculating also increase simultaneously.

$$x = \{y_1, y_2, \ldots, y_n\} \tag{1}$$

$$path = \{S, y_1, y_2, \ldots, y_n, T\} \tag{2}$$

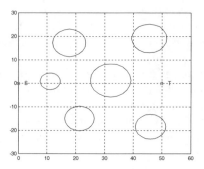

Fig. 1. Flight space diagram

2.3   Route Performance Evaluation

Before determining the aircraft route, it is necessary to make sure that each performance index of the route. It mainly includes the cost of fuel and the cost of threat. The goal of the route is to make a minimum total cost. This article uses the following formula to evaluate the route performance [7]:

$$\min J = \sum_i^n [k_1 w_{ti} + k_2 w_{fi}] \tag{3}$$

In the formula, J is the route index function, n is the number of waypoints, and w_{ti} is the fuel cost of i-th segment route, which is proportional to the voyage of L_i, k_1, k_2 represent weight coefficient, which we can use to make a choice preferentially according to the UAV's requirements; w_{fi} is the threat cost of the i-th segment route. The threat cost of each waypoint is:

$$w_{fi} = \rho / d_{min} \tag{4}$$

In the formula, ρ represents the avoiding threats coefficient. The larger ρ is, the safer unmanned aerial vehicle is. d_{min} represents the minimum distance between route and all threats. If the aircraft entered the no-fly zone, w_{fi} will be infinity. When taking (4) into (3), we can get the following cost function:

$$\min J = \sum_i^n [k_1 L_i + k_2 \rho / d_{min}] \tag{5}$$

3 UAV Route Planning Combined with GA Algorithm

3.1 Improved CS Algorithm (ICS)

Cuckoo Search algorithm is by the scholars of the University of Cambridge, UK, Yang and Deb, to simulate the behavior that Cuckoos search nest to lay eggs. Then, they found a new Search algorithm, namely the Cuckoo Search algorithm [12]. Because of this algorithm is simple, efficient and easy to implement, it is widely used in various fields. The algorithm simulates the natural process that Cuckoo search nest to lay eggs, where the parameters of handling problem are coded to form the bird's nest. Population consists of lots of bird's nests. The individuals of population choose bird's nest by Levy flight of Cuckoo and determine to abandon the bird's nest to update population according to a certain probability. To iterate until it gets the final optimization results.

The updating formula of path and position that Cuckoo searches nest is as follows:

$$x_i^{(t+1)} = x_i^{(t)} + \partial^{\oplus} L(\lambda), \quad i = 1, 2, \cdots, n \tag{6}$$

In the formula, $x_i^{(t)}$ represents the position of the i-th one in the t-th generation of nests, $\oplus$ represents dot product, ∂ represents Step Length, $L(\lambda)$ represents path that Levy randomly searches. $L \sim u = t^{-\tau}, 1 < \tau \leq 3$.

The (6) after refining:

$$x_i^{(t+1)} = x_i^{(t)} + \partial . * step^{(t)} . * \left(x_i^{(t)} - best^{(t)} \right), \quad i = 1, 2, \cdots, n \tag{7}$$

In the formula, $step^{(t)}$ represents the t-th generation step size that Levy flight generated. $best^{(t)}$ is the best solution to the t-th generation. ∂ in basic cuckoo search algorithm is a constant, it is not conducive to rapid convergence and accuracy of the algorithm. So do the following adjustments:

$$\partial = \partial_{max} - \frac{\partial_{max} - \partial_{min}}{itermax} \times t \tag{8}$$

In the formula, ∂_{max} represents the maximum step size, ∂_{min} represents the minimum step size, $itermax$ represents the maximum number of iterations, t represents the current number of iterations.

3.2 CS Algorithm Combined with Genetic Algorithm Operators

Genetic algorithm is one of the earliest intelligent algorithms, which simulates the survival of the fittest in nature. With the aid of heredity and variation operation, it can solve the problem and get an approximate optimal solution. Scholars study it successfully. Its application fields are also quite extensive. Intelligent algorithms are easy to fall into local optimum prevalent drawbacks commonly. In order to make the cuckoo search algorithm out of local optimal solution, to speed up the search efficiency, it combines the crossover and mutation operator of genetic algorithm.

3.3 Algorithm Steps

Step 1: Initialize parameters and randomly initialize solutions.

Step 2: Using the formula (5) to undertake an evaluation of all the nests to obtain *fmin* and *best*.

Step 3: With the formula (7) and (8), generating a new nest and obtaining *fmin*1 and *best*1 by evaluated the fitness value. If *fmin*1 <*fmin*, replacing the original optimal value and optimal solution with *fmin*1 and *fbest*1, otherwise being unchanged.

Step 4: Using the ideas of genetic algorithms to transform the nest.

Step5: Obtain *fmin*2 and *best*2 by evaluating fitness value again. If *fmin*2 <*fmin*, replacing the original optimal value and the optimal solution with *fbest*2 and *fmin*2, otherwise unchanged.

Step 6: Determine whether the cycle ends, if it is, outputting waypoint information, otherwise, skipping to step 3.

4 The Simulation Examples and Analysis

In order to verify the performance of the proposed algorithm, using standard cuckoo search algorithm and improved cuckoo search algorithm to perform three times experiments respectively, using MATLAB R2010a to write Codes, and all instances run in Celeron processor (R) dual-core CPU T3100, 1.90 GHz, 2 GB memory for PC If the area UAV flight was 50 km × 50 km electronic maps. Parameters are set to: population size is 25, the total number of iterations is 500 or 100; Weight coefficient $k_1 = 1$, $k_2 = 1$, the rate of abandoning colonies $pa = 0.25$, the ratio of crossing $pc = 0.85$, aversion coefficient $\rho = 30$, $\partial_{max} = 0.1$, $\partial_{min} = 0.01$. Each experiment independently operates 10 times.

- Experiment 1: six threat points, nine control points.
  Unmanned aerial vehicle route planning based on standard Cuckoo search algorithm runs 10 times and gets navigation route as shown in Fig. 2. The change of the optimal solution is shown in Fig. 3. The improved algorithms for navigation route as shown in Fig. 4.
  From Figs. 2 and 4 shows, when the number of threat points and control points are less, with standard cuckoo search algorithm and improved the cuckoo search algorithm, obtaining a resemble route planning renderings by running 10 times. The difference of route cost is small, stability is better, and navigation route is smooth, and above shows the feasibility of the algorithm. However, the advantages of the improved algorithm is not obvious, in order to illustrate the efficiency of the improved algorithm preferably, at this time, increasing the number of control points and threat point, continue to experiment two and three (Fig. 7).
- Experiment 2: number of control points for 24 (Fig. 9).
- Experiment 3: randomly generate 50 points.
  Here for all adopts approximate circle represents a threatened area, radius for the effective distance of a threatened area, circle for the threatened area. A threat of Figs. 2 and 6 point coordinate is (11.1 0.8), (17.7 17.2), (21.3 −15), (32.1 1.2),

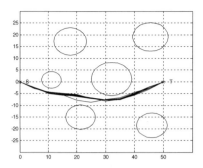

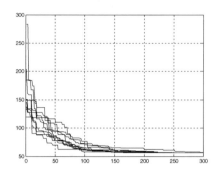

Fig. 2. Standard CS algorithm for the route plan

Fig. 3. Change curve of the standard CS

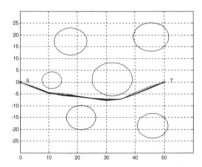

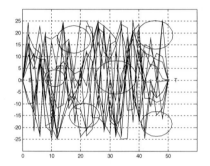

Fig. 4. ICS algorithm for the route plan

Fig. 5. Standard CS algorithm for the route plan

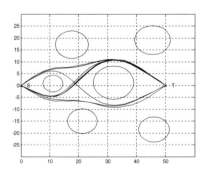

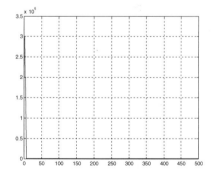

Fig. 6. ICS algorithm of route plan

Fig. 7. Change curve of the ICS

728   C. Xie and H. Zheng

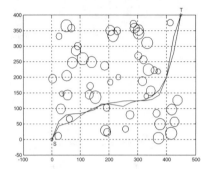

Fig. 8. ICS for route plan under complex environment

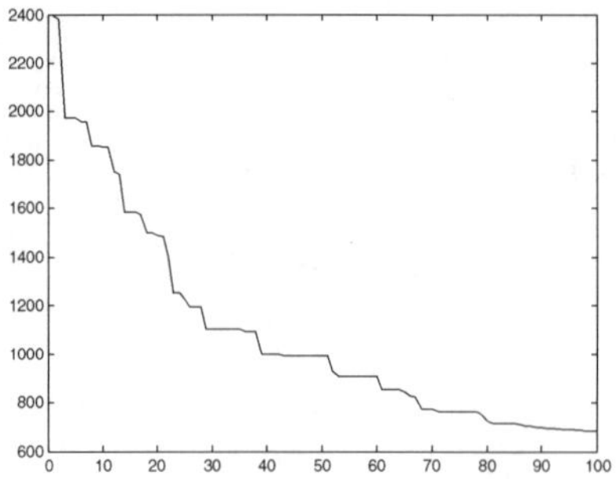

Fig. 9. Change curve of ICS algorithm under complex environment

(45.5 19.1), (46 −18.5), Threat radius is (3.5, 5.8, 5.2, 7.04, 6.11, 5.3), Fig. 8 threat the coordinates and radius are randomly generated. From Fig. 5, we can easily to find that, when the control points increase, the standard cuckoo search algorithm could not find any feasible and safe route by running 10 times, while the improved cuckoo search algorithm can find the optimal solution absolutely every time, and improve the success rate of path planning. In Fig. 8, when the threat environment is more complex, the improved algorithm still can find safe and feasible route, but this moment, the success rate of route planning only is 20 %. All the simulation experiment results show that in a complex environment, the algorithm designed in this paper for unmanned aerial vehicle route planning is effective. Its run time is under 10 s. Planning route is smoother, which also is advantageous to the unmanned aerial vehicle flight.

5 Conclusion

UAV Route planning is a challenging problem, especially when constrained environment is more complex. Due to the use of deterministic search algorithms within a reasonable time is difficult to obtain the optimal solution, in this article, a new kind of cuckoo search algorithm analyzes and studies the route planning problem of unmanned aerial vehicle, combined with the crossover and mutation operator of genetic algorithm, and modify the representation of a fixed step. All simulation results show that, when the number of threat points and control points are more, improved cuckoo search algorithm is faster, the success rate of which is higher. The method can find a safe and feasible route, but the success is expected to be improved. This will be the direction in the future research.

Acknowledgment. This work is supported by the Project of Guangxi High School Science Foundation under Grant no. KY2015YB539

References

1. Latombe, J.: Robot Motion Planning. Springer, New York (1990)
2. Richards, N.D., Sharma, M., Ward, D.G.: A hybrid A* automaton approach to on-line path planning with obstacle avoidance. In: AIAA-2004-6229 (2004)
3. Liu, Z., Shi, J.G., Gao, X.G.: Application of Voronoi diagram in flight path planning. Acta Aeronautica et Astronautica Sinica **29**, 16–18 (2008)
4. Jennings, A.L., Ordonez, R., Ceccarelli, N.: Dynamic programming applied to UAV way point path planning in wind. In: Proceedings of IEEE International Symposium on Computer-Aided Control System Design, pp. 215–220 (2008)
5. Yu, J.-X., Zhou, C.-L., Liu, D.-P.: Based on improved genetic algorithm for UAV route planning and simulation. Comput. Simul. **30**(12), 17–20 (2013)
6. Zheng, R., Feng, Z.-M., Lu, M.-Q.: Application of particle genetic algorithm to plan planning of unmanned aerial vehicle. Comput. Simul. **28**(6), 88–91 (2011)
7. Qiu, X.-H., Qiu, Y.-C.: Application of ant algorithm to path planning of unmanned aerial vehicle. Comput. Simul. **27**(9), 102–105 (2010)
8. Hua, S.-S.: Research and simulation of UAV route planning optimization method. Comput. Simul. **30**(4), 45–48 (2013)
9. Turker, T., Sahingoz, O.K., Yilmaz, G.: 2D path planning for UAVs in Radar threatening environment using simulated annealing algorithm. In: 2015 International Conference on Unmanned Aircraft Systems (ICUAS), pp. 56–61 (2015)
10. Chi, T.-Y., Ming, Y., Kuo, S.-Y., Liao, C.C.: Civil UAV path planning algorithm for considering connection with cellular data network. In: 2012 IEEE 12th International Conference on Computer and Information Technology, pp. 327–330 (2012)
11. Zhang, B., Liu, W., Mao, Z.: Cooperative and geometric learning algorithm (CGLA) for path planning of UAVs with limited information. Automatica **50**, 809–820 (2014)
12. Yang, X.S., Deb, S.: Cuckoo search via Levy flights. In: Proceedings of World Congress on Nature and Biologically Inspired Computing, pp. 210–214. IEEE Publications, India (2009)

Dual-System Water Cycle Algorithm for Constrained Engineering Optimization Problems

Qifang Luo[1], Chunming Wen[1], Shilei Qiao[1],
and Yongquan Zhou[1,2(✉)]

[1] College of Information Science and Engineering,
Guangxi University for Nationalities, Nanning 530006, China
l.qf@163.com
[2] Guangxi High School Key Laboratory of Complex System
and Computational Intelligence, Nanning 530006, China

Abstract. In this paper presents an improved version of the water cycle algorithm (WCA) based on a dual cycle system, together referred to as the dual-system water cycle algorithm (DS-WCA). The DS-WCA makes the WCA faster and more robust. The new processes of inland and ocean cycles are applied to increase the diversity of the population and accelerate the convergence speed, respectively. We evaluate the ability of the DS-WCA to solve four engineering design problems. Simulations indicate that the proposed algorithm is able to obtain optimized or near-optimized solutions in all cases. Compared to other state-of-the art evolutionary algorithms, the DS-WCA performs significantly better in terms of the quality, speed, and stability of the final solutions.

Keywords: Water cycle algorithm · Dual-system water cycle algorithm · Constraint optimization · Engineering design · Optimal design

1 Introduction

Currently, because evolutionary algorithms can solve problems more easily than traditional optimization algorithms, evolutionary algorithms are widely applied in different fields, such as numerical computing and engineering optimization. Increasingly more modern evolutionary algorithms inspired by natural or animal behavior are emerging, and they are becoming increasingly popular. The water cycle algorithm (WCA) is one such evolutionary algorithm, based on the observations of the water cycle and how rivers and streams flow downhill towards the sea in the real world. The water cycle algorithm (WCA) was proposed by Eskandar et al. [1]. This algorithm gradually aroused attention and has been increasingly applied to different areas of study. For example, Sadollah et al. [2] applied the WCA to solve multi-objective optimization problems. Zhang et al. [3] applied the WCA to solve practical engineering optimization problems. Sadollah et al. [4] proposed an ER-WCA (water cycle algorithm with evaporation rate) for solving constrained and unconstrained optimization problems [5–7].

© Springer International Publishing Switzerland 2016
D.-S. Huang et al. (Eds.): ICIC 2016, Part I, LNCS 9771, pp. 730–741, 2016.
DOI: 10.1007/978-3-319-42291-6_73

However, the problems of slow convergence speed and low accuracy still exist. In fact, the water cycle system includes two parts in nature—the outer cycle and the inner cycle. The former is also called the cycle between the land and sea, and the latter includes inland and ocean cycles [8]. This paper presents an improved version of the water cycle algorithm for constrained engineering optimization problems, i.e., the so-called dual-system water cycle algorithm (DS-WCA). The proposed algorithm is applied to two engineering problems to assess its efficiency. The simulation results showed that the proposed algorithm is effective and more robust than the traditional WCA.

2 The Basic WCA Algorithms

The WCA simulates the flow of rivers and streams towards the sea according to water cycle processes. The starting point is the process of rainfall. We chose the best individual (best raindrop) as the sea. Then, some other good raindrops are chosen as rivers, while the rest are considered to be streams.

2.1 Initialization

To begin the algorithm, an initial population is generated randomly as follows:

$$Population = \begin{bmatrix} Sea \\ River_1 \\ River_2 \\ River_3 \\ \vdots \\ Stream_{Nsr+1} \\ Stream_{Nsr+2} \\ Stream_{Nsr+3} \\ \vdots \\ Stream_{N_{pop}} \end{bmatrix} = \begin{bmatrix} x_{1,1} & x_{1,2} & x_{1,3} & \cdots & x_{1,N} \\ x_{2,1} & x_{2,2} & x_{2,3} & \cdots & x_{2,N} \\ \vdots & \vdots & \vdots & \vdots & \vdots \\ x_{N_{pop},1} & x_{N_{pop},2} & x_{N_{pop},3} & \cdots & x_{N_{pop},N} \end{bmatrix} \tag{1}$$

where N_{pop} and N are the number of individuals in a population and the number of design dimensions, respectively. In the first step, N_{pop} streams are created. Some of N_{sr} (good individuals) are selected as the sea and rivers. The rest of the population is calculated using the following equation:

$$N_{sr} = Number\ of\ Rivers + \underbrace{1}_{Sea} \tag{2}$$

$$N_{stream} = N_{pop} - N_{sr} \tag{3}$$

Depending on flow magnitude, the amount of water entering a river or sea varies from stream to stream. In the WCA, the designated streams for each river and sea are calculated as:

$$NS_n = round\left\{ \left| \frac{f(River_n)}{\sum_{i=1}^{N_{sr}} f(River_i)} \right| \times N_{Stream} \right\}, \quad n = 1, 2, \ldots N_{sr} \quad (4)$$

where NS_n is the number of streams that flow to the specific rivers and f is the evaluation function in the algorithm.

2.2   Flow and Exchange

As in nature, the streams are created from the raindrops and join each other to form new rivers. All rivers and streams flow to the sea (best optimal point). A stream flows to the river along the connecting line between them, using a randomly chosen distance, given as follows:

$$X \in (0, C \times d) \quad C > 1 \quad (5)$$

where C is a value between 1 and 2. The current distance between the stream and river is represented as d. The value of X in Eq. (5) corresponds to a distributed random number between 0 and $(C \times d)$.

If the value of C is greater than one, the streams are able to flow in different directions towards the rivers. This concept may also be used in rivers flowing to the sea. Therefore, the new position for streams and rivers may be given as

$$X_{Stream}^{i+1} = X_{Stream}^i + rand \times C \times (X_{River}^i - X_{Stream}^i) \quad (6)$$

$$X_{River}^{i+1} = X_{River}^i + rand \times C \times (X_{Sea} - X_{River}^i) \quad (7)$$

If the solution of a stream is better than its connecting river, then the positions of the river and stream are exchanged. Similarly, if the solution of a river is better than the sea, the positions of the river and sea are exchanged.

2.3   The Rainfall Process

The rainfall process is one of the most important operators that can prevent the algorithm from rapid convergence. In the WCA, this assumption is proposed to avoid getting trapped in local optima. First, we check whether the stream or river is close enough to the sea to make the rainfall process occur; thus, the rainfall condition may be given as

$$if \left| X_{sea} - X_{River}^i \right| < eps \quad i = 1, 2, 3, \ldots, N_{sr}.$$

where eps is a very small constant in Matlab.

In the rainfall process, the new raindrops form streams in different locations. The following equation is used:

$$X^{new}_{Stream} = LB + rand \times (UB - LB) \tag{8}$$

where LB and UB are the lower and upper bounds defined by the given problem, respectively. Again, the best newly formed raindrop is considered as a river flowing to the sea. The rest of the new raindrops are assumed to form new streams, which flow to the rivers.

The schematic view of the WCA is illustrated in Fig. 1; the empty shapes refer to the new positions found by streams and rivers.

The basic process of WCA includes the following:

(1) Initial raindrops (population) and the division of streams, rivers, and sea.
(2) The streams flow to the rivers and the rivers flow to the sea.
(3) The positions of stream and river, river and sea are exchanged.
(4) New streams or rivers are formed by the process of rainfall.
(5) Check the convergence criteria. If the stopping criterion is satisfied, the algorithm will be stopped; otherwise, return to (2).

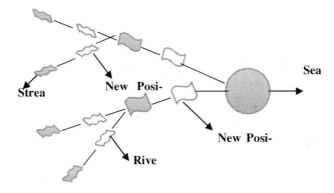

Fig. 1. Schematic view of the water cycle system

3 Dual-System Water Cycle Algorithms

This paper presents an improved version of the water cycle algorithm for constrained engineering optimization problems, i.e., the so-called dual-system (outer cycle system and inner cycle system) water cycle algorithm (DS-WCA).

3.1 Background of the Dual-System Water Cycle

In fact, the water cycle system includes two parts of the outer cycle system and the inner cycle system in nature. The former is also called the cycle between the land and sea, the latter includes the inland cycle and ocean cycle [8]. As we can see in Fig. 2, the streams form by the processes of rain and ice melting, then, the streams flow to a specific river and the river flows to the sea. Finally, evaporation and condensation will form rain again. This is the process of the outer cycle system. The confluence between rivers, and the process of evaporation and rain on the Earth's surface form the inland cycle. Secondly, the evaporation and rain on the ocean's surface form the ocean cycle, which is a process of the inner cycle system. Based on the above-mentioned system, we present an improved version of the water cycle algorithm for constrained engineering optimization problems, i.e., the so-called dual-system water cycle algorithm.

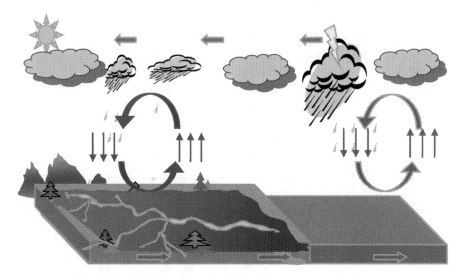

Fig. 2. Simplified diagram of the dual-system water cycle

Evolutionary algorithms can be considered as an efficient method with which to solve constrained optimization problems. Two important characteristics of meta-heuristics are exploration and exploitation [9]. Exploration guarantees that the algorithm can explore the search space more efficiently, often by randomization. This is the essential step that guarantees that the system can jump out of any local optima and generate new solutions as diversely as possible. Exploitation intends to use the information from the current best solutions. This process searches around the neighborhood of the current best solutions and selects the best candidates. Similarly, there are two different parts in the DSWCA. The outer cycle system is similar to the process of exploration, and the inner cycle system is similar to the process of exploitation. In the early stages of optimization, the outer cycle system can search for the optimal value more efficiently and avoid the algorithm falling into the local optimum. In the later stages of the optimization, the inner cycle system can search around the neighborhood of the current best solution and determine the solution more accurately by generating new rivers, which increases the diversity of the population.

3.2   The Principle and Framework of the Dual-System Water Cycle

3.2.1   The Outer Cycle Process

The outer cycle is still in accordance with the WCA, i.e., a stream flows to a river and this river flows to the sea; the new positions for streams and rivers can be calculated using Eqs. (6) and (7). If the solution given by a stream is better than its connecting river, the positions of the river and stream are exchanged.

The rainfall condition may be given as

$$if \left| X_{sea} - X_{River}^i \right| < eps \quad i = 1, 2, 3, \ldots, N_{sr}.$$

where *eps* is a very small constant in Matlab. In the rainfall process, the following equation is used:

$$X_{Stream}^{new} = LB + rand \times (UB - LB) \tag{9}$$

$$X_{River}^{new} = LB + rand \times (UB - LB) \tag{10}$$

where *LB* and *UB* are the lower and upper bounds defined by the given problem, respectively.

3.2.2   The Inner Cycle Process

In the inner cycle, we define a new operation—confluence between rivers. The process of confluence between rivers is very common in nature; this process may be given as

$$X_{River}^{i+1} = X_{River}^i + rand \times C \times (X_{River}^i - X_{River}^k) \quad k = 1, 2, 3, \cdots N_{sr}, \quad k \neq i. \tag{11}$$

Similarly, the rainfall condition may be given as

$$if \left| X_{sea} - X_{River}^i \right| < eps \quad i = 1, 2, 3, \cdots, N_{sr}.$$

where *eps* is a very small constant in Matlab.

In the inner cycle, the rainfall process occurs on the surface of the sea; hence, this process of the ocean cycle may be given as

$$X_{River}^{new} = X_{sea} + rand \times \left(\frac{1}{N_{iter}} \right) \tag{12}$$

$$X_{Stream}^{new} = X_{sea} + rand \times \left(\frac{1}{N_{iter}} \right) \tag{13}$$

where N_{iter} is the current number of iterations, *rand* is a uniformly distributed random number between 0 and 1, and *C* is a value between 1 and 2.

3.2.3   The Conditions of the Outer and Inner Cycles

$$water\, cycle \begin{cases} inner\, cycle & if \max(c\_sr)<0 \quad and \quad |X_{River}-sea|<eps \\ outer\, cycle & otherwise \end{cases} \quad (14)$$

In the above equation, c_sr is an array formed by constrained conditions of the optimization problems, and eps is a very small constant in Matlab.

3.2.4   The DS-WCA Framework
According to the above description and analysis, the steps of the DS-WCA are summarized as follows:

Step 1: Choose the initial parameters of the DS-WCA: N_{pop}, N_{sr} and maximum iterations.
Step 2: Generate a random initial population and form the initial streams, rivers, and sea using Eqs. (1), (2) and (3).
Step 3: Determine the intensity of flow for the rivers and sea using Eq. (4).
Step 4: Judge the algorithm in terms of entering the inner cycle system or outer cycle system by Eq. (13). If entering into the inner cycle, begin Step 5; otherwise, begin Step 8.
Step 5: The streams flow to the rivers and the rivers flow to the sea by Eqs. (6) and (7).
Step 6: Exchange the positions of streams, rivers, and sea according to their fitness value and constraints.
Step 7: If the rainfall condition is satisfied, the raining process will occur using Eqs. (9) and (10). Otherwise, begin Step 10.
Step 8: The process of confluence between rivers is given by Eq. (11).
Step 9: If the rainfall condition is satisfied, the raining process will occur using Eqs. (12) and (13). Otherwise, begin Step 10.
Step 10: Check the convergence criteria. If the stopping criterion is satisfied, the algorithm will be stopped; otherwise, return to Step 4.

4   Simulations and Comparisons

In this section, the proposed DS-WCA has been applied for solving four engineering design problems [10]; then, the performance of the DS-WCA was compared with other optimizers. The task of optimizing each of the test functions was executed using 25 independent runs. The initial parameters for DS-WCA, N and N_{sr}, ε were chosen as 100, 8, and 0.0001.

(1) Pressure vessel design: A cylindrical pressure vessel capped at both ends by hemispherical heads is presented in Fig. 3. The total cost, including a combination of single $60°$ welding cost, material, and forming cost, is minimized. The involved

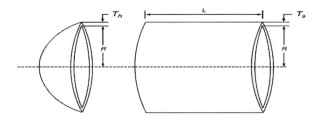

Fig. 3. Schematic view of the pressure vessel problem

variables are the thickness of the shell $T_s(x_1)$, thickness of the head $T_h(x_2)$, the inner radius $R(x_3)$, and the length of the cylindrical section of the vessel $L(x_4)$. The optimization problem can be formulated as follows:

$$
\begin{aligned}
Minimize \quad & f(x) = 0.6224x_1x_3x_4 + 1.7781x_2x_3^2 + 3.1661xx + 19.84x_1^2x_4 \\
Subject\ to \quad & g_1(x) = -x_1 + 0.0193x_3 \leq 0 \\
& g_2(x) = -x_2 + 0.00954x_3 \leq 0 \\
& g_3(x) = -\pi x_3^2 x_4 - \tfrac{4}{3}\pi x_3^3 + 1296000 \leq 0 \\
& g_4(x) = x_4 - 240 \leq 0 \\
& 0 \leq x_1, x_2 \leq 100 \\
& 10 \leq x_3, x_4 \leq 200
\end{aligned}
$$

Table 1 shows the comparisons of the best solutions obtained by the proposed DS-WCA and WCA (Table 2).

Table 1. Comparison of the best solution given by previous for the pressure vessel problem

Solution and constraint	WCA	DS-WCA
x_1	0.7958	0.7782
x_2	0.3933	0.3846
x_3	41.2310	40.3196
x_4	187.6901	200.0000
$g_1(x)$	−0.0000	−0.0000
$g_2(x)$	−0.0000	−0.0000
$g_3(x)$	−0.0143	−0.0001
$g_4(x)$	−52.3099	−40.0000
$f(x)$	5916.09144	5885.33277

(2) Tension/compression spring design problem: The tension/compression spring design problem has the objective of minimizing the weight ($f(x)$) of a tension/compression spring (as shown in Fig. 4) subject to constraints on minimum deflection (g_1), shear stress (g_2), surge frequency (g_3), and limits on outside diameter (g_4). The design variables involved are the wire diameter $d(x_1)$, the mean coil diameter $D(x_2)$, and number N of active coils (x_3). The problem can be expressed as follows (Table 3):

Table 2. Comparison of the statistical results of various algorithms for the pressure vessel problem

Author [s]	Best	Mean	Worst	Std
Aragón et al. [12]	6390.554	7694.067	6737.065	3.57E+02
Bernardino et al. [11]	6059.855	7388.16	6545.126	1.24E+02
Huang et al. [13]	6059.73	6085.23	6371.05	4.30E+01
Mezura-Montes and Coello [14]	6059.75	6850	7332.88	4.26E+02
Mezura-Montes and Hernández-Ocana [15]	6059.73	6081.78	6150.13	6.72E+01
Zahara and Kao [16]	5930.31	5946.79	5960.06	9.16E+00
He and Wang [17]	606108	6147.13	6363.8	8.65E+01
Parsopoulos and Vrahatis [18]	6544.27	9032.55	11638.2	9.96E+02
Gandomi et al. [19]	6059.71	6179.13	6318.95	1.37E+02
Eskandar et al. [1]	5916.09	6414.12	7308.27	4.20E+02
Our result	**5885.33**	5885.71	5889.14	9.773E−01

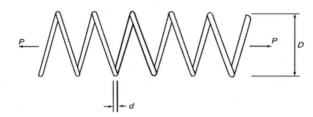

Fig. 4. Schematic view of the tension/compression spring problem

$$\text{Minimize} \quad f(x) = (x_3 + 2)x_2 x_1^2$$
$$\text{Subject to} \quad g_1(x) = 1 - \frac{x_2^3 x_3}{71785 x_1^4} \leq 0$$
$$g_2(x) = \frac{4x_2^2 - x_1 x_2}{12566(x_2 x_1^3 - x_1^4)} + \frac{1}{5108 x_1^2} - 1 \leq 0$$
$$g_3(x) = 1 - \frac{140.45 x_1}{x_2^2 x_3} \leq 0$$
$$g_4(x) = \frac{x_1 + x_2}{1.5} - 1 \leq 0$$
$$0.05 \leq x_1 \leq 1, \quad 0.25 \leq x_2 \leq 1.3, \quad 2 \leq x_3 \leq 15$$

Table 3. Comparison of the best solution given by previous studies for the tension/compression spring problem

Solution and constraint	WCA	DS-WCA
x_1	0.0500	0.0517
x_2	0.3167	0.3574
x_3	14.0171	11.2478
$g_1(x)$	−0.0000	−0.0000
$g_2(x)$	−0.0000	−0.0000
$g_3(x)$	−3.9687	−4.0552
$g_4(x)$	−0.7550	−0.7272
$f(x)$	0.0127186832	0.0126652484

Table 4. Comparison of statistical results for various algorithms for the tension/compression spring problem

Author [s]	Best	Mean	Worst	Std.
Coello Coello [21]	0.012704	0.012769	0.012822	3.94E−05
Coello and Montes [20]	0.012681	0.012742	0.012973	5.90E−05
Bernardino et al. [10]	0.012674	0.012730	0.012924	5.20E−04
Dos Santos Coelho [24]	0.012669	0.013854	0.018127	1.34E−03
Liu et al. [25]	0.012857	0.019555	0.071802	1.17E−02
Lampinen [23]	0.012670	0.012703	0.012790	2.70E−05
Ray and Liew [22]	0.012669	0.012922	0.016717	5.90E−04
Eskandar et al. [1]	0.012718	0.013054	0.014825	1.10E−04
Our result	**0.012665**	0.012665	0.012665	4.21E−08

As seen in Table 4, in terms of the best results and standard deviations, the proposed method is superior to the other optimization methods.

5 Conclusions

This paper presents an improved version of the water cycle algorithm for constrained optimization problems, i.e., the so-called dual-system water cycle algorithm (DS-WCA). In this dual-system, the outer cycle system is similar to the process of exploration, and the inner cycle system is similar to the process of exploitation. In the early stages of optimization, the outer cycle system can search for the optimal value more efficiently and avoid the algorithm falling into the local optimum. In the later stages of the optimization, the inner cycle system can search around the neighborhood of the current best solution and more accurately find the solution. In the present study, the dual-system water cycle algorithm has been employed to solve constrained optimization problems. The DS-WCA was validated using four engineering design problems. The simulations indicate that the proposed algorithm generally offers better solutions than the WCA and other optimizers compared in this research, in addition to its stability in almost every problem.

Acknowledgment. This work is supported by National Science Foundation of China under Grant No. 61563008

References

1. Eskandar, H., Sadollah, A., Bahreininejad, A., Hamdi, M.: Water cycle algorithm–a novel metaheuristic optimization method for solving constrained engineering optimization problems. Comput. Struct. **110**, 151–166 (2012)
2. Sadollah, A., Eskandar, H., Bahreininejad, A., Kim, J.H.: Water cycle algorithm for solving multi-objective optimization problems. Soft. Comput. **19**, 2587–2603 (2015)

3. Zhang, C., Liao, G.W., Li, L.J.: Optimizations of space truss structures using WCA algorithm. Prog. Steel Build. Struct. **1**(16), 35–38 (2014)
4. Sadollah, A., Eskandar, H., Bahreininejad, A., Kim, J.H.: Water cycle algorithm with evaporation rate for solving constrained and unconstrained optimization problems. Appl. Soft Comput. **30**, 58–71 (2015)
5. Eskandar, H., Sadollah, A., Bahreininejad, A., Lumpur, K.: Weight optimization of truss structures using water cycle algorithm. Int. J. Optim. Civil Eng. **3**(1), 115–129 (2013)
6. Sadollah, A., Eskandar, H., Kim, J.H.: Water cycle algorithm for solving constrained multi-objective optimization problems. Appl. Soft Comput. **27**, 279–298 (2015)
7. Sadollah, A., Eskandar, H., Bahreininejad, A., Kim, J.H.: Water cycle, mine blast and improved mine blast algorithms for discrete sizing optimization of truss structures. Comput. Struct. **149**, 1–16 (2015)
8. Randall, D.A., Dazlich, D.A.: Diurnal variability of the hydrologic cycle in a general circulation model. J. Atmos. Sci. **48**(1), 40–62 (1991)
9. Yang, X.-S.: Harmony search as a metaheuristic algorithm. In: Geem, Z.W. (ed.) Music-Inspired Harmony Search Algorithm. SCI, vol. 191, pp. 1–14. Springer, Heidelberg (2009)
10. Bernardino, H.S., Barbosa, I.J.C., Lemonge, A., Fonseca, L.G.: A new hybrid AIS-GA for constrained optimization problems in mechanical engineering. In: Evolutionary Computation, 2008, IEEE World Congress on Computational Intelligence, pp. 1455–1462. IEEE (2008)
11. Bernardino, H.S., Barbosa, H.J., Lemonge, A.C.: Constraint handling in genetic algorithms via artificial immune systems. In: Late-Breaking Paper at Genetic and Evolutionary Computation Conference, GECCO (2006)
12. Aragón, V.S., Esquivel, S.C., Coello, C.A.C.: A modified version of a T-Cell Algorithm for constrained optimization problems. Int. J. Numer. Meth. Eng. **84**(3), 351–378 (2010)
13. Huang, F.Z., Wang, L., He, Q.: An effective co-evolutionary differential evolution for constrained optimization. Appl. Math. Comput. **186**(1), 340–356 (2007)
14. Mezura-Montes, E., Coello, C.A.C.: An empirical study about the usefulness of evolution strategies to solve constrained optimization problems. Int. J. Gen. Syst. **37**(4), 443–473 (2008)
15. Mezura-Montes, E., Hernández-Ocana, B.: Bacterial foraging for engineering design problems: preliminary results. In: Memorias del 4o Congreso Nacional de Computacion Evolutiva (COMCEV 2008), CIMAT, Gto. Mexico, October 2008
16. Zahara, E., Kao, Y.T.: Hybrid Nelder-Mead simplex search and particle swarm optimization for constrained engineering design problems. Expert Syst. Appl. **36**(2), 3880–3886 (2009)
17. He, Q., Wang, L.: An effective co-evolutionary particle swarm optimization for constrained engineering design problems. Eng. Appl. Artif. Intell. **20**(1), 89–99 (2007)
18. Parsopoulos, K.E., Vrahatis, M.N.: Unified particle swarm optimization for solving constrained engineering optimization problems. In: Wang, L., Chen, K., Ong, Y. (eds.) ICNC 2005. LNCS, vol. 3612, pp. 582–591. Springer, Heidelberg (2005)
19. Gandomi, A.H., Yang, X.S., Alavi, A.H., Talatahari, S.: Bat algorithm for constrained optimization tasks. Neural Comput. Appl. **22**(6), 1239–1255 (2013)
20. Coello, C.A.C., Montes, E.M.: Constraint-handling in genetic algorithms through the use of dominance-based tournament selection. Adv. Eng. Inform. **16**(3), 193–203 (2002)
21. Coello Coello, C.A.: Constraint-handling using an evolutionary multi-objective optimization technique. Civil Eng. Syst. **17**(4), 319–346 (2000)
22. Ray, T., Liew, K.M.: Society and civilization: an optimization algorithm based on the simulation of social behavior. IEEE Trans. Evol. Comput. **7**(4), 386–396 (2003)

23. Lampinen, J.: A constraint handling approach for the differential evolution algorithm. In: Proceedings of the World on Congress on Computational Intelligence, vol. 2, pp. 1468–1473. IEEE (2002)
24. Dos Santos Coelho, L.: Gaussian quantum-behaved particle swarm optimization approaches for constrained engineering design problems. Expert Syst. Appl. **37**(2), 1676–1683 (2010)
25. Liu, H., Cai, Z., Wang, Y.: Hybridizing particle swarm optimization with differential evolution for constrained numerical and engineering optimization. Appl. Soft Comput. **10** (2), 629–640 (2010)

Accurate Prediction of Protein Hot Spots Residues Based on Gentle AdaBoost Algorithm

Zhen Sun[1], Jun Zhang[1(✉)], Chun-Hou Zheng[1], Bing Wang[2],
and Peng Chen[3]

[1] School of Electronic Engineering and Automation, Anhui University, 111#,
Jiulong Road, Economic Development Zone, Hefei 230601, Anhui, China
wwwzhangjun@163.com
[2] School of Electrical and Information Engineering,
Anhui University of Technology, Ma Anshan 243032, China
[3] Institute of Health Sciences, Anhui University, Hefei 230601, Anhui, China

Abstract. Hot spots are critical for protein interactions. Since the experimental method to measure protein hot spots are very cost and time-consuming, computational method is an option to predict hot spots in protein interfaces. In this work, we use Gentle AdaBoost to identify hot spot without feature selection. For all algorithm, ASEdb and BID are used as separate training and test dataset. We extract sequential and structural information of protein, which constitutes 178 dimensional feature vectors. Comparing with other algorithms, our proposed algorithm obtained satisfactory experimental results either on training dataset or test dataset, which yeilds F1 score of 0.79 and 0.69 on training dataset and test dataset, respectively.

Keywords: Hot spots · Gentle AdaBoost · Protein interaction

1 Introduction

Protein-protein interactions are central to most biological processes [1–3]. It has been found that the free energy change protein-protein interactions. Only a small fraction of interface residues which contribute a large portion of the binding energy are called hot spots [4]. In the binding interface, hot sports are clustered in tightly packed regions and surrounded by residues that are less important energetically [5, 6]. When the residues is mutated to alanine, an experimental method can be used to identify a hot spot if the change in its binding free energy is larger than a certain threshold. Several databases focusing on changes in the binding free energy from single amino acid mutations are existing. Alanine Scanning Energetics database (ASEdb) [7] is created by collecting hot spots data identified by alanine scanning mutagenesie experiments. The Building Interface Database (BID) [8] stores protein interaction information which come from literature and is checked by experiment.

Since alanine scanning mutagenesis experiments are cost and time-consuming, many researchers use computational method to predict hot spots [5]. Among them

© Springer International Publishing Switzerland 2016
D.-S. Huang et al. (Eds.): ICIC 2016, Part I, LNCS 9771, pp. 742–749, 2016.
DOI: 10.1007/978-3-319-42291-6_74

including energy function-based physical models, evolutionary conservation-based methods, docking-based methods and machine learning methods. Recently, support vector machines (SVMs) were often used to identify hot spot. Cho et al. [9] identified hot spots using sequence, structure and molecular interaction. Xia et al. [10] combined protrusion-based features with solvent accessibility to predict hot spots. The two methods called KFC2a and KFC2b, were made by Zhu et al. [11] based on mainly solvent accessible surface area and local plasticity.

In spite of many algorithms identify hot spots in protein interfaces successfully, most of these them focus on individual feature. However, if more valuable features were developed, a better classification algorithm is more needed. Usually, classifier ensemble is an option. In this paper, Gentle AdaBoost [12, 13] was used to identify hot spots in protein interfaces. A wide variety of features from a combination of protein sequence, structure information, physicochemical information were obtained. The used data come from ASEdb and BID benchmark datasets. Additional interface residues between hot spots and non-hot spots will be deleted from the training data set. Then we evaluate the performance of the proposed method on test set. The experimental results show that our method outperforms the state-of-the-art hot spot predictions methods.

2 Materials and Methods

2.1 Datasets

Two state-of-the-art hot spot data sets are used in this work. ASEdb and BID database are used as experimental data set. In this study, ASEdb is used as training data and BID database is used as independent test set. For all dataset, CATH query system was used to eliminate the redundancy in the data set. The sequence identity less than 35 % and the SSAP score less than or equal to 80 is deemed to redundancy. 15 protein complexes without redundancy data is shown in Table 1. Among them we eliminate protein chains which does not correspond with Consurf-DB files. In this study, a hot spots has been defined as the interface residue which corresponding binding free energy is not less than 2.0 kcal/mol. According to the method of Tuncbag et al. [14], non-hot spot residues is defined as the interface residue with binding free energy less than 0.4 kcal/mol. Other residues was eliminated from training set.

Table 1. Training set of protein structures

PDB	First molecule	Second molecule
1a4y	Angiogenin	Ribonuclease inhibitor
1a22	Human growth hormone	Human growth hormone binding protein
1ahw	Immunoglobulin Fab 5G9	Tissue factor
1brs	Barnase	Barstar
1bxi	Colicin E9 Immunity Im9	Colicin E9 DNase
1cbw	BPTI Typsin inhibitor	Chymotrypsin
1dan	Blood coagulation factor VIIA	Tissue factor

(Continued)

744   Z. Sun et al.

Table 1. *(Continued)*

PDB	First molecule	Second molecule
1dvf	Idiotopic antibody FV D1.3	Anti-idiotopic antibody FV E5.2
1fc2	Fc fragment	Fragment B of protein A
1fcc	Fc(IGG1)	Protein G
1gc1	Envelope protein GP120	CD4
1jrh	Antibody A6	Interferon-gamma receptor
1vfb	Mouse monoclonal antibody D1.3	Hen egg lysozyme
2ptc	BPTI	Trypsin
3hfm	Hen Egg Lysozyme	Ig FAB fragment HyHEL-10

The BID database is used to test the proposed methods. In the database, the relative disruptive effect of alanine mutation is labeled as "strong", "intermediate", "weak", or "insignificant". In our study, that mutations are considered as hot spots, which have been labeled "strong", another been considered as non-hot spots.

2.2 Features Description

Total 178 features used by APIS, KFC2 and DBSI were employed in this work. There features can be summarized into five categories: Physicochemical features, Features about protein structure, Features related to neighbors of the target residue and others.

Physicochemical features of an amino acid residue consist ten features: number of atoms, number of electrostatic charge, number of potential hydrogen bonds, hydrophobicity, hydrophilicity, propensity, isoelectric point, mass, expected number of contacts with 14 Å sphere, and electron-ion interaction potential.

Structure feature contains protrusion index, depth index, accessible surface area (ASA), relative ASA (RASA). The values of these feature by PSAIA can be calculated. Four useful residue attributes including total mean (mean value of all atom values), side-chain mean (mean value of all side-chain atom values), maximum (highest of all atom values) and minimum (lowest of all atom values) is come from DI and PI; five residue attributes from ASA and RASA: total (sum of all atom value), backbone (sum of all backbone atom values), side-chain (sum of all side-chain atom values), polar (sum of all oxygen, nitrogen atom values) and non-polar (sum of all carbon atom values). The relative changes in ASA, DI and PI between the complex and monomer state of the residues were also computed as follows:

$$RcASA = \frac{\text{ASA in Monomer} - \text{ASA in Complex}}{\text{ASA in Monomer}} \qquad (1)$$

$$RcDI = \frac{\text{DI in Complex} - \text{DI in Monomer}}{\text{DI in Complex}} \qquad (2)$$

$$\mathrm{RcPI} = \frac{\mathrm{PI \ in \ Monomer} - \mathrm{PI \ in \ Complex}}{\mathrm{PI \ in \ Monomer}} \qquad (3)$$

In a process of protein-protein interaction, the environmental factor of a target interface residue is so important that as there we defining environmental features of a target residue with two distances cutoffs 4.0 and 5.0 Å. On this occasion, we figured the number of residues and the number of atoms, where is around the side chain of the target residue. The number of rotatable single bonds within the side chain and the total hydrophobicity of residues around the side chain of the target residue are calculated. Otherwise computed the weighted rotatable single bond number by number of rotatable single bonds divided by the number of atoms in the side chain.

Biochemical contact features were computed and as Darnell et al. [16]. Others, we identified non-covalent interactions within protein complex by the molecular modeling package WHAT-IF. Atomic contacts, hydrogen bonds and salt bridges, the three kinds of non-covalent interactions were calculated.

An atomic contact were recorded when the distance between the van der Waals surfaces of two atoms less than 0.25 Å. Using three categories: polar, non-polar and generic, atomic contacts could be classified. Only contacts across the interface were tabulated.

We used an optimized hydrogen bond network model to identify hydrogen bond by WHAT-IF program. Based on the strength of the hydrogen bond assign hydrogen bond scores between 0 and 1. The hydrogen bond score of a residue is considered to be the sum of the hydrogen bond scores of its atoms.

We distinguish between different salt bridges using the distance between the centers of an acidic oxygen and a basic nitrogen. The salt bridge feature of a residue is the number of aslt bridges it makes with its binging partner.

Additionally, the evolution rate for each residue was obtained using the Rate4Site algorithm, which is implemented in the ConSurf-DB server (Table 2).

Table 2. The composition of each group

The number of features[a]	The type of features
18	Physicochemical features
55	Features about solvent accessibility
58	The neighbors features of the target residue
47	others

3 Method of Gentle AdaBoost

Boosting is a widely used ensemble method which combining the many "weak" classifiers to produce a powerful "committee." Giving the training dataset consisting n data point and its label pair $(x_1, y_1), \ldots, (x_n, y_n)$, x_i express as a feature vector and $y_i = -1$ or 1 express class label. We define $F(x) = \sum_1^M C_m f_m(x)$, $f_m(x)$ is a classifier producing value plus or minus 1 and C_m are constants; the corresponding prediction is sign $(F(x))$. When a training sample was misclassified, Adaboost algorithm will give a higher weight to that sample. Finally, we define the final classifier as a linear combination of the classifiers from each stage.

The simple boosting procedure has been developed from in the PAC-learning framework. The text expressed that a weak learner could always improve its performance by training two additional classifiers on filtered versions of the input data stream. A weak learner is an algorithm who producing a two-class classifier with high probability better than a coin flip. After learning an initial classifier h_1 on the first N training points:

1. h_2 is learned on a new sample of N points, half of which are misclassified by h_1.
2. h_3 is learned on N points for which h_1 and h_2 disagree.
3. The boosted classifier is h_b = Majority Vote(h_1, h_2, h_3).

In this work, we use an additive model called Gentle AdaBoost, the weak learner is CART [15]. Here we give brief description of the algorithm.

Gentle AdaBoost

---

1. **Start with weight** $w_i = \frac{1}{N}, i = 1, 2, ..., N, F(x) = 0$.
2. **Repeat for** $m = 1, 2, ..., M$:
   (a) **Fit the regression function** $f_m(x)$ **by weighted least-squares of** y_i **to** x_i **with weights** w_i.
   (b) **Update** $F(x) \leftarrow F(x) + f_m(x)$.
   (c) **Update** $w_i \leftarrow w_i + w_i \exp(-y_i f_m(x_i))$ **and renormalize** $\sum_i w_i = 1$.
3. **Output the classifier** $\text{sign}[F(x)] = \text{sign}[\sum_{m=1}^{M} f_m(x)]$.

Here N represent number of sample; M represent iterative number, value as 200; w_i is weight.

3.1 Measurements of Prediction Performance

Our data set is a typical unbalance classification problem. So, the widely-used F1 score was used to evaluate model performance in our work. The F1 score is defined as a function of Precision and Recall. Precision measures the fraction of positive hot spot identifying that are correct. Recall indicates how many total hot spots are accurately predicted. Accuracy express how many total hot spots and non-hot spots are accurately identified. If a predicted hot spot is an true hot spot, these measures is defined as true positives (TP). If a predicted hot spot is not a true hot spot, the count is defined as false positives (FP). If a residue is decided as a non-hot spot and it is a true non-hot spot, these measures is defined as true negatives (TN). Other residue is decided as non-hot spots and an actual hot spots defined as false negatives (FN). These evaluation measures are defined as follows:

$$Pre = \frac{TP}{TP + FP} \tag{4}$$

$$Rec = \frac{TP}{TP + Fn} \tag{5}$$

$$Acc = \frac{TP + TN}{TP + TN + FP + FN} \tag{6}$$

$$F1 = \frac{2 * Rec * Pre}{Rec + Pre} \tag{7}$$

4 Results and Discussion

4.1 Prediction for the Train Set

We used Sequential Forward Selection as a feature selection method. However, we do not obtained satisfactory results. When all feature was used for our model, Gentle AdaBoost predicted hot spot residues with the highest accuracy for the training data set (Accuracy 0.83).

4.2 Cross-Validation for the Training Data Set

Firstly, we use 10-fold cross-validation for training data set to evaluate Gentle Ada-Boost performance. The experimental results are shown in Table 3. We compared Gentle AdaBoost with other models by assessing residues using other methods in the same training data set. It can be found that Gentle AdaBoost has highest values of Accuracy. Only Rec and F1 lower than KFC2a [10] while better than others methods.

Table 3. The experimental result on cross-validation data for training data set

Method	Acc	Pre	Rec	F1	Mcc
Robetta	0.77	0.89	0.51	0.65	0.54
FOLDEF	0.7	0.91	0.31	0.46	0.41
KFC	0.74	0.75	0.55	0.64	0.45
MINERVA	0.81	**0.93**	0.58	0.72	0.62
KFC2a	**0.83**	0.78	**0.84**	**0.8**	**0.67**
KFC2b	0.81	0.82	0.74	0.75	0.62
APIS	0.75	0.7	0.74	0.71	0.5
Gentle AdaBoost	**0.83**	0.81	0.77	0.79	0.66

The highest values are highlighted in bold.

4.3 Prediction for the Independent Test Set

For further validating the performance of Gentle AdaBoost, Independent test set come from BID was used as test set in our model. The effect of each model is measured by four metrics: Accuracy, Precision, Recall and F1 score. The detail has been showed in Table 4.

Table 4. Comparison of different hot spot prediction methods in the independent test set

Method	Acc	Pre	Recall	F1	ΔF1
Robetta	0.72	0.52	0.33	0.41	✕
FOLDEF	0.69	0.48	0.26	0.34	-0.07
KFC	0.69	0.48	0.31	0.38	-0.03
MINERVA	0.76	0.65	0.44	0.52	+0.11
KFC2a	0.73	0.55	**0.74**	0.63	+0.22
KFC2b	0.77	0.64	0.55	0.6	+0.19
APIS	0.75	0.57	0.72	0.64	+0.23
Gentle AdaBoost	**0.77**	**0.82**	0.59	**0.69**	**+0.28**

It can be shown that Gentle AdaBoost substantially outperforms the other methods in three performance metrics (Acc, Rec and F1 score). Furthermore, the F1-score of Gentle AdaBoost is 0.69, while those of the existing methods fall in the range of 0.34–0.64. Comparing with APIS [10] and KFC2 [11], Gentle AdaBoost has achieve better performance.

The highest values are highlighted in bold. ΔF1 is change in F1 score for a model when compared with Robetta [17].

5 Conclusion

In this paper, Gentle AdaBoost algorithm was used to predict protein hot spots residues. The experimental dataset was extracted from ASEdb and BID database. ASEdb was used as training data and BID database was used as test dataset. Comparing with other algorithm, whether in training data or test data, Gentile AdaBoost algorithm achieve better prediction performance on four criteria. It can be conclude that our proposed algorithm has potential advantage to predict unbalance protein hot spots. We further need to validate the proposed algorithm with more large hotspots dataset.

Acknowledgments. This work was supported by National Natural Science Foundation of China under grant nos. 61271098 and 61032007, and Provincial Natural Science Research Program of Higher Education Institutions of Anhui Province under grant no. KJ2012A005.

References

1. Chothia, C., Janin, J.: Principles of protein-protein recognition. Nature **256**(5520), 705–708 (1975)
2. Janin, J.: Principles of protein-protein recognition from structure to thermodynamics. Biochimie **77**(7), 497–505 (1995)
3. Kann, M.G.: Protein interactions and disease: computational approaches to uncover the etiology of diseases. Briefings Bioinform. **8**(5), 333–346 (2007)
4. Moreira, I.S., Fernandes, P.A., Ramos, M.J.: Hot spots—a review of the protein–protein interface determinant amino-acid residues. Proteins: Struct., Funct., Bioinf. **68**(4), 803–812 (2007)
5. Bogan, A.A., Thorn, K.S.: Anatomy of hot spots in protein interfaces. J. Mol. Biol. **280**(1), 1–9 (1998)
6. Wells, J.A.: Systematic mutational analyses of protein-protein interfaces. Methods Enzymol. **202**, 390–411 (1991)
7. Thorn, K.S., Bogan, A.A.: ASEdb: a database of alanine mutations and their effects on the free energy of binding in protein interactions. Bioinformatics **17**(3), 284–285 (2001)
8. Fischer, T., et al.: The binding interface database (BID): a compilation of amino acid hot spots in protein interfaces. Bioinformatics **19**(11), 1453–1454 (2003)
9. Cho, K-i, Kim, D., Lee, D.: A feature-based approach to modeling protein–protein interaction hot spots. Nucleic Acids Res. **37**(8), 2672–2687 (2009)
10. Xia, J.F., Zhao, X.M., Song, J., Huang, D.S.: APIS: accurate prediction of hot spots in protein interfaces by combining protrusion index with solvent accessibility. BMC Bioinformatics **11**, 174 (2010)
11. Zhu, X., Mitchell, J.C.: KFC2: a knowledge-based hot spot prediction method based on interface solvation, atomic density, and plasticity features. Proteins: Struct. Funct. Bioinf. **79**(9), 2671–2683 (2011)
12. Demirkır, C., Sankur, B.: Face detection using look-up table based gentle AdaBoost. In: Kanade, T., Jain, A., Ratha, N.K. (eds.) AVBPA 2005. LNCS, vol. 3546, pp. 339–345. Springer, Heidelberg (2005)
13. Friedman, J., Hastie, T., Tibshirani, R.: Additive logistic regression: a statistical view of boosting (with discussion and a rejoinder by the authors). Ann. Stat. **28**(2), 337–407 (2000)
14. Tuncbag, N., Gursoy, A., Keskin, O.: Identification of computational hot spots in protein interfaces: combining solvent accessibility and inter-residue potentials improves the accuracy. Bioinformatics **25**(12), 1513–1520 (2009)
15. Breiman, L., et al.: Classification and Regression Trees. CRC Press, Boca Raton (1984)
16. Darnell, S.J., Page, D., Mitchell, J.C.: An automated decision-tree approach to predicting protein interaction hot spots. Proteins-Struct. Funct. Bioinf. **68**(4), 813–823 (2007)
17. Keskin, O., et al.: Protein-protein interactions: structurally conserved residues at protein-protein interfaces. Biophys. J. **86**(1), 267a (2004)

Is There a Relationship Between Neighborhoods of Minority Class Instances and the Performance of Classification Methods?

Asdrúbal López-Chau[1(✉)], Farid García-Lamont[2], and Jair Cervantes[2]

[1] Universidad Autónoma del Estado de México,
CU UAEM Zumpango, Camino, Viejo a Jilotzingo s/n Col. Valle Hermoso,
5600 Zumpango, Estado de México, Mexico
alchau@uaemex.mx
[2] Universidad Autónoma del Estado de México, CU UAEM Texcoco,
Av. Jardín Zumpango s/n, Fracc. El Tejocote, 56259 Texcoco, Mexico

Abstract. The performance of classification methods is notably damaged with imbalanced data sets. Although some studies to analyze this behavior have realized before, most of the conclusions obtained from experiments correspond to synthetic data sets. In this paper, we study the relationship between the performance of five classification methods and neighbors of minority class instances. According to the results of experiments, we found strong empiric evidence that the type of neighborhoods of minority class instances affect classification accuracy. Indeed, we observe that the type of neighborhood is more important than the imbalance rate. In order to validate the results, we use ten real-world imbalanced data sets, and measure AUC ROC and True Positive Rates.

Keywords: Imbalanced classification · Nearest neighbors · Minority class · SMOTE

1 Introduction

Imbalanced data sets contain a large number of instances of one category identified as the majority class, and just few instances of the opposite type, the minority class. Currently, there are many applications that generate this type of data, for example, medical diagnosis [1–3], fraud detection in telecommunications [4] and agriculture [5]. By far, the underlying concept hidden in the minority class instances is the most important [6], but also the hardest to capture by classification methods.

Classic classification methods were designed based on the hypothesis that data sets are balanced. Therefore, in most cases these methods ignore the minority class, and just focus on predicting correctly the instances of the majority class [6, 7].

In the literature, most of research on the imbalance problem has focused on three main topics [8]. The first one is the proposal of new methods to face the imbalance problem [9–13]. The second is about measuring the performance of classification

© Springer International Publishing Switzerland 2016
D.-S. Huang et al. (Eds.): ICIC 2016, Part I, LNCS 9771, pp. 750–761, 2016.
DOI: 10.1007/978-3-319-42291-6_75

methods with imbalanced data sets [14–17], most of these sets are synthetic or from specific domains. The third topic groups the works that study the complexity of data sets [18, 19]. These researches are very valuable to understand why classification methods fail with imbalanced data sets. Japkowicz and Stephen [18] realized a set of experiments to establish possible relationships between concept complexity, size of the training set and class imbalance level. Conclusions from such experiments suggest that data complexity hinders the performance of classifiers. Prati et al. [20] and Batista et al. [15] show that class overlapping has a more negative impact than imbalance itself. Both, [18, 20], use synthetic data sets. Batista et al. [15] utilized data sets from UCI repository to measure. In [19], Weiss presents a summary of results about the main features of data sets that cause low precision accuracy of classification methods in this domain.

A common approach in previous works is they characterize the complexity of entire data sets. However, the analysis of data sets at different levels is necessary to understand more deeply the imbalance problem. Recently, Smith et al. [21] presented a study in which data sets are analyzed at instance level. They examine the examples that are frequently misclassified by nine classification methods. The conclusion is similar to the found in other works: class overlap causes instance hardness. One of the hardness measures utilized in [21] is k-nearest neighbors.

We observe that an important number of algorithms designed to alleviate the imbalance problem are based in the k-nearest neighbors algorithm, see for example [10, 13, 21–26]. Therefore, we wonder if there exists a relationship between the performance of classification methods and the neighborhoods of instances. For the case of imbalanced data sets, we focus on the minority class instances, which usually contain the information about the most important concept in a data set.

Ten real-world data sets publicly available were used in our experiments. For each data set, we trained and tested five classifiers (Neural Network (NN), K-nearest neighbor classification (kNNC), Support Vector Machine (SVM), C4.5 and Naive Bayes (NB)). We measured AUC ROC and True Positive Rates (TPR) achieved. Based on this, we analyze the relationship between performance of classification methods and the neighborhoods of minority class instances. Surprisingly, we found empiric evidence such relationship exits.

The rest of this paper is organized as follows. In Sect. 2, a brief explanation of classification methods and two of the main measures of the performance of classifiers is given. The Sect. 3 shows four main types of neighborhoods of instances, and there we explain the methodology used to explore data sets. In the Sect. 4, the results of applying the methodology on ten real-world data sets are presented. Finally, in the Sect. 5, a number of observations and future research paths are included.

2 Preliminaries

2.1 Related Works

The effect of imbalance on classifiers has been studied in last years. Japkowicz and Stephen [18] tested Support Vector Machine, Multilayer Perceptron and C4.5 classifiers with imbalanced data sets. According to results of experiments, the performance of classifiers is severely damaged by the following factors: (1) the degree of class imbalance; (2) the complexity of the concept represented by the data; (3) the overall size of the training set. However, these conclusions are useful for synthetic data sets.

In [14, 27], a set of experiments have been conducted to compare the performance of classification methods with imbalanced data sets. Some observations from these and other works are the following: Between-class imbalance damage the performance of classification methods [19], within-class imbalance has a more negative impact than between-class imbalance [28, 18], noise in the data and rare cases are harmful for classifiers [7, 29]. Meanwhile, Lopez et al. [30] study some problems related to data intrinsic characteristics. Their conclusions can be summarized as follows: the presence of small clusters of minority class instances injures the predictions of classification methods; class overlapping makes very hard or even impossible the distinction between the minority and majority classes [30]; noisy and borderline instances cause more difficulties than imbalance to classification algorithms. A common characteristic among previous works is that the analysis of performances is realized at data set level. In our study, the approach is at instance level.

2.2   Classification Methods

In the literature, some popular classification methods are Support Vector Machine (SVM), Neural Networks (NN), Decision Trees (DT), Naive Bayes (NB) and k-nearest neighbor classification (kNNC).

SVM is one of the newest classification methods. It solves a quadratic programming problem to find an optimal separating hyper plane that classifies the instances. Usually, SVM uses a kernel (nonlinear function) to work in a high dimensional feature space, where the instances are easy to classify. Brown [31], Tang [32] and others authors claim that SVM is severely affected by the imbalance.

NN is a classification method whose structure resembles to biological neurons in the human brain. The structure of NN consists of an input layer, one or more hidden layers and an output one. Each layer contains nodes interconnected with the modes of the next and the previous layers. The performance of NNs is also damaged when they are applied to imbalanced data sets.

DTs partition recursively the input space, producing regions with low impurity. The metric for measure the purity of partitions is information gain for the C4.5 classifier.

NB is a probabilistic classification method that assumes independency between values of attributes. It uses the Bayes theorem to build a model to predict the class of previously unseen instances.

The kNNC predicts the category of a previously unseen instance based on the predominant class of the K nearest neighbors of the instance. The main issue of kNNC is the choice of parameter k. A low value of it affects the sensitivity to noise; a large value of k can affect the prediction accuracy due to the presence of instances of several classes.

2.3   Metrics for Evaluation of Classifiers

Henceforth, X will represent a data set, containing examples of class +1 (minority) or −1 (majority). The structure of X is as follows: $X = (x_i, y_i)|x_i \in R^d, \quad y_i \in \{+1, -1\}, i = 1, \ldots, N$ where d is the number of attributes in X; x_i is an instance, y_i is the class (label) of the instances x_i, and N = |X| is the number of instances in X.

The performance of a classifier C built from X is usually tested using unseen examples. The errors of the predictions committed by C, occur when $(y_k = C(x_k)) = y_k$, where y_k is the real class of x_k.

	True class	
	Positive	Negative
Prediction Positive	TP	FP
Negative	FN	TN

Fig. 1. Confusion matrix

Several measures of C can be obtained using a confusion matrix or contingency table [36]. Figure 1 shows the entries of a confusion matrix for two classes. The meaning of each element is the following:

TP (True Positive examples) is the number of examples of class +1, which are correctly predicted. TN (True Negative examples) is the number of examples of class −1, which are correctly predicted. FP (False Positive examples) is the number of examples of class −1, which are predicted as positive (+1). FN (False Negative examples) is the number of examples of class +1, which are predicted as negative (−1).

Instead of computing classification accuracy, which measures only the percentage of correctly classified examples, the two main measures used in this paper are True Positive Rate (TPR) which is the percentage of positive instances correctly classified; and AUC, which provides a single measure of a classifier's performance for evaluating which model is better [33]. These two measures are computed as follows:

$$TP_{rate} = TP/(TP + FN) \qquad (1)$$

$$AUC = 0.5 * (1 + TP_{rate} - FPR) \qquad (2)$$

With FPR = FP/(FP + TN). The area under the ROC curve (AUC or ROC Area) reduces ROC performance to a single scalar value representing expected performance [34]. In spite of controversial opinions, the AUC is often used as a valid measure to compare classifiers; the greater AUC, the better performance of classifiers.

3   Proposed Approach

K-nearest neighbors is an algorithm which has been applied as a tool for supporting many other methods. We use it to explore the distribution of classes within the neighborhood of minority class instances. For each minority class instance x_i, we compute its K-nearest neighbors, and count the number of instances of each class. Based on this counting, we assign a category to the neighborhood of x_i. The possible categories are the following:

Instance with neighborhood type A. Both, instance x_i and all of its k-nearest neighbors share the same class. We infer that x_i is part of a main concept in the data set.

Instance with neighborhood type B. Instance x_i has different class from all of its k-nearest neighbors. We infer that this instance is noise.

Instance with neighborhood type C. Most neighbors are minority class instances, whereas the rest are not. We interpret that x_i is near of decision boundaries.

Instance with neighborhood type D. Most neighbors are majority class instances, whereas the rest are not. This type of instances are interpreted as belonging to weak concepts.

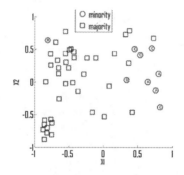

Fig. 2. Illustration of the four types of neighborhoods (A,B, C and D) for the minority class instances. The number of nearest neighbors used for this data set is K = 3.

As a way to illustrate the types of neighborhoods of the minority class instances, we present a synthetic data set in Fig. 2. The minority class instances are represented with circles, and the type of neighborhood with a letter (A, B, C or D).

Instances marked with letter B, are usually treated as noise by classifiers. Instances marked with A are the easiest to classify correctly. Instances C and D are borderline instances; however, the latter type usually correspond to weak concepts.

Analyzing the Neighborhoods. The exploration is realized using the following procedure:

1. Compute the K-nearest neighbors of each minority class instance x_i in data set.
2. Count the neighbors of each class to determine the type of neighborhood of x_i.
3. Repeat N times (a) and (b):
   (a) Separate data set into training and testing sets.
   (b) Train a Classifier C, and count the number of mis-classifications for the minority class instances committed by C.

Step 3 in previous procedure is applied to detect which instances are predicted incorrectly. Using N = 30 or greater values, produces similar values.

The number of nearest neighbors affects the analysis; however, using a standard value of K = 5 is useful in most cases. This value is widely used in many methods, it was first proposed in [10].

4   Experiments and Discussion

Most of experiments reported in the literature about the imbalance problem are realized using synthetic data sets. Commonly, these data sets are designed based on some assumptions, for example, minority class instances create small clusters and disjunctions [18]. Nevertheless, data from real-world applications are too different, as we will see in this section. Therefore, the conclusions obtained from such experiments are difficult be extended to real-world scenarios.

In our case, we use ten publicly available imbalanced data sets, they can be downloaded from [35]. Table 1 shows the main characteristics of them. Column name Alias is the short name that is used to show the results. IR means the imbalance ratio.

Five classification methods were tested, these are the following: SVM, NN, C45 (decision tree), NB and kNNC. The implementation of these methods is the provided by the Weka [36] environment. Ten cross-validation was used to obtain the reported results.

The experiments were conducted on a computer with the following features: 2.6 GHz Intel Core i5 processor, OS X 10.11.3, 8 GB RAM. The procedure explained before was implemented in the Java language.

In order to explore the types of neighborhoods, the value for K was set to five for each data set. Table 2 shows the number of instances in each type of neighborhood.

Table 1. Data sets used in experiments

Data set	Alias	Num. attributes	Num. majority	Num. minority	Imbalance ratio
Vowel0	Vowel	13	898	90	9
Glass2	Glass	9	197	17	11
Cleveland	Cleveland	13	160	13	12
Ecoli4	Ecoli 4	7	316	20	15
Page-blocks-1-3 vs 4	Page-blocks	10	444	28	15
Ecoli-0-1-3-7 vs 2-6	Ecoli 0	7	274	7	39
winequality-red-3 vs 5	Winequality	11	681	10	68
Poker-8 vs 6	Poker	10	1,460	17	85
kddcup-rootkit-imap vs back	Kddcup	41	2,203	22	100
Abalone	Abalone	8	4,142	32	129

The five classification methods were trained and tested with these ten data sets. True Positive Rate (TPR) for the minority class instances was measured, these are presented in Fig. 3. The "ideal classifier" shown in this figure is the hypothetical one that does not commit any prediction errors for the previously unseen instances. The sets in which the classifiers achieved the greatest values of TPR were the following: Vowel, Page-bocks and Kddcup. These are considered the "easiest" data sets. The "hardest" data sets (lowest TPR values) were Glass, Winequality and Abalone.

Table 2. Number of instances in each category of neighborhood

Dataset	Number of instances in each type of neighborhood			
	A	B	C	D
Vowel	**81**	0	7	2
Glass	0	6	2	**9**
Cleveland	0	6	0	**7**
Ecoli	**10**	3	5	2
Page-blocks	**11**	2	8	7
Ecoli	0	2	**5**	0
Winequality	0	**9**	0	1
Poker	0	**12**	0	5
Kddcup	**20**	0	2	0
Abalone	0	**30**	0	2

Another important measure to evaluate the general performance of classifiers is AUC ROC. The greater value of this measure the better the classifier. The AUC ROC achieved by the five classification methods are presented in Table 3. In this table, the value of the parameter tuned for each classifier is also included.

The three best averages of AUC ROC were achieved with Kddcup, Vowel and Page-blocks data sets. The worst averages of AUC ROC were obtained with Poker, Winequality and Abalone data sets. Both measures, TPR and average of AUC ROC agree with the data sets that are "easy" or "hard" to classify for the five classification methods.

Once the performances of the classification methods is known, we compare them with the types of neighborhoods, which are summarized in Table 2. Some observations are the following:

The hardest data sets for classifiers are those which contain instances with neighborhoods type B and D. In these cases, minority class instances do not form clusters.

Two of the ten datasets (Vowel and Kddcup) contain a greater number of instances with neighborhood type A than instances with other types of neighborhoods. Hence, these data sets should be "easy" for the classification methods.

In three of the ten datasets (Poker, Winequality and Abalone) the number of instances with neighborhood type B is large, and there exists some minority class

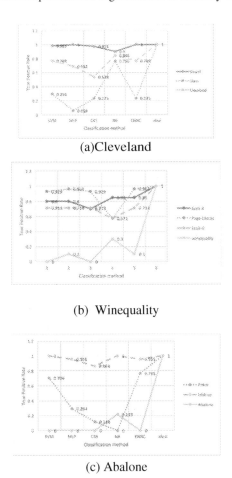

(a)Cleveland

(b) Winequality

(c) Abalone

Fig. 3. True positive rates (TPR). The note indicates the data set name in which the performance of classification methods is the worst.

instances with neighborhood type D. Therefore, most the instances of these data sets seem noise. This makes hard the task for classification methods.

Five of the ten datasets data sets have many borderline instances (type C), and instances with neighborhoods type D. These suggest the existence of weak concepts. Furthermore, the number of instances with neighborhood A is zero. This hinders the performance of classifiers.

Table 4 shows a comparative between the predominant type of region of each data set, against the TPR achieved by each classification method. Analyzing this table, one can observe that data sets in which the predominant neighborhood of minority class instances is type A, the TPR is high. On the other hand, if the predominant neighborhood is type B, the TPR is very low. In the other two cases (neighborhoods of type C and D), the performance is moderately damaged.

Table 3. AUC ROC achieved

Dataset	SVM (gamma)	MLP (Learning rate)	C4.5 (C)	NB	kNNC (K)
Vowel0	0.983 (0.526)	**1.000** (0.300)	0.974 (0.250)	0.980	**1.000** (1.000)
Glass2	0.629 (9.293)	0.776 (0.300)	**0.812** (0.250)	0.715	0.570 (1.000)
Cleveland	0.875 (0.100)	0.768 (1.000)	0.824 (0.594)	0.912	**0.907** (2.000)
Ecoli4	0.894 (0.723)	**0.986** (0.100)	0.796 (0.100)	0.991	0.901 (2.000)
Page-blocks	0.962 (0.933)	**0.999** (0.100)	0.961 (0.100)	0.959	0.979 (4.000)
Ecoli 0	0.857 (1.433)	**0.946** (0.100)	0.884 (0.199)	0.903	0.839 (2.000)
Winequality	0.500 (1.000)	**0.801** (0.300)	0.499 (0.100)	0.558	0.701 (4.000)
Poker	0.853 (1.000)	0.646 (0.100)	0.550 (0.199)	0.456	**0.882** (1.000)
kddcup	**1.000** (0.500)	**1.000** (0.100)	**1.000** (0.250)	**1.000**	**1.000** (5.000)
Abalone19	0.500 (0.100)	**0.800** (0.300)	0.481 (0.100)	0.691	0.587 (5.000)

Table 4. Comparison between TPR and types of predominant neighborhood

Dataset	Neighborhood	SVM	MLP	C45	NB	KNNC
Vowel	A	0.983	**1.000**	0.974	0.9000	1.000
Glass	D	0.294	0.059	0.235	**0.765**	0.235
Cleveland	D	0.769	**0.962**	0.538	0.846	0.769
Ecoli 4	A	0.800	0.800	0.700	**0.850**	**0.850**
Page-blocks	A	0.929	**0.964**	0.929	0.571	**0.964**
Ecoli 0	C	**0.714**	**0.714**	**0.714**	0.571	**0.714**
Winequality	B	0.000	**0.100**	0.000	0.003	0.001
Poker	B	0.706	0.294	0.118	0.000	**0.765**
kddc	A	**1.000**	0.955	0.864	1.000	0.955
Abalone	B	0.000	0.000	0.000	**0.219**	0.000

5 Conclusions

It is well-known that classification methods do not perform well with imbalanced data sets. In spite of many experiments have been conducted in others works, most of them are focused to compare the performance of several classification methods using synthetic data sets. In this paper, we take a very different approach. First, we detect the type of neighborhood for each minority class instances. Then, we train and tested five classification methods with ten real-world data sets, and measure the True Positive Rate and AUC ROC.

We found that in real-world data sets, many minority class instances seem noise, i.e., most of their neighbors have opposite class. In this situation, the performance of classifiers is highly hindered. Data sets with large number of instances with all their neighbors belonging to the minority class, are the easiest for classification methods. Experiments with a greater number of data sets are necessary to understand more deeply these relationships.

Future research paths include to design methods to predict the performances of classification methods on data sets with different complexities.

References

1. Esfandiari, N., Babavalian, M.R., Moghadam, A.M.E., Tabar, V.K.: Review: knowledge discovery in medicine: current issue and future trend. Expert Syst. Appl. **41**(9), 4434–4463 (2014). doi:10.1016/j.eswa.2014.01.011
2. Wang, S.L., Zhu, Y.H., Jia, W., Huang, D.S.: Robust classification method of tumor subtype by using correlation filters. IEEE/ACM Trans. Comput. Biol. Bioinform. **9**(2), 580–591 (2012)
3. Zhao, X., Liu, K., Zhu, G., He, F., Duval, B., Richer, J., Huang, D., Jiang, C., Hao, J., Chen, L.: Identifying cancer-related microRNAs based on gene expression data. Bioinformatics **31** (8), 1226–1234 (2015). doi:10.1093/bioinformatics/btu811
4. Hilas, C.S., Mastorocostas, P.A.: An application of supervised and unsupervised learning approaches to telecommunications fraud detection. Knowl.-Based Syst. **21**(7), 721–726 (2008). http://www.sciencedirect.com/science/article/pii/S0950705108000786
5. Hu, R., Jia, W., Ling, H., Huang, D.: Multiscale distance matrix for fast plant leaf recognition. IEEE Trans. Image Process. **21**(11), 4667–4672 (2012)
6. He, H., Garcia, E.A.: Learning from imbalanced data. IEEE Trans. Knowl. Data Eng. **21**(9), 1263–1284 (2009). doi:10.1109/TKDE.2008.239
7. Hulse, J.V., Khoshgoftaar, T.: Knowledge discovery from imbalanced and noisy data. Data Knowl. Eng. **68**(12), 1513–1542 (2009)
8. García, V., Sánchez, J.S., Mollineda, R.A.: Exploring the performance of resampling strategies for the class imbalance problem. In: García-Pedrajas, N., Herrera, F., Fyfe, C., Benítez, J.M., Ali, M. (eds.) IEA/AIE 2010, Part I. LNCS, vol. 6096, pp. 541–549. Springer, Heidelberg (2010)
9. Garcia, S., Herrera, F.: Evolutionary training set selection to optimize c4.5 in imbalanced problems. In: Eighth International Conference on Hybrid Intelligent Systems, 2008, HIS 2008, pp. 567–572 (2008)
10. Chawla, N.V., Bowyer, K.W., Hall, L.O., Kegelmeyer, W.P.: SMOTE: synthetic minority over-sampling technique. J. Artif. Int. Res. **16**(1), 321–357 (2002). http://dl.acm.org/citation.cfm?id=1622407.1622416
11. Akbani, R., Kwek, S.S., Japkowicz, N.: Applying support vector machines to imbalanced datasets. In: Boulicaut, J.-F., Esposito, F., Giannotti, F., Pedreschi, D. (eds.) ECML 2004. LNCS (LNAI), vol. 3201, pp. 39–50. Springer, Heidelberg (2004)
12. Raskutti, B., Kowalczyk, A.: Extreme re-balancing for svms: a case study. SIGKDD Explor. Newsl. **6**(1), 60–69 (2004). doi:10.1145/1007730.1007739
13. Han, H., Wang, W.-Y., Mao, B.-H.: Borderline-SMOTE: a new over-sampling method in imbalanced data sets learning. In: Huang, D.-S., Zhang, X.-P., Huang, G.-B. (eds.) ICIC 2005. LNCS, vol. 3644, pp. 878–887. Springer, Heidelberg (2005). doi:10.1007/11538059_91
14. Drummond, C., Holte, R.C.: C4.5, class imbalance, and cost sensitivity: why under-sampling beats over-sampling. In: Proceedings of the International Conference Machine Learning, Workshop Learning from Imbalanced Data Sets, pp. 1–8 (2003)
15. Batista, G.E., Prati, R.C., Monard, M.C.: A study of the behavior of several methods for balancing machine learning training data. SIGKDD Explor. Newsl. **6**(1), 20–29 (2004). doi:10.1145/1007730.1007735

16. Huang, Y.M., Hung, C.M., Jiau, H.C.: Evaluation of neural networks and data mining methods on a credit assessment task for class imbalance problem. Nonlinear Anal.: Real World Appl. **7**(4), 720–747 (2006)
17. Liu, X.Y., Zhou, Z.H.: The influence of class imbalance on cost-sensitive learning: an empirical study. In: Sixth International Conference on Data Mining, 2006, ICDM 2006, pp. 970–974 (2006)
18. Japkowicz, N., Stephen, S.: The class imbalance problem: a systematic study. Intell. Data Anal. **6**(5), 429 (2002)
19. Weiss, G.M.: Mining with rarity: a unifying framework. SIGKDD Explor. Newsl. **6**(1), 7–19 (2004). doi:10.1145/1007730.1007734
20. Prati, R.C., Batista, G.E., Monard, M.C.: Class imbalances versus class overlapping: an analysis of a learning system behavior. In: Monroy, R., Arroyo-Figueroa, G., Sucar, L., Sossa, H. (eds.) MICAI 2004. LNCS (LNAI), vol. 2972, pp. 312–321. Springer, Heidelberg (2004)
21. Smith, M.R., Martinez, T., Giraud-Carrier, C.: An instance level analysis of data complexity. Mach. Learn. **95**(2), 225–256 (2014)
22. Kriminger, E., Principe, J., Lakshminarayan, C.: Nearest neighbor distributions for imbalanced classification. In: The 2012 International Joint Conference on Neural Networks (IJCNN), pp. 1–5 (2012)
23. Barua, S., Islam, M., Murase, K.: A novel synthetic minority oversampling technique for imbalanced data set learning. In: Lu, B.-L., Zhang, L., Kwok, J. (eds.) ICONIP 2011, Part II. LNCS, vol. 7063, pp. 735–744. Springer, Heidelberg (2011)
24. Bunkhumpornpat, C., Sinapiromsaran, K., Lursinsap, C.: Safe-level-SMOTE: safe-level-synthetic minority over-sampling technique for handling the class imbalanced problem. In: Theeramunkong, T., Kijsirikul, B., Cercone, N., Ho, T.-B. (eds.) PAKDD 2009. LNCS, vol. 5476, pp. 475–482. Springer, Heidelberg (2009)
25. Sáez, J.A., Luengo, J., Stefanowski, J., Herrera, F.: Smote-IPF: addressing the noisy and borderline examples problem in imbalanced classification by a re-sampling method with filtering. Inf. Sci. **291**, 184–203 (2015). http://www.sciencedirect.com/science/article/pii/S0020025514008561
26. Padmaja, T., Dhulipalla, N., Bapi, R., Radha Krishna, P.: Unbalanced data classification using extreme outlier elimination and sampling techniques for fraud detection. In: International Conference on Advanced Computing and Communications, 2007, ADCOM 2007, pp. 511–516 (2007)
27. Luengo, J., Fernandez, A., Herrera, F., Herrera, F.: Addressing data-complexity for imbalanced data-sets: a preliminary study on the use of preprocessing for c4.5. In: Ninth International Conference on Intelligent Systems Design and Applications, 2009. ISDA 2009, pp. 523–528 (2009)
28. Gong, R., Huang, S.H.: A Kolmogorov Smirnov statistic based segmentation approach to learning from imbalanced datasets: with application in property refinance prediction. Expert Syst. Appl. **39**(6), 6192–6200 (2012). http://www.sciencedirect.com/science/article/pii/S0957417411016824
29. Seiffert, C., Khoshgoftaar, T., Van Hulse, J., Folleco, A.: An empirical study of the classification performance of learners on imbalanced and noisy software quality data. In: IEEE International Conference on Information Reuse and Integration, 2007, IRI 2007, pp. 651–658. IEEE, Aug 2007
30. López, V., Fernández, A., García, S., Palade, V., Herrera, F.: An insight into classification with imbalanced data: empirical results and current trends on using data intrinsic characteristics. Inf. Sci. **250**, 113–141 (2013). http://www.sciencedirect.com/science/article/pii/S0020025513005124

31. Brown, I., Mues, C.: An experimental comparison of classification algorithms for imbalanced credit scoring data sets. Expert Syst. Appl. **39**(3), 3446–3453 (2012). http://www.sciencedirect.com/science/article/pii/S095741741101342X
32. Tang, Y., Zhang, Y.Q., Chawla, N., Krasser, S.: SVMS modeling for highly imbalanced classification. IEEE Trans. Syst. Man Cybern. Part B: Cybern. **39**(1), 281–288 (2009)
33. Kohavi, R., Provost, F.: Glossary of terms. Mach. Learn. **30**(2–3), 271–274 (1998). http://dl.acm.org/citation.cfm?id=288808.288815
34. Fawcett, T.: An introduction to ROC analysis. Pattern Recogn. Lett. **27**(8), 861–874 (2006). doi:10.1016/j.patrec.2005.10.010
35. Alcalá-Fdez, J., Fernandez, A., Luengo, J., Derrac, J., García, S., Sánchez, L., Herrera, F.: KEEL data-mining software tool: data set repository, integration of algorithms and experimental analysis framework. J. Multiple-valued Log. Soft Comput. **17**, 255–287 (2011)
36. Hall, M., Frank, E., Holmes, G., Pfahringer, B., Reutemann, P., Witten, I.H.: The WEKA data mining software: an update. SIGKDD Explor. Newsl. **11**(1), 10–18 (2009). http://doi.acm.org/10.1145/1656274.1656278

A Hybrid Particle Swarm Optimization Embedded Trust Region Method

Jun He[1(✉)], Fei Han[1], and Shou-Bao Su[2]

[1] School of Computer Science and Communication Engineering, Jiangsu
University, Zhenjiang, Jiangsu 212013, China
{2211408046,hanfei}@ujs.edu.cn
[2] School of Computers Engineering, Jinling Institute of Technology,
Nanjing, Jiangsu 211169, China
showbo@jit.edu.cn

Abstract. As one of swarm intelligence algorithms, particle swarm optimization (PSO) has good global search ability, but the main disadvantage is that it is easy to fall into the local minima, and the convergence accuracy is restricted. Trust region method is an important numerical method for solving nonlinear optimization problems with reliability, stability and strong convergence. In this paper, a hybrid PSO Embedded trust region method is proposed to improve the search ability of the swarm. The algorithm effectively combines the global search of particle swarm optimization with the fast and precise local search capability of the trust region method. The experiment results show that it has much better accuracy and convergence to the global optimal solution.

Keywords: Hybrid particle swarm optimization · Trust region method · Numerical experiment

1 Introduction

Particle swarm optimization (PSO) [1] was firstly proposed by Kennedy and Eberhart in 1995. The basic idea is inspired by modeling and simulating the results of the study on the behavior of flocks of birds in the early stage. Particle swarm optimization algorithm is easy to fall into the local optimum, and the convergence accuracy is not high. In the past 20 years, many researchers have conducted research in-depth, and put forward some improved PSOs, and achieved better results than traditional PSO.

In [2] an improved particle swarm optimization algorithm with inertia weight was proposed to balance the global search ability and local search ability of the algorithm. The greater the inertia weight, the stronger the global search ability, the smaller the inertia weight, the stronger the local search ability. In literature [3, 4], Clerc introduced the concept of shrinkage factor, and the method described a selection of the inertia weight w, the acceleration constants c1 and c2 to ensure the convergence of the algorithm. In [5] the author presented a particle swarm optimizer with passive congregation to improve the performance of standard PSO. Passive congregation is an important biological force preserving swarm integrity. By introducing passive congregation to PSO, information can be transferred among individuals of the swarm. In [6] Sugantan

© Springer International Publishing Switzerland 2016
D.-S. Huang et al. (Eds.): ICIC 2016, Part I, LNCS 9771, pp. 762–771, 2016.
DOI: 10.1007/978-3-319-42291-6_76

proposed an improved particle swarm optimization based on the neighborhood region. The basic idea is that at the beginning, the neighborhood of each individual is itself. With the growth of the evolution generation, the neighborhood is also expanding to the whole population. In order to avoid premature convergence of particle swarm optimization, Riget [7] adopted two operators 'Attractive' and 'Repulsive' to ensure population diversity so as to improve the efficiency of the algorithm. In [8–10], a new hybrid optimization algorithm that combines the PSO algorithm with gradient-based local search algorithms to achieve faster convergence and better accuracy of final solution without getting trapped in local minima was presented. These proposed methods improve the search ability of basic PSO. Aiming at the above-mentioned problems, in [11], an improved PSO algorithm, named HARPSOGS, combining ARPSO with the steepest gradient descent method, was proposed. Gradient search makes the swarm converge to local minima quickly. The gradient descent method is one dimensional line search method, and the basic idea is firstly to determine the search direction, find an acceptable point, and then adjust the search direction according to a certain strategy, repeat. However, it often leads to the failure of the algorithm because of the large step size, especially when the problem is ill-conditioned.

These improved algorithms above have improved the search ability, but still have great randomness in nature with the result that increases the convergence time and reduces the convergence accuracy. The trust region method has a fast local convergence, an ideal global convergence, and a super linear convergence. Another significant advantage of the trust region algorithm is its stable numerical performance, and it is suitable for solving the ill-conditioned optimization problem. In this paper, an improved hybrid particle swarm optimization called HARPSOTR is proposed to overcome the problems. The algorithm effectively combines the global search ability of particle swarm optimization with the global search ability and precise local search capability of the trust region method [12–15]. The experiment results show that it can be more accurate and fast convergence to the global optimal solution, compared with other traditional improved particle swarm optimization algorithm.

2 Basic Particle Swarm Optimization and Trust Region Method

2.1 Basic Particle Swarm Optimization

Particle swarm optimization simulates the foraging behavior of bird clusters by means of collective association between birds to reach the best position. In PSO, a set of potential solutions to the problem is called a population. Each potential solution is called a particle (particle). Each particle is represented by its geometric position and velocity vector. In the process of evolution, each particle has an information exchange with other particles, based on its own experience and the best "experience" of neighboring particles to determine their flight. Assume that the dimension of the search space is D, and the swarm is $S = (X_1, X_2, X_3, \ldots, X_{Np})$; each particle represents a position in the D dimension; the position of the i-th particle in the search space can be denoted as $X_i = (x_{i1}, x_{i2}, \ldots, x_{iD})$, $i = 1, 2, \ldots$, Np, where Np is the number of all particles.

The best position of the i-th particle being searched until now is called pbest which is expressed as $P_i = (p_{i1}, p_{i2}, \ldots, p_{iD})$. The best position of the all particles is called gbest which is denoted as $P_g = (p_{g1}, p_{g2}, \ldots, p_{gD})$. The velocity of the i-th particle is expressed as $V_i = (v_{i1}, v_{i2}, \ldots, v_{iD})$. According to literature [1], the basic PSO was described as:

$$V_i(t+1) = V_i(t) + c1 * rand() * (P_i(t) - X_i(t)) + c2 * rand() * (P_g(t) - X_i(t)) \quad (1)$$

$$X_i(t+1) = X_i(t) + V_i(t+1) \quad (2)$$

where c1, c2 are the acceleration constants with positive values; rand() is a random number ranged from 0 to 1. In order to obtain better performance, adaptive particle swarm optimization (APSO) algorithm was proposed [2], and the corresponding velocity update of particles was denoted as follows:

$$V_i(t+1) = W(t) * V_i(t) + c1 * rand() * (P_i(t) - X_i(t)) + c2 * rand() * (P_g(t) - X_i(t)) \quad (3)$$

Where W(t) can be computed by the following equation:

$$W(t) = W_{max} - t * (W_{max} - W_{min}) / N_{pso} \quad (4)$$

In Eq. (4), W_{max}, W_{min} and N_{pso} are the initial inertial weight, the final inertial weight and the maximum iterations, respectively.

In literature [6], attractive and repulsive particle swarm optimization (ARPSO) was proposed to ensure population diversity so as to improve the efficiency of the algorithm. The velocity-update formula is described as:

$$V_i(t+1) = W * V_i(t) + dir[c1 * rand() * (P_i(t) - X_i(t)) + c2 * rand() * (P_g(t) - X_i(t))] \quad (5)$$

Where

$$dir = \begin{cases} -1, & dir > 0, diversity < d_{low} \\ 1, & dir < 0, diversity > d_{high} \end{cases} \quad (6)$$

And a function was defined to calculate the diversity of the swarm as follows:

$$diversity(S) = \frac{1}{|N_p| \cdot |L|} \cdot \sum_{i=1}^{|S|} \sqrt{\sum_{j=1}^{N} (p_{ij} - \bar{p}_j)^2} \quad (7)$$

Where $|L|$ is the length of the maximum radius of the search space, p_{ij} is the j-th component of the i-th particle and p_j is the j-th component of the average over all particles.

2.2 Trust Region Method

The trust region method was first proposed by Powell in 1870s [12]. In [13] Yuan proved the convergences of trust region methods. Considering the following unconstrained optimization problem

$$\min_{x \in R^n} f(x) \tag{8}$$

Where $f(x)$ is a continuous differential function in R^n. At k-th iteration, it calculates a trial step d_k by solving the following trust region sub-problem

$$\min \varphi_k(d) = g_k^T d + \frac{1}{2} d^T B_k d \tag{9}$$

$$s.t. \|d\| \leq \Delta_k \tag{10}$$

Where $g_k = \nabla f(x_k)$ is the gradient vector at the current approximation iterate x_k, B_k is a $n \times n$ symmetric matrix which may be the exact Hessian $H(x_k)$ or the quasi-Newton approximation and $\Delta_k > 0$ is the trust region radius. In this paper, the notation $\|\cdot\|$ denotes the Euclidean norm in R^n. Let d_k be the solution of (8). The predicted reduction is defined by the reduction of the approximate model, that is, $\Pr ed_k = \varphi_k(0) - \varphi_k(d_k)$, and the actual reduction is defined by $Ared_k = f(x_k) - f(x_k + d_k)$. The ratio between these two reductions is defined by $\rho_k = \frac{Ared_k}{\Pr ed_{k\infty}}$, and it plays a key role to decide whether the trial step is acceptable and to adjust the new trust region radius.

Basic algorithm:
Step1: Given $x_0, radius\ r_1, 0 < \mu, \eta < 1$, accuracy ε, $k := 1$;
Step2: Calculate $f(x_k), \nabla f(x_k)$. If $\|\nabla f(x_k)\| \leq \varepsilon$, end, x_k is the best solution, otherwise, calculate B_k;
Step3: Solving sub-problem: $\min \varphi_k(d) = f_k + g_k^T d + \frac{1}{2} d^T B_k d \quad s.t. \|d\| \leq \Delta_k$, obtaining d_k;
Step4: Calculate ρ_k, if $\rho_k \leq \mu$, $x_{k+1} = x_k$, otherwise, $x_{k+1} = x_k + d_k$.
Step5: Update r_k, if $\rho_k \leq \mu$, $r_{k+1} = \frac{1}{2} r_k$, if $\mu < \rho_k < \eta$, $r_{k+1} = r_k$, otherwise $r_{k+1} = 2r_k$.
Step6: $k := k + 1$, go to step2.

The traditional trust region method needs to resolve the subproblem when the trial step fails, and it is difficult to solve the problem of the new iterative point and takes much time to do repetitive calculations. In order to solve the problems, many scholars put forward some improved trust region methods. In literature [14], a self-adaptive trust region method is proposed. The trust region radius is updated at a variable rate according to the ratio between the actual reduction and the predicted reduction of the objective function, rather than by simply enlarging or reducing the original trust region radius at a constant rate. In literature [15], a new algorithm is proposed for nonlinear optimization that employs both trust region techniques and line searches. Unlike traditional trust region methods, the algorithm does not resolve the subproblem if the trial step results in an increase in the objective function, but instead performs a backtracking line search from the failed point. The traditional trust region method is based on quadratic model. In literature [16], an adaptive trust region method based on the conic model for unconstrained optimization problems is proposed.

3 The Proposed Hybrid PSO

Since ARPSO keeps reasonable population diversity effectively by adaptively attracting and repelling each other to avoid falling into local minima and has better performance. However, it is still a stochastic evolutionary algorithm as other PSOs and may converge to local minima or take more time to find the global optimal solution. To improve the search performance of ARPSO, we must improve the efficiency of local search and the ability to maintain population diversity. As a deterministic search algorithm, trust region method has fast and efficient local search capability and global convergence. It applies quadric model to correct the step size, which makes the target function decrease more efficiently than the linear search method, such as gradient descent. On one hand, when ARPSO strap into local minima, it is used to get rid of local minima to guide the particles to the global minimum, because the trust region method has global convergence. On the other hand, it improves the convergence and search efficiency in the local search. Therefore, ARPSO and the trust region method are complementary each other on improving the convergence accuracy and rate of the PSO. In this paper, an improved hybrid PSO, called HARPSOTR, combining the deterministic search method based on trust region with the stochastic particle swarm optimization algorithm to improve the search ability of the population, is proposed. The detailed steps are as follows:

> Step 1: Initialize velocity and position of all particles randomly and all parameters;
> Step 2: Calculate pbest of each particle, gbest of all particles;
> Step 3: Each particle update its velocity and position by the following formulas:

$$X' = X(t) + V_{arpso} \tag{11}$$

$$X'' = X_{tr} \tag{12}$$

$$X_{i+1}(t+1) = \begin{cases} X' & if \quad f(X') < f(X'') \\ X'' & else \end{cases} \tag{13}$$

> Where V_{arpso} is the velocity update of the i-th particle obtained by Eq. (5), X_{tr} is the position update of the i-th particle obtained by Ex. (8); $f(X)$ is the fitness function;
> Step 4: if count $= \beta$, expanding or reducing radius, update gbest by Ex. (8);
> Where β is a constant and used to record the number of times gbest has no changes.
> Step 5: return to step 2 until the maximum number of iterations or the predetermined convergence accuracy is reached.

4   Experiment Results and Discussion

In this section, the performance of the proposed hybrid PSO algorithm is compared with the standard PSO, APSO, PSOPC, ARPSO and CPSO. Test functions are shown in Table 1. In this experiment, six high dimensional functions are chosen to test the search efficiency of the improved PSO. Ellipsoid (F1) and Quadric (F2) are continuous differentiable unimodal functions, which have only one minimum point. Ackley (F3), Griewangk (F4), Rosenbrock (F5) and Rastrigin (F6) are multimodal functions, and F4 has a lot of local minima and only a global minimum. PSO is difficult to find the minimum points. The minimum local area of F5 is very flat which makes it difficult to converge to the minimum for the particle swarm algorithm based on stochastic search. F6 is highly multimodal and has many local minima.

All the programs are carried out in MATLAB 2012a environment on an Intel Core (TM) i3-2120 3.30 GHZ CPU. The population size for all PSOs is 40 in all experiments. And the acceleration constants c1 and c2 for SPSO, PSOPC, HARPSOTR and CPSO all are set as 1.4962. The constants c1 and c2 both are set as 2.0 and 1.5 for APSO and ARPSO. The parameters, d_{low} and d_{high} are selected as 5e-6 and 0.25, respectively, in ARPSO and HARPSOTR. The constant β is set as 12 for HARPSOTR. The inertia weight w decreases linearly from 0.9 to 0.4 for all PSOs according to the literature [2]. All the results shown in this paper are the mean values of 20 trails.

Table 1. Test functions used in the experiment

Test function	Expression	Search space	Global minima
Ellipsoid(F1)	$\sum_{i=1}^{n} i \cdot x_i^2$	$(-100, 100)^n$	0
Quadric(F2)	$\sum_{i=1}^{n} (\sum_{j=1}^{i} x_j)^2$	$(-100, 100)^n$	0
Ackley(F3)	$-a \cdot e^{-b\sqrt{\frac{\sum_{i=1}^{n} x_i^2}{n}}} - e^{\frac{\sum_{i=1}^{n} \cos(c \cdot x_i)}{n}} + a + e^1$	$(-32, 32)^n$	0
Griewangk(F4)	$1 + \sum_{i=1}^{n} x_i^2 / 4000 - \prod_{i=1}^{n} \cos\left(\frac{x_i}{\sqrt{i}}\right)$	$(-100, 100)^n$	0
Rosenbrock(F5)	$\sum_{i=2}^{n} 100 \cdot \left(x_i - x_{i-1}^2\right)^2 - (1 - x_i)^2$	$(-100, 100)^n$	0
Rastrigin(F6)	$10n + \sum_{i=1}^{n} (x_i^2 - 10\cos(2\pi x_i))$	$(-5.12, 5.12)^n$	0

Figure 1 shows the mean convergence curve of different PSOs on all six functions with ten dimensions. For the function F1, HARSOTR and PSOPC have a better search performance than other PSOs, and HARPSOTR has a faster convergence than PSOPC. From the Fig. 1(b), obviously, HARPSOTR converges much more accuracy than other PSOs, because F2 is a quadric function, and trust region method is based on quadric model. For the function F3, compared with F1 and F2, the convergence rate and accuracy is much low. However, HARPSOTR has a better performance than all other PSOs, though ARPSO may have better accuracy after more than 5000 iterations. From the Fig. 1(d), it can be seen that all PSOs are caught in the local minima quickly on the function F4, though HARPSO has slightly better than other PSOs. From the Fig. 1(e), the convergence accuracy of the HARPSOTR has been decreasing continuously during the whole search process and the proposed hybrid PSO converges much more accurately in all PSOs. For the function F6, in the experiment, HARPSOTR cannot find the global minimum sometimes. In Fig. 1(f), it shows most of the results.

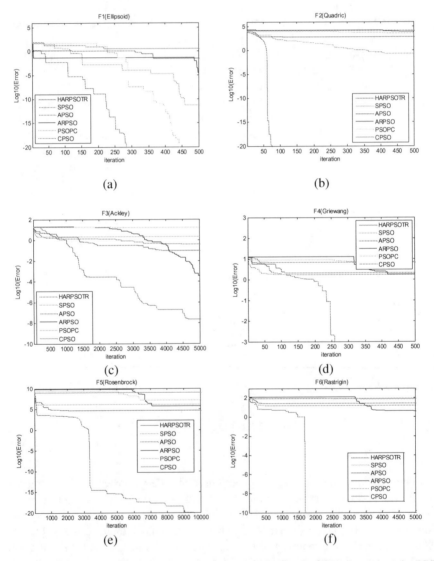

Fig. 1. Best solution versus iteration number for the six test functions by using six PSOs (a) Ellipsoid (b) Quadric (c) Ackley (d) Griewangk (e) Rosenbrock (f) Rastrigin

Table 2. Mean best solution for the six test functions by using six PSOs

Test functions	Dimension	SPSO	APSO	ARPSO	PSOPC	CPSO	HARPSOGS	HARPSOTR
Ellipsoid F1	10	1.1317e−13	14.8081	0.0251	8.3897e−20	3.4430	-	4.9110e−52
	20	3.3141e−34	7.5587	9.1065e−20	5.0863e−23	39.5338	-	1.2116e−40
	30	2.9837e−38	30.7443	6.2949e−17	1.3271e−21	41.9863	-	5.3875e−60
Quadric F2	10	42.2678	4.2300e+03	8.1094e+03	6.3468e+03	5.5704e+03	-	1.0497e−29
	20	3.1721e+04	1.0340e+04	1.0340e+03	2.9591e+04	2.5043e+04	-	6.6630e−21
	30	7.2290e+03	2.8474e+04	1.1597e+05	6.8389e+04	5.5968e+04	-	1.1710e−26
Ackley F3	10	2.2994	0.8793	0.0018	16.5853	1.9017	-	1.7472e−06
	20	9.1444	8.7933	0.0255	18.6405	8.6724	-	3.2557e−06
	30	12.0488	12.8253	0.8735	18.7432	13.0073	-	3.5123e−05
Griewangk F4	10	25.3574	58.0362	63.5619	74.2862	69.7222	2.5535e−14	0.2501
	20	73.1461	1.4340ee+02	2.1471e+02	1.8974e+02	2.0832e+02	1.0856e−13	0.0468
	30	1.5542e+02	2.9911e+02	3.5861e+02	3.2580e+02	3.6770e+02	2.2123e−9	0.0168
Rosenbrock F5	10	9.8041e+05	1.2882e+07	4.0711e+6	7.1383e+07	7.2693e+08	5.4010e−10	1.2229e−17
	20	1.8648e+10	1.1879e+10	8.3954e+08	3.8151e+09	3.1529e+09	1.1348e−09	1.7904e−15
	30	7.5275e+11	3.6750e+09	4.7616e+09	9.1057e+09	7.0218e+09	5.6649e−07	4.2930e−20
Rastrigin F6	10	22.2122	48.6746	5.2649	66.7179	80.3944	2.117e−012	0
	20	78.2352	1.5679e+02	52.0200	1.8432e+02	1.8093e+02	6.8801e−11	13.7304
	30	1.5212e+02	2.4621e+02	3.3281e+02	3.2834e+02	3.5350e+02	9.9437e−10	58.5033

Table 2 shows mean best solution for the six test functions on different dimensions by using six PSOs. It can be found that the proposed hybrid PSO has better convergence accuracy and rate than other PSOs in most cases. Although HARPSOGS obtains the mean convergence accuracy than HARPSOTR on the function F4 and F6, it requires more iterations and the proposed algorithm sometimes can find the global minimum. In general, the proposed hybrid PSO is superior to HARPSOGS and other PSOs.

5   Conclusions

In this paper, a hybrid PSO algorithm, embedded trust region method, is proposed to improve the search ability of ARPSO. As the initial population of PSO is randomly selected, the optimization results have a certain degree of randomness. In order to avoid falling into local optimal solution, research the current optimum solution by trust region method, which is contributed to the evolution of PSO again. The experiments show the proposed method could effectively avoid premature convergence and had better convergence performance than other PSOs. Because trust region method needs to calculate Hessian matrix or the Quasi-Newton approximation, it takes much time to find the successor point. And the selection of radius is very important to the result. In literature [14–16], some methods about how to reduce the computational complexity and criteria for the selection of radius are presented. The selection of the constant is also significant to the result. In the future research work, the influence of parameters r (radius) and on the performance of the proposed PSO will be analyzed and we will continue to improve the shortcomings of the algorithm and use this promising hybrid algorithm for more complex problems.

Acknowledgements. This work was supported by the National Natural Science Foundation of China (Nos.61271385, 61572241, 61375121).

References

1. James, K., Eberhart, R.: Particle swarm optimization. In: Proceedings of IEEE International Conference on Neural Networks, vol. 4 (1995)
2. Shi, Y., Eberhart, R.: A modified particle swarm optimizer. Comput. Intell. **6**(1), 69–73 (1998)
3. Clerc, M.: The swarm and the queen: towards a deterministic and adaptive particle swarm optimization. In: Proceedings of the 1999 Congress on Evolutionary Computation CEC 1999 (1999)
4. Corne, D., Dorigo, M., Glover, F.: New Ideas in Optimization, pp. 11–32. McGraw-Hill Ltd., UK (1999)
5. He, S., Wu, Q.H., Wen, J.Y., et al.: A particle swarm optimizer with passive congregation. Biosystems **78**(1), 135–147 (2004)
6. Sugantan, P.N.: Particle swarm optimization with neighborhood operator. In: Proceedings of the 1999 Congress on Evolutionary Computation Piscataway, NJ, pp. 1958–1962. IEEE Service Center (1999)

7. Riget, J., Vesterstrom, J.M.: A diversity-guided particle swarm optimizer - the ARPSO. EVAlife Technical report (2002)
8. Sugino, R.: Numerical performance of PSO algorithm using the gradient method. Theoret. Appl. Mech. **57**, 461–468 (2009)
9. Noel, M.: A new gradient based particle swarm optimization algorithm for accurate computation of global minimum. Appl. Soft Comput. **12**(1), 353–359 (2012)
10. Mathew, M.: A new gradient based particle swarm optimization algorithm for accurate computation of global minimum. Appl. Soft Comput. **12**(1), 353–359 (2012)
11. Liu, Q., Han, F.: A hybrid attractive and repulsive particle swarm optimization based on gradient search. In: Huang, D.-S., Jo, K.-H., Zhou, Y.-Q., Han, K. (eds.) ICIC 2013. LNCS, vol. 7996, pp. 155–162. Springer, Heidelberg (2013)
12. Powell, M.J.D.: A new algorithm for unconstrained optimization. Nonlinear Programming, pp. 31–65. Academic Press, New York (1970)
13. Yuan, Y.X.: Convergences of trust region methods. Chin. Numer. Math. **16**, 333–346 (1994)
14. Hei, L.: A self-adaptive trust region algorithm. J. Comput. Math. **21**(2), 229–236 (2003)
15. Nodedal, J., Yuan, Y.X.: Combining trust region and line search techniques. In: Yuan, Y.X. (ed.) Advances in Nonlinear Programming, pp. 153–175. Springer, US (1998)
16. Di, S., Sun, W.: Trust region method for conic model to solve unconstrained optimization problems. Optim. Methods Softw. **6**, 237–263 (1996)

Machine Learning and Data Analysis for Medical and Engineering Applications

Inferring Disease-Related Domain Using Network-Based Method

Zhongwen Zhang[1,2,3], Peng Chen[4], Jun Zhang[5],
and Bing Wang[1,2,3(✉)]

[1] School of Electronics and Information Engineering,
Tongji University, Shanghai, China
14zhangzhongwen@tongji.edu.cn,
wangbing@ustc.edu
[2] The Advanced Research Institute of Intelligent Sensing Network,
Tongji University, Shanghai, China
[3] The Key Laboratory of Embedded System and Service Computing,
Tongji University, Shanghai, China
[4] Institute of Health Sciences, Anhui University, Hefei, Anhui, China
pchen.ustcl0@gmail.com
[5] College of Electrical Engineering and Automation, Anhui University,
Hefei, Anhui, China
wwwzhangjun@gmail.com

Abstract. Domain-domain interaction (DDI) network analysis has been widely applied in the investigation of the mechanisms of diseases. This project focus on mapping the melanoma-related mutations to domain and domain-domain interaction (DDI) level and use potential correlation method to find the human domains that have never been found to have anything to do with the melanoma-related study. Firstly, we extract melanoma-related mutations from COSMIC database, then map the mutations to protein database UniProt and domain database Pfam to find the melanoma-related domains; Secondly, we get the melanoma-related DDI information and the human DDI information from the DDI database iPfam; Thirdly, we construct the melanoma-related DDI network and human DDI network and then combine two approaches to use potential correlation method to analyze the two DDI networks. Finally, we find that among all the human domains who have potential correlation relationship with melanoma, there are 27 human domains that have never been found to have anything to do with melanoma in the existing literature and study. The result shows the effectiveness of our method and further study based on these human domains may lead to some new methods to cure melanoma.

Keywords: DDI · Melanoma · Potential correlation method

1 Introduction

Melanoma, also known as malignant melanoma, is the main cause of death of patients with skin cancer. The disease accounts for 80 % of deaths from skin cancer with an estimated 76,690 new patients and 9,480 deaths expected in the United States in

© Springer International Publishing Switzerland 2016
D.-S. Huang et al. (Eds.): ICIC 2016, Part I, LNCS 9771, pp. 775–783, 2016.
DOI: 10.1007/978-3-319-42291-6_77

2013 [1]. As such, the 5-year survival rate for patients with advanced melanoma has historically been below 15 % [2]. Rapid advances in next generation sequencing (NGS) technologies have made comprehensive characterization of cancer genomes feasible [3–5]. A large number of whole exome sequencing (WES) and whole genome sequencing (WGS) studies have already been performed in melanoma [6–10]. Many melanoma related databases were established. These data gives the researchers a wide platform to know more about the melanoma and find the more effective ways to beat melanoma.

Domain-domain interactions (DDIs) play fundamental roles in biology cellular systems such as molecule transportation transcription regulation, DNA replication, signal transduction. DDIs have been widely used in the investigation of the causal mechanisms of diseases and disease comorbidity [11]. During the last decade, rapid progress in high-throughput experimental techniques, especially yeast two-hybrid system, has greatly accelerated the generation of DDI data [12, 13]. With the available massive amount of DDI data, researchers can better interpret and evaluate the data. So far, the quantity and quality of DDIs in human, have enabled investigators to construct reliable interactomes that serve as references in biomedical research, especially in studying molecular mechanisms of complex diseases. Specifically in humans, DDI data has been widely applied to identify and prioritize disease candidate genes, understand the relationship between disease genes, explore network properties of disease genes, and reconstruct disease-specific subnetworks.

In this study, we take the potential correlation among biological molecules analysis method as the main research method. By constructing the melanoma-related (mr) DDI network and human DDI network and using potential correlation method to analyze the mr DDI network and human DDI network, we find the human domains that have interaction relationship with mr domains. According to the potential correlation method, we have fully confidence to infer that these human domains, which have potential correlation relationship with melanoma, could be a great help for the treatment of melanoma. So further study based on these human domains may found the new methods to cure melanoma.

2   Materials and Methods

2.1   Melanoma Mutation Data

We downloaded the mr mutation data from the COSMIC database (version: CosmicMutantExport_v70.tsv) (http://www.sanger.ac.uk/genetics/CGP/cosmic) [14]. As the data file in COSMIC contains all cancer mutations and only melanoma-related mutations are needed for this project, so we extract all the mr mutations from the COSMIC database. For the current analysis, we primarily focus on somatic single nucleotide variants (SNV) since they constitute the largest fraction of oncogenic drivers identified in melanoma. For these SNVs, we collected all nonsilent parts, including missense, nonsense mutations, as well as accompanying information such as tumor name, tumor type, and gene name. We generally refer to these SNVs as mutations in this article. Finally, the number of mr mutation which caused nosilent mutations is 279,760.

2.2 Domain Data

All domain data is extracted from PFAM Database [15], which provides a large collection of high quality protein domain families (pfam-A) and low quality protein domain families (pfam-B). Here, only pfam-A were used in this study for data quality issue. Currently there are 14,831 domain families in PFAM database (Version: 27.0: ftp://ftp.ebi.ac.uk/pub/databases/Pfam/releases/Pfam27.0/).

Melanoma-related domain data: Based on the mutation records in COSMIC which have corresponding amino acid change, there are 58,645 mutations can be found corresponding proteins in the protein database UniproKB [16]. As the data presented for each entry in Pfam is based on the protein database UniProt, and the proteins in Pfam has the detail information about all the domains of that protein. Thus all the mr domains can be obtained. Finally, 1,494 mr domain families can find mr mutations.

Human domain data: As the protein database UniproKB has all the human proteins extracted from literature and curator-evaluated computational analysis, and the protein domain database Pfam has detail information of proteins and their functional domains, we get all human domains from Pfam. Finally, 5,763 human-related domain families are found. Until now, the mr domains and all the human domains are all prepared for further analysis.

2.3 DDI Data and DDI Network

Since the domain definition adopted in this project come from the widely popular Pfam database, the DDI databases iPfam (http://ipfam.sanger.ac.uk/) based on Pfam is used [17] (version:/homodomain interaction.csv: ftp://selab.janelia.org/pub/ipfam/Current_Release/homodomain_interaction.csv).

In the above step, the mr domains and all the human domains are ready for further analysis. The iPfam database has the detail information of the domain interactions. We can get the DDI information of mr domains and the human domains. Finally, there are 604 mr domains that have interactions within mr domains, and there are 1,952 human-related domain which have interactions within human-related domains.

Now we have the detail information of mr DDI and the human DDI, so we can construct the DDI network of the mr mutations domains and the human domains. In a DDI network, a node denotes a domain encoded by a gene and an edge denotes an interaction between two domains. Thus, the Cytoscape, an Open Source Platform for Complex Network Analysis and Visualization, is used to construct and analyze the mr DDI network and the human DDI network [18].

2.4 DDI Network Topological Measures and Statistical Analysis

In this project, we combine two approaches to find the human domains which have potential correlation relationship with melanoma, the defined of potential correlation relationship is as

$$C_{pot} = f(v, w) : v \in V, w \notin V \tag{1}$$

where $f(v, w)$ means node v interacted with w and V means mr domains, so node w has potential correlation relationship with V. The first approach is as follows: firstly we analyze the mr DDI network and try to find the most important domains in the mr DDI network. As these 'important' domains play an essential role in the mr DDI network, we have fully confidence to believe that the human domains which have interaction relationship with these important mr domains may also play an important role in the mr process. Then we try to find, in the human DDI network except mr domains, the human domains that interact with mr domains. These human domains have potential correlation relationship with melanoma and further study based on these human domains may lead to new methods to cure melanoma.

We combined five common network topological measures, i.e., Closeness, Eccentricity, Stress, Centroid and Betweenness centrality, to examine network topological properties in the DDI network, with the help of these six common network topological measures, we can find the 'important' domains in the mr DDI network. We briefly describe the measurement here, more details are provided in the [19]. For the DDI network G = (V,E), V represent the node and E represent the interaction between nodes. For a node in a DDI network, Closeness measures the number of links of the shortest path traveling from the node to another node, the defined of Closeness is

$$C_{clo}(v) = \frac{1}{\sum_{w \in V} dist(v, w)} \tag{2}$$

dist(v,w) is the distance of the shortest path from node v to node w; Betweenness measures the sum of the fractions that the node is a member of the set of shortest paths that connect all the pairs of nodes in the network, the defined of Betweenness is

$$C_{bwt}(v) = \sum_{s \neq v \in V} \sum_{t \neq v \in V} \delta_{st}(v) \tag{3}$$

where $\delta_{st}(v) = \frac{\sigma_{st}(v)}{\sigma_{st}}$ and σ_{st} is the number of shortest paths between node s and node t; Eccentricity is the maximum non-infinite length of a shortest path between node n and another node in the network, the defined of Eccentricity is

$$C_{ecc}(v) = \frac{1}{\max\{ dist(v, w) : w \in V \}}; \tag{4}$$

Stress is calculated by measuring the number of shortest paths passing through a node, the defined of Stress is

$$C_{str}(v) = \sum_{s \neq v \in V} \sum_{t \neq v \in V} \sigma_{st}(v) \tag{5}$$

$\sigma_{st}(v)$ is the number of shortest paths between node s and node t passing through the node v; Centroid is computed by focusing the calculus on couples of nodes (v, w) and systematically counting the nodes that are closer to node v or to node w, the defined of Centroid is

$$C_{cen}(v) = \min\{f(v, w) : w \in V\{v\}\} \tag{6}$$

where $f(v, w) = \gamma_v(w) - \gamma_w(v)$, $\gamma_v(w)$ is the number of node closer to node v than to node w. We used an Cytoscape APP CentiScaPe2.1, to calculate these network topological measures. As Centroid, Closeness together with Eccentricity, and Betweenness, together with Stress, can respectively imply the importance of a node in the network, so we combine these three measures to get the 'important' domains in the melanoma-related DDI network.

The second approach is to calculate how many human domains except melanoma-related domains are interacted with the mr domains and then select the domain that has three or more interaction partners with mr domains for further study. These human domains are not found to have directly relationship with melanoma, but they have direct interaction relationship with mr domains, so these human domains have potential correlation relationship with melanoma, we have fully confidence in that these human domains may also have relationship with melanoma and further study based on these human domains may lead to new methods to cure melanoma (Fig. 1).

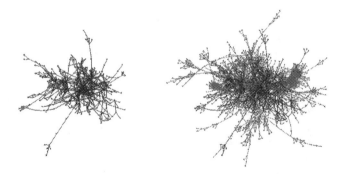

Fig. 1. The left one is the melanoma-related DDI network and the right one is the human-related DDI network. For this project, we try to find in the human-related DDI network except the melanoma-related part, the human-related domains which have two or more interaction relationship with melanoma-related domains.

After the two approaches, we have two lists of human domains which have the potential correlation relationship with melanoma, then we discard some domains that have already been found to play an important role in mr process through the existing literature and the description of these domains. The rest are those that haven't been found to have close relationship with mr activities. According to the potential correlation method, we find that these human domains may also play an important role in mr activities and further study based on these human domains need to be done.

3   Result and Discussion

3.1   Data Summary for Mr Domain and Human-Related Domain Analysis

Based on the mr mutation records in COSMIC, which have corresponding amino acid change, there are 58,645 mutations that can find corresponding proteins in UniproKB. Among the 58,645 mr mutations, there are 5,203 genes involved. Among 58,645 mr mutations, 23685 are located in domain regions, 34,960 are located in non-domain regions, and 1,494 mr domain families can find mr mutations. According to the DDI information from the iPfam database and the mr domains, there are 844 DDI pairs acting within the mr domains, and 1,058 DDI pairs acting between the mr domains and the non mr domains.

As the protein database UniproKB has all the human proteins extracted from literature and curator-evaluated computational analysis, and the database Pfam has detail information of proteins and their functional domains, we download all the human proteins from UniproKB and then use these human proteins to extract all the human domains from Pfam. After this, a list of all human proteins including 20,204 proteins is extracted from UniproKB. Then we get all human domains from Pfam. Finally, 5,763 human-related domain families are found. According to the DDI information from the iPfam database and the human-related domains, there are 3,820 DDI pairs acting within the human-related domains.

3.2   DDI Network Analysis

In this project, we combine two approaches together to find the human domains that have potential correlation relationship with melanoma.

The first approach is to find the 'important' melanoma-related domains in the mr DDI network, and then find, in the human DDI network except mr domains, the human domains that have potential correlation relationship with melanoma. We use the Cytoscape App Centiscape2.1, to analyze the mr DDI network. Centiscape2.1 can calculate Closeness, Stress, Betweenness, Centroid Value and Eccentricity of the network. As Centroid, Closeness together with Eccentricity, and Betweenness, together with Stress, can respectively imply the importance of a node in the network, we select the top 20 mr domains of these five index from the result of Centiscape2.1. Then we extract the mr domains that appear in at least two columns among the five index. The result list shows the important mr domains in the mr DDI network. Then we find in the human-related domains, except mr domains, the human domains that have interactions with at least two 'important' mr domains. The result shows the human domains that has potential correlation relationship with mr domains. And according to the potential correlation method, these human domains maybe very important to the mr study.

Approach two is to find in the human-related domains except the mr domains, the domain that has at least two interaction relationships with mr domains. There are 177 human-related domains that interact with mr domains. Domains with more than three interaction partners are selected and verified. Noting that there are 8 human domains that have only 2 interaction partners with mr domains, but they also appear in the first approach's 'important' list, so they are also counted in. Finally, there are 83 human-related domains that have more than three interaction partners with mr domains.

As the result of approach one is included in the result of approach two, but have more confidence compared with approach two. So we finally get a list of 83 human-related domains that have potential correlation relationship with melanoma. As the existing literature or researches may have already studied the relationship among these human domains and melanoma, we check out the existing literature or research and the description of these human domains, and just keep the human domains that have never been found to have anything to do with melanoma. After verifying, 56 human-related domains were found to have close relationship with melanoma and 27 human-related domains are not find to have anything to do with melanoma. Among the 27 human-related domains, there are 16 (Table 1) domains have relationship with disease, so we highly believe that they may also have relationship with melanoma, and there are 11 (Table 2) domains haven't found anything with disease, but as they have interaction relationship with mr domains, so they have closed relationship with melanoma, and further study based on these domains may find new ways to cure melanoma.

Table 1. Human-related domains that have potential correlation relationship with melanoma and have been found to have relationship with diseases.

Domain ID	Domain name	#Ref	Domain ID	Domain name	#Ref
PF01247	Ribosomal L35Ae	[20]	PF00861	Ribosomal L18p	[28]
PF08763	Ca_chan_IQ	[21]	PF00572	Ribosome L13	[29]
PF12026	DUF3513	[22]	PF02221	E1_DerP2_DerF2	[30]
PF13774	Longin	[23]	PF01507	PAPS_reduct	[31]
PF01115	F_actin_cap_B	[24]	PF03297	Ribosomal S25	[32]
PF11601	Shal-type	[25]	PF03464	eRF1_2	[33]
PF10163	EnY2	[26]	PF12424	ATP_Ca_trans_C	[34]
PF01929	Ribosomal L14e	[27]	PF15036	Interleukin-34	[35]

Table 2. Human-related domains that have correlation relationship with melanoma and have not been found to have any relationship with diseases.

Domain ID	Domain name	Domain ID	Domain name
PF01775	Ribosomal L18ae	PF00831	Ribosomal L29
PF00828	Ribosomal L18e	PF01655	Ribosomal L32e
PF03494	beta-APP	PF14492	EFG_II
PF09596	MamL-1	PF13621	Cupin 8
PF01020	Ribosomal L40e	PF00736	EF1 guanine nucleotide exchange
PF01781	Ribosomal L38e		

4   Conclusion

This project focus on finding new methods to cure melanoma by mapping the melanoma-related mutations to domain and domain-domain interaction (DDI) level and use potential correlation method to find the human domains that have never been found to have anything to do with the melanoma-related study. With the help of two network analyze method, we get a list of 83 human domains that have potential correlation relationship with melanoma. We check out the existing literature or research and the description of these human domains. After verifying, 56 human-related domains were found to have close relationship with melanoma. This proves the effectivity of the potential correlation method, so for the rest 27 human-related domains that were not find to have anything to do with melanoma, according to the principle of the potential correlation, further study based on these human domains may lead to some new method to cure melanoma.

Acknowledgement. This work was supported by the National Science Foundation of China (Nos. 61472282, 61300058 and 61271098) and Anhui Provincial Natural Science Foundation (No.1508085MF129).

References

1. Siegel, R., Naishadham, D., Jemal, A.: Cancer statistics, 2013. CA Cancer J. Clin. **63**, 11–30 (2013)
2. Dutton-Regester, K., Hayward, N.: Reviewing the somatic genetics of melanoma: from current to future analytical approaches. Pigment Cell Melanoma Res. **25**, 144–154 (2012)
3. Meyerson, M., Gabriel, S., Getz, G.: Advances in understanding cancer genomes through second-generation sequencing. Nat. Rev. Genet. **11**(10), 685–696 (2010)
4. Reis-Filho, J.S.: Next-generation sequencing. Breast Cancer Res. **11**(Suppl. 3), S12 (2009)
5. Lawrence, M.S., Stojanov, P., Polak, P., et al.: Mutational heterogeneity in cancer and the search for new cancer-associated genes. Nature **499**(7457), 214–218 (2013)
6. Kunz, M., Dannemann, M., Kelso, J.: High-throughput sequencing of the melanoma genome. Exp. Dermatol. **22**(1), 10–17 (2013)
7. Pleasance, E.D., Cheetham, R.K., Stephens, P.J., et al.: A comprehensive catalogue of somatic mutations from a human cancer genome. Nature **463**(7278), 191–196 (2010)
8. Wei, X., Walia, V., Lin, J.C., et al.: Exome sequencing identifies GRIN2A as frequently mutated in melanoma. Nat. Genet. **43**(5), 442–446 (2011)
9. Yokoyama, S., Woods, S.L., Boyle, G.M., et al.: A novel recurrent mutation in MITF predisposes to familial and sporadic melanoma. Nature **480**(7375), 99–103 (2011)
10. Berger, M.F., Hodis, E., Heffernan, T.P., et al.: Melanoma genome sequencing reveals frequent PREX2 mutations. Nature **485**(7399), 502–506 (2012)
11. Wang, B., Sun, W., Zhang, J., et al.: Current status of machine learning-based methods for identifying protein-protein interaction sites. Curr. Bioinform. **8**(2), 177–182 (2013)
12. Zhu, L., You, Z.H., Huang, D.S., et al.: t-LSE: a novel robust geometric approach for modeling protein-protein interaction networks. PLoS ONE **8**(4), e58368 (2013)
13. Krogan, N.J., Cagney, G., Yu, H., et al.: Global landscape of protein complexes in the yeast Saccharomyces cerevisiae. Nature **440**(7084), 637–643 (2006)

14. Forbes, S.A., Beare, D., Gunasekaran, P., et al.: COSMIC: exploring the world's knowledge of somatic mutations in human cancer. Nucleic Acids Res. **43**(D1), D805–D811 (2015)
15. Finn R.D, Bateman, A., Clements, J., et al.: Pfam: the protein families database. Nucleic acids research, 2013: gkt1223
16. UniProt Consortium and others. UniProt: a hub for protein information. Nucleic Acids Res. gku989 (2014)
17. Finn, R.D., Miller, B.L., Clements, J., et al.: iPfam: a database of protein family and domain interactions found in the Protein Data Bank. Nucleic Acids Res. **42**(D1), D364–D373 (2014)
18. Shannon, P., Markiel, A., Ozier, O., et al.: Cytoscape: a software environment for integrated models of biomolecular interaction networks. Genome Res. **13**(11), 2498–2504 (2003)
19. Scardoni, G., Tosadori, G., Faizan, M., Spoto, F., Fabbri, F., Laudanna, C.: Biological network analysis with CentiScaPe: centralities and experimental dataset integration[J]. F1000Res. **3** (2014)
20. Monji, M., Senju, S., Nakatsura, T., et al.: Head and neck cancer antigens recognized by the humoral immune system. Biochem. Biophys. Res. Commun. **294**(3), 734–741 (2002)
21. Das, A., et al.: Role of voltage-gated T-type calcium channels in the viability of human melanoma. Universitat de Lleida (2012)
22. Titz, B., Low, T., Komisopoulou, E., et al.: The proximal signaling network of the BCR-ABL1 oncogene shows a modular organization. Oncogene **29**(44), 5895–5910 (2010)
23. Williams, K.C., McNeilly, R.E., Coppolino, M.G.: SNAP23, Syntaxin4, and vesicle-associated membrane protein 7 (VAMP7) mediate trafficking of membrane type 1–matrix metalloproteinase (MT1-MMP) during invadopodium formation and tumor cell invasion. Mol. Biol. Cell **25**(13), 2061–2070 (2014)
24. Huang, C.M., Elmets, C.A., van Kampen, K.R., et al.: Prospective highlights of functional skin proteomics. Mass Spectrom. Rev. **24**(5), 647–660 (2005)
25. Wulff, H., Castle, N.A., Pardo, L.A.: Voltage-gated potassium channels as therapeutic targets. Nat. Rev. Drug Discov. **8**(12), 982–1001 (2009)
26. Bedi, U.: Regulation of H2B monoubiquitination pathway in breast cancer. Niedersächsische Staats-und Universitätsbibliothek Göttingen (2014)
27. Huang, X.P., Zhao, C.X., Li, Q.J., et al.: Alteration of RPL14 in squamous cell carcinomas and preneoplastic lesions of the esophagus. Gene **366**(1), 161–168 (2006)
28. Allen-Vercoe, E., Holt, R., Moore, R., et al.: Detection of fusobacterium in a gastrointestinal sample to diagnose gastrointestinal cancer: U.S. Patent Application 13/877,421. 2011-10-4
29. Amsterdam, A., Sadler, K.C., Lai, K., et al.: Many ribosomal protein genes are cancer genes in zebrafish. PLoS Biol. **2**(5), e139 (2004)
30. Gay, N.J., Gangloff, M.: Structure and function of Toll receptors and their ligands. Annu. Rev. Biochem. **76**, 141–165 (2007)
31. Jain, A.K., Jain, S., Rana, A.C.: Metabolic enzyme considerations in cancer therapy. Malays. J. Med. Sci.: MJMS **14**(1), 10 (2007)
32. Zhang, X., Wang, W., Wang, H., et al.: Identification of ribosomal protein S25 (RPS25)–MDM2-p53 regulatory feedback loop. Oncogene **32**(22), 2782–2791 (2013)
33. McGill, G.G., Horstmann, M., Widlund, H.R., et al.: Bcl2 regulation by the melanocyte master regulator Mitf modulates lineage survival and melanoma cell viability. Cell **109**(6), 707–718 (2002)
34. Sennoune, S.R., Luo, D., Martínez-Zaguilán, R.: Plasmalemmal vacuolar-type H+-ATPase in cancer biology. Cell Biochem. Biophys. **40**(2), 185–206 (2004)
35. Gunawardhana, S., Zins, K., Lucas, T., et al.: Novel CSF-1 receptor ligand IL-34 modulates macrophage-breast cancer cell crosstalk. Cancer Res. **74**(19 Suppl.), 1160 (2014)

Training Neural Networks as Experimental Models: Classifying Biomedical Datasets for Sickle Cell Disease

Mohammed Khalaf[1(✉)], Abir Jaafar Hussain[1], Dhiya Al-Jumeily[1],
Robert Keight[1], Russell Keenan[2], Paul Fergus[1], Haya Al-Askar[3],
Andy Shaw[1], and Ibrahim Olatunji Idowu[1]

[1] Faculty of Engineering and Technology, Liverpool John Moores University,
Byrom Street, Liverpool L3 3AF, UK
M.I.Khalaf@2014.ljmu.ac.uk,
{a.hussain,d.aljumeily,p.fergus,a.shaw}@ljmu.ac.uk,
R.Keight@2015.ljmu.ac.uk
[2] Liverpool Paediatric Haemophilia Centre, Haematology Treatment Centre,
Alder Hey Children's Hospital,
Eaton Road, West Derby, Liverpool L12 2AP, UK
Russell.keenan@alderhey.nhs.uk
[3] College of Computer Engineering and Science, Sattam Bin Abdulaziz
University, Al-Kharj, Kingdom of Saudi Arabia

Abstract. This paper discusses the use of various type of neural network architectures for the classification of medical data. Extensive research has indicated that neural networks generate significant improvements when used for the pre-processing of medical time-series data signals and have assisted in obtaining high accuracy in the classification of medical data. Up to date, most of hospitals and healthcare sectors in the United Kingdom are using manual approach for analysing patient input for sickle cell disease, which depends on clinician's experience that can lead to time consuming and stress to patients. The results obtained from a range of models during our experiments have shown that the proposed Back-propagation trained feed-forward neural network classifier generated significantly better outcomes over the other range of classifiers. Using the Receiver Operating Characteristic curve, experiments results showed the following outcomes for our models, in order of best to worst: Back-propagation trained feed-forward neural net classifier: 0.989, Functional Link Neural Network: 0.972, in comparison to the Radial basis neural Network Classifiers with areas of 0.875, and the Voted Perception classifier: 0.766. A Linear Neural Network was used as baseline classifier to illustrate the importance of the previous models, producing an area of 0.849, followed by a random guessing model with an area of 0.524.

Keywords: Neural network architectures · Sickle cell disease · Real datasets · The area under curve · Receiver operating characteristic curve · e-Health

© Springer International Publishing Switzerland 2016
D.-S. Huang et al. (Eds.): ICIC 2016, Part I, LNCS 9771, pp. 784–795, 2016.
DOI: 10.1007/978-3-319-42291-6_78

1 Introduction

Sickle cell disease (SCD) is considered long-term disease that facing medical sector, affects human from early childhood. It has a high severe impact on the life expectancy and the patient's quality of life due to red blood cell (RBCs) abnormality. This disease affects patients who suffer from genetic blood disorders, where the haemoglobin behaves abnormal in the blood. SCD is phenotypically complex, with various medical outcomes ranging from early childhood mortality to a nearly unrecognised condition [1]. With respect to sickle cell disease, the most significant symptoms that could show effects on patients are fatigue, shortness of breath, dizziness, and headaches. The World health Organisation (WHO) reported that 5 % of the population carries the hae-moglobin disorder around the world, mostly, sickle cell disease and thalassemia [2].

These days, the significance improvement in medical/science data have provide a new solution to improve healthcare outcomes and services. Recent research has demonstrated the positive effects of a drug called hydroxyurea/hydroxycarbamide in terms of modifying the disease phenotype [3]. The clinical procedure to manage SCD modifying therapy is time consuming and difficult for medical staff. In order to curtail the significant medical variability presented by such difficult crisis, healthcare consultants need to improve adherence to therapy, which is regularly poor and subsequently results in fewer benefits and elevated risks to patients. Up to the present, the new trend of machine learning is becoming essential for the analysis of data within the medical domain. Machine learning provides a number of services for prognostic and diagnostic issues in a range of clinical societies. This kind of techniques are being utilized in order to analyse the significance of healthcare factors in association with their integrations for prognosis, for instance, to overall patient management, for providing therapy and sup-port, and the most important to predict the disease progression [4].

Neural Networks (NNs) is widely-used classification methods for medical domains [5]. This type of machine learning approach can identify groups of attributes in order to recognise the specific kind of diseases and illnesses. The main reason behind that is due to their characteristics of parallel processing, self-organization, non-linearity, and the most important point is self-learning [6]. In this paper, the applications of NN approaches for health datasets classification will be discussed. The performance of four types of NN algorithms, comprising the feed-forward neural net classifier (BPXNC), the Functional Link Neural Network (FLNN), Radial basis neural Network Classifiers (RBNC), and Voted Perception classifier (VPC), will be considered. The networks will be applied to classify the amount of medication dosage for those who suffer from SCD, the results of which will be presented and discussed.

The structure of this paper is organised as follows. Section 2 discusses various NN architectures while Sect. 3 shows the clinical care pathway. The methodology for classifying sickle cell datasets is shown in Sect. 4, while Sect. 5 shows the result section. Section 6 illustrates the conclusion.

2   Neural Networks

Neural networks (NNs) are considered a form of biologically inspired algorithm (BIA), based on the constellations of connected elements observed in the biological brain, namely networks of specialised cells called neurons [7]. NNs are inspired software programs that were designed in order to simulate the way that mankind brain processes any type of information. NNs attempts to model actual systems depending on the information provided to it. This kind of machine learning involves hundreds of single unit artificial neurons and is considered a powerful computational and mathematical data model that is cable of representing complex input and output connections. The main motivation behind developing an NNs is the capacity to perform intelligent tasks, which are performed in a manner similar to the operation of the human brain. Furthermore, the power of NNs comes from connecting the neurons in a one particular network that have ability to represent non-linear and linear relationships.

NNs for computational modelling need a number of neurons so that they can be connected together to form a network. In this context, neurons are organised in layers and have processing units, which takes one or more inputs to generate an output. In this case, at each neurone, all inputs have to be connected with a weight that modifies the strength of each input. As a result, neurons will simply collect all the inputs together to calculate an output. The weights in each NNs are trained using different types of learning algorithm, for example supervised and unsupervised learning. In order that the network promotes the most important features within training process, learning algorithms are utilised to update the weights of the NNs using mathematical equations.

The main backbone of using weights is to check the output if it is too high, then the weights should be lowered by with certain amount to be fit to the output for all the input instance. On the other hand, if the predicated output is too low, then the weights need to be increment by the set amount. The hidden layer learns to provide a representation for the inputs. In each NNs, one or more hidden layers can be applied. The process of constructing such an architecture is referred to as learning in NNs [8]. There are four different NN classifiers used in these experiments:

A. **The back-propagation trained feed-forward neural network classifier (BPXNC):** This type of NNs is trained to set a number of input data through making adjustment of the complete weights [9]. In this context, the information t collected from imputes is fed forward to improve the weights between neurons. This classifier then will be able to read the output and inputs value within the training datasets in order to decrease the differentiation between observed and predicted values [10].

B. **Functional Link Neural networks (FLNN):** FLNN has single layer of neurons capable to handle non-separable classes through by increasing the dimensionality of the whole input spaces [11]. It is called a flat classifier. This feature provides such an excellent support that makes the learning algorithm in the network less complex. We have selected this model in the purpose of enhancing the performance of network.

C. **Voted Perception classifier (VPC):** This algorithm is based on the perceptron models of frank and Rosenblatt. The classifiers can work with the input data that is considered linearly separable. It uses kernel functions to display very high dimensional spaces. The VPC is much more efficient and simple to implement. VPC is trained using supervised learning algorithms. The training process implements many full sweeps by the training data. In this model, the classification are performed for a new objective by permitting the ensemble of perceptron's to make vote in the NN on the label of each test point [12].

D. **The Radial basis neural Network Classifiers (RBNC):** RBNC is considered popular models in NN architectures, which is often utilised in classification problem and complex pattern recognition. Typically, this model has an input layer, a one hidden layer but without activation function, and output layer. The mapping procedure can be modified for RBNC through adjusted the weights values in the output layer [9].

3   Clinical Care Pathway

Clinical care pathways are considered one of the major methods to manage and provide the best quality in clinical concerning the standardisation of care procedures. It is obvious that their implementation improves outcomes and decreases the variability in medical practice [13]. In this context, this type of processes refers to how medical consultants can treat the SCD patients according to their conditions. It promotes efficient and organised patient care depend on practice. The major goal of this study is to explore and develop methodology for the datasets that are collected from the local hospital, which can increase the efficiency of the data analysis procedures. Healthcare professional provides 50 mg of hydroxyurea drug/liquid at the beginning until reached 1700 mg according to the patient's condition. Typically, clinicians reduce the amount of dosage medication when there is any toxicity appear in patient body, especially in liver or kidneys part. It is very important to check if patient takes their medication regularly of not. This will have high impacts on patient conditions through increasing the level of Hb and MCV rates. Figure 1 demonstrates the full flowchart that healthcare professionals use to analyse patients' blood test and provide accurate amount of medication.

4   Methodology

In this paper, the previous work in the field of machine learning have been forced to check and predict the severe crises of SCD, rather than using optimal predication techniques to give correct amounts of hydroxyurea in modifying the disease phenotype [9]. Currently, there is no standardisation of disease modifying therapy management. Through using the proposed computerised comprehensive system, the goal is to provide an optimised and reproducible standard of care in various medical settings across the UK. The main aim of this research is to use recent advanced NN approaches, in order to provide an excellent support to the clinicians by providing specific medication for each individual conditions. In this context, and due to the pattern of the SCD

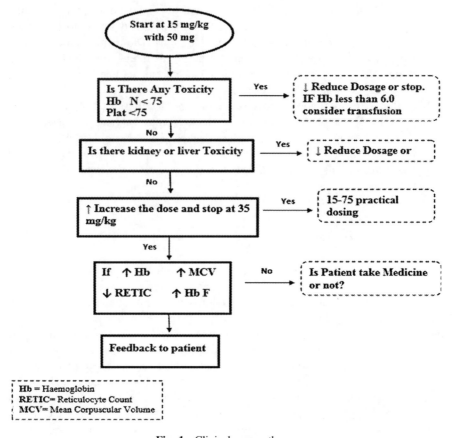

Fig. 1. Clinical care pathway

dataset, we attempt to propose classification of the patient's datasets records at an earlier stage, according to how much of a dose the patient will need to take. This can potentially lower the costs, avoid unnecessary admission to hospitals, mitigate patient illness before proceed to the worse condition, improve patient welfare, and unnecessary interventions.

NN algorithms can be implemented to figure strong integrated classifiers, including previous patient cases that have been gathered from the local hospital for SCD in the city of Liverpool, UK, over the last ten years using 1168 patient records.

4.1   Data Collection

The real datasets that are used in this study for SCD were collected within a ten-year period. Each patient sample covers 13 instances deemed vital factors for classifying the SCD datasets as showed in Table 1 [14]. In order to collect these datasets, the local hospital has supported this study with many patient records to obtain better accuracy

Table 1. SCD datasets features.

No	Types of attributes
1	Weight
2	Haemoglobin (Hb)
3	Mean corpuscular volume (MCV)
4	Platelets (PLTS)
5	Neutrophils (white blood cell NEUT)
6	Reticulocyte count (RETIC A)
7	Reticulocyte count (RETIC %)
8	Alanine aminotransferase (ALT)
9	Body bio blood (BIO)
10	Hb F
11	Bilirubin (BILI)
12	Lactate dehydrogenase (LDH)

and services. The resulting dataset comprised 1168 sample points, with a single target variable describing the hydroxyurea medication dosage in milligrams. To facilitate our classification study, the target dosage was discretised into 6 bins, denoted classes 1 through 6. It is formed through dividing the output range (in Milligrams) into membership intervals of equal size: Class 1: $[148 \leq Y < 410 \text{ mg}]$, Class 2: $[410 \leq Y < 659 \text{ mg}]$, Class 3: $[659 \leq Y \leq 919 \text{ mg}]$, Class 4: $[919 \leq Y < 1200 \text{ mg}]$, Class 5: $[1200 \leq Y < 1430 \text{ mg}]$, Class 6: $[1430 \leq Y \leq 1700 \text{ mg}]$.

4.2 Experimental Setup

The experimental setup covers the design of the test environment used in our experiments, the models tested, and the configuration of each model. Finally, the performance evaluation metrics utilised to measure the results of the NN algorithms are presented for the SCD datasets.

This study is composed of trained models using four types of integrated NN approaches: Back-propagation trained feed-forward neural net classifier, Functional Link Neural Network, the Radial basis neural Network Classifiers, and the Voted Perception classifier. These models are appropriate to act as comparators of high performance [15]. The linear model used includes a linear transformation function with a single layer neural network at each class output unit. To obtain performance estimates for the respective models, we ran each simulation 50 times and calculated the mean of the responses. The full set of models used in the experiments are described in Table 2. Finally, the random oracle model is used to establish random case performance through the assignment of random responses for each class.

We used a competing models to the classification task. In addition, to apply ROM that provide a baseline indicator to show the performance yielded by random guessing [16]. Furthermore, we presented a linear model to discover the variance in performance presented between this weak classifier and the non-linear classifiers, such as NNs.

Table 2. Classification models description.

Models	Description	Architecture	Training algorithm	Role
ROM	Random oracle model	Pseudorandom number generator	N/A	Random guessing baseline
BPXNC	Feed-forward neural network algorithm	Context units: one context unit for each output unit	This classifier is trained properly to map a set of input data in order to make iterative modification for the whole weights	Non-linear comparison model
FLNN	Functional link neural network	Units: 13-30-3, tansig activations	Gradient descent with momentum and adaptive learning rate backpropagation	Test model
RBNC	Feed-forward neural net including N sigmoid neurons	13 inputs, 3 outputs	The classifier has radial basis units with only 1 hidden layer	Non-linear comparison model
VPC	Combines an ensemble of perceptron's through voting procedure	Units: 13-30-3, tansig activations	This classifier performed by permitting the ensemble of perceptron's to vote in the neural network on the label of each test point	Non-linear comparison model
LNN	Linear combiner network	Units: 13-3, linear activations	Batch training with weight and bias learning rules	Linear comparison model

Table 3. Performance metric calculation.

Metric name	Calculation
Sensitivity	TP/(TP + FN)
Specificity	TN/(TN + FP)
Precision	TP/(TP + FP)
F1 score	2 * (Precision * Recall)/(Precision + Recall)
Youden's J statistic (J Score)	Sensitivity + Specificity − 1
Accuracy	(TP + TN)/(TP + FN + TN + FP)
Area under ROC curve (AUC)	0 <= Area under the ROC Curve <= 1

The integration of random control baselines, weak and strong, offers an experimental frame through which to gauge the relative performance.

In our study, the classifier models evaluation involves of both in-sample (training), and out-of-sample (testing) diagnostics, involving sensitivity, specificity, precision, the F1 score, Youden's J statistic, and the classification accuracy calculated as shown in Table 3. Moreover, the classifiers were categorised using the Area under the Curve (AUC) and the Receiver Operating Characteristic (ROC) plots, where the classification capability across all operating points was determined.

5 Results

In this section, we analyse the results from the various experiments that have been applies in this study as listed in Tables 4 and 5, presenting outcomes for training and testing of the models, respectively. We run further performance visualisations through the use of AUC comparison plots (Figs. 2 and 3) and the ROC plots as shown in Figs. 4 and 5. In order to make compassion among classifiers, the BPXNC outperformed the rest of models as demonstrated in Table 5, illustrating capability both in generalising and in fitting the training data to unseen examples. The calculated means of AUCs for the BPXNC model, obtained for six classes during training yielded an area of 0.997 (ideal), in comparison to 0.989 over the test sample. Classes one to six were found to illustrate optimal consistent generalisation and performance from the training to the test sets for this classifier. It was discovered that the FLNN model was able to yield an average AUC of 0.972, outperforming the RBNC classifier and showing an overall rank of second place. The RBNC model produced an average AUC of 0.875, ranking third overall, outperforming the lower ranking models by a reasonable margin. All three of the top performing models, the BPXNC, FLNN, and RBNC, obtained nearly ideal AUCs and represent viable candidates for future use. These models produced exceptional results in terms of both training and generalisation. We found that the VPC classifiers, constituted the next performance level obtained, with test AUCs of 0.766 respectively.

Table 4. The performance of classifiers (training phase).

Model	Sensitivity	Specificity	Precision	F1	J	Accuracy	AUC
ROM	0.493	0.542	0.179	0.253	0.0351	0.543	0.488
BPXNC	0.983	0.988	0.93	0.954	0.97	0.986	0.997
VPC	0.743	0.73	0.337	0.45	0.474	0.733	0.781
RBNC	0.83	0.865	0.62	0.702	0.695	0.86	0.907
FLNN	0.954	0.964	0.809	0.873	0.918	0.963	0.985
LNN	0.854	0.827	0.524	0.642	0.68	0.835	0.87

The LNN was able to learn the non-linear components in the data and so produced such an excellent classification results. The average test of AUCs for this model ranged 0.849, which is seen to demonstrate performance significantly below that of the other models.

Table 5. The performance of classifiers (testing phase).

Model	Sensitivity	Specificity	Precision	F1	J	Accuracy	AUC
ROM	0.548	0.547	0.197	0.268	0.0949	0.539	0.524
BPXNC	0.978	0.98	0.91	0.942	0.958	0.98	0.989
VPC	0.729	0.745	0.383	0.496	0.475	0.749	0.766
RBNC	0.835	0.842	0.526	0.639	0.678	0.845	0.875
FLNN	0.95	0.966	0.803	0.862	0.916	0.962	0.972
LNN	0.824	0.848	0.534	0.641	0.672	0.84	0.849

The ROM random guessing model is as expected produce a weak classification as shown in the ROC plots for all classes (see Figs. 1 and 2), illustrating by contrast the consequence of the results from the other trained classifiers. The average AUC for this model yield 0.524, which considered the lowest outcomes among other models. This process of guessing yields both train and test set AUCs lower than the LNN trained model baseline.

In order to explore our datasets in 2- dimensional, the data representation was explored to investigate if there are any regularities that might be uncovered within its structure. Exploratory analysis is considered such a significant step in the machine learning approaches, permitting the human advisor to obtain an intuition of the data as well as the potential learnability of the data. The results from the whole data exploration can be utilised to lead the modelling point, since a key component learnability is known to be a task of the correspondence between the kind of demonstration it is supplied with and the learning algorithm. To undertake an exploration of the utilised data in these experiments, we computed summary statistics, followed by visualisation methods including Principal Component Analysis (PCA) and t-distributed Stochastic Neighbourhood Embedding (tSNE) as reported in Figs. 6 and 7, respectively. Results from the visualisation procedures reveal that some discernible structure is present within the data. The PCA plot indicates that there are potential clusters of values present within the data. The tSNE plot demonstrates that the data can be geometrically

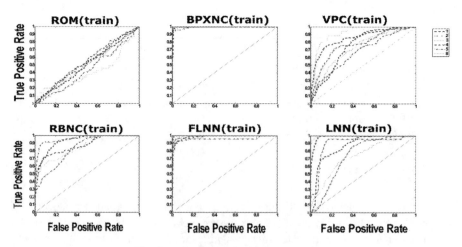

Fig. 2. ROC curve (Train) for classifiers

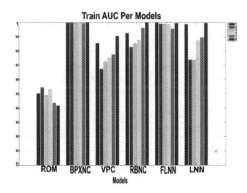

Fig. 3. ROC curve (Testing) for classifiers

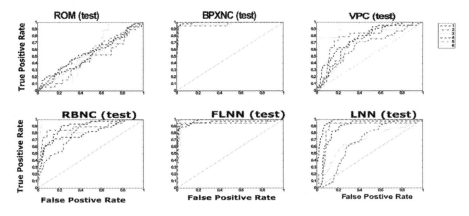

Fig. 4. Train AUC per model (Color figure online)

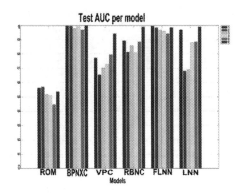

Fig. 5. Testing AUC per model (Color figure online)

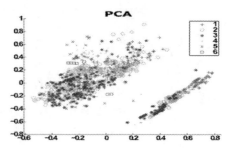

Fig. 6. PCA plot (Color figure online)

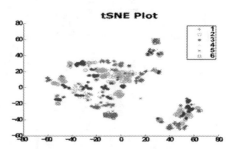

Fig. 7. tSNE plot (Color figure online)

separated when allowing different intervals of dosage level. Moreover, the exploratory data analysis illustrates no noticeable defects that might call into question the outcomes of consequent analysis.

6   Conclusion

This study presents an empirical investigation into the use of various neural network algorithms to classify the level of dosage for SCD medication. In this paper, various model architectures are used for analysing the medical datasets obtained from SCD patients. The main purpose of this research is to examine the effectiveness of these models in terms of training and testing setting, investigating if such architectures could enhance classification results. It was found through experimental investigation, including the usage of SCD datasets and approaches such as BPXNC, the FLNN, RBNC, and VPC that the analysis of medical datasets is viable and produces precise results. The results gained from a range of models during our experiments have demonstrated that the proposed Back-propagation trained feed-forward neural network classifier produced the best results with the AUC 0.989 in compare with other models. It is found that the outcomes are based on a considerable data sample, containing more than 1100 sample examples, which supports the significance of the findings. It is recommended that further research will be used to make confirmation on our findings, where a large number of data could be utilised also to advance the performance of the results. In this circumstance, we recommend however that a machine learning algorithms, for instance, support vector machine and k-nearest neighbour classifiers could be used to increase the scale and scope of this research.

References

1. Sebastiani, P., Ramoni, M.F., Nolan, V., Baldwin, C.T., Steinberg, M.H.: Genetic dissection and prognostic modeling of overt stroke in sickle cell anemia. Nat. Genet. **37**, 435–440 (2005)
2. Weatherall, D.J.: The inherited diseases of hemoglobin are an emerging global health burden. Blood **115**, 4331–4336 (2010)
3. Kosaryan, M., Karami, H., Zafari, M., Yaghobi, N.: Report on patients with non transfusion-dependent β-thalassemia major being treated with hydroxyurea attending the Thalassemia Research Center, Sari, Mazandaran Province, Islamic Republic of Iran in 2013. Hemoglobin **38**, 115–118 (2014)
4. Magoulas, G.D., Prentza, A.: Machine learning in medical applications. In: Paliouras, G., Karkaletsis, V., Spyropoulos, C.D. (eds.) ACAI 1999. LNCS (LNAI), vol. 2049, pp. 300–307. Springer, Heidelberg (2001)
5. Al-Shayea, Q.K.: Artificial neural networks in medical diagnosis. Int. J. Comput. Sci. Issues **8**, 150–154 (2011)
6. Liu, B., Wang, M., Yu, L., Liu, Z., Yu, H.: Study of feature classification methods in BCI based on neural networks. In: 27th Annual International Conference of the Engineering in Medicine and Biology Society, IEEE-EMBS 2005, pp. 2932–2935. IEEE (2005)
7. Brabazon, A., O'Neill, M.: Biologically Inspired Algorithms for Financial Modelling. Springer Science & Business Media, Berlin (2006)
8. Karayiannis, N., Venetsanopoulos, A.N.: Artificial Neural Networks: Learning Algorithms, Performance Evaluation, and Applications. Springer Science & Business Media, Berlin (2013)
9. Idowu, I.O., Fergus, P., Hussain, A., Dobbins, C., Askar, H.A.: Advance artificial neural network classification techniques using EHG for detecting preterm births. In: 2014 Eighth International Conference on Complex, Intelligent and Software Intensive Systems (CISIS), pp. 95–100 (2014)
10. Fergus, P., De-Shuang, H., Hamdan, H.: Prediction of intrapartum hypoxia from cardiotocography data using machine learning. In: Applied Computing in Medicine and Health-Emerging Topics in Computer Science and Applied Computing, pp. 125–146 (2016)
11. Dehuri, S., Cho, S.-B.: A comprehensive survey on functional link neural networks and an adaptive PSO–BP learning for CFLNN. Neural Comput. Appl. **19**, 187–205 (2010)
12. Hussain, A.J., Fergus, P., Al-Askar, H., Al-Jumeily, D., Jager, F.: Dynamic neural network architecture inspired by the immune algorithm to predict preterm deliveries in pregnant women. Neurocomputing **151**, 963–974 (2015)
13. Panella, M., Marchisio, S., Stanislao, F.: Reducing clinical variations with clinical pathways: do pathways work? Int. J. Qual. Health Care **15**, 509–521 (2003)
14. Khalaf, M., Hussain, A.J., Al-Jumeily, D., Fergus, P., Keenan, R., Radi, N.: A framework to support ubiquitous healthcare monitoring and diagnostic for sickle cell disease. In: Huang, D.-S., Jo, K.-H., Hussain, A. (eds.) ICIC 2015. LNCS, vol. 9226, pp. 665–675. Springer, Heidelberg (2015)
15. Duch, W.: Towards comprehensive foundations of computational intelligence. In: Duch, W., Mańdziuk, J. (eds.) Challenges for Computational Intelligence. Studies in Computational Intelligence, vol. 63, pp. 261–316. Springer, Berlin (2007)
16. Jia, X.-Y., Li, B., Liu, Y.-M.: Random oracle model. Ruanjian Xuebao/J. Softw. **23**, 140-151 (2012)

Multi-agent Systems for Dynamic Forensic Investigation

Phillip Kendrick[1](✉), Abir Jaafar Hussain[1], and Natalia Criado[2]

[1] Department of Computer Science, Liverpool John Moores University, Liverpool, UK
P.G.Kendrick@2012.ljmu.ac.uk, A.Hussain@ljmu.ac.uk
[2] Department of Informatics, King's College London, London, UK
Natalia.Criado@kcl.ac.uk

Abstract. In recent years Multi-Agent Systems have proven to be a useful paradigm for areas where inconsistency and uncertainty are the norm. Network security environments suffer from these problems and could benefit from a Multi-Agent model for dynamic forensic investigations. Building upon previous solutions that lack the necessary levels of scalability and autonomy, we present a decentralised model for collecting and analysing network security data to attain higher levels of accuracy and efficiency. The main contributions of the paper are: (i) a Multi-Agent model for the dynamic organisation of agents participating in forensic investigations; (ii) an agent architecture endowed with mechanisms for collecting and analysing network data; (iii) a protocol for allowing agents to coordinate and make collective decisions on the maliciousness of suspicious activity; and (iv) a simulator tool to test the proposed decentralised model, agents and communication protocol under a wide range of circumstances and scenarios.

Keywords: Forensic investigation · Multi-agent system · Simulator · Cyber security

1 Introduction

Providing effective cyber security will require efficient and scalable solutions to meet the ever increasing number of needs found within modern expanded networks. Typical security solutions utilise a number of specialised devices (e.g., Intrusion Detection Systems (IDS) [1, 2], firewalls and forensic tool-kits [3]) often requiring high performance hardware to offset the cost of processing large amounts of data. These solutions often suffer from two problems that we will address in this paper: information overflow, which occurs when systems inefficiently collect all of the available information for bulk processing; and the failure to detect advanced stealthy attacks, which occurs when systems fail to observe the necessary information required for accurate attack analysis. By combining Multi-Agent Systems (MAS) [4] with intelligent information gathering techniques, improvements can be made by taking advantage of automated forensic investigations to proactively gather the necessary data about suspicious activity[1].

A MAS, at its most basic level, can be defined as a collection of intelligent agents. Agents are independent pieces of software, often capable of working together to solve

---

[1] Suspicious activity is defined as any activity that does not appear to fit the norm of the network.

© Springer International Publishing Switzerland 2016
D.-S. Huang et al. (Eds.): ICIC 2016, Part I, LNCS 9771, pp. 796–807, 2016.
DOI: 10.1007/978-3-319-42291-6_79

problems that could not be solved by a single agent. The following properties are characteristic of an intelligent agent [4]: (1) Autonomy - The agent's ability to act independently without any external human operator interaction. (2) Reactivity - The agent's ability to sense environmental changes and react to the situation. (3) Proactivity - The agent's ability to choose actions to achieve goals. (4) Adaptability - The agent's ability to change goals in response to unforeseen circumstances. (5) Communication & Coordination - The agent's ability to communicate with other agents to perform more complex tasks together.

IDS are deployed in an attempt to solve the problem of detecting malicious actors who perform actions without authorisation. Problems such as the costly requirements for high performance hardware [5], the inability to monitor all relevant information sources [6], the inefficiency of having to process large amounts of information flowing through a network [5], as well as structural vulnerabilities with centralised technologies make the current era of IDS ineffective.

In this paper, intrusion detection and forensic data collection techniques are combined with MAS for more effective and efficient detection in response to cyber attacks. This is performed by gathering the relevant evidence. In particular, we present a Multi-Agent model for the dynamic organisation of agents participating in forensic investigations; an agent architecture endowed with mechanisms for collecting and analysing network data; a protocol for allowing agents to coordinate and make collective decisions on the maliciousness of suspicious activity; and a simulator tool to test the proposed decentralised model, agent architecture and protocol under a wide range of circumstances and scenarios.

The remainder of this paper is organised as follows. Section 2 contains an overview of previous research in the area of Network Security and MAS. Section 3 provides details of the proposed model including a formal definition of the agent architecture and the communication protocol. Section 4 describes the simulation tool, the parameters for customising the experiments and the simulation results. Finally in Sect. 5 a conclusion, discussion of potential applications of the proposed model and future work is given.

2 Related Research

Shakarian et al. [7] described a cyber attribution system [8, 9] that takes into consideration different data sources and uses MAS to reason about the origin of an attack through the use of agent reasoning. The system uses information gathered about the attack as well as information gathered from a wide range of military sources to reason in-depth about the attribution of an attack. A highlighted danger of relying on external sources of information is the trustworthiness of the source, which must be taken into consideration. This use of external information provided an effective way to gain extra contextual information for detected attacks but was heavily reliant on previously collected and catalogued information from military sources.

Haack et al. [10] developed a hierarchical MAS model for monitoring and reporting data within the security environment. The system was composed of a number of agent types, each with a specific task to perform, such as event monitor, alert and report for system operators. The flow of information consisted of a high level policy created by a system operator which would be disseminated to the lower level agents responsible for monitoring networked components. Alerts would be generated by the monitor agents and aggregated by a higher level agent to make decisions about potential security events. The structure of this system is inherently centralised as the information, which is transferred up and down an agent pipeline, gets processed by one dedicated agent rather than having decisions made locally by the individual agents.

Jahanbin et al. [11] proposed a MAS framework for forensic information gathering which uses three types of agents for data collection, data analysis and alert generation. The authors note how the MAS paradigm is well suited to the task of forensic data collection as agents can be dispatched to areas of the network to perform collection and analysis of evidence such as log files. This system is structurally similar to Haack et al. [10] with layered agents passing information up the agent pipeline to a central agent for decision making. This central agent structure is similar to an IDS as it collects information and then makes a judgement based on that information, however, if some information is missing, the system would continue processing new information rather than actively searching missing data.

Baig [12] performed a survey of the current application of MAS in a number of critical infrastructure fields including intrusion detection. Emphasis was placed on system resilience so that if the system was attacked, resulting in some agents being forced off-line, the remaining agents should reorganise themselves to continue operation. Having agents specifically designed to adapt to network changes (e.g., hosts being turned off, firewalls restricting access to a subnet or intentional compromise) was shown to be a critical consideration, especially in the security environment.

Mees [13] designed a MAS to detect Advanced Persistent Threats by using external data sources to lookup the origins of suspicious connections. Within the framework three agents were described: a consultation agent to evaluate the location of the IP address, an analysis agent to compare the suspect connection with previously seen traffic patterns and a third agent to attempt to distinguish between human and robot connections by performing task-specific analysis. By using agents in this way to gather external information from data sources located beyond the local network perimeter, the agents were able to gather extra information that might not have been available to traditional IDSs which, for security, do not usually make external connections. This system utilises agents capable of performing multiple tasks which doesn't take advantage of having a greater number of more specialised agents for improve scalability. Our model uses a greater number of specialised agents to encourage competition between agents where multiple actions could be taken at any given time.

3   Dynamic Forensic Investigation Model

To address the problems of threat detection in networking environments, we propose a Multi-Agent model that combines the traditional tasks of an IDS using forensic collection capabilities to intelligently respond to cyber attacks. By bestowing both forensic information gathering and IDS-like information analysis capabilities onto agents, the proposed model is capable of automatically following the relevant lines of digital investigation after the detection of a suspicious activity.

The current era of cyber security solutions relies on the bulk collection and processing of data to detect instances of malicious behaviour. This centralised approach often leads to an overflow of information resulting in performance bottlenecks which inhibits the accurate analysis of suspicious activity. This can be improved upon by using multiple agents for data collection and analysis, which will be located in multiple physical areas on the network, thereby reducing the amount of information flowing to one single node. The total amount of information that must be processed for a particular suspicious activity can be reduced by only actively collecting extra data when there is some evidence to suggest that the data is relevant. Instead of collecting all possible data at all times, agents should collect the minimal amount to detect specific attacks and then, if an agent suspects an attack is occurring, start to collect more data specific to that activity. This is different to the current era of IDS technologies which typically collect all available information at all times regardless of the type of suspicious activity. By performing data collection and analysis tasks using a number of agents, and only utilising their functionalities when there is cause to do so, the resulting system will be both more accurate and efficient since only necessary tasks will be performed.

Rather than having a central repository of all information collected as seen in previous systems in [10, 11, 14–17], in the proposed system, the information is locally stored within the individual agents. Furthermore, agents are capable of performing one data collection action using information as input when interacting with a data source[2] to obtain more information (output) about a suspicious activity. By distributing data collection tasks among the agents in this way, the processing can be kept locally near to the sources of information rather than having a potentially vulnerable central system that could be the target of attacks. This organisation in itself provides many security and performance benefits over current centralised models, for example, agents are replaceable, hence if one agent goes off-line due to an attack or poor network conditions, the remaining agents could reorganise themselves around the information gap to continue collecting data.

3.1   Model Overview

The model is formally defined as a set of agents ($G = \{g_1, ..., g_i\}$), where each agent (g_i) can perform a data collection action for gathering information about suspicious activities. This information can be formalised by a set of features F representing different attributes or characteristics of a given activity; e.g., the IP address of a given connection, VPN usage, etc. Each feature ($f \in F$) has a domain (D_f) containing all its possible values.

A data collection action is any forensic collection task that retrieves data from some data source; e.g., the collection of Domain Name Service (DNS) logs or a list of currently connected IP addresses. Data collection actions are represented using conditions and effects. Conditions must be satisfied to execute the action; i.e., conditions represent the

---

[2] A data source is defined to be a source of information that exists, this may be an external data source such as a DNS server or a local source such as connection logs.

information that needs to be known in order to perform a data collection action. The effects represent the new information that will be known once the data collection action has been executed successfully.

Definition 1. *A data collection action is defined as a tuple* $\langle C, e \rangle$
   C is the action conditions; i.e., a set of pairs (f, v) where $f \in F$ and $v \in D_f$;
   $e \in F$ is the action effect; i.e., a feature whose value will be determined by the action.

Once an agent executes its data collection action it performs an analysis of the available information for suspicious activity; e.g., analysing the geographical origin of the collected IP addresses to identify malicious connections. More formally, a data analysis action is defined as follows:

Definition 2. *Given a set of pairs (f, v) representing the available information about a suspicious activity, we define a data analysis action as a function returning a value between $[0, 1]$ representing the probability of the suspicious activity being malicious.*

Note that the effects of a data collection action are used as conditions for other data collection actions, allowing agents to cooperate in *extended data collection process*. In particular, only those agents whose action conditions are satisfied are included in the process (i.e., those agents capable of collecting more information by using the information already collected as input). Our model includes a protocol to allow agents to coordinate when they participate in extended data collection process. The benefit of using this model is that a line of investigation only takes place when there is previous evidence (i.e., as a satisfied condition) to suggest that more useful data may be found. Traditional IDSs do not consider the context of previous information to tailor its future actions and target areas that has evidence suggesting more data might be found. The proposed model creates a chain of agent knowledge that is built up as each agent performs its own data collection which is later used to analyse the attack as a whole. Informally, we say that the conditions of a data collection action are satisfied when all the pairs feature value in the action condition are also contained by the available information.

Definition 3. *Given a set I formed by pairs (f, v) representing the available information about a suspicious activity, and a data collection action $\langle C, e \rangle$ we define that action conditions are satisfied iff for all $(f, v) \in C$, $(f, v) \in I$; and not satisfied otherwise.*

At the beginning of a security event, only a small amount of information is known about the attack so agents will perform actions that will provide context to the situation. Then, agents with a small number of conditions will first perform their data collection task, simpler data collection actions will have fewer conditions on their execution. As more knowledge is acquired, more complex agents with multiple conditions will be able to perform their data collection tasks while agents that have not had their conditions satisfied will not be able to participate. This will improve the efficiency of intrusion detection by allowing agents to logically avoid certain collection tasks until such a time when there is reasonable evidence to suggest that further useful data will be found. For example, assuming that there are 100 available actions, with 50 actions that are intended to run on a remote network and 50 that are intended to run on the local network. By performing the simple action of determining whether the attacker is located either remotely or locally, the agent could reduce the number of possible actions that would need to be performed. Performing remote analysis when the attacker has been identified as being on the local network would be computationally wasteful. Using this principle we will achieve greater levels of efficiency by only collecting and analysing pieces of information that are relevant to the context of the situation.

3.2 Agent Architecture

Agents in the proposed model consist of a set of modules that provide the base functionality, each of the following modules: monitoring, intrusion flagging, data collection, local analysis, communication and global analysis module, are described in detail below.

Monitoring Module. The monitoring module takes sampling tasks (e.g., monitoring data such as incoming IP address, VPN usage, etc.), distributed across a number of agents, to monitor for different network activities. The agents can monitor a host, service or network point and the observed data will be stored locally to be used in future anomaly analysis.

Intrusion Flagging Module. The intrusion flagging module is used to analyse the monitored data obtained by the previous module against the agent's local database to make a preliminary decision on whether the data is suspicious or not. This is done through the use of either signature or anomaly detection. An *escalation threshold* ($\pi \in [0, 1]$) is used to decide whether the information deviates from the normal patterns of the system. This threshold can be set relatively low as the decision on the suspiciousness of the event in question is only used to decide whether an extended data collection process should take place. In comparison to current IDS technologies, in which the first layer of anomaly analysis is often the only layer used in the decision making process, our model allows different agents to cooperate to further investigate events.

Data Collection Module. The data collection module is used to interact with the data source that allows the agent to obtain more information about a suspicious activity. As aforementioned, a data collection action requires some known information (i.e., conditions) to obtain some new information (i.e., effect). Note that some data collection actions (e.g., primitive information gathering actions) may not require any conditions to produce an output.

Local Analysis Module. The output received from the data collection module will be analysed by the local analysis module, which uses the agents local database to classify the information as either malicious or not. As aforementioned, the local analysis will calculate the probability of the suspicious activity being malicious. In particular, the agent builds a *local report* containing its identity, the output of the data collection module, and its local analysis value. This local report will be used to extend the available information about the suspicious activity.

Definition 4. *A local report defined as a tuple* $\langle g, (f, v), p \rangle$ *where:*

$g \in G$ *is the agent's identifier;*

(f, v) is a pair feature value corresponding to the output of the data collection action performed by agent g;

$p \in [0, 1]$ *is the agent's analysis of the suspicious activity; i.e., the probability of the suspicious activity being malicious.*

During the local analysis, information collected from sources will be analysed using the agent's local database of results collected from random sampling and anomaly analysis. This will produce an analysis of the information, also considering external factors such as the validity of the information retrieved, the integrity of the source or the integrity of the agent itself. As an example, consider an agent that collects information from a mail server. As part of the agent's tool-kit, it has the ability to check third-party bug tracking databases to detect whether there are any known vulnerabilities in the server. Consider the case that the agent finds a critical vulnerability with the data source it is assigned to. When the agent comes to take part in an extended collection task and finds the event to be innocuous with some degree of certainty, it could choose to reduce the degree of certainty to a lower value due to the existence of a vulnerability. This is done to adapt to changes in the environment and to take into account other factors that existing systems would normally not consider.

Communications Module. The communications module is used to send and receive data from other agents to perform extended data collection processes. The communications module will send the available information to other agents to be added to the existing information by performing their own local data collection task. In particular, both the available information and the newly collected information (local report) will be combined into a *global report*.

Definition 5. *A global report R is defined as a set of local reports* $\{\langle g_1, (f_1, v_1), p_1 \rangle, \ldots, \langle g_n, (f_n, v_n), p_n \rangle\}$ *containing the information collected by different agents participating in a same extended data collection process.*

Extended data collection occurred when each agent performs its local data collection task, aggregates the newly collected data with the current global report and then sends it to the next agent.

Global Analysis Module. The global analysis module as input a global report and makes a decision about the maliciousness of the suspicious activity under investigation.

Definition 6. *Given a global report G representing the local decisions made by the agents participating in an extend data collection process, the global decision is a function returning a value between [0, 1] representing the collective judgement about the maliciousness of the investigated activity.*

Various methods could be used to aggregate the results, for example, the average or highest number of decisions could be used to arrive at different outcomes. In MAS, voting [18] is used to allow agents to make group decisions based on the individual experiences of each agent, this is used within the proposed system to aggregate the individual local decisions produced by each agent into a global decision about the maliciousness of the event. An example voting system might take the highest number of votes for either malicious or innocuous and set the global decision to that value. A threshold (*termination threshold*) is used to determine if the global decision has reached the required level of certainty to justify the end of the extended data collection process or if further investigation is needed.

3.3   Interaction Protocol

To allow agents participating in an extended data collection to coordinate our model includes an interaction protocol (depicted in Fig. 1) The protocol is formed by five main phases: (i) request for participants; (ii) proposals from available participants; (iii) participant selection; (iv) inform summary; (v) inform result, described as follows.

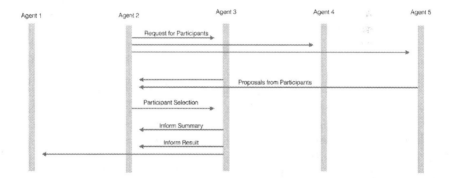

Fig. 1. An example information flow between 5 agents. Agent 2 is the current agent. Agent 2 requests participants to perform extended data collection, Agents 3 and 5 respond while Agent 4 cannot participate so ignores the request. Agent 2 selects Agent 3 to be the next agent to perform its data collection task. After the data collection task, Agent 3 sends back a summary to Agent 2, which after making the final decision, sends the result to all previous agents (Agents 1 and 2) to end the extended data collection process.

Request for Participants. Once an agent has performed its data collection and analysis tasks, the agent must add its local report to the global report and then send it onto the next agent for further information collection. The communication module is used to facilitate this.

The first step of the interaction protocol is aimed at requesting help from other agents that can participate in the data collection process. In particular, the set of pairs feature value are extracted from the global report and then broadcast (by the initiating agent) to the other agents. Given a global report $\langle g_1, (f_1, v_1), p_1 \rangle, \ldots, \langle g_n, (f_n, v_n), p_n \rangle$ a request is formalised as a set $\{(f_1, v_1), \ldots, (f_n, v_n)\}$ containing the available information about a suspicious activity.

Proposals from Participants. Any agents whose data collection action is satisfied by the information contained can respond by indicating their availability to participate in the extended data collection task.

Participant Selection. It is possible that several agents respond to the initial request indicating that they can work with the available data. The initiator must decide which agent will be selected to continue with the data collection process. In particular, the initiator will send the whole global report to this selected agent.

Unlike in other MAS solutions, the proposed model does not include a central repository of agents which can be queried to find the most suitable agent for a given task. This improves scalability but requires a system to allow agents to find each other. The agents will maintain a local database of agents that they have previously worked with. Deciding which agent should be selected as the preferred agent will affect the performance of the system as a whole. If the most optimal agent is selected for the task most of the time, the search process will improve as less time is spent performing data collection by unreliable agents. There are a number of ways in which the preferred agent can be identified based on what is important in a given situation. If accuracy is especially important for the current event the agents may select the agent that most often votes correctly, this will produce result in a more accurate search but could come at the cost of efficiency. If time is an important factor during some event the agent may choose the fastest performing agent to collect information quickly, this will produce a result faster than the previous but could potentially result in a less certain decision. While analysis of factors such as these could be done to determine the optimal preferred agent selection algorithm, events within the security environment can often be unpredictable and allowing the agents to choose the preferred agent at run time could produce a more adaptable solution.

For example, in a Distributed Denial of Service (DDoS) attack, selecting the preferred agent based on the speed at which agents are capable of collecting information could not result in a timely collection if one part of the network is under attack. However, selecting agents based on their geographical location (e.g., using the agents that are not under attack), would improve the efficiency of the response.

Inform Summary. To endow agents with information to select the preferred agent for a given task the inform summary step sends back its performance summary to the agents predecessor. This summary will be logged by the initiator agent for use in the preferred agent selection process. The parameters sent in the summary will include the agent's local decision about the maliciousness of the event, as well as, performance variables such as the time taken to perform the collection task, the importance of the data collected and the computational cost of performing the collection.

Inform Result. Once a final decision has been reached, the final decision will be sent to all of the participating agents, this can then be used by the agents to review its method for selecting the preferred agent (e.g., the preference can be increased for those agents with local decisions in-line with the final decision).

4   Simulation Tool

The proposed model has been implemented as a simulator tool for future testing. Agents are initialised with a set of binary features to simulate information about the security environment. The agents are displayed graphically as nodes and the path of the extended data collection process as it is passed between the agents as edges (see Fig. 2).

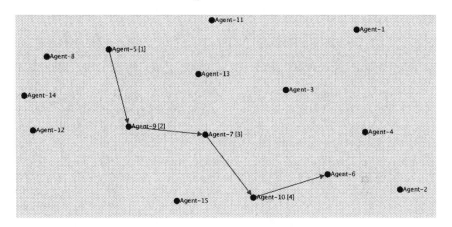

Fig. 2. An example simulation using 15 agents, 5 of which participated in the extended collection while the rest did not have their conditions satisfied so were unable to run. Nodes represent individual agents, blue arrows as well as the number in the square brackets show the order in which they ran. (Color figure online)

The simulator consists of a number of parameters that can be configured to control all aspects of the environment and the agents. Environmental parameters such as number of runs and number of iterations control the number of times the simulation should be repeated (reinitialising the agent settings each time) and the number of simulated events, both malicious and innocuous, that will occur per run. Agent variables controlling the number of agents, as well as the number of conditions assigned to each agent, can be configured to simulate scenarios using a small or large number of agents with a variable number of conditions that must first be satisfied before the agent can perform its collection action. The analysis variable is used to model the agent's ability to correctly detect whether a simulated event was malicious or innocuous. This analysis value is compared to a random assigned real number value between the [0, 1] interval to simulate noise. The higher the analysis value, the more certain an agent will be that its decision is correct. In addition, various thresholds such as for deciding when an agent should submit its local decision can be configured to set the level of certainty the agent must have before voting whether the data is malicious. Controlling this variable can be done adaptively and will be the focus of future work. Likewise the global decision threshold can be configured to test the effectiveness of different voting algorithms when calculating the agent's final global decision about the event as a whole.

Three types of log files are created during execution of the simulator, the first of which is an action level account showing the initial values for each agent as well as whether their local decision was correct or not. The second file includes an iteration level report which shows the agents involved during each simulated event, whether the global decision was correct or not and the statistics to measure the agents performance (e.g., true positive rate, false positive rate, sensitivity, specificity, etc.). The final log is saved after each run and stores averaged values for the previously mentioned agent performance statistics as well as any simulator threshold values.

5   Discussion

In this paper we have presented a multi-agent model for forensic investigation with potential applications in domains where devices are often distributed across a wide area and information is scattered between local and remote networks. Besides that, Network Security could benefit from this adaptive model as the environment is unpredictable and long term goals cannot be planned for. For example, in intrusion detection scenarios the best way to identify an attack or attacker is often heavily dependent on the attacker's actions and so cannot be defined in advance. Our model could also be used over a number of decentralised Internet of Things devices where resources are scarce. Agents could be installed onto low-end hardware since each agent only has to maintain a local database of logged data specific to their task. Finally, our model could be applied to scenarios that require adaptive information gathering or exploration tasks, as our agents agents are endowed with mechanisms to organise themselves to perform tasks without top-down or fixed instruction. Our model only requires that tasks be broken down into individual actions that can be distributed among the agents.

Future work will be focused on exploring the most optimal configurations under different scenarios and accuracy, efficiency and reliability conditions. Another direction of research involves extending the proposed model with mechanisms to take automated action in response to the detected threats.

References

1. Mukherjee, B., Heberlein, L.T., Levitt, K.N.: Network intrusion detection. IEEE Netw. **8**(3), 26–41 (1994)
2. Verwoerd, T., Hunt, R.: Intrusion detection techniques and approaches. Comput. Commun. **25**(15), 1356–1365 (2002)
3. Clint, M.R., Reith, M., Carr, C., Gunsch, G.: An examination of digital forensic models. Int. J. Digit. Evid. **1**(3), 1–12 (2002)
4. Woolridge, M.: An introduction to multiagent systems, 2nd edn. Wiley, Hoboken (2011)
5. Liao, H.-J., Lin, C.-H.R., Lin, Y.-C., Tung, K.-Y.: Intrusion detection system: a comprehensive review. J. Netw. Comput. Appl. **36**(1), 16–24 (2012)
6. Corey, V., Peterman, C., Shearin, S., Greenberg, M.S., Van Bokkelen, J.: Network forensics analysis. IEEE Internet Comput. **6**(6), 60–66 (2002)

7. Shakarian, P., Simari, G.I., Moores, G., Parsons, S.: Cyber attribution: an argumentation-based approach. In: Jajodia, S., Shakarian, P., Subrahmanian, V.S., Swarup, V., Wang, C. (eds.) Cyber Warfare, pp. 151–171. Springer, Berlin (2015)

8. Shakarian, P., Simari, G.I., Moores, G., Parsons, S., Falappa, M.A.: An argumentation-based framework to address the attribution problem in cyber-warfare. CoRR, abs/1404.6699 (2014)

9. Shakarian, P., Simari, G.I., Falappa, M.A.: Belief revision in structured probabilistic argumentation. In: Beierle, C., Meghini, C. (eds.) FoIKS 2014. LNCS, vol. 8367, pp. 324–343. Springer, Heidelberg (2014)

10. Haack, J.N., Fink, G.A., Maiden, W.M., McKinnon, A.D., Templeton, S.J., Fulp, E.W.: Ant-based cyber security. In: Proceedings of - 2011 8th International Conference on Information Technol. New Generations, ITNG 2011, pp. 918–926 (2010)

11. Jahanbin, A., Ghafarian, A., Seno, S.A.H., Nikookar, S.: A computer forensics approach based on autonomous intelligent multi-agent system. Int. J. Database Theory Appl. 6(5), 1–12 (2013)

12. Baig, Z.A.: Multi-agent systems for protecting critical infrastructures: a survey. J. Netw. Comput. Appl. 35(3), 1151–1161 (2012)

13. Mees, W.: Multi-agent anomaly-based APT detection. In: Proceedings of Information Systems Technology Panel Symposium, pp. 1–10 (2012)

14. Seresht, N.A., Azmi, R.: MAIS-IDS: a distributed intrusion detection system using multi-agent AIS approach. Eng. Appl. Artif. Intell. 35, 286–298 (2014)

15. Alkhateeb, F., Al Maghayreh, E., Aljawarneh, S.: A multi agent-based system for securing university campus: Design and architecture. In: 2010 International Conference on Intelligent Systems, Modelling and Simulation, pp. 75–79. IEEE, January 2010

16. Orfila, A., Carbo, J., Ribagorda, A.: Intrusion detection effectiveness improvement by a multi-agent system. Int. J. Comput. Sci. Appl. 2(1), 1–6 (2005)

17. Helmer, G., Wong, J.S.K., Honavar, V., Miller, L., Wang, Y.: Lightweight agents for intrusion detection. J. Syst. Softw. 67(2), 109–122 (2003)

18. Russell, S., Norvig, P.: Artificial Intelligence: A Modern Approach. Prentice-Hall, Englewood Cliffs, 25, 27 (1995)

A Genetic Analytics Approach for Risk Variant Identification to Support Intervention Strategies for People Susceptible to Polygenic Obesity and Overweight

C. Aday Curbelo Montañez[1]([✉]), P. Fergus[1], A. Hussain[1],
D. Al-Jumeily[1], B. Abdulaimma[1], and Haya Al-Askar[2]

[1] Applied Computing Research Group, Faculty of Engineering and Technology,
Liverpool John Moores University, Byrom Street, Liverpool L3 3AF, UK
{C.A.Curbelomontanez, B.T.Abdulaimma}@2015.ljmu.ac.uk,
{P.Fergus, A.Hussain, D.Aljumeily}@ljmu.ac.uk
[2] College of Computer Engineering and Science,
Sattam Bin Abdulaziz University, Al-Kharj, Kingdom of Saudi Arabia
sun_2258@hotmail.com

Abstract. Obesity is a growing epidemic that has increased steadily over the past several decades. It affects significant parts of the global population and this has resulted in obesity being high on the political agenda in many countries. It represents one of the most difficult clinical and public health challenges worldwide. While eating healthy and exercising regularly are obvious ways to combat obesity, there is a need to understand the underlying genetic constructs and pathways that lead to the manifestation of obesity and their susceptibility metrics in specific individuals. In particular, the interpretation of genetic profiles will allow for the identification of Deoxyribonucleic Acid variations, known as Single Nucleotide Polymorphism, associated with traits directly linked to obesity and validated with Genome-Wide Association Studies. Using a robust data science methodology, this paper uses a subset of the TwinsUK dataset that contains genetic data from extremely obese individuals with a BMI $\geq$ 40, to identify significant obesity traits for potential use in genetic screening for disease risk prediction. The paper posits an approach for methodical risk variant identification to support intervention strategies that will help mitigate long-term adverse health outcomes in people susceptible to obesity and overweight.

Keywords: Genetics · SNPs · Obesity · Data science · R · Bioconductor · gwascat

1 Introduction

According to the World Health Organization (WHO)[1], the occurrence of obesity and overweight worldwide has doubled since 1980. More recently, the figures suggest that in 2014 more than 1.9 billion adults were overweight and 600 million were obese.

---

[1] http://www.who.int/.

© Springer International Publishing Switzerland 2016
D.-S. Huang et al. (Eds.): ICIC 2016, Part I, LNCS 9771, pp. 808–819, 2016.
DOI: 10.1007/978-3-319-42291-6_80

The condition was initially recognized as a disease in 1948 by the WHO [1] and since then its prevalence has continued to increase making it a global phenomenon and one of the main contributors to poor health. Today it is considered one of the most difficult clinical and public health challenges worldwide [2–4]. It is the leading cause of Type 2 Diabetes, cardiovascular disease, premature death, hypertension, osteoarthritis, stroke and certain cancers [2, 3, 5]. Consequently, it is high on the political agenda of many countries. While North America is considered the most obese continent, the UK is ranked as one of the most obese nations in Europe[2].

The predisposition to obesity in humans is referred to as polygenic obesity and is considered a complex and multifactorial disease that manifests itself through interactions between genetic, behavioural and environmental factors. While obesity tends to exist within families, the way it is inherited does not correspond to known patterns. We know that it is dependent on environmental factors and numerous studies have shown that an individual's predisposition to obesity is more similar among genetically related individuals than those that are not. The phenotypes associated with obesity exhibit significant additive heritability (the proportion of the variability of a trait that is attributable to genetic factors) [6]. In the particular case of Body Mass Index (BMI), family and twin studies have shown that between 40 % and 70 % of the inter-individual variation in susceptibility to obesity can be attributed to genetic differences in the population [7–9]. The remaining percentage is associated with other factors, such as lifestyle. It is clear that the prediction of obesity based solely on genetic information will never be completely accurate. Thus, the ideal genetic predictor, should complement, but never replace traditional predictors.

Building on existing works and tools for genetic analysis, this paper explores polygenic obesity and associated risk factors. Using a robust data science methodology, data extracted from Genome-Wide Association Studies (GWAS) is utilised to identify significant Single Nucleotide Polymorphism's (SNPs) associated with obesity traits. This study serves as a genetic analytics framework for the interpretation of genetic profiles for the discovery of significant biomarkers and their associated genes for use in early detection and intervention strategies.

2   Background

The central dogma of molecular biology describes how cells replicate their Deoxyribonucleic Acid (DNA) in daughter cells via cell division. It also focuses on how proteins are synthesized from DNA, according to the flows of information from DNA to ribonucleic Acid (RNA) and RNA to protein. Genes encode amino acid chains that compose the proteins, which have specific functions in the organism [10, 11]. Understanding the function of each gene, and the proteins they synthesize can help to identify the cause(s) of different diseases. This has the real potential to find better ways to treat and prevent diseases through early detection and prevention strategies, personalized drugs and tailored therapies.

---

[2] http://www.who.int/.

In the context of polygenic obesity, this disease is complex and does not exhibit a typical Mendelian pattern of transmission. There is evidence derived from GWAS that suggest single nucleotide polymorphisms (SNPs) in certain genes are associated with obesity risk factors and BMI. Examples of these genes include those associated with fat mass and obesity (FTO) and the Melanocortin 4 receptor gene (MC4R) [12–15]. Additional studies have reported that certain genes, including those mentioned, have a strong link with energy consumption in the nervous system when the hypothalamus part of the brain is stimulated [16].

While in [17], a meta-analysis of 340,000 people was conducted and 97 Genome-wide Significance (GWS) loci associated with BMI were reported, of which 56 were novel. At least 52 of these have been previously identified and are reported to be associated with a predisposition to obesity. However, their predictive capability is poor, because it only explains a fraction of the total variance. Therefore, it cannot compete with the predictability of traditional risk factors such as, obesity in parents and children. Nonetheless, some of these loci are currently being used in direct-to-consumer (DTC) personal genomic profiles to estimate the risk of obesity in the lives of individuals [15].

The predictive capability of loci is however likely to improve as the cost of genetic sequencing decreases. In 2003, the cost was $2.7 billion, while in 2013 it was $5000 [18]. This has resulted in a plethora of sequencing methods designed to determine the order of nucleotide bases that make up a molecule of DNA or RNA [19]. The first generation of sequencing, known as Sanger sequencing, made the Human Genome Project possible. More recently, these have been replaced with next-generation sequencing (NGS) techniques that provide cheaper alternatives [20, 21]. In particular, they allow for faster discovery of genes and regulatory elements associated with diseases, thus significantly increasing the clinical diagnosis performed with the nucleic acid sequence.

This is further supported by the fact that a considerable number of base pairs (nucleotides) in the human genome are identical for all humans [22]. Hence, in genetic association studies, bases where there is variation between humans are only considered. These differences are commonly referred to as SNPs. Initially, studies were commissioned to investigate the basic genetics of common obesity based on candidate gene approaches and linkage analyses. However, other types of more effective studies utilizing hypothesis-free methodologies have been used in obesity studies to identify many obesity related loci. These studies are termed GWAS and permit the analysis of a larger number of genetic variants for association with traits of interest.

However, despite these advances, sequencing the whole genome is still expensive; therefore, researchers tend to focus on coding regions of the human genome known as exons. It is estimated that 85 % of disease-causing mutations are located in coding and functional regions of the genome, so whole-exome sequencing (WES) methods are a potential alternative to unravel the cause of rare genetic disorders [23].

This paper adopts this approach were 68 exonerated sequenced samples from extremely obese people are used and compared with a publicly available and manually curated collection of published GWASs.

3   Methodology

In this paper, we compare the genetic profiles of extremely obese people, obtained from the UK10K_OBESITY_TWINSUK REL-2013-04-20 dataset hosted by Sanger Institute[3], with GWAS results indexed in the National Human Genome Research Institute (NHGRI) Catalog[4]. The dataset represents 68 cases of extremely obese twins. These samples are a subset of the TwinsUK cohort samples that have a BMI $\geq$ 40 and were exome sequenced by the UK10K Obesity group [24, 25].

The UK10K_OBESITY_TWINSUK REL-2013-04-20 dataset contains SNPs and associated metadata variables that include chromosome, position, identifier or reference (REF) and alternate (ALT) bases. The samples in the VCF format were sequenced using GRCh37 and accessed in NHGRI Catalog data (hg19 genome build) based on this assembly.

The primary focus in this study is to identify the frequency of SNPs or their variants of interest in the 68 examined profiles (only Europeans considered) and which of these are indexed in the NHGRI Catalog. This mapping allows us to link SNP-traits in our samples with risk alleles indexed in the NHGRI Catalog. This process results in a set of traits that best represent phenotypes related to obesity.

3.1   Data Pre-processing

The dataset contains 5,827,333 observations (records) and 4 variables that describe the chromosome number, position, SNP ID and REF, and ALT bases which are combined to represent the genotype, as shown in Table 1. These are the variables used in this study to identify SNPs and risk alleles previously reported in GWAS and indexed in the NHGRI Catalog. Column names were edited for simplicity and convenience.

In the initial data pre-processing stages observations with missing values (NA values) were removed including those observations with no SNP IDs. Duplicate SNPs in all the samples were summed and utilised as a feature to represent SNP frequency. This initial data pre-processing step reduced the number of observations from 5,827,333 to 361,899.

3.2   NHGRI Catalog Comparison

Using metadata from the NHGRI Catalog ("ebicat37") the SNP ID and the Genotype contained in the samples were linked with SNPs and risk alleles reported in the Catalog. This was achieved by using the RSID and Genotype values in the samples to index SNPS and STRONGEST.SNP.RISK.ALLELE in the NHGRI Catalog. Those SNPs with risk alleles marked as "?" in the Catalog where discarded from our samples.

This resulted in 1,304 matching SNPs from our samples. However, the potential risk alleles identified in the NHGRI Catalog were obtained in studies conducted using

---

[3] http://www.sanger.ac.uk/.

[4] https://www.ebi.ac.uk/.

Table 1. Columns extracted from the UKTwins dataset and used in this study

#Chrom	Position	RSID	Genotype
1	14907	Rs79585140	A/G
1	63671	Rs116440577	G/A
1	63697	Rs201582574	T/C
1	69511	Rs75062661	A/G

people from different populations and ethnic backgrounds. Therefore, since our samples belonged to UK twins, only those variants identified in studies conducted in European populations were used. This was achieved using a filtering process to select SNPs where the column INITIAL.SAMPLE.DESCRIPTION contained the word "European". This reduced the number of SNPs further to 1,041–11 variables were included as shown in Table 2.

Table 2. Variables selected after European filtering process.

Variables
#chrom
position
rsid
genotype
SNPCount
STRONGEST.SNP.RISK.ALLELE
DISEASE.TRAIT
P.VALUE
REPORTED.GENE.S.
INITIAL.SAMPLE.DESCRIPTION
JOURNAL

Following the identification of risk variants associated with European populations, traits linked with obesity were selected as shown in Table 3.

Figure 1 shows the 1041 SNPs contained in our dataset that are indexed in the NHGRI Catalog. The x axis represents the chromosomes, while the y value shows each of the SNPs – the strongest associations have the smallest P-values, therefore negative logarithms will be the greatest. So while all of the SNPs in the study are considered to be statistically significant [26], SNPs with higher y values are deemed to be the most significant. In other words, all the SNPs represented in Fig. 1 have been significantly associated with obesity-related traits in European populations, according to GWAS publications included in the NHGRI Catalog. The SNPs highlighted in green correspond

Table 3. Obesity traits selected.

Variables	Abbreviations
Body mass index	BMI
Visceral adipose tissue adjusted for BMI	VATaBMI
Obesity	O
Obesity(early onset extreme)	OEOE
Weight	W
Weight loss	WL
Birth weight	BW
Height	H
Fat body mass	FBM
Eating disorders	ED
Waist-hip ratio	WHR
Waist Circumference	WC

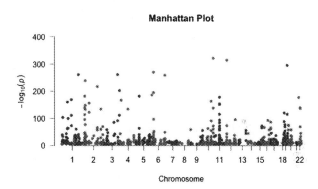

Fig. 1. Manhattan plot with obesity traits highlighted in green. (Color figure online)

to the obesity-related traits previously listed in Table 3. The $-\log_{10}(p)$ represents the statistical significance of the association reported in the GWAS.

Obesity-related SNPs where identified in chromosomes 1, 2, 3, 4, 5, 6, 9, 10, 11, 12, 15, 16, 17, 19, 20 and 22 and highlighted in green. However, the chromosomes of specific interest are 1, 3 and 19 which each have four obesity SNPs. It should be noted that not all the SNPs included in the three chromosomes are significant as will be discussed in more detail in the results section. Additionally, the absence of obesity-related traits in chromosomes 7, 8, 13, 14, 18 and 21 can also be noticed. Chromosomes X and Y were not included in the representation as there were no identifiable SNPs indexed from the NHGRI Catalog for these chromosomes in our data.

Using our 1041 SNPs, we filtered the dataset to only include those SNPs related to obesity – this reduced the dataset to 39 obesity related SNPs. Figure 2 describes the 39 obesity related SNPs that appear in the initial 68 samples of extremely obese subjects.

Figure 2 represents the SNPs and their obesity associated traits for each chromosome – colours show the frequency of the variants. The full trait names described on the y axis are listed in Table 3. The SNPs that are repeated more than 45 times are labelled in Fig. 2. In some cases, SNPs overlap as they are related to the same traits in the same chromosome. The trait height was identified nineteen times in different chromosomes. Obesity was also found in 5 chromosomes. The trait obesity (early onset extreme). Waist-hip ratio and obesity (early onset extreme) traits are represented in pink and labelled as they were repeated 68 and 64 times respectively. Surprisingly, one of the SNPs associated with body mass index, rs13107325 (chromosome 4), was only repeated in six samples and for that reason it was represented as an orange-green colour.

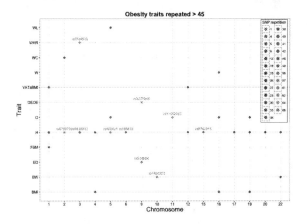

Fig. 2. Obesity related traits and their frequency in different chromosomes (Color figure online)

4   Results

In this study we investigated the occurrence of SNPs in 68 samples of cases with extreme obesity. The results in Table 4 provide a brief description of the obesity related SNPs that were identified in the samples. The table identifies the chromosome number, position, SNP id, SNP frequency, obesity trait, p-value and reported genes. From all the traits identified, the most prominent, given its association with 19 identified SNPs, is *height*. However, this trait would not represent a risk as in most cases it corresponded to the REF base on the genotype, not the ALT. Nonetheless, we do not have any clinical data to confirm this.

Our findings show that the SNP associated with the waist-hip ratio trait, *rs6784615*, was present across all samples, followed by the SNP *rs2275848*, associated with obesity (early onset extreme) which is repeated 64 times. Other outstanding SNPs such as *rs1801253*, *rs11042023* and *rs514024* were linked to birth weight, obesity and eating disorders and were present in 61, 61 and 56 of the samples respectively.

The SNP *rs7498665* in chromosome 16 has been associated to several obesity-related traits according to the information included in the NHGRI Catalog.

Table 4. Obesity related SNPs identified in the 68 extremely obese samples.

#Chrom	Position	SNP ID	Freq	Trait	P.Value	Reported Genes
1	149906413	rs11205303	42	Height	4e-23	MTMR11
1	150727539	rs2230061	40	Fat body mass	4e-08	CTSS
1	184023529	rs1046934	39	Height	2e-31	TSEN15
1	3649562	rs9662633	5	Visceral adipose tissue adjusted for BMI	6e-06	TP73
2	232378231	rs6750795	57	Height	2e-08	C2orf52
2	27801759	rs1919128	30	Waist Circumference - Triglycerides (WC-TG)	2e-09	C2orf16
3	52506426	rs6784615	68	Waist-hip ratio	4e-10	NISCH, STAB1
3	56667682	rs9835332	49	Height	5e-13	C3orf63
3	172165727	rs572169	41	Height	3e-18	GHSR
3	172165727	rs572169	41	Height	1e-12	GHSR
4	1701317	rs2247341	42	Height	2e-11	SLBP/FGFR3
4	103188709	rs13107325	6	Body mass index	2e-13	SLC39A8
5	176517326	rs422421	62	Height	1e-12	FGFR4/NSD1
5	75003678	rs2307111	41	Obesity	3e-12	C5orf37
5	159820931	rs10515808	12	Weight loss (gastric bypass surgery)	4e-07	C1QTNF2
6	32050067	rs185819	49	Height	3e-08	HLA class III
9	95887320	rs2275848	64	Obesity (early onset extreme)	1e-06	NINJ1
9	130504070	rs514024	56	Eating disorders (purging via substances)	5e-06	PKN3, SET, WDR34, ZDHHC1, ZER1
10	115805056	rs1801253	62	Birth weight	4e-09	ADRB1
11	8662516	rs11042023	62	Obesity	1e-11	RPL27A
12	124399550	rs1316952	18	Visceral adipose tissue adjusted for BMI	5e-06	DNAH10
12	56740682	rs2066807	5	Height	1e-13	STAT2
15	74336633	rs5742915	45	Height	1e-15	PML
15	89388905	rs16942341	4	Height	4e-27	ACAN
16	28883241	rs7498665	39	Obesity	3e-13	SH2B1
16	28883241	rs7498665	39	Body mass index	5e-11	SH2B1
16	28883241	rs7498665	39	Body mass index	3e-10	ATP2A1, SH2B1
16	28883241	rs7498665	39	Obesity	5e-12	SH2B1

16	28883241	rs7498665	39	Weight	1e-09	ATP2A1, SH2B1
17	43216281	rs4986172	33	Height	2e-16	ACBD4
17	47390014	rs2072153	29	Height	4e-08	ZNF652
19	41937095	rs17318596	36	Height	5e-16	ATP5SL
19	46202172	rs2287019	27	Body mass index	2e-16	GIPR, QPCTL
19	46181392	rs1800437	25	Obesity	3e-14	GIPR
19	17283303	rs2279008	25	Height	3e-08	MYO9B
20	34025756	rs143384	39	Height	1e-58	GDF5
20	32333181	rs7274811	26	Height	6e-22	ZNF341
22	42336172	rs5758511	30	Birth weight	3e-06	CENPM
22	40820151	rs5757949	29	Height	4e-06	MKL1, SGSM3

This particular SNP has been identified in 36 of our samples and can be related to body mass index, obesity or weight traits. It has also been associated with genes, such as SH2B1 and ATP2A1.

The SNP *rs572169* is listed twice as it can be observed in Table 4. In the NHGRI Catalog the same SNP was reported twice with different p-values from different studies. After performing the comparison between the 68 samples of extremely obese people and the NHGRI Catalog, both results were included in the table. The SNP *rs572169* is not the only one identified with different p-values. *rs7498665* was also included in the catalog and reported in different studies. This time, *rs7498665* was reported within different genes: SH2B1 or ATP2A1, SH2B1, as well as different p-values.

Although this study has focused on risk variants associated with obesity traits, other SNPs with other traits within the 10 most repeated have also been identified. In addition to *rs6784615,* the SNP *rs37370* was also present in all samples which is linked to *Asymmetrical dimethylarginine levels.* Other SNPs were associated with *psoriasis, parkinson's disease, osteosarcoma, Insulin-like growth factors, coronary heart disease, myocardial infarction (early onset), inflammatory bowel disease, Ulcerative colitis, blood metabolite levels, Type 2 diabetes* as well as some type of cancer among other diseases. In future works these SNPs and their association with obesity, including causation factors will be studied further.

5   Discussion

Understanding obesity as a complex disease is an arduous task. In an attempt to better understand how obesity is developed, we sought to identify the main genetic variants present in samples of people with extreme obesity. This was achieved using information contained in the NHGRI Catalog as a reference for the identification of risk variants in our samples. The methodology proposed in this paper investigated a solid foundation for identifying SNPs and their associated risks and in this sense provides a foundation for much more complex genetic analytics.

However, the study has a number of limitations, the major one being the size of the dataset – 68 samples are not representative of a population. However, we did analyse samples belonging to twins in the UK which is part of a more limited and specific population group.

The lack of contextual information about the samples also proved to be an issue. Although we knew that the samples belonged to British twins with extreme obesity (BMI > 40), we did not have any other information regarding their age, gender, lifestyle or clinical history, which greatly limited the scope of this study. For this reason, using manipulation and data visualization techniques, the objective of this study was to identify statistically significant obesity related variants from GWAS. This has provided a robust foundation for future research directions, particularly in the area of detection and prediction and the use of advanced machine learning algorithms, which will be the focus of our future work.

We also compared whole-exome sequencing data (WES data) against whole genome results data (GWAS data) and this means that we could be ignoring other regions of the DNA that could provide more insights to our study which will be considered in future work.

Starting from an initial number of 5827333 observations, we managed to reduce the dataset using standard data science tools and techniques, such as R^5 and Bioconductor[6] allowing us to identify 39 SNPs associated with obesity risks listed in Table 4. Among all the SNPs associated with obesity the most prominent one in the 68 samples was rs6784615. This SNP is associated with waist-hip ratio (WHR), a trait commonly used when it comes to quantifying obesity [15].

6 Conclusions

The identification of biomarkers contained in the entire genome, or more specifically, those found in the protein-coding regions or exons, has helped us to identify genetic variations associated with particular traits in extremely obese samples.

This study used a manually curated and publicly available dataset with results from GWAS for the identification of obesity-related traits in 68 extremely obese samples. Metadata from the NHGRI Catalog was accessed using the R/Bioconductor package *gwascat* and the hg19 genome build *ebicat37* data. The dataset and the samples were analysed using rigorous data science techniques, which led to a significant reduction in the initial number of observations and the identification of SNPs that could help to explain the development of extreme obesity. Missing values and no relational data items were removed from the study. The variables *rsid* and *genotype* as well as *SNPS* and *STRONGEST.SNP.RISK.ALLELE* were the key for the identification of SNP-obesity related traits in our samples.

---

[5] https://www.r-project.org.

[6] https://www.bioconductor.org/.

While the results show specific variants related to obesity traits present in all the samples, they are unlikely by themselves to be causative variants. The potential for variants identified in GWAS to predict the risk of complex diseases is problematic. At the moment, currently known variants do not fully explain the risk of disease occurrence to be of clinical use in predictive systems. However, the presence of specific obesity-related SNPs in extremely obese individuals could help us to gain a better understanding about what SNPs we need to look for. Of course, we need to consider the inclusion of contextual and clinical data and study the interaction between SNPs to gain a better understanding of obesity aetiology. The results do however raise further questions. First, it would be useful to see if the identified SNPs interact conjointly to cause extreme obesity. Second, it would also be useful to explore machine learning methods, such as support vector machines and other advanced artificial neural network architectures to study the non-linearity interaction between SNPs. While this study focused on UK twins, it would be interesting to look at other population groups as well.

References

1. James, W.P.T.: WHO recognition of the global obesity epidemic. Int. J. Obes. (Lond.) **32** (Suppl. 7), S120–S126 (2008)
2. Walley, A.J., Blakemore, A.I.F., Froguel, P.: Genetics of obesity and the prediction of risk for health. Hum. Mol. Genet. **15**, 124–130 (2006)
3. Yang, W., Kelly, T., He, J.: Genetic epidemiology of obesity. Epidemiol. Rev. **29**, 49–61 (2007)
4. Ryley, A.: Children's BMI, overweight and obesity (Chapter 11). In: Health Survey for England 2012, England (2013)
5. Vallgarda, S., Nielsen, M.E.J., Hartlev, M., Sandoe, P.: Backward- and forward-looking responsibility for obesity: policies from WHO, the EU and England. Eur. J. Publ. Health **25**, 1–4 (2015)
6. Butte, N.F., Cai, G., Cole, S., Comuzzie, A.G.: Viva la Familia study: genetic and environmental contributions to childhood obesity and its comorbidities in the Hispanic population. Am. J. Clin. Nutr. **84**, 646–654 (2006)
7. Wardle, J., Carnell, S., Haworth, C., Plomin, R.: Evidence for a strong genetic influence on childhood adiposity despite the force of the obesogenic environment. Am. J. Clin. Nutr. **87**, 398–404 (2008)
8. Zaitlen, N., Kraft, P., Patterson, N., Pasaniuc, B., et al.: Using extended genealogy to estimate components of heritability for 23 quantitative and dichotomous traits. PLoS Genet. **9**, e1003520 (2013)
9. Moustafa, J.S.E.S., Froguel, P.: From obesity genetics to the future of personalized obesity therapy. Nat. Rev. Endocrinol. **9**, 402–413 (2013)
10. Alberts, B., Johnson, A., Lewis, J., Morgan, D., et al.: Molecular Biology of the Cell. Garland Science, Taylor & Francis Group, New York (2015)
11. Watson, J.D., Gann, A., Baker, T.A., Levine, M., et al.: Molecular Biology of the Gene. Pearson. Cold Spring Harbor Laboratory Press, New York (2014)
12. Wang, K., Li, W., Zhang, C.K., Wang, Z., et al.: A genome-wide association study on obesity and obesity-related traits. PLoS ONE **6**, e18939 (2011)
13. Xi, B., Wang, C., Wu, L., Zhang, M., et al.: Influence of physical inactivity on associations between single nucleotide polymorphisms and genetic predisposition to childhood obesity. Am. J. Epidemiol. **173**, 1256–1262 (2011)

14. Corella, D., Ortega-Azorín, C., Sorlí, J.V., Covas, M.I., et al.: Statistical and biological gene-lifestyle interactions of MC4R and FTO with diet and physical activity on obesity: new effects on alcohol consumption. PLoS ONE **7**, e52344 (2012)
15. Loos, R.J.F.: Genetic determinants of common obesity and their value in prediction. Best Pract. Res. Clin. Endocrinol. Metab. **26**, 211–226 (2012)
16. Willer, C.J., Speliotes, E.K., Loos, R.J.F., Li, S., et al.: Six new loci associated with body mass index highlight a neuronal influence on body weight regulation. Nat. Genet. **41**, 25–34 (2009)
17. Locke, A.E., Kahali, B., Berndt, S.I., et al.: Genetic studies of body mass index yield new insights for obesity biology. Nature **518**, 197–206 (2015)
18. Shaer, O., Nov, O., Okerlund, J., Balestra, M., et al.: Informing the design of direct-to-consumer interactive personal genomics reports. J. Med. Internet Res. **17**, e146 (2015)
19. Grada, A., Weinbrecht, K.: Next-generation sequencing: methodology and application. J. Invest. Dermatol. **133**, 11–14 (2013)
20. Rabbani, B., Tekin, M., Mahdieh, N.: The promise of whole-exome sequencing in medical genetics. J. Hum. Genet. **59**, 5–15 (2014)
21. Shendure, J., Ji, H.: Next-generation DNA sequencing. Nat. Biotechnol. **26**, 1135–1145 (2008)
22. Fall, T., Ingelsson, E.: Genome-wide association studies of obesity and metabolic syndrome. Mol. Cell. Endocrinol. **382**, 740–757 (2014)
23. Majewski, J., Schwartzentruber, J., Lalonde, E., Montpetit, A., et al.: What can exome sequencing do for you? J. Med. Genet. **48**, 580–589 (2011)
24. Moayyeri, A., Hammond, C.J., Valdes, A.M., Spector, T.D.: Cohort profile: TwinsUK and healthy ageing twin study. Int. J. Epidemiol. **42**, 76–85 (2013)
25. Moayyeri, A., Hammond, C.J., Hart, D.J., Spector, T.D.: The UK adult twin registry (TwinsUK Resource). Twin Res. Hum. Genet. **16**, 144–149 (2013)
26. Welter, D., MacArthur, J., Morales, J., Burdett, T., et al.: The NHGRI GWAS catalog, a curated resource of SNP-trait associations. Nucl. Acids Res. **42**, D1001–D1006 (2014)

A Dynamic, Modular Intelligent-Agent Framework for Astronomical Light Curve Analysis and Classification

Paul R. McWhirter[1,2(✉)], Sean Wright[1], Iain A. Steele[2], Dhiya Al-Jumeily[1], Abir Hussain[1], and Paul Fergus[1]

[1] Applied Computing Research Group, Faculty of Engineering and Technology, Liverpool John Moores University, Byrom Street, Liverpool L3 3AF, UK
P.R.McWhirter@2014.ljmu.ac.uk, S.Wright@2012.ljmu.ac.uk,
{D.Aljumeily,A.Hussain,P.Fergus}@ljmu.ac.uk
[2] Astrophysics Research Institute, IC2, Liverpool Science Park, Liverpool John Moores University, 146 Brownlow Hill, Liverpool L3 5RF, UK
I.A.Steele@ljmu.ac.uk

Abstract. Modern time-domain astronomy is capable of collecting a staggeringly large amount of data on millions of objects in real time. This makes it almost impossible for objects to be identified manually. Therefore the production of methods and systems for the automated classification of time-domain astronomical objects is of great importance. The Liverpool Telescope has a number of wide-field image gathering instruments mounted upon its structure. Utilizing a database established by a pre-processing operation upon these images, containing millions of candidate variable stars with multiple time-varying magnitude observations, we applied a method designed to extract time-translation invariant features from the time-series light curves. These efforts were met with limited success due to noise and uneven sampling within the time-series data. Additionally, finely surveying these light curves is a processing intensive task. Fortunately, these algorithms are capable of multi-threaded implementations based on available resources. Therefore we propose a new system designed to utilize multiple intelligent agents that distribute the data analysis across multiple machines whilst simultaneously a powerful intelligence service operates to constrain the light curves and eliminate false signals due to noise and local alias periods.

Keywords: Data analysis methods · Big data mining · Machine learning · Uneven time-series analysis · Light curve analysis · Variable stars · Binary stars · Harmonic regression · Harmonic feature extraction · Multi-agent systems · Period detection

1 Introduction

Astronomy is entering a period of unprecedented data gathering capability. Advances in observational, storage and data processing technologies have allowed for extended sky surveys such as the Sloan Digital Sky Survey (SDSS) to be conducted and exploited [1]. Within the next decade a number of even larger surveys are also planned. Technology is now at a point where it has become possible to gather data on wide regions of the sky repeatedly over variable time periods [2]. This data can be analyzed for periodic

© Springer International Publishing Switzerland 2016
D.-S. Huang et al. (Eds.): ICIC 2016, Part I, LNCS 9771, pp. 820–831, 2016.
DOI: 10.1007/978-3-319-42291-6_81

structure which can then be used to create physical models by fitting weighted regression learning algorithms. These methods can provide us with valuable knowledge about the presence and classification of astronomical objects that are periodically changing in time as well as identifying transient phenomena [3].

Time domain astronomy is a research area characterized by the large datasets generated by sky surveys [4]. This time-series data contains information on the temporal component of measurements and the whole time-series contains observations across multiple epochs. In Time-Domain Astronomy, it is common for these observations to have a significantly uneven distribution in time with inconsistent intervals between observations [5]. As a result, astronomy maintains a demand for data processing techniques capable of the automated processing of time-series data on individual objects that can contain observations over the space of days followed by no additional observations for a period of months. In this paper we propose a new theoretical platform for the analysis of this vast quantity of time-domain astronomy data by introducing intelligent-agents. A typical agent is a type of computer system that is embedded in a type of environment that is capable of conducting an autonomous action within that environment in order to meet its objectives. An Intelligent Agent on the other hand is an extension of this approach with the ability to make decisions and adapt to its changing environment.

The rest of this paper is structured as follows. In Sect. 2, the background of time-domain astronomy is discussed with reference made to the numerous classifiable objects. Section 3 introduces the Small Telescopes Installed at the Liverpool Telescope (STILT) instruments, wide field imaging devices and the pre-processing pipeline used to construct an Structured Query Language (SQL) time-series database from the raw images. In Sect. 4, the feature extraction is discussed through using light curve model fitting resulting in the extraction of important, magnitude and phase independent, features. In Sect. 5, a system utilizing intelligent-agents is proposed for the successful processing and classification of numerous light curves. The final conclusions and proposals of future work are provided within Sect. 6.

2   Background

Astronomical time-series data is generated through the production of wide-angle images of the sky. By identifying objects in multiple images with different observation times, information on the change of the brightness of these objects can be determined. The resulting brightness-over-time data for each individual object is called the objects light curve [2, 6].

Many astronomical objects exhibit brightness variability due to a large number of differing physical processes that uniquely influence an object's light curve. Therefore, the light curve can be used in the classification of variable objects based on the signature of these physical processes and the detection of unknown candidate objects or even unknown variability phenomena [7]. The first major type of variable astronomical phenomenon is Variable Stars [8]. Variable stars are unstable stars and undergo periods of pulsation where they grow and contract in size [9]. These size oscillations produce changes to the stars temperature and brightness resulting in a measurable change upon the light curves [2, 6]. The light curves of these pulsating stars can be used to produce descriptive features. Models produced from these features can then be used to identify the class of candidate variable objects and the period of their oscillations.

A second important type of variable object is the eclipsing binary. In these systems, two or more stars are in close proximity to each other and execute orbits around a common gravitational center-point. The close proximity of the stars often means that they cannot be distinguished on an image and appear as a single source of light. Variations in these objects are caused by the plane of the orbit aligning with the view from Earth. As a result, one star periodically passes in front of another resulting in a change in the brightness of the source of light in the astronomical images.

Finally, there are also transient events that result in harder-to-predict phenomena [3]. Flare Stars are stars that can undergo occasional outbursts due to magnetic and plasma processes within their atmospheres. These events can be repetitive but not usually with the degree of periodicity of variable stars. For purely transient events, two of the most studied examples are Novae and Supernovae caused by the cataclysmic eruption of stellar material, producing some of the brightness objects in the known universe as the victim star is destroyed or badly disrupted during the event.

3   The STILT Dataset

The Small Telescopes Installed at the Liverpool Telescope (STILT) dataset is a wide field object SQL database. It contains 1.24 billion separate object observations of 27.74 million independent stellar objects. It was generated through the pre-processing of observational images gathered by the STILT instruments [2]. The STILT instruments consist of three cameras with varying field of views mounted directly to the body of the main Liverpool Telescope aimed co-parallel with the main telescope's field of view. The first instrument, SkycamA is capable of imaging the entire sky from La Palma. It is primarily used for monitoring the status of weather but it can be of use in the detection of bright transient objects. This camera does not contribute any observations to the STILT database. The next camera is named SkycamT and is responsible for most of the observations. It is a single CCD camera capable of detecting light across the visible spectrum with a wide-angle lens with a field of view of 21 by 21 degrees and a magnitude limit of +12. Finally, the remainder of the database is constructed from observations by the SkycamZ instrument. This instrument contains a CCD camera which is also capable of detecting light from across the visible spectrum attached to a small telescope with a field of view of only one by one degree but with a greatly increased magnitude limit of +18. The database contains time-series data on the magnitude of detected objects over a period of time from March 2009 to March 2012 [2]. As the Skycam images are centered on the view of the main Liverpool Telescope, observations of specific objects are only recorded when they are within the field of view of the camera as the telescope is focused within the vicinity of the objects. This results in time-series with uneven length gaps between observations, increasing the difficulty of identifying variations in the magnitude of the observations.

4   Feature Extraction

We begin our analysis using a methodology proposed by Debosscher et al. in 2007 and improved upon by Richards et al. in 2011 [3, 10]. The goal is to describe the time-series data for each object as a set of harmonic features that are invariant to the objects mean magnitude and time-translation phase allowing direct comparison between objects. The whole SkycamT database used in this investigation is 180 GB in size with 20 GB of indexes for faster query response times. For each object, a set methodology is applied to generate an associated feature vector. The database is queried for all observations of a specific object. The returned table has its magnitude, modified Julian date and magnitude error columns retrieved. The identification of the dominant periodic oscillation within the object's time-series is then required. There are a number of possible algorithms that can be deployed on uneven time-series.

Phase dispersion minimization [11] and the String Length Lafler-Kinman (SLLK) statistic can identify how well-aligned data points are placed in phase-space across a sample range of periods [12]. This is accomplished through computing the distance between each data point in phase space. An extension to this idea of aligning data within phase space is a recently proposed periodogram based on the Blum-Kiefer-Rosenblatt (BKR) statistical independence test [13]. Instead of utilizing the alignment of the data such as in the string-length methods, a rank correlation test is performed. There also exist Information Theoretic approaches such as slotted Correntropy and the improved Correntropy Kernelized Periodogram (CKP) [14] that have proven to be very effective and are a focus for future initiatives on the Skycam database [6]. However, for this initial investigation a Lomb-Scargle Periodogram (LSP) is utilized to identify the primary periodic signal within the data.

The LSP uses a least-squares spectral analysis. It is a method of estimating the frequency spectrum of time-series data by the fitting of multiple sinusoids to the data using least-squares regression [3, 5]. This method is performed over a frequency range resulting in the statistic normalized power that has a larger value if the fitted sinusoid has a lower chi-squared error with a candidate frequency. It operates over a frequency range with a finite set of candidate frequencies separated by intervals. As the objects in the database can have low and high period variations, this interval is set as constant to produce a uniform sample across the full frequency range. The lowest frequency is the longest period that can be expected to be detected by the periodogram. It is defined as the reciprocal of the difference between the maximum and minimum modified Julian Date of the object's observations named the total observation time t_{tot}. The maximum frequency is related to the minimum periods that can be found from an object's data. Hypothetically scanning down to the minimum possible periods for variable stars is recommended. However, for some pulsating white dwarf stars, this can be as low as 1–2 min [10]. In previous methods, a Pseudo-Nyquist frequency was proposed for the determining of a maximum frequency for unevenly sampled data which approximates the Nyquist frequency by taking the mean of the individual time intervals between the observations of an object. This equation is shown in Eq. 1.

$$f_{nyq} = 0.5 \left\langle \frac{1}{\Delta T} \right\rangle \tag{1}$$

where f_{nyq} is the Pseudo-Nyquist frequency and ΔT is the time intervals between observations of an object. This frequency is determined by the mean intervals between observations in an uneven time-series. Gaps in observations are not considered uneven observations and instead are just considered times with a lack of observations and do not contribute to the calculation of the Pseudo-Nyquist frequency. Despite this distinction, a globally accepted definition of a 'gap' and an 'uneven sample' was not identified. Therefore we attempt to approach this problem to provide a good solution. Theoretically, the time intervals between observations could all be considered gaps. This is not really possible as the start and stop times of Skycam observations are unlikely to be an integer number of minutes (the interval between exposures). Ignoring this, theoretically the sampling rate could be the Nyquist frequency of the evenly sampled exposure intervals which is half of the reciprocal of a minute, 720 cycles per day, equivalent to a period of 2 min. Currently we have defined a gap as an interval greater than two standard deviations for this method. This is unlikely to be an acceptable final answer as the standard deviation has limited meaning on non-normal distributions. Future experiments will investigate this question in order to provide a more concrete definition.

Finally, the frequency step between candidate frequencies must be determined to produce a frequency spectrum with a finite number of frequencies. Both Debosscher et al. and Richards et al. make use of a frequency step of 0.1 divided by the total observation time as defined above [3, 10]. In this method the frequency step is defined as shown in Eq. 2.

$$f_{step} = \frac{1}{ovsm \times t_{tot}} \tag{2}$$

where f_{step} is the frequency step and *ovsm* is the oversampling factor. When the oversampling factor is set to 10 the frequency step equates to that used in the previous methods [3, 10]. Our early experiments suggest this might be too fine a frequency grid for the STILT data as noisy peaks seem to be produced for some objects. Interestingly, for the light source Algol, an eclipsing binary system, this oversampling factor was required. This remains a challenge in the development of this method. The frequency associated with the maximum power from the LSP is recorded as the primary frequency. Upon the determination of the candidate period, weighted linear regression is performed to fit a four-harmonic sinusoid model using the period detected by the periodogram. This model is then subtracted from the time-series in a process called pre-whitening. This is done as to eliminate any periodic activity within the time-series based on the dominant period detected by the periodogram. This pre-whitened time-series is then used to identify a second period independent of the first dominant period. A third period is then identified using the same method. A harmonic best-fit is computed on the original data by weighted linear regression using a model with twenty six coefficients is utilized as shown in Eq. 3.

$$y(t) = ct + \sum_{i=1}^{3} \sum_{j=1}^{4} \left\{ a_{ij} \sin\left(2\pi j f_i t\right) + b_{ij} \cos\left(2\pi j f_i t\right) \right\} + b_0 \tag{3}$$

where the b_0 parameter is the mean magnitude of the light curve and the c parameter is the linear trend of the time-series. The frequencies f_i and the coefficients a_{ij} and b_{ij} are retained along with the linear trend c and provide a good description of the light curve. These coefficients are not yet time-translation invariant and must be transformed into better descriptors of the light curve. This is accomplished by transforming the Fourier coefficients into a set of amplitudes A_{ij} and phases PH_{ij}. This computation is performed through the use of trigonometric identities [10]. This results in the production of twenty eight features that are time-translation invariant. These features include the slope of the linear trend, the three frequencies used in the final harmonic model, the twelve amplitude coefficients and eleven phase coefficients and the ratio of data variance (called variance ratio) between the variance before the pre-whitening of the harmonic model of the primary period and after. The LSP is known to strongly identify periodicities for variable stars that are highly sinusoidal such as Mira-class variables. But it can also struggle with less sinusoidal light curves such as eclipsing binaries occasionally missing the period of offering a multiple of the correct period instead of the true period.

Figure 1 demonstrates the model produced by the described method for the SkycamT data collected on the star Mira showing a clear sinusoidal oscillation. The period of Mira has been widely reported as 332 days, verified by surveys such as HIPPARCOS [15]. Despite the discussed weaknesses, if the periodogram returns a result similar to the stars correct period, the linear regression can produce an accurate model. The model is not a perfect fit but should be sufficient to generate features within the ranges expected of Mira class variables.

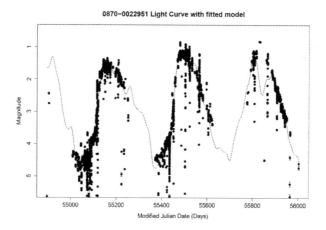

Fig. 1. The light curve of the star Mira with a harmonic fit with a period of 316 days compared to the accepted value of 332 days. Despite this, the harmonic model is a close match to the data.

A good solution for improving the accuracy of the period search is to incorporate a more powerful period finding algorithm. The artificial data points have allowed the harmonic fits the degree of freedom required to fit the unevenly sampled magnitude data resulting in powerful fits such as that demonstrated by Fig. 2. Unfortunately this has also resulted in situations where the model deviates drastically in the non-sampled regions due to 'noisy fringes' in the data. The linear regression is resilient to noise and can evaluate the signals within very poor data. However, coupled with the uneven sampling rate, the noise can result in the linear regression entering a region devoid of data whilst fitting for a steeper or shallower gradient due to the noise of these last data points. As a result the amplitude of the sinusoids can peak beyond a physically realistic state.

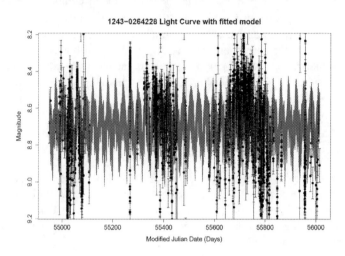

Fig. 2. A harmonic fit on the star 1243-0264228. By allowing the model to vary within the non-sampled spaces, a superior harmonic fit can be constructed

Finally, the performance of the analysis system is of great concern. Fortunately, the weighted linear regression is implemented using a very efficient normal equation method. Therefore the LSP which is $O(N^2)$ in processing complexity is the primary processing component of this analysis method. This order is a result of needing to run every observation of an object over a high resolution frequency spectrum to extract the dominate periods otherwise important harmonic variations may be missed. The frequency spectrum can contain tens of thousands of candidate periods which are completely independent allowing for parallelized implementation.

5   Proposed Solution

In order to significantly improve performance, we propose a system that makes use of multiple intelligent agents to subdivide the processing tasks between multiple clusters whilst simultaneously a new 'intelligence service' will constantly monitor the models being generated by the individual agents. This intelligence system will learn stochastically as models are continuously generated for the light curves of different objects. We use intelligent agents for this task as they are capable of continuously making decisions based on the quantity of data being processed and the system resources available [16]. The proposed framework will be dynamic and modular as well as scalable in nature. The modular aspect of the framework will allow a user the ability to develop tasks as well as the type of strategies used when replicating an agent. This dynamic aspect of the framework allows it to be scalable by using replication.

Due to the modularity of the framework, a user has the ability to define their tasks. Depending on the tasks and the environment that will be operated; the agent will either know in advance when it is first started or at which time in the future. There are two distinct scenarios in which these agents can operate. In the first scenario, an agent is processing some calculations in a static SQL Database. The agent will know in advance the size of the tables required for the tasks by conducting a count to retrieve the size and it can compute the amount of agents and the tasks that needs to be processed on each agent depending on the user's pre-defined discretion. In the second scenario, an agent is processing in-frequent amounts of data bursts, sometimes small amounts of data and sometimes large amounts of data in a timely manner. The agent will have to monitor its resources and depending on the threshold defined by the user the agent can determine if it can complete the tasks in the required time [16]. When an agent has decided it can either not complete the tasks or has reached a point that it no longer can complete its task due to the load then it will replicate itself. This is the dynamic and scalable aspect of the framework because the agent has to make a decision that it cannot complete its tasks and has to offload some tasks to another agent. This decision is made from what we call 'the dynamic strategy' by looking at the resources of the host machine and comparing them with the threshold defined by the user. This is only used for the replication phase.

Once an agent has requested a replication then that agent will only run the first task and the newly replicated agent will run the remaining tasks. When that agent has decided that it can no longer complete the tasks in a suitable time period then it will replicate itself again. Every time an agent starts to struggle due to the load then it will replicate itself until each task is running on a single agent. Previously replication has mainly been used for fault tolerance [17–19]. However, we are using replication for performance and scalability increase. Some tasks could possibly be synchronous and some tasks may be asynchronous. This means that tasks could possibly rely on results computed from previous tasks and some tasks may be independent and can run in parallel. We define two types of task, synchronous tasks that rely on the result from a previous task or asynchronous tasks that rely on a type of data either from a previous tasks result or based on the source data.

The system will contain two types of agents, 'The Task Agent' and 'The Resource Management Agent/Resource Controller'. When an agent wants to replicate itself, it will contact all Resource Management agents on the network to see if any resources are available for it to replicate and responses will be send back informing the replicating agent of the amount of usable resources available on each of the machine. If an appropriate amount of resources are available then the agent will request the resource management agent to replicate the task agent. When replicating, the resource management agent will spawn a container for the replicated agent to run within with a specific amount of resources that the user has defined for the task [20]. We choose to containerize the agent for two reasons. Firstly, containerizing the agent allows us to isolate the agent from other agents on the machine while also having the capability to deploy the agent almost instantly due to their minimal runtime requirements. Secondly, we can configure the container to only use a specific amount of resources, for example a predefined number of CPU Cores and percentage of usage for each CPU, allowing for more control over the resources allocated to the replicated agent [21]. Figure 3 demonstrates the structure of this replication sequence.

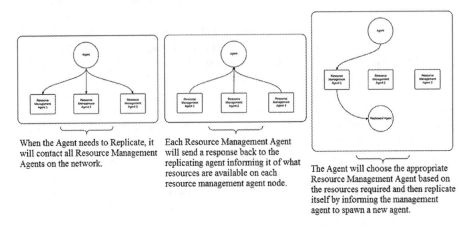

When the Agent needs to Replicate, it will contact all Resource Management Agents on the network.

Each Resource Management Agent will send a response back to the replicating agent informing it of what resources are available on each resource management agent node.

The Agent will choose the appropriate Resource Management Agent based on the resources required and then replicate itself by informing the management agent to spawn a new agent.

Fig. 3. The sequence of operations involved in agent replication

The proposed system for the processing of astronomical light curves will be using the static Skycam database. This means that no new entries are ever added to the database and the data is left standalone. We will use a static based strategy as we know that the database is not going to be updated. This strategy will count the number of objects in the database and then based on the quantity of resources allocated to each agent, a decision that the programmer has made when building their strategy, will either allow a single agent to conduct all the tasks utilizing the total resources (as shown in Fig. 4) or each machine running an instance of the Resource Management Agent will spawn task agents to run the tasks for multiple objects simultaneously.

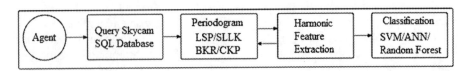

Fig. 4. A single agent is capable of running every task.

Whilst the multi-agents distribute the processing tasks generating the harmonic best-fit models, a second intelligence service operates independently. This service monitors the outbound models and determines whether they are realistic. This service with have a confidence threshold which dictates how well the model has fit the time-series data based on its trained state. If it has a low confidence it attempts to modify the model based on previous patterns that have been discovered within the multiple objects light curves processed previously. The intelligence service is always operating using supervised learning trained using the known light curves of well sampled objects to continuously improve its prediction. This system is envisaged to use a form of neural network containing multiple models based on the different light curve profiles discovered. Both Recurrent and Convolutional neural networks have been shown to be potent at predicting time-series [22]. Additionally, these methods can be extended into deep learning through the addition of more layers if the light curves require the production of more powerful features. In the event of a harmonic fit model receiving a low confidence, the intelligence service will attempt to construct a new harmonic model using neural networks trained on the previous high confidence models through the stochastic learning process. We are hopeful that the combination of scalable processing and accurate time-series predictions will lead to high performance processing of the STILT database through the generation of robust features by supervised and unsupervised learning for future multi-class classification analysis.

6   Conclusion and Future Work

The weighted linear regression harmonic best-fit models can be used to produce time-translation invariant features from uneven time-series. The STILT database contains many objects with sufficient noise and uneven sampling to result in poor or physically unrealistic harmonic models. The LSP can produce multiples of the correct period and occasionally it misses the periodic signal completely. Replacing the LSP with a more powerful and less limiting algorithm might alleviate this problem. The proposed solution seeks to introduce scalability and multithreading for the high performance processing of the STILT database. This is accomplished through the use of a multiple intelligent agent platform. This platform is capable of distributing the processing tasks across multiple available machines as required based on the processing workload and available resources. Finally, a new intelligence service using more powerful machine learning algorithms such as Recurrent and Convolutional neural networks can regulate the models generated by the harmonic best-fit producing consistent results despite the degree of freedom. Our future work will involve the incorporation of the proposed methods into a newly developed data analytics platform. Following this, the models produced can be evaluated through testing previously classified variable objects in the STILT dataset as well as sourcing external datasets for comparative results. These efforts allow the production of robust light curve features that are well placed for future incorporation into a powerful multi-class classification system.

Acknowledgment. This work was funded through a Liverpool John Moores University scholarship in partial fulfilment of the requirements for the degree of Doctor of Philosophy. The Liverpool Telescope is operated on the island of La Palma by Liverpool John Moores University in the Spanish Observatorio del Roque de los Muchachos of the Instituto de Astrofísica de Canarias with financial support from the UK Science and Technology Facilities Council [23].

References

1. York, D.G., Adelman, J., et al.: The sloan digital sky survey: technical summary. Astron. J. **120**(3), 1579–2000 (2000)
2. Mawson, N.R., Steele, I.A., Smith, R.J.: STILT: system design and performance. Astron. Nachr. **334**(7), 729–737 (2013)
3. Richards, J.W., Starr, D.L., et al.: On machine-learned classification of variable stars with sparse and noisy time-series data. Astrophys. J. **733**(1), 10–32 (2011)
4. Vaughan, S.: Random time series in astronomy. Philos. Trans. Roy. Soc. **371**, 20110549 (2011)
5. Scargle, J.D.: Studies in astronomical time series analysis. II. statistical aspects of spectral analysis of unevenly spaced data. Astrophys. J. **263**, 835–853 (1982)
6. Huijse, P., Estévez, P.A., et al.: An information theoretic algorithm for finding periodicities in stellar light curves. IEEE Trans. Sig. Process. **60**(10), 5135–5145 (2012)
7. Protopapas, P., Giammarco, J.M., et al.: Finding outlier light curves in catalogues of periodic variable stars. Roy. Astron. Soc. Mon. Not. **369**, 677–696 (2006)
8. Eyer, L., Mowlavi, N.: Variable stars across the observational HR diagram. J. Phys.: Conf. Ser. **118**(1), 012010 (2008)
9. Percy, J.: Understanding Variable Stars. Cambridge University Press, Cambridge (2007)
10. Debosscher, J., Sarro, L.M., et al.: Automated supervised classification of variable stars I. Methodol. Astron. Astrophys. **475**, 1159–1183 (2007)
11. Stellingwerf, R.F.: Period determination using phase dispersion minimization. Astrophys. J. **224**, 953–960 (1978)
12. Clarke, D.: String/rope length methods using the Lafler-Kinman statistic. Astron. Astrophys. **2**(386), 763–774 (2002)
13. Zucker, S.: Detection of periodicity based on independence tests - II. Improved serial independence measure. Mon. Not. Lett. Roy. Astron. Soc. **1**(457), 118–121 (2016)
14. Liu, W., Pokharel, P.P., Principe, J.C.: Correntropy: a localized similarity measure. In: International Joint Conference on Neural Networks, IJCNN 2006, Vancouver, BC (2006)
15. Bedding, T.R., Zulstra, A.A.: HIPPARCOS period-luminosity relations for Mira and semiregular variables. Astrophys. J. **506**, 47–50 (1998)
16. Padgham, L., Winikoff, M.: Developing Intelligent Agent Systems a Practical Guide. Wiley, Hoboken (2004)
17. Almeida, A.D.L., Aknine, S., et al.: Plan-based replication for fault-tolerant multi-agent systems. In: 20th International Parallel and Distributed Processing Symposium, IPDPS 2006, Rhodes Island (2006)
18. Guessoum, Z., Faci, N., Briot, J.-P.: Adaptive replication of large-scale multi-agent systems – towards a fault-tolerant multi-agent platform. In: Garcia, A., Choren, R., Lucena, C., Giorgini, P., Holvoet, T., Romanovsky, A. (eds.) SELMAS 2005. LNCS, vol. 3914, pp. 238–253. Springer, Heidelberg (2006)

19. Sylvain, D., Guessoum, Z., Ziane, M.: Adaptive replication in fault-tolerant multi-agent systems. In: 2011 IEEE/WIC/ACM International Conference on Web Intelligence and Intelligent Agent Technology (WI-IAT), Lyon (2011)
20. Docker: What is Docker? Understand how docker works and how you can use it (2016). https://www.docker.com/what-docker
21. Docker: Docker Run Reference (2016). https://docs.docker.com/engine/reference/run/
22. Langkvist, M., Karlsson, L., Loutfi, A.: A review of unsupervised feature learning and deep learning for time-series modeling. Pattern Recogn. Lett. **1**(42), 11–24 (2014)
23. Steele, I.A., Smith, R.J., et al.: The Liverpool telescope: performance and first results. In: Society of Photo-Optical Instrumentation Engineers (SPIE) Conference Series (2004)

A Smart Health Monitoring Technology

Carl Chalmers[✉], William Hurst, Michael Mackay, and Paul Fergus

Liverpool John Moores University, Byrom Street, Liverpool L3 3AF, UK
{C.Chalmers,W.Hurst,M.I.Mackay,P.Fergus}@ljmu.ac.uk

Abstract. With the implementation of the Advanced Metering Infrastructure (AMI), comes the opportunity to gain valuable insights into an individual's daily habits, patterns and routines. A vital part of the AMI is the smart meter. It enables the monitoring of a consumer's electricity usage with a high degree of accuracy. Each device reports and records a consumer's energy usage readings at regular intervals. This facilitates the identification of emerging abnormal behaviours and trends, which can provide operative monitoring for people living alone with various health conditions. Through profiling, the detection of sudden changes in behaviour is made possible, based on the daily activities a patient is expected to undertake during a 24-h period. As such, this paper presents the development of a system which detects accurately the granular differences in energy usage which are the result of a change in an individual's health state. Such a process provides accurate monitoring for people living with self-limiting conditions and enables an early intervention practice (EIP) when a patient's condition is deteriorating. The results in this paper focus on one particular behavioural trend, the detection of sleep disturbances; which is related to various illnesses, such as depression and Alzheimer's. The results demonstrate that it is possible to detect sleep pattern changes to an accuracy of 95.96 % with 0.943 for sensitivity, 0.975 for specificity and an overall error of 0.040 when using the VPC Neural Network classifier. This type of behavioral detection can be used to provide a partial assessment of a patient's wellbeing.

Keywords: Smart grids · Advanced metering infrastructure · Smart meters · Profiling · Assistive technologies · Early intervention practice · Customer access devices · Activates of daily living

1   Introduction

Many countries around the world are currently undertaking a large-scale implementation of smart meters and their associated infrastructures. These smart energy devices offer vast advancements to the traditional energy grid. However, in order to maximise their true potential, different applications need to be considered beyond the traditional uses of electricity generation, distribution and consumption.

The smart metering infrastructure provides new possibilities for a variety of different applications that where unachievable using the traditional grid topology. Specifically, smart meters, enable detailed around the clock monitoring of energy usage. This granular data captures detailed habits and routines through the users' interactions with electrical

© Springer International Publishing Switzerland 2016
D.-S. Huang et al. (Eds.): ICIC 2016, Part I, LNCS 9771, pp. 832–842, 2016.
DOI: 10.1007/978-3-319-42291-6_82

devices and appliances. Each smart meter accurately records the electrical load for a given property at 30-min intervals. However, this reading frequency can be reduced to obtain even finer readings. This data enables the detection and identification of sudden changes in patient behaviour.

Each year, the number of people in the UK living with self-limiting conditions, such as dementia, Parkinson's disease and mental health problems, is increasing [1]. This is largely due to an aging population and improvements in diagnosis and medical treatments. The number of populace living with dementia worldwide [2] is currently estimated at 35.6 million and this number is set double by 2030 and more than triple by 2050. Additionally, 1 in 4 people currently experience a type of mental health problem each year [3]. Supporting these sufferers places a considerable strain on organisations, such as the National Health Service (NHS), local councils, front line social services and carers/relatives [4]. In monetary terms, dementia alone costs the NHS over £17 billion a year [5]; exacerbated by the cost of depression patients, which is predicted to increase to 1.45 million in the UK; adding a further £2.96 billion cost to essential services by 2026. This figure excludes other mental health conditions, such as Anxiety disorders, Schizophrenic disorders, bipolar related conditions, eating disorders and personality disorders.

Effective around the clock monitoring of these conditions can be a considerable challenge and often leads to patients having to reside in care homes and other accommodation. In addition, the need to detect accurately, sudden or worsening changes in a patient's condition, is vital for early intervention.

Community mental health groups, crisis and home resolution teams, assistive outreach teams and early psychosis teams all play a key role in preventing costly inpatient admissions. If any changes are not dealt with early, the prognosis is often worse and, as a result, costs for treatment will undoubtedly be higher [6]. An early intervention approach has been shown to reduce the severity of symptoms, improve relapse rates and significantly decrease the use of inpatient care. Evidence suggests that a comprehensive implementation of Early Intervention Practice (EIP) in England could save up to £40 million a year in psychosis services alone. Being able to detect deteriorating conditions in dementia patents earlier, enables physicians to better diagnose and identify stage progression for the disease. This enables earlier intervention for the illness before cognitive deficits affect or worsen mental capacity; supporting the individual and their family in adapting to the illness simultaneously.

This paper explores the idea of utilising smart meter data to detect abnormal changes in behaviour using a supervised classification approach. Section 2 provides a background into smart meters and patient behaviours. Section 2.1 presents an insight into the behavioural trends that smart meters can detect at different reading intervals. Additionally, we present a case study using sample smart meter data to detect sleep disturbances. Section 3.2 discusses the proposed methodology and system framework, while Sect. 4 presents the results. The paper is concluded in Sect. 5.

2   Background

Smart grids fundamentally change the way in which electricity is generated, distributed and monitored. This enhanced communication removes the traditional need for energy usage readings to be collected manually. Instead, a robust automatic reporting system with greater granularity of readings is offered [7].

2.1   Smart Meters

Smart meters the foundation of any future smart electricity grid and provide consumers with highly reliable, flexible and accurate metering services. Most significantly, they provide real-time energy usage readings at granular intervals [8]. They obtain information from the end users' load devices and measure the energy consumption. Smart meters are able to store 13 months of historical energy usage, which allows for the creation of detailed energy usage profiles [9]. Currently readings are taken at 30 min intervals. However, as these meters become more sophisticated, they are able to measure household power consumption at even more granular time-scales [10]. Smart meters are able to report energy usage as low as 10-s intervals, even though this is not currently widely deployed due to the significant amount of data it would generate [11]. This is of particular benefit for health monitoring applications, where changes in behaviour need to be identified at the granular level. Figure 1 highlights, the additional information that can obtained from increasing the reading frequency.

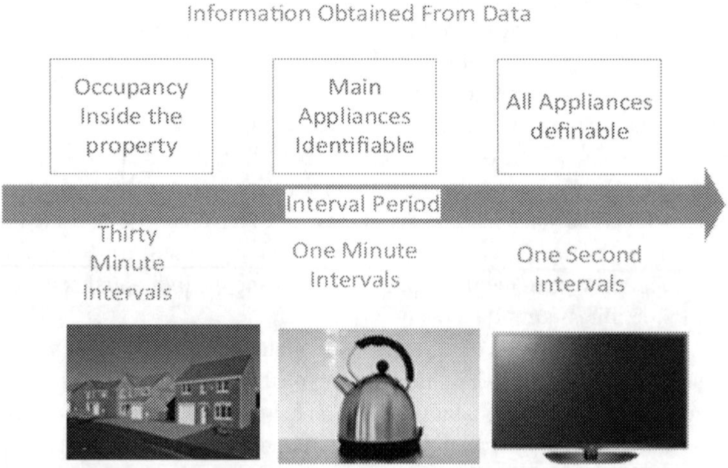

Fig. 1.   Information obtained by increasing interval reading.

Smart meters utilise ZigBee smart energy. The UK Department of Energy & Climate Change (DECC) has announced Smart metering equipment technical specifications (SMETS) 2, which cites the use of ZigBee Smart Energy 1.x. [12]. Consumer Access Devices (CAD), will access updated consumption and tariff information directly from their smart meter; a CAD can request updates of electricity information every 10 s and gas information every 30 min.

2.2 Patient Behaviors

Being able to detected and predict these changes in patterns of behavior requires a detailed understanding of the symptoms that are anticipated for each medical condition.

For example, there are a common set of features for conditions such as Alzheimer's disease and other dementias. These include agitation, anxiety, depression, apathy, delusions, sleep and appetite disturbance, elation, irritability, disinhibition and hallucinations [13]. The severity of each of these conditions alters with disease progression. Likewise, severe depression exhibits many similar behavioural symptoms as dementia, e.g. memory problems and social disengagement. Additionally, depression can cause physical complications, such as chronic joint pain, limb pain, back pain, gastrointestinal problems, tiredness, sleep disturbances, psychomotor activity changes and appetite changes [14]. These changes are reflected in how the sufferer interacts with people, their environment and their electric devices. Being able to detect any erratic or sudden behaviour change early, caters for better intervention and can lead to an early diagnosis in psychosis. Each individual is different and, as such, present their own set of symptoms and warning signs; however, one or more warning signs are likely to be evident, for example:

Memory problems.
Severe decline of social relationships.
Dropping out of activities - or out of life in general.
Odd or bizarre behaviour.
Deterioration of personal hygiene.
Hyperactivity or inactivity, or alternating between the two.
Severe sleep disturbances.
Significantly decreased activity.

Understanding these often subtle changes can lead to reduced hospital admissions through the implantation of an early intervention care plan. This is especially true for dementia suffers where the understanding of a patient's ability to undertake normal Activates of Daily Living (ADL) is an important component of the overall assessment of a patient's wellbeing. For that reason, the following section discusses how smart meter data can be utilised to detect specific patterns and device interactions.

3 Approach

In order to simulate the real-time data gathering capabilities of a smart meter, an energy monitor has been installed in a patient's premise to gather energy usage data. Energy usage is collected in real-time. This establishes the patient's routine and identifies any noteworthy trends in device utilisation.

3.1 Case Study

Figure 2 shows an overview of the electricity monitor, shown in (a). The blue current sensor transformer clip (CT) is fastened around the live cable shown in (b) to measure the electrical load. The second white sensor, which is the Optical Pulse Sensor, shown in (c) works by sensing the LED pulse output from the utility meter.

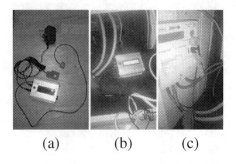

(a)   (b)   (c)

Fig. 2. Energy monitor.

Each pulse corresponds to a certain amount of energy passing through the meter. By counting these pulses, a kWh value can be calculated. Due to smart meters not being generally available until 2020 energy monitors can be used to simulate smart meter data collection capabilities. The data obtained from the energy monitor is logged to an SQL database as demonstrated in Fig. 3. The proposed system interfaces with the database directly.

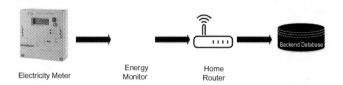

Electricity Meter   Energy   Home   Backend Database
                         Monitor   Router

Fig. 3. End to end data collection.

Increasing the reading frequencies facilitates the identification of individual device utilisation this is demonstrated in Fig. 4. The y-axis highlights the energy usage in Watts, while the x-axis shows the reading time.

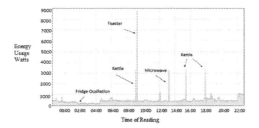

Fig. 4. Real time energy readings.

Obtaining energy readings at 1 to 10 s intervals constructs individual energy signatures for each device. This is achieved by identifying the amount of energy being consumed, as demonstrated in Fig. 5; where, as before, the y-axis displays energy usage in Watts and the x-axis shows the reading time. This allows background noise from certain devices, such as the fridge oscillation, air conditioning and standby electricity usage, to be filtered out to leave clear usage signatures for important devices, such as a kettle, lights and cooking equipment.

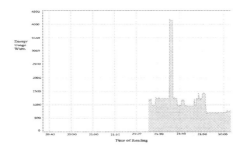

Fig. 5. Energy signature for a kettle.

As previously highlighted, changes in sleep patterns are a common feature in several mental-health medical conditions. Being able to recognise changes and patterns in behaviour and, in particular, sleep disturbances, facilitates the prediction of future cognitive and non-cognitive changes. Often, dementia sufferers in hospital are admitted due to other poor health caused by other illnesses [15]. These are often caused by patient inactivity which ultimately leads to additional health complications. Detecting these behaviours early enables the prevention of additional complications.

3.2   Methodology

The data processing components of the system operates in two modes; firstly, the training mode, which involves collecting and processing energy usage data to train the data classifiers. Secondly, the prediction mode identifies both normal and abnormal behaviours using the trained classifiers from mode 1.

Mode 1 – Training: When the system is in training mode normal and abnormal data is collected from the data store. Normal data refers to a patient's usual behavioural routines in a household. Abnormal data relates to a deviation from expected patterns of behaviour. Data features are extracted from the data set in order to train the classifiers to be able to detect abnormal patterns in a dataset.

Mode 2 – Prediction: Using the trained classifiers the system automatically detects both normal and abnormal patient behaviour in real-time. Where appropriate the system alerts the patients support network to a potential problem if detected.

In the first instance the system alerts the patient to check in, by performing specific device interaction. This reduces any possible false alarms and verifies that that the patient requires no further assistance. The system identifies if interaction has taken place; if this is not the case an alert is communicated to a third-party health care practitioner as shown if Fig. 6 both of these modes of operation or subsequently detailed in this section.

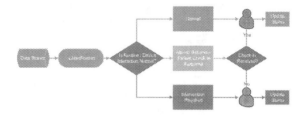

Fig. 6. Mode two.

4   Evaluation

For the following analysis, one year's worth of energy usage readings for 8 different smart meter users was selected. The 8 consumers were selected as a sub group of the population as they accurately represent the population as a whole, this approach is more practical for the initial data analysis. Out of the 8 consumers selected 4 have normal readings and 4 have abnormal readings. The subjects with normal readings were classified as having no energy usage readings greater than 2 Kwh between the hours of 1:30 and 4:00 for the entire year period. Abnormal subjects were classified where they exceeded 2 Kwh between the hours of 1:30 and 4:00 on 3 or more occasions in a one-year period. All households in the experiment have one occupant to ensure accurate results. Initially 7 features per consumer were derived for each 24-h period totalling 8760 results for each of the following features: General supply min; general supply max; general supply median; general supply standard deviation; general supply mean; off peak max and off peak mean.

4.1   Dimensionality Reduction

In order to ascertain the optimum features and the greatest variance for the classification, Principle Component Analysis (PCA) is undertaken on the features created from the variables in the dataset (with each variable being a smart meter) this process is shown in Fig. 7 during the feature selection step. General supply min, general supply max, general supply median, general supply standard deviation, general supply mean, off peak max and off peak mean, were devised from each variable to establish a total set of 28 features. Using this methodology helps to ensure the identification of the most useful features and a reduction of 67 % in the number of features was obtained (left 9 out of the original 28).

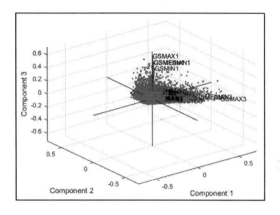

Fig. 7. PCA analysis.

4.2   Neural Networks Classification

In order to perform the classification of the data a selection of classifiers where used these include: back-propagation trained feed-forward neural network classifier (BPXNC), levenberg-marquardt trained feed-forward neural net classifier (LMNC), automatic neural network classifier (NEURC), radial basis function neural network classifier (RBNC), trainable linear perceptron classifier (PERLC), voted perception classifier (VPC) and the random neural network classifier (RNNC) [16]. Each of these classifiers where selected for their ability to learn how to recognise abnormal values in a dataset [17]. They also employ a supervised learning approach, which is a key part of the approach. Table 1 presents the confusion matrix for the NEURC classifier as an example.

Table 1.   Confusion matrix for NEURC

True Labels	Estimated Labels		
	1	**2**	**Totals**
1	4133	246	4379
2	107	4272	4379
Totals	424-	4518	8758

The classifiers' performance is calculated using confusion matrices to assess the success of the classification or Area under the Curve (AUC), sensitivity, specificity and error [18]. The results are calculated mathematically, using the following formulae, where *TP* refers to True Positive, *TN* implies True Negative and *FP* and *FN* refer to False Positive and False Negative respectively.

$$Sensitivity = TP/(TP + FN) \tag{1}$$

$$Specificity = TN/(TN + FP) \tag{2}$$

$$Accuracy = (TP + TN)/(TP + FP + FN + TN) \tag{3}$$

The NEURC is the most accurate; able to classify 95.96 % of the data correctly with an error of 0.040. For the NEURC classifier 4133 out of 4379 normal behaviours are correctly classified. The same classifier is able to assess 4272 out of 4379 abnormal behaviours are accurately classified on average. All of the results from the classification experiments are depicted in Table 2.

Table 2. Classification results

Classifiers	AUC (%)	Sensitivity	Specificity	Error
VPC	84.81	0.844	0.851	0.1519
RNNC	93.26	0.914	0.950	0.0674
PERLC	64.01	0.301	0.978	0.3599
BPXNC	89.50	0.887	0.902	0.105
LMNC	92.57	0.898	0.953	0.074
NEURC	95.96	0.943	0.975	0.040
RBNC	93.33	0.898	0.968	0.107

4.3 Discussion

It is clear from the results in Table 2, that the classifiers are able to detect accurately both the normal and abnormal behaviours in the data set. Figure 8(a) shows a visualisation of two features in a scatter plot. However this type of visualisation makes it challenging to see the division of the abnormal behaviour from the normal patterns.

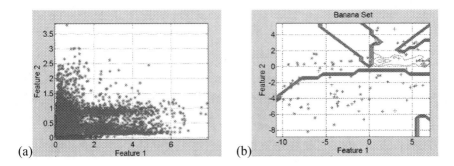

Fig. 8. (a) Visualisation of two features (b) NEURC classification visualisation.

Figure 8(b) shows a visualisation if the NEURC classification process, overlaid on the Banana Plot; where the contour lines represent a conception of the data separation process during calculation.

Using the above techniques, support the findings that neural networks can be used to detect abnormal behaviour in smart meter datasets for health care monitoring. Using this approach, our system is able to perform an analysis of real-time datasets to detect when a user's behaviour is changing as a result of illness. The neurc classifier provides an accurate monitoring algorithm for monitoring people living with self-limiting conditions requiring an enable early intervention practice (EIP).

5   Conclusion

The implementation of smart meters and their associated infrastructures addresses many of the traditional constraints imposed by the current energy generation and distribution infrastructure. It allows detailed monitoring and consumer energy usage profiling. This leads to more efficient energy usage, planning and fault tolerance. Being able to collect and analyse sufficient amounts of usage data makes it possible to identify reoccurring patterns and trends in a patient's behaviour.

This can be utilised to address many problems, not just in the field of electricity and generation but also in health monitoring. In this paper, we have discussed the need to understand patient's behaviours based on their conditions and how these behaviours can be reflected in their energy usage. Identifying these changes and patterns facilitate the possibility for independent living and early intervention practice, reducing the demand for hospital care and enhanced resource planning.

The case study put forward in this research investigates the detection of sleep disturbances, which is one of the many key indicators of an onset or change in condition. We have demonstrated how this can be achieved by analysing smart meter data using our proposed methodology and by implementing neural network classifiers to detect abnormal energy usage.

References

1. Department of Health. Report Long-term conditions compendium of Information: 3rd edition. https://www.gov.uk/government/publications/long-term-conditions-compendium-of-information-third-edition

2. Chan, M.: Dementia A public health priority, World Health Organization. http://www.globalaging.org/agingwatch/Articles/Dementia%20a%20public%20health%20priority.pdf

3. Mental Health Foundation. The Fundamental Facts The latest facts and figures on mental health.   http://www.mentalhealth.org.uk/content/assets/PDF/publications/fundamental_facts_2007.pdf?view=Standard

4. Prince, M., Knapp, M., Guerchet, M., McCrone, P., Prina, M., Comas-Herrera, A., Wittenberg, R., Adelaja, B., Hu, B., King, D., Rehill, A., Salimkumar, D.: Dementia UK: Second Edition - Overview. Alzheimer's Society (2014)

5. McCrone, P., Dhanasiri, S., Patel, A., Knapp, M., Lawton-Smith, S.: Paying the price the cost of mental health care in England to 2026. King's Fund

6. Mental Health Network NHS CONFEDERATION. Early intervention in psychosis services. http://www.nhsconfed.org/resources/2011/05/early-intervention-in-psychosis-services

7. Popa, M.: Data collecting from smart meters in an advanced metering infrastructure. In: INES 2011 15th International Conference on Intelligent Engineering Systems, pp. 137–142 (2012)

8. Wang, L., Devabhaktuni, V., Gudi, N.: Smart meters for power grid – challenges, issues, advantages and status. In: 2011 IEEE/PES Power Systems Conference and Exposition (PSCE), pp. 1–7 (2011)

9. Benzi, F., Anglani, N., Bassi, E., Frosini, L.: Electricity smart meters interfacing the households. IEE Trans. Ind. Electron. **58**(10), 4487–4494 (2011)

10. McKenna, E., Richardson, I., Thomson, M.: Smart meter data: balancing consumer privacy concerns with legitimate applications. Energy Policy **41**, 807–814 (2012)

11. Vojdani, A.: Smart integration. IEEE Power Energy Mag. **71**(9), 71–79 (2008)

12. Department of Energy and Climate Change. Smart Meters, Smart Data, Smart Growth. https://www.gov.uk/government/publications/smart-meters-smart-data-smart-growth

13. Mega, M.S., Cummings, J.L., Fiorello, T., Gornbein, J.: The spectrum of behavioural changes in Alzheimer's disease. Neurology **46**, 130–135 (1996)

14. Trivedi, M.H.: The link between depression and physical symptoms. Primary Care Companion J. Clin. Psychiatry **6**, 12–16 (2004)

15. Selikson, S., Damus, K., Hameramn, D.: Risk factors associated with immobility. J. Am. Geriatr. Soc. **36**, 707–712 (2015)

16. Cho, H.J., Lavretsky, H., Olmstead, R., Levin, M.J., Oxman, M.N., Irwin, M.R.: Sleep disturbance and depression recurrence in community-dwelling older adults: a prospective study. Am. J. Psychiatry (2008)

17. Marom, N., Rokach, L., Shmilovici, A.: Using the confusion matrix for improving ensemble classifiers. In: Proceedings of the Twenty-Sixth IEEE Convention of Electrical and Electronics Engineers in Israel, pp. 000555–000559 (2010)

18. Aljaaf, A.J., Al-Jumeily, D., Hussain, A.J., Fergus, P., Al-Jumaily, M., Radi, N.: Applied machine learning classifiers for medical applications: clarifying the behavioural patterns using a variety of datasets. In: 2015 International Conference on Systems, Signals and Image Processing (IWSSIP) (2015)

A Framework on a Computer Assisted and Systematic Methodology for Detection of Chronic Lower Back Pain Using Artificial Intelligence and Computer Graphics Technologies

Ala S. Al Kafri[1(✉)], Sud Sudirman[1], Abir J. Hussain[1], Paul Fergus[1], Dhiya Al-Jumeily[1], Mohammed Al-Jumaily[2], and Haya Al-Askar[3]

[1] Faculty of Engineering and Technology, Liverpool John Moores University,
Byrom Street, Liverpool L3 3AF, UK
a.s.alkafri@2015.ljmu.ac.uk,
{s.sudirman,a.hussain,p.fergus,d.aljumeily}@ljmu.ac.uk
[2] Consultant Neurosurgeon and Spine Surgeon,
Dr Sulaiman Al Habib Hospital, Dubai Healthcare City, Dubai, UAE
maljumaily@yahoo.fr
[3] College of Computer Engineering and Science, Sattam Bin Abdulaziz University,
Al-Kharj, Kingdom of Saudi Arabia
sun_2258@hotmail.com

Abstract. Back pain is one of the major musculoskeletal pain problems that can affect many people and is considered as one of the main causes of disability all over the world. Lower back pain, which is the most common type of back pain, is estimated to affect at least 60 % to 80 % of the adult population in the United Kingdom at some time in their lives. Some of those patients develop a more serious condition namely Chronic Lower Back Pain in which physicians must carry out a more involved diagnostic procedure to determine its cause. In most cases, this procedure involves a long and laborious task by the physicians to visually identify abnormalities from the patient's Magnetic Resonance Images. Limited technological advances have been made in the past decades to support this process. This paper presents a comprehensive literature review on these technological advances and presents a framework of a methodology for diagnosing and predicting Chronic Lower Back Pain. This framework will combine current state-of-the-art computing technologies including those in the area of artificial intelligence, physics modelling, and computer graphics, and is argued to be able to improve the diagnosis process.

Keywords: Computer aided/assisted diagnosis · Chronic Lower Back Pain · Artificial intelligence · Physics modelling · Computer graphics

1 Introduction

Back pain is one of the major musculoskeletal pain problems which affected many people and it is considered as one of the main causes of disability all over the world [1].

© Springer International Publishing Switzerland 2016
D.-S. Huang et al. (Eds.): ICIC 2016, Part I, LNCS 9771, pp. 843–854, 2016.
DOI: 10.1007/978-3-319-42291-6_83

The Pain Community Centre [2] indicated that in the United Kingdom (UK), 2.5 million people have back pain every day of the year. The survey also found that back problems are the leading cause of disability with nearly 119 million days per year lost. The survey also found that one in eight unemployed people give back pain as the reason for unemployment. Statistically, an individual who has been off sick with back pain for a month has a 20 % chance to still being off work a year later [3]. The percentage of people who return to see their general practitioner (GP) with back pain within 3 months is more than 29 % [4].

There are two types of back pain, upper and lower ones. Lower back pain is more common than the former and is estimated to affect at least 60 % to 80 % of the adult population in the UK at some time in their lives. While most of them will have resolution of their back pain with simple measures such as using simple analgesia and exercise, a small proportion of them develop a more chronic condition [5, 6]. Lumbar spine is the lower back area in the spinal column which contains five vertebrae labelled L1 to L5 [7, 8]. Figure 1 describes the lumbar spine and its parts which are the area affected by Chronic Lower Back Pain (CLBP) [9]. Magnetic Resonance Image (MRI) is mainly used to diagnose patients with CLBP or those with symptoms consistent with radiculopathy or spinal column stenosis [10]. Physicians perform the diagnosis normally by studying the MR images through visual inspection of the data. CLBP can be caused by a number of factors including fractures, lumbar disc degeneration, lumbar disc herniation, or infection in the nerve roots. If they suspect disc herniation as a possible cause of the pain, they would utilize axial view of the MRI to help form their decision [11]. In the case of vertebrae infection or fracture, MRI is also the best choice for diagnosis because it allows displaying the full infected area including the bone marrow to differentiate it from more serious cases such as crushed vertebrae [12].

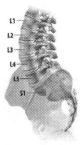

Fig. 1. Lumbar Spine that contain the vertebrae from L1 to L5 [13].

Visual observation and analysis of MR images could take up much of a physician time and effort. Moreover, it can increase the probability of misdiagnosis. As a result, physicians would opt to use a Computer Aided Diagnosis (CAD) to help this task. There are a number of CAD systems that can be used for various clinical purposes ranging from a CAD system for detecting colonic polyp and breast cancer in mammography, to another for detecting prostate cancer using MR images [11]. Despite the availability of these systems, physicians still have to overcome a number of technical challenges due to the wide range of imaging characteristics and resolutions [14] as well as due to the limitation of the algorithms employed to highlight areas of interest.

On the opposite end, there are also some progresses in the rehabilitation mechanism of CLBP patients. For example, a lower back pain rehabilitation system is proposed in [16] using a wireless sensor technology which helps the patients and physiotherapists carry out the rehabilitation exercises. In addition to the problem of diagnosing the cause of CLBP and rehabilitation of the patients, the importance of a reliable prevention mechanism was highlighted in [15] which stated that an accurate means of identifying patients at high risk for chronic disabling pain could lead to more cost-effective care. Furthermore, it can be argued that CLBP is suffered by patients who have history of untreated non-chronic LBP. Therefore, it is imperative that there should be a computer aided system in place to help physicians in their tasks in identifying potential problems that might occur in the future based on existing physiology of the lumbar spine and the patient's characteristics. There has been limited progress in this regard, including one by Neubert et al. [17] who claimed that 3-dimensional MR images have the potential to help physicians to detect and monitor the spine disorder at an early stage. Consequently, we argue that this is one of the most promising areas of research in which computer technologies can play significant part in solving the problem. A number of research works to further the technology in this regard have been made. Ahn [18] developed an interactive computerised simulation of a virtual model of human cervical spine which incorporates physics based modelling and implemented using a physics engine library. Physics engine libraries are traditionally used to develop gaming, robotics, or flight simulations but more recently they are used by researchers for medical purposes [19]. Furthermore, our initial review of the literature reveals that there are some progress in the modelling of the lumbar spine as a 3D computer model as well as mathematical/physics model [20–22]. It is believed that these advances in physics modelling coupled with computer technologies can help solve the problem of future prediction of CLBP. This initial review of the various techniques to detect flaws in lumbar spine and 3D modelling of lumbar spine has highlighted the significance of identifying and understanding a problem space that associated with computer assisted and systematic methodology. We have identified two main research issues associated with this and are proposing two solutions that address them in this paper. These solutions will improve the speed and accuracy of the physicians' and radiologists' tasks in diagnosing and managing CLBP patients. The remainder of this paper is organised as follows. Section 2 presents the literature review of existing techniques that helps diagnosis and management of CLBP. The framework of the proposed system is presented in Sect. 3. Section 4 contains the discussion and analysis and Sect. 5 presents the paper's conclusion.

2   Review of Existing Techniques in Diagnosing and Management of CLBP

There are two main techniques that help in diagnosing and managing CLBP that will be explained in this section.

2.1   Lumbar Disc Herniation Detection Using Computer Vision and Artificial Intelligence Techniques

Clinical studies have indicated that morphological characteristics of lumbar discs and their signal intensity on a patient's MRI image have close relationship to the clinical outcome [23]. To this end, computer vision and artificial intelligence algorithms can be utilised to exploit these facts by analysing the MR images, calculating appropriate image features (or feature descriptors), and classifying them to decide if any particular regions in the image belong to problematic areas. Image features can be considered as a set of important information derived from an image or a subset of an image that can uniquely describe the image contents. This information is extremely important in computer vision as it can be used to label or mark specific locations of the image and can be used in comparing various images. There are two types of image features which are namely global and local features [24]. Local features are computed at different locations in the image using only small support area of around the location point. As such, even when the other parts of the image undergo changes, as long as the support area remains the same, local features would more likely not be affected. This is one of the strong points of local features over global features because they are robust to occlusion. Examples of local features are corners, edges, and texture descriptors. On the other hand, global features are derived from the entire image that resulted in their ability in generalising the entire image into one single feature vector. One example of global features is image code, which is a compressed form of the image using an appropriate coding technique that preserves the high level information of the image contents. Alternatively, global features could be constructed from a collection of local features such as shape descriptors, contours descriptors, texture descriptors, etc. Image analysis and comparison are performed by means of classifying its features. This is done by comparing the features from the test image in question with those from training data. A brute force approach for comparing two sets of image features would compare every feature in one set to every feature in the other and keeping track of the "best so far" match. This results in a heavy computational complexity in the order of $O(N^2)$ where N is the number of feature in each image. A number of algorithms have been proposed to improve the computational complexity, including the popular kd-tree technique [25]. This technique uses exact nearest neighbour search and works very well for low dimensional data but quickly loses its effectiveness as dimensionality increases. The popularity of the kd-tree technique has seen a number of derivatives that further improve the algorithm including [26, 27]. The success of a more recent matching technique called Fast Library for

Approximate Nearest Neighbour (FLANN) [28] is another example how the computational complexity of image feature comparison can be further reduced to allow near real-time execution.

The uniqueness of each proposed algorithm in this category often lies in the choice of features and matching algorithms as well as novel application of existing approaches to new or untested problems. This research reviews a number of algorithms that are proposed to identify regions in the MRI that are responsible for CLBP that explained below.

Muja and Lowe [28] proposed a visualization and quantitative analysis framework using image segmentation technique to derive six features that are extracted from patients MR images, which were found to have close relationship with Lumbar Disc Herniation score. The six features include the distribution of the protruded disc, the ratio between the protruded part and the dural sacs, and its relative signal intensity. Alomari et al. [11, 29] proposed a probabilistic model for automatic herniation detection that incorporates appearance and shape features of the lumbar intervertebral discs. The technique models the shape of the disc using both the T1-weighted and T2-weighted co-registered sagittal views for building a 2 dimensional (2D) feature image. The disc shape feature is modelled using Active Shape Model algorithm while the appearance is modelled using the normalized pixel intensity. These feature-pairs are then classified using Gibbs-based classifier. The paper reported that 91 % accuracy is achieved in detecting the herniation. A vertebrae detection and labelling algorithm of lumbar MR images is proposed in [14]. The paper firstly converts the 2D MR images to 3D before using them as an input to the detection algorithm. This detection algorithm is a combination of two detectors namely Deformable Part Model (DPM) [30] and inference using dynamic programming on chain [31]. A computational method to diagnose Lumbar Spinal Stenosis (LSS) from the patient's Magnetic Resonance Myelography (MRM) and MRI is proposed in [32]. LSS is a medical condition in which the spinal canal narrows and compresses the spine. In this paper, an image segmentation process is first carried out as a pre-processing step to identify the affected dural sac area in the input images. It then produces the relevant image features based on the inter and intra context information of the segments and use them to detect the presence of LSS [32]. Detection of problematic areas in medical images is not the only application of computer vision and AI in medicine. One evidence for this can be seen in [33]. In the paper, an image processing algorithm is used, not for detection, but to improve the clarity and quality of 3D MRI and computed tomography (CT) images so that they can be viewed without using a disparity device.

2.2   Three-Dimensional Geometrical and Physics Modelling of Lumbar Spine for Future Prediction of CLBP

Three-dimensional surface modelling of lumbar spine has been carried out and widely published in the literature. A number of physics models of the lumbar spine have also been developed [34]. However, little of these have been used to help physicians in future prediction of CLBP. This section describes existing related techniques and technologies in this area.

Starting with the type of input data used to generate the 3D model, a study that compares the quality of 3D-surface model generated from both CT scan and MRI is proposed [20]. The research interestingly concluded that CT scan is better than MRI scan in producing adequate surface registration for image segmentation and generation of a 3D-surface model [20]. Most geometrical and dynamic physics modelling of lumbar spine in the literature are carried out using Finite Element Modelling (FEM). One of the earliest techniques that uses FEM is detailed in [22, 35]. A more recent technique to create 3D geometrical and mechanical model of lumbar spine with FEM is proposed by Nabhani and Wake [36] which modelled the L4 and L5 vertebrae. The paper reports large stress concentrations in the superior and inferior facet region and on the central surfaces of the vertebral body and in the cortical shell of the vertebrae. The software package used to reconstruct the vertebrae model is I-DEAS Master Series. Noailly et al. also used FEM to model the L3, L4 and L5 vertebrae [37]. The paper, however, offers an inconclusive finding about model validation through comparison of computed global behaviours with experimental results. A number of researches studied the effect of body movement and the application of external pressure on the generated lumbar model. Feipel et al. studied the kinematic behaviour of the lumbar spine during walking including the effect of walking speed on the lumbar motion (translation, rotation, and bending) patterns [38]. The study concluded that walking velocity affects the range of the lumbar motion but not the sagittal plane motion. Another study by Papadakis [39] shows that there is a statistically significant difference between gait variability in a group of people with lumbar spinal stenosis and a healthy group of individuals. Receiver Operating Characteristic (ROC) is used in this study to measure the method value and to find the cut-off value. In addition, to finding the condition in the day of measurement the Oswestry Low Back Pain was used [39]. The effect of external pressure on the shape of lumbar spine is studied in [40]. The paper concludes that a posterior-to-anterior (PA) force which is applied during MRI scan at a single lumbar spinous process causes motion of the entire lumbar region. The findings of these works suggest that there are many factors that can determine the shape of the lumbar spine at any one time and they may affect both the resulting 3D geometry model as well as physics model of the spine. With the advancement in computer graphics and computer game technology, realistic simulation of real life objects ranging from racing cars kinematics to projectiles trajectory and to character movements is becoming a reality. It is therefore sensible to consider these new technologies for more serious applications such as computer aided diagnosis. A study on the appropriateness of using extensible physics engines for medical simulation purposes is given in [19]. A review and survey of recent techniques which use game engines in simulating clinical training is given in [41].

3   Discussion and Analysis

Based upon the review of the literature it can be concluded that there are significant gaps that need to be bridged between the relevant computing technologies and their application as an aid tool in the diagnosis and management of CLBP patients. The aim of this research is to develop a computer assisted and systematic methodology for detection and prediction of potential sources of chronic lower back pain using these technologies. Therefore, we hypothesise that the bridging of these gaps, by employing relevant state-of-the-art computing technologies in computer aided system for diagnosis and management of CLBP patients, would improve the efficiency and accuracy of the medical process to diagnose and manage CLBP patients. Previous researchers [5, 42] have highlighted the importance of identifying and understanding the problem area that are related to the medical decision assisted systems. Moreover, the mechanism of artificial intelligence and computer graphics technologies which could be applied in the lower back pain detection and modelling process is another factor which needs to be identified carefully. After extensively reviewing existing research work, we found two important issues that need to be addressed. First, how we can use current advances in computer technologies to help physicians diagnose the cause of CLBP. Second, how we can use current advances in computer technologies to help physicians predict future occurrence of CLBP in their patients.

In this research we are proposing a framework for a novel methodology which can help physicians in their efforts to diagnose and manage CLBP cases. The methodology consists from two parts: The first part is a lumbar disc herniation detection using computer vision and artificial intelligence (AI) techniques. In this case, the system takes MRI of CLBP patient's lumbar spines as inputs and produces highlights of the lumbar disc if herniation is detected. The development of this part would include (1) the development of novel image features suitable for differentiating herniated and normal discs from their image appearances and (2) finding a suitable and best state-of-the-art Artificial Intelligence technique for classifying the features by analysing and comparing their performances. The second part is a 3D geometrical and physics modelling of lumbar spine for analysing the source of CLBP. The model would take into account the patient's characteristics such as height, weight, gender, and age as well as the current state of his/her lumbar spine as derived from the patient's MR images. The model will be used to provide dynamic and interactive 3D visual representation of the patient's lumbar spine to the physician and can simulate the positions in which pain is generated. In addition, the same approach could be used to predict the progression of the degenerative process of the disc.

4   Framework of the Proposed System

The proposed system will have two functionalities. Firstly, the system will be used to detect lumbar disc herniation using computer vision and artificial intelligence techniques. In this functionality, the patient's contrast weighted MR images is segmented to remove any irrelevant parts from the image and keep only the disc, vertebrae and spinal cord canal. The system then calculates the disc height [43], the distance between adjacent vertebrae, the distance between disc and spinal cord [44], and the feature descriptors for the disc shape [42]. The system also requires additional inputs such as the patient's age, gender, height and weight. These inputs are used to determine the expected values for the disc height, the distance between adjacent vertebrae, the distance between disc and thecal sac, and the feature descriptors for the disc shape, when abnormalities do not occur. The system then applies a knowledge-based or artificial intelligence algorithm to compare the two sets (the calculated set and the determined set) to decide whether disc herniation occurred. This process is illustrated in Fig. 2.

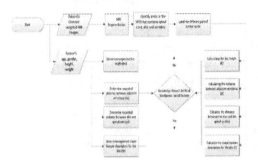

Fig. 2. Steps for disc herniation detection system

There are a number of algorithms that could be used in the knowledge-based/artificial intelligence system. Our approach is to experiment with a number of classifiers and image feature descriptors and perform the training and classification process using combinations of them. The best pair of classifier and feature descriptors will be chosen based on their accuracy. To illustrate the training and classification process, we use one of the most popular and widely used classifiers namely the Haar Cascade Classifier [45]. Haar Cascade classifier can be trained using thousands of sample images and utilises the sliding window technique to locally compute the feature descriptors. The training will use contrast weighted MR images as inputs as well as labelled affected region that have been done manually. Contrast weighted images are used to emphasis different types of tissues within the same MR images. This process is further illustrated in Fig. 3.

Fig. 3. System training

The trained system will then be able to produce labelled images of the affected areas if disc herniation is detected in the input MR images, as shown in Fig. 4. For the second functionality, the system will be used to give the physician the ability to predict the future occurrence of disc herniation. This functionality uses the same first step as the first one that is the segmentation of the patient's contrast weighted MR images to remove irrelevant parts. A 3D model of the patient's lumbar spine is then developed using the segmented MR images. A physics model is then attached to this 3D model to govern the kinematics of the lumbar spine using one of the physics engine. Similar to the first functionality, the system also takes patient's age, gender, height, and weight as the second set of input. This set of input is then used to adjust the parameters of the physics model to adapt it to the patient's characteristics. This will then be used for either the visualisation of the lumbar spine and prediction of future occurrence of disc herniation. This process is illustrated in Fig. 5.

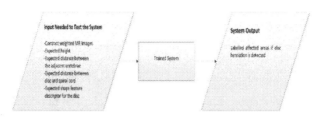

Fig. 4. System testing

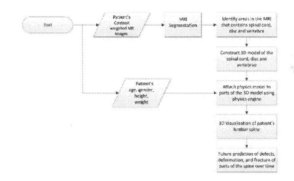

Fig. 5. Steps for lumbar spine visualisation and prediction system

5   Conclusion

The progress in image processing, computer vision, artificial intelligence, computer graphics and physics simulation is moving rapidly in the past decade. However, they have not been utilized in any significant way to improve Computer Aided Diagnostic technique in particular the way Chronic Lower Back Pain cases are diagnosed and managed. Most physicians still rely on a long and laborious task to visually identify abnormalities from the patient's Magnetic Resonance Images. Furthermore, currently there is no solution to assist physicians to understand how patients' physiological characteristics, posture and position affect the way pain at lower back spine is generated. This paper proposed a framework for a novel methodology that utilises state-of-the-art computing technologies, which can be used as a computer aided diagnosis tool, to help physicians in their efforts to diagnose and manage CLBP cases. The methodology consists from two parts namely, (a) lumbar disc herniation detection using computer vision and artificial intelligence (AI) techniques, and (b) 3D geometrical and physics modelling of lumbar spine for visualising lumbar spine and prediction of future occurrence of disc herniation. This proposed system will be able to solve some of the existing problems with the current Chronic Lower Back Pain diagnosis and management procedures.

References

1. McCamey, K., Evans, P.: Low back pain. Prim. Care **34**(1), 71–82 (2007)
2. Pain Community Centre. http://www.paincommunitycentre.org/article/low-back-pain-problem#ref6
3. Waddell, G.: The Back Pain Revolution, 2nd edn. Churchill Livingstone (Elsevier), Edinburgh (2004)
4. Croft, P.R., Macfarlane, G.J., Papageorgiou, A.C., Thomas, E., Silman, A.J.: Outcome of low back pain in general practice: a prospective study. Br. Med. J. **316**(7141), 1356 (1998)
5. Waddell, G., Burton, A.K.: Occupational health guidelines for the management of low back pain at work: evidence review. Occup. Med. (Chic. Ill) **51**(2), 124–135 (2001)
6. Burton, K., Kendall, N.: Musculoskeletal disorders. BMJ **348**, bmj.g1076 (2014)
7. Ellis, R.M.: Back pain. BMJ **310**(6989), 1220 (1995)
8. Raj, P., Nolte, H., Stanton-Hicks, M.: Anatomy of the spine. In: Raj, P., Klinder, T., Stanton-Hicks, M. (eds.) Illustrated Manual of Regional Anesthesia, pp. 3–7. Springer, Heidelberg (1988)
9. Proximal, P.P., Supplement, P.P., Powers, C.M., Bolgla, L.A., Callaghan, M.J., Collins, N., Sheehan, F.T.: Patellofemoral pain: proximal, distal, and local factors–2nd international research retreat. J. Orthop. Sports Phys. Ther. **42**(6), 1–55 (2012)
10. Methods, R., Practices, B.: Appropriateness of care: use of MRI in the investigation of patient low back pain executive summary, pp. 1–29 (2015)
11. Raja'S, A., Corso, J.J., Chaudhary, V., Dhillon, G.: Automatic diagnosis of lumbar disc herniation with shape and appearance features from MRI. In: Proceedings of SPIE Medical Imaging, p. 76241A (2010)
12. David, A.L.: 9 Spine. Imaging for students, D, pp. 187–206 (2012)
13. The Healthy Spine | Spinal Simplicity. http://www.spinalsimplicity.com/the-healthy-spine/
14. Lootus, M., Kadir, T., Zisserman, A.: Vertebrae detection and labelling in lumbar MR Images. In: Yao, J., Klinder, T., Li, S. (eds.) Computational Methods and Clinical Applications for Spine Imaging. Lecture Notes in Computational Vision and Biomechanics, vol. 17, pp. 219–230. Springer, Switzerland (2014)
15. Turner, J.A., Shortreed, S.M., Saunders, K.W., Leresche, L., Berlin, J.A., Von Korff, M.: Optimizing prediction of back pain outcomes. Pain **154**(8), 1391–1401 (2013)
16. Su, W.-C., Yeh, S.-C., Lee, S.-H., Huang, H.-C.: A virtual reality lower-back pain rehabilitation approach: system design and user acceptance analysis. In: Antona, M., Stephanidis, C. (eds.) UAHCI 2015. LNCS (LNAI and LNBI), vol. 9177, pp. 374–382. Springer, Heidelberg (2015)
17. Neubert, A., Fripp, J., Engstrom, C., Schwarz, R., Lauer, L., Salvado, O., Crozier, S.: Automated detection, 3D segmentation and analysis of high resolution spine MR images using statistical shape models. Phys. Med. Biol. **57**(24), 8357 (2012)
18. Ahn, H.S.: A virtual model of the human cervical spine for physics-based simulation and applications. The University of Memphis (2005)
19. Nourian, S., Shen, X., Georganas, N.D.: Role of extensible physics engine in surgery simulations. In: 2005 IEEE International Workshop on Haptic Audio Visual Environments and their Applications (2005)

20. Hoad, C.L., Martel, A.L., Kerslake, R., Grevitt, M.: A 3D MRI sequence for computer assisted surgery of the lumbar spine. Phys. Med. Biol. **46**(8), N213 (2001)
21. Morais, S.T.: Development of a biomechanical spine model for dynamic analysis, Universidade do Minho (2011)
22. Shirazi-Adl, A., Ahmed, A.M., Shrivastava, S.C.: A finite element study of a lumbar motion segment subjected to pure sagittal plane moments. J. Biomech. **19**(4), 331–350 (1986)
23. Alomari, R.S., Corso, J.J., Chaudhary, V., Dhillon, G.: Lumbar spine disc herniation diagnosis with a joint shape model. Clin. Appl. Spine Imaging **17**, 87–98 (2014)
24. Lisin, D.A., Mattar, M.A., Blaschko, M.B., Benfield, M.C., Learned-miller, E.G.: Combining Local and Global Image Features for Object Class Recognition. In: CVPR Workshops (2005)
25. Freidman, J.H., Bentley, J.L., Finkel, R.A.: An algorithm for finding best matches in logarithmic expected time. ACM Trans. Math. Softw. **3**(3), 209–226 (1977)
26. Arya, S., Mount, D.M., Netanyahu, N.S., Silverman, R., Wu, A.Y.: An optimal algorithm for approximate nearest neighbor searching in fixed dimensions. In: Proceedings of 5th ACM-SIAM Symposium on Discrete Algorithms, vol. 1, no. 212, pp. 573–582 (1994)
27. Beis, J.S., Lowe, D.G.: Shape indexing using approximate nearest-neighbour search in high-dimensional spaces. In: Proceedings of IEEE Computer Society Conference on Computer Vision and Pattern Recognition, pp. 1000–1006 (1997)
28. Muja, M., Lowe, D.G.: Scalable nearest neighbour methods for high dimensional data. IEEE Trans. Pattern Anal. Mach. Intell. **36**(11), 1–14 (2014)
29. Alomari, R.S., Corso, J.J., Chaudhary, V., Dhillon, G.: Lumbar spine disc herniation diagnosis with a joint shape model. In: Yao, J., Klinder, T., Li, S. (eds.) Computational Methods and Clinical Applications for Spine Imaging. Lecture Notes in Computational Vision and Biomechanics, vol. 17, pp. 87–98. Springer, Switzerland (2014)
30. Felzenszwalb, P.F., Girshick, R.B., McAllester, D., Ramanan, D.: Object detection with discriminatively trained part-based models. Pattern Anal. Mach. Intell. IEEE Trans. **32**(9), 1627–1645 (2010)
31. Pedro, D.P.H., Felzenszwalb, F.: Pictorial structures for object recognition. Int. J. Comput. Vis. **61**, 55–79 (2004)
32. Koh, J., Alomari, R.S., Dhillon, G.: Lumbar spinal stenosis CAD from clinical MRM and MRI based on inter-and intra-context Features with a two-level classifier **7963**, 796304–796308 (2011)
33. Kamogawa, J., Kato, O.: Virtual Anatomy of Spinal Disorders by 3-D MRI/CT Fusion Imaging (2010). (no. Table 1)
34. Huynh, K., Gibson, I., Gao, Z.: Development of a Detailed Human Spine Model with Haptic Interface, pp. 165–195 (2012). cdn.intechweb.org
35. Lavaste, F., Skalli, W., Robin, S., Roy-Camille, R., Mazel, C.: Three-dimensional geometrical and mechanical modelling of the lumbar spine. J. Biomech. **25**(10), 1153–1164 (1992)
36. Nabhani, F., Wake, M.: Computer modelling and stress analysis of the lumbar spine. J. Mater. Process. Technol. **127**(1), 40–47 (2002)
37. Noailly, J., Wilke, H.-J., Planell, J.A., Lacroix, D.: How does the geometry affect the internal biomechanics of a lumbar spine bi-segment finite element model? Consequences on the validation process. J. Biomech. **40**(11), 2414–2425 (2007)
38. Feipel, V., De Mesmaeker, T., Klein, P., Rooze, M.: Three-dimensional kinematics of the lumbar spine during treadmill walking at different speeds. Eur. Spine J. **10**(1), 16–22 (2001)
39. Papadakis, N.C., Christakis, D.G., Tzagarakis, G.N., Chlouverakis, G.I., Kampanis, N.A., Stergiopoulos, K.N., Katonis, P.G.: Gait variability measurements in lumbar spinal stenosis patients: part A. Comparison with healthy subjects. Physiol. Meas. **30**(11), 1171–1186 (2009)

40. Kulig, K., Landel, R.F., Powers, C.M.: Assessment of lumbar spine kinematics using dynamic MRI: a proposed mechanism of sagittal plane motion induced by manual posterior-to-anterior mobilization. J. Orthop. Sport. Phys. Ther. **34**(2), 57–64 (2004)
41. Marks, S., Windsor, J., Wünsche, B.: Evaluation of game engines for simulated clinical training. In: New Zealand Computer Science Research Student Conference (NZCSRSC) (2008)
42. Alomari, R.S., Corso, J.J., Chaudhary, V., Dhillon, G.: Automatic diagnosis of lumbar disc herniation with shape and appearance features from MRI. Prog. Biomed. Opt. Imaging **11**, 76241A (2010)
43. Ghosh, S., Alomari, R.S., Chaudhary, V., Dhillon, G.: Computer-aided diagnosis for lumbar MRI using heterogeneous classifiers. In: Proceedings of International Symposium Biomedical Imaging, pp. 1179–1182 (2011)
44. Jordan, J., Konstantinou, K., O'Dowd, J.: Herniated lumbar disc. BMJ Clin. Evid. (2009)
45. Viola, P., Jones, M.: Rapid object detection using a boosted cascade of simple features. In: Proceedings of 2001 IEEE Conference on Computer Vision and Pattern Recognition CVPR 2001, vol. 1, C, pp. 511–518 (2001)

Partially Synthesised Dataset to Improve Prediction Accuracy

Ahmed J. Aljaaf[1]([✉]), Dhiya Al-Jumeily[1], Abir J. Hussain[1],
Paul Fergus[1], Mohammed Al-Jumaily[2], and Hani Hamdan[3]

[1] Applied Computing Research Group, Liverpool John Moores University,
Byrom Street, Liverpool L3 3AF, UK
A.J.Kaky@2013.ljmu.ac.uk,
{d.aljumeily,a.hussain,p.fergus}@ljmu.ac.uk
[2] Department of Neurosurgery, Dr. Sulaiman al Habib Hospital,
Dubai Healthcare City, Dubai, UAE
[3] Département Signal & Statistiques,
CentraleSupélec, Châtenay-Malabry, France
Hani.Hamdan@centralesupelec.fr

Abstract. The real world data sources, such as statistical agencies, library databanks and research institutes are the major data sources for researchers. Using this type of data involves several advantages including, the improvement of credibility and validity of the experiment and more importantly, it is related to a real world problems and typically unbiased. However, this type of data is most likely unavailable or inaccessible for everyone due to the following reasons. First, privacy and confidentiality concerns, since the data must to be protected on legal and ethical basis. Second, collecting real world data is costly and time consuming. Third, the data may be unavailable, particularly in the newly arises research subjects. Therefore, many studies have attributed the use of fully and/or partially synthesised data instead of real world data due to simplicity of creation, requires a relatively small amount of time and sufficient quantity can be generated to fit the requirements. In this context, this study introduces the use of partially synthesised data to improve the prediction of heart diseases from risk factors. We are proposing the generation of partially synthetic data from agreed principles using rule-based method, in which an extra risk factor will be added to the real-world data. In the conducted experiment, more than 85 % of the data was derived from observed values (i.e., real-world data), while the remaining data has been synthetically generated using a rule-based method and in accordance with the World Health Organisation criteria. The analysis revealed an improvement of the variance in the data using the first two principal components of partially synthesised data. A further evaluation has been conducted using five popular supervised machine-learning classifiers. In which, partially synthesised data considerably improves the prediction of heart diseases. Where the majority of classifiers have approximately doubled their predictive performance using an extra risk factor.

Keywords: Partially synthesised data · Prediction · Heart diseases · Machine learning · Rule-based method

© Springer International Publishing Switzerland 2016
D.-S. Huang et al. (Eds.): ICIC 2016, Part I, LNCS 9771, pp. 855–866, 2016.
DOI: 10.1007/978-3-319-42291-6_84

1 Introduction

There is growing interest from external researchers for access to data records collected by statistical agencies, organisations and research institutes. However, the privacy of individuals and confidentiality of data must be protected on legal and ethical grounds. Meanwhile, there is a demand to release a sufficient detail of data to maintain the reality and validity of statistical inference on the target population. To satisfy these desires, one method is to restrict data for approved analyses by authorised individuals. A second method is to release synthetic data rather than observed values, which typically conducted by a statistical disclosure control (SDC) technique [1]. The term of synthetic data has emerged since 1993 by Rubin [2]. The main aim was to protect the privacy and confidentiality of personal information through releasing synthetically produced data rather than actual data [2]. In general, a synthetic data can be created by a computer program using a random number generator or a formula that derived from real-world data [4]. There are two approaches of generating synthetic data, fully synthesis and partially synthesis data. Under the first approach, all data attributes are synthesised and no real data are released, while a subset of data attributes is synthesised under the partially synthesis approach [3, 5].

The real world data sources, such as statistical agencies, library databanks, research institutes and random generation procedures, are the major sources for researchers. Using this type of data involves a range of advantages. First, the data is relevant to real world problems, which enables reliable estimation of the usefulness of the results. Second, it improves the credibility and validity of the experiment. More importantly, this type of data is typically unbiased [4]. However, many studies showed that the use of synthetically generated data instead of real-world data is attributed to several factors including; (a) the difficulty of using real-world data because of the privacy policies. (b) The available quantities of the real-world data may not be sufficient for the purposes of the experiment. (c) The collection of real-world data might be inapplicable, costly or time consuming. (d) The real-world data might be unavailable, particularly in the newly arises research subjects [3, 4]. Moreover, Synthetic data have a considerable advantages including; the simplicity of generation, requires relatively small amount of time in comparison with a real-world data collection, a sufficient quantity can be generated to fit the requirements with the diversity and relevance that can mimic the real-world data [4].

This study introduces a new method of creating synthetic data. We are proposing the generation of synthetic data from agreed principles using rule-based method. This new method has been proposed with the aim of improving prediction accuracy. In particular the prediction of heart diseases. We are targeting the improvement of heart diseases prediction through adding an extra risk factor. This risk factor will be synthetically generated in accordance with the World Health Organisation (WHO) criteria for classification of adults underweight, overweight and obesity according to BMI [20]. The experiment will be conducted using partially synthesised data, where more than 85 % of the data have been extracted from real-world data, while less than 15 % has been synthetically generated using rule-based method. The real-world part of data consists of six risk factors extracted from the Cleveland Clinic Foundation heart disease dataset, which available online at [16]. The synthesised part of data consists of one additional risk factor, which synthetically generated based on agreed principles. The Cleveland Clinic Foundation heart disease dataset was intensively used in the majority of studies that addressed the early prediction of heart diseases. These studies have used a full range of data attributes. In contrast, we are extracting only the risk factors, which represents the real-world part of data in this study. An adequate review of studies that targeted the prediction of heart diseases can be found in [6].

The researchers gave considerable attention to the prediction of heart diseases. Where the early prediction of heart disease has a significant influence on patient safety, as it can contributes to an effective and successful treatment before any severe degradation of cardiac output [6]. Heart diseases is a public health problem with high societal and economic burdens. It is considered the main cause of frequent hospitalisations in individuals 65 years of age or older, and slightly less than 5 million Americans suffer from heart diseases [8]. Heart diseases can occur because of many potential causes, some are illnesses in their own right, while others are secondarily to another underlying diseases [7, 8]. The commonest cause of heart failure is coronary disease by 62 % compared to other risk factors such as hypertension, valvular disease, myocarditis, diabetes, alcohol excess, obesity and smoking [8, 9]. In general, heart diseases can be used to describe a condition in which the heart is unable to pump a sufficient amount of blood around the body [7].

In this paper, we aimed to (a) review the latest studies that addressed the aspect of synthetic data generation, (b) describe our proposed method of synthetic data generation using rule-based method and in accordance with agreed principles, (c) inspect the feasibility of our method and adding an extra risk factor using principal component analysis, (d) evaluate the utilisation of an extra risk factor to improve the prediction of heart diseases using five popular supervised machine-learning classifiers, and finally, (e) highlight the results and study contributions.

2   Synthesised Data in Real-World Applications

Although the focus and the requirement are quite different in each field, the use of synthetic data has become an appealing alternative in many diverse scientific disciplines, including performance analysis, software testing, privacy protection and synthetic oversampling. Macia et al. [10] have proposed the use of fully synthesised data to investigate the performance of machine learning classifiers. As they have stated, the use of synthetically generated datasets can offer a controlled environment to analyse the performance of machine learning classifiers and therefore provide a better understanding of their behaviours. In the same context, Sojoudi and Doyle [11] have used a synthetic data generated by an electrical circuit model to investigate the performance of three methods, namely thresholding the correlation matrix, graphical lasso and Chow-Liu algorithm. These methods have been used as an alternative to identify the direct interactions between brain regions. They noticed that the first two methods (i.e., thresholding the correlation and graphical lasso algorithm) are susceptible to errors.

In area of software evaluation, Whiting et al. [12] contributed in creating fully synthesised data to test visual analytics applications. Their main aim was to enable tool developers to determine the effectiveness of their software within an acceptable time frame. Similarly, Babaee and her colleague [13] have introduced the use of synthetic 2D X-ray images to validate medical image processing applications. Initially, a model of an organ is created using modelling software, then the model converted to computerised tomography image (CT) through assigning a proper Hounsfield unit to each voxel. As they have reported, this method may provide a ground truth data for researchers to validate their proposed medical image processing methods.

In another study, the researchers were targeting imbalanced learning problems. They have used a Kernel density estimation method to construct partially synthesised oversampling approach to address imbalanced class distribution in a particular data set. The experiment and evaluation was conducted using different medical related datasets with promising results [14]. Finally, Park and his partners have introduced a non-parametric systematic data for privacy protection of healthcare data. As they have claimed, their proposed method synthesises artificial records while maintaining the statistical features of the original records to the maximum extent possible. Using different data mining and statistical analysis methods, they have concluded that the synthetic dataset delivers results that largely similar to the original dataset [15].

3 Materials and Methods

3.1 Dataset

The experiment conducted using a partially synthesised data, which consists of two parts. First, real-world observations (i.e., risk factors), which represents 85.72 % of the data. This part includes six risk factors extracted from the Cleveland Clinic Foundation heart disease dataset, which is available online at [16]. These risk factors are patient's age, gender, resting blood pressure, serum cholesterol, fasting blood sugar and maximum heart rate. This study selects these risk factors according to sophisticated researches in cardiovascular disease. As presented in the Framingham heart study, which predicting risk factors of cardiovascular disease for 30 years, patient's age, gender, blood pressure, serum cholesterol and blood sugar are considered standard risk factors [17]. Another long-term follow-up study on healthy individuals aged 25–74 years found that a high resting heart rate is an independent risk factor for coronary artery disease incidence or mortality among white and black individuals [18]. This part of data (i.e., real-world observations) consists of 297 consistent instances and without missing values. The output class includes four labels, which are no risk, low risk, moderate risk and high risk of developing heart diseases.

Around two-thirds of the data are for male individuals, which is 201 instances. Mean age of individuals in the data set is 54 years. Figure 1 presents the age distribution, which clearly shows that the risk of developing heart disease starts approximately in the fourth decade of life. The risk is then about to double across every ten years to reach its peak in the sixth decade of life. Aging as shown by many studies poses the largest risk factor for cardiovascular diseases, where aging is associated with changes in cardiovascular tissues, which leads to the loss of arterial elasticity and increase arterial thickening and stiffness. These changes may subsequently contribute to hypertension, stroke, and arterial fibrillation [19].

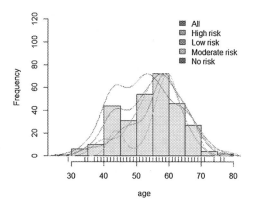

Fig. 1. Distribution of age (Color figure online)

The second part is an additional risk factor (i.e., body mass index BMI), which represents 14.28 % of the data and has been synthetically generated in accordance with the World Health Organisation (WHO) criteria for classification of adults underweight, overweight and obesity according to BMI [20]. This study consider adding BMI as an additional risk factor because; (a) it is neither been collected with the original data nor been involved with the same data for inference. (b) The increase in BMI could dramatically increase the prospect of heart diseases. A study conducted in the USA showed that 30 % reduction in the proportion of obese people would prevent approximately 44 thousand cases of heart diseases each year [7]. Moreover, being obese has been shown to double the risk of heart diseases [7, 8]. Finally, (c) adding an additional risk factor would potentially improve the early prediction of heart diseases.

The second part (i.e., synthetic data) has been generated using rule-based method, in which, we are modelling the creation of WHO in the form of IF-THEN statements. These statements are implemented to generate the synthetic part of data. The classification of WHO identifies a principal cut-off points to categorise individuals according to their BMI, which is a manner of labelling someone as underweight, normal, overweight or obese. BMI is calculated by dividing an individual's weight in kilograms by the square of his/her height in metres [21]. As mentioned by WHO's global atlas on cardiovascular disease prevention and control, risks of coronary heart disease, raised blood pressure, type 2 diabetes and ischaemic stroke increase steadily with an increasing BMI [21]. Accordingly, the main aim of this paper is to involve BMI as an additional risk factor with the real world observation in order to improve the prediction accuracy of developing heart diseases. The following figure illustrates the distribution of the class labels in accordance with the cut-off points of the WHO criteria as a first step toward generating synthetic data from agreed principles.

The first step was identifying the ranges of each class label. As shown in Fig. 2, no risk class label assigned to normal range of BMI, which ranges from 19 and 25 according to WHO creation. Then low risk class label, which refers to overweight or pre-obese, represents BMI ranges between 25 and 30 according to WHO creation. Followed by moderate risk class label that corresponds to obese type one and finally high risk class label for obese type two and three. An overlapping area is maintained over the class labels,

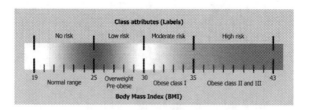

Fig. 2. The distribution of class labels in accordance with BMI cut-off points

in which every class label is sharing a single unit with the following class label. For example, unit 25 of BMI is mutual between no risk and low risk class labels, and so on for the remaining class labels. This overlapping intended to simulate the real-world data disturbance and preserves statistical analysis from bias. The second step in the data generation process is to transform this explanation into conditional rules (i.e., IF-THEN statement). These rules then converted into a computer program that passes through the real-world dataset and generates BMI based on WHO creation and corresponding to a particular class label. The following algorithm demonstrates a second step of the data generation process, in which, we are using one SWITCH statement rather than a series of IF-THEN statement to express and translate the WHO criteria as a set of rules.

```
Input: a labelled set of real data
Output: a partially synthesised data
  Begin
  For each row r in the data file do
    Read class value clv,
    Switch clv do
      Case clv : no risk
        Normal range of body mass index BMI
        BMI = random number between (19, 25)
      Case clv : low risk
        A range from overweight to pre-obese of BMI
        BMI = random number between (25, 30)
      Case clv : moderate risk
        A range of obese class one.
        BMI = random number between (30, 35)
        Case clv : high risk
          A range of obese class two and three
          BMI = random number between (35, 43)
    End switch
    Assign BMI value to the corresponding clv
    End for
  End
```

Algorithm 1. Partially synthesised data algorithm

Finally, let's consider A is patient's age, G is patient's gender, R is resting blood pressure, SC is serum cholesterol, FBS is fasting blood sugar and MHR is maximum heart rate. These risk factors are extracted for real world dataset. After adding the synthesised body mass index BMI, the partially synthesised dataset can be represented as a set $S = \{ <A_1, G_1, R_1, SC_1, FBS_1, MHR_1, BMI_1 > , \ldots, < A_n, G_n, R_n, SC_n, FBS_n, MHR_n, BMI_n > \}$, where n is the size of the set S, which is the total number of instances. The Class attribute consists of four labels to classify patients with heart diseases into four risk levels and represented as $C = \{c_1, \ldots, c_m\}$, in this study $m = 4$, which are no risk, low risk, moderate risk, high risk.

3.2   Statistical Analysis

This section inspects the feasibility of involving the BMI as another risk factor for improving the prediction of heart diseases. A principal component analysis (PCA) is employed and the data is normalised. This experiment used a partially synthesised dataset, in which 85.72 % of the data obtained from real-world data, while the remaining 14.28 % have been synthetically generated using rule-based method. This combination of data belongs to different measurement scales including dichotomous (i.e., binary values) and continuous values. Two out of seven attributes are binary values, which represent 28.57 % of the data attributes. These attributes are patient's gender and fasting blood sugar. They were reported as 1 for male and 120 mg/dl or more of blood sugar, while zero for female and less than 120 mg/dl of blood sugar. The remaining five attributes are belonging to continuous values, but they have a different kind of distribution. For example, the mean and standard deviation of blood pressure attribute are 131.69 and 17.76, respectively, whereas they are 247.35 and 51.99 for serum cholesterol attribute. The mean of age attribute is 54.54 years.

These diverse ranges of means and standard deviations need to be transformed into a normal distribution before conducting statistical analysis, in which all the attributes will have the same mean or standard deviation. Therefore, the z-score normalisation method has been used to transform and unify the ranges of all attributes within the dataset. The z-score normalisation method re-scales the data to generate a new range dataset with zero mean and one standard deviation. The z-score normalisation method is mathematically represented as follows.

$$\bar{v}_i = \frac{v_i - \mu_x}{\sigma_x}$$

Where $\bar{v}_i$ is the normalised value, v_i is the original value, ith column of row attribute X, and μ_x and σ_x are the mean and standard deviation of row attribute X, respectively [22].

Although the principal component analysis (PCA) is commonly applied in data science for reducing the number of dimensions, particularly with high dimensional data, the main purposes of PCA is to identify patterns in data to highlight commonalities and variations. Therefore, PCA has been applied to describe the hidden structure of the data and investigate whether adding another risk factor maximising the amount of variance within the data or not. Where, maximising the amount of variance using the fewest number possible of principal components would be an ideal scenario of PCA, which reflect positively on the prediction accuracy. Figure 3 shows a score plot of the first principal component versus the second principal component of both real and partially synthesised data.

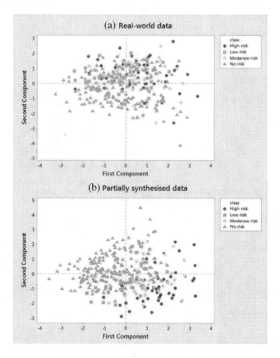

Fig. 3. First principal component vs. second principal component

In Fig. 3 A (i.e., real-world data), the score plot of the first two components shows a highly overlapping area among the class labels. Despite the cumulative proportion of eigenvalues of these two components should reveal an obvious groupings of data points, it appears difficult to aggregate the largest amount possible of data points that belong to a certain class label within a clear group. This is the leading cause at showing unsatisfactory results in the prediction of heart diseases. In contrast, with 14.28 % synthesised data, the score plot of the first two components shows an improvement of the variance in the data. Figure 3 B (i.e., partially synthesised data), demonstrates a more separate distribution in the data points, which indicate that the cumulative proportion of eigenvalues of these two components reveals an obvious groupings of data points. This will reflect positively on detecting clusters and improving predictions. Next section investigates this matter further through applying different machine learning methods.

3.3   Results

This section utilises five popular supervised machine-learning classifiers to assess both real and partially synthesised dataset, particularly evaluating the value of employing an extra risk factor to improve the prediction of heart diseases. The targeted classifiers are NaiveBayes (NB), Multilayer Perceptron Neural network (MLP), Support Vector Machine (SVM), Logistic Regression (LoR) and C4.5 Decision Tree (DT). This diversity of classifiers would clearly reveal whether an improvement in prediction of heart diseases from risk factors was accomplished using partially synthesised data or not. Two experiments have been conducted. In the first experiment, we have examined the sensitivity, specificity, mean absolute error and prediction accuracy of the targeted classifiers using the real-world data. In the second experiment, we re-examined these classifiers using partially synthesised data. In both experiments, we have used k-folds cross validation methods. In which, the data are partitioned into k equal subsets with almost the same proportions of different class labels. Of the k subsets, a single subset retained for testing, while the remaining k-1 subsets used for training. The cross validation is then repeated k times, until each subset applied exactly once for testing. Finally, the results averaged to estimate a model predictive performance. In this section, k = 10.

Table 1. The evaluation of real-world data (i.e., risk factors)

Classifiers	Sensitivity	Specificity	Mean absolute error	Accuracy
NB	0.53	0.68	0.278	53.87
MLP	0.54	0.68	0.269	54.54
SVM	0.53	0.48	0.319	53.87
LoR	0.55	0.66	0.272	55.89
DT	0.51	0.70	0.271	51.85

Table 1 shows the overall predictive performance of the machine learning classifiers using real-world data. Results indicate a considerably low overall predictive performance, where the prediction accuracy ranges between 51 % and 56 %. Almost all the predictive models show convergent ranges of sensitivity, specificity and mean absolute error. This confirms the analysis results through the score plot of the first two components in Fig. 3A. Although unacceptable predictive performance (i.e., below average), Logistic Regression has achieved highest sensitivity and accuracy, followed by Multilayer Perceptron with 54 % of sensitivity and accuracy. NaiveBayes has overcome Support Vector Machine with 2 % of specificity; however, they have presented an identical sensitivity and accuracy. Despite Decision Tree comes at the end of the list with 51 % of sensitivity and accuracy, it has recorded the best specificity. The majority of predictive models showed approximately 0.27 of mean absolute error, except Support Vector Machine model, which recorded a slightly higher error rate.

Table 2 introduces the overall predictive performance using partially synthesised data. The majority of classifiers have achieved impressive overall results with more than 90 % of sensitivity, specificity and prediction accuracy. These results clearly

Table 2. The evaluation of partially synthesised data (i.e., adding extra risk factor)

Classifiers	Sensitivity	Specificity	Mean absolute error	Accuracy
NB	0.91	0.95	0.064	91.58
MLP	0.90	0.97	0.052	90.90
SVM	0.87	0.92	0.260	87.20
LoR	0.93	0.98	0.037	93.93
DT	0.91	0.96	0.048	91.58

demonstrate that the use of partially synthesised data (i.e., adding an extra risk factor) has had a significant impact on prediction accuracy. Logistic regression was also the leading model with 93 % of sensitivity, accuracy and 98 % of specificity. Although Multilayer Perceptron achieved the second highest specificity, its sensitivity and accuracy were slightly lower than NaiveBayes and Decision Tree that recorded similar sensitivity and accuracy. Support Vector Machine registered the lowest overall predictive performance. In contrast to the first experiment, the mean absolute error considerably dropped for the majority of models, whereas the overall predictive results substantially increased for all the predictive models.

3.4   Discussion

The conducted experiment highlighted the involvement of an additional risk factor, which synthetically generated using rule-based method and according to the standards of WHO, with a set of risk factors that extracted from real-world data to predict heart disease. Despite the lack of an agreed principle to indicate the accepted ratio of synthesised data in a particular data set, it seems that synthesising less than 15 % of the data will not have a serious impact on the quality of statistical inference. In contrast, PCA has been conducted to investigate the hidden structure of the data, which reveals an improvement of the variance in the data using the first two principal components of partially synthesised data. This has been confirmed using various machine learning methods, where partially synthesised data significantly improves the prediction accuracy of heart diseases. The majority of classifiers have approximately doubled their predictive ability using the BMI as an extra risk factor.

This study holds two contributions. The main contribution shows the idea of generating synthetic data from agreed principles. A rule-based method has been used for this purpose. This strategy can be generalised into many other research areas. In particular the researches that aim to use fully and/or partially synthesised data in certain scientific discipline. For example, improving predictions, software testing and evaluation, security aspects and so on. A reliable implementation of synthetic data generation using rule-based method requires a predefined criteria. Where these criteria need to be agreed worldwide in order to generate a valid set of data with a minimum possibility of bias. An expert knowledge is also under an obligation to express and translate these criteria into a set of rules. Where this study used the criteria of WHO with the aim of using an extra risk factor to improve the prediction of heart diseases. The second contribution is the participation of an additional risk factor to improve the prediction accuracy. This has been considered with several specialised studies [7, 8]. In contrast to the first contribution, the method of using extra risk factor to improve the prediction accuracy cannot be generalised as a new way to improve the prediction of certain diseases. It is entirely restricted to this study.

4 Conclusion

This paper presents the idea of generating synthetic data from agreed principles. The main aim was the improvement of the prediction of heart diseases from risk factors. Partially synthesised data have been used, in which more than 85 % of the data extracted from real-world data, while the remaining was synthetically generated using rule-based method and in accordance with the criteria of World Health Organisation. A statistical analysis has shown an improvement in the variance of data after adding an extra risk factor. A further investigation has been conducted utilising five well-known supervised machine learning methods. The classifiers have approximately doubled their predictive performance using an extra risk factor, which confirms the statistical analysis result.

References

1. Loong, B.: Topics and applications in synthetic data. Doctoral dissertation, Harvard University. (2012)
2. Rnbin, D.B.: Discussion statistical disclosure limitation. J. Official Stat. **9**(3), 461–468 (1993)
3. Jeske, D.R., Samadi, B., Lin, P.J., Ye, L., Cox, S., Xiao, R., Younglove, T., Ly, M., Holt, D., Rich, R.: Generation of synthetic data sets for evaluating the accuracy of knowledge discovery systems. In: Eleventh ACM SIGKDD International Conference on Knowledge Discovery in Data Mining, pp. 756–762 (2005)
4. Hall, N.G., Posner, M.E.: The generation of experimental data for computational testing in optimization. In: Bartz-Beielstein, T., Chiarandini, M., Paquete, L., Preuss, M. (eds.) Experimental Methods for the Analysis of Optimization Algorithms, pp. 73–101. Springer, Heidelberg (2010)
5. Sakshaug, J.W.: Synthetic data for small area estimation. Doctoral dissertation, The University of Michigan (2011)
6. Aljaaf, A.J., Al-Jumeily, D., Hussain, A.J., Dawson, T., Fergus, P., Al-Jumaily, M.: Predicting the likelihood of heart failure with a multi level risk assessment using decision tree. In: Third International Conference on Technological Advances in Electrical, Electronics and Computer Engineering (TAEECE), pp. 101–106. IEEE, Beirut (2015)
7. The European Society of Cardiology: Heart failure: preventing disease and death worldwide (2016). http://www.escardio.org/communities/HFA/Documents/whfa-whitepaper.pdf. Accessed 2 Feb 2016
8. Roger, V.L.: The heart failure epidemic. Int. J. Environ. Res. Public Health **7**(4), 1807–1830 (2010)
9. Scottish Intercollegiate Guidelines Network (SIGN): Management of chronic heart failure: a national clinical guideline (2016). http://sign.ac.uk/pdf/sign97.pdf. Accessed 5 Feb 2016

10. Macia, N., Bernado-Mansilla, E., Orriols-Puig, A.: Preliminary approach on synthetic data sets generation based on class separability measure. In: 19th International Conference on Pattern Recognition (ICPR), pp. 1–4. IEEE (2008)
11. Sojoudi, S., Doyle, J.: Study of the brain functional network using synthetic data. In: 52nd Annual Allerton Conference on Communication, Control, and Computing (Allerton), pp. 350–357. IEEE (2014)
12. Whiting, M.A., Haack, J., Varley, C.: Creating realistic, scenario-based synthetic data for test and evaluation of information analytics software. In: Proceedings of the 2008 Workshop on Beyond Time and Errors: Novel Evaluation Methods for Information Visualization, Florence Italy (2008)
13. Babaee, M., Nilchi, A.R.N.: Synthetic data generation for X-ray imaging. In: 21st Iranian Conference on in Biomedical Engineering (ICBME), pp. 190–194. IEEE (2014)
14. Tang, B., He, H.: KernelADASYN: Kernel based adaptive synthetic data generation for imbalanced learning. In: IEEE Congress on Evolutionary Computation (CEC), pp. 664–671. IEEE (2015)
15. Park, Y., Ghosh, J., Shankar, M.: Perturbed Gibbs samplers for generating large-scale privacy-safe synthetic health data. In: IEEE International Conference on Healthcare Informatics (ICHI), pp. 493–498. IEEE (2013)
16. The Cleveland Clinic Foundation: Heart Disease Data Set (2016). http://archive.ics.uci.edu/ml/datasets/Heart+Disease. Accessed 3 Feb 2016
17. Pencina, M.J., D'Agostino, R.B., Larson, M.G., Massaro, J.M., Vasan, R.S.: Predicting the 30-year risk of cardiovascular disease: the Framingham heart study. Circulation **119**, 3078–3084 (2009)
18. Gillum, R.F., Makuc, D.M., Feldman, J.J.: Pulse rate, coronary heart disease, and death: the NHANES I epidemiologic follow-up study. Am. Heart J. **121**, 172–177 (1991)
19. North, B.J., Sinclair, D.A.: The Intersection between aging and cardiovascular disease. Circ. Res. **110**, 1097–1108 (2012)
20. World Health Organisation: The International Classification of adult underweight, overweight and obesity according to BMI (2016). http://apps.who.int/bmi/index.jsp?introPage=intro_3.html. Accessed 5 Feb 2016
21. The World Health Organization: Global Atlas on cardiovascular disease prevention and control (2016). http://www.who.int. Accessed 3 Feb 2016
22. Al Shalabi, L., Shaaban, Z.: Normalization as a preprocessing engine for data mining and the approach of preference matrix. In: the International Conference on Dependability of Computer Systems (DepCos-RELCOMEX 2006), pp. 207–214. IEEE (2006)

Author Index

Printed in the United States
By Bookmasters